ADVANCES IN WATER DESALINATION

WILEY SERIES ON ADVANCES IN WATER DESALINATION

NOAM LIOR, Series Editor

ADVANCES IN WATER DESALINATION

Edited by

Noam Lior
University of Pennsylvania

WILEY

A JOHN WILEY & SONS, INC., PUBLICATION

Cover Images: (large photo) © Airyelf/iStockphoto; (circle 1) © Blanka Boskov/iStockphooto; (circle 2) Photograph shows the 127 million m^3/year RO desalination plant in Hadera, Israel. Courtesy of IDE Technologies, the builder and operator of the plant; (circle 3) Photograph shows the 23,500 m^3/day per unit MSF desalination plant in Al-Jobail, Saudi Arabia. Courtesy of Sasakura Engineering Ltd., the builder of the plant; (circle 4) Line art depicting filtration.

Published by John Wiley & Sons, Inc., Hoboken, New Jersey
Published simultaneously in Canada

Library of Congress Cataloging-in-Publication Data:

Advances in water desalination / edited by Noam Lior.
p. cm.
Includes bibliographical references and index.
ISBN 978-0-470-05459-8 (hardback)
1. Saline water conversion. I. Lior, Noam.
TD479.A36 2012
628.1′67–dc23

2012006128

Printed in the United States of America

10 9 8 7 6 5 4 3 2 1

CONTENTS

*Corresponding author.

PREFACE

This volume contains a wide spectrum of principal and timely information about (1) advances in fundamentals of desalination analysis and design when taking into consideration the increasing concerns about environmental and fuel cost effects on the processes, (2) an evaluation of the state of the art of pretreatment and desalination technologies considering environmental and performance aspects, (3) a critical comprehensive survey of the economic aspects of water desalination, (4) a review of advances in hollow-fiber reverse-osmosis membrane modules, (5) an introduction and review of the emerging adsorption desalination process, and (6) a comprehensive review of advanced instrumentation, measurements, control, and automation in the MSF (Multi-Stage Flash) and RO (Reverse Osmosis) desalination processes.

Perhaps the current main leading challenge in water desalination is its sustainability. The first three chapters in the book address two of the three sustainability pillars: environmental impact and economics. Economics became a growing concern due to the rapidly increasing and wildly fluctuating prices of energy, which is becoming a more dominant fraction of the produced water cost.

I note with great sorrow that the author of Chapter 1, Dr. Professor Yehya El-Sayed, has passed away before the printing of this book. As one of the world's leading and well-acknowledged thermodynamicists, he brought the science to engineering practice in general and to water desalination in particular, especially in his seminal work on applications of exergy and exergo-economic analysis to this field. His life-work and this chapter demonstrates his foresight in dealing scientifically with water desalination sustainability, and will remain a permanent tribute to his memory.

I would like to acknowledge the essential contributions of the chapter authors who shared with us their precious knowledge and experience, of the book series Editorial Board members who are international leading desalination experts, Ms. Miriam Balaban, Dr. Professor Mohamed Ali Darwish, Dr. Professor Osamu Miyatake, Dr. Professor Shichang Wang and Dr. Mark Wilf who provided valuable guidance and review, and of Dr. Arza Seidel of John Wiley & Sons who has patiently and professionally overseen the creation of this book series and volume.

Some material for Chapter 1 is not included in the book (in its various formats) and may be downloaded at http://booksupport.wiley.com. For more information about Wiley products, including a variety of print and electronic formats, visit www.wiley.com.

Professor NOAM LIOR

University of Pennsylvania
Philadelphia, PA 19104-6315, USA
lior@seas.upenn.edu
Editor-in-Chief
Philadelphia, 17 March 2012

INTRODUCTION TO THE BOOK SERIES ADVANCES IN WATER DESALINATION

Rapidly increasing scarcity of water usable for drinking, irrigation, industry, and general sanitation, caused by rising use and pollution of existing fresh water sources, has created an enormous rise (lately of around 12%/year) in water desalination. Water desalination consists of separation processes that produce new fresh water from seawater and other water sources which are too saline for use. Large commercial scale desalination began in 1965 and had a worldwide capacity of only about 8000 m^3/day in 1970. It now produces about 72 million m^3/day of desalted water by about 16,000 facilities worldwide. Within 10 years, production is forecasted to triple with an expected investment of around $60 billion.

Water desalination is accomplished by a variety of different technologies, which are gradually changing to reduce capital costs, energy consumption and environmental impacts. It consumes large amounts of energy and materials, and has an associated important and increasingly recognized impact on the environment. Research and development, improved construction, operation, cost allocation in multi–purpose plants, and financing methods, and education and information exchange must continue to be advanced to reduce the cost of the water produced and improve process sustainability.

Advances in Water Desalination is designed to meet the knowledge needs in this rapidly advancing field. One book volume is published per year, and contains 5–7 invited, high quality timely reviews, each treating in depth a specific aspect of the desalination and related water treatment field and the chapters are written and reviewed by top experts in the field. All aspects are addressed and include science, technology, economics, commercialization, environmental and social impacts, and sustainability.

The series will be useful for desalination practitioners in industry and business, scientists and researchers, and students.

The series is advised and directed by an international Editorial Board of desalination and water experts from academia and industry.

I am grateful to Dr. Arza Seidel of John Wiley & Sons who has patiently and professionally overseen the creation of this book series.

Professor Noam Lior

University of Pennsylvania
Philadelphia, PA 19104-6315, USA
lior@seas.upenn.edu
Editor-in-Chief
Philadelphia, 17 March 2012

ABOUT THE AUTHORS

Prof. Gary Amy is Director of the Water Desalination and Reuse Research Center and Named Professor of Environmental Science and Engineering at the King Abdullah University of Science and Technology (KAUST) in the Kingdom of Saudi Arabia.

Prof. Amy's research focuses on membrane technology, innovative adsorbents, ozone/advanced oxidation, river bank filtration and soil aquifer treatment, natural organic matter and disinfection by-products, and organic and inorganic micropollutants.

Dr. Amitzur Ze'ev Barak born 1938 in Israel, got both his B.Sc. in Mechanical & Energy Engineering/Nuclear Engineering [1960] and his Doctor of Sciences-in-Technology, Civil Engineering/Hydro-Sciences [1974] at the Technion, the Israel Institute of Technology. Joined the desalination community [1962] as the Research Desalination Engineer at the IDE-Israel Desalination Engineering Ltd. Coinventor and development-manager of two low-temperature evaporative desalination processes—LTMVC [mechanical vapor compression, 1964] and LTMED [multieffect-distillation, 1969]. For these activities, he received the Israeli "Prime-Minister Award for Applied Research" [1976].

Manager of the Thermal Desalination R&D Department at IDE Ltd [1968–1974].

Manager of the "Joint US-Israel Desalination Program," and director of all the Israeli governmental R&D activities on desalination [1976–1981].

Senior staff engineer for planning at the Israeli Atomic Energy Commission [1982–2003]. Since 2003, Professor of Chemical Engineering, Civil Engineering, and Mechanical Engineering at the Ariel University Center of Samaria, Israel. Consultant to the IAEA (International Atomic Energy Agency), the CERN, and dozen other entities on energy and desalination. Published over 60 papers and has six patents on desalination and solar energy.

Anutosh Chakraborty received his B.Sc. Eng. from BUET, Bangladesh, in 1997. He obtained his M. Engg. and Ph.D. degrees from the National University of Singapore (NUS) in 2001 and 2005, respectively. He worked as a JSPS Fellow at the interdisciplinary Graduate School of Engineering Sciences of Kyushu University, Japan. At present, he is working at the School of Mechanical and Aerospace Engineering, Nanyang Technological University (NTU), Singapore, as an Assistant Professor. His research interests focus on micro/nanoscale transport phenomena, thin-film thermoelectric device; adsorption thermodynamics, adsorption cooling, gas storage, and desalination; and CO_2-based cooling system. At present, Dr. Chakraborty has published about 100 articles in peer-reviewed journals and international conference proceedings and holds six patents.

Dr. Ali El Nashar is a mechanical engineer with specialization in the fields of energy and desalination and with a special interest in solar desalination and power generation. He received his Ph.D. degree in nuclear engineering from the Queen Mary College, University of London, UK, in 1968. His work experience covers applied research and development work at both academic and industrial institutions. He has been involved in teaching and research at several academic institutions in Egypt, UK, and USA, among them are the University of Alexandria and University of Mansoura in Egypt; the Queen Mary College, London University and Lanchester Polytechnic in the United Kingdom; and the Clemson University and Florida Institute of Technology in the United States. He has worked as the manager of cogeneration and desalination department at the Abu Dhabi Water & Electricity Authority (ADWEA) from 1982 to 2002, where his department participated in the commissioning of new desalination and power plants as well as monitoring the performance of existing plants operated by the ADWEA. He was also in charge of the solar desalination research program in the ADWEA during this period, where he supervised the installation, commissioning, and testing of the solar desalination demonstration plant in Umm Al Nar, which was designed and operated as a part of a joint research program with Japan's New Energy Development Organization (NEDO). He has been a member of several professional organizations, including the ASME, IDA, and ISE, and the editor of the IDA, Energy, and ISE. He has consulted for a number of international organizations, including the United Nations Environmental Program (UNEP); Arab Agency for Industrial Development

(AAID); Technology International, Inc. (USA); CH2M-Hill, Inc. (USA); Science Applications, Inc. (USA); Dow Chemical Europe (Switzerland); and Industrial Center for Water & Energy Systems (ICWES), Abu Dhabi. Dr. El-Nashar has more than 50 published papers, reports, and book chapters in his field of interest.

Dr. Yehia El-Sayed 1928–2010. Yehia El-Sayed was born in Alexandria, Egypt, on September 13, 1928. He received his bachelor's degree from Alexandria University and his doctorate in Mechanical Engineering from Manchester University in England. He taught and conducted research at Assiut University (Egypt), Kansas State University, Dartmouth College, Glasgow University (Scotland), Tripoli University (Libya), and the Massachusetts Institute of Technology. His legacy persists in the thousands of students and colleagues whose careers and intellectual development he has influenced. He was a recognized international authority in desalination, thermodynamics, and thermoeconomics. He authored two books and numerous scientific papers. A Life Fellow of the American Society of Mechanical Engineering, he was a two-time recipient of ASME's prestigious Edward F. Obert Award, in addition to a Best Paper Award from the International Desalination Association. Dr. El-Sayed's contributions brought the fundamentals of science to usefulness in engineering practice across the spectrum of energy conversions systems, providing principles for optimizing their technical and economic efficiency.

Editor-in-Chief's note: Yehia El-Sayed submitted his chapter but regrettably passed away before the publication of this book. His wisdom, kindness, and friendship will be missed by the desalination and thermodynamics scientific communities, including me.

Prof. Maria D Kennedy Ph.D., is the Professor of Water Treatment Technologies at UNESCO-IHE. She is a board member of the European Desalination Society. She has 18 years of research experience and currently specializes in research and development in the field of membrane technology. Her research areas of interest include membrane fouling (indices), scaling and cleaning, and modeling of membrane systems. She has been involved in international training projects in Israel (West Bank), Jordan, Oman, St. Maarten, and Yemen in the field of desalination and water reuse.

Dr. Atsuo Kumano is a professional engineer Japan, and in charge of technical matters in the Desalination Membrane Department in Toyobo Co., Ltd. Dr. Kumano's research and development focus on membrane technology and its engineering, for water treatment membranes such as reverse osmosis membrane for seawater desalination, and wastewater treatment including hollow fibre configuration module analysis.

Dr. Kumano holds a Ph.D. in Chemical Science and Engineering from Kobe University, Japan, 2011; an M.S. in Environmental Engineering from Osaka University, Japan, 1983; and a B.S. in Environmental Engineering from Osaka University, Japan, 1981.

Dr. Sabine Lattemann is a part-time Research Scientist at the Water Desalination and Reuse Center (WDRC) of the King Abdullah University of Science and Technology (KAUST) in the Kingdom of Saudi Arabia.

Sabine has over 10 years of experience in environmental impact assessment (EIA) studies. Her main areas of interest include the desalination of seawater, offshore wind energy development projects, and maritime shipping impacts. From 2007 to 2010, Sabine worked on the topic of environmental impacts and life cycle assessment of seawater desalination plants within the European research project "MEDINA." From 2004 to 2007, she chaired the environmental working group of the World Health Organization Project "Desalination for safe water supply."

Sabine holds a Postgraduate Diploma in Marine Science from Otago University (New Zealand), an M.Sc. in Marine Environmental Science from the University of Oldenburg (Germany), and a doctorate degree from the UNESCO-IHE Institute for Water Education and Delft University of Technology (The Netherlands).

Dr. Noam Lior is a Professor of Mechanical Engineering and Applied Mechanics at the University of Pennsylvania, where he is also a member of the Graduate Group of International Studies, Lauder Institute of Management and International Studies (MA/MBA program); of the Institute for Environmental Science; and of the Initiative for Global Environmental Leadership (IGEL) at the Wharton Business School. He did his Ph.D. work on water desalination at the Seawater Conversion Laboratory of the University of California, Berkeley, and thus started active research, teaching, and consulting in this field in 1966. His editorships include the following.

Editor-in-Chief:

Advances in Water Desalination book series, John Wiley, since 2006.
Energy, The International Journal, 1998–2009.

Board of Editors Member:

Desalination, The International Journal of Desalting and Water Purification, since 1988;

Energy Conversion and Management Journal, since 1994;

Desalination and Water Treatment—Science and Engineering, journal, Desalination Publications–International Science Services, since 2008;

Frontiers of Energy and Power Engineering, Springer, since 2008;

The Energy Bulletin, an international quarterly published by the International Sustainable Energy Development Center (ISEDC, under UNESCO auspices), Moscow, Russian Federation, since 2011;

Thermal Science and Engineering Journal (Japan), 1999–2008;

The ASME Journal of Solar Energy Engineering, 1983–1989;

The International Desalination & Water Reuse Quarterly, 1997–2003.

He has more than 350 technical publications, many of which are in the energy and desalination fields, and is the editor of the book *Measurements and Control in Water Desalination* (Elsevier, 1986).

Kim Choon Ng is working as a Professor at the Mechanical Engineering Department of the National University of Singapore. Professor Kim Choon specializes in the design of thermally driven adsorption cycles for desalination and cooling, with the objective of achieving a specific energy consumption of less than 1.5 kWh per cubic meter. The newly patented cycle of AD + MED desalination plant has the highest water production rates to date, producing potable water from either seawater or brackish-water using only low temperature waste heat. The novelties of the AD + MED cycle are that (i) it can operate with MED stages at temperatures below the ambient conditions with seven to nine stages, (ii) it has almost no major moving parts, (iii) it has minimal fouling because the temperature of heat source is from 50 to 80 °C, and (iv) it is environmental friendly. In addition, he employs the highly efficient ozone microbubble systems for the pretreatment of the feed water. His main research interests are adsorption thermodynamics, adsorption desalination and cooling, and microbubble treatment of wastewater with ozone. He has published more than 250 articles in peer-reviewed journals and international conference proceedings. He has edited three books and holds 10 patents.

Bidyut Baran Saha obtained his B.Sc. (Hons.) and M.Sc. degrees from Dhaka University of Bangladesh in 1987 and 1990, respectively. He received his Ph.D. in 1997 from Tokyo University of Agriculture and Technology, Japan. He worked as an Associate Professor at the interdisciplinary Graduate School of Engineering Sciences of Kyushu University until 2008. He worked as a Senior Research Fellow at the Mechanical Engineering Department of the National University of Singapore before joining the Mechanical Engineering Department of Kyushu University in 2010 as a Professor. He also holds a Professor position at the Thermophysical Properties Division of the International Institute for Carbon-Neutral Energy Research (WPI-I^2CNER), Kyushu University. His main research interests are thermally powered sorption systems, adsorption desalination, heat transfer enhancement, and energy efficiency assessment. He has published more than 200 articles in peer-reviewed journals and international conference proceedings. He has edited three books and holds seven patents. Recently, he served as the Guest Editor for the *Heat Transfer Engineering* journal for a special issue on the Recent Developments of Adsorption Technologies for Energy Efficiency and Environmental Sustainability. He is also serving as the Managing Guest Editor of *Applied Thermal Engineering*. He worked as the General Chairman for the Innovative Materials for the Processes in Energy Systems (IMRES) for Fuel Cells, Heat Pumps and Sorption Systems, 2010 Singapore, and will organize IMPRES2013 at Fukuoka, Japan.

Sergio G Salinas Rodriguez Ph.D., M.Sc, is a Lecturer in Water Treatment Technology at UNESCO-IHE. He has six years of experience in research and in integrated membrane systems. He has performed research in the fields of fouling indices and organic matter characterization for seawater reverse osmosis systems.

Prof. Jan C Schippers Ph.D., M.Sc, is a member of the Water Supply Group at UNESCO-IHE and a professor at the Wageningen University. He has extensive professional experience in drinking and industrial water supply projects in Morocco, Qatar, Libya, Gabon, Cape Verde, Namibia, Uzbekistan, Chile, France, and many other countries. Prof. Schippers is the past president of the European Desalination Society and the chairman of scientific and program committees of numerous international conferences and workshops of the IWA and EDS.

Dr. Corrado Sommariva is a consultant of international reputation. He is presently the Managing Director of the ILF Consulting Engineers, Middle East, and the head of the worldwide desalination activities of the ILF. Dr. Sommariva has experience in both thermal, reverse osmosis and wastewater system and has served in all the major desalination developments in the Middle East in various roles. Dr. Sommariva has a Ph.D. in Chemical Engineering from Genoa University and a diploma in Management from Leicester University.

Dr. Sommariva has served in the IDA board in the past 12 years. He has served as the first VP in 2003–2004. Furthermore, Dr. Sommariva served in the European Desalination Society (EDS) board for the past 14 years and has served as the president in the year 2004–2005.

Within his main activities in IDA, Dr. Sommariva served as chairman of the affiliate committee and started the humanitarian outreach initiative that has culminated with the establishment of the humanitarian committee in the IDA.

Dr. Sommariva has been the Chairman of the WHO committee for the establishment of safe drinking water from desalination and the Technical Co-Chair of the IDA World Congress in Dubai. He is an honorary Professor at the Genoa and L'Aquila Universities, where he holds regular courses on desalination and water-reuse-related matters. Dr. Sommariva also holds regular courses with the IDA and the Bushnaq academy.

Dr. Sommariva has published over 50 papers on desalination covering leading edge research and economics and two books on desalination management and economics and project financing.

Starting from a very technical background, he has worked for the past 20 years on desalination in various roles. He joined the ILF in 2009 after working nine years with Mott MacDonald, where he has been leading the desalination and water treatment group as the Managing Director of Generation, Middle East.

Kyaw Thu received his Ph.D. from the National University of Singapore (NUS), Singapore, in 2010 and B.E. (Mechanical Engineering) from the Yangon Technological University (Y.T.U.), Myanmar, in 2004. At present, he is working as a Research Scientist at the Water Desalination and Reuse Center, King Abdullah University of Science and Technology (KAUST). His research areas include adsorption science, theoretical and experimental analysis of thermally activated adsorption and absorption cycles for cooling and desalination, heat and mass transfer and energy efficiency of HVAC systems, Combined Heat and Power (CHP) Cycles, and solar thermal engineering.

CHAPTER 1

Water Desalination Revisited in Changing Physical and Economic Environments

YEHIA M. EL-SAYED*

*Dr. El-Sayed has regrettably passed away prior to the publication of this chapter. Final proofreading and some updating were done by the Editor. A tribute to his life was published as Testimonial, Yehia M. El-Sayed, *Energy* **36** 2315 (2011).

Advances in Water Desalination, First Edition. Edited by Noam Lior.

1.1 INTRODUCTION

The topic of water desalination is revisited because of the negative impact of the rising oil price index on the economic environment and the adverse effects of the increasing carbon footprint on the physical environment. In this introductory chapter, these negative factors are discussed with respect to their impact on past and present desalination methods. The impact of these factors on the design and operation practices of desalination and energy-intensive systems in general is highlighted. The energy analysis methodologies developed during the last two decades, including the methodology discussed in the present study, are summarized. General references on the subject matter are listed in the Further Reading section at the end of this chapter.

The software mentioned in this chapter may be downloaded at http://booksupport.wiley.com.

1.1.1 Past and Present Desalination

Interest in water desalination began in the late 1950s and early 1960s when the price of oil was only $3 per barrel (bl). A number of desalting processes and systems were considered that sought to minimize the cost of water production. For seawater, the leading methods were multistage flash distillation, vapor compression and freezing. Other processes, such as electrodialysis and reverse osmosis, lagged somewhat behind. Balancing the cost of the resources utilized in fueling a system and the resources utilized in making its devices favored moderate efficiency devices. For example, multistage flash distillation (MSF) in a cogeneration system used a maximum temperature of around 190°F (~80°C) in 8–12 stages. Cost allocated to water was as low as $0.3/m^3$. Environmental constraints were virtually absent.

As the oil price index increased to $25/bl, the number of the stages of conventional MSF increased to about 20 and the cost allocated to water rose to about $1/m^3$. At the same time, the awareness and concern regarding increased CO_2 emissions also increased.

Present desalination methods are facing a continuing increase in oil prices and a continuing increase of CO_2 content in the air. This creates a serious concern to designers and operators of desalination plants, power plants, and energy-intensive plants in general. Innovative ideas, along with expanded R&D in certain directions, will be essential to boost prevailing technological advances to achieve higher-efficiency devices at lower cost.

Unfortunately, if the efficiencies of these devices are not high enough and their costs are not low enough, then promoting conservation may be necessary in order to reduce demand, followed by undesirable rationing.

1.1.2 The Emerged Concern

Early traditional approaches to the synthesis and design of energy-intensive systems relied on the intuition of experienced engineers and designers. Modest concern was given to fuel consumption, and no concern was given to the environment or to waste management.

The continuing rise in oil prices and the continuing increase in the carbon footprint did, indeed, create a concern. Today the concern is at its peak, fueled by an increase in world population looking for a higher standard of living.

The concern regarding the environment did rise to a global level and did pose a difficult challenge for the designers and operators of energy-intensive systems. Cost-effective fuel conservation became a focus of attention in the design and in the operation of these systems. The design aspects became a complex multidisciplinary process requiring specialized knowledge in each discipline. The operation aspects became more responsive to any missmanagement of energy, emissions, and waste disposal. Many research and development (R&D) projects emerged to target a new generation of energy systems to meet the challenge at both the producer end and the consumer end.

There was an increased demand for improved methods of system analysis to achieve lower cost and higher efficiency, to facilitate the work of system designers. The methods of improved energy analysis influenced the design and the manufacture of energy conversion devices. Devices are now designed for the system as a whole rather than being selected from lines of preexisting components. Manufacture models are developed for the devices to reduce overall cost. The low cost of "number crunching" has enhanced the development of energy-intensive analysis.

Almost all methods developed involve optimization and seek innovation through energy-intensive analysis. Common tools are modeling and computational algorithms. However, the tendency for models to involve assumptions and view the same system from different perspective has created variations in the quality and reliability of the developed models. It is, therefore, important that models be verified and also that both designers and operators be aware of the purpose of each model and its limitations.

1.1.3 The Emerged Energy Analysis Methodologies

The interaction between cost and efficiency has always been recognized qualitatively. However, the interest in formulating the interaction was first highlighted in connection with seawater distillation in the 1960s to gain insight into the interaction between the surface of separation requirement and energy requirement. The first landmark of the work on thermoeconomics [1] dealt with seawater desalination processes. Further development followed in 1970 [2,3]. Professor Tribus coined the word *thermoeconomics*. Professor Gaggioli [4,5] generated interest in extending the development to all kinds of energy-intensive systems.

Since then the interest spread nationally and internationally by a large number of investigators, and the development is still continuing. Various schools of thought regarding optimal system design have evolved in the last 30 years with the following common objectives:

- Increasing the ability to pinpoint and quantify energy inefficiencies.
- Providing further insight into possible improvements in system design and operation.
- Automation of certain aspects of the search for improvement.

Investigators differ with respect to the techniques of managing system complexity. Four techniques may be identified, all of which allow changes in system structure directly or indirectly:

- Construct an internal system economy as a system decomposition strategy. Most of the work by these techniques falls under the heading of either *thermoeconomics* or *exergoeconomics* [6–8].
- Consider a composite heat exchange profile of all heat exchange processes to identify where to add or reject heat and to produce and/or supply work appropriately. All work performed using this technique is termed "pinch technology" [9].
- Let the computer automate the analysis by supplying it with a large database of devices and their characteristics. All the work performed using this technique is classified as *expert systems* or *artificial intelligence* [10].
- Consider evolutionary techniques based on the survival-of-the-fittest theory [11,12] to identify the desired system.

The author recommended the references listed in Sections 1.F.1–1.F.7 at the end of this chapter as useful readings for the preceding material.

1.2 THE METHODOLOGY USED IN THIS STUDY

The methodology discussed in this chapter, termed *thermoeconomics*, begins with simple thermodynamic computations of a given system configuration on a trajectory leading to an optimal design via multidisciplinary computations involving the disciplines of design, manufacture, and economics, in addition to thermodynamics.

In a typical thermodynamic model, the cost factor is absent. Decision variables are mainly efficiency parameters of the processes involved, along with a few parameters such as pressure, temperature, and composition. The computations target fuel consumption, overall system efficiency, and duty parameters of the system devices. Evaluating cost involves input resources from the disciplines of design and manufacture in a prevailing economic environment. This, in turn, requires formulated communications among the participating disciplines.

Thermoeconomic analysis targets minimized production costs and is based on three main principles:

- Improved thermodynamic analysis, through the concept of exergy, to add transparency to the distribution of lost work (exergy destructions) throughout a system configuration.
- Improved costing analysis, by quantifying the manufacturing and operating costs of the devices of a system, to add transparency to the interaction between cost and efficiency.
- Enhanced optimization, via reasonable simplifying assumptions, to reach improved design points for alternative and evolving system configurations.

1.2.1 Improved Thermodynamic Analysis

Improved thermodynamic analysis extends the conventional thermodynamic computations to include the second law of thermodynamics *quantitatively* rather than *qualitatively*. The extended computations are simply entropy balance computations in addition to property computations and the conventional mass, energy, and momentum balances. Entropy is conserved in an ideal process and is created in a real process. The ideal adiabatic work of a compressor or a turbine (isentropic), for example, is obtained when the entropy remains constant. Actual adiabatic work is associated with entropy creation. The adiabatic efficiency relates the actual work to the ideal. The process inefficiency (irreversibility) measured as a lost work potential $= T_0\,S^{\mathrm{c}}$, where T_0 is an ultimate sink temperature.

The main advantage of extended computations is that they enable assignment of fuel consumption to each process in a system. *Fuel* here means the input energy resource often applied at one location within the system boundaries. The energy resource may be fossil fuel, power, heat, solar, wind, or any other driving resource.

Thus, the manner in which a fuel is utilized throughout a system is revealed. Processes of high fuel consumption are identified. Means of fuel saving are inspired by a structural change of the system or/and by a design point change. New avenues of research and development are discovered.

It is important to note that engineers previously did not recognize the need to perform entropy balances. They could perform the thermodynamic analysis using property computations, and efficiency-related variables of a process such as pressure or heat loss, adiabatic efficiency, and heat exchange effectiveness. They missed the advantage of the distribution of fuel consumption throughout a given system.

A more complete picture of efficiencies and inefficiencies is obtained by using a general potential work function known as *exergy*. For simple chemical systems, this represents the maximum useful work relative to a dead-state environment defined by pressure P_0, temperature T_0, and composition $\{X_{c0}\}$. Exergy also represents the minimum amount of work needed to create the system from the dead-state environment.

1.2.1.1 The Exergy Function The exergy function is a general potential work function for simple chemical systems. The function evolved from the work of Carnot and Clausius, and is due to Gibbs [13]. The function is expressed as follows:

$$E^{s} = U + P_0V - T_0S - \sum \mu_{c0}N_c \tag{1.1}$$

Here, E^{s} is the maximum work that could be obtained from a sample of matter of energy U, volume V, number of moles (or mass) of each matter species N_c when the sample of matter is allowed to come to equilibrium with an environment of pressure P_0, temperature T_0, and chemical potential μ_{c0} for each species N_c. The same expression measures the least work required to create such a sample of matter from same environment. A form useful to second-law computations for systems in the steady state is

$$E^{f} = H - T_0S - \sum \mu_{c0}N_i \tag{1.2a}$$

where E^{f} is flow exergy. For convenience, it is often expressed as the sum of two changes: (1) a change under constant composition $\{X_c\}$ from the state at P and T to a state at a reference point between P_0 and T_0 and (2) a change under constant P_0 and T_0 from composition $\{X_c\}$ to a state at reference $\{X_{c0}\}$. The state at $P_0 + T_0 + \{X_{c0}\}$ defines the reference dead-state environment for computing exergy

$$E^{f} = (H - H^{0}) - T_0(S - S^{\circ}) + \sum(\mu_c - \mu_{c0})N_c \tag{1.2b}$$

where $(H^{0} - T_0S^{\circ})_{P_0,T_0,X_c} = (\sum \mu_cN_c)_{P_0 \cdot T_0}$ is used.

All special forms of potential workfunctions such as Carnot work, Keenam's availability, Helmholz free energy, and Gibbs free energy are obtainable from ideal interaction between a simple chemical system and large dead state environment using mass, energy and entropy balances as given by El-Sayed [14].

Section 1.A.3 (in the end-of-chapter Appendix) gives some useful forms of flow exergy in terms of measurable parameters and discusses the selection of the dead-state environment(s). Two or more dead-state environments may be used whenever there is no interest in their relative work potential. A known equilibrium chemical reaction may be introduced to establish the equivalent equilibrium composition of a missing species in a selected dead state environment.

1.2.2 Improved Costing Analysis

Most engineering activities seek the extreme of an objective function, which is usually a multicriterion function. Some criteria can be quantified in terms of monetary values such as fuel, equipment, and maintenance costs. Others involve nonunique assumptions regarding quantification of economic factors such as environmental impact, reliability, safety, and public health. In the design phase of an energy system, however, concern peaks around two criteria—*fuel* and *equipment*—without violating other desired criteria. A closer look at the interaction between fuel and equipment (products of specified materials and shapes) now follows to establish an improved costing analysis along with the improved thermodynamic analysis—in other words, to establish a thermoeconomic analysis.

Even when the objective function focuses on fuel and equipment only as costs, the analysis becomes multidisciplinary in nature. At least four disciplines of knowledge participate in information exchange: thermodynamics, design, manufacture, and economics. A communication protocol has to be established among the participating disciplines to provide cost with a rational basis.

Unfortunately, bidding information and some engineering practices for estimating the capital costs of major energy conversion devices are not helpful in the improvement of system design. The estimations are often oversimplified by a duty parameter for a group of devices such as a simple gas turbine unit costs of $500/kW. Such costs are not responsive to efficiency changes. The obvious way to recover missed information is to communicate with designers and manufacturers or to apply their practices encoded by suitable mathematical models.

1.2.2.1 The Quantification of the Manufacturing and Operation Resources for a Device Any energy conversion device requires two resources: those needed to manufacture it, R_{manuf}, and those needed to operate it R_{operate}. These two resources increase with the device duty (capacity and pressure–temperature severity) and are in conflict with the device performing efficiency (one or more efficiency parameters). Since both resources are expensive, their minimum sum is sought.

1.2.2.1.1 The Manufacturing Resources The leading manufacturing activities are materials, R&D, design, and construction. Exergy destruction associated with the performed activities of these activities are difficult to trace back or evaluate. The capital cost of a device **Z** in monetary units is an indicator of the performed activities, if not the best indicator. The capital cost, in turn, may be expressed by one or more characterizing parameters and their unit-dimensional costs:

$$Z = \Sigma c_{ai} A_i + k \tag{1.3}$$

Usually one characterizing surface A_i of unit surface cost c_{ai} is an adequate quantification of Z. A_i is evaluated by an updated design model. The unit cost c_{ai} is a manufacturing cost evaluated by an updated manufacture model. The rate of the manufacturing resources then becomes

$$R_{\text{manuf}} = \mathbf{Z} = c_z c_a (V_{\text{manuf}}) A(V_{\text{design}}) \tag{1.4}$$

where **Z** is the capital cost rate and c_z is the capital recovery rate.

1.2.2.1.2 Operating Resources The primary operation resources are related to fueling and other maintenance materials and activities. The *fueling resource* is what the device pulls or draws from the fueling supply point. In other words, it is simply the *exergy destruction* performed by the device. Engineers, however, use efficiency parameters (pressure loss ratio, adiabatic efficiency, effectiveness, etc.) to account for exergy destruction. All devices destroy exergy for their operation, depending on their performance efficiency. Only ideal devices (operating at 100% efficiency), which do not exist, have zero exergy destruction when performing their duties. The rates of operating resources that do not go to the products are directly quantified by the rates of exergy destruction. In monetary units, the operating resources can be expressed as

$$R_{\text{operate}} = c_d D(\{V_{\text{duty}}\}, \{V_{\text{efficiency}}\}) \tag{1.5}$$

where D is the rate of exergy destruction of a device depending on its duty and efficiency and c_d is the cost of its exergy destruction; c_d depends on the cost of the fuel feeding the system and on the position of the device within the system configuration. The objective function J_i of a device i to minimize at the device level is

$$\begin{aligned} J_i &= R_{\text{manuf}} + R_{\text{operate}} \\ &= c_{zi} c_{ai} (V_{\text{manufacture}}) A_i (V_{\text{design}}) + c_{di} D_i (\{V_{\text{duty}}\}, \{V_{\text{efficiency}}\}) \end{aligned} \tag{1.6}$$

1.2.2.2 Correlating the Manufacturing Resources of a Device in Terms of Thermodynamic Variables Communication between the thermodynamic and the design models makes it possible to express A_i as a minimized surface $A_{i\,\min}(\{V_{\text{duty}}\}, \{V_{\text{efficiency}}\})$, and communication between the design and the manufacture models allows one to express $c_{ai} = Z_{\min}(V_{\text{manuf}})/A(V_{\text{design}})$ as a minimized unit surface price $c_{a\,\min}(\{V_{\text{duty}}\}, \{V_{\text{efficiency}}\})$.

State-of-the-art or updated design and manufacture models are sought for major system devices. A conventional thermodynamic model delivers to each device its respective $\{V_{\text{duty}}\}, \{V_{\text{efficiency}}\}$ obtained from one feasible system solution. The design model of the device minimizes the characterizing surface of the device by adjusting the design dimensions of the design model that represent its design degrees of freedom. The minimized surface A_{min} is sent to the manufacture model to minimize the manufacturing cost of the device design blueprint by adjusting the decision variables of the manufacture model, which represent its manufacturing degrees of freedom. The minimized unit surface cost $c_{\text{a min}}$ is the minimized manufacturing cost/A_{min}.

This process is repeated over a range of feasible system solutions of interest to optimal system design. A matrix of rows representing feasible system solutions as related to a device and of columns representing thermodynamic duty and efficiency variables, design decision variables, and manufacture decision variables allows the manufacturing cost of a device in terms of design and manufacturing variables to be correlated in terms of thermodynamic variables.

A device objective function in terms of thermodynamic variables can be expressed as

$$\begin{aligned} J_i &= R_{\text{manuf}} + R_{\text{operate}} \\ &= c_z c_{\text{a}i\ \text{min}}(\{V_{\text{duty}}\}, \{V_{\text{efficiency}}\}) A_{i\ \text{min}}(\{V_{\text{duty}}\}, \{V_{\text{efficiency}}\}) \\ &\quad + c_{\text{d}i} D_i(\{V_{\text{duty}}\}, \{V_{\text{efficiency}}\}) \end{aligned} \tag{1.7}$$

where $c_{\text{a}i\text{mn}}, A_{i\ \text{min}}$, and D_i are all functions of $\{V_{\text{duty}}\}$ and $\{V_{\text{efficiency}}\}$, tending, in general, to increase with duty, and are at conflict with efficiency.

Communication between the system thermodynamic model and the design models of its devices has been applied to a fair number of any conversion devices as given in Section 1.A.1. An example of such communication for forced-convection heat exchangers, in which the manufacturing cost of a heat exchanger is expressed in terms of thermodynamic variables, is given in Section 1.A.2.

However, the communication between design and manufacture is still lagging. The unit surface manufacture cost is derived, at the moment, from published cost information rather than by manufacturing models. The communication between design and manufacture models of devices is still being formulated.

1.2.3 Enhanced Optimization

1.2.3.1 Two Simplifying Assumptions The optimization of an energy system configuration is most expedient when the system devices are optimized one by one with respect to the decision variables of the system. Improved thermodynamic and costing analyses have two basic features that qualify a system for device-by-device optimization:

- The assignment of fuel consumption to each device of the system establishes the operating costs of the system devices.

- Most of the decision variables are efficiency parameters whose major impact is on the local manufacturing costs of their respective devices.

Two simplifying assumptions are introduced to allow device-by-device optimization with respect to efficiency decisions as explained in the following paragraphs:

- An average exergy destruction cost applies to all devices.
- Efficiency decisions are local to their devices followed by a correction for their effect on other devices.

1.2.3.2 The Conditions of Device-by-Device Optimization The objective function of a device is expressed in Equation (1.7). The objective function of a system configuration, in terms of $\{V_{\text{duty}}, V_{\text{efficiency}}\}$, given a sizing parameter for the production rate and having one fueling resource, is

$$\begin{aligned}\text{Minimize } J_s &= c_F F + \sum_{i=1}^{n} \underline{Z_T} + C_R \\ &= c_F F + \sum_{i=1}^{n} \underline{Z_i} + C_R \\ &= c_F F + \sum_{i=1}^{n} C_{zi} Z_i + C_R \\ &= c_F F(\{V_{duty}, V_{efficiency}\}) + \sum_{i=1}^{n} c_{zi} c_{ai} A_i(\{V_{duty}, V_{efficiency}\}) + C_R \end{aligned} \tag{1.8}$$

where F is fuel rate; $\mathbf{Z}_{\text{T}}$ the total capital cost recovery rate; $\underline{\mathbf{Z}_i}$ the capital cost recovery rate of a device; n, the number of devices; and Z_i, the capital cost of each device represented by one characterizing dimension A_i. C_{R} is a constant remainder cost as far as the system design is concerned. When a design becomes a project, C_{R} may become a variable with respect to other non-system-design decisions.

To express the cost objective function of a system [Eq. (1.8)] in terms of the functions of the manufacturing and operating resources of its devices [Eq. (1.7)], the following condition must apply to a device i after dropping the constant C_{R}:

$$\frac{\partial J_s}{\partial Y_j} = \frac{\partial J_i}{\partial Y_j} = 0 \tag{1.9}$$

where Y_j is a system decision variable, J_{s} is the objective function of the system, and J_i is that function of a device i in the system:

$$\begin{aligned}\frac{\partial J_{\text{s}}}{\partial Y_{\text{j}}} &= c_{\text{f}}\left(\frac{\partial E_{\text{F}}}{\partial D_i}\right)\left(\frac{\partial D_i}{\partial Y_j}\right) + \left(\frac{\partial Z_{\text{T}}}{\partial Z_i}\right)\left(\frac{\partial Z_i}{\partial Y_j}\right) = c_{\text{f}}\text{K}_{e\,\text{ji}}\frac{\partial D_i}{\partial Y_j} + K_{zji}\left(\frac{\partial Z_i}{\partial Y_j}\right) \\ &= \frac{\partial J_i}{\partial Y_j}\end{aligned} \tag{1.10}$$

where

$$c_F F = c_f E_F \tag{1.11a}$$

$$K_{eji} = \left(\frac{\partial E_F}{\partial D_i}\right) \text{by a small change in } Y_j \tag{1.11b}$$

$$K_{zji} = \left(\frac{\partial Z_T}{\partial Z_i}\right) \text{by a small change in } Y_j \tag{1.11c}$$

IF K_{eji} and K_{zji} are independent of Y_j or at least weak functions of Y_j, then Equation (1.9) gives the objective function of a device as follows:

$$J_i = c_f K_{eji} D_i + K_{zji} Z_i \tag{1.12a}$$

$$= c_{di} D_i + c_{zi} c_{ai} A_i \tag{1.12b}$$

Then $c_{di} = c_f K_{eji}$, and the capital cost rate is modified by K_{zji}.

The condition that a device can be self-optimized in conformity with the objective function of its system is that K_{eji} and K_{zji} can be treated as constants.

The major effects of most efficiency decision variables on their respective devices ($K_{eji} = K_{e\,ii}$), converging to the condition of Equation (1.9) with $K_{zii} = 1$. They are denoted as local Y_L. Few efficiency decisions have their major effect on more than one device such as heat exchange effectiveness of two heat exchangers in series. These are identified as global Y_G. Their values $\{K_{eji}$ and $K_{zji}\}$ will continue to change, leading to random fluctuations of the system objective function with no sign of convergence. A slower optimization routine, often gradient-based, has to be used for these few global decisions. Because most efficiency decision variables are designated as local, it is worthwhile to utilize the piecewise optimization of the system devices, to gain insight into possible improvements and to ensure rapid optimization.

1.2.3.3 *The Form of $A_{i\,\min}$ and D_i of a Device* A suitable form to express $A_{i\,\min}$ and D_i in terms $\{V_{i\,\text{duty}}\}$ and $\{V_{i\,\text{efficiency}}\}$, particularly for optimization, is a form extracted from geometric programming:

$$A_{i\,\min} = k_a \prod_{j=1}^{n} (V_{i\,duty})_j^{d_a} (V_{i\,\text{efficiency}})_j^{e_a} \tag{1.13}$$

$$D_i = k_d \prod_{j=1}^{n} (V_{i\,\text{duty}})_j^{d_d} (V_{i\,efficiency})_j^{e_d} \tag{1.14}$$

where k_a and k_d are constants; n is the number of correlating variables, and d_a, e_a, d_d, and e_d are exponents. For the local decisions

$$J_i = c_f K_{ei} D_i(Y_{Li}) + K_{zi} c_{zi} c_{ai} A_i(Y_{Li}) \tag{1.15}$$

where the exergy destruction price $c_{di} = c_f K_{ei}$ and $K_{ei} = \delta E_F/\delta D_i$ through a change δY_{Li} and is always a positive quantity. K_{ei} converges to a constant, and K^{zi} converges to 1.

Equation (1.15) boils down, as far as the optimization of Y_{Li} is concerned, to a generalized form of a Kelvin optimality equation:

$$J_i = k_e Y_{Li}{}^{ne} + k_z Y_{Li}{}^{nz} \tag{1.16}$$

where k_e and k_z are lumped energy and the capital factors, considered weak functions of Y_{Li}, and n_e and n_z are exponents of opposite signs. The Kelvin optimality equation has the exponents 1 and -1. If k_e and k_z were precisely constants, then the optimum is reached in one system computation by the analytical solution

$$Y_{Li\,opt} = \left[\frac{-(k_z n_z)}{(k_e n_e)}\right]^{1/(n_e - n_z)} \tag{1.17}$$

1.2.3.4 Convergence to System Optimum The decisions idealized as local are not in complete isolation from the rest of the system. They influence the duties passed over from their devices, as mass rates, heat rates, or power, to other devices. The effect of these duties on cost within the range of system optimization is linear. To allow for this mild variation to adjust and converge to the system optimum, system computations are repeated using the analytical solutions of Equation (1.17) as an updating equation.

Substituting D_i and A_i for k_e and k_z, we obtain the updating equation for convergence:

$$Y_{Li\,new} = Y_{Li\,old}\left[\frac{(-n_m/n_e)(c_{zi} c_{ai} A_i)}{c_{di} D_i}\right]^{1/(n_e - n_m)} \tag{1.18}$$

Equation (1.18) happens to converge to a system's optimum in seconds (four to six iterations).

1.2.3.5 Optimization of System Devices by One Average Exergy Destruction Price According to Equation (1.15), each device i has its own exergy destruction price c_{di}. With K_{zi} converging to 1, we obtain

$$\sum c_{di} D_i = c_f E_f = c_f\left(\sum D_i + \sum D_j + \sum E_p\right) \tag{1.19}$$

where $\{E_p, E_f)$ are exergies of feeds and products, $\{D\}$ are exergy destruction by the devices, and $\{E_j\}$ exergy of wasted streams and c_f is fuel price per unit exergy. Then, introducting an average c_{da} such that

$$c_{da}\sum D_i = \sum c_{di} D_i = c_f E_f = c_f\left(\sum D_i + \sum D_j + \sum E_p\right)$$

we obtain

$$c_{da} = c_f(1 + \sum D_j / \sum D_i + \sum E_p / \sum D_i) \tag{1.20}$$

A slightly higher c_d than c_{da} often improves further the desired objective function.

1.2.3.6 Global Decision Variables Few decision variables belong to the system as a whole and are considered global. Operating pressure and temperature levels of a system are examples of global decisions. Occasionally a local decision such as a temperature difference has a global effect. Devices are not decomposed with respect to these decisions. A nonlinear programming algorithm may be invoked to solve for the optimum of these decisions simultaneously. If the range of variation of global decisions is narrow, manual search may be sufficient. For automated optimization, a simplified gradient-based method that ignores cross second derivatives may also be sufficient. This simplified method avoids singular matrices, which block solutions and often occur in systems of process-oriented description. It also converges, if guided to differentiate between a maximum and a minimum, as shown by the following updating equations for a global decision Y_G:

$$Y_{G\,\text{new}} = Y_{G\,\text{old}} \pm \Delta Y \tag{1.21a}$$

$$\Delta Y = \text{ABS}\left[\frac{\delta Y}{(g_2 - g_1)(-g_1)}\right] \tag{1.21b}$$

$$g_1 = \frac{(J_1 - J_0)}{\delta Y} \tag{1.21c}$$

$$g_2 = \frac{(J_2 - J_1)}{\delta Y} \tag{1.21d}$$

$$\delta Y = Y_{G1} - Y_{G0} = Y_{G2} - Y_{G1} \tag{1.21e}$$

The updating equation [Eq. (1.21)] requires three system computations to obtain three neighboring values of the objective function assuming, for example, Y_{G0}, $Y_{G0} + \delta Y$ and $Y_{GO} + 2\delta Y$ for each global decision. After $\{\Delta Y\}$ of the simultaneous solution has been obtained, the $\pm$ sign is then assigned to guide the change in the favored direction because zero gradient represents both maximum and minimum.

References listed in Section 1.F.8 at the end of this chapter are additional useful readings for the preceding Section 1.2.

1.3 THE SCOPE OF ANALYSIS

1.3.1 Desalination Related to Physical and Economic Environments

Desalted water is either coproduced with power production where the combined system is fossil-fuel-driven or self-produced, driven indirectly by fossil fuel by engines or by power from the grid. Most grid power is fossil-fuel-driven. The remaining grid power is driven by renewable sources of energy or by nuclear energy.

When desalted water is fossil-fuel-driven, two streams are to be dumped in the environment: an exhaust gas stream and a concentrated brine stream. When the exhaust is dumped in air CO_2 emission occurs. When concentrated brine is dumped back into the sea, marine life is damaged; and when dumped underground, the salinity of the underground water rises fast because of the limited amount of underground water. Dumping waste directly in the physical environment is the cheapest way to dispose of waste, but at the expense of the environment.

When desalted water is driven by solar, wind, or tidal energy, only the brine stream needs to be dumped. Exhaust gases are absent as well as CO_2 emission. Thus, in terms of CO_2 emission, renewable-energy-driven desalination systems are the most ecofriendly.[1]

For fossil-fuel-driven desalination systems, the higher the efficiency of the system, the lower the fuel burning and hence the CO_2 emission for the same produced product(s). This pattern continues until cost loses its competitiveness in the market as a limit to the reduction of CO_2 emission. The economic environment imposes the limit.

In view of the points discussed above, a number of desalination systems will be evaluated in terms of efficiency, cost, and CO_2 emission, assuming that direct dumping of concentrated brine is tolerated.

The avoidance of direct brine dumping will be treated by going to zero liquid discharge where more desalted water is obtained and solid salts can be safely transported isolated dumping locations. Predumping treatment is another option to safe dumping but is not considered in this study.

The idea of generating power by the concentration difference between concentrated brine and seawater will be investigated as a source of power though it does avoid the effect of direct dumping.

1.3.2 The Systems Considered

Systems with nine different configuration types, each intended for a specified purposes, are considered here. Four configuration types are fossil-fuel-driven burning natural gas, two are grid-power-driven, two are solar-driven, and one is concentrated-brine-driven. The purpose is to capture ideas that may help meet the challenges of diminishing fossil-fuel resources, increased CO_2 emissions, and hazardous-waste dumping.

The methodology of analysis is explained in Section 1.2. Accordingly, each system is described with respect to its working fluids and their thermodynamic properties and by its devices and their thermodynamic decision variables. The decision variables are used to solve mass balance, energy balance, and exergy

[1] More accurate assessment of energy and environmental impact should include calculation of embodied energy and emissions, i.e. the energy and emissions associated with the system construction. These might be rather high when renewable energy such as solar, wind, marine or osmotic is used, because of the relatively large quantity of hardware needed. These were not included in this chapter, which does not diminish the value of its conclusions, especially since embodied values are often small relative to operational ones. The Editor-in Chief.

balance equations leading to a feasible solution with the lower number of iterative loops. A characterizing surface of heat transfer, mass transfer, or momentum transfer is identified for each device. The cost of the device is rated per unit manufacturing cost of the characterizing surface. Decision variables are changed manually to minimize a cost objective function of the system.

The flow diagrams of the systems considered for their purposes are Figures 1.1–1.9. Figure 1.1 shows the gas turbine/multistage flash distillation (GT/MSF) cogeneration system with 100 MW power. Figure 1.2 shows the simple combined cycle (SCC) at 100 MW power, with compressor pressure 135 psia and firing temperature 1600°F. Figure 1.3 shows the vapor compression (VC) system of 10 migd (million imperial gallons per day) water. Figure 1.4 shows VC at the same capacity but with zero liquid discharge. Figure 1.5 depicts the reverse-osmosis (RO) system in one and two stages of 10 migd water. The two-stage system is a standby system in case one stage fails to deliver potable product water. Figure 1.6 shows an RO of the same capacity but for zero liquid

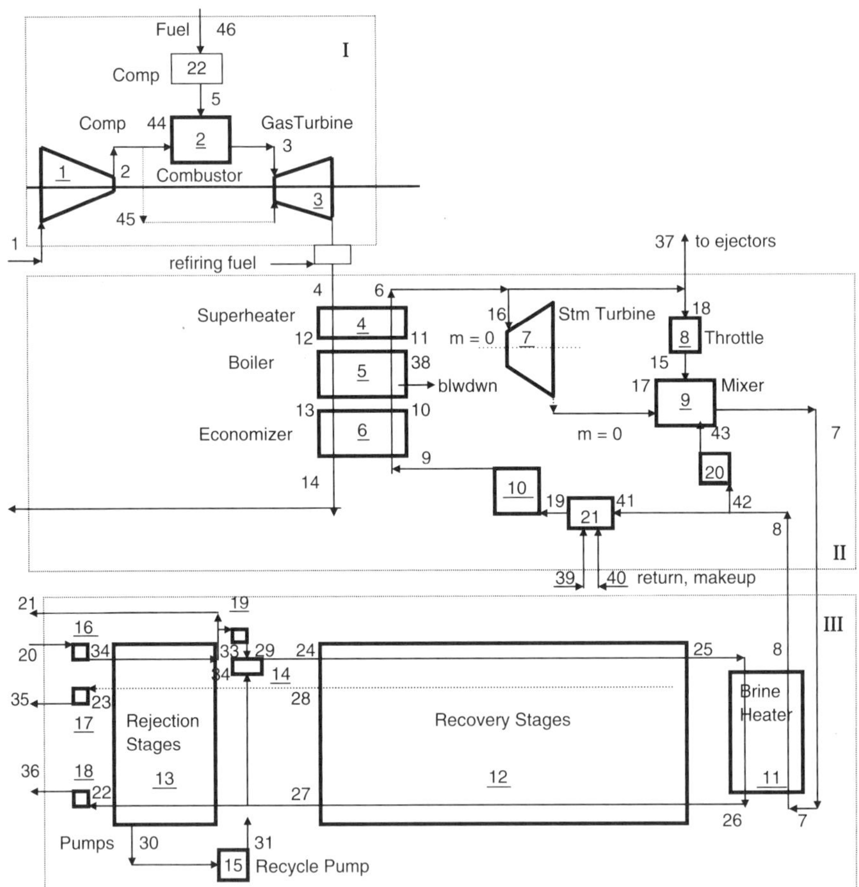

Figure 1.1 Gas turbine/multistage flash distillation cogeneration system.

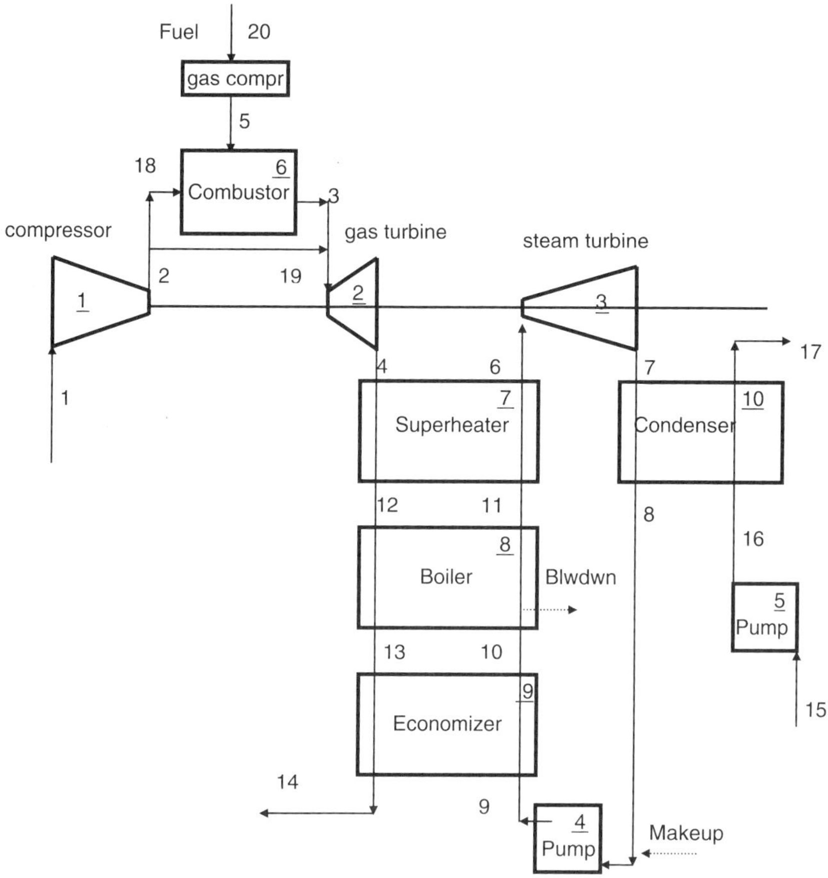

Figure 1.2 Simple combined cycle SCC.

discharge. Figure 1.7 shows a 0.2-usmgd solar photovoltaic/reverse-osmosis (PV/RO) system for small communities of about 1000 people. Figure 1.8 shows a 1-usmgd solar photovoltaic/electrodialysis (PV/ED) system for partial recovery of irrigation drainage. Figure 1.9 represents a concentrated-brine-driven system for power generation Delta-Xs-Power (osmosis power). A concentrated brine stream of 10 usmgd is assumed.

For the systems in Figures 1.1–1.6, the imperial gallon was used. For the systems in Figures 1.7–1.9, the us gallon was used. (The imperial gallon is 1.2 US gallons.)

For the systems in Figures 1.1–1.3, the optimization is automated for the efficiency decision variables. The design models of the devices of these systems have many design degrees of freedom to generate preformulated design-based costing equations for the devices. This, in turn, allows for automated computation of the minimized characterizing surfaces. Minimization of the cost objective functions the devices is enhanced given the unit surface costs and the unit exergy destruction costs of the devices.

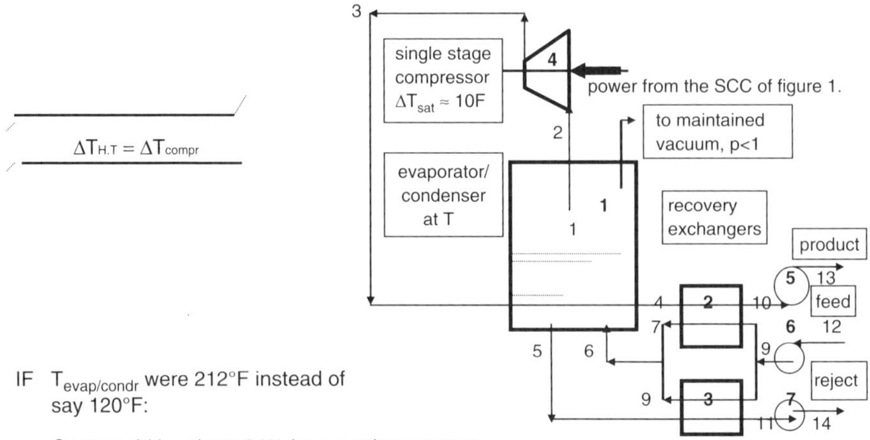

IF $T_{evap/condr}$ were 212°F instead of say 120°F:

- Cost would be about 14% less, setting a target cost 0.75–0.80 $/ton
  Compressor Capacity would increase 15 times.
- Venting substitutes vacuum.

Figure 1.3 Vapor compression distiller VC.

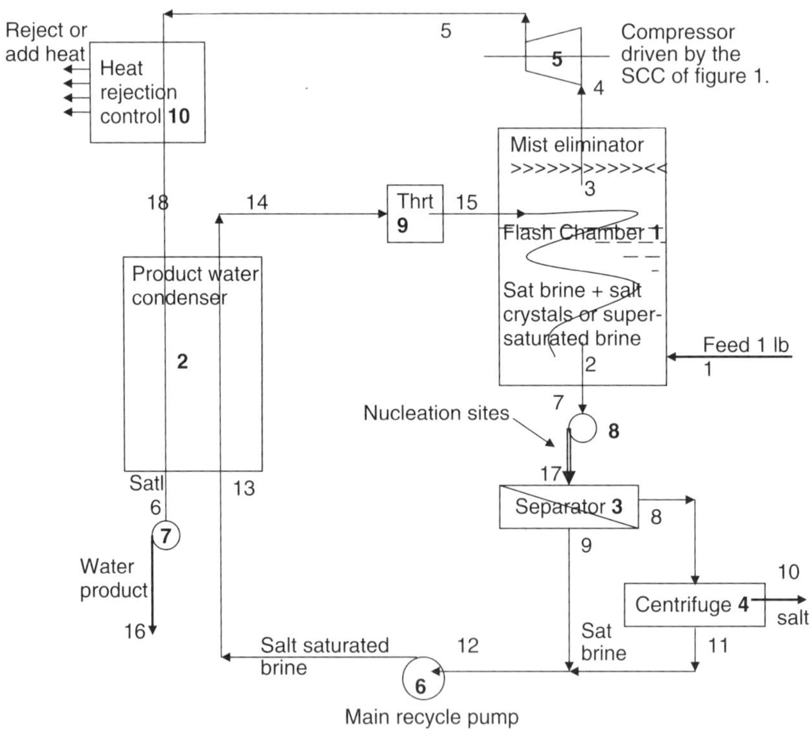

Figure 1.4 Vapor compression distillation system with zero liquid discharge.

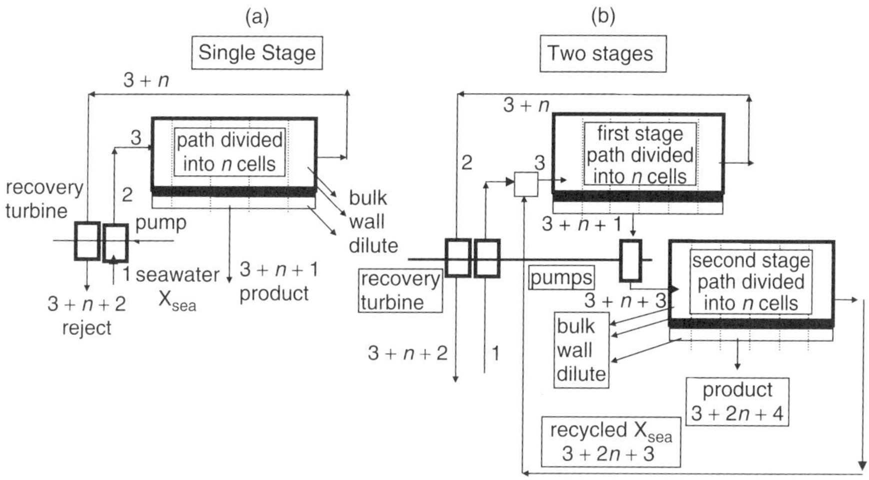

Figure 1.5 Reverse-osmosis desalting system RO: (a) single-stage; (b) two-stage.

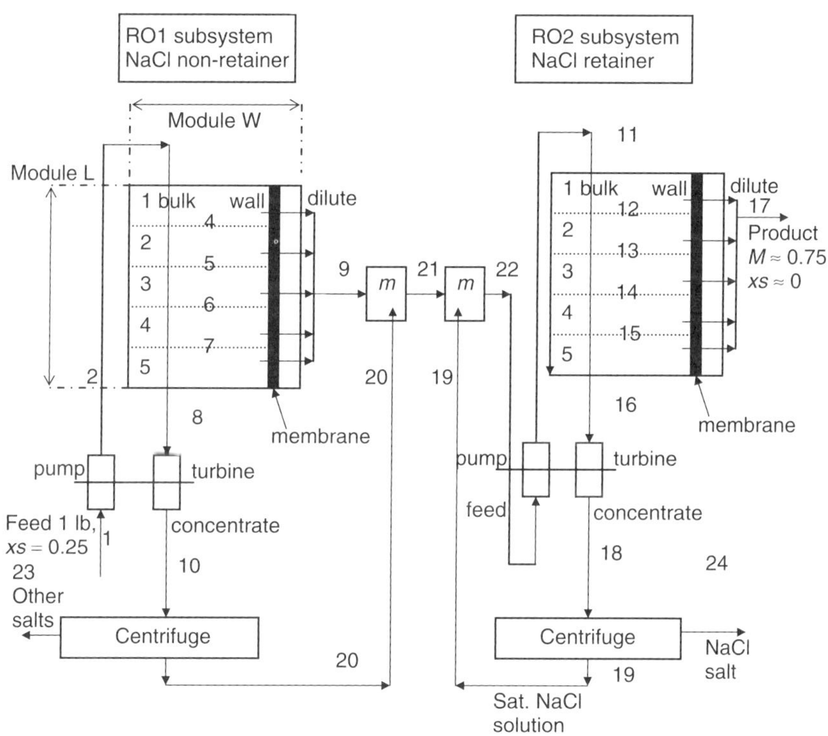

Figure 1.6 Reverse osmosis with zero liquid discharge.

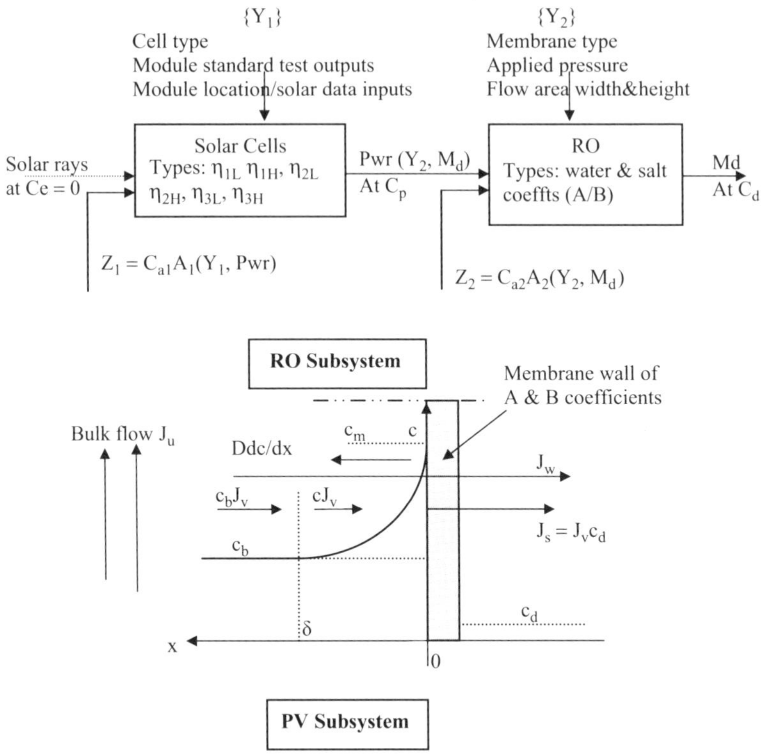

Type	Lower η°		Higher η°	
	η	\$/pW	η	\$/pW
1) Amorphous Silicon (a-Si:H):	6%	3.4	15%	1.7
2) Czochralski (Cz-Si):	12%	4.6	17%	3.4
3) Multi-junction, 500X concentrator	17%	2.9	22%	2.4

Figure 1.7 The Photovoltaic/Reverse osmosis desalination system PV/RO.

For the remaining systems the optimization is manual since there are not enough design options (design degrees of freedom), unlike those of the devices of systems 1–3.

Each system has two design points: a reference design point and an improved design point, by automated or manual optimization. Two economic environments are also considered. One represents an oil price index of $25/bl (barrel) and the other one represents $100/bl. Although it is difficult to predict the price structure under rising oil prices, a simple prediction is assumed. Fuel, power, and steam costs at a rate of 100/bl are set at 4 times those at $25/bl. Capital cost of devices and the cost products are set at a lower rate of 2 times.

The GT/MSF system is designed for two products (power and water) while being driven by a single fuel resource. Often, the decision variables of the system permit one product rate as a decision variable. A power of 100 MW is selected as

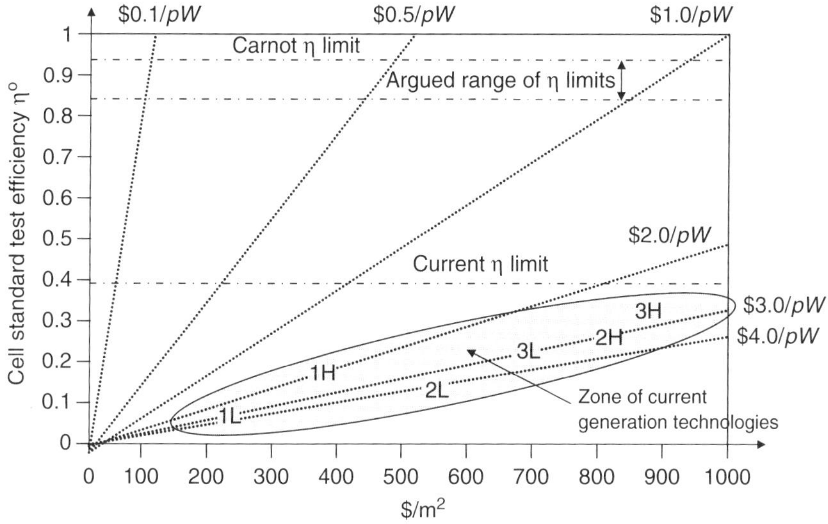

Figure 1.8 Current competitiveness of PV/RO system [18]. pW: peak watts.

the decision product rate. The objective of maximizing profitability is considered for this particular system, in which computed water product rates ranged from 8 to 10 migd.

For cogeneration systems such as the GT/MSF system, various assumptions have been proposed to allocate the production cost to water and power. The assumptions are all logical but the allocations differ significantly. The allocation assumed here uses the capital cost of each subsystem as belonging to that subsystem. The fuel is allocated in proportion to the exergy destructions and waste streams of each subsystem. This gives a lower bound to the cost of water since no devices or their exergy destructions are shared by the two subsystems. Higher water cost is obtained if the exergy destruction of combustion is shared.

For the GT/MSF and SCC systems a default power load profile that varies from 20 to 100 MW with a load factor of 0.583 is assumed. Both ideal and actual control features are considered. Ideal control assumes design efficiency at all load fractions (implying variable geometry devices). The actual control considered keeps the rate of airflow to the gas turbine compressor at the design value while increasing the air/fuel ratio. A quadratic equation for system efficiency as function of load fraction is assumed for design efficiency at maximum load and 20% efficiency at minimum load. Both GT/MSF and SCC systems are run without and with night products to evaluate the effects of improved load factor. Both systems considered include an RO subsystem for the night product. Two time periods are identified. The first lasts from midnight to 6 A.M. where a power of 80 ± 0.5 MW is available. The second lasts from 7 P.M. to 11 P.M. where a power of 40 MW ± 0.5 MW is available. For the SCC system, a water electrolysis subsystem producing H_2 and O_2 as night products is also considered.

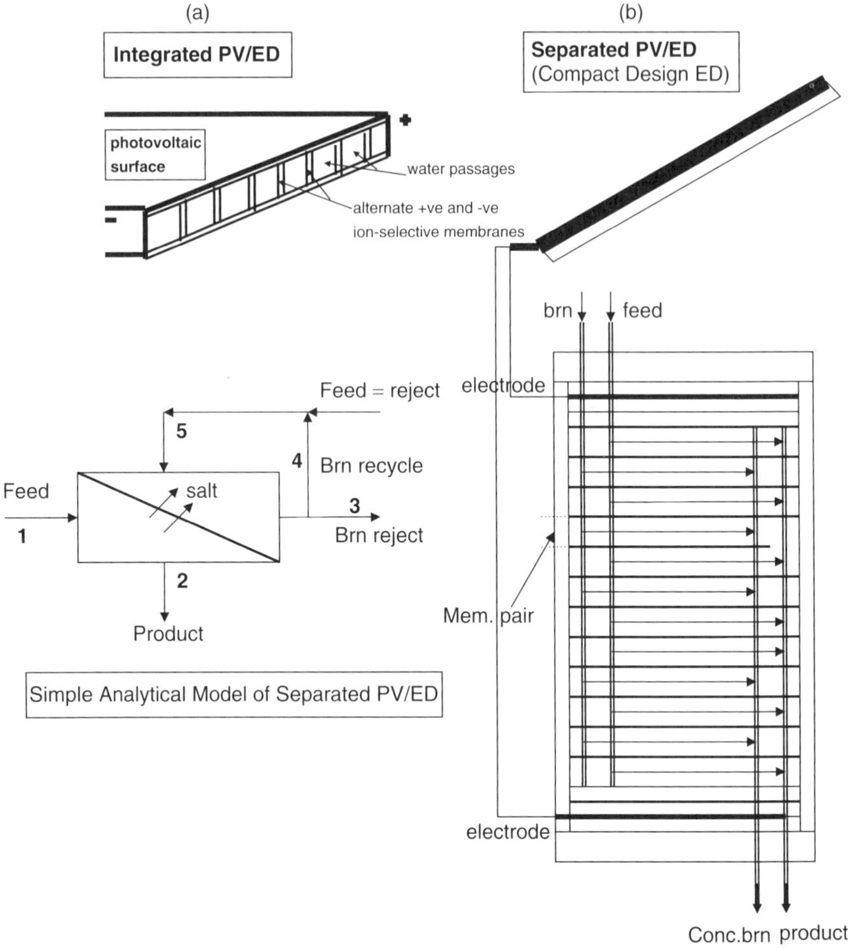

Figure 1.9 Photovoltaic/electrodialysis (PV/ED) system: (a) integrated PV/ED; (b) separated PV/ED (compact design ED).

For the SCC-driven VC and RO desalting systems, two types of rejected stream disposal are assumed: a conventional brine discharge and a salt discharge (zero liquid discharge).

For the PV/RO and PV/ED solar systems, a solar intensity profile at 30° north latitude is assumed.

For the Δ-X_{salt}-Power system (osmosis power), sodium chloride ideal solution is assumed. Salt content 0.04 is assumed for sea and ≤ 0.25 salt content is assumed for the driving brine.

Sample runs of the various systems considered are given in Tables 1.1–1.7. Table 1.1 lists data obtained for GT/MSF cogeneration runs; Table 1.2, SCC power runs; Table 1.3, SCC/VC runs; Table 1.4, SCC/RO runs; Table 1.5, PV/RO runs; Table 1.6, PV/ED runs; Table 1.7, osmosis power of Δ-X_{salt}.

Table 1.1 Gas Turbine/Multistage Flash Distillation Systems

	Reference System						Improved by Higher Profitability					
	Design Steady-Load Operation		Variable Load Operation No 2nd Product		Night RO Desalter Operation		Design Steady-Load Operation		Variable Load Operation No 2nd Product		Night RO Desalter Operation	
	Oil Index, $		Oil Index, $		Oil Index, $		Oil Index, $		Oil Index, $		Oil Index, $	
Parameter	25	100	25	100	25	100	25	100	25	100	25	100
Power, MW	100	100	58.3	58.3	86.5	86.5	100	100	58.3	58.3	99	99
Water, migd	8.04	8.04	8.04	8.04	8.04	8.04	10.1	9.25	10.1	9.25	10.1	9.25
Water, t/h	1,496	1,496	1,496	1,496	1,496	1,496	1,871	1,721	1,871	1,721	1,871	1,721
Fuel, MWt	350	350	224	224	308	308	338	306	216	194	297	266
To water	64.9	64.9	41.7	41.7	57.2	57.2	66.8	55	43	33	59	46
To power	285	285	183	183	251	251	271	251	173	161	238	221
1st law eff	0.301	0.301	0.266	0.266	0.326	0.326	0.313	0.334	0.276	0.303	0.338	0.375
2nd law eff	0.355	0.355	0.289	0.289	0.355	0.355	0.365	0.391	0.295	0.329	0.368	0.405
CO_2, t/h	64.1	64.1	63.8	63.8	64	64	62	56.1	61.6	55.8	61.8	56
To water	11.9	11.9	11.8	11.8	11.9	11.9	12	10	12.2	10	12.2	10
To power	52.2	52.2	51.9	51.9	52.1	52.1	50	46	49.4	45.8	49.6	46
lb/lb fuel	2.6	2.6	4.04	4.04	2.9	2.9	2.6	2.6	4.04	4.07	2.95	2.98
lb/kWhe	1.14	1.14	1.96	1.96	1.33	1.33	1.1	1.01	1.86	1.73	1.1	1.02
lb/tn water	17.5	17.5	17.4	17.4	17.5	17.5	14.1	12.8	14.3	12.8	14.3	12.8
Capital, $M	154	308	154	308	154+	308+	191	410	191	410	191+	410+
Capital, $/h	1,925	3,854	1,925	3,854	1,925+	3,854+	2,388	5,125	2,388	5,125	2,388+	5,125+
Fuel, $/h	3,497	13,987	2,196	8,785	3,013	12,125	3,379	12,247	2,115	7,694	2,924	10,620
Production, $/h	5,424	17,840	4,171	12,830	5,593	17,562	5,764	17,366	4,547	12,981	6,052	16,599

(Continued)

Table 1.1 (*Continued*)

	Reference System						Improved by Higher Profitability					
	Design Steady-Load Operation		Variable Load Operation No 2nd Product		Night RO Desalter Operation		Design Steady-Load Operation		Variable Load Operation No 2nd Product		Night RO Desalter Operation	
	Oil Index, $		Oil Index, $		Oil Index, $		Oil Index, $		Oil Index, $		Oil Index, $	
Parameter	25	100	25	100	25	100	25	100	25	100	25	100
Revenue, $/h	6,687	24,965	4,570	15,556	11,882	30,181	7,199	25,538	5,065	16,142	12,377	30,767
Profit, $/h	1,263	7,124	399	2,726	6,180	12,619	1,434	8,172	518	3,162	6,325	14,169
Design case allocation												
C_{wtr}, $/t	1.02	2.91	0.859	2.264	0.96	2.678	1.058	2.88	0.925	2.558	1.01	2.987
C_{pwr}, $/kWh	0.038	0.134	0.048	0.158	0.04	0.138	0.037	0.123	0.047	0.143	0.039	0.123
C_{da}, $/kWh	0.023	0.091					0.035	0.12				
Some decisions												
MSF stages	18	18	18	18	18	18	22	28	22	28	22	28
Pinch	50	50	50	50	50	50	9	11	9	11	9	11
Efficiency levels	0.85	0.85	9.85	0.85	0.85	0.85	0.9	0.92	0.9	0.92	0.9	0.92
Op_{pen} MW	—	—	26.1	26.1	8.04	8.04	—	—	25.1	22.8	7.7	7.05
$/h	—	—	260	1,042	80	320	—	—	251	913	77	281
Cap+, $/h	—	—	—	—	696	1,392	—	—	—	—	696	1,392
Prd+, t/h	—	—	—	—	5,625	5,625	—	—	—	—	5,625	5,625

Table 1.2 Combined Cycle Power Systems

Parameter	Reference System						Improved by Lower Production Cost					
	Design Steady-Load Operation		Variable Load Operation, No 2nd Product		Night Operation of RO and EL		Design Steady-Load Operation		Variable Load Operation, No 2nd Product		Night Operation of RO and EL	
	Oil Index, $		Oil Index, $		Oil Index, $		Oil Index, $		Oil Index, $		Oil Index, $	
	25	100	25	100	25	100	25	100	25	100	25	100
Power, MW	100	100	58.3	58.3	86.5	86.5	100	100	58.3	58.3	86.5	86.5
Fuel, MW	248.5	248.5	164.5	164.5	220.9	220.9	246.8	227.4	163.3	150.5	219.3	202.1
CO_2, t/h	45.5	45.5	28.9	28.9	39.7	39.7	45.2	41.7	28.7	26.5	39.4	36.3
lb/lb fuel	2.7	2.7	2.6	2.6	2.65	2.65	2.7	2.7	2.6	2.59	2.65	2.64
lb/kWhe	1.002	1.002	1.09	1.09	1.01	1.01	0.99	0.916	1.08	0.948	1.002	0.924
1ST law Efficiency	0.402	0.402	0.355	0.355	0.391	0.391		0.44	0.357	0.388	0.394	0.428
2nd law Efficiency	0.461	0.461	0.385	0.385	0.425	0.425		0.51	0.388	0.421	0.428	0.465
Capital, $M	92.3	184.7	92.3	184.7	92.3+	184.7+	82	189.4	82	189.4	82+	189.4+
Capital, $/h	1,155	2,310	1,155	2,310	1, 155+	2, 310+	1,032	2,369	1,023	2,369	1, 023+	2, 369+
Fuel, $/h	2,485	9,943	1,645	6,581	2,209	8,837	2,468	9,096	1,633	6,021	2,193	8,085
Production, $/h	3,641	12,253	2,800	8,891			3,500	11,465	2,656	8,390		
Revenue, $/h	4,500	20,000	2,625	11,667			4,500	20,000	2,625	11,667		
Profit, $/h	859	7,747	−175	2,776			1,000	8,535	−31	3,277		
Operation												
F_{pen} MW			19.5	19.5	6.03	6.03			19.3	17.8	5.98	5.5
F_{pen}, $/h			195	781	60	241			193	714	59	220
RO at night												
Production, $/h					4,060	12,542					3,922	11,150
Revenue, $/h					9,938	26,292					9,938	26,292

(*Continued*)

Table 1.2 ***(Continued)***

Parameter	Reference System						Improved by Lower Production Cost					
	Design Steady-Load Operation		Variable Load Operation, No 2nd Product		Night Operation of RO and EL		Design Steady-Load Operation		Variable Load Operation, No 2nd Product		Night Operation of RO and EL	
	Oil Index, $		Oil Index, $		Oil Index, $		Oil Index, $		Oil Index, $		Oil Index, $	
	25	100	25	100	25	100	25	100	25	100	25	100
Profit, $/h					5,877	13,752					6,016	15,142
NightCap, $/h					696	1,392					696	1,392
Production, t/h					5,625	5,625					5,625	5,625
C_{prd}, $/t					0.24	0.74					0.24	0.74
EL at night												
Production, $/h					3,404	11,277					3,266	10,533
Revenue, $/h					5,860	18,136					5,860	18,136
Profit, $/h					2,456	6,910					2,594	7,603
NightCap, $/h					39.8	79.6					39.8	79.6
P_{rd}, t/h					3.23	3.23					3.23	3.23
C_{prd}, $/t					228	889					228	889
Cp, $/kWh	0.0364	0.1225					0.035	0.115				
Cd, $/kWhe	0.021	0.091					0.03	0.12				
Some decisions												
Pinch	25						5	5				
Dth10	10						4	4				
Efficiency, compressor	0.84						0.87	0.91				
Efficiency, GT	0.9						0.89	0.92				
Efficiency, stream	0.85						0.82	0.87				
Efficiency, pump	0.7						0.92	0.94				

Table 1.3 Sample Combined Cycle/Vapor Compression (CC/VC) System

	Conventional Brine Discharge				Zero Liquid Discharge			
	Reference System		Improved by Lower Production Cost		Reference System		Improved by Lower Production Cost	
Parameter	Oil Price Index, $25/bl	Oil Price Index, $100/bl	Oil Price Index, 25$/bl	Oil Price Index, $100/bl	Oil Price Index, $25/bl	Oil Price Index, $100/bl	Oil Price Index, $25/bl	Oil Price Index, $100/bl
Water, migd	10	10	10	10	10	10	10	10
t/h	1,894	1,894	1,894	1,894	1,894	1,894	1,894	1,894
Power, MW	43.958	43.958	17.455	17.455	77.199	77.199	77.199	77.199
KWh/t	23.21	23.21	9.216	9.216	40.76	40.76	40.76	40.76
Fuel, MW	109.265	109.265	43.9956	39.704	191.895	191.895	190.526	175.563
CO_2, t/h	20.02	20.02	7.895	7.272	35.162	35.162	34.905	32.155
lb/lb fuel	2.7	2.7	2.7	2.7	2.7	2.7	2.7	2.7
lb/kWhe	1.002	1.002	0.995	0.916	1.002	1.002	0.994	0.961
lb/ton water	23.24	23.24	9.17	8.44	40.84	40.84	40.54	37.35
1st-law efficiency	0.4023	0.4023	0.405	0.4396	0.4023	0.4023	0.4052	0.4395
2nd-law efficiency	0.4611	0.4611	0.4617	0.5016	0.4611	0.4611	0.4619	0.5016
Water Side								
Capital, $M	39.4	78.8	104.7	209.5	112.7	225.5	112.7	255.5
Capital, $/h	492.6	985.3	1,309.7	2,619.4	1,410	2820	1,410	2,820
Power side								
Capital, $M	40.06	81.2	14.4	33.2	71.3	142.6	63.7	146.3
$/h	508.0	1,016	1,180.9	415.0	891.7	1,783.5	797.3	1,829
Fuel, $/h	1,092.6	4,370	510.8	1,588.16	1,919	7,675.8	1,905.3	7,022.5
Production C_{wtr}, $/h	2,093.2	6,371.8	1,921.5	4,622.6	4,220.5	12,279	4,112.3	11,671
Production C_{pwr}, $/h	1,600.6	5,386.5	611.8	2,003.1	2,810.6	9,459	2,702.5	8,851.8
C_{wtr} $/t	1.1	3.36	1.01	2.44	3.02	8.8	2.94	8.36

(Continued)

Table 1.3 ***(Continued)***

Parameter	Conventional Brine Discharge				Zero Liquid Discharge			
	Reference System		Improved by Lower Production Cost		Reference System		Improved by Lower Production Cost	
	Oil Price Index, $25/bl	Oil Price Index, $100/bl	Oil Price Index, 25$/bl	Oil Price Index, $100/bl	Oil Price Index, $25/bl	Oil Price Index, $100/bl	Oil Price Index, $25/bl	Oil Price Index, $100/bl
C_{pwr}, $/kWh	0.0364	0.1225	0.035	0.1147	0.0364	0.1225	0.035	0.1146
Sales								
C_{wtr}, $/t	1.3	2.6	1.3	2.6	1.3	2.6	1.3	2.6
C_{pwr}, $/kWh	0.045	0.20	0.045	0.2	0.45	0.2	0.45	0.2
C_d, $/kWhe	0.021	0.091	0.03	0.12	—	0.091	—	0.12
Leading Decisions								
	VC	CC	VC	CC	VC	CC	VC	CC
	Dtv = 12	Pinch = 25	Dtv = 4.5	Pinch = 5	Psctn = 0.7	Pinch = 25	Psctn = 0.7	Pinch = 5
	Effc = 0.7	Dt10 = 10	Effc = 0.89	Dtcon = 4	Pd = 1.9	Dt10 = 10	Pd = 1.9	Dtcon = 4
	Effp = 0.7	Effc = 0.84	Effp = 0.89	Effc = 0.87	Dtflash = 10	Effc = 0.84	Dtflash = 10	Effc = 0.915
		Effgt = 0.90		Effgt = 0.89	Dtsat = 10	Effgt = 0.90	Dtsat = 10	Effgt = 0.92
		Effstt = 0.85		Effstt = 0.82	Effc = 0.85	Effstt = 0.85	Effc = 0.85	Effstt = 0.87
		Effpmp = 0.7		Effpmp = 0.89		Effpmp = 0.7		Effpmp = 0.93

Notation: Effc, Effgt, Effp, Effsst—compressor, gas turbine, steam turbine efficiencies; Dtv, Dt10, Dtcon, Dtsat: respective temperature differeneces, °F; Pd, Psctn: respective pressures, psia.

Table 1.4 **Simple Combined Cycle/Reverse-Osmosis System**[a]

Parameter	Conventional Brine Discharge				Zero Liquid Discharge			
	Reference System		Improved by Lower Production Cost		Reference System		Improved by Lower Production Cost	
	Oil Price Index, $20/bl	Oil Price Index, $100/bl	Oil Price Index, 20$/bl	Oil Price Index, $100/bl	Oil Price Index, $20/bl	Oil Price Index, $100/bl	Oil Price Index, $20/bl	Oil Price Index, $100/bl
Water, migd	10	10	10	10	10	10	10	10
Water, t/h	1,894	1,894	1,894	1,894	1,894	1,894	1,894	1,894
Power, MW	12.174	12.174	5.613	5.613	110.46	110.46	110.46	110.46
Power, KWh/t	6.376	6 376	2.919	2.919	58.32	58.32	58.32	58.32
Fuel, MW	30.262	30.262	13.865	12.771	274.57	274.57	272.59	261.191
CO_2, t/h	5.545	5.545	2.54	2.339	50.3	50.3	49.9	46
lb/lb fuel	2.7	2.7	2.7	2.7	2.7	2.7	2.7	2.7
lb/kWhe	1.002	1.002	0.995	0.916	1.002	1.002	0.994	0.916
Ib/ton water	6.44	6.44	2.95	2.72	58.43	58.43	57.96	53.43
1st-law eff	0.4023	0.4023	0.4049	0.4395	0.4023	0.4023	0.4052	0.4397
2nd-law eff	0.4611	0.4611	0.4615	0.5015	0.4611	0.4611	0.4620	0.5018
Water Side								
Capital M$	6.28	13.06	12.834	26.55	0.506	1.012	0.506	1.012
$/h	229	465.1	470.2	951.5	18.98	38.0	18.98	38.0
Power Side								
Capital, $M	11.2	22.5	4.6	10.7	102	204	91.2	209.2
Capital, $/h	140.9	281.7	58.4	134.0	1,275.7	2,551	1,140	2,615.8
Fuel, $/h	302.6	1,210.5	138.6	510.8	2,745.7	10,983	2,726	10,048
$C_{prd\ wtr}$, $/h	672.9	1,957.3	667.2	1,596.2	4,040	13,572	3,885	12,701
$C_{prd\ pwr}$, $/h	443.3	1,492.2	197	644.8	4,021	13,634	3,866	12,663

(*Continued*)

Table 1.4 ***(Continued)***

Parameter	Conventional Brine Discharge				Zero Liquid Discharge			
	Reference System		Improved by Lower Production Cost		Reference System		Improved by Lower Production Cost	
	Oil Price Index, \$20/bl	Oil Price Index, \$100/bl	Oil Price Index, 20\$/bl	Oil Price Index, \$100/bl	Oil Price Index, \$20/bl	Oil Price Index, \$100/bl	Oil Price Index, \$20/bl	Oil Price Index, \$100/bl
C_{wtr}, \$/t	0.355	1.03	0.353	0.84	2.13	7.16	2.05	6.7
C_{pwr}, \$/kWh	0.0364	0.1225	0.0351	0.1148	0.0364	0.1225	0.0349	0.1146
Sales								
C_{wtr}, \$/t	1.3	2.6	1.3	2.6	1.3	2.6	1.3	2.6
C_{pwr}, \$/kWh	0.045	0.2	0.045	0.2	0.045	0.2	0.045	0.2
C_d, \$/kWhe	0.021	0.091	0.03	0.12				0.12
Leading Decisions								
	RO	CC	RO	CC	RO	CC	RO	CC
—	$P = 1{,}500$	Pinch = 25	$P = 1{,}000$	Pinch = 5	$P = 1{,}000$, 5,000	Pinch = 25	$P = 1{,}000$, 5,000	Pinch = 5
—	Effp/t = 0.75	Dt10 = 10	Effp/t = 0.88	Dtcon = 4	Effp/t = 0.9	Dt10 = 10	Effp/t = 0.9	Dtcon = 4
—	$A = 0.02$	Effc = 0.84	$A = 0.035$	Effc = 0.916	$A = 0.01$, 0.02	Effc = 0.84	$A = 0.01$, 0.02	Effc = 0.915
—	$B = 0.001$	Effgt = 0.90	$B = 0.0005$	Effgt = 0.92	$B = 0.05$, 0.0003	Effgt = 0.9	$B = 0.05$, 0.0003	Effgt = 0.92
—	$W = 40$	Effstt = 0.85	$W = 40$	Effstt = 0.87	$X_{s1} = 0.04$, 0.041,0.03	Effstt = 0.85	$X_{s1} = 0.04$, 0.041,0.03	Effstt = 0.87
—	$H = 0.05$	Effpm = 0.7	$H = 0.02$	Effpmp = 0.89	$X_{s2} = 0.26$, 0.271, 0.0005	Effpmp = 0.7	$X_{s2} = 0.26$, 0.271, 0.0005	Effpmp = 0.94

[a] See notations footnote in Table 1.3 and Nomenclature Section (at end of chapter) for abbreviation and symbol definitions.

Table 1.5 Reverse-Osmosis/Photovoltaic Systems

Parameter	Runs 1	2	3	4	5	6	7	8	9
Input									
A_{wtr}, lb/(h · ft^2 · psi)	0.04	0.04	0.04	0.04	0.04	0.04	0.01	0.002	0.01
B_{salt}, ft/h	0.0005	0.0005	0.0005	0.0005	0.0005	0.0005	0.0005	0.0003	0.0008
Top pressure, psia	900	900	900	900	900	900	1,250	1,800	600
H_{brn} flow, in.	0.06	0.06	0.06	0.01	0.06	0.01	0.05	0.015	0.015
PV cell type	1	2	3	1	1	3	1	1	1
Cell lab efficiency	0.13	0.14	0.19	0.13	0.09	0.19	0.13	0.13	0.13
\$/peak, W	0.308	6.15	4.34	2.50	5.67	3.53	3.08	3.08	3.08
Solar flux, kW/m^2	0.65	0.65	0.65	0.8	0.5	0.8	0.65	0.65	0.65
Operation days per year	365	365	365	365	240	365	365	365	365
Eff pump/turbine, each	0.9	0.9	0.9	0.9	0.8	0.9	0.9	0.9	0.9
X_{salt} feed, lb/lbm	0.04	0.04	0.04	0.04	0.04	0.04	0.04	0.04	0.01
$X_{salt\ reject}$, lb/lbm	0.07	0.07	0.07	0.07	0.07	0.07	0.07	0.07	0.04
Field	0.85	0.85	0.85	0.85	0.60	0.85	0.85	0.85	0.85
Output									
Product salt, ppm	467	467	487	201	467	201	484	489	494
RO surface, ft^2	14,809	14,809	14,809	9,491	14,826	9,491	17,892	45,857	35,209
PV surface, ft^2	17,109	15,096	11,665	15,885	42,412	10,830	23,767	37,820	10,177
Power, kW	129	120	129	148	174	148	179	286	77
EO dissipation, kW	102	102	102	120	147	120	152	258	49
PV dissipation, kW	902	780	573	1,030	1,793	655	1,253	1,993	536
RO process efficiency	0.32	0.32	0.32	0.28	0.24	0.28	0.23	0.15	0.19
RO mmbr number	1,593	1,593	1,593	4,780	1,591	4,780	1,549	1,732	1,255
Pressure, φ_p	1.2	1.2	1.2	1.6	1.2	1.6	1.8	3.6	2.0
ΔP_{brn}, psi	3.1	3.1	3.1	402	3.1	402	6.5	606	197
Average H_{mass}, ft/h	0.197	0.197	0.197	0.776	0.197	0.776	0.210	0.345	0.293
\$25/bl economy (0.05,0.5)									
Capital, \$M	0.612	1.017	0.779	0.509	1.212	0.664	0.813	1.536	0.732
PV unit cost \$/ft^2	23.7	53.7	49.1	23.7	23.7	49.1	23.7	23.7	23.7
C_{pwr} \$/kWh (0.05)	0.107	0.215	0.152	0.087	0.723	0.123	0.107	0.107	0.107
C_{wtr} \$/m^3 (0.5)	0.673	0.940	0.783	0.508	1.624	0.610	0.862	1.849	1.126
Competitiveness	Near	—	Near	Yes	—	Near	Near	—	—
\$100/bl economy (0.2,1)	—	—	—	—	—	—	—	—	—
Capital, \$M	1.224	2.035	1.558	1.018	2.424	1.328	1.625	3.071	1.462
PV unit cost, \$/ft^2	47.4	107.4	98.1	47.4	47.4	98.1	47.4	47.4	47.4
C_{pwr}, \$/kWh (0.2)	0.215	0.430	0.303	0.175	0.602	0.247	0.215	0.215	2.15
Water cost, \$/m^3 (1)	1.347	1.879	1.566	1.016	3.248	1.220	1.723	3.698	2.252
Competitiveness	Almost	—	Near	Yes	—	Almost	Near	—	—

Table 1.6 Photovoltaic/Electrodialysis Systems

	Runs								
Parameter	1	2	3	4	5	6	7	8	9
Inputs									
Feed, ppm	5,000	5,000	5,000	5,000	5,000	5,000	2,000	10,000	10,000
Product, ppm	500	500	500	500	500	500	500	300	500
ED desalting efficiency	0.3	0.3	0.3	0.6	0.1	0.5	0.3	0.3	0.6
Applied voltage	120	120	120	120	60	120	60	120	60
Current densityAmp/m^2	100	100	100	100	50	100	50	100	50
Brine recycle ratio	0.8	0.8	0.8	0.0	0.85	0.8	0.0	0.8	0.0
PV cell type	11	22	33	11	11	33	11	11	11
Cell lab efficiency	0.1275	0.145	0.187	0.135	0.09	0.187	0.1275	0.1275	0.135
$/peak W	3.077	6.154	4.344	2.361	5.667	3.529	3.077	3.077	2.906
Solar flux, kW/m^2	0.65	0.65	0.65	0.8	0.5	0.8	0.65	0.65	0.65
Operation days per year	365	365	365	365	240	365	365	365	365
Field desalination-efficiency	0.85	0.85	0.85	0.9	0.60	0.85	0.85	0.85	0.9
Output									
W_{theor}, Btu/lb produced	0.19	0.19	0.19	0.12	0.21	0.19	0.18	0.25	0.17
ED surface, m^2	55.4	55.4	55.4	17.9	725.6	33.2	207.8	71.6	96.7
PV surface, m^2	2,406.6	2,123.4	1,640.8	602.4	14,473.2	801.9	2,257.3	3,106.4	996.2
Power, kW	195.53	195.53	195.53	63.79	638.52	117.61	183.41	252.39	85.70
ED dissipation, kW	202	202	202	29	652	124	150	270	38
PV dissipation, kW	1,365	1,181	867	417	6,585	522	1,289	1,762	560
ED overall efficiency	0.299	0.299	0.299	0.593	0.100	0.497	0.299	0.299	0.595
ED membrane number	55	55	55	18	726	33	208	72	97
ΔP flow, psi	0.3	0.3	0.3	0.3	0.3	0.3	0.3	0.3	0.3
Reject brine X_{salt} ppm	27,511	27,511	27,511	9,502	35,015	27,511	3,501	58,515	1,905
Brine X_{salt}, ppm	23,009	23,009	23,009	5,000	30,513	23,009	2,000	48,812	10,000
$25/bl economy (0.05,0.5)	—	—	—	—	—	—	—	—	—
Captl $M	0.62	1.23	0.87	0.16	3.78	0.43	0.60	0.80	0.27
PV unit cost $/m^2	255	578	528	255	255	528	255	255	255
C_{pwr}, $/kWh (0.05)	0.107	0.215	0.152	0.082	0.301	0.123	0.107	0.107	0.102
C_{wtr}, $/m^3 (0.5)	0.044	0.089	0.063	0.011	0.406	0.031	0.042	0.057	0.018
Competitiveness	Near	—	—	Yes	—	Near	Near	Almost	Almost
$100/bl economy (0.2,1)	—	—	—	—	—	—	—	—	—
Capital, $M	1.24	2.47	1.75	0.31	7.56	0.85	1.20	1.60	0.53
PV unit cost, $/m^2	510	1,156	1,056	510	510	1,056	510	510	510
C_{pwr}, $/kWh (0.2)	0.215	0.430	0.303	0.165	0.602	0.247	0.215	0.215	2.03
Water cost $/m^3 (1)	0.089	0.177	0.125	0.022	0.811	0.061	0.083	0.114	0.037
Competiteness	Near	—	—	Yes	—	Near	Near	Yes	Yes

Table 1.7 Osmosis Power

Parameter	Runs								
	1	2	3	4	5	6	7	8	9
Input									
A_{wtr} lb/(h · ft^2 · psi)	0.04	0.04	0.03	0.03	0.02	0.02	0.01	0.01	0.05
B_{salt}, ft/h	0.0005	0.0	0.0006	0.0	0.0005	0.0	0.001	0.0	0.0003
X_{salt} high, lb/lbm	0.1	0.1	0.07	0.07	0.15	0.15	0.2	0.2	0.25
X_{salt} reject, lb/lbm	0.06	0.06	0.05	0.05	0.06	0.06	0.06	0.08	0.06
$H_{c_{high}}$, in.	0.6	0.6	0.4	0.4	0.2	0.2	0.5	0.5	0.6
$H_{c_{low}}$, in.	0.5	0.5	0.5	0.5	0.3	0.3	0.4	0.4	0.5
Output									
X_{leak}, ppm	421	0	673	0	841	0	3356	0	202
A_{wall}, M · ft^2	41.2	45.6	110.8	118.4	75.6	55.0	81.8	66.0	13.8
A_{bulk}M · ft^2	39.6	39.6	107.4	107.4	50.0	50.0	78.6	61.3	13.2
Net power, MW	2.5	2.5	0.5	0.4	7.5	7.4	13.1	15.1	17.2
Ideal power, MW	34.3	34.3	6.2	6.2	68	68	80.1	72.9	749.2
∂X_{plzn}, high, lb/lbm	1.2	1.2	0.042	0.042	0.405	0.405	0.842	0.720	1.425
∂X_{plzn}, low, lb/lbm	0.042	0.180	0.012	0.068	1.178	0.093	0.024	0.096	0.023
$25/bl economy									
Capital wall, $M	494.1	546.9	1,329.1	1,420.2	907.4	650.8	981.7	793.2	166.1
Capital bulk, $M	475.4	475.6	1,288.4	1,289.3	604.9	605.4	937.7	735.1	158.2
Capital wall, $/kWh	4.45	5.55	24.29	27.88	2.42	2.56	2.01	2.33	0.22
Capital bulk, $/kWh	4.09	4.09	22.75	22.75	2.07	2.07	1.75	1.95	0.19
Capital ideal, $/kWh	3.44	3.44	19.11	19.11	1.73	1.73	1.47	1.62	0.161
$100/bl economy									
Capital wall, $M	988.1	1,093.8	2,658.3	2,840.4	1,814.8	1319.6	1963.3	1,584.4	332.3
Capital bulk, $M	950.8	951.8	2,576.8	2,578.6	1,209.8	1210.8	1875.3	1,470.1	316.5
Capital wall, $/kWh	8.9	11.09	48.57	55.76	4.87	5.12	4.01	4.67	0.44
Capital bulk, $/kWh	8.18	8.18	45.50	45.50	4.13	4.13	3.51	3.85	0.37
Capital ideal, $/kWh	6.87	6.87	38.22	38.22	3.74	3.74	2.95	3.24	0.31

The results have been presented in tabular rather than graphical format because of the large number of decision variables. Thermodynamic decision variables are no less than 15 for any system, and the thermodynamic decisions trigger both design and manufacture decisions because of the design and manufacture degrees of freedom involved. Graphs are more transparent in presenting results for cases of one or two decision variables. The number of two and three-dimensional relations explodes, however, with a large number of decision variables.

1.4 THE ANALYZED SYSTEMS IN DETAIL

1.4.1 Gas Turbine/Multistage Flash Distillation Cogeneration Systems

1.4.1.1 Flow Diagram Figure 1.1 is a flow diagram showing a steam turbine both idle and refiring active under less than design power load. Ejector steam and blowdown are allowed for. Airblade cooling for higher firing temperature is allowed for but not employed. The system has 63 thermodynamic decision variables, 24 of which can be manipulated to improve its cost objective function of higher profitability.

1.4.1.2 Major Features of the Results

- A night product raises the load factor from 0.583 to 0.865 and produces 5625 t/h desalted water <500 ppm, assuming an RO power requirement of 5 kWh/t. Profitability is reduced in the absence of a night product and is raised to 5–10 times the design steady-state value in the presence of a night product, provided all night product is salable.
- The CO_2 emission is slightly reduced from 64 to 56 t/h with higher profitability since burning fuel by refiring is essential to maintain the steady production of the MSF distiller. Auxiliary boilers and the throttling of high-pressure steam are alternatives that maintain the steady production of MSF distiller with the same weak effect on CO_2 emission.
- The cost of water remains around \$1/t and the cost of power around \$0.038/kWh for \$25/bl economy. For \$100/bl economy the costs are $\sim$ \$3/t for water and $\sim$ \$0.12/kWh for power.
- The first- and second-law efficiencies are raised from their design values of 0.3 and 0.355 to 0.37 and 0.405, respectively, with higher profitability.
- Automated optimization changes all the 24 manipulated decision variables. For example, the number of MSF stages is raised for its design value of 18 to 28, the pinch point is reduced from 50°F to around 10°F, and the level of adiabatic efficiencies of compressor, gas turbine, and steam turbine is raised from 0.85 to 0.92 for higher profitability.
- The improved design points differ for each of the two economies considered, namely, \$20/bl and \$100/bl oil.

- The fuel penalty of actual control compared to ideal control (the design efficiency remains constant at all load fractions) for the reference system is 26 MW with no night product. This corresponds to \$260/h for \$20/barrel economy and to \$1042/h for \$100/bl economy. The penalty for higher profitability is only about 7 MW with a night product of desalted water. This corresponds to \$70/h for a \$25/bl economy and to \$280/h for \$100/bl economy

1.4.2 The Simple Combined Cycle Systems

1.4.2.1 Flow Diagram Air-cooling (see Fig. 1.2) of blades is allowed for but not employed since the firing temperature is only up to 1600°F. The steam turbine expands steam to condensing temperature of 100°F. The power to be delivered after system needs have been satisfied is set to 100 MW. The number of thermodynamic decision variables is 34, 18 of which manipulated to lower production cost.

1.4.2.2 Major Features of the Results

- A night product raises the load factor from 0.583 to 0.865 and produces 5625 t/h desalted water <500 ppm during installation of a night RO desalter of currently attainable power requirement of 5 kWh/t and produces 0.36 t/h H_2 and 2.88 t/h O_2 during installation of a night water electrolyzer of power requirement 78.25 MWh/t H_2, which represents ΔG_f of the water content divided by an efficiency factor of 0.42 for a direct-current intensity of 1 A/cm^2. Profitability becomes a loss in the absence of night products and is raised to 5–10 times the design steady-state value in the presence of night products, provided all night products are salable.
- The CO_2 emission design value is 45 t/h (compared to 64 t/h of the GT/MSF case). The emission is reduced to 40 and 36 t/h with improved lower production cost.
- The cost of power is ~ \$0.035/kWh for 25\$/barral economy and ~ \$0.12/kWh in \$100/bl economy.
- The first- and second-law efficiencies are raised from their design values of 0.4 and 0.46 to 0.43 and 0.465, respectively, for lower production cost.
- The fuel penalty of actual control compared to ideal control is 19.5 MW with no night product. This corresponds to \$195/h for \$25/bl economy and to \$781/h for a \$100/bl economy. The penalty for lower production cost is only ~5.5 MW, with a night product of desalted water or of H_2 and O_2. This corresponds to \$59/h for a \$25/bl economy and to \$220/h for \$100/bl economy.
- The cost of night product of water is a \$0.24/t, and that of H_2 and O_2 is \$228/t for a \$25/bl economy. For a \$100/bl economy, the corresponding costs are \$0.74/t and \$889/t.

1.4.3 Vapor Compression Systems Driven by the Figure 1.2 Simple Combined Cycle

1.4.3.1 Flow Diagrams Figure 1.3 shows a vapor compression (VC) distiller with 16 thermodynamic decision variables, three of which are manipulated to lower production cost. The operating temperature is maintained at 160°F.

Figure 1.4 shows its version of Figure 1.3 for zero liquid discharge (ZLD) with 19 decision variables, none of which are manipulated for lower production cost. Compressor efficiency is set at 85%, with a of pressure ratio of 2.73 and a suction pressure of 0.7 psia. This low suction pressure calls for more than 10 compressors of reasonable inlet area operating in parallel.

Both VC versions are driven by the simple combined cycle shown in Figure 1.2, with 34 decision variables, 18 of which are manipulated to lower production cost.

1.4.3.2 Major Features of the Results

- No night product is introduced because the production is at steady state, producing 10 migd water by VC desalter.
- The reference design of the VC distiller has a capital cost of $39M for compressor and heat exchange surfaces and requires 44 MW power. The driving simple combined cycle (SCC) requires, in turn, 109 MW fuel. For a $25/bl economy, the cost of water is $1.1/t. The cost of power is as delivered by the combined cycle $0.0364/kWh. For a $100/bl economy, the cost of water is $3.36/t and the cost of power as delivered by the combined cycle is $0.1225/kWh.
- The improved design by lower production cost has a capital cost of $210M and requires 17.5 MW power. The driving SCC requires, in turn, 40 MW fuel. For a $25/bl economy, the cost of water is $1.1/t and the cost of power as delivered by the combined cycle is $0.035/kWh. For a $100/bl economy, the cost of water is $2.44/t and the cost of power as delivered by the combined cycle is $0.115/kWh.
- The concept of zero liquid discharge is considered using single-stage VC. The result, so far, has not been cost-effective. Power requirement is almost doubled (77 MW), and so is CO_2 emission (35 t/h). The unit cost of water almost tripled. However, the high pressure of the RO case is avoided but large parallel compressors operating under vacuum are needed. Section 1.A.4 explains the large power requirement via the theoretical work of separation extended to zero liquid discharge.

1.4.4 Reverse Osmosis Desalination Systems Driven by the Figure 1.2 Simple Combined Cycle

1.4.4.1 Flow Diagrams Figure 1.5 shows conventional single- (a) and two-stage (b) reverse-osmosis desalination systems. The two-stage system is only a standby system in case product water at concentrations of <500 ppm cannot be obtained by the single-stage system. The standby option, however, was not required. The single-stage system has 18 thermodynamic decision variables, 12 of which are manipulated to minimize production cost with product quality <500 ppm.

The design model of the RO process [15], which is based on numerous equations [the model is shown in the following diagram, and the equations include Eqs. (1.22)–(1.31)], and does not have many design degrees of freedom; moreover, solutions with quality for product water >500 ppm have to be excluded:

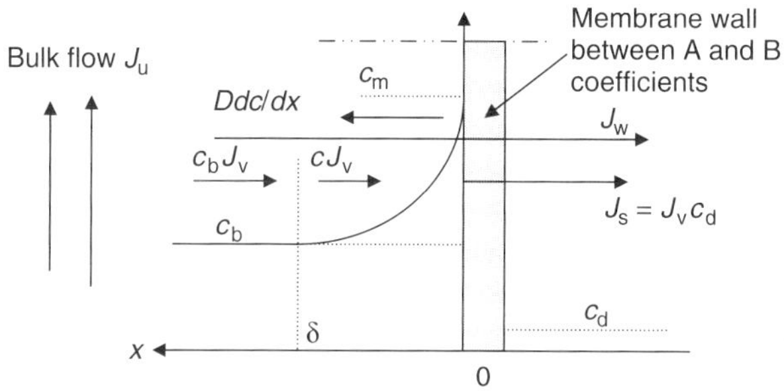

$$J_w = A(\Delta P - \Delta\Pi) = A[(P_m - P_d) - (\Pi_m - \Pi_d)]$$
$$= A[(P_m - P_d) - \varphi RT(c_m - c_d)] \quad (1.22a)$$

$$J_s = B(\Delta c) = B(c_m - c_d) \quad (1.23a)$$

$$J_v = h_m\left[\ln\left(\frac{c_m - c_d}{c_b - c_d}\right)\right] \quad (1.24a)$$

where

$$J_v(c - c_d) - Ddc/dx = 0 \quad (1.25)$$

$$\int_{x=0}^{\delta} J_v dx = \int_{c_m}^{c_b} D\mathrm{d}c/(c - c_d) \quad (1.26a)$$

$$J_v\delta/D = \frac{J_v}{h_m} = \ln\left[\left(\frac{c_m - c_d}{c_b - c_d}\right)\right] \quad (1.26b)$$

$$\delta = D/h_m \quad (1.27)$$

$$J_w = J_v(\rho - c_d) \quad (1.28)$$

$$J_s = J_v c_d \quad (1.29)$$

$$S_{RO} = \frac{M_d}{J_w + J_s} \quad (1.30)$$

$$Z_{RO} = C_{aRO}S_{RO} \quad (1.31)$$

where C_{aRO} is the membrane cost per unit membrane surface area. A multiplier >1 accommodates care of the balance of plants (default 1.5).

Equations (1.22a)–(1.24a), (1.28), and (1.29) are five equations in five unknowns J_w, J_s, J_v, c_m and c_d. The number of equations and unknowns can be reduced to 3 by substituting Equations (1.28) and (1.29) in Equations (1.22a)–(1.24a), and the equations and the unknowns may be reduced to 3:

$$J_w = A[(P_m - P_d) - \varphi RT(c_m - c_d)] \tag{1.22b}$$

$$= B(c_m - c_d)\frac{(\rho - c_d)}{c_d} \tag{1.23b}$$

$$= h_m(\rho - c_d)\left[\ln\left(\frac{c_m - c_d}{c_b - c_d}\right)\right] \tag{1.24b}$$

The unknowns are J_w, c_m, and c_d.

The following equations are needed to compute pressure losses and mass transfer coefficient h_m and hence concentration boundary-layer thickness d. The equations are based on dimensional analysis. Note that the assumption of single-thickness boundary layer is often used in the analysis of RO membranes. Baker [16, p. 176] uses $\delta = 20$ μm. This assumption is relaxed by dimensional-analysis-based equations, where the thickness responds to geometry changes of flow conduit.

$$f = 16/\mathrm{N_{RE}} \quad \text{for fully developed laminar flow} \tag{1.32}$$

$$= 0.078\mathrm{N_{RE}}^{-0.25} \quad \text{for fully developed turbulent flow}, N_{RE} < 30,000 \tag{1.33}$$

$$N_{SH} = 1.85\left(\frac{N_{PE}}{N_V}\right)^{1/3} \quad \text{to asymptotic value of } N_{SH}N_{PE} = 48 \tag{1.34}$$

for laminar flow between plates(narrow passages)

$$= 1.62\left(\frac{N_{PE}}{N_V}\right)^{1/3} \quad \text{to asymptotic value of } N_{SH}N_{PE} = 16 \tag{1.35}$$

for laminar in tubes, $N_{Re} < 1000$

$$= 0.04N_{RE}^{0.75}N_{SC}^{\ 0.33} \tag{1.36}$$

for fully developed turbulent flow, $N_{RE} > 10,000$

$$N_{RE} = \text{Reynolds number} \quad = Ud\frac{\mu}{\rho} \tag{1.37}$$

$$N_{PE} = \text{Peclet number} \quad = V\frac{d}{D} \tag{1.38}$$

$$N_{SH} = \text{Sherwood number} \quad = h_m \frac{d}{D} \tag{1.39}$$

$$N_{SC} = \text{Schmidt number} \quad = \mu / \frac{\rho}{D} \tag{1.40}$$

$$N_v = \text{permeated/flow velocity geometry} = J_v \frac{l}{J_u/d} \tag{1.41}$$

J_v = velocity normal to flow(x direction)

J_u = velocity in flow direction

l = length in brine flow direction

d = equivalent diameter of the brine flow area (hydraulic diameter)

$$d = 4WHn_1/2/(W + H)/n_2 \tag{1.42}$$

n_1 = blockage factor of brine flow area

n_2 = wetted perimeter increase factor

The factors n_1 and n_2 depend on the thickness and the shape of feed spacers. The factors produce an equivalent reduced height H_e of an empty conduit. The values assigned to n_1 and n_2 in this model are 0.25 and 2.0, respectively. An empty conduit has $n_1 = n_2 = 1$.

The diffusion coefficient computed by Wilke–Chang equation [17, Reid p. 598] $= 3.55 \cdot 10^{-6}$ cm^2/s. The diffusion coefficient quoted from Baker [16, p. 176] $= 10 \cdot 10^{-6}$ cm^2/s, is used in this model. Density ρ and viscosity μ are computed by transport properties routine of reference [14].

A membrane relation that guarantees an allowable product salt content (e.g. <500 ppm) may be derived by introducing a membrane dimensionless number N_m as follows. First, the wall osmosis pressure difference $(\Pi_m - \Pi_d)$ may be written in terms of wall and product salt concentrations (mass or mole per unit volume):

$$(\Pi_m - \Pi_d) = \varphi RT(c_m - c_d) \tag{1.43}$$

where $\varphi = 1$ for ideal solution and > 1 to accommodate deviations from ideal solution. In this study, ideal solution is assumed.

Applied pressure $(P_m - P_d)$ may be written in a similar way by introducing φ_P:

$$(P_m - P_d) = \varphi_P RT(c_m - c_d) \tag{1.44}$$

$$\frac{J_w}{J_s} = A(\varphi_P - \varphi)R\frac{T}{B} = N_M \tag{1.45}$$

Equation (1.45) is realized in this model development and is believed to be important in RO membrane design. It presents the combined influence of membrane coefficients and applied pressure (A, B, and P). The invested pressure above osmosis is measured by $\varphi_P - \varphi$, where $\varphi = 1$ for ideal solutions.

The following are some features of N_M:

- For seawater, $N_M > 1500$ seems to guarantee product <500 ppm.
- The larger the N_M, the lower the membrane surface requirement for a given product rate.
- N_M is independent of concentration polarization.

A unit of 200 usgpd is considered (surface ~1 m^2). Its flow path is divided into five sections. The feed concentration is equal to the bulk concentration to first section. All the parameters of interest along the flow path of each section are computed. For each section, the permeated flux J_w and membrane concentrations c_m and c_d are obtained by solving Equations (1.22b)–(1.24b) simultaneously. The unit modules and their pressure shells can be arranged in numerous ways to manage flow pressure losses and shell diameter. A module of 30 units in parallel is selected. A shell contains three modules in series. About 18 shells provide the required product rate of 415 m^3/d (0.11 usmgd). This satisfies the domestic water demands of a community of about 1000 persons (0.3 m^3 per day per person).

Figure 1.6 depicts a version of a RO with zero liquid discharge of two stages. The number of stages depends on the solubility limits of the various salt species. In the absence of membranes selective to specific salt species, Figure 1.6 assumes a hypothetical version of two stages: (1) a retainer for species other than sodium chloride and (2) a retainer to sodium chloride. The system has 27 decision variables, 12 of which are manipulated for production cost minimization.

1.4.4.2 Major Features of the Results Please refer to Table 1.4 for sample runs

- No night product is introduced because the production is at steady state, producing 10 migd water by an RO desalter
- The reference design of the RO desalter has a membrane surface of 402,452 ft^2 and requires 12.17 MW power (6.376 kWh/t water product). The driving simple combined cycle (SCC) requires, in turn, 30.3 MW fuel. For a \$25/bl economy, the cost of water is \$0.35/t. The cost of power is as delivered by the combined cycle \$0.0364/kWh. For a \$100/barrel economy, the cost of water is \$1.03/t and the cost of power as delivered by the combined cycle is \$0.1225/kWh.
- The improved design by lower production cost has a membrane surface of 806,080 ft^2 and requires 5.6 MW power (2.919 kWh/t water product). The driving simple combined cycle requires, in turn, 13.86 MW fuel. For a \$25/barrel economy, the cost of water is \$0.35/t and the cost of power as delivered by the combined cycle is \$0.035/kWh. For a \$100/bl economy, the cost of water is \$0.84/t and the cost of power as delivered by the combined cycle is \$0.1148/kWh.

- For RO, the reference system main decision variables are water coefficient 0.02 lb/(h.ft^2·psi), salt coefficient 0.015 ft/h, and pressure 1500 psia. For CC, the main decisions are pinch 25°F, compressor efficiency 0.85, and turbine efficiency 0.9.
- The concept of zero liquid discharge is considered using two stages in series. The result, so far, has not been cost-effective. The power requirement increased about 8 times, and CO_2 emission almost doubled. Moreover, high-pressure membranes (5000 psia), probably ceramic, need to be developed. The large power requirement via the theoretical work of separation extended to zero liquid discharge is discussed in Section 1.A.4.

1.4.5 Photovoltaic/Reverse-Osmosis (PV/RO) Solar Systems

1.4.5.1 Flow Diagram Figure 1.7 shows the flow diagram of the solar desalter of small distributed 0.2 migd desalted water for communities of about 10,000 people along with the main variables of the RO subsystem and of the solar subsystem. Figure 1.8 shows the future potential of this particular solar desalting system. Figure 1.9 shows two photovoltaic/electrodialysis (PV/ED) configurations.

1.4.5.2 Major Features of Results Please refer to Table 1.5 for sample runs

- All runs use inputs of attainable membrane water and salt coefficients
- The first eight runs assume seawater feed of 0.04 salt mass-fraction and reject around 0.07 salt mass fraction. The last assume brackish-water feed of 0.01 salt mass fraction and reject 0.04 mass fraction.
- Solar intensity ranged from 0.5 to 0.8 kW/m^2, averaging 0.65 kW/m^2.
- All runs assume operation of 365 days per year except run 5, which assumes 240 days per year.
- All runs assume loss of cell field efficiency to 0.85 of that laboratory evaluation.
- For a \$25/bl oil price index economic environment, a competitive power cost of \$0.05/kWh and a competitive water cost using RO of 0.5 \$/m^3 can be assumed. One run is competitive. A few runs are nearly competitive, utilizing the advantage of no CO_2 emission.
- For \$100/bl oil price index economic environment, the competitive power cost is \$0.2/kWh (4 times that of the \$25/bl index) and a competitive water cost using RO of \$0.1/m^3 (2 times that of the \$25/bl index) can be assumed. Competitiveness increases because the cost of a material product escalates at a lower rate than does power cost.
- Figure 1.8 indicates a promising potential future for PV/RO desalination technology [18].

1.4.6 Photovoltaic/Electrodialysis Solar System

The separated PV/ED is the configuration (Fig. 1.9b) considered for analysis. The feed is assumed to have zero exergy. A simple analysis model is assumed that does not reveal the distributions of salt and water through flow passages because of insufficient information on ion exchange membranes.

1.4.6.1 Major Features of the Results Please refer to Table 1.6 for sample runs

- Efficiencies are used as decisions rather than being computed whenever the available characteristics of the active surface (membrane or solar cell) do not permit computiation of the efficiency.
- All runs use inputs of attainable or near attainable efficiencies. Ion exchange membrane efficiencies ranged from 0.1 to 0.6. Pump efficiency was set at 0.8. DC–AC conversion efficiency was set at 0.95.
- The first six runs assume feed of 5000 ppm salt content. One run assumes feed of 2000 ppm and one run assumes feed of 10,000 ppm.
- The brine recycle ratio is 0.8 for most runs. For one run the ratio is 0.85. For three runs the ratio was set to zero.
- Solar intensity ranged from 0.5 to 0.8 kW/m^2 and averaged 0.65 kW/m^2.
- All runs assume operation of 365 days per year except run 5, which assumes 240 days per year.
- The first three runs compare the three types of solar cells, each at its upper efficiency.
- Eight runs assume loss of cell field efficiency to 0.85 of that laboratory standard test values. One run limits the loss to 0.9 the standard test value.
- For \$25/bl oil price index economic environment, a competitive power cost of \$0.05/kWh and a competitive water cost using RO \$0.5/m^3 can be assumed. One run is competitive. A few runs are nearly competitive, utilizing the advantage of no CO_2 emission.
- For \$100/bl oil price index economic environment, the competitive power cost is \$0.2/kWh (4 times that of the \$25/bl index) and a competitive water cost using ED of \$0.1/m^3 (2 times that of the \$25/bl index) can be assumed. Three runs show competitiveness. Competitiveness increases because the cost of a material product escalates at a lower rate than power cost.

1.4.7 Osmosis Power Systems

1.4.7.1 Flow Diagram Figure 1.10 shows the flow diagram of a single-stage osmosis power system utilizing the chemical exergy difference between two streams of brines of different salt concentration. The less concentrated brine is assumed to be seawater of salt mass fraction 0.04. Water mobility through water-selective membranes is assumed. The stage is divided into 10 cells. The major controlling variables of a cell are given in Figure 1.10b.

However, the issue of tapping power from the chemical exergy difference between two brine streams of different salt concentrations by an electrodialysis device using the mobility of salt ions through ion exchange selective membranes,

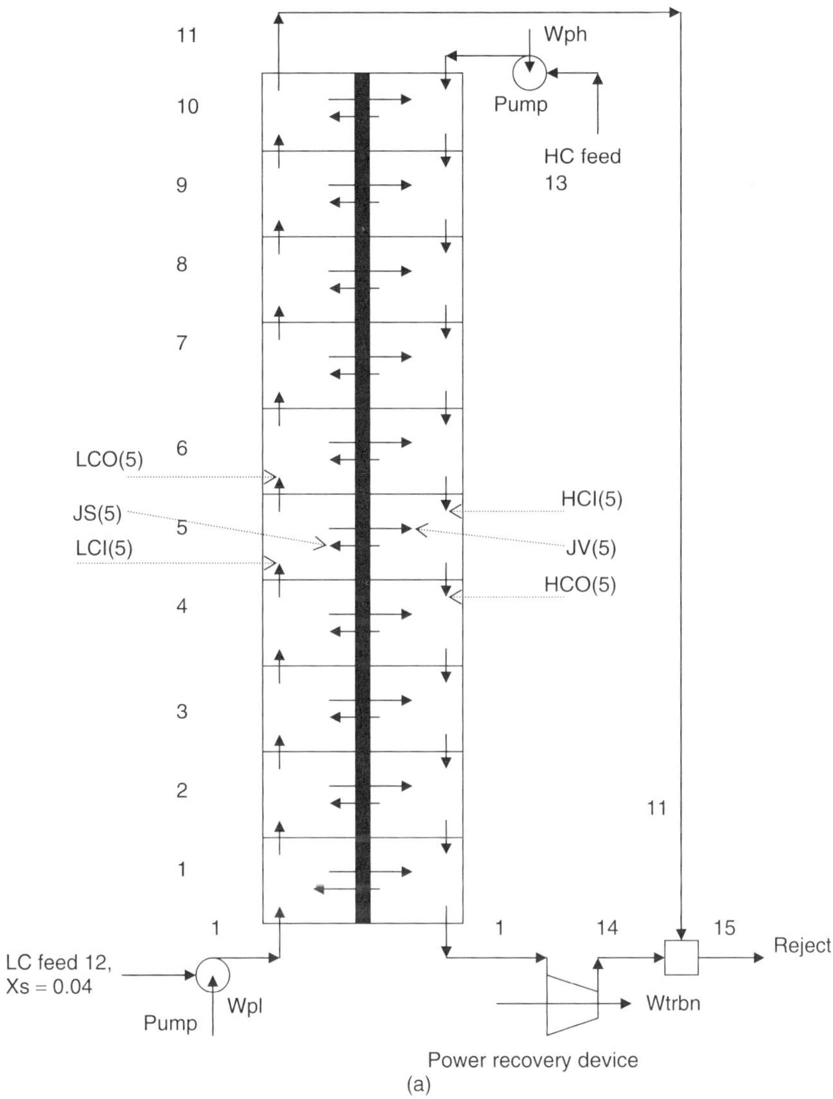

Figure 1.10 The osmosis power system: (a) flow diagram; (b) variables controlling the cell [19].

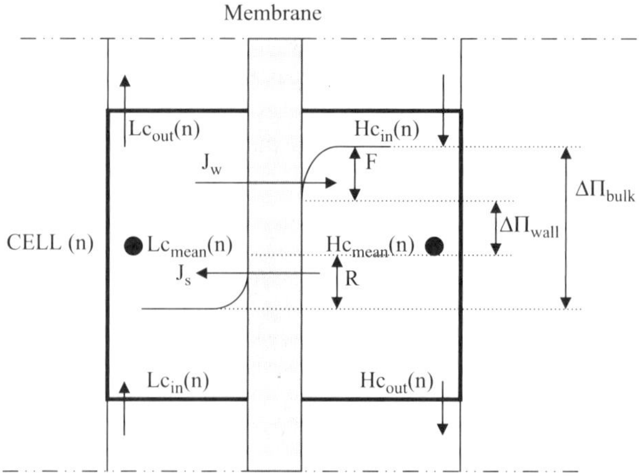

(b)

Figure 1.10 *(Continued)*

remains to be addressed. Consider the following equations:

$$\Delta\Pi_{\text{bulk}} = \varphi RT\left(\rho_{c_{\text{high bulk}}} XS_{c_{\text{high bulk}}} - \rho_{c_{\text{low bulk}}} XS_{c_{\text{low bulk}}}\right) \tag{1.46}$$

$$\Delta\Pi_{\text{wall}} = \Delta\Pi_{\text{bulk}} - F - R \tag{1.47}$$

where F is the fall in the salt content of the high-concentration fluid due to dilution by the diffusing pure water flux J_{w} to the high-concentration side of the membrane causing a polarization effect $\partial X_{\text{plzn high}}$ and R is the rise in the salt content of the low concentration fluid due to loss of diffusing pure water flux J_{w} to the high concentration side of the membrane and the diffusing salt flux J_{s} to the low concentration side, causing a polarization effect $\partial X_{\text{plzn low}}$. Also, $J_{\text{s}} = 0$ for membrane salt permeability coefficient $B \approx 0$:

$$\text{Work recovered} = A(\Delta\Pi_{\text{wall}} - \Delta P)\Delta P \tag{1.48}$$

$$\text{Theoretical work recovery } W_{\text{theor}} = \frac{A}{4}\Delta\Pi^2_{\text{bulk}} \tag{1.49}$$

1.4.7.2 Major Features of the Results

Please refer to Table 1.7 for sample runs

- Table 1.7 shows a sample result of inputs and outputs of the osmosis power systems of a concentrated feed stream of 10 migd of salt mass fractions 0.07–0.25 relative to seawater with a salt mass fraction of 0.04 for the economies of \$25 and \$100 per barrel of oil.

- Power from conventional salt content of rejected brine (0.06–0.08 lb_{salt}/lb_{mix}) due to concentration difference is not cost-effective. A small amount of power < 0.5 MW is gained by large membrane surface $110M \cdot ft^2$. For brine with a salt mass fraction of 0.25, a power as high as 20 MW is gained by a membrane surface around 12 M ft^2. Only for a brine of salt content 0.1 and higher does power recovery begin to make sense.
- For a \$25/bl (oil price index) economy and brine of salt mass fraction <0.1, power cost is as high as \$25/kWh. The power cost for a corresponding ideal process is \$19/kWh. For brine of salt mass fraction 0.25, power cost is reduced to \$0.2–0.4/kWh. The power cost for a corresponding ideal process is \$0.16/kWh. For the \$100/bl, economy, the respective costs are doubled.
- The salt mass fraction of brine from oilwells can be as high as 0.25. Power generation by this brine, when combined with seawater via membranes, can be cost-effective in some situations.

1.4.8 Future Competitiveness of Combined Desalination Systems

1.4.8.1 Prediction Criteria Efficiency, dumped waste and product cost are three major criteria that identify the most fit desalination system in the future. Overall system efficiency, CO_2 emission per unit product, and cost/unit product are considered for the systems analyzed. Dumped brine is assumed to be tolerated because zero liquid discharge is still far from being cost-effective.

1.4.8.2 Predicted Competitiveness GT/MSF systems show that the case of variable power demand cogeneration counteracts most of the advantage of cogeneration. This widely used cogeneration is likely to loose attractiveness in the future. The advantage is, however, maintained for base power load cogeneration.

SCC systems for power generation show that power-driven night products of low storage cost improves the plant load factor and raises its profitability provided the products are in short supply. The management of power generation by organized night products may gain competitive advantage in the future.

The CC/RO desalting systems show that the attractiveness of power driven desalting systems is likely to surpass that of distillation because of higher efficiency, lower emissions and lower product cost. The CC/VC desalting systems come second to CC/RO. The lower operation pressure and the higher biofouling resistance are advantages, but the handling of large specific volumes is a disadvantage. The development of strong light material for high-speed low-pressure-ratio compressors, or the development of scale-free VC operation at atmospheric pressure, reduce the disadvantage as well as the gap between the product cost by RO and VC. If the disadvantage is reduced, the power-driven VC will gain also achieve future promise for zero-liquid-discharge (ZLD) desalination.

The PV/RO system has zero CO_2 emission and zero fossil fuel consumption but does not avoid dumping of concentrated brine in the physical environment. Their future attractiveness is on the rise. The PV/ED system comes second to PV/RO if it undergoes sufficient development.

The osmosis system for power production is only a possibility, not a reality, and requires the availability of brine near saturation to combine with seawater to obtain power at acceptable cost. If developed, it can be useful for eliminating product water of oilfields located near a sea.

References listed in Section 1.F.9 at the end of this chapter are additional useful readings for the preceding Section 1.4.

1.5 RECOMMENDED RESEARCH DIRECTIONS

1.5.1 Avoiding CO_2 Emissions

- Desalination systems driven by renewable energy sources, particularly solar, are recommended if CO_2 emission is to be avoided.
- Competitiveness requires high-efficiency desalting systems and high-efficiency solar conversion systems. RO is in the lead for high efficiency, particularly for seawater. ED can be in the lead for lower salt content sources. Photovoltaic desalination is in the lead for higher-efficiency conversion to power.
- Competitiveness is increased by:
    - RO of higher water and lower salt permeability and lower cost per unit surface
    - ED of higher current density and lower electric resistance and lower cost per unit surface
    - PV solar cells of higher standard test efficiency and higher field efficiency and lower cost per unit surface

1.5.2 Reducing CO_2 Emissions

- If fossil fuels have to be used, the solutions to lowering CO_2 emissions is to develop higher efficiency energy conversion devices and/or produce more products for the same emissions.
- Competitiveness is increased by
    - Cogeneration of power and desalted water by base-load power plants
    - Producing night low-storage-cost products to improve the load factors of power plants and of variable-load (non-base-load) cogeneration plants.

1.5.3 Desalination of Zero Liquid Discharge

- Adequate understanding of the feed saturation limits, their sequence, their dynamics of salt release and the separation of their solids are essential for the idea of zero liquid discharge.
- Membrane desalting is inherently high efficiency and cost-effective due to its avoidance of water phase change.
    - Membranes should be designed to discriminate between salt species with respect to their solubility limits. Ideally two types are needed; one membrane retains all species except sodium chloride, and the other one retains sodium chloride and stands pressures as high as 5000 psia.

  - A doping method for longer supersaturation time is needed to avoid the clogging of membrane passages.

- Vapor compression desalting pays an energy penalty for the presence of phase change but has the advantages of low-pressure operation, high biofouling resistance, and less severe clogging problems. These advantages are also desirable for zero liquid discharge
  - The vapor compressor should be made of a strong lightweight composite to run efficiently at the desired high speed with lower stresses.
  - Vapor compression intake at atmospheric pressure helps further increase the compressor efficiency and reduce its cost if scaling can be avoided.

1.6 CONCLUSIONS

Solar desalination of conventional concentration ratios has a high potential to replace fossil fuel and avoids its CO_2 emission for the production of desalted water. Supporting research to improve photovoltaic conversion efficiency and the driven desalting efficiency is worthwhile. Reverse osmosis and electrodialysis are in the lead for higher desalting efficiency.

Zero liquid discharge desalination to avoid environmental damage due to dumping of rejected brine is far from being cost effective. Further research is needed.

Reverse osmosis driven by a simple combined cycle to produce water produces much lower CO_2 emission than conventional power distillation cogeneration systems. RO driven by high-firing cooled-blade combined cycle further reduces CO_2 emission.

Power plants and cogeneration plants that burn fossil fuel and operate under variable power demand can benefit from night products of low storage cost in short supply. Night products improve the plant load factor, produce more products for the same CO_2 emission and raise profitability.

Increasing the efficiencies of conventional energy conversion devices and reducing their costs have their limits in meeting the challenge of rising fuel prices and rising CO_2 emission. New processes using new materials and new devices need to be discovered.

1.7 THE SOFTWARE PROGRAMS DEVELOPED BY THE AUTHOR FOR SYSTEM ANALYSIS

(The software discussed may be downloaded at http://booksupport.wiley.com).

1.7.1 Four Programs Developed and Their Entries

- *DesRvst*: handles the systems of the first six configurations. These are the systems of: GT/MSF, SCC, VC/SCC, VCZLD, RO/SCC, and ROZLD. Each system has its own results of states, processes and costs displayable

or printable. An overall energy balance and exergy balance verifies the consistency of the results. The program has the following entries:
  - GT/MSF cogeneration systems
  - Simple combined cycle systems with night RO desalter
  - Simple combined cycle systems with night electrolyzer
  - RO rejecting brine driven by simple combined cycle
  - RO rejecting salt driven by simple combined cycle
  - VC rejecting brine driven by simple combined cycle
  - VC rejecting salt driven by simple combined cycle
  - Change economic environment (default $25/bl)
  - Change power demand profile
  - Change units IP/SI (default IP)
  - Change zero exergy state (default sea)

- *SOLRO*: handles PV/RO systems and has the following entries:
  - Design solar-panel-driven RO system
  - Run sample of up to 10 systems
  - Display all results
  - Print all results
  - Print selected results
  - Change units IP/SI (default IP)
  - Change zero exergy state (default sea)
  - Change economic environment (default $25/bl)
  - Review software description
  - Terminate

- *SOLED* handles PV/ED systems and has the following entries:
  - Design solar-panels-driven ED system
  - Run sample of up to 10 systems
  - Display all results
  - Print all results
  - Print selected results
  - Change units IP/SI (default IP)
  - Change zero exergy state (default sea)
  - Change economic environment (default $25/bl)
  - Review software description
  - Terminate

- *Osmosis* handles osmosis power systems and has the following entries:
  - Compute one-stage osmosis given hs and dP/P

- Design one-stage osmosis given flow passages
- Display all results
- Print all results
- Print selected results
- Run sample of 10 systems given *hs* and *dP/P*
- Print sample runs
- Change Units IP/SI (default IP)
- Change zero exergy state (default sea)
- Change economic environment (default $25/bl)
- Review software description
- Terminate

1.7.2 Major Ingredients of Each Program

These ingredients are as follows:

- System description in terms of working fluids, devices, and thermodynamic decision variables
- A routine for changing decision variables manually
- Computational routine in communication with fluid thermodynamic and transport properties and with the connectivity of devices to compute a solution per a selected reference unit mass. Connectivity may be explicit (handling several configurations) or embedded in the computational routine (handling one configuration at a time). The speed of computation depends on system connectivity, system decision variables, and the complexity of the controlling equations.
- A routine to compute parameters of interest, once a solution is obtained, such as heat, power and mass rates; efficiencies; exergy destructions; costs and objective function
- For the systems shown in Figures 1.1–1.3, an optimization routine that automates the optimization of the system

1.7.3 The Software

The software developed for system analysis contains the following properties:

- Installation of the software in the user's computer is automatic.
- The software contains:
  - The chapter explained in slides
  - Executable versions of the four programs
  - The source code of their master programs
  - Sample source codes of property and process programs

APPENDIX

Some Useful Equations and Facts for Water Desalination Modeling as Employed in this Chapter [14].

1.A.1 Brief Description of the Thermodynamic Model of a System and the Design Models of Its Main Components

1.A.1.1 Thermodynamic Model The model has a database of fluid properties and elementary processes that are the building blocks of a fair number of power generation and cogeneration systems. The fluid property database contains the equations essential for computation of the thermodynamic and transport properties of H_2O, NH_3, R_{12}, and NH_3/H_2O mixtures, seven ideal gases (O_2, N_2, H_2O, CO_2, SO_2, CO and H_2) and their combination which can cover air, gas mixtures, combustion gases dry or wet; and seven liquids (lubricating oil, ethylene glycol, glycerin, kerosene, sodium, bismuth, mercury, and seawater/brines). Refrigerants R142a and R153b were also included. The process database contains 22 elementary processes that allow the description of a large number of systems. The main elementary processes handle expansion, compression, heat exchange, mixing, combustion, and throttling. Few processes are simple combinations of the elementary processes, such as a multistage process. Few are purely computational, performing tasks such as splitting, merging, and tearing. The performance of a main elementary process is described by its overall efficiency and loading parameters. More than one set of the essential input parameters is allowed by the thermodynamic model to enhance system computation with the fewest iteration loops.

The model is also used to express the exergy destruction of a device in terms of the device efficiency and loading parameters. To compute the exponents $\{n_e\}$ of a device that correlates its exergy destruction D in terms of its efficiency and loading parameters, the process model of the device is run with different input variables covering the range of interest to its system. The computed exergy destruction D is listed versus its correlating parameters. A curve-fitting procedure gives the value of $\{n_e\}$ applicable over the range of variation considered.

1.A.1.2 Sample Design Models The purpose of the following design models is to provide a rational basis for the cost of their devices. For this purpose *all* design models target the evaluation of a *dominating flow passage surface* for which a unit cost gives a fair prediction of the estimated device cost. The design models represent some of the current design practices and not necessarily the best ones. They also need to be updated to accommodate changes in design practices.

1.A.1.2.1 The Axial Air Compressor The basic features are axial, two dimensional analysis at the mean radius, subsonic, 50% reaction, diffusion factor <0.45, and ideal gas properties. Blade geometry is kept constant. All stages except the final one experience the same temperature rise. Tip blade speed, axial velocity, root/tip radius ratio, and work factors are kept constant at 1150 ft/s^{-1} (250 m/s^{-1}), 500 ft/s^{-1} (150 m/s^{-1}), 0.5, and. 0.98–0.83 (0.83 after the third stage), respectively.

A polytropic efficiency is assumed, velocity triangles computed, and the stage efficiency is evaluated from cascade tests corresponding to the blade geometry. Computations are iterated until polytropic and stage efficiencies are matched. Mass rate, pressure ratio, and temperature rise per stage are varied, and the number of stages, total surface of fixed and moving blades, adiabatic efficiency, speed, and recommended solidity are computed. An arbitrary value of solidity can also be entered as input. The total surface of the moving and fixed blades is correlated in terms of air mass flow rate, pressure ratio, and efficiency ratio $\eta/(1-\eta)$. Ambient conditions are assumed for air at compressor inlet. A version of the model accommodates low-pressure ratio *axial and radial steam compressors*.

1.A.1.2.2 The Gas Turbine The basic features are axial, uncooled blades, two-dimensional analysis at mean radius, subsonic, 50% reaction, and ideal gas properties. Blade geometry is kept constant. Loading and flow coefficients ψ,ϕ, mean blade speed, and inlet temperature are kept constant at 1.4, 0.8, 1115 ft/s (240 m/s) and 1600°F (870°C) respectively. The inlet temperature implies uncooled expansion. The first two values seem to minimize the needed total surface of blades. Stage efficiencies of nozzle and rotor blades are assumed, velocity triangles are computed, and stage and tip clearance losses are evaluated from cascade tests corresponding to the blade geometry. Computations are iterated until the assumed efficiencies and the losses are matched. Mass rate, pressure ratio, and speed are varied, and the number of stages, total surface of nozzle and rotor blades, adiabatic efficiency, and recommended solidity are computed. An arbitrary value of solidity can be entered as input. The model does not guarantee that the speed matches that of the compressor. The total blade surface of the fixed and moving blades is correlated in terms of gas flow rate, expansion ratio, and efficiency parameter $\eta/(1-\eta)$. Gas pressure at exit is assumed to be ambient.

1.A.1.2.3 The Steam Turbine The steam turbine is similar to the gas turbine except for a few differences. Actual steam properties were used to compute the specific heat and the isentropic index instead of the constant values assumed in the case of air and combustion gases. Inlet temperature and pressure and exit pressure instead of the pressure ratio became inputs. Exit pressure was changed to cover both condensing and backpressure turbines. In some cases the blade heights were too short and high rotational speeds were entered to reduce mean diameter and increase blade height. The total surface of blades did not change with the change in speed. The total surface of the blades is correlated in terms of steam mass rate, (T/P) at inlet, exit pressure, and efficiency ratio $\eta/(1-\eta)$. Impulse stages are not included.

1.A.1.2.4 Centrifugal Pumps The basic features are centrifugal, axial flow at inlet and radial flow at exit, with velocity head recovery. Loading (head) and flow coefficients Ψ, Φ, number of impeller blades, root/eye radius ratio, velocity exiting casing, specific volume, and maximum head per stage are kept constant at 1.4, 0.8, 7, 0.4, 6 ft/s (1.8 m/s), 0.016 cu ft/lb (0.001 m^3/kg), and 500 ft (150 m), respectively. Velocity triangles and flow passages are computed given specific speed.

Mass flow rate, head and specific speed are varied. Speed, surface of impeller, diffuser surface, and efficiency are computed. Specific speed is changed such that surface is minimized. One costing equation did not fit all cases. One equation was used for low flow rates and high-pressure heads (feed pumps) and one for large flow rates and low-pressure heads (circulating pumps). Extending flow rates to > 500 lb/s need to be implemented. Impeller surface is correlated in terms mass rate, pressure head, and efficiency ratio $\eta/(1-\eta)$.

1.A.1.2.5 Gas Turbine Combustor The basic features are annular tube and burning natural gas. Inlet and exit temperatures, air/fuel mass ratio, and number of cans are kept constant at 1600°F (870°C), 500°F (260°C), at values of 75, and 7, respectively. Air mass rate, pressure, and pressure loss are varied and the combustor surface computed. (60 m/s). The maximum velocity was set at 200 ft/s. Combustion intensity varied at ~80 kW $(ft^3 \cdot atm)^{-1}$ (2000 kW $(m^3 \cdot atm)^{-1}$). The combustor surface is correlated in terms air mass rate, inlet pressure, and pressure loss.

1.A.1.2.6 Heat Exchangers The basic features are forced convection heat exchange, single- and two-phase fluids, and three generic types of exchangers (double-tube, fin-plate, and shell-and-tube). For the shell-and-tube type, flow may be counter or crosscounter, tubes may be plain or finned on the outside, and shell may be cylindrical or duct-type. In two-phase procedures, more than one equation is used for film coefficients and friction factor multipliers. Pressure losses were based on the worst-case multiplier. The plate-fin type consists of layers of plates with straight parallel fins on each side of each plate. The fins on one side are perpendicular to those of the other side. Two sets of layers may be connected in series to allow for mixing. A surface geometry is selected. For shell and tube geometry, tube length, diameter, and pitches, and shell diameter or width and depth are entered. For the plate-fin geometry, the number of plates, fins per inch on either side, and their heights and thickness are entered. Two groups of boundary parameters can be entered: mass rates and temperature and pressures at all inlets and exits or mass rates and inlet pressures and temperatures and effectiveness. With both entries, film coefficients of heat transfer and pressure drops are computed. The heat exchange surface is computed in either of two ways: surface by geometry or surface $= Q/U\,\Delta T$. With the first entry computations are iterated until the two areas are matched and the two pressure drops are accommodated. With the second one, only the surface iteration is needed. The pressure drops are output parameters. The iterations are both manual and automated to minimize the heat exchange surface. This program, evolved in parallel with the thermodynamic model.

The superheater, the boiler, and the economizer of the heat recovery steam generator assumed duct- type shell and tubes with outside circular fins. Fin geometry on the outside of the steam generator tubes and fouling factors in heat exchange are kept constant. The brine heater and the flash stages are assumed to have plain tubes. The brine heater is assumed to have a cylindrical shell. The flash stages are assumed to have a duct-type shell. A constant temperature drop is assumed for all the stages and a chamber at a temperature of 150°F (65°C) is assumed to represent all the stages. A heat transfer temperature difference correction is introduced for the rejection stages. Two types of air preheater are considered: shell-and tube with circular fins on the outside or a plate-fin type. For the evaporator/condenser a vertical shell-and-tube type with plain tubes is assumed. The heating steam condenses in the tubes, and the liquid is sprayed on the outside at a rate 10 times the vaporized liquid. In all the exchangers fouling factors are kept constant. Inlet parameters are varied. The temperature profile is first computed and checked for crossings and pinch point. Rate of heat exchange, effective temperature difference of heat transfer, and heat exchange surface are computed among other detailed heat transfer outputs. Pressure drops, if outputs, are computed. Given the inlet pressure temperature and mass rate for the two fluids, all heat exchange surfaces are correlated in terms the rate of heat exchange, a temperature difference (terminal or logarithmic mean) and hot and cold side pressure losses. In a flash stage, any temperature drop induced by flashing is used instead of a pressure loss. The effect of pressure and temperature levels is accounted for in the unit cost (severity of operation).

1.A.1.2.7 Radiant Heat Exchange in Boiler A simple model is assumed. The basic features are square vertical duct type, forming water walls combined with reflectors and backed with insulation. The water boils in the tubes, and the vapor formed is separated in an upper drum. The total radiation exchange between the entire gas volume and the walls is based on the mean beam length. The absorption bands of H_2O and CO_2 of the hot gases are factored in. The effect of temperature variation along the duct is accounted for by dividing the gas path into five sections, each with a uniform temperature. The gas is assumed to enter at the adiabatic flame temperature. The effect of convective film coefficients and wall resistance is included. The height of the duct and its width were observed to control both the gas-side and the steam-side pressure losses, beside the heat exchange surface. The gas-side pressure loss was not significant because of the significant change in gas temperature and the associated lowering of gas velocities. The tube-side pressure drop progressively increased as the height is increased relative to the width with negligible effect on the heat transfer coefficient. The heat exchange is expressed by an overall heat transfer coefficient ranging from 30 to 50 Btu $(\mathrm{h} \cdot \mathrm{ft}^2 \cdot \mathrm{F})^{-1}$ [0.170–0.284 kW $(\mathrm{m}^2 \cdot \mathrm{K})^{-1}$], while convection-induced heat exchange was lower by only one order of magnitude. The surface of the wall tubes is correlated in terms of the rate of heat transfer and the conventional logarithmic mean temperature difference. An equivalent temperature driving force $\Delta T_r = (T_{gas}/T_{flame})^4 - (T_{steam}/T_{flame})^4$ was used in earlier applications with a different correlation ($A = 0.39Q\,\Delta T_r^{-2}$).

1.A.1.2.8 Curve-Fitting Costing Equations and Exergy Destruction Various mathematical procedures are available for curve fitting by minimizing the deviations around a fit. The number of the surfaces A or exergy destructions D generated should be much larger than the correlating parameters, which usually vary from 2 to 4. One simple procedure is to use sets of number equal to the number of the correlating parameters plus one, to obtain the coefficient k and exponents $\{n\}$. The ratios of the computed A generated by a set to the corresponding one generated by the design model are computed. The process is repeated with different sets until a set is found where the ratios deviate the least from one. The same rule of thumb applies to D. The correlating parameters of the costing equations and their range of applicability are given in Section 1.A.2.

Deviations in curve fitting were within ±10%, and in very few cases of a wide range of applicability deviations of ±20% were found. It is important to note that improved correlations depend on improving the quality of models and reducing their range of applicability.

References listed in Section 1.F.10 at the end of this chapter are additional useful readings for the preceding Section 1.A.1.

1.A.2 The Capital and Fuel Costing Equations of some common Devices (Tables 1.A.1 and 1.A.2)

For illustration, let the device be a forced-convection heat exchanger. It is assumed to be the superheater, component 7 of the heat recovery steam generator of the simple combined cycle of Figure 2. A duct shell-and-finned tube type is assumed. The fins are assumed circular on the outside that is, on the gas side. The design model of heat exchangers described in Secion 1.A.2 is used.

The boundary parameters P, T, $\{x\}$, M at inlets and exits of the exchanger as embedded in the system at a design point for the system are used. The exchanger physical surface and its geometry are defined by length, diameter, spacing, number, material, material thickness and fin geometry of the tubes. These parameters are usually more than sufficient to allow for adjustment to match the computed surface and pressure drops by film coefficients and friction factors for the given heat load and its temperature profile. Any extra design degrees of freedom are used to minimize the surface and/or to satisfy reliable design practices. The design process is thus a matching/minimizing process.

The minimized surface as a function of performance is generated by repeating this design process for different boundary parameters within a range relevant to the optimization of the system. A specific geometry of minimized surface is obtained for each set of boundary parameters. The surface is then expressed by an appropriate set of performance parameters such as heat loads, mass rates, heat exchange temperature differences, effectiveness, and pressure losses. In this example, the surface of the fins and tubes is expressed in terms of the heat load, the logarithmic mean temperature difference and pressure losses on the shell side and on the tube side. The following form is used:

$$A = kQ^{n1} \Delta T_m{}^{n2} \Delta P_t^{n3} \Delta P_s{}^{n4} \tag{1.A.1}$$

where A is converted to a costing equation by

$$Z = c_a A \tag{1.A.2}$$

Table 1.A.1 Generated Capital Costing Equations and Their Local Objectives

	Costing Equation $Z = c_a A$, $A = k x_1^{n1} x_2^{n2} x_3^{n3} x_4^{n4}$			Local Objective $J(Y) = c_d D + c_z Z = K e Y^{ne} + k_z Y^{nz}$			
Device	c_a \$k/ft^2 /(m^2)	k, IP/(SI)	$x_1^{n1} x_2^{n2} x_3^{n3} x_4^{n4}$ ranges of $\{x\}$,(IP/SI)	Y	n_e	n_z	Dissipation
1 Axial compressor	50 (538)	0.15 (0.0063)	$M^1\,Pr^{0.45}\,e^{0.45}$ 50–1000, 5–15, 2.3–11.5 (25–455, 5–15, 2.3, 11.5)	e	−0.95	0.45	D_{PT}
2 Gas turbine	50 (538)	0.32 (0.0135)	$M^1 Pr^{-0.5} e^{0.85}$ 50–1000, 5–15, 4–19 (25–455, 5–15, 4–19)	e	−0.8	0.85	D_{PT}
3 Steam turbine	50 (533)	0.9 (1.978)	$M^1 (T_i/P_i)^{0.05} P_e^{-0.75} e^{0.9}$ 25–100, 1.5–30, 1–150, 4–19 (11–45, 120–2400, 0.0071–1.03, 4–19	e	−0.8	0.90	D_{PT}
4 Pump feed	3 (32)	0.0025 (0.000435)	$M^{0.55} \Delta P^{0.55} e^{1.05}$ 5–70, 14–900, 1.8–9 (2–32, 100–6200, 1.8–9)	e	−1	1.05	D_{PT}
5 Pump cooling-water	3 (32)	0.0063 (0.00183)	$M^1 \Delta P^{0.1} e^{0.7}$ 100–500, 2–25, 4–14 (45–230, 14–170, 4–14)	e	−1	0.7	D_{PT}
6 Fan/blower	3 (32)	0.063 (0.0183)	$M^1 \Delta P^{0.1} e^{0.7}$ 100–500, 0.1–0.6, 2–9 (45 = 230, 0.7–4, 2–9)	e	−1	1.05	D_{PT}
7 Combustor	0.2 (2.15)	5.85 (0.261)	$M^{0.5} P^{0.24} dp^{-0.75}$ 400–900, 50–200, 0.01–0.3 (180–410, 0.34–1.38, 0.01–0.3)	dP	1	−0.75	D_P

(Continued)

Table 1.A.1 (*Continued*)

Device	Costing Equation $Z = c_a A$ $A = kx_1^{n1}x_2^{n2}x_3^{n3}x_4^{n4}$			Local Objective $J(Y) = c_d D + c_z Z = KeY^{ne} + k_z Y^{nz}$			
	c_a \$k/ft^2 /(m^2)	k, IP/(SI)	$x_1^{n1}x_2^{n2}x_3^{n3}x_4^{n4}$ ranges of $\{x\}$,(IP/SI)	Y	n_e	n_z	Dissipation
8 Superheater, convective	0.03 (0.32)	340 (32.48)	$Q^1 \Delta T_m^{-1} dP_t^{-0.15} dP_s^{-0.14}$	ΔT_m	1	−1	D_T
			10–15, 100–200, 6–13, 0.06–0.44	dP_t	1	−0.15	D_{PT}
			(10–15, 55–110, 40–90, 0.4–3)	dP_s	1	−0.14	D_{Ps}
9 Boiler, convective	0.03 (0.32)	340 (32.48)	$Q^1 \Delta T_m^{-1} dP_t^{-0.33} dP_s^{-0.26}$	ΔT_m	0.45	−1	D_T
			(25–55, 75–200, 6–13, 0.06–0.44	dP_t	1	−0.33	D_{PT}
			25–55, 40–110, 40–90, 0.4–3	dP_s	1	−0.26	D_{Ps}
10 Economizer	0.03 (0.32)	310 (29.89)	$Q^1 \Delta T_m^{-1} dP_t^{-0.16} dP_s^{-0.125}$	ΔT_m	0.45	−1	D_T
			15–40, 70–105, 6–26, 0.007–0.56	dP_t	1	−0.16	D_{Pt}
			15–40, 38–60, 40–180, 0.5–4	dP_s	1	−0.125	D_{Ps}
11 Brine	0.04 (0.43)	3.3 (0.367)	$Q^1 \Delta T_t^{-0.7} dP_t^{-0.08} dP_s^{-0.04}$	ΔT_t	0.9	−0.7	D_T
Feed heater			40–185, 5–15, 0.1–7, 0.001–1.3	dP_t	1	−0.08	D_{PT}
condenser			(40–185, 2.5–80, 0.7–50, 0.007–9	dP_s	1	−0.04	D_{Ps}
12 MSF	0.04 (0.43)	10 (1.6)	$Q^1 \Delta T_n^{-0.75} \Delta T_t^{-0.5} dP_t^{-0.1}$	ΔT_n	1.5	−0.75	D_T
			14–110, 3–10, 3–12, 0.2–10	ΔT_t	1	−0.5	D_T
			14–110, 1.7–6, 1.7–7, 13–70	dP_t	1	−0.1	D_{Pt}
13 Radiant boiler	0.08 (0.86)	100 (16.73)	$Q^1 \Delta T_m^{-1}$ 50–600, 0.1–1	$dP_t = 0.0004A - 25$ $dP_t = 0.000037A - 1.72$	No tradeoffs; dependent on surface		

14 Air Preheater plate-fin	0.008 (0.086)	37,000 (3496)	$Q^1 \Delta T_m^{-2} dP_h^{-0.3} dP_c^{-0.3}$	ΔT_m	1	−2	D_T
			10–100, 50–150, 0.03–1.5, 0.03–1.5	dP_h	1	−0.3	D_{Ph}
			(10–100, 28–83, 0.2–10, 0.2–10)	dP_c	1	−0.3	D_{Pc}
15 Air Preheater shell-and-tube	0.03 (0.32)	2750 (235)	$Q^1 \Delta T_m^{-1.5} dP_t^{-0.3} dP_s^{-0.2}$	ΔT_m	1	−1.5	D_T
			10–100, 50–150, 0.03–1.5, 0.03–1.5	dP_t	1	−0.3	D_{Ph}
			(10–100, 28–83, 0.2–10, 0.2–10)	dP_s	1	−0.2	D_{Pc}
16 Throttle valve	0.75 (8.07)	0.45 (0.989)	$M^1 (T_i/P_i)^{0.05} P_e^{-0.75}$	No tradeoff			
			5–20, 1.5–5, 0.5–100				
			(2–9, 120–400, 0.003–0.7)				
17 Mixing chamber	30 (1060)	1 (1)	V^1	No tradeoff			
18 c_a pressure factor	—	0.191 (0.85)	$P^{0.3}$	No tradeoff			
19 Evaporater condenser	0.04	6.2	$Q^1 \Delta T_m^{-1} dP_t^{-0.01} dP^{-0.1}$	ΔT_m	1	−1	D_T
			150–800, 4–40, 0.01–0.05, 0.01–0.04	dP_t	1	−0.01	D_{Ph}
			(50–800, 2–22, 0.06–0.35, 0.06–0.3	dP_s	1	−0.1	D_{Pc}
20 Vapor Compressor, radial	9 (96.9)	0.0018 (0.000076)	$(v\ M)^1\ Pr^1\ e^{0.7}$	e	−0.95	0.7	D_{PT}
			M50–1000, 1.1–2, 2.3–11.5				
			(M22–455, 1.1–2, 2.3–11.5)				
21 *hx* general approximation	0.04	5	$Q^1 \Delta T_m^{-1} dP_t^{-0.15} dP_s^{-0.15}$	ΔT_m	−1	0.5	D_T
			15–100, 4–40, 0.05–1, 0.03–0.4	dP_t	1	−0.15	D_{Ph}
			(15–100, 2–22, 0.3–7, 0.2–2.3)	dP_s	1	−0.15	D_{Pc}

Notations: In columns 2–4, Imperial (UK) values are presented First, Followed by SI (intemational; metric) values in parentheses. Imperial (IP) units: c_a, \$*k*/ft^2; A, ft^2; M, lb/s; Q, kW (range MW); P , P_e, psia; T_i, R; ΔT, °F;
$\Delta P, dP$, psi; V, ft^3/s. SI units: c_a, \$k/m^2; A, m^2; M, kg/s; Q, kW (range MW); P_i, P_e, MPa; T_i, K; ΔT, °C;
$\Delta P, dP$, kPa; V, m^3/s. Nomenclature: D = dissipation (exergy destruction), kW; $D = D_P + D_T + D_C$, $D_{PT} = D_P + D_T$; c_d = unit cost of exergy destruction, \$/kWh, $e = \eta/(1-\eta)$; Pr = pressure ratio. General notes: Effect of pressure on unit costs is assumed 1 for pressures < 250 psia; cost of a steam ejector is assumed double that of the throttle valve.

Table 1.A.2 Generated and Gathered Off-Design Performance Equations

Component	Equations
	By Design Models
6) Combustor	$\Delta P = \Delta P_d (M_g/M_{gd})^{1.75}$
7) Superheater	$\eta = \eta_d (M_h/M_{hd})^{0.2} (M_c/M_{cd})^{-0.15}$
	$\Delta P_h = \Delta P_{hd} (M_h/M_{hd})^{1.75}$
	$\Delta P_c = \Delta P_{cd} (M_c/M_{cd})^{1.8}$
8) Boiler	$\eta = \eta_d (M_h/M_{hd})^{-0.05} (M_c/M_{cd})^{0.01}$
	$\Delta P_h = \Delta P_{hd} (M_h/M_{hd})^{1.75}$
	$\Delta P_c = \Delta P_{cd} (M_c/M_{cd})^{1.75}$
9) Economizer	$\eta = \eta_d (M_h/M_{hd})^{0.15} (M_c/M_{cd})^{-0.05}$
	$\Delta P_h = \Delta P_{hd} (M_h/M_{hd})^{1.75}$
	$\Delta P_c = \Delta P_{cd} (M_c/M_{cd})^{1.1}$
10) Condenser	$\eta = \eta_d (M_h/M_{hd})^{-0.05} (M_c/M_{cd})^{-0.35}$
	$\Delta P_h = \Delta P_{hd} (M_h/M_{hd})^{1.41}$
	$\Delta P_c = \Delta P_{cd} (M_c/M_{cd})^{0.6}$
	By Generalized Correlations
11) Compressor (axial)	$M_r = M/M_d$
	$\eta_r = a_1 + a_2 M_r + a_3 M_r^2$
	$M_r \geq 0.5, \eta_r = 0.9$
	$\eta = \eta_d \eta_r / \eta_{rd}$
	$PR = PR_d M_r$
12) Adjustable IGV	$a_1 = -0.7508, a_2 = 3.2414, a_3 = -1.5906$
Adjustable IGV + 5 stators	$a_1 = 0.3337, a_2 = 1.0917, a_3 = -0.5254$
Gas turbine	$PR_r = PR/PR_d, M_r = N_M/N_{Md}$
	η_r or $M_r = (a_1 + a_2 PR_r + a_3 PR_r^2)$
	$N_M = M/P(T)^{0.5}$, a correlating flow number
	$\eta = \eta_d \eta_r / \eta_{rd}$
	$\eta_{rd} = 0.9, a_1 = 0.6164, a_2 = 0.6179, a_3 = -0.3343$
	For $PR_r \geq 0.53, M_r = 1$
	For $PR_r < 0.53, M_r$ has
	$a_1 = 0.1228, a_2 = 2.8283, a_3 = -2.2145, T_{firing} = (M_r N_{Md} P_i / M_{air})^2$ to match rpm (r/min)
Steam turbine	$M_r = M/M_d, PR_r = PR/PR_d$
	$\eta/\eta_d = A_1 + A_2 M_r + A_3 M_r^2$
	$A_i = a_{i1} + a_{i2} PR_r + a_{i3} PR_r^2$
	Reaction turbine $a_{ij} =$
	0.247917, 0.128125, −0.0101042
	1.23125, −0.221875, 0.0215625
	−0.479167, 0.09375, −0.0114583
	Impulse turbine $a_{ij} =$
	0.425833, 0.001875, 0.00302083
	0.882500, 0.066875, −0.01031250
	−0.308333, −0.068750, 0.00729167
Feed pump	$M_r = M/M_d, \Phi_r = 0.1$
	$\Psi_r = 0.52, \eta_r = 0.85$
	$\Phi = \Phi_r M_r, \Phi \leq 0.15$
	$\Psi = 0.595 - 0.3\Phi - 4\Phi^2$
	$\eta = (14\Phi - 56.9\Phi^2)/\eta_r \eta_d$
	$P = P_d \Psi / \Psi_r$
Cooling-water pump	$M_r = M/M_d, \Phi_r = 0.13,$
	$\Psi_r = 0.3, \quad \eta_r = 0.9$
	$\Phi = \Phi_r M_r, \Phi \leq 0.18$
	$\Psi = 0.55 - 1.83\Phi - 0.667\Phi^2$
	$\eta = (12.167\Phi - 43.3\Phi^2)/\eta_r \eta_d$
	$P = P_d \Psi / \Psi_r$

The unit cost c_a depends on the type of material and the manufacture process and is also dependent on time and location. It is expressed thermodynamically as function of pressure, temperature and composition (severity of operation). In this example, c_a is assumed per unit total surface of fins and tubes. Ten minimized surfaces were generated by changing inlet P, T and M; the allowed pressure losses and effectiveness were also included. Heat load, exit conditions, and logarithmic mean temperature difference are recorded. The parameters that remain fixed are the fin geometry, tube thickness, tube arrangement (staggered), fouling factors, flow directions (gas horizontal, steam with gravity). In this particular example, the effect of gravity on pressure losses is negligible. Table 1.A.3 lists the recorded parameters of the 10 minimized surfaces and the quality of the correlation.

The constant k and the four exponents n_1, n_2, n_3 and n_4 of Equation (1.A.1) are computed by using the surfaces of five cases simultaneously. These five cases are selected randomly from the total number of cases. The computed constant and exponents that best fit the surfaces of all the cases is selected. The simultaneous solution involves the inverse of a 4×4 matrix. When the matrix determinant is relatively small, unreasonable exponents are obtained and have to be rejected. Also, some selections may give rise to singular solutions and fail to give any values altogether. There are, however, many sets that give solutions. There is also opportunity to round off the best-fit exponents along with a modified value of the constant k such that the quality of the fit is not changed. The best fit is selected by comparing various sets. No formal regression approach is used to seek the best fit.

The constant and exponents obtained were $k = 30.71, n_1 = 1, n_2 = -1$, $n_3 = -0.15$, and $n_4 = -0.14$. The equation is applicable in the range $Q = 8$ to 66 MW, $\Delta T_m = 38$ to 130°C, $\Delta P_t = 20$ to 90 kPa, and $\Delta P_s = 0.2$ to 1.2 kPa with average scatter $\pm 8\%, \max +10\%$. Inside tube surfaces covered the range 110 to 975 m^2.

1.A.3 Some Useful Forms of Flow Exergy Expressions

1.A.3.1 Equations

$$E = H - T_0 S - \sum \mu_{i0} X_i \tag{1.A.3}$$

where E is the flow exergy per unit matter. Then we can, either use

$$[H_{0d} - T_0 S_{0d}]_{T_0, P_0, \{Xi_0\}} = \sum \mu_{i0} X_i \tag{1.A.4}$$

or introduce

$$[H_0 - T_0 S_0]_{T_0, P_0, \{Xi\}} = \sum \mu_i X_i \tag{1.A.5}$$

Equation (1.A.4) uses the dead-state enthalpy and entropy directly by the subscript od. Equation (1.A.5) introduces an intermediate state at T_0, P_0 without changing

Table 1.A.3 The Superheater Performance/Design Correlating Matrix Minimized Surfaces Versus Thermodynamic and Geometric Parameters

Run	A_{tube} m^2	Q, MW	ΔT_{lm}, °C	η	ΔP_t, kPa	ΔP_s, kPa	L_{tube}, m	d_o, cm	W_{sh}, m	Pitches 1, 2, cm		A_{fin}/A_i	N_{tube}	N_{pass}
1	486	15.76	66	0.921	42	0.462	20.4	2.5	11.9	5	4.52	11.8	364	1
2	915	66.80	128	0.609	41	0.475	5.8	2.5	52.1	5	4.52	11.8	2,397	2
3	620	17.32	49	0.883	42	0.544	29.6	2.5	8.8	5	4.52	11.8	321	1
4	897	31.50	66	0.921	48	0.627	20.4	2.5	20.4	5	4.52	11.8	673	1
5	856	34.66	66	0.921	37	1.192	16.8	2.5	20.1	5	4.52	11.8	776	1
6	976	34.28	39	0.921	82	0.903	12.2	5	15.5	10	9.04	19.7	1,258	8
7	188	7.88	66	0.921	90	0.834	85.3	7.6	0.91	15	13.6	27.8	10	1
8	276	8.67	66	0.921	90	0.227	45.7	3.8	3.7	7.6	6.78	15.7	57	1
9	355	9.52	66	0.919	21	0.234	29.6	3.8	6.4	7.66	0.78	15.7	114	1
10	112	9.52	126	0.400	83	0.965	34.1	7.6	2.1	15	13.6	27.8	15	1
						Scatter of the Correlating Costing Equation								
Run			1	2	3	4	5	6	7	8		9	10	
A_{eqn}/A_{table}			0.965	1.10	1.08	0.98	1.06	0.92	1.02	0.92		0.976	1.08	

composition using the subscript o. In fact, any intermediate state convenient for property computations can be introduced. Using Equation (1.A.5), we obtain

$$E = (H - H_0) - T_0(S - S_0) + \sum(\mu_i - \mu_{i0})X_i \tag{1.A.6}$$

where the thermal mechanical part of exergy is

$$E^{\mathrm{tm}} = (H - H_0) - T_0(S - S_0) \tag{1.A.7}$$

and the chemical part is

$$E^{\mathrm{c}} = \sum(\mu_i - \mu_{i0})X_i \tag{1.A.8a}$$

$$= \mathrm{R}T_0 \sum X_i \ln\left(\frac{a_i}{a_{i0}}\right) \tag{1.A.8b}$$

where

$$a_i = \gamma_i X_i = \frac{f_i}{f^0} \tag{1.A.9}$$

- *For Ideal-Gas Mixtures*. The Thermal Mechanical component may be further divided into the Thermal part:

$$E^{\mathrm{t}} = C_{\mathrm{p}}(T - T_0)\left(\frac{1 - T_0}{T_{\mathrm{m}}}\right) \tag{1.A.10}$$

where

$$T_{\mathrm{m}} = \frac{T - T_0}{\ln(T/T_0)} \tag{1.A.11}$$

and the Mechanical part

$$E^{\mathrm{m}} = RT_0 \ln\left(\frac{P}{P_0}\right) \tag{1.A.12}$$

In terms of mole fractions, the Chemical part becomes

$$E^{\mathrm{c}} = RT_0 \sum X_i \ln\left(\frac{X_i}{X_{i0}}\right) \tag{1.A.13}$$

- *For Nonideal Mixture Excess Gibbs Function*

$$G_x = RT \sum N_i \ln \gamma_i \tag{1.A.14}$$

Differentiation gives

$$\gamma_i = \frac{(\partial G_x/\partial N_i)}{(RT)} \tag{1.A.15}$$

$$H_x = -T^2 \times \partial \left(\frac{G_x}{T}\right) / \partial T \tag{1.A.16}$$

$$S_x = (H_x - G_x)/T \tag{1.A.17}$$

$$V_x = \partial G_x / \partial P \tag{1.A.18}$$

- *Changes in Terms of Measurables*

$$dh = C_\mathrm{p} dT + [V - T(\partial V/\partial T)_\mathrm{P}) dP] \tag{1.A.19}$$

$$dS = C_\mathrm{p} \frac{dT}{T} - \left(\frac{\partial V}{\partial T}\right)_\mathrm{P} dP \tag{1.A.20}$$

$$f^{\mathrm{vi}} = \phi_i Y_i P \tag{1.A.21}$$

$$f^{\mathrm{li}} = \gamma_i X_i P_{\mathrm{si}} \phi_{\mathrm{si}} F \tag{1.A.22}$$

$$F = \exp\left(\int \frac{V^{\mathrm{li}} dP}{RT}\right) \tag{1.A.23}$$

$$f^{\mathrm{vi}} = f^{\mathrm{li}} \tag{1.A.24}$$

- *Two-Component Mixture, the Gibbs Excess Function is*

$$G_x = X_1 X_2 [A + B(X_1 - X_2)] \tag{1.A.25}$$

$$\ln \gamma_1 = \frac{[(A + 3B){X_2}^2 - 4B{X_2}^3]}{RT} \tag{1.A.26}$$

$$\ln \gamma_2 = \frac{[(A - 3B){X_1}^2 + 4B{X_1}^3]}{RT} \tag{1.A.27}$$

where A and B are particular constants for the two components

- *Using More than One Dead-State Composition*

$$\sum (\mu_i - \mu_{i0}) X_i = \sum_{i1} (\mu_i - \mu_{i0}) X_i + \sum_{i2} (\mu_i - \mu_{\mathrm{ir}}) X_i + \sum_{i2} (\mu_{\mathrm{ir}} - \mu_{i0}) X_i \tag{1.A.28}$$

Another dead state, denoted here with the subscript ir (e.g., sea), is assumed for species (e.g., salts) having traces in the usually assumed dead state denoted here with the subscript io (e.g., air) but is relatively abundant in ir. The last term is a constant of no interest in exergy change beyond the dead state ir.

- *Introducing Known Intermediate Chemical Changes*

$$\Delta G_R = \sum_R (\mu_i - \mu_{i0}) X_{iR} - \sum_P (\mu_i - \mu_{i0}) X_{iP} \tag{1.A.29}$$

Let μ_j be a reactant (e.g., a hydrocarbon fuel) of minute equilibrium mole fraction in the assumed dead state (e.g., air); then μ_{j0} is determined by

$$(\mu_j - \mu_{j0}) X_j = \Delta G_R + \sum_P (\mu_i - \mu_{i0}) X_{iP} - \sum_R (\mu_i - \mu_{i0}) X_{iR, i \neq j} \tag{1.A.30}$$

1.A.3.2 Balances

1.A.3.2.1 Exergy Balance

$$\sum_{in} E_b = \sum_{out} E_b + \sum D \tag{1.A.31}$$

where

$$D = \text{exergy destruction} \tag{1.A.32}$$

$$E_b = E^q + E^w + E^f \tag{1.A.33}$$

$$E^q = Q(1 - T_0/T_b) \tag{1.A.34}$$

$$E^w = W_s \tag{1.A.35}$$

$$E^f = ME$$

1.A.3.2.2 Entropy Balance

$$\sum_{out} S_b - \sum_{in} S_b - S^{cr} \tag{1.A.36}$$

where

$$S_b = S^m + S^q \tag{1.A.37}$$

$$S^q = Q/T_b \tag{1.A.38}$$

$$S^m = MS \tag{1.A.39}$$

$$D = T_0 S^{cr} \tag{1.A.40}$$

1.A.3.2.3 A Note on the Dead State Environment An absolute dead state of zero exergy does not exist, but a reference one can be set. Arbitrary reference states have long been used. In thermodynamic properties zero enthalpy and entropy differ for different working fluids. In chemical reactions, elements are selected as reference to compute the energy and free energy of formation of compounds.

A reference dead state for zero exergy is defined by a pressure P_0, a temperature T_0 and a set of chemical species of composition $\{X_{io}\}$ suitable for analyzing the utilization of energy in a particular situation. The composition $\{X_{i0}\}$ is preferred to resemble a natural state in which the chemical species of interest are not traces in order to establish chemical exergies. Atmospheric air is an appropriate dead state to use for work with a number of gases, including combustion products, although the pure species may be used as reference. Seawater is appropriate for work with desalination. Bauxite is appropriate when dealing with the purification of aluminum. More than one dead state may be assigned as shown by equation (26), so long the potential work between the two dead states is not of immediate interest.

A selected dead state implies a large environment of constant values for P_0, T_0 and $\{X_{i0}\}$. In most natural environments, P_0 and $\{X_{i0}\}$ remain more or less constant but T_0 may exhibit daily and seasonal variations. When the change has significant effect on the value of exergy, exergy analysis is repeated as function of time periods of different dead state temperatures.

1.A.4 Theoretical Separation Work Extended to Zero Liquid Discharge

Theoretical work of separation may shed light on the thermodynamics and partly on the economics of zero liquid discharge. Conventional desalination approximately doubles the dissolved salt concentration of the feed. In other words, conventional desalination produces half of the feed as does desalted water product. Zero liquid discharge produces all the water in the feed as desalted water product. A theoretical work about double that of conventional desalination would hence be encouraging to the idea of zero liquid discharge.

The thermodynamic computations here assume sodium chloride as the ideal solution. Deviations from ideal solution are expected as concentration increases and the theoretical work is expected to be even higher.

Assuming an ideal NaCl solution, let X = mass fraction and x = mole fraction, $T = 80^\circ$F. We can then calculate the theoretical power at infinite feed per unit product as follows:

$$w_{\min} = \left(\frac{\partial \mathrm{G}}{\partial \mathrm{M_w}}\right)_{\mathrm{feed}=\infty} = \mathrm{RTln}\left(\frac{1}{x_{\mathrm{w}}}\right) \qquad (1.A.41a)$$

X	x	$w_{\min}$ Btu/lb pure, feed $= \infty(x_b \approx x_f)$	$RT\ln(1/x_w) \approx RTx_s (x_s \leq 0.3)$
0.95	0.854	115	50.90
0.75	0.480	39.0	28.61
0.50	0.235	16.0	14.00
0.28	0.1069	6.74	6.37
0.27	0.1022	6.42	6.09
0.25	0.0930	5.82	5.54
0.15	0.0515	3.15	3.07
0.07	0.0226	1.37	1.35
0.04	0.0127	0.76	0.76
0.035	0.0114	0.68	0.68

The theoretical work with finite feed per unit product is obtained by integrating equation (1.A.41a) from initial x_f to final concentration x_b:

$$w_{\min} = \frac{1}{M_w}\int_{x_f}^{x_b} \frac{\partial G}{\partial M_w} dM_w = \frac{RT}{M_w}\int_{x_f}^{x_b} \ln\left(\frac{1}{x_w}\right) dM_w$$

Integrating and noting that $dM_w = d(M_f(x_b - x_f)/x_b) = M_f x_f dx_b / x_b^2$ and letting $\ln(1/x_w) = x_b$ we obtain

$$w_{\min} \approx \frac{RTx_b x_f}{(x_b - x_f)} \ln\left(\frac{x_b}{x_f}\right) \qquad \text{(given } M_f \text{ and } x_f\text{)} \qquad (1.A.42)$$

The theoretical work of separation taking in consideration the saturation limits of the salt species are a consequence of water separation by equation 1.A.42 for an ideal concentrator and of water and salt separation [obtained by Eq. (1.A.42)] in an ideal crystallizer.

The theoretical work taking in consideration the saturation salt content of NaCl at $80^\circ F = 0.27 lb_s/lb_{soln}$, is obtained as follows. From feed X_f to $X_{sat} = 0.27$, Equation (1.A.42) applies, producing M_{d1}. The work of salt separation in an actual process requires a departure from saturation. Letting the departure for estimating the theoretical work be zero, meaning an infinite feed of equation (1.A.41a), we obtain

$$w_{s\min} \approx RTx_{sat} \qquad (1.A.41b)$$

The separated water and separated salt occur at the same x_{sat}; this means that

$$M_s/(M_{d2} + M_s) = x_{sat}$$

where M_{d2} is determined by the separated salt and vice versa.

For a given feed M_f

$$W_{\min,Mf} \approx M_{d1} RTx_{sat} \frac{x_f}{(x_{sat} - x_f)} \ln\frac{x_{sat}}{x_f} + M_{d2} RTx_{sat} \qquad (1.A.43a)$$

where

$$M_{d1} = M_f(x_{sat} - x_f)/x_{sat}$$
$$M_{d2} = M_s(1 - x_{sat})/x_{sat}$$

Per unit product water

$$w_{min} = \frac{W_{min,Mf}}{M_f - M_f x_f} \tag{1.A.43b}$$

Four initial and final concentrations lb_s/lb_{soln} of NaCl aqueous solution are of interest to saline water desalination:

X_f	x_f	X_{sat}	x_{sat}	X_b	x_b	
0.035	0.0114	—	—	0.07	0.0226	represents conventional seawater desalting
0.035	0.0114	0.270	0.1022	1.00	1.00	represents ZLD seawater desalting
0.070	0.0226	0.270	0.1022	1.00	1.00	represents ZLD retrofit seawater desalting
0.250	0.0930	0.270	0.1022	1.00	1.00	represents ZLD oil field water desalting

At $T = 80°F$	Per lb feed eqn (3)			Per/lb product Eq. (1.A.43b)	
M_{d1}	W_{min}	M_{d2}	$W_{s,min}$	w_{min}	
1	0.94	0	0	$0.94/1.00 = 0.94$	usual seawater desalting
0.8704	1.68	0.0945	6.092	$2.03/0.965 = 2.1$	ZLD seawater desalting
0.7407	2.61	0.189	6.092	$3.09/0.930 = 3.3$	ZLD retrofit seawater
0.0741	5.81	0.675	6.092	$4.54/0.750 = 6.1$	ZLD oil field water

The preceding theoretical analysis shows the trend of desalination theoretical energy requirement in terms of the salt content of the feed and the reject assuming

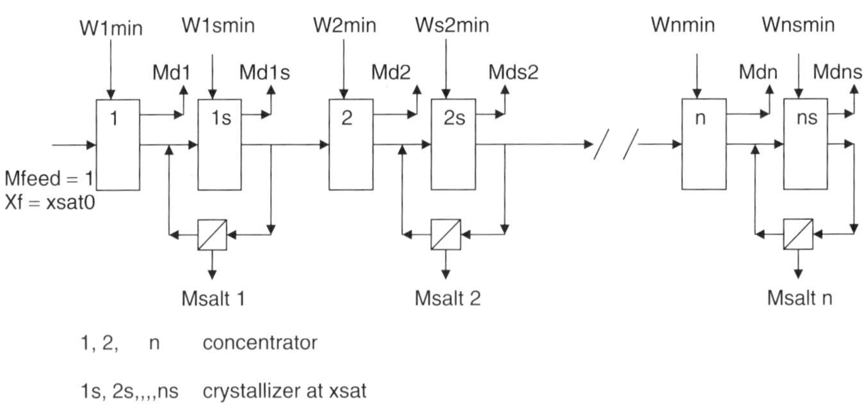

Figure 1.A.1 Theoretical processing of a feed of n salt saturation limits.

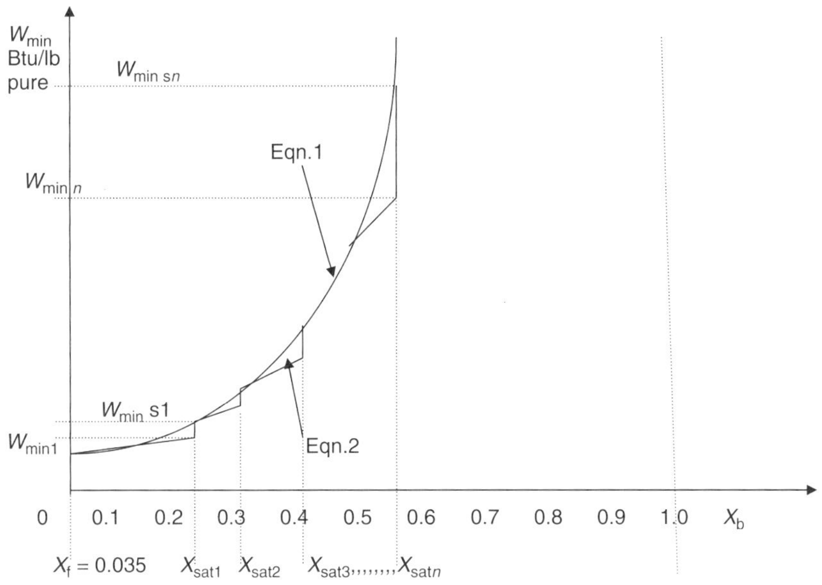

Figure 1.A.2 Feed of n Saturation limits on $w_{\text{theor}}-x_{\text{sat}}$ coordinates.

sodium chloride aqueous solution as feed. Zero-liquid-discharge desalting is 2 times the conventional seawater theoretical requirement for seawater salt content as feed. It is 3 times the conventional reject brine salt content as feed, and it is 6 times that with oilfield product–water salt content as feed. Without improved desalting efficiency, a similar trend is expected for actual work requirement and for cost since the cost of energy is a major part of the desalination processes. Improved desalting efficiency may be sufficient to justify competitive zero liquid discharge of seawater or its conventional reject brine. For produced water as feed, a net environmental benefit is also needed to justify competitiveness.

The theoretical work described above is derived assuming sodium chloride aqueous solution as an ideal solution. A factor >1 is often used to accommodate the deviation from ideality.

Any saline water contains more than one salt species. Depending on composition and the solubility limits of the individual salts in water, a sequence of precipitations peculiar to the saline water is expected.

Let n be the number of saturation limits occurring one after the other (consecutively), and let the feed be denoted by $x_{\text{sat}}(0)$ of $M_{\text{d}}(0) = 0$. The minimum work per unit feed becomes

$$W_{\text{min,fd}=1} = \sum_{i=0}^{i=n} M_{\text{d}}(i)RT \frac{x_{\text{sat}}(i)\, x_{\text{sat}}(i-1)}{x_{\text{sat}}(i) - x_{sat}(i-1)} \ln\left[\frac{x_{\text{sat}}(\text{i})}{x_{\text{sat}}(i-1)}\right] \qquad (1.A.43c)$$

where

$$M_d(i) = \sum_{i=1}^{i=n-1} \left(1 - M_d(i)\frac{x_{sat}(i) - x_{sat}(i-1)}{x_{sat}(i)}\right) \quad (1.A.44)$$

Figure 1.A.1 shows the theoretical processing of a saline water of n saturation limits by a sequence of concentrators where the work defined by Equation (1.A.42) applies, and for crystallizers where the work by equation (1.A.41a) applies at saturation limits. Figure 1.A.2 shows the same information on work-concentration coordinates.

Please refer to Tables 1.A.4–1.A.7. for selected useful properties

Table 1.A.4 Constant-Pressure Specific Heats of Ideal Gases

Gas	$C_p(\tau)$; $\tau = T\text{K}/100$		Temperature Range, K	Maximum error, %
O_2	$C_{po} = 37.432 + 0.020102\tau^{1.5}$	$-178.57\tau^{-1.5} + 236.88\tau^{-2}$	300–3500	0.30
N_2	$C_{po} = 39.060 - 512.79\tau^{1.5}$	$+1072.7\tau^{-2} - 820.40\tau^{-3}$	300–3500	0.43
H_2O	$C_{po} = 143.05 - 183.54\tau^{0.2.5}$	$+82.75\tau^{0.5} - 3.6989\tau$	300–3500	0.43
CO_2	$C_{po} = -3.7357 + 30.529\tau^{0.5}$	$-4.1034\tau + 0.024198\tau^2$	300–3500	0.19
CO	$C_{po} = 69.145 - 0.70463\tau^{0.75}$	$-200.77\tau^{-0.5} + 176.76\tau^{-0.75}$	300–3500	0.42
H_2	$C_{po} = 56.565 - 702.74\tau^{-0.75}$	$+1165.0\tau^{-1} - 560.7\tau^{-1.5}$	300–3500	0.60
OH	$C_{po} = 81.546 - 59.35\tau^{0.25}$	$+17.329\tau^{0.75} - 4.2660\tau$	300–3500	0.43
NO	$C_{po} = 59.283 - 1.7096\tau^{0.5}$	$-70.613\tau^{-0.5} + 74.889\tau^{-1.5}$	300–3500	0.34
NO_2	$C_{po} = 46.045 + 216.10\tau^{-.5}$	$-363.66\tau^{-0.75} + 232.55\tau^{-2}$	300–3500	0.26
CH_4	$C_{po} = -672.87 + 439.74\tau^{0.25}$	$-24.875\tau^{07.5} + 323.88\tau^{-0.5}$	300–2000	0.15
C_2H_4	$C_{po} = -95.395 + 123.15\tau^{0.5}$	$-35.641\tau^{07.5} + 182.77\tau^{-3}$	300–2000	0.07
C_2H_6	$C_{po} = 6.895 + 17.26\tau$	$-0.6402\tau^2 + 0.00728\tau^3$	300–1500	0.83
C_3H_8	$C_{po} = -4.042 + 30.46\tau$	$-1.571\tau^2 + 0.03171\tau^3$	300–1500	0.40
C_4H_{10}	$C_{po} = 3.945 + 37.12\tau$	$-1.833\tau^2 + 0.03498\tau^3$	300–1500	0.54

Table 1.A.5 Critical Constants

Substance	Formula	Molecular Weight	Temprerature, K	Pressure, MPa	Volume, m^3/kmol
Ammonia	NH_3	17.03	405.5	11.28	0.0724
Argon	Ar	39.948	151	4.86	0.0749
Bromine	Br_2	159.808	584	10.34	0.1355
Carbon dioxide	CO_2	44.01	304.2	7.39	0.0943
Carbon monoxide	CO	28.011	133	3.50	0.0930
Chlorine	Cl_2	70.906	417	7.71	0.1242
Deuterium	D_2	4.00	38.4	1.66	—
Helium	He	4.003	5.3	0.23	0.0578
Hydrogen	H_2	2.016	33.3	1.30	0.0649
Krypton	Kr	83.80	209.4	5.50	0.0924
Neon	Ne	20.183	44.5	2.73	0.0417

Table 1.A.5 (*Continued*)

Substance	Formula	Molecular Weight	Temprerature, K	Pressure, MPa	Volume, m^3/kmol
Nitrogen	N_2	28.013	126.2	3.39	0.0899
Nitrous oxide	N_2O	44.013	309.7	7.27	0.0961
Oxygen	O_2	31.999	154.8	5.08	0.0780
Sulfur dioxide	SO_2	64.063	430.7	7.88	0.1217
Water	H_2O	18.015	647.3	22.09	0.0568
Xenon	Xe	131.30	289.8	5.88	0.1188
Benzene	C_6H_6	78.115	562	4.92	0.2603
n-Butane	C_4H_{10}	58.124	425.2	3.80	0.2547
Carbon tetrachloride	CCl_4	153.82	556.4	4.56	0.2759
Chloroform	$CHCl_3$	119.38	536.6	5.47	0.2403
Dichlorodifluoromethane	CCl_2F_2	120.91	384.7	4.01	0.2179
Dichlorofluoromethane	CCl_2F	102.92	451.7	5.17	0.1973
Ethane	C_2H_6	30.070	305.5	4.88	0.1480
Ethyl alcohol	C_2H_5OH	46.070	516	6.38	0.1673
Ethylene	C_2H_4	28.054	282.4	5.12	0.1242
n-Hexane	C_6H_{14}	86.178	507.9	3.03	0.3677
Methane	CH_4	16.043	191.1	4.64	0.0993
Methyl alcohol	CH_3OH	32.042	513.2	7.95	0.1180
Propane	C_3H_8	44.097	370	4.26	0.1998
Propene	C_3H_6	42.081	365	4.62	0.1810
Propyne	C_3H_4	40.065	401	5.35	—
Trichlorofluoromethane	CCl_3F	137.37	471.2	4.38	0.2478

Table 1.A.6 Enthalpy and Gibbs Free Energy of Formation and Absolute Entropy of Some Substances at 25°C and 0.1 MPa

Substance	Formula	Molecular Weight	State	H_f, kJ/kmol	G_f, kJ/.kmol	S, kJ/ (kmol.K)
Carbon monoxide	CO	28.011	gas	−110,529	−137,150	197.653
Carbon dioxide	CO_2	44.011	gas	−393,522	−394,374	213.795
Water	H_2O	18.015	gas	−241,827	−228,583	188.833
Water	H_2O	18,015	liquid	−285,838	−237,178	70.049
Methane	CH_4	16.043	gas	−74,873	−50,751	186.256
Acetylene	C_2H_2	26.038	gas	+226,731	+209,234	200.958
Ethene	C_2H_4	28.054	gas	+52,283	+68,207	219.548
Ethane	C_2H_6	30.070	gas	−84,667	−32,777	229.602
Propane	C_3H_8	44.097	gas	−103,847	−23,316	270.019
Butane	C_4H_{10}	58.124	gas	−126,148	−16,914	310.227
Octane	C_8H_{18}	114.23	gas	−208,447	+16,859	466.835
Octane	C_8H_{18}	114.23	liquid	−249,952	+6,940	360.896
Carbon (graphite)	C	12.011	solid	0	0	5.795

Table 1.A.7 Logarithms to Base *e* of the Equilibrium Constant K^a

T, K	$H_2 = 2H$	$O_2 = 2O$	$N_2 = 2N$	$H_2O = H_2 + 0.5O_2$	$H_2O = 0.5H_2 + OH$	$CO_2 = CO + 0.5O_2$	$0.5N_2 + 0.5O_2 = NO$
298	−164.005	−186.975	−367.480	−92.208	−106.208	−103.762	−35.052
500	−92.827	−105.630	−213.372	−52.691	−60.281	−57.616	−20.295
1000	−39.803	−45.150	−99.127	−23.163	−26.034	−23.529	−9.388
1200	−30.874	−35.005	−80.011	−18.182	−20.283	−17.871	−7.569
1400	−24.463	−27.742	−66.329	−14.609	−16.099	−13.842	−6.270
1600	−19.837	−22.285	−56.055	−11.921	−13.066	−10.830	−5.294
1800	−15.866	−18.030	−48.051	−9.826	−10.657	−8.497	−4.536
2000	−12.840	−14.622	−41.645	−8.145	−8.728	−6.635	−3.931
2200	−10.353	−11.827	−36.391	−6.768	−7.148	−5.120	−3.433
2400	−8.276	−9.497	−32.011	−5.619	−5.832	−3.860	−3.019
2600	−6.517	−7.521	−28.304	−4.648	−4.719	−2.801	−2.671
2800	−5.002	−5.826	−25.117	−3.812	−3.763	−1.894	−2.372
3000	−3.685	−4.357	−22.359	−3.086	−2.937	−1.111	−2.114
3200	−2.534	−3.072	−19.937	−2.451	−2.212	−0.429	−1.888
3400	−1.516	−1.935	−17.800	−1.891	−1.576	0.169	−1.690
3600	−0.609	−0.926	−15.898	−1.392	−1.088	0.701	−1.513
3800	0.202	−0.019	−14.199	−0.945	−0.501	1.176	−1.356
4000	0.934	0.796	−12.660	−0.542	−0.044	1.599	−1.216
4500	2.486	2.513	−9.414	0.312	0.920	2.490	−0.921
5000	3.725	3.895	−6.807	0.996	1.689	3.197	−0.686
5500	4.743	5.023	−4.666	1.560	2.318	3.771	−0.497
6000	5.590	5.963	−2.865	2.032	2.843	4.245	−0.341

[a]For the reaction $n_aA + n_bB = n_cC + n_dD$, the equilibrium constant is defined as $K = (a_C{}^{nc} a_D{}^{nd})/(a_A{}^{na} a_B{}^{nb})$.

Symbols (some additional symbols definitions are in the text where they are mentioned, and in the associated software)

A	membrane water permeability coefficient lb/(h · ft^2· psi) [or m/(h.bar)]; surface area, A_{min} surface area minimized by design degrees of freedom
B	membrane salt permeation coefficient ft/h (or m/h)
c	salt concentration per unit volume, lb/ft^3 (or kg/m^3); c_b of bulk; c_m of high-pressure-side membrane wall; c_d of diluted product
c_d	unit cost of exergy destruction \$/kWh, c_{da}, system average
C_a	cost per unit characterizing surface (area)
C_{aED}	cost per unit surface of ion exchange electrodialysis membranes
C_{aPV}	solar cell module cost per unit surface
C°_{aPV}	solar cell module cost per unit surface under standard test conditions
C_{aRO}	cost per unit membrane surface under reverse-osmosis (R conditions)
C_f	fuel price per kWh exergy
C_F	fuel price per kWh higher heating value
C_{pwr}	unit power cost, \$/kWh
C_{wtr}	production cost per unit product, \$/m^3

C_z	capital recovery rate \$/\$·y: C_{zM} for membranes, C_{zRO} for remaining RO devices, C_{zPV} for solar cells
Cap	capital cost, Cap^+, added capital
d	Diameter or equivalent diameter, ft
D	diffusion coefficient, m^2/s; dissipation, kW (exergy destruction rate)
$D^{th}10$	temperature difference in the condenser, F
E	exergy; E^s, of system; E^f, of flow
f	friction factor
F	fuel; $F_{penalty}$ for fuel penalty, F_{idealc} for fuel of ideal control, Np_{fuel} for fuel of night product, $F_{refired}$ for fuel of refiring
F_{ms}	PV cost multiplier accounting for added cost from module to system
h_m	RO mass transfer coefficient, ft/h (or m/h)
hhv	fuel higher heating value
H	enthalpy; height of flow passage, in. (or mm); H_b, of RO membrane feed side; H_d, of its product low-pressure, side; H_{hc}, H_{lc}, osmosis high- and low-concentration sides, respectively
J	objective function, \$/h; mass flux, lb/(h·ft^2) [or kg/(h · m^2)] J_w, of pure water; J_s of salt; J_v, volume flux, ft/h (or m/h)
LF	power load factor
mgd	million gallons per day; migpd, Imperial; usmgd, American
M_d	PV/RO water product rate usmgd (m^3/d)
N	dimensionless number; N_V, velocity/geometry number; N_M, a membrane number [Eq. 4.24]
OP	variable-load operation, OP_{pen}, operation cost penalty
P	Applied pressure, psia (or bar); P_0, dead-state pressure
P_w	Power, kW; P_{load}, power required; P_{dlvrd}, power delivered
ppm	RO product water salt content, parts per million; ED feed parts per million
P_{zn}	RO concentration polarization ratio x_{sm}/x_{sb} (salt at membrane wall/bulk)
R	universal gas constant; resources, R_{manuf}, manufacturing resources of a device, $R_{operate}$ operating resources of a device
R_{eff}	Ratio of PV expected solar cell field efficiency to standard test conditions top cell efficiency η/η°
R_{sol}	the ratio of design solar intensity to the standard test intensity sol/sol$^\circ$
S	entropy
S_{PV}	solar cell surface, m^2; S°_{PV}, of standard test conditions
S_{RO}	RO membrane surface, m^2
Sol	design solar intensity, kW/m^2; Sol$^\circ$: of standard test conditions (1kW/m^2 = 1 sun)
T	temperature; T_o dead-state temperature
V	device decision variables, $V_{efficiency}$, thermodynamic; V_{design} design $V_{manufacture}$ manufacture
X	dependent variable, X_{duty}, for a device

x_s	Salt mass fraction; x_{sf}, of feed, x_{sj}, of reject brine; x_{sm}, at membrane wall; x_{sb}, of bulk flow; x_{sd}, of product water
Z	capital cost, \$; Z_{RO}, for RO subsystem; Z_{PV}, for PV module; Z_{PVS} for solar subsystem; Z_{VC}, for VC subsystem; Z_{ED}, for electrodialysis subsystem; Z_{GT}, for gas turbine subsystem; Z_{MSF}, multistage flash distillation subsystem
Z	capital cost rate
δ	concentration boundary-layer thickness, μm (10^{-6}m)
ΔP_h	pressure loss, psi (or kPa), high-pressure side of RO membrane
ϕ	departure from ideal solution; 1 for ideal solution; ϕ_P, of RO applied pressure
η	efficiency; η_{RO}, of separation process; η_{PV}, of solar-to-power conversion; $\eta^o{}_{PV}$, of standard test conditions; η_{pmp}, adiabatic efficiency of pressurizing pump; η_{trb}, adiabatic efficiency of work recovery turbine or device
ρ	density
μ	viscosity; chemical potential; μ_{c0}, chemical potential of species at dead state P_0 and T_0

Abbreviations

ED	electrodialysis desalter
EL	water electrolysis system
GT/MSF	gas turbine/multistage flash distillation cogeneration system
PV	photovoltaic (solar cells)
RO	reverse-osmosis desalter
SCC	simple combined cycle
VC	vapor compression (distiller)
ZLD	zero liquid discharge

SELECTED REFERENCES FOR SECTION 1.1–1.3

1. Tribus M, Evans R, *The Thermoeconomics of Sea Water Conversion*, UCLA Report 62–63, Aug. 1962.
2. El-Sayed Y, Evans R, Thermoeconomics and the design of heat systems, *J. Eng. Power* 27–35 (Jan. 1970).
3. El-Sayed Y, Aplenc A, Application of the thermoeconomic approach to the analysis and optimization of a vapor-compression desalting system, *J. Eng. Power* 17–26 (Jan. 1970).
4. Gaggioli R, ed., *Thermodynamics: Second Law Analysis*, ACS Symp. Series 122, *American Chemical Society*, 1980.
5. Gaggioli R, ed., *Efficiency and Costing*, ACS Symp. Series 235, 1983.
6. El-Sayed Y, A second-law-based optimization, Parts 1 and 2, *J. Eng. Gas Turb. Power* **118**:693–703 (Oct. 1996).

7. Torres C, Serra L, Valero A, Lozano M, The productive structure and thermoeconomic theories of system optimization, *Proc. ASME Adv. Energy Syst. Div*. **36**:429–436 (Nov. 1996).
8. Lazaretto A, Tsatsaronis G, On the quest for objective equations in exergy costing, *Proc. ASME*. **37**:197–210 (1997).
9. Linhoff B, The use of pinch analysis to knock down capital costs and emissions, *Chem. Eng. Progress*, 32–57 (Aug. 1994).
10. Sciubba E, Melli R, *Artificial Intelligence in Thermal Systems Design*, Nova Scientific Publishers, 1998.
11. Dimopoulos GG, Christos A, Frangopoulos CA, Optimization of energy systems based on evolutionary and social metaphors, *Energy* **33**:171–179 (2008).
12. Bouvy C, Lucas K, Multicriterial optimization of communal energy supply concepts, *Proc. 19th Int. ECOS Conf*., Greece, 2006, Vol. 1, pp. 543–551.
13. Gibbs JW, (1978), *On the Equilibrium of Heterogeneous Substances, The Collected Work*, Yale Univ. Press, 1978, (orig. publ. 1928, Vol. 1, p. 77).
14. El-Sayed Y, *The Thermoeconomics of Energy Conversions*, Elsevier (UK), 2003.
15. El-Sayed YM, The rising potential of competitive solar desalination, *Desalination* **216**:314–324 (2007).
16. Baker RW, *Membrane Technology and Applications*. Wiley, Hoboken, NJ, 2004.
17. Reid R, Pruasnitz J, Poling B, *The Properties of Gases and Liquids*, 4th ed., McGraw-Hill, 1989.
18. Green MA, *Third Generation of Photovoltaics, Advanced Solar Energy Conversion* Springer-Verlag, 2003.
19. Mulder M, *Basic Principles of Membrane Technology*, Kluwer Academic, 2003.
20. Sehra A, Bettner J, Cohen A, Design of a high performance axial compressor for utility gas turbines, *J. Turbomach*. 114(2):277–286 (1992).
21. Cohen H, Rogers G, Saravenamuttoo H, *Gas Turbine Theory*, 3rd ed., Wiley, 1987.
22. Hegetschweiler H, Bartlet R, Predicting performance of large steam turbine-generator units, *Trans. Am. Soc. Mech. Eng. ASME*. 79:1085–1114 (1957).
23. Sabersky R, Acosta A, Hauptmann E, *Fluid Flow, a First Course in Fluid Mechanics*, 3rd ed., Macmillan, 1989.

FURTHER READING

1.F.1 International Symposia on Energy Analysis

Pleskov YV, *Solar Energy Conversion: A Photoelectrochemical Approach*, Springer-Verlag, 1992.

Luque A, Hegedus S, ed., *Handbook of Photovoltaic Science and Engineering*, Wiley, Hoboken, NJ, 2003.

Moran M, Sciubba E, eds., *Proc. 4th Int. Symp. Second Law Analysis of Thermal Systems*, Rome, May 25–29, 1987.

Cai, Ruixian, Moran M, eds., *Proc. Int. Symp. Thermodynamic Analysis and Improvement of Energy Systems*, (TAIES'89), Beijing, June 5–8, 1989.

Stecco S, Moran M, eds., *Proc. Florence World Energy Research Symp.* (FLOWERS'90), Florence, Italy, May 28–June 1, 1990.

Proc. Int. Conf. Analysis of Thermal and Energy Systems (Athens'91), Athens, 1991.

Valero A, Tsatsaronis G, editors: *Proc. Int. Symp. Efficiency, Costs, Optimization, and Simulation of Energy Systems* (ECOS'92), Zaragoza, Spain, June 15–18, 1992.

Proc. Int. Symp. Energy Systems and Ecology (Ensec'93), Krakow, Poland, July 5–9, 1994.

Proc. Florence World Energy Research Symp. (FLOWERS'94), Florence, July 6–8, 1994.

Proc. Int. Symp. Efficiency, Costs, Optimization, Simulation and Environmental Impact of Energy Systems (ECOS'95), Istanbul, July 9–14, 1995.

Proc. Int. Symp. Efficiency, Costs, Optimization, Simulation and Environmental Impact of Energy Systems (ECOS'96), Stockholm, June 25–27, 1996.

Proc. Int. Symp. Thermodynamic Analysis and Improvement of Energy Systems, (TAIES'97), Beijing, June 10–13, 1997.

Proc. Int. Symp. Efficiency, Costs, Optimization, Simulation and Environmental Aspects of Energy Systems and Processes (ECOS'98), Nancy, France, July 8–10, 1998.

Proc. Int. Symp. Efficiency, Costs, Optimization, Simulation and Environmental Aspects of Energy Systems (ECOS'99), Tokyo, June 8–10, 1999.

Proc. Int. Symp. Efficiency, Costs, Optimization, and Simulation of Energy Systems (ECOS'00), Univ. Twente, The Netherlands, July 5–7 2000.

Proc. Int. Symp. Efficiency, Costs, Optimization, Simulation and Environmental Impact of Energy Systems (ECOS'01), Istanbul, July 4–6, 2001.

Proc. Int. Symp. Efficiency, Costs, Optimization, and Simulation of Energy Systems (ECOS'02), Berlin, July 3–5, 2002.

Proc. Int. Symp. Efficiency, Costs, Optimization, and Simulation of Energy Systems (ECOS'03), Denmark, June 30–July 2, 2003.

Proc. 17th Int. Conf. Efficiency, Cost, Optimization, Simulation and Environmental Impact of Energy and Process Systems (ECOS'2004), Guanajauto, Mexico, July 7–9, 2004.

Proc. 18th Int. Conf. Efficiency, Cost, Optimization, Simulation and Environmental Impact of Energy Systems (ECOS'2005), Trondheim, Norway, June 20–22, 2005.

Proc. 19th Int. Conf. Efficiency, Cost, Optimization, Simulation and Environmental Impact of Energy Systems (ECOS'2006), Aghia Pelagia, Crete, Greece, July 12–14, 2006.

Proc. 20th Int. Conf. Efficiency, Cost, Optimization, Simulation and Environmental Impact of Energy Systems (ECOS'2008), Krakow Gliwicw, Poland, June 24–27, 2008.

The 22nd International Conference on Efficiency, Cost, Optimization, Simulation and Environmental Impact of Energy Systems (ECOS'2009) held in Foz do Iguaçú, Brazil, 31 August -- 3 September 2009.

The 23rd International Conference on Efficiency, Cost, Optimization, Simulation and Environmental Impact of Energy Systems (ECOS'2010) held in Lausanne, Switzerland, 14-17 June 2010.

The 24th International Conference on Efficiency, Cost, Optimization, Simulation and Environmental Impact of Energy Systems (ECOS'2011) held in Novi Sad, Serbia, 4-7 July 2011.

1.F.2 Selected International Symposia on Desalination

Proc. IDA World Congress Desalination and Water Sciences, Abu Dhabi, UAE, Nov. 18–24, 1995.

Proc. IDA World Congress Desalination and Water Reuse, Madrid, Oct. 6–9, 1997.

Proc. European Conf. Desalination and the Environment, Las Palmas, Gran Canaria, 1999.

Proc. The Int. Conf. Seawater Desalination Technologies on the Threshold of the New Millennium, Kuwait, Nov. 4–7, 2000.

Proc. IDA Int. Conference Desalination, Singapore, March 21–22, 2001.

Proc. IDA World Congress Desalination and Water Reuse, Bahrain, March 8–13, 2002.

Proc. IDA World Congress Desalination, and Water Reuse, Paradise Island, Bahamas, Sept. 28–Oct. 3, 2003.

Proc. Int. Conf. Water Security for Future Generations,, Changchun, Jilin Province, China, July 26–31, 2004.

Proc. IDA World Congress Desalination and Water Reuse, Singapore, Sept. 11–16, 2005.

Proc. ADST 2006 Int. Conf. Desalination Technologies and Water Reuse, Alexandria, Egypt, May 6–8, 2006.

Proc. AMTA Membrane Technologies Conf., Las Vegas, NV (USA), July 23–26, 2007.

Proc. IDA World Congress Desalination and Water Reuse, Maspalomas, Gran Canaria, Spain, Oct. 21–26, 2007.

IDA World Congress on Desalination and water Reuse, Dubai, November 7-12, 2009.

IDA World Congress on Desalination and water Reuse, Perth, Australia, September 4-9, 2011.

1.F.3 Books on Thermodynamics

Gaggioli RA, ed., *Thermodynamics: Second Law Analysis*, ACS Symp. Series Vol. 122, American Chemical Society, 1980.

Moran MJ, *Availability Analysis, A Guide to Efficient Energy Use*, Prentice Hall, 1982.

Kotas TJ, *The Exergy Method of Thermal Plant Analysis*, Butterworth, 1984.

Reid R, Pruasnitz J, Poling B, *The Properties of Gases and Liquids*, 4th ed., McGraw-Hill, 1989.

Szargut J, Morris D, Steward F, *Exergy Analysis of Thermal, Chemical and Metallurgical Processes*, Hemisphere Publishing, 1985.

Van Wylen GJ, Sonntag RE, *Fundamentals of Classical Thermodynamics*, Wiley 1996.

1.F.4 Books on Optimization and Equation Solvers

Wilde DJ, Beightler CS, *Foundations of Optimization*, Prentice-Hall, 1967.

Intriligator M, *Mathematical Optimization and Economic Theory*, Prentice-Hall, 1971.

Wismer D, Chattergy R, *Introduction to Nonlinear Optimization, a Problem Solving Approach*, North-Holland Series in System Science and Engineering, 1978.

Chapra S, Canale R, *Numerical Methods for Engineers*, 2nd ed., McGraw-Hill, 1988.

1.F.5 Books on Design of Energy Conversion Devices

Hottel H, Sarofim A, *Radiative Transfer*, McGraw-Hill, 1967.

Siegel R, Howell J, *Thermal Radiation Heat Transfer*, 2nd ed., Hemisphere Publishing, 1981.

Rohsenow W, Hartnett J, Ganic E, *Handbook of Heat Transfer Applications*, 2nd ed., McGraw Hill Book Company.

Cohen H, Rogers G, Saravenamuttoo H, *Gas Turbine Theory*, 3rd ed., Wiley, 1987.

Sabersky R, Acosta A, Hauptmann E, *Fluid Flow, a First Course in Fluid Mechanics*, 3rd ed., Macmillan, 1989.

1.F.6 Books on Optimal Design

Edgerton R, *Available Energy and Environmental Economics*, Lexington Books.

Gaggioli R, ed., *Efficiency and Costing*, ACS Symp. Series Vol. 235, American Chemical Society, 1983.

Papalambros PY, Wilde DJ, *Principles of Optimal Design, Modeling and Computation*, Cambridge Univ. Press, 1988.

Edger TF, Himmelblau DM, *Optimization of Chemical Processes*, McGraw-Hill, 1988.

Bejan A, Tsatsaronis G, Moran M, *Thermal Design and Optimization*, Wiley, 1996.

El-Sayed YM, *The Thermoeconomics of Energy Conversions*, Elsevier, 2003.

1.F.7 Books on Emerging Technologies (Fuel/Solar Cells and Selective Membranes)

Kordesch K, Simader G, *Fuel Cells and Their Applications*, VCH, 1996.

Minh NQ, Takahashi T, *Science and Technology of Ceramic Fuel Cells*, Elsevier, 1995.

Appleby AJ, ed., *Fuel Cells: Trends in Research and Applications*, Hemisphere Publishing, 1987.

Mazer JA, *Solar Cells: An Introduction to Crystalline Photovoltaic Technology*, Kluwer Academic, 1997.

Pleskov YV, *Solar Energy Conversion: A Photoelectrochemical Approach*, Springer-Verlag, 1990.

Mulder M, *Basic Principles of Membrane Technology*, 2nd ed., Kluwer Academic 1996.

Baker RW, *Membrane Technology and Applications*, Wiley, 2004.

Luque A, Hegedus S, eds., *Handbook of Photovoltaic Science and Engineering*, Hegedus, editors, Wiley, 2003.

Green MA, *Third Generation Photovoltaics, Advanced Solar Energy Conversion*, Springer, 2003.

1.F.8 General Additional Reading for Section 1.2

Bejan A, Tsatsaronis G, Moran M, *Thermal Design and Optimization*, Wiley, 1996.

Chapra SC, Canale RP, *Numerical Methods for Engineers*, McGraw-Hill, 1988.

DeGarmo E, Sullivan W, Bontadelli J, Engineering Economy, 9th ed., Macmillan, 1992.

Dunbar WR, Lior N, Gaggioli R, Combining fuel cells with fuel-fired power plants for improved exergy efficiency, *Energy* 16:(10)1259–1274 (1991).

Dunbar WR, Lior N, Gaggioli R, The component equations of energy and exergy, *ASME J. Energy Resources Technol*. 114:75–83 (1992).

Dunbar WR, Lior N, Gaggioli R, The effect of the fuel-cell unit size on the efficiency of a fuel-cell-topped Rankine Cycle, *ASME J. Energy Resources Technol*. 115:105–107 (1993).

Dunbar WR, Lior N, Sources of combustion irreversibility, *Comb. Sci. Technol*. 103:41–6 (1994).

Edger TF, Himmelblau DM, *Optimization of Chemical Processes*, McGraw-Hill, 1988.

Edgerton R, *Available Energy and Environmental Economics*, Lexington Books, 1982.

Evans RB, *A Proof that Essergy is the Only Consistent Measure of Potential Work*, PhD thesis, Thayer School of Engineering, Dartmouth College, Hanover, NH, 1969.

Incropera F, DeWitt D, Fundamentals of Heat and Mass Transfer, 3rd ed., Wiley, 1990.

Keenan JH, *Thermodynamics*, Wiley, 1941.

Kotas TJ, *The Exergy Method of Thermal Plant Analysis*, Butterworth, 1984.

Lior N, Sarmiento-Darkin W, Al-Sharqawi HS, The exergy fields in transport processes: Their calculation and use, *Energy* 31:553–578 (2006).

Means RS, *Mechanical, Building, and Electrical Cost Data*, Construction Publishers and Consultants, Kingston, MA, 1998.

Modern Cost Engineering: Methods and Data, compiled by *Chemical Engineering*, McGraw-Hill, 1979.

Moran MJ, *Availability Analysis, A Guide to Efficient Energy Use*, Prentice-Hall, 1982.

Papalambros PY, Wilde DJ, *Principles of Optimal Design, Modeling and Computation*, Cambridge Univ. Press, 1988.

Rant Z, Exergie, ein neues wort fur technische arbeitsfahigkeit, *Forsch. Ing. Wissen*. **22**(1):25–32 (1956).

Szargut J, Morris D, Steward F, *Exergy Analysis of Thermal, Chemical and Metallurgical Processes*, Hemisphere Publishing, 1985.

Van Wylen RE, Sonntag GJ, *Fundamentals of Classical Thermodynamics*, Wiley, 1996.

Wilde DJ, Beightler CS, *Foundations of Optimization*, Prentice-Hall 1967.

1.F.9 General Additional Reading for Section 1.4

Archer Enterprises, *Report on the Commercial Electrolytic Production of Hydrogen*, Geneva, NY.

Darwish MA, Al-Najem NM, Lior N, Towards sustainable seawater desalting in the Gulf area, *Desalination* Vol. 235, 5887, (2009).

Echeverria A, Desalination plant in Tenerife, San Loranzo Valley, *Proc. IDA World Congress Desalination and Water Reuse*, Madrid, Oct. 6–9, 1997, Vol. II, pp. 675–694.

El-Nashar AM, Validating the simulation program "Soldes" with actual performance data from an operating solar desalination plant, *Proc. IDA World Congress Desalination and Water Reuse*, Madrid, Oct. 6–9, 1997, Vol. 1, pp. 107–142.

El-Sayed YM, Desirable enforcements to optimal system design, *Proc. 19th Int. Conf. Efficiency, Cost, Optimization, Simulation and Environmental Impact of Energy Systems* (ECOS'06), Aghia Pelagia, Crete, Greece, July 12–14, 2006, Vol. 1, pp. 483–492.

Maloney K, Dopp R, *Highly Efficient Hydrogen Generation via Water Electrolysis Using Nanometal Electrodes*, Energy Research Laboratory, QuantumSphere Inc., Santa Ana, CA.

Torres del Corral M, (del) Pino, MP, Lodos MG, Rodrigous M, Physico-chemical and electrodialysis reversal treatment to reclaim wastewater from a sewage treatment plant, Proce. IDA World Congress Desalination and Water Reuse, Madrid, Oct. 6–9, 1997, Vol. II, pp. 695–711.

Van Wylen RE, Sonntag GJ, *Fundamentals of Classical Thermodynamics*, Wiley, 1996.

Yongqing Wang, Noam Lior, Fuel allocation in a combined steam-injected gas turbine and thermal seawater desalination system, *Desalination* Vol. 214, 306326, (2007).

1.F.10 Literature on Design Models

El-Sayed Y, On the feasibility of large vapor compression distillation units, *Desalination* 107:13–27, (1996).

El-Sayed Y, On the development of large vapor compression distillation units, *Proc. Int. Desalination Assoc. Desalination Seminar*, Cairo, Sept. 6–8, 1997.

El-Sayed Y, Gaggioli R, The integration of synthesis and optimization for conceptual designs of energy systems, *J. Energy Resources Technol*. 110:109–113 (1988).

Hoyt H, Sarofim A, *Radiative Transfer*, McGraw-Hill, 1967.

Rohsenow W, Hartnett J, Ganic E, *Handbook of Heat Transfer Applications*, 2nd ed., McGraw-Hill, (1985).

Siegel R, and Howell J, *Thermal Radiation Heat Transfer*, 2nd ed., Hemisphere Publishing, 1981.

CHAPTER 2

Environmental and Performance Aspects of Pretreatment and Desalination Technologies

SABINE LATTEMANN*, SERGIO G. SALINAS RODRIGUEZ, MARIA D. KENNEDY, JAN C. SCHIPPERS, and GARY L. AMY

*Corresponding author

Advances in Water Desalination, First Edition. Edited by Noam Lior.

2.1 INTRODUCTION

Seawater desalination is a rapidly growing coastal industry. The combined production capacity of all seawater desalination plants worldwide has increased by 30% over the last few years: from 28 million cubic meters per day (Mm^3/d) in 2007—which is the equivalent of the average discharge of the River Seine in Paris—to more than 40.4 Mm^3/d in 2009. For the 'pioneering' countries in the Middle East, which have been depending on desalinated water for decades, the driving factors for desalination were scarce freshwater resources coupled with abundant oil resources to engage in energy-intensive and costly thermal desalination projects. For the newly emerging desalination markets, the preferred process is seawater reverse osmosis (SWRO), and the driving factors are more diverse, including demographic growth, economic development, urbanization, droughts, climate change, and declining conventional water resources in terms of quality and quantity caused by overuse, pollution, or salinization of surface and groundwaters. As the effort and cost to supply water from conventional sources has been rising in many parts of the world and the costs of seawater desalination have been declining over the years, desalination has also become economically more competitive.

While market analysts agree that seawater desalination capacities will continue to grow rapidly in the future, the controversial debate on the extent to which this development may affect the environment continues. Seawater desalination is an energy-intensive process, consumes considerable amounts of resources in the form of chemicals and materials, and may have negative effects on the marine environment due to the discharges of concentrate waste waters and residual chemicals into the sea. The growing number of desalination plants worldwide and the increasing size of single facilities emphasises the need for greener desalination technologies and more sustainable desalination projects.

Two complementing approaches are the development and implementation of best available technology (BAT) standards and best practice guidelines for environmental impact assessment (EIA) studies. While BAT is a technology-based approach, which favors state-of-the-art technologies that reduce resource consumption and waste emissions, EIA aims at minimizing impacts at a site- and project-specific level through environmental monitoring, evaluation of impacts, and mitigation where necessary.

This chapter attempts to determine whether the present desalination technologies can be considered as green and sustainable solutions, as recently claimed for some projects. To follow up on the issue the present state of the art in terms of desalination capacities and desalination technologies will be discussed. Section 2.2 provides an overview of the worldwide desalination capacity and discusses regional trends. The figures show that desalination is developing into a coast-based industry, which may have harmful effects on the environment if not well-designed and managed. The effects on the environment will generally depend on the location and size of a desalination project and the technology used. Section 2.3 briefly outlines the state of the art of SWRO processes with particular focus on the different intake, pretreatment, and outfall designs. Our intention is not to reproduce textbook materials but to highlight chemical and energy use aspects and to evaluate different intake and pretreatment systems in terms of their operational performance and environmental footprint. While the chemical use of a desalination plant depends mainly on the pretreatment system and hence on feedwater quality, the actual desalination process accounts for most of the energy use of a desalination plant.

Section 2.4 summarizes the key environmental concerns of desalination plants. These are the impingement and entrainment of organisms caused by the intakes, the impacts on the marine environment caused by the discharge of concentrate and chemical residues, and energy consumption that causes emissions of greenhouse gases and air pollutants. Section 2.5 attempts to determine whether desalination can be considered a green technology and evaluates approaches for impact mitigation. A concept for BAT for seawater desalination technologies will be proposed, in combination with a methodological approach for the EIA of desalination projects.

The scope of EIA studies is outlined, including environmental monitoring, toxicity, and hydrodynamic modeling studies. In Section 2.6, the usefulness of multicriteria analysis (MCA) as a decision support tool for EIAs is explored and used to compare different intake and pretreatment options for seawater reverse osmosis plants.

2.2 GLOBAL DESALINATION CAPACITY

The use of desalination has increased in many parts of the world. According to the International Desalination Association's (IDA) worldwide desalting plant inventory, the production capacity of all desalination plants worldwide increased from about ~44.1 Mm^3/d in 2007 to ~62.5 Mm^3/d in 2010, which represents a 40% increase. The values include facilities that treat brine [defined as water containing >50,000 ppm total dissolved solids (TDS)], seawater (20,000–50,000 ppm), brackish water (3000–20,000 ppm), river water (500–3000 ppm), pure water (<500 ppm), or waste water; these data have been obtained either during construction, online, or presumed online [1,2].

Much of the growth of the desalination market takes place in the seawater sector, although waste water desalination and brackish water desalination will presumably become more important in the future. In some regions, such as California and Israel, waste water desalination even preceded the implementation of seawater desalination projects. At present (i.e., 2010), 5% of the total capacity of 62.5 Mm^3/d is produced from waste water (Fig. 2.1, top left); 17% of the present capacity is produced from brackish water sources, mainly brackish groundwater. A limited number of plants are being located in estuarine sites, such as the Thames Gateway desalination plant in East London with a capacity of 150,000 m^3/d. The plant withdraws brackish water with a maximum salt content of 11,000 mg/L during low tide and therefore requires only about half the energy (1.7 kWh/m^3) of SWRO plants. However, the tidal and seasonal variability of the raw water with regard to dissolved and particulate organic matter requires a complex pretreatment consisting of coagulation, flocculation, clarification, media filtration, and ultrafiltration [3,4]. The lower energy demand is a main benefit of estuarine sites; however, the pretreatment challenge may be the reason why only a limited number of projects have been implemented to date.

At present, ~65% of the capacity is produced from seawater. Seawater is hence the predominant source water for desalination and accounts for a worldwide water production of 40.4 Mm^3/d – a value that is comparable to the average discharge of the Seine River (43.2 Mm^3/d or 500 m^3/s) before disemboguing into the English Channel.

2.2.1 Capacity by Process Type and Source Water Type

For all source water types reverse osmosis (RO) is the prevalent desalination process. It accounts for more than 60% (38.7 Mm^3/d) of the global capacity (Fig. 2.1, top right); 33% or 20.9 Mm^3/d of the global capacity is produced by distillation

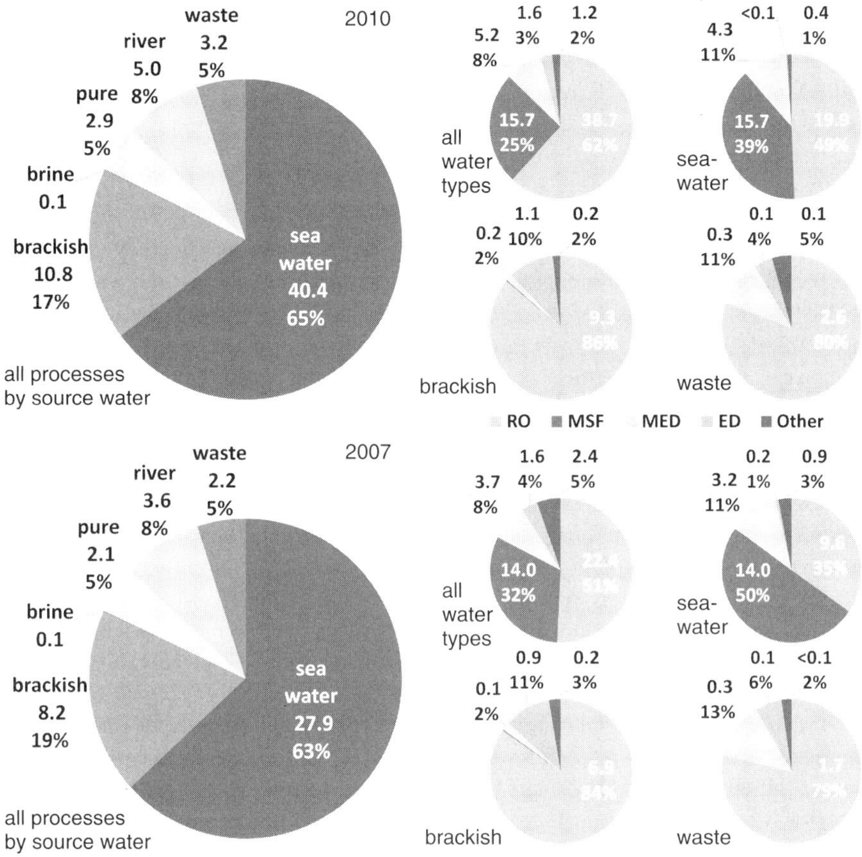

Figure 2.1 Global desalination capacity in 2010 (top) and 2007 (bottom) in Mm^3/d and percent by source-water type and by process. [RO—reverse osmosis; MSF—multi-stage flash (distillation); MED—multi-effect distillation; ED—electrodialysis] (primary data obtained from Refs. 1 and 2). A colored graph is published in Ref. 27.

plants, either multi-stage flash (MSF) or multi-effect distillation (MED) plants, with relative market shares of 25% (15.7 Mm^3/d) and 8% (5.2 Mm^3/d), respectively. A minor desalination process is the membrane-based electrodialysis (ED) process with about 3% market share (1.6 Mm^3/d). Other processes, such as electrodionization (EDI) and nanofiltration (NF) account for another 2% (1.2 Mm^3/d) of the global desalination capacity.

The picture changes if one distinguishes between the different source-water types. For seawater, SWRO and *thermal* processes each account for half of the global production (19.9 Mm^3/d and 20.0 Mm^3/d). MSF is still the predominant *thermal* process, accounting for 39% of the global production. On the contrary, RO is the dominant process for brackish water (86%, 9.3 Mm^3/d) and waste water (80%, 2.6 Mm^3/d) desalination, whereas distillation plays only a minor role for these source waters.

2.2.2 Capacity by Region and Source Water Type

Approximately 44% (27.2 Mm^3/d) of the global desalination production takes place in the Middle East, mainly in the Gulf countries (24.9 Mm^3/d). At present, ~17% (10.2 Mm^3/d) of the global capacity is located in the Americas, mainly in North America (7.8 Mm^3/d); 14% each in Asia (9.0 Mm^3/d) and in Europe (8.7 Mm^3/d), mainly in southern Europe (6.9 Mm^3/d); 8% (5.2 Mm^3/d) in Africa, mainly in northern Africa (4.4 Mm^3/d); and 3% (2.1 Mm^3/d) in the Pacific region, almost entirely in Australia (Fig. 2.2).

Seawater desalination accounts for two-thirds or more of the capacity in the Caribbean (91%), Africa (91%, 83%), the Middle East (87%), Australia (76%), and southern Europe (66%), and is the predominant process in most remaining regions except for North America; northern Europe; and Japan, Korea, and Taiwan. Finally, 54% (21.6 Mm^3/d) of the seawater desalination capacity is located in the six Gulf cooperation council (GCC) states: in Saudi Arabia, the United Arab Emirates, Kuwait, Bahrain, Qatar, and Oman. Although the combined capacity of these countries increased by 26% or 4.5 Mm^3/d between 2007 and 2010, the region's share of the global capacity dropped by 7% from 61% to 54%. Except for Oman, the desalination plants of the GCC states are located in either the Arabian Gulf or in the Red Sea. If one includes the installed capacities of other riparian countries in the Red Sea and the Mediterranean Sea, the cumulative capacity in these three enclosed sea areas exceeds 75% of the global seawater desalination capacity.

North America is the only region where brackish water desalination is the predominant process, with a capacity of 3.5 Mm^3/d, which represents 44% of the regional desalination capacity and almost a third (32%) of the global brackish water desalination capacity (Fig. 2.2). Of the global brackish water desalination capacity 21% is located in the six GCC states (2.3 Mm^3/d) and 13% (1.4 Mm^3/d) in southern Europe. The remaining shares of the other regions range between 1% and 6%.

River water desalination occurs mainly in North America, which accounts for almost half (47%) of the global river water desalination capacity (2.3 Mm^3/d). Japan, Korea, and Taiwan (0.6 Mm^3/d, 12%), southern Europe (0.6 Mm^3/d, 11%), and northern Europe (0.5 Mm^3/d, 10%) each represent ~10% of the global river water desalination capacity. In northern Europe, river water is the most important source water for desalination, closely followed by seawater. Japan, Korea, and Taiwan are the only countries where pure-water production is the most important process, accounting for 45% (1.4 Mm^3/d) of the regional and 47% of the global pure-water desalination capacity.

Waste Water purification occurs primarily in East Asia with 0.8 Mm^3/d or 26% of the global waste water desalination capacity, followed by North America (0.6 Mm^3/d, 20%) and the GCC states (0.5 Mm^3/d, 15%). Southern Europe (0.3 Mm^3/d, 10%); Japan, Korea, and Taiwan (0.3 Mm^3/d, 9%); and Australia (0.3 Mm^3/d, 9%) each represent ~10% of the global waste water desalination capacity.

In conclusion, about 65% or 40.4 Mm^3/d of the worldwide desalination capacity was produced from seawater sources in 2010. The global seawater desalination

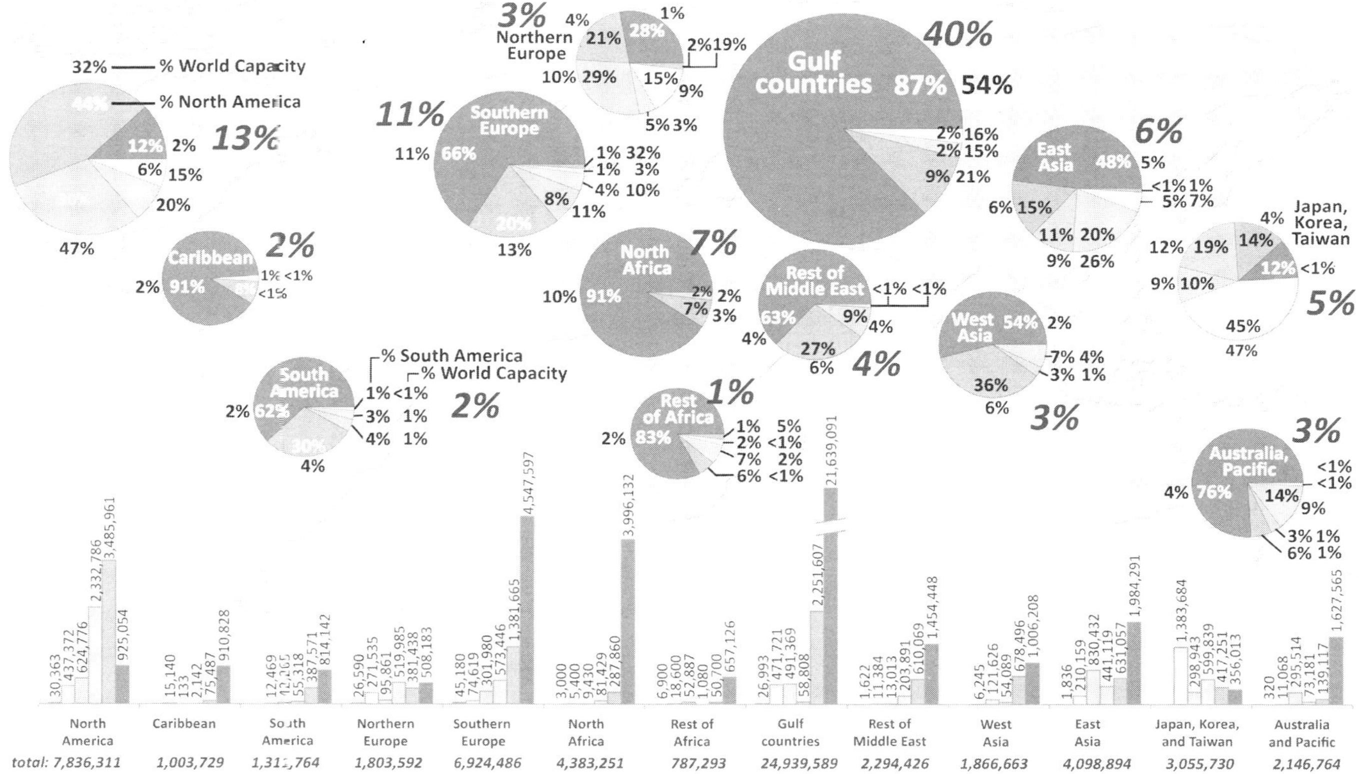

Figure 2.2 Global desalination capacity by region in m^3/d (bar diagrams, bottom) and percent (pie diagrams, top). The pie diagrams and the percentages inside the pie diagram show the different source-water types used for desalination in a given region. The figures outside the pie diagram denote the share of the global capacity. The figures in italics refer to the total capacity in a region (bottom), respectively the percentage of the global capacity. For example, the total desalination capacity in North America is 7,836,311 m^3/d or 13% of the global desalination capacity (including all source-water types). Brackish water desalination is the predominant process and accounts for a production of 3,485,961 m^3/d. This is equivalent to 44% of the total desalination capacity in North America, or 32% of the global brackish water desalination capacity (Primary data obtained from Ref. 2.) A colored graph is published in Ref. 27.

capacity increased by 45% compared to the year 2007 (27.9 Mm^3/d), which represents a 2% increase in the global share. Of the desalinated seawater, 50% is produced by thermal processes. The MSF distillation process is reserved almost exclusively for the desalination of seawater, mainly in the Gulf countries. However, the RO process is rapidly gaining ground. It accounted for about one-third (35%) of the global seawater desalination capacity in 2007, but its market share increased to almost half (49%) of the global seawater desalination capacity in 2010. While thermal plants still predominate in the oil-rich countries of the Middle East, SWRO is usually the preferred process where cheap fossil energy or low value heat is not available. Consequently, SWRO has been the first choice in many countries outside the Middle East for years. It can be expected that this trend will soon be reflected in the global figures when RO capacities surpass the global thermal desalination capacity in the near future. Although brackish water and waste water treatment methods offer a great future potential, desalination of seawater will remain the dominant desalination process for years to come. With coastal population densities on the increase in many parts of the world—half of the world's population already lives within 200 Km of the ocean, and 70% of the world's metropolises (largest mega cities) are located near coasts [5]—the development potential for seawater desalination facilities is huge. However, as the need for desalination accelerates in many parts of the world, concerns are raised over potential negative impacts of desalination on the environment.

As seawater desalination accounts for most of the production and is the focus of this chapter, the term *desalination* is used as a synonym for seawater desalination in this publication. Because of the potential cumulative impacts of desalination activity on the marine environment, it is also of greater interest to consider capacities by *sea area* rather than by world region. The installed capacity in the Gulf, the Red Sea, and the Mediterranean Sea is therefore described in the following sections, using primary data from the International Desalination Association (IDA) [1].

Figures 2.3–2.5 show all sites in the Gulf, in the Red Sea and in the Mediterranean Sea with a cumulative MSF/MED/RO capacity of $\geq$1000 m^3/d, specifically identifying all sites $\geq 100,000$ m^3/d (Gulf and Red Sea) and $\geq$50,000 m^3/d (Mediterranean Sea) by name and capacity. The total capacity (triangles) of each riparian state and the installed capacity in the sea region is given (first published in Ref. 6, updated after publication of Ref. 1).

2.2.2.1 *The Arabian Gulf* In terms of sea areas, the largest number of desalination plants can be found in the Gulf, with a total seawater desalination capacity of approximately 12.1 Mm^3/d—or ~44% of the worldwide daily production (Fig. 2.3). The major producers of desalinated water in the Gulf (and worldwide) are Saudi Arabia (representing 25% of the worldwide seawater desalination capacity, with 11% located in the Gulf region, 12% in the Red Sea region, and 2% in unknown locations), the United Arab Emirates (representing 23% of the worldwide seawater desalination capacity), and Kuwait (6%). Thermal desalination processes

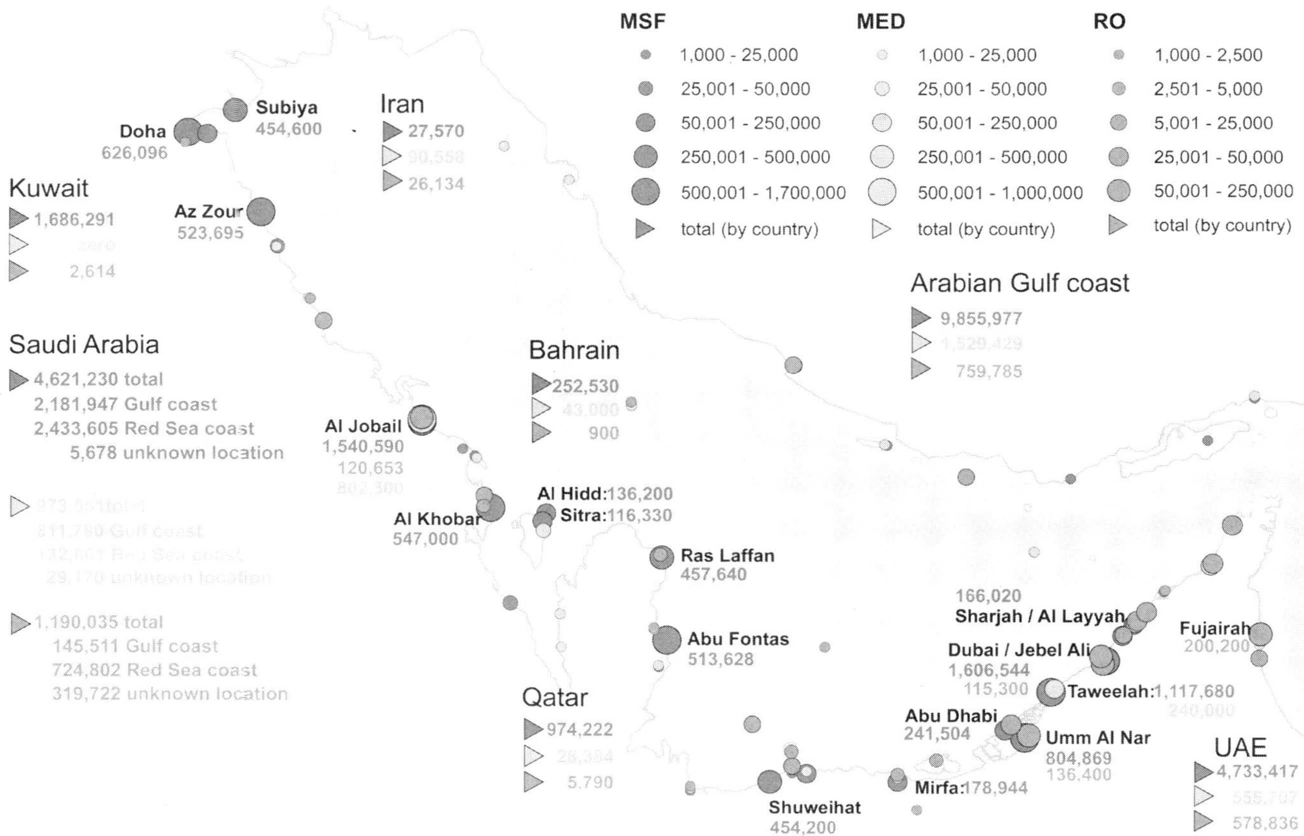

Figure 2.3 Cumulative MSF, MED, and SWRO capacities in the Gulf in cubic meters per day (m^3/d). A colored graph is published in Ref. 27.

dominate in the Gulf region, as water and electricity are typically generated by large cogeneration plants, which use low-value steam as a heat source and electricity from power plants for desalination. About 81% of the desalinated water in the Gulf region is produced by MSF and 13% by MED plants, and only a minor amount by SWRO (6%).

2.2.2.2 The Red Sea In the Red Sea area, desalination plants have a total production capacity of 3.6 Mm^3/d (representing 13% of the worldwide desalination capacity, Fig. 2.4). As in the Gulf, most of the water is produced in large cogeneration plants (72%), mainly on the Saudi Arabian coast in the locations of Yanbu, Rabigh, Jeddah, Shoaiba, and Assir. The world's largest desalination complex with a total water production of 1.6 Mm^3/d is located in Shoaiba. Saudi Arabia accounts for 92% of the desalinated water production in the Red Sea region, with 78% (2.6 Mm^3/d) of the national production coming from thermal plants, whereas Egypt accounts for only 7% of the desalinated water production from the Red Sea, of which 90% (0.2 Mm^3/d) is produced by smaller RO plants, mainly on the Sinai Peninsula and in the tourist resorts along the Red Sea coast.

2.2.2.3 The Mediterranean Sea In the Mediterranean, the total water production from seawater is about 4.0 Mm^3/d (representing 14% of the worldwide desalination capacity, Fig. 2.5). Spain, with about 8% of the world's total production (2.2 Mm^3/d), is the third largest producer of desalinated water globally and the largest in the region. However, about 25% of the Spanish capacity is located on the Canary Islands in the Atlantic Ocean, and "only" ~65% in the Mediterranean. (The Spanish AGUA program further augmented the water supply on the Mediterranean coast by desalination, increasing the capacity from 1.4 Mm^3/d (2005) to over 2.7 Mm^3/d [1000 Hm^3/y (year)] by 2010.) The government program, which also included water use efficiency and reuse measures, was introduced to avert another main water supply project, that is, diversion of the Ebro river to southern Spain.

While thermal processes (MSF and MED) dominate in the Gulf and Red Sea, the main process in the Mediterranean is seawater RO (SWRO). In 2002, both SWRO and distillation plants still held about equal market shares in the Mediterranean [7]. Today, SWRO accounts for 70% of the production in the Mediterranean and 99% of the Spanish production in the Mediterranean. Distillation plants are still found in Libya, Algeria, and Italy (in decreasing order of priority), but new plants in these countries are also often SWRO plants. For instance, a tremendous expansion of capacities is currently taking place in Algeria, northern Africa's fastest-growing desalination market, where the first large SWRO plant (200,000 m^3/d) opened in February 2008. It is the first in a series of other large projects which will increase the country's desalination capacity to 4 Mm^3/d by 2020.

On the Mediterranean coast of Israel, three large SWRO plants are currently in operation—the Ashkelon plant, with a capacity of approximately 300,000 m^3/d; the Hadera plant, with approximately 350,000 m^3/d; and the Palmachin plant, with approximately 93,000 m^3/d. In 2008, the Israeli government approved an emergency

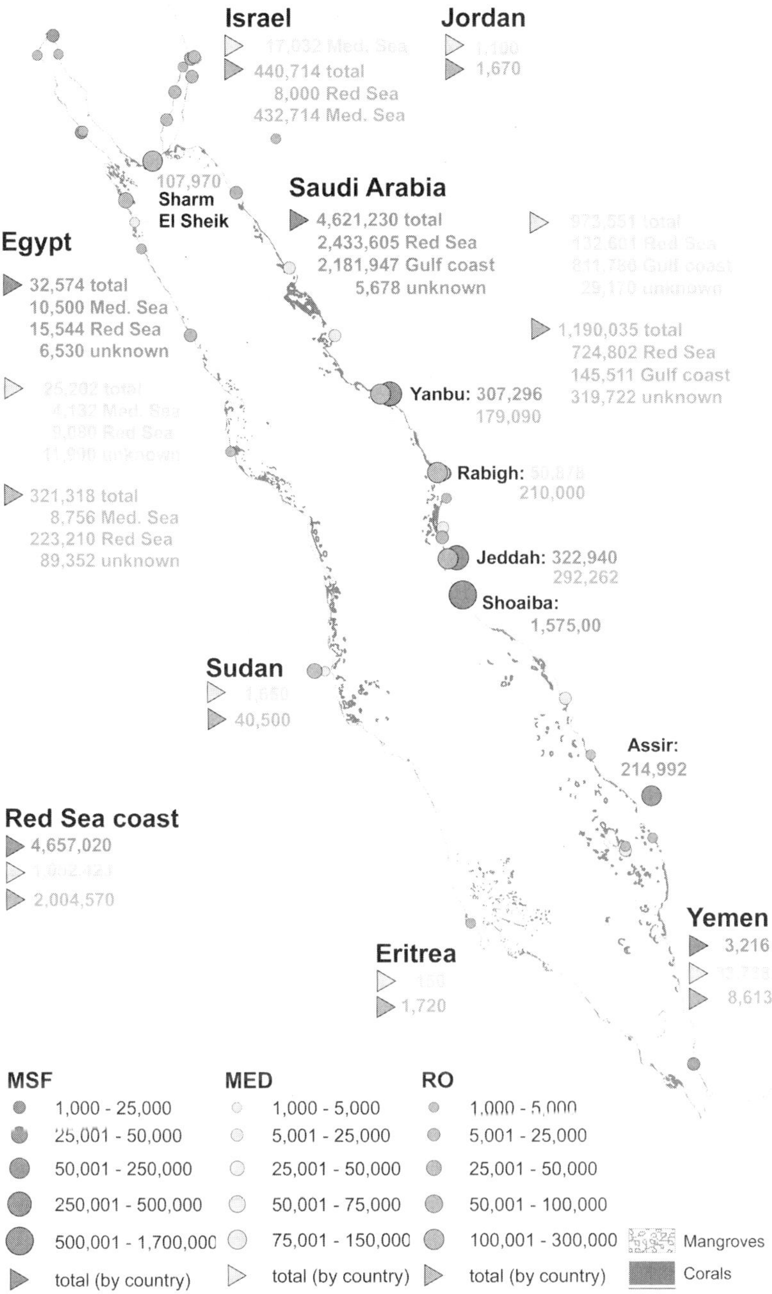

Figure 2.4 Cumulative MSF, MED, and SWRO capacities in the Red Sea in m^3/d. A colored graph is published in Ref. 27

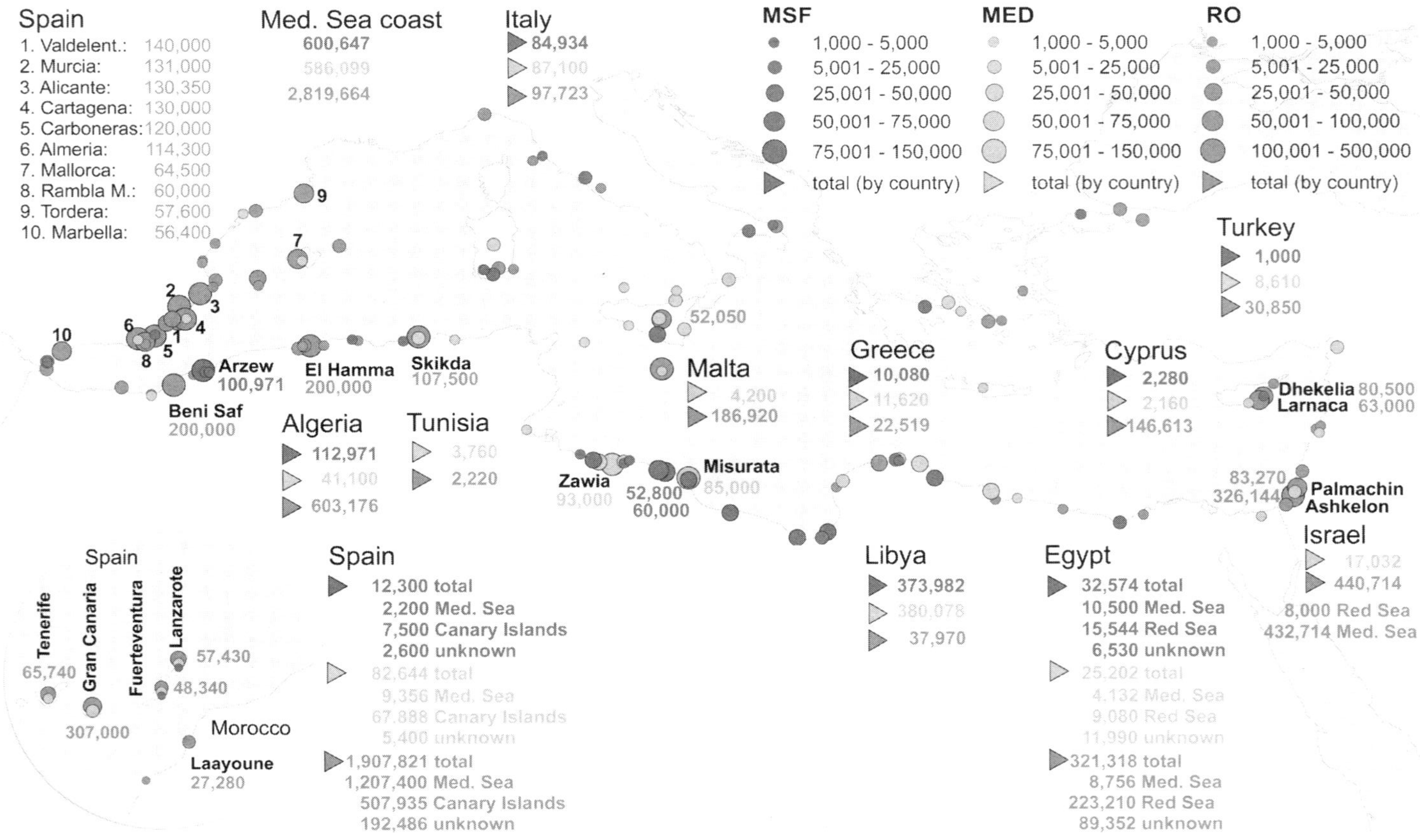

Figure 2.5 Cumulative MSF, MED, and SWRO capacities in the Mediterranean Sea in m^3/d. A colored graph is published in Ref. 27

program to address the country's growing water shortage, which is expected to raise the target for desalinated water production to 2.1–2.7 Mm^3/d by 2020 depending on water demand and other alternatives [9]. Furthermore, it is planned to sharply increase the use of the country's brackish water resources, from around 16,500 m^3/d to 220,000–274,000 m^3/d [10]. Other measures include more water-efficient practices, fixed water quotas, greater enforcement of water restrictions, and upgrading of waste water treatment capacities in order to increase recycling of waste water from presently 75% to 95% within 5 years [8].

2.2.2.4 Other Sea Regions While seawater desalination is already a well-established technology in the abovementioned sea regions, the era of large-scale desalination projects is about to start in other parts of the world. In California, a potential for 15–20 seawater desalination projects with a combined capacity of 1.7 Mm^3/d is expected for 2030. These would increase the share of desalination to 6% of California's 2000 urban water use. The two largest and most advanced projects are located in the cities of Carlsbad and Huntington Beach with a proposed capacity of 200,000 m^3/d each [11].

In Australia, the first large SWRO plant with a capacity of 144,000 m^3/d became operational in Perth in 2006, followed by a second project of similar size in the Gold Coast in 2008, and a third project in Sydney in 2010 with a capacity of 250,000 m^3/d. Further projects are under development in Adelaide, the Upper Spencer Gulf, near Perth, and in Karratha, with capacities ranging between 120,000 and 140,000 m^3/d, and in Victoria and Queensland with capacities of 400,000-450,000 m^3/d [8,12].

A third impressive example is China. Desalination capacity is currently estimated at around 366,000 m^3/d, which may increase by the factor of 100 to 36 Mm^3/d until 2020. It is expected that most of the investment will go into the four northeastern coastal provinces of Tianjin, Hebei, Liaoning, and Shandong, where total water shortage reached 16.6–25.5 billion m^3/y in 2010. Besides desalination of seawater, waste water reclamation is a serious option under consideration [8].

In conclusion, about three-quarters of the global seawater desalination capacity is located in the Mediterranean region and the Middle East, specifically in the sea areas of the Gulf, the Red Sea, and the Mediterranean Sea. Small semienclosed seas may be understood as self-contained ecosystems [13], which are prone to pollution and anthropogenic impacts. The response of the affected ecosystem depends on the magnitude of the impact and the sensitivity of the system to pollution and disturbance.

An outstanding example is the Gulf, with the world's highest density of large desalination plants. Very few similar areas exist that face such a high concentration of disturbance, and the prognosis for the Gulf continuing to provide abundant natural resources is poor [14]. Seawater desalination plants are included in the list of major sources of land-based marine pollution in the Gulf by the United Nations Environment Program (UNEP) and the Regional Organization for the Protection of the Marine Environment (ROPME) [15,16]. The problem for the Gulf, not only with regard to desalination, is the limited cross-border collaboration and even domestic collaboration among government agencies. Even where each project receives environmental assessment or attention, each is treated more or less in isolation, rarely in combination. However, the accumulative effects of an increasing number of projects in such a small, biologically interacting sea such as the Gulf exacerbates the overall deterioration [14].

On the contrary, desalination plants in Australia and California are unlikely to produce measurable effects on the Pacific Oceans; however, local coastal habitats may be affected by the discharges, either by individual projects or cumulatively by several projects along a coastline.

2.3 STATE OF THE ART OF THE TECHNOLOGY

All desalination processes have been extensively described in the literature [e.g., 17–20]. The intention of this chapter is not to reproduce these textbook materials but to describe the different desalination processes with regard to the use of resources, environmental considerations, and performance. While the design of the intake, pretreatment, desalination process, and outfall largely determine the impacts on the marine environment, energy demand and air quality impacts depend mainly on the process type.

2.3.1 Seawater Intake

Seawater desalination plants can receive feedwater through either a *surface* water intake or a *subsurface* intake embedded in the seafloor or beach sediments. Surface intakes include the nearshore intakes of most distillation plants, which are often located directly at the shoreline, and the submerged intakes that are more common for large SWRO projects and are typically located further offshore and in greater water depths.

For SWRO plants, different types of subsurface intakes with either vertical or horizontal collectors have also been used successfully [19,21–25] [Fig. 2.6]. Vertical wells and horizontal radial wells are both embedded in the permeable *onshore* sediments, whereas horizontal drains are drilled horizontally from a central point on land into the *offshore* sediments. Where the natural sediments are not sufficiently permeable, *onshore* or *offshore* infiltration galleries can be constructed by replacing the natural sediments by a more permeable medium into which the perforated pipes are embedded.

2.3.1.1 Surface Intake Surface intakes are usually equipped with a combination of screens to reduce the amount of debris and the number of organisms that are absorbed by the plant with the feedwater. Most advances in screen design stem from the power industry. State-of-the-art intake systems [19] can effectively reduce the impingement of aquatic organisms against screens and the entrainment of organisms into the plant. Passive screens with a large surface area and no moving parts are commonly used for large SWRO plants. The screens are operated

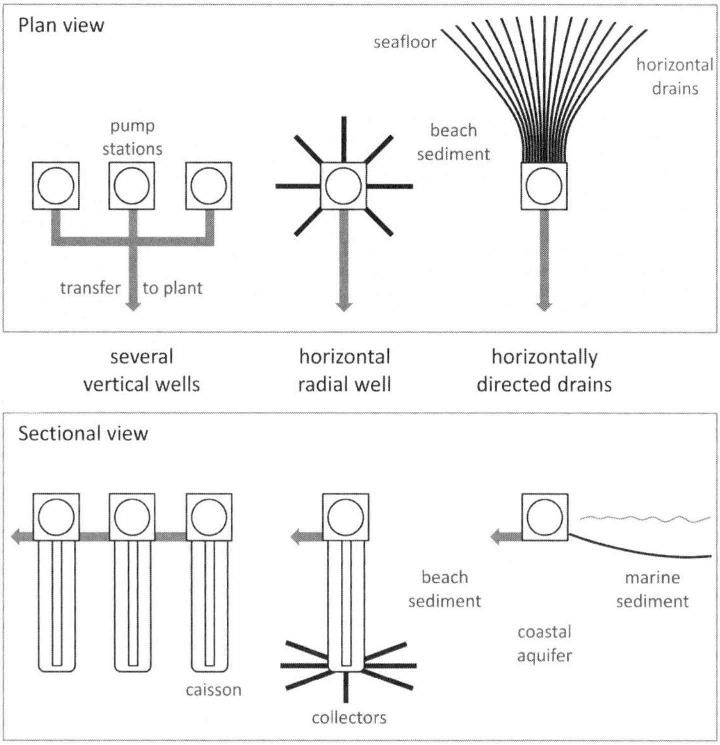

Figure 2.6 Subsurface intakes: vertical wells, horizontal radial well, and horizontal drains.

with a very low velocity to reduce impingement of marine organisms and can be backflushed with compressed air to remove debris.

The prevailing intakes for SWRO plants are offshore submerged intakes, where the intake is located at some distance from the coast and in greater water depth. As marine life is usually less abundant in these areas, submerged intakes often produce a better feedwater quality with lower contents of suspended solids and microorganisms than do nearshore intakes. Depending on the seafloor bathymetry, this may require a distance of several hundred meters from the shore and a water depth of 10–20 m.

The seawater transmission pipeline from an offshore submerged intake to the shore can be placed either on or below the seabed, using open-trench or tunneling techniques. Intakes located directly at the shoreline, which is more common for distillation plants but also seen in some SWRO plants, are often protected by a jetty or breakwater basin in order to reduce wave action and to allow suspended material to settle.

2.3.1.2 Subsurface Intake Because of the limited output capacity of beachwells and because of the lower recovery of SWRO systems compared to brackish water reverse osmosis (BWRO) systems, a large number of single wells would be required for a large SWRO plant, which is difficult to realize [18]. Beachwells are therefore usually considered only for small SWRO plants with capacities ≤20,000 m^3/d. A feasible alternative to open intakes for large seawater desalination plants are horizontal drains, such as the Neodren system, which is installed from an onshore site by horizontal directional drilling into the seafloor sediments. Sufficient flow rates can be realized for all plant sizes depending on the number of drains installed. The technology is, for example, used in the San Pedro del Pinatar plant in Spain, with a capacity of 65,000 m^3/d [26; Table 4 in Ref. 27], and was also considered in the planning studies for the Barcelona plant (200,000 m^3/d) [28] and other large plants in Algeria with capacities of ≤500,000 m^3/d [Table 5 in Ref. 27; 29] but not selected in the final design. For a 200,000-m^3/d plant, a maximum batch of 25 drains would be required, spaced at a distance of 2–3 m onshore (50–75 m total) and spreading out 300 m offshore in a fanlike arrangement.

In Long Beach, California, an ocean floor demonstration system is currently being tested which combines offshore infiltration galleries for the intake and a discharge gallery for the outfall (see also Section 2.3.5.2). Both consist of perforated laterals placed under the ocean floor to collect or to discharge the water. The infiltration rates are 2 and 4 L/m^2 per minute, and the discharge rates are 5.3–6.9 L/m^2 per minute. The intake in combination with 100 μM and 5 μM cartridge filters were found to achieve sufficient DOC and turbidity removal to be used as feedwater to a SWRO plant. The performance was comparable to effluent produced by a microfiltration system [24].

As subsurface intakes use natural sediments for prefiltration, they can be described as a 'natural treatment system' or 'biofilter'. Favorable conditions for subsurface intakes are geologic formations with a high transmissivity and a certain sediment thickness, whereas unfavorable conditions include sediments with high volumes of mud and a low degree of 'flushing' [30]. Biofilters often produce a better feedwater quality than open intakes as the water is typically characterized by lower and less variable amounts of organic carbon, suspended solids, nutrients, and microorganisms, and hence by a lower fouling potential [21]. This considerably reduces the pretreatment requirements in the 'engineered pretreatment system' following the biofilter. Further pretreatment is usually limited to acid and/or antiscalant addition followed by cartridge filtration. Extensive field experience shows that SWRO systems treating wellwater, with cartridge filtration as the only filtration step, can be operateded successfully over the years [18].

Shallow beachwells sometimes contain significant amounts of suspended particles [18]. Moreover, water from beachwells is often anaerobic or anoxic and may contain hydrogen sulfide, as well as iron(II) and manganese(II) depending on the geology. Aeration may lead to the precipitation of ferric hydroxide, which causes turbidity, and the formation of manganese dioxide deposits on the membranes over time. In some SWRO plants operating on wellwater, iron(II) and manganese(II) were initially absent but increased over time, e.g., in Malta. This appearance of insoluble salts in the feedwater is a risk of beachwell intakes, which may necessitate the installation of granular media filters in case feedwater conditions should deteriorate over the lifetime of the project. Moreover, the water composition of wellwater with regard to sparingly soluble salts (barium and strontium sulfates and silicates) may differ substantially from that of surface seawater, which may necessitate the use of an antiscalant.

2.3.2 Pretreatment Processes and Chemical Use

The main operational concerns and reasons for pretreatment in SWRO plants (Fig. 2.7) are particulate fouling by suspended particles, biofouling by microorganisms, organic fouling by dissolved organic matter, scaling by sparingly soluble inorganic compounds, and oxidation and halogenation by residual chlorine. The type and amount of pretreatment depends on the intake water quality. An overview on the main pretreatment alternatives for SWRO plants is given in the following subsections.

2.3.2.1 Minimal Pretreatment As subsurface intakes such as beachwells and offshore horizontal drains are biofilters that can reduce both organic and biofouling on the SWRO membranes, the following engineered pretreatment system is often minimal. The conventional steps of chlorination–dechlorination and coagulation–flocculation are seldom required, and systems are operated with a single filter only, using selective pretreatment against scaling where required and depending on the feedwater composition [Tables 3 and 4 in Ref. 27]. In San Pedro del Pinatar (Spain), for example, two "twin plants" with a capacity of 65,000 m^3/d are being operated. One plant is equipped with an open intake followed by a two-stage dual media filter and uses about 4 mg/L of coagulant ($FeCl_3$), whereas the other uses a subsurface intake followed by a sand filter and operates without coagulant. Both plants use about 1 mg/L of antiscalant and no chlorine [31]. A smaller beachwell plant in Javea, Spain (26,000 m^3/d) operates without any pretreatment chemicals and requires acid and caustic cleaning of the membranes only about once a year [32].

2.3.2.2 Conventional Pretreatment As surface intakes must accommodate more variable water quality due to seasonal weather conditions and algae blooms, pretreatment is generally more complex and extensive than for subsurface intakes. The conventional pretreatment for SWRO plants with surface intakes includes shock chlorination to control marine growth in the intake system, followed by coagulation–flocculation and filtration to remove suspended solids and colloids, and dechlorination prior to the RO units. Sometimes additional screening, sedimentation or flotation is included as an initial pretreatment step (Fig. 2.7).

Chlorine is usually added intermittently to the feedwater in doses of $\leq$10 mg/L to control biogrowth on the intake screens, inside the intake pipe, and in the pretreatment line, with one to three dosing points along that line. Chlorination can be carried out daily, weekly, or biweekly, depending on the site. Some plants also operate more successfully without chlorination, which breaks natural organic matter into biodegradable compounds that may increase biofouling in SWRO elements. The water is dechlorinated ahead of the SWRO units with a reducing agent, sodium bisulfite (SBS), usually in a dosage of 3 times the residual chlorine concentration

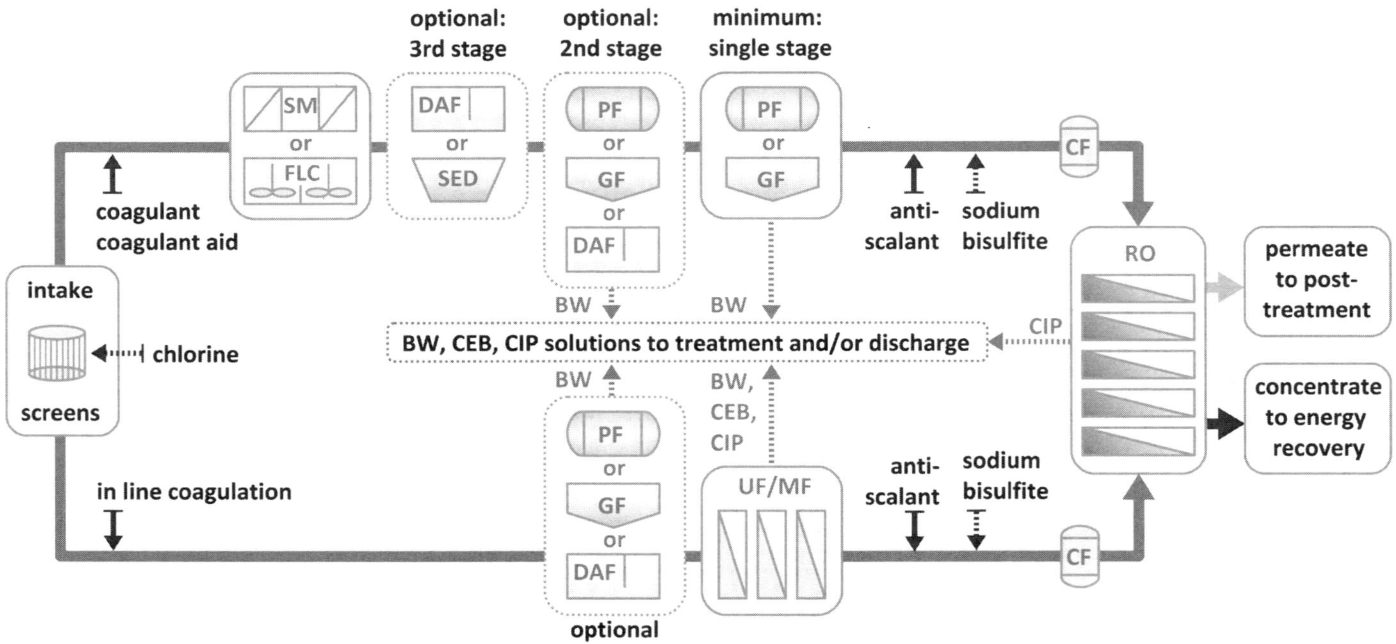

Figure 2.7 Simplified flow-chart of a SWRO plant with conventional pretreatment (top) or UF/MF pretreatment (bottom): (SM—static mixer; FLC—flocculation chambers; DAF—dissolved air flotation; SED—sedimentation; PF—pressurized media filter; GF—gravity media filter; CF—cartridge filter; BW—backwash; CEB—chemically enhanced backwash; CIP—cleaning in place); dotted lines show intermittent flows/doses (adapted from data in Refs. 18 and 33-35).

to remove any residual oxidants from the RO feedwater, which might damage the membranes [Table 2 in Ref. 27].

Coagulation–flocculation is a combined process. *Coagulation* is the destabilization of the particle surface charge of small and colloid particles, which is followed by *flocculation*, i.e., the formation of larger flocs. The dosing of primary coagulants, typically ferric chloride ($FeCl_3$) or ferrous sulfate ($FeSO_4$), depends on the feedwater quality and ranges between 0.2 and 20 mg/L [Table 5 in Ref. 27]. Doses of 1–6 mg/L are assumed to be the rule and 20 mg/L the exception. The resultant formation of metal hydroxides is very rapid, but effective coagulation requires intensive mixing to bring the hydroxides in contact with the colloid particles. This is achieved downstream of the injection point by static mixers or flocculation chambers that induce more gentle mixing. The process can be further enhanced by the addition of a coagulant aid, which is typically a longchain organic polymer. Cationic polymers can also be used as primary coagulants [18].

Filtration is either performed in pressurized vessels or gravity concrete chambers. The filters contain a single or dual medium, usually anthracite and sand, are arranged in a single- or two-stage configuration, and are backwashed with either filtrate or concentrate. The backwash water is discharged into the sea or dewatered and the sludge transported to a landfill. The backwash frequency and the amounts of sludge produced depend on the amount of suspended solids in the feedwater and the required coagulant dose. Two-stage dual-medium filtration is usually effective in producing a consistent feed quality, even under fluctuating water conditions [36]. Sedimentation or dissolved air flotation (DAF) can precede single- or dual-stage filtration in locations with treatment-recalcitrant feedwater [33]. In the Gulf, some new SWRO plants have incorporated DAF followed by UF or media filters to deal with extended periods of red tides [37]. In many SWRO plants, a single-stage pressurized or gravity filter was found to be sufficient [Table 5 in Ref. 27].

Scaling by inorganic salts, metal oxides, and hydroxides does not present a problem with the majority of seawater feeds as a result of the relatively low recovery rate ($\leq$50%) of SWRO plants, the high ionic strength of seawater, and the low concentration of bicarbonate ions in seawater [36]. Nevertheless, most SWRO plants use 1–2 mg/L of antiscalants or 20–50 mg/L of acid to avoid the risk of scale formation on the RO membranes. Some plants also operate without any antiscalant, e.g., on the Cayman Islands or in the Mediterranean. Laboratory studies suggest that calcium carbonate scaling—the main scalant in SWRO—should not occur since the induction time of calcium carbonate is considerably longer than the residence time of the water in the membranes [38]. This result raises questions about the current practice of antiscalant use in SWRO, which should be further evaluated by pilot studies and in full-scale SWRO plants. If a second stage is necessary for boron removal, which requires that the pH be raised to ~10, antiscalants are needed to prevent the formation of magnesium hydroxide [$Mg(OH)_2$]. SWRO plants with a single stage often use acid only, while plants with a second RO stage often use polymer antiscalant to maintain a higher pH for boron removal.

2.3.2.3 Ultrafiltration (UF) and Microfiltration (MF) An emerging alternative to conventional pretreatment in SWRO is UF/MF pretreatment. The number of full-scale SWRO plants equipped with UF/MF pretreatment, however, is still fairly low compared to the many pilot studies that have demonstrated a good performance of UF/MF in SWRO, and the manifold benefits that have been cited in the literature.

One proclaimed benefit is the consistently high feedwater quality that is produced regardless of source-water fluctuations, which generally results in a reduced fouling potential of the SWRO membranes, a lower cleaning frequency, and hence a lower chemical use, as well as a longer membrane life and therefore savings in material and energy use. Moreover, a better feed quality offers the potential to operate at a higher flux, so that the total membrane surface area can be reduced. The pretreatment system would be smaller, too, because less feedwater needs to be treated. It is therefore estimated that UF/MF pretreatment can save about one-third in plant area size [18] and about one-third in the membrane replacement rate [39]. Despite these benefits, however, conventional pretreatment is still often chosen over UF/MF for new SWRO projects. One possible reason could be the still limited performance data of full-scale UF/MF-SWRO plants [40].

UF/MF was once considered as a possibly "chemical-free" alternative to conventional pretreatment that may eliminate the need for coagulant dosing with good feedwater quality [41] and the need for a continuous presence of free chlorine [36,42]. However, the use of a disinfectant, either continuously or intermittently, and the use of inline coagulation, which is the dosing of a coagulant without a sedimentation or filtration step, seem to be common practice in many UF/MF systems in order to improve the process performance and filtrate quality. Moreover, UF/MF membranes usually require intermittent chemically enhanced backwashing (CEB) and cleaning in place (CIP).

Performance data on more than 40 UF/MF-SWRO systems could be obtained from the more recent literature (years 2000–2009). Of these, only 16 provided information on full-scale *operational* plants [Table 6 in Ref. 27], the remaining were *pilot systems* [Table 7 in Ref. 27]. Examining the available literature led to the following conclusions concerning chemical use in UF/MF pretreatment for SWRO.

The majority of UF/MF-SWRO systems showed a reliable and often superior performance to conventional pretreatment systems in terms of *turbidity* and *microorganisms* removal. The favorable performance, however, must partly be attributed to the use of chemicals similar in type and dosage to chemical use in conventional pretreatment. However, some UF/MF systems performed less well compared to conventional pretreatment in terms of dissolved organics removal. MF/UF pretreatment may therefore not always be an adequate solution for the prevention of organic fouling [43], unless coagulation prior to UF/MF is used, which can improve the adsorption and removal of organics as fine particulates. Coagulants may not be needed in every UF/MF system. The need and the optimum dosage should be established by pilot testing [39], as overdosing may also result in operational problems of the UF/MF membranes [44] and of the SWRO membranes [45], as well as in unnecessary costs and environmental impacts.

The reported coagulant levels in this review ranged between 0.2 and 10 mg/L in UF/MF pilot plants and between 0.3 and 10 mg/L in UF/MF full-scale plants, with values of 0.2–2 mg/L assumed to be the norm and 10 mg/L the exception [Tables 6 and 7 in Ref. 27]. This compares to ranges of 0.2–20 mg/L in conventional pretreatment, where ranges of 1–6 mg/L are assumed to be the norm and 20 mg/L the exception [Table 5 in Ref. 27].

Chlorination levels in several UF/MF systems ranged between 0.3 and 2 mg/L, which is similar to conventional pretreatment where low-level continuous chlorination of 1–2 mg/L is applied. However, many conventional SWRO plants now use intermittent chlorination. For example, an intermittent dose of 10 mg/L for one hour per day equals a continuous dosage of "only" 0.4 mg/L. As chlorination intervals in conventional pretreatment are often even longer, e.g., once per week, chlorine use may be even lower. Many UF/MF systems additionally employ CEBs with considerable chlorine doses, which may compare to an additional chlorine level of 0.5–2 mg/L if the chemicals were used continuously. For instance, a UF-SWRO plant in Asia reported feedwater chlorination in a dosage of 1 mg/L and chlorination of every backwash once every hour in a dosage of 15 mg/L [40].

In conclusion, the reported chemical use in UF/MF pretreatment does not seem to live up to the expectations that were initially placed into the technology of being a chemical-free process. However, total chemical use in UF/MF is in some cases claimed to be significantly lower or minimal compared to conventional levels [40,41,46]. Pearce [39] assumes that UF/MF systems can be operated at a coagulant dose concentration of 43% of conventional pretreatment (i.e., 0.3 mg/L instead of 0.7 mg/L as Fe) and that chlorine use can be reduced to intermittent chlorine cleaning. Indeed, a lower consumption of chemicals in UF/MF seems to be feasible.

For example, a lifecycle analysis and aspects of operability led to the selection of a UF system over a classic two-stage filtration system for a desalination project south of Perth with a capacity of 140,000 m^3/d, which will presumably operate without coagulation [47]. Our literature review indicates that coagulant use seems to be lower in UF/MF pretreatment than in conventional pretreatment, whereas chlorine use appears to be higher in some UF/MF-SWRO plants, due to a combination of continuous chlorination and intermittent shock chlorination during CEB and CIP. Some UF/MF systems were found to be liable to biofouling without continuous or intermittent chlorination of the UF feed, CEB or CIP [48,49].

The overall chemical use of an UF/MF SWRO system depends on the filtration time and the backwashing, CEB, and CIP intervals. One option to postpone backwashing and CEB might be additional pretreatment prior to membrane filtration, such as a beachwell or a sand filter. Other options are to lower the flux, which will increase the total membrane area to be installed [50], or to operate without coagulant pretreatment but more frequent cleaning [39]. A well-designed system operated on good feedwater quality may thus, indeed, reduce chemical use. However, an integrated membrane system does not seem to be superior to conventional pretreatment in terms of chemical use when inline coagulation, chlorination and frequent CEB is employed. In that case, the only remaining benefit may stem from the reduced cleaning frequency (CIP intervals) of the SWRO membranes due to the improved quality of the feedwater, which may cause a reduction in chemical use of the overall system.

2.3.3 Comparing Pretreatment Processes with Respect to Organic and Particulate Foulant Reductions in SWRO Systems

This section studies various seawater pretreatment systems regarding to particulate and organic matter removal. To study organic matter, liquid chromatography with online dissolved organic carbon detection (LC-OCD) and fluorescence excitation emission were considered, while for particulate matter the modified fouling index at constant flux was employed.

The results from six locations are presented. These locations present various pretreatment options found in seawater reverse-osmosis systems: coagulation and dual-medium filtration, microfiltration, beachwells, and ultrafiltration. The plant names and locations are not disclosed, as requested by plant operators.

2.3.3.1 Plant Description The locations summarized in Table 2.1 were studied. The locations include conventional and membrane pretreatment, and different intakes as beachwells and a system termed the *underocean floor*. The studied period corresponds to July 2007–August 2009.

Table 2.1 Summary of Seawater Reverse Osmosis (SWRO) Locations and Plants

Location	Intake	Pretreatment	RO unit
A (northern Mediterranean water)	Direct intake	UF (0.01 μM)	$R = 40\%$, $J = 15$ L/(m^2·h)
B (US—Long Beach, CA)		Beachwell	
D (eastern Mediterranean water)	Submerged pipe	Coagulation—DMF	$R = 40\%$
S (northwestern Mediterranean water)	Submerged pump ($L = 2.5$ km)	UF (0.02 μM)	$R = 45\%$, $J = 15$ L/(m^2·h); 7 elements per vessel, 8-in. module; never cleaned
U (northern Pacific Ocean	Underocean floor	Underocean floor	
Z (North Sea water)	Submerged ($L = 100$ m)	Strainer–UF (~300 kDa)	$R = 40\%$, $J = 15$ L/(m^2·h) 6 elements/vessel, 8-in. module never cleaned

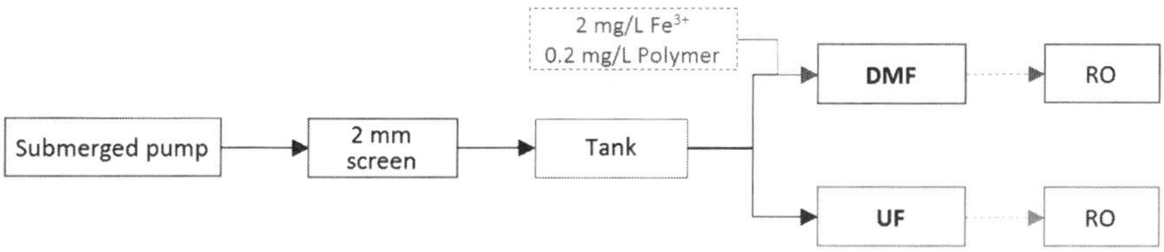

Figure 2.8 Scheme of site A.

Site A The pilot plant consists of two parallel RO pretreatment lines: ultrafiltration (UF) and coagulation combined with dual-medium filtration (Coag + DMF) (Fig. 2.8).

The water intake is a submerged (2 m) pump close to the coastline (1 m distance). Water is pumped (12 m height) for ~40 m length into a plastic tank for sand particles to settle and for algae removal through a 2-mm screen. Water is pumped again ~30 m distance to the pilot plant installation. Seawater is received in a tank (0.8 m^3) and then distributed to the DMF line and to the UF line. There is no pH correction or acid addition in any part of the plant.

The RO units consist of one 4-in. module working at 20% recovery. For the prediction model estimation, typical full-scale conditions were applied.

Iron chloride ($FeCl_3$) at a dose of 2 mg Fe^{3+}/L was employed. Coagulant is dosed in front of the DMF by dosing pumps. A cationic polymer (0.15–0.2 mg/L) is added as well before DMF. The DMF consist of a cylinder (~40 cm diameter) containing two media layers (anthracite and sand). Table 2.2 summarizes the characteristics of the unit.

The pH of the water after Coag + DMF decreases from the raw-water value (~8.21) to 7.8–8.0. The residence time from the raw seawater tank inside the installation to the DMF effluent sampling point is around 25 min.

The pilot plant has one UF module working permanently at a constant flux. The residence time from the raw seawater tank inside the pilot plant to the UF

Table 2.2 Dual-Medium Filtration Unit Description of Site A

Parameter	Value	Comment
Flow rate	~0.9 m^3/h	Flow from top to bottom
Media	anthracite (80 cm), sand (80 cm)	Anthracite is the upper layer
Cleaning	$\Delta P = 400$ mbar	Backwash with air and water; it takes 1 h before new water is produced
Total height of DMF	3.5 m	—
SDI_{15}	~3.5	Typical value after 10 h, including coagulation

Table 2.3 Ultrafiltration Unit Description of Site A

Parameter	Value	Comment
Operation	Constant flux (1.6 m^3/h)	Typical pressure ~70 kPa
Flux	57 L(m^2·h)	
Nominal pore size	0.01 µm	Loose UF
Material	PVDF	
Brand	MEMCOR CMF-S S10V	Siemens
Backwash	2 min	Air scour and backwash with permeate water
CEB	Every 40 cycles	Every ~10 h
Membrane area	27.9 m^2	
Filtration	Outside to inside	

permeate sampling point is around 3 min. The description of the UF is presented in Table 2.3.

Site B The beachwell has a sand depth of 0.8–1 m and a filtration rate of 0.4 m/h.

Site D The raw-water turbidity ranges from 0.5 to 5 nephelometric turbidity units (NTU), with total organic carbon content (TOC) ranging between 0.7 and 1.5 mg/L, and elemental carbon content (EC) is ~56.5 mS/cm. The pretreatment of the plant consists of coagulation with ferric sulfate + single-stage granular filtration, with the effluent of this step fed to the RO units (FILMTEC SW30HR) after cartridge filtration.

Site S This site is located 15 km from Barcelona. It consists of two separated/independent treatment lines (W and T, Fig. 2.9). Pilot plant W consists of ultrafiltration followed by reverse osmosis at 51% recovery, and pilot plant T consists of coagulation + dissolved air flotation followed by ultrafiltration and then reverse osmosis at 45% recovery. A second treatment line exists with two-stage DMF that at the moment does not feed RO. The pilot plants receive water from an open intake (submerged pipe) located 2.5 km from the coast and 25 m below the

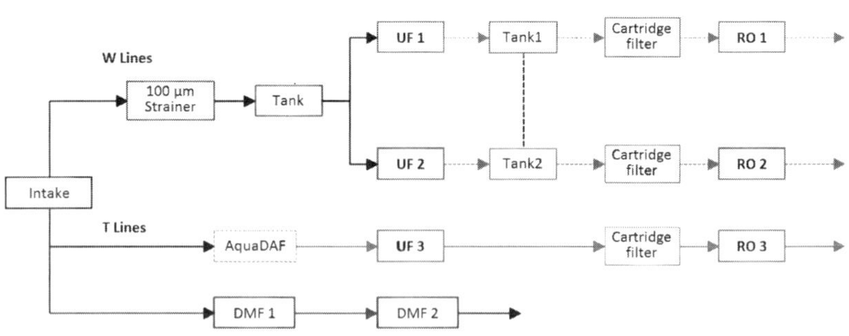

Figure 2.9 Scheme of site S.

water surface. The pipe is cleaned by chlorination (frequency not disclosed). The average suspended solids of the raw water was about 3.5 mg/L, with an average silt density index (SDI) value of 5.

The two site S lines are described as follows:

W Lines. Before the raw water is being fed to the UF, it passes through a 100 μm-strainer. The ultrafiltration units (UF1 and UF2) operate at constant flux at ~60 L/(m^2·h). Backwash is applied at double the operation flow with air scour every 30 min, consisting of 10 s air scour, 15 s backwash with UF permeate, and 45 s forward flush with raw water (Table 2.4).

The RO systems consist of two units working in parallel (RO1 and RO2). Each unit has six 4-in. SW30-4040HR elements and operate at 52% recovery. The RO2 unit has a hybrid configuration (high-rejection modules in the front of the pressure vessel and high-production modules at the end), while RO1 has a standard configuration. Both RO units operate at constant pressure (70 bar). Bisulfite and antiscalant are added in front of the RO (Permatreat, phosphonated). Since the start of operation of the plant (April 2009), the RO membranes were never cleaned. The RO production capacity is ~0.76 m^3/h.

T Lines. The AquaDAF unit was not necessary during the testing period, due to good raw seawater quality. The UF3 modules are Zenon membranes that operate in a submerged mode (Table 2.5). There are two dual-medium filters (DMFs) that operate in series. In DMF1 there are two layers of sand and pumice and in DMF2 sand and anthracite. DMF1 was not in operation during the testing period because the raw seawater quality was good. Coagulation is used in combination with DMF. During the testing period the dose was 1.5 mg/L $FeCl_3$. The RO3 unit is one pressure vessel containing seven 8-in. elements. The RO operates at 45% recovery.

Site U This system (see Fig. 2.10) is based on the design criteria associated with slow sand filtration systems. The century-old slow sand filtration concept has been utilized around the world and now offers the opportunity to be applied in

Table 2.4 UF1/UF2 Unit Description of Site S

Parameter	Value	Comment
Operation	Constant flux (1.9 m^3/h)	Typical pressure 0.7 bar
Flux	~58 L/(m^2·h)	—
Nominal pore size	0.03 μm	—
Material	PVDF	—
Brand	SFP2660	OM Excell—Dow
Backwash	1.25 min	With air scour, permeate water
CEB	Every 24 h	~48 cycles
Membrane area	33 m^2	—
Filtration	Outside to inside	—

Table 2.5 UF3 Unit Description of Site S

Parameter	Value	Comment
Flux	~50 L/(m^2·h)	—
Nominal pore size	0.02 μm	—
Material	PVDF	—
Brand	ZeeWeed	Zenon
Membrane area	55 m^2	—
Filtration	Outside to inside	—

Figure 2.10 Site U: underocean floor intake–discharge system at Long Beach, California.

an innovative manner for seawater desalination systems. By incorporating slow sand filtration (loading rate of <2 L/(m^2·h)] into the seawater collection process, a natural, biological filtration process reduces organic and suspended solids loading on the desalination plant. Therefore, additional pretreatment is not required, thus reducing costs and improving the desalination process.

Advantages of the underocean floor seawater intake system over open ocean intakes or desalination pretreatment processes are: (1) the flow rate and operation of the under ocean floor intake system is unaffected by wave action and tidal forces; (2) it is virtually maintenance-free, eliminating operation and maintenance costs; (3) it requires no backwashing, cleaning, treatment, recharging, and/or rehabilitation; and (4) it serves the dual role of an intake and pretreatment component in an environmentally sensitive manner.

Site Z The pilot plant is located in Noord-Beveland in The Netherlands. The plant (Fig. 2.11) makes use of coagulation and ultrafiltration as a pretreatment to the reverse-osmosis units. Before coagulation, water pH is reduced to 6.5 and then coagulant [poly(aluminium chloride)] (PACl) is added (0.5 mg/L as Al^{3+}), and mixing occurs mechanically and hydraulically in the mixing tank. Average conditions for the UF operation are described in Table 2.6. The operational conditions in the plant such as coagulant dose, coagulant type, and pH correction are under experimentation and thus have changed over time.

2.3.3.2 Pretreatment for Particulate and NOM Removal

Site A: Dual-Media Filtration versus Microfiltration The Site A plant has two parallel treatment trains. The first treatment line consists of pH correction (6.8 with sulfuric acid), coagulation with ferric chloride + polymer addition, and dual-medium filtration (anthracite and sand). The second treatment train consists of pH correction and microfiltration [0.1-μm PVDF membranes operating at 50 L/(m^2·h)]. Table 2.7 shows that the DOC concentration is ~1.2 mg/L at the intake of the plant. SUVA values are in all cases <2, suggesting that nonorganic matter (NOM) consists mostly of nonhumics with low hydrophobicity and low molecular weight [51].

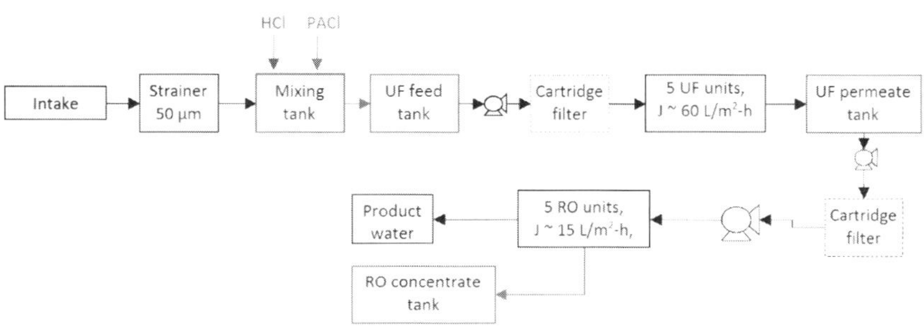

Figure 2.11 Scheme of site Z.

Table 2.6 UF Unit Description of Site Z

Parameter	Value	Comment
Operation	Constant pressure	It could work at constant flux for short periods
Flux	~60 L/(m^2·h)	After cleaning
Nominal pore size	~300 kDa	—
Material	PES	—
Brand	SeaGuard	NORIT filtration
Backwash	Every 45 min	—
CEB	1-2 x/day	—
Membrane area	37 m^2	per module
Filtration	inside to outside	—

Table 2.7 Site A Raw-Water DOC (mg/L) and SUVA Values [L/(mg·m)]

Pretreatment Train	Raw Water	Coag + DMF Out	MF Out
SUVA	0.78 ± 0.12	0.69 ± 0.12	0.59 ± 0.16
DOC	1.19 ± 0.32	0.77 ± 0.03	0.85 ± 0.04

The LC-OCD results are presented in Figure 2.12 for both pretreatment trains including the percentage removal of the organic matter fractions. By comparing coagulation + DMF and MF, it can be seen that the former is more effective in removing organic matter: 35% DOC removal for Coag + DMF compared with 28% DOC removal for MF. In both treatment trains, the biopolymers are significantly removed (47% Coag + DMF and 36% MF). DMF combined with inline coagulation is more effective than MF without coagulant addition. As mentioned in the plant description, the pilot plant has two pretreatments in parallel with treatments for

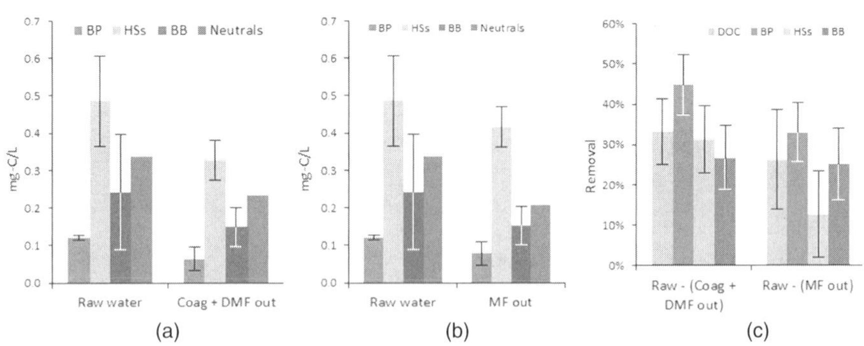

Figure 2.12 Site A LC-OCD results: Coag+DMF (a) and microfiltration (b,c).

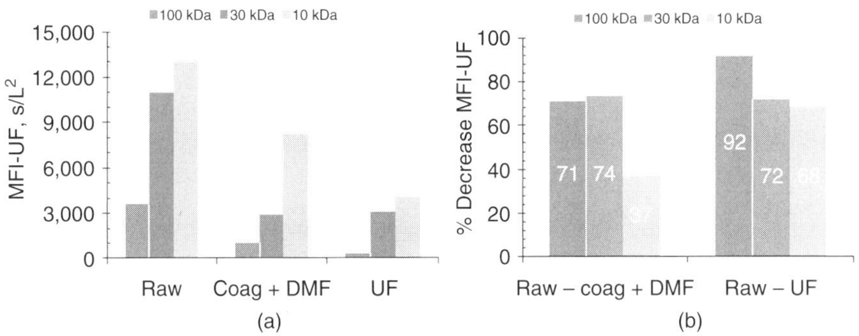

Figure 2.13 Site A MFI-UF values (a) and percentage removal (b) for raw seawater, Coag + DMF effluent, and UF permeate with 100, 30 and 10 kDa membranes at 250 L/(m^2·h).

the same raw water. Both pretreatments were compared by measuring the MFI-UF values before and after treatment of the units and with various membranes (100, 30, and 10 kDa) at 250 L/(m^2·h). The results are presented in Figure 2.13.

The raw seawater has an electrical conductivity equal to 57.1 mS/cm at pH 8.21, and during the testing period the water temperature was 19°C. The MFI-UF value for the raw water is comparable with North Sea water. The higher value is obtained with the smaller MWCO (10 kDa) as more particles/colloids are retained by the membrane. The MFI-UF value for 30 kDa is between 100 and 10 kDa and closer to the latter result. In general the same trend is observed for DMF and UF water, with the values for the UF permeate lower than those for the DMF effluent. The percentages in MFI-UF removal for each pretreatment line are presented in Figure 2.13b. For all the MWCOs, water passage through UF results in further decrease of MFI-UF values in comparison with Coag + DMF. For UF, MFI-UF values were 92%, 72%, and 68% for 100, 30 and 10 kDa, respectively. For Coag + DMF the reductions in MFI-UF values were 71%, 74%, and 37% for 100, 30, and 10 kDa, respectively.

Figure 2.14 shows the additional increase in MFI-UF by changing the MWCO for both pretreatment lines. The change from 100 to 30 kDa produces the higher increase in MFI-UF value. The additional increases in MFI-UF for DMF and UF are presented in Figure 2.13.

Site B: Beachwell The beachwell has a sand depth of 0.8 1 m and a filtration rate of 0.4 m/h. The DOC content at site B averages 0.94 mg/L with a SUVA value of $\sim 0.70L/(mg{\cdot}\mathrm{m})$. After passage through the beachwells the DOC is reduced to 0.74 mg/L (see Table 2.8).

The results of LC-OCD analysis are present in Figure 2.15. Humic substances are the main fraction for the raw water and for the beachwell effluent. The removal effectiveness of beachwells shows that the DOC removal is ~21%, with the biopolymer fraction removed by ~70%. This is a significant reduction in organic matter with size >20 kDa.

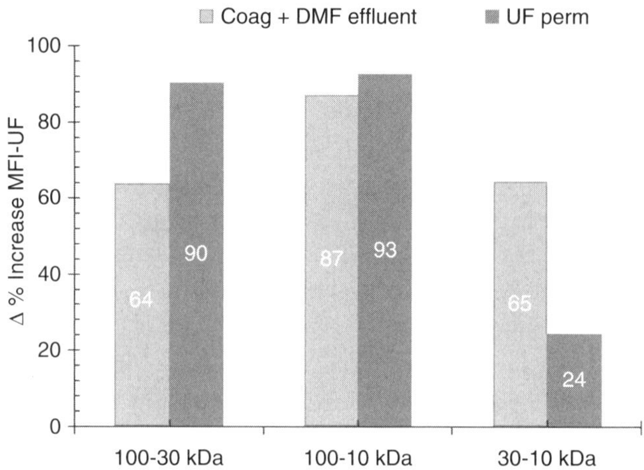

Figure 2.14 Percentage additional increase in site A MFI-UF values between 100–30-, 100–10-, and 30–10-kDa membranes for raw seawater and Coag + DMF effluent and raw water and UF permeate with 100-, 30-, and 10-kDa membranes at 250 L/(m^2·h).

In case of particulate fouling potential, the measured MFI values are presented in Table 2.9. A reduction of 50% was found with a 30-kDa membrane while a reduction of 35% was found for 10-kDa membrane. This might suggest that the beachwells reject more particles with membranes >30 kDa.

Site D: Coagulation + Dual-Medium Filtration The raw-water turbidity ranges from 0.5 to 5 NTU, with TOC 0.7–1.5 mg/L, and EC ~56.5 mS/cm. The pretreatment of the plant consists of coagulation with ferric sulfate and single-stage granular filtration, with the effluent of this step fed to the RO units after cartridge filtration. The LC-OCD results show that the raw water consists mainly consists of humic substances (~50%), biopolymers (~10%), building blocks and neutrals (~20% each; Fig. 2.16).

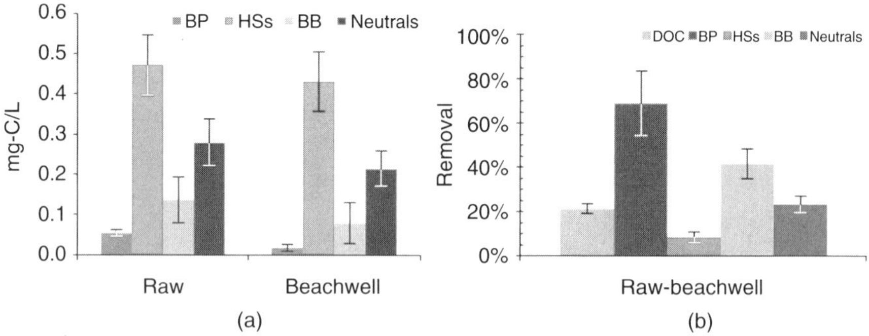

Figure 2.15 Site B LC-OCD results.

Table 2.8 Site B Raw-Water DOC [(mg·C/L)] and SUVA Values [L/(mg·m)]

Pretreatment Train	Raw Water	Beachwell
SUVA	0.70 ± 0.05	0.61 ± 0.28
DOC	0.94 ± 0.09	0.74 ± 0.06

Table 2.9 MFI-UF Measured at Constant Pressure (2 bar)

Water Type	MFI 30 kDa	Removal	MFI 10 kDa	Removal
Raw water	4980	—	11,140	—
Beachwell	2570	~50%	7,300	~35%

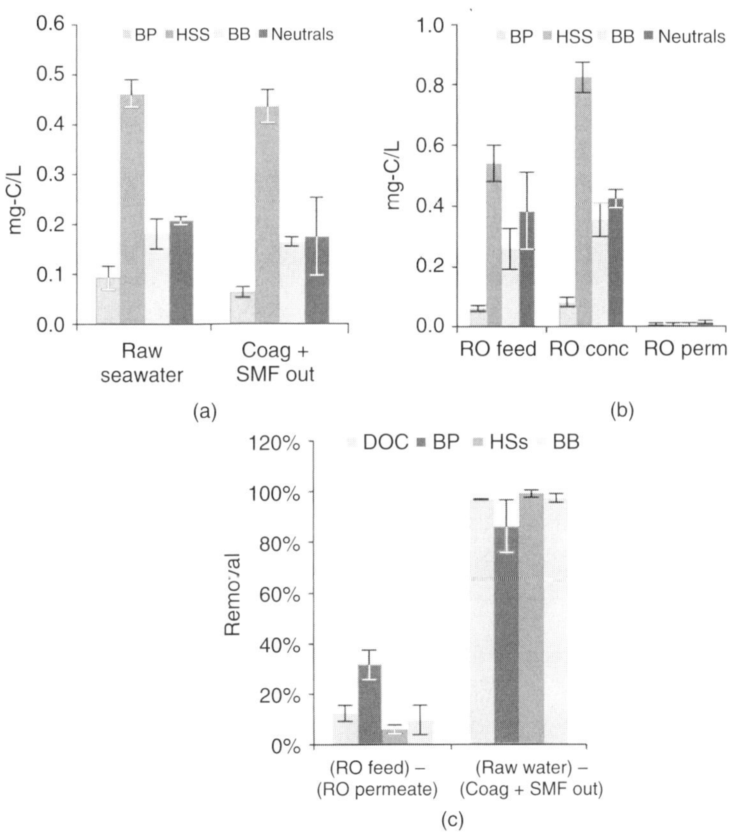

Figure 2.16 Site D LC-OCD results: pretreatment (a), RO system (b,c).

Table 2.10 MFI-UF Measured at 250 L/m²h

Water Type	MFI 100 kDa	Removal	MFI 10 kDa	Removal
Raw water	400	—	1800	—
Coag + SMF out	205	~50%	850	~52%

In terms of organic matter, coagulation + single-stage media filtration removes 12% of DOC where the major removed fraction is the biopolymers (~32%). The passage of water through the RO membranes (FILMTEC SW30HR) removes >98% of the organic carbon. The MFI results are presented in Table 2.10. For the 100-kDa membrane, as well as for the 10 kDa membrane, the MFI value decreased by around 50%.

Site S: Ultrafiltration From both pilot plants the MFI-UF values were measured at constant flux with a 10-kDa polyethersulfone (PES) membrane. Initially for the raw water the MFI-UF was measured with a 30-kDa membrane. The value obtained was 1050 s/L². As this value was observed to be low in comparison with other seawaters, i.e., ~4500 s/L² with a 100 kDa for North Sea water, it was decided to perform the profiling of the pilot plants with a 10-kDa membrane. The results are presented in Figure 2.17.

The MFI-UF values for UF1 and UF3 are close: 850 and 780 s/L², respectively. According to the manufacturers, the UF1 has a nominal pore size of 0.03 μm and UF3 a nominal pore size of 0.02 μm. This suggests that a tighter UF is more suitable for pretreatment to remove particles and colloids. The DMF2 value was high at ~1950 s/L². Unfortunately, the operational data was not disclosed. Iron chloride was added at a concentration of 1 mg/L as $FeCl_3$ in front of the DMF2. The percentage in decrease of MFI-UF values (Figure 2.17 right) before and after pretreatment was 65%, 67%, 68% and 19% for UF1, UF2, UF3 and DMF2, respectively.

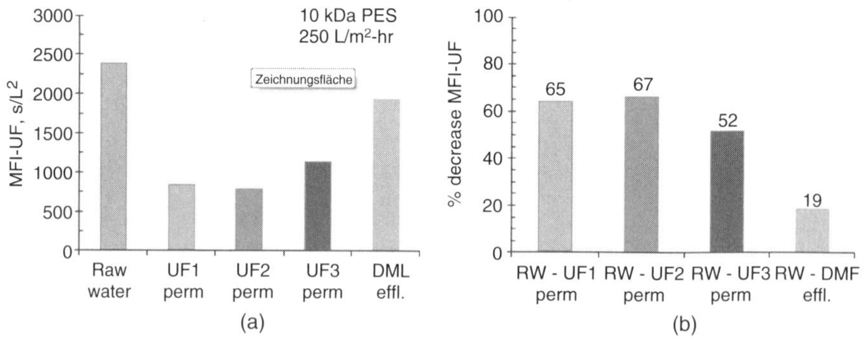

Figure 2.17 Site S – MFI-UF values (left) and MFI-UF percentage reduction (right) for W and T lines.

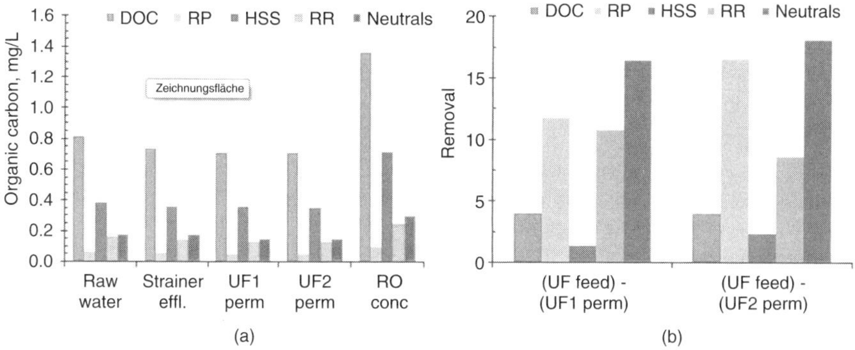

Figure 2.18 Site S LC-OCD results (a) and removal (b).

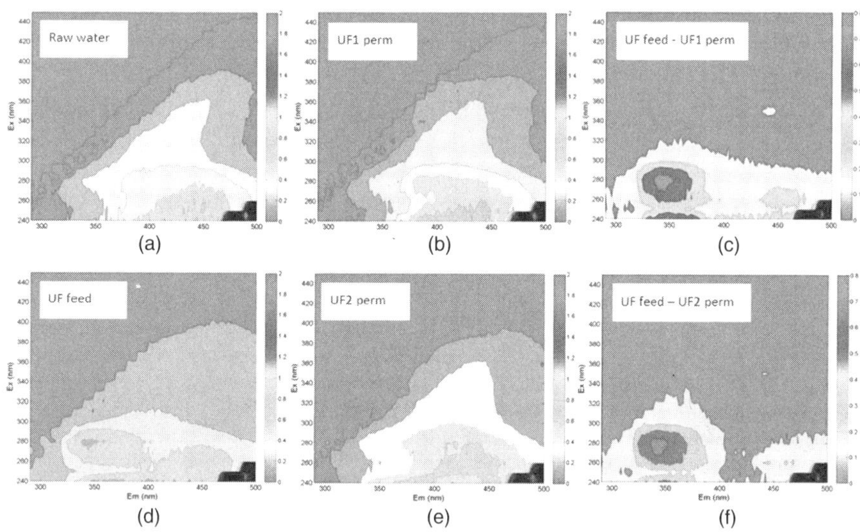

Figure 2.19 F-EEMs Raw water, UF1 perm, UF feed-UF1 perm (a–c) and UF feed, UF2 perm and UF feed-UF2 perm (d–f).

The LC-OCD results from the plant are presented in Figure 2.18. The DOC removal achieved by the UF units was around 4%. Biopolymers were removed by ~14% on average.

The fluorescence-EEM spectra for some points along the treatment plant are presented in Figure 2.19. The fluorescence index for this location is FI = 1.71, which represents that the organic matter fluorophores are mainly autochthonous (microbially derived). Protein-like (Fig. 2.19d) and humic-like (Fig. 2.19e,f) organic matter fluorophores are present in the water samples. According to the differential EEMs, both UF units mainly remove mainly protein-like material.

Site U: Underocean Floor The LC-OCD results obtained from the system are presented in Figure 2.20. It can be observed that there is a significant removal of biopolymers (~75%) with this intake system. An additional 5 μm cartridge filter provided ~13% extra biopolymer reduction.

The fluorescence-EEM results are presented in Figure 2.21. Protein-like material and humic-like fluorophores were observed. The figure shows that protein like material and humic-like material were removed mainly by the underocean floor (UOF) intake. The fluorescence index for the raw water was FI = 1.81, indicating that the organic matter fluorophores are mainly autochthonous (microbially derived).

Site Z: Coagulation + Ultrafiltration MFI-UF plant profiling with a 100-kDa membrane for several dates during the period 2009–2010 is presented in Table 2.11. The MFI values were measured at 250 L/(m^2·h). The samples from 5/10/10 (i.e., May 10, 2010) are significantly higher than the ones in the previous year. This significant increase was correlated with the algae bloom and thus increase in biopolymers (TEP) concentration in the raw water during the testing period.

Although the raw-water values varied in time, in all the testing dates the MFI decrease after the UF was between 89 and 97%. Figure 2.22 shows the MFI-UF

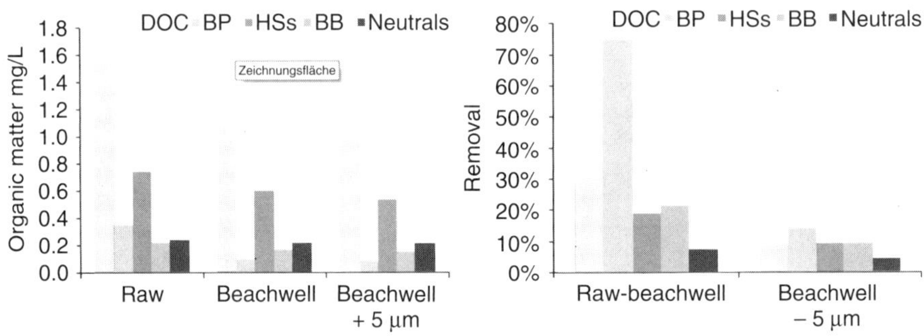

Figure 2.20 Site U LC-OCD results (a) and removal after beachwell (b).

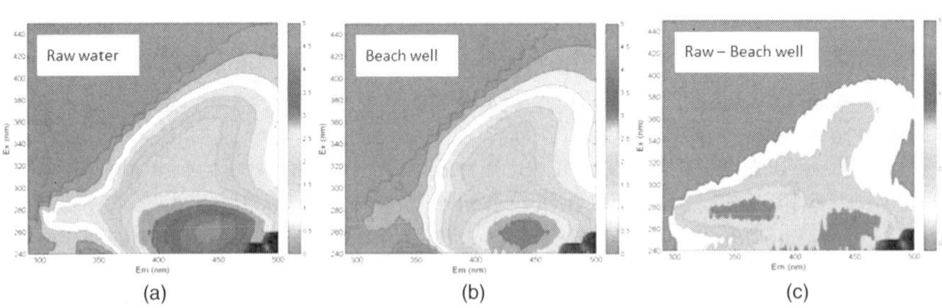

Figure 2.21 Site U: F-EEM for raw water (a), Beachwell (b), and differential (c).

Table 2.11 Site Z MFI-UF (100 kDa) Values (in s/L^2) and Percentage Removal

Date	Raw Water	UF Feed	UF Permeability	Removal, %
4/23/09	4,310	2,935	190	94
4/28/09	4,840	4,295	125	97
6/16/09	3,800	3,650	395	89
7/2/09	2,950	2,285	203	91
7/6/09	2,840	2,450	200	92
5/10/10	25,340	17,190	980	94

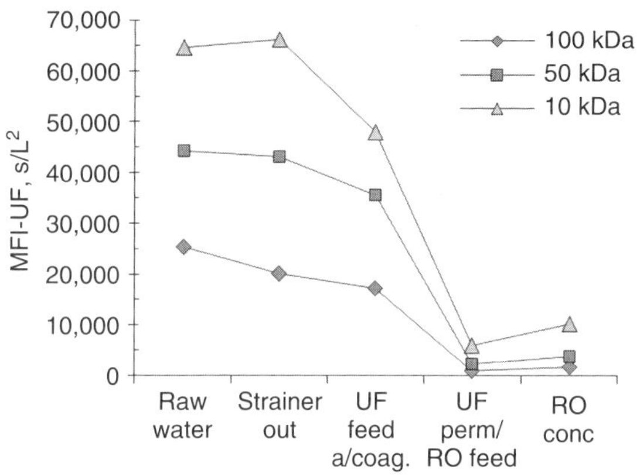

Figure 2.22 Site Z: MFI-UF plant profiling measured with 100-, 50-, and 10-kDa membranes at 250 L/(m^2·h).

values measured with 100-, 50-, and 10-kDa membranes at 250 L/(m^2·h) along a SWRO plant treating water from the North Sea for the higher foulant period. The plant is located in The Netherlands. The percentages in reduction of MFI values after water passage through the ultrafiltration units were 94.3%, 93.4%, and 87.6% for membranes of 100, 50, and 10 kDa, respectively.

The results from fluorescence spectrometry for a plant treating water from the North Sea are presented in Figure 2.23. In this case, the fluorescence intensities are presented according to representative points in the spectra. Humic substances (humic, fulvic, and marine–humic-type materials) predominate the spectra with higher fluorescence intensities than in the amino-acid-like materials.

From Figure 2.24, the AA-like tyrosine peak was removed by 23% and the AA-like tryptophan peak was removed by 22% after the UF units. Coagulation (strainer effluent UF feed) removed 15% AA-like material and 6%, 9% and 6% for fulvic, humic, and marine–humic material, respectively. The fluorescence index, FI ~1.75, indicates that the organic matter fluorophores are mainly autochthonous, that is, microbially derived. The average LC-OCD results for the sampling campaigns are presented in Figure 2.24. Biopolymers were removed mainly by the UF units (~50%).

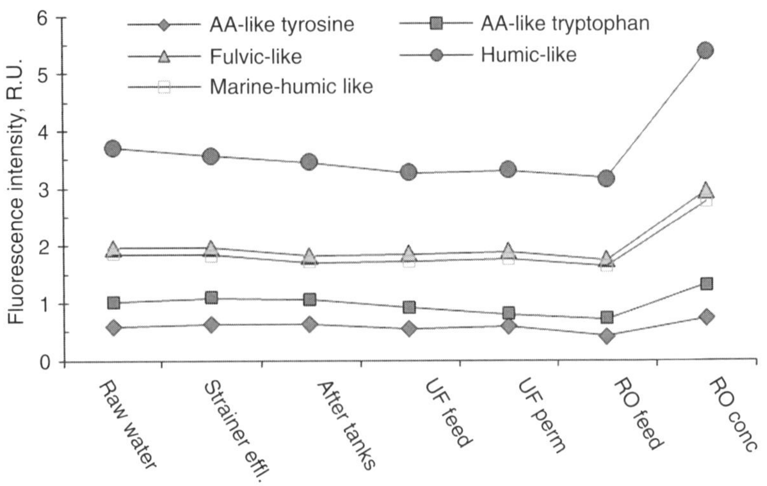

Figure 2.23 Site Z - Fluorescence intensities (RU) along the treatment plant.

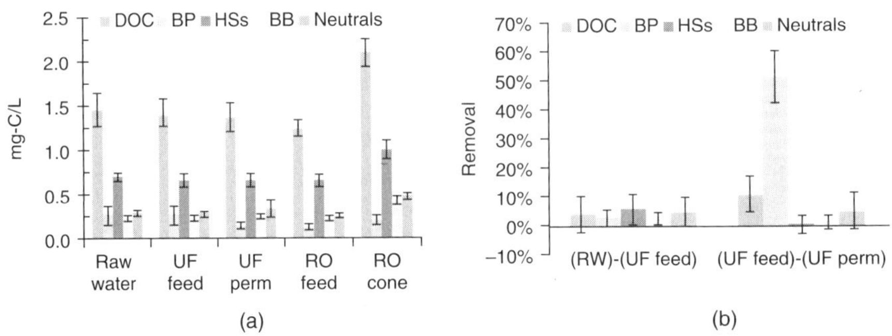

Figure 2.24 Site Z organic matter fractions (a) and removal (b).

2.3.3.3 Summary and Conclusions Seawater organic matter averaged 1 (mg·C)/L; where humic substances represent ~50% biopolymers ~10%, and building blocks and neutrals the remaining 40%.

Table 2.12 summarizes of the pretreatment NOM removal achieved in the different locations.

The removal of biopolymers (~70%) via beachwell and UOF pretreatment was roughly twice that achieved in conventional and membrane pretreatment systems. For site C (estuarine water), coagulation + continous sand filtration removed 12% DOC and 17% biopolymers. The UF units removed nearly 70% of the biopolymers that were fed to the membranes.

Table 2.12 Summary of Pretreatment MFI and NOM Removal Efficiencies

Site	Pretreatment	DOC, %	Biopolymers, %	Humics, %	MFI, with 10 kDa, %
A	Coag+DMF	35	47	30	40
A	MF	26	36	8	68
B	Beachwell	21	70	9	35
D	Coag+SMF	12	32	6	52
S	UF	4	15	1	65
U	Underocean floor	30	75	19	—
Z	UF	8	51	1	>90

2.3.4 Desalination Process and Energy Use

The total energy demand of a desalination plant depends on a range of factors such as recovery, pretreatment type (e.g., conventional vs. UF/MF), the type of SWRO membranes used (e.g., low-energy membranes), the efficiency of pumps and motors, the type and efficiency of the energy recovery system installed (if any), and environmental conditions (e.g., feedwater temperature). Energy demand also depends on the product-water specifications. Employing a second SWRO pass for boron removal will increase the energy demand of the process. Table 2.13 summarizes the typical energy requirements of the main desalination processes and compares them to other water supply options.

Table 2.13 Energy Consumption Due to Desalination, and Conventional Water Supply Options

Method	Electrical Energy, kWh/m^3	Main Energy Form	Thermal Energy, MJ/m^3	Performance Ratio, kg/2326 kJ	Reference(s)
BWRO	0.5–3.0	Electrical	—	—	19
SWRO	2.5–7.0	Electrical	—	—	19
MSF	3.0–5.0	Steam/thermal	250–330	7.0–9.0	19
MED-TVC	1.5–2.5	Steam/thermal	145–390	8.0–14	19
Surface water treatment	0.2–0.4	Electrical	—	—	11,41
Waste water reclamation	0.5–1.0	Electrical	—	—	11
Long-distance transport[a]	1.6–2.8,[b] 12.0[c]	Electrical	—	—	—

[a] Depends on the transport distance and the elevation gap between source and destination; e.g., normal distribution costs are around ~0.6 kWh/m^3 [41], based on UK experience.

[b] Power required to convey surface water to San Diego, Los Angeles, and Orange County [11].

[c] Power required if water were to be conveyed to Perth via the Kimberley pipeline project [52].

Modern SWRO plants can achieve a *specific* energy demand of <2.5 kWh/m^3 and a total energy demand <3.5 kWh/m^3 by using state-of-the-art equipment (such as pressure exchangers, variable-frequency pumps, and low-pressure membranes) and under favorable conditions (i.e., a low fouling potential, temperature >15°C, salinity <35).[1]

The real energy demand may be higher under less favorable conditions. For example, the calculated *specific* energy demand of a state-of-the-art facility with a feed salinity of 40 and a temperature of 20°C (typical for eastern Mediterranean seawater), a total recovery of 41%, and equipped with the most efficient energy recovery system, is approximately 3.8 kWh/m^3 (Table 2.14). An additional 0.2–0.8 kWh/m^3 is required for pretreatment, waste water, and sludge treatment (depending on the feedwater quality), administration buildings and laboratories, and posttreatment and drinking-water pumping to supply network, which leads to a *total* energy consumption of about 4–4.6 kWh/m^3 [33,53].

For instance, the Spanish hydrological plan assumes a *total* energy use of 4 kWh/m^3 under the assumption that plants are equipped with state-of-the-art technologies [54], which is similar to the energy demands reported for other large SWRO projects in Israel (3.9 kWh/m^3 [55]) and California (4.5 kWh/m^3 [56]). Older or smaller SWRO plants without energy recovery may use up to 5 kWh/m^3 at 50% recovery. Values given for more recent large SWRO projects that also include the *transfer* of water to the supply grid ranged between 4.2 and 5.3 kWh/m^3 [57–59] (for further examples, see also Table 5 in Ref. 27).

Table 2.14 Energy Consumption of SWRO with Different Pretreatments[a]

Pretreatment Type	Total Energy Demand, kWh/m^3	Specific Energy Demand[b], kWh/m^3	Energy Demand for Pretreatment[c], kWh/m^3	Energy Demand for Waste Water and Sludge Treatment, kWh/m^3
FL, SM, 1 GF	4.01	3.79	0.035	0.019
MF/UF	4.24	3.81	0.215	0.042
DAF + 2 filters	4.37	3.78	0.395	0.024
DAF + MF/UF	4.64	3.83	0.580	0.052

[a]Per m^3 of *product* water at 20°C, a feed salinity of 40, a total recovery of 41%, and using work exchangers.

[b]First and second RO passes, including cleaning operations.

[c]Without seawater extraction, screening, and pumping.

Notations: Other types of energy consumption (internal pumping, auxiliaries, administration buildings, laboratories, posttreatment, water transfer to supply network) account for a difference of 0.1–0.2 kWh/m^3 between the total energy demand and the sum of the given specific and single demands. (FL—flocculation; SM—static mixer; GF—gravity filter; DAF—dissolved air flotation; UF—ultrafiltration; MF—microfiltration).

Source: Ludwig [35].

[1]For the definition of salinity, see footnote 3 in Section 2.4.2.1.

The Affordable Desalination Collaboration operated a demonstration plant in California over 2 years using state-of-the-art, off-the-shelf technology and set a world record in *specific* energy consumption of 1.58 kWh/m^3 with a low-energy membrane operated at 42% recovery. However, this result is currently not realistic in full-scale applications as it was achieved at the expense of permeate water quality and process recovery. The *specific* energy demand of SWRO plants usually increases with recovery, but the *total* energy demand decreases with the recovery rate as less feedwater must be pumped and treated to obtain the same volume of permeate at a higher recovery. Optimizing the energy demand of the whole process is a complex undertaking, as the single subsystems of a SWRO plant, particularly its pretreatment and first and second passes, are closely interrelated [35]. At the most affordable point for a single-stage 190,000 m^3/d plant, a total treatment energy in the range of 2.75–2.98 kWh/m^3 was demonstrated [60].

A report published by the US National Research Council estimates that the practical upper limit of energy savings in RO may be about 15% from current levels, assuming a system operating at 40% recovery, using a 95% energy recovery device and a seawater RO membrane with twice the permeability of today's best membranes. Improvements in module design appear to have the greatest potential for reducing the overall energy costs, unless a breakthrough in an alternate technology to RO is achieved [19].

The energy demand of the pretreatment (Table 2.15) is lowest for flocculation with a static mixer and one- or two-stage gravity filtration (0.015–0.02 kWh per m^3 *filtrate* water). It increases if the static mixer is replaced by a flocculation basin (0.10–0.12 kWh/m^3), and if additional flotation or sedimentation steps are added (0.14–0.16 kWh/m^3) [35]. The energy demand of a more extensive conventional pretreatment is comparable to the energy demand that is generally given for UF/MF pretreatment in the literature, which is 0.10–0.20 kWh/m^3 [35,36,50,61]. However, plant operators give lower energy demands of only 0.03–0.09 kWh per m^3 filtrate water. The most energy intensive option would be UF/MF pretreatment with additional pretreatment such as flotation, with an estimated energy demand of 0.25 kWh/m^3 [35]. Table 2.14 shows the pretreatment energy demand of a plant per m^3 *product* water operated at 41% total recovery. Depending on the feedwater quality, to which the pretreatment is customized, the energy demand may account for more than 10% of the overall energy demand of the plant [35].

When comparing energy requirements, the whole process should be taken into account. For instance, the overall energy costs in a SWRO system with UF/MF pretreatment may be lower than that for conventional pretreatment due to a lower energy consumption in the SWRO stage. A better feedwater quality results in lower SWRO fouling and a reduction in RO pressure drop caused by fouling [62].

2.3.5 Outfalls

The most widely used method of concentrate disposal is surface water discharge. It is a relatively low-energy, low-technology, and low-cost solution, assuming that the length of the pipeline is reasonable and the concentrate does not need further

Table 2.15 Energy Consumption Levels of Different SWRO Pretreatments[a]

Pretreatment	Energy Consumption			Reference(s)
	kWh/m^3 Filtrate	kWh/m^3 Product[b] (for 40% Recovery)	kWh/m^3 Product[b] (for 50% Recovery)	
FL, SM, 1 filter,[c]	0.015	0.037	0.030	35
FL, SM, 2 filters[c]	0.020	0.049	0.040	35
FL, FB, 1 filter[c]	0.100	0.244	0.200	35
FL, FB, 2 filters[c]	0.120	0.293	0.240	35
SED, 1 filter[c]	0.140	0.341	0.280	35
SED, 2 filters[c]	0.150	0.366	0.300	35
DAF, 1 filter[c]	0.150	0.366	0.300	35
DAF, 2 filters[c]	0.160	0.390	0.320	35
MF/UF low[c]	0.100	0.244	0.200	35
MF/UF high[c]	0.200	0.488	0.400	35
DAF + MF/UF[c]	0.250	0.610	0.500	35
NORIT XIGA, SEAGUARD[d] (inside-out, pressurized, PES/PVP, 0.2–0.4 bar TMP)	0.030	0.073	0.060	50,63
Inge Multibore[e] (inside-out, pressurized, PES, 0.25 bar TMP)	0.050	0.122	0.100	48
Dow SFP[f] (outside-in, pressurized, PVDF, air scour, 0.5 bar TMP)	0.090	0.220	0.180	64,65
Zenon ZeeWeed (inside-out, submerged/vacuum, PVDF, air scour, 0.1–0.35 bar)	No information			46,66,67

[a]The values presented here reflect the specific demand of the pretreatment only without intake and initial screening.

[b]To obtain the energy demand normalized to 1 m^3 of *RO permeate* (*product*) water, the values given for 1 m^3 of *pretreated* (*filtrate*) water are divided by the recovery rate of the plant (i.e., 0.4 for 40% recovery).

[c]Without intake/screening, assuming a feed pressure of 1 bar for initial pretreatment.

[d]Value of 0.03 kWh/m^3 for the UF excluding the intake. The intake in a full-scale SWRO plant with this pretreatment accounts for a site-specific energy demand of 0.08 kWh/m^3 and screening for <0.01 kWh/m^3, amounting to a total energy demand of 0.12 kWh/m^3 in this case.

[e]Value of 0.05 kWh/m^3 for a pilot plant and including ultrafiltration, backwash, and CEB. The intake (1.2 km offshore) accounts for an additional 0.02 kWh/m^3 [68]. Pilot plants typically have a higher specific energy demand than full-scale plants, which can be assumed to have a specific energy demand of ~0.01 kWh/m^3 for the UF at 0.4 bar transmembrane pressure [69].

[f]UF energy demand in two operational plants including backwash and air scrub but without intake.

Notations: FL—flocculation; SM—static mixer; FB—flocculation basin; SED—sedimentation; DAF—dissolved air flotation; UF—ultrafiltration; MF—microfiltration; PES—polyethersulfone; PVP—polyvinylpyrollidone; PVDF—poly(vinylidene fluoride); TMP—transmembrane pressure.

treatment. However, it has the potential for negative impacts on aquatic organisms, but the implementation of suitable mitigation measures, such as a good site location, an advanced outfall design with diffusers, or the predilution with other discharges such as power plant cooling water can likely minimize most potential negative environmental effects.

The discharge design primarily influences the mixing behavior in the near-field region, which extends up to a few hundred meters from the outfall location. In that region, a velocity discontinuity between the effluent and the ambient flow arises from the initial momentum flux and the buoyancy flux of the effluent. It causes turbulent mixing, which leads to an entrainment of seawater and thereby decreases differences in salinity, temperature, or residual chemicals between the effluent and ambient water bodies. Ambient currents may deflect the jet trajectory, inducing higher dilution, whereas ambient density stratification has a negative effect on vertical spreading. Boundary interactions can occur, for example, at the water surface, the seabed, or pycnoclines. They generally define the transition from near-field to far-field mixing processes. The far field can extend up to several kilometers and is dominated by ambient processes, such as passive diffusion, which cause a further slow mixing of the plume [70].

2.3.5.1 Diffusers The use of multiport diffusers can effectively increase the mixing process of the concentrate in the discharge site by increasing the volume of seawater in contact with the concentrate and by creating turbulent mixing conditions. A number of factors affect the dilution potential of diffusers, including the exit velocity and the volume of the concentrate, the depth of nozzles below the sea surface, the vertical angle of nozzles, and the number and spacing of nozzles [71]. The concentrate typically exits the diffuser nozzle at a high velocity and is directed in an upward slope toward the sea surface. With such a design, a salinity level of one unit above background levels can be achieved at the edge of the regulatory mixing zone.

Two broad categories of concentrate outlet structures can be distinguished: *rosette-style diffusers*, which consist of several outlets risers above the seafloor with a small number of nozzles attached to each riser, and *pipeline-style diffusers*, which consist of nozzles arranged along a pipe instead of a rosette (Fig. 2.25). All large Australian SWRO projects, including those for Victoria [71], Sydney [72], Perth [73], and Gold Coast plants [74], use or are proposed to use either a rosette- or pipeline-style diffuser systems.

2.3.5.2 Subsurface Discharge Brine disposal can also take place via a subsurface discharge structure. In coastal areas, beachwells or infiltration galleries beneath the beach or seafloor can be used to induce mixing in the groundwater table to slowly dissipate the plume into the surf zone.

In Long Beach, California, an ocean floor demonstration system has been tested that combines seabed drains for the intake and a discharge gallery for the outfall, both located in the seafloor sediments [24] (see also Section 2.3.1.2).

A discharge gallery has also been in use at the Marina Coast Water District desalination plant (1000 m^3/d) in California for 10 years, which is one of the first

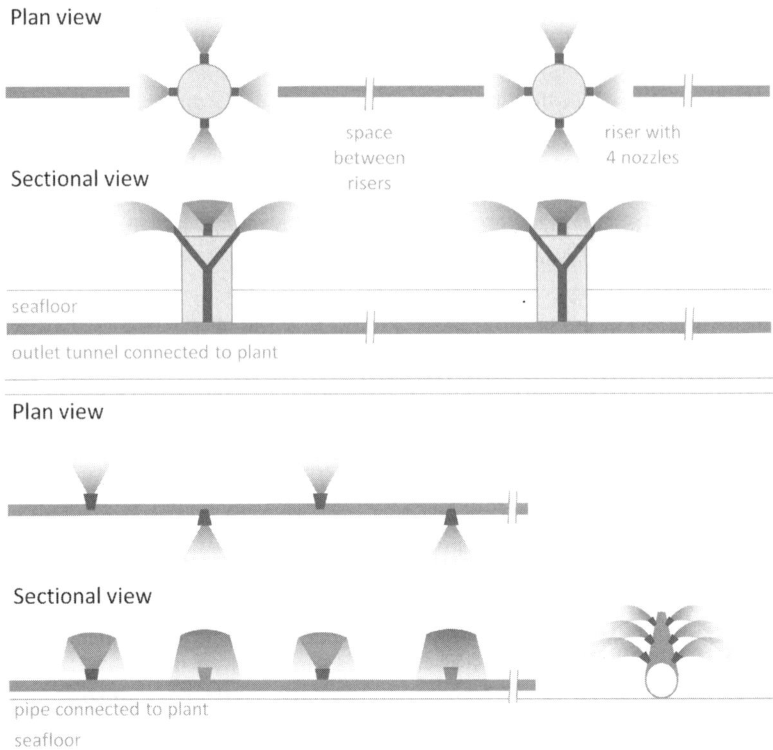

Figure 2.25 Rosette-style (a) and pipeline-style diffusers (b) for concentrate discharge.

plants to use such a system for brine disposal. In contrast to the demonstration facility in Long Beach, the discharge gallery is located in the beach sediments, where the concentrate is diluted through mixing with natural groundwater, and is subsequently dissipated into the surf zone. A long-term monitoring program concluded that there was not a detectable increase in salinity of the receiving waters due to brine discharge [22].

Subsurface outfalls are considered to be an effective way of minimizing the environmental impacts of concentrate discharge, at least in some locations where suitable hydro-geological conditions exist. However, they are more feasible for smaller SWRO plants and limited experience and monitoring data limits their implementation to date.

Another, more mature, technology is deep-well injection of concentrate into a deep geologic formation, usually inland, and isolated from drinking-water aquifers. It is generally used for larger flows, due to high development costs [19]. Other options for brine disposal include sewer discharge, evaporation ponds, land application, or zero liquid discharge (ZLD). They are used mainly where surface discharge is not possible, e.g., for inland BWRO plants, but rarely for SWRO plants [75].

2.3.5.3 Codischarge with Cooling Water

Co-location of MSF/MED distillation plants with power plants is common practice, and some large SWRO plants either have been or have been proposed to be co-located with power plants. Examples include the Carlsbad and Huntington Beach SWRO plants in southern California [56,76], the Tampa Bay SWRO plant in Florida [77], and the Ashkelon and Hadera SWRO plants in Israel [78,79]. The main environmental benefits of co-locating SWRO plants to power plants are [80]:

- The existing intake and outfall structures are utilized, reducing construction impacts.
- The impacts of land use and landscaping are reduced, as the facility is constructed in an industrial area, and does not require additional power transmission lines.
- If the intake water is taken from the cooling water discharge conduits of the power plant, the required energy demand of the SWRO process can be reduced by 5–8% because of a higher membrane permeability at higher water temperature (mostly relevant for ambient seawater temperatures $<20^\circ C$).[2]
- If cooling water is reused as feedwater to the desalination process, the total amount of feedwater intake is reduced, limiting the impingement and entrainment effects to the level found at the existing power plant.
- The concentrate can be blended with the cooling water before discharge, which significantly reduces the salinity of the concentrate before disposal.

2.3.5.4 Codischarge with Waste Water

Another option for codischarge exists with waste water treatment plant effluents, as proposed for two small SWRO plants in California (Santa Barbara and Santa Cruz [22,81]) and Europe's largest SWRO plant in Barcelona [28]. The main advantage is that the salinity of the concentrate is very effectively diluted. A dilution ratio of 1:1 is sufficient to reduce the salinity of the concentrate to ambient seawater salinity levels, as the SWRO concentrate usually has twice the ambient salinity. However, there are several issues associated with the practice of blending SWRO concentrate with waste water treatment plant effluents, and the discharge through an existing waste water treatment plant outfall has found a limited application to date.

One consideration is the potential for whole effluent toxicity of the blended discharge that may result from an ion imbalance of the blend of the two waste streams [75]. As waste water effluent has a freshwater origin with a different ratio of key ions than does seawater, the ion imbalance may be responsible for the observed toxic effects [82]. Furthermore, residual contaminants in the waste water may have negative effects on marine life. However, both effects are attributed to the waste water and not to the concentrate from the desalination process. Consequently, they may occur wherever waste water treatment plant effluents are discharged into the sea.

[2]As higher feed temperature results in a higher degree of salt passage, it may also result in higher energy consumption if a second RO stage has to be implemented [18].

Another consideration is that waste water may be considered as a resource, which should not be wasted to the ocean in water-scarce areas, as recycling is usually preferable to disposal in the concept of waste management. It must also be determined whether a new desalination project is necessary or could be reduced in size if the existing waste water sources were reused to full potential instead of being discharged into the ocean. Another argument for waste water reuse using desalination technologies is the elimination of a waste product. Effluents from conventional waste water treatment plants still contain diverse contaminants, including nutrients, metals, or micropollutants such as pharmaceutical and personal-care products, which are burdensome for many rivers, estuaries, and coastal seas. Purifying and reusing waste water not only produces a new water supply but also eliminates waste water discharge if the remaining waste stream from the process is treated using ZLD technologies and disposed of appropriately.

Water reuse is practiced in many parts of the world, but the use of desalination technologies in water reuse has been limited so far (see Section 2.2). The world's largest facility treating waste water with an output capacity of 310,000 m^3/d is located in Sulaybia, Kuwait. It uses UF followed by RO to treat secondary effluent waste water. The energy demand for waste water desalination is lower than that for SWRO because of the considerably lower salt content of the water, which is another environmental benefit. An expansion of waste water desalination is therefore expected in the future. A second advantage is that most of the water is already available where it is most needed, i.e., near urban areas, avoiding long transport distances. The purified water can be used for industrial purposes, landscaping activities in urban areas, or aquifer recharge. From a technical logistic perspective, the product can even comply with WHO drinking-water standards [82,83], but direct potable reuse, such as that as practiced in Windhoek, Namibia, has found a very limited application to date.

2.4 POTENTIAL ENVIRONMENTAL IMPACTS

The main environmental concerns of new desalination projects usually revolve around a few key issues (Fig. 2.26). This section briefly describes three key concerns (for a detailed analysis, see Ref. 27): impingement/entrainment caused by intakes, marine pollution caused by concentrate and chemical discharges, and emissions of greenhouse gases and air pollutants due to energy use.

2.4.1 Impingement and Entrainment

Open seawater intakes usually result in the loss of eggs and larvae of fish and invertebrate species, spores from algae and seagrasses, phytoplankton and zooplankton, as well as smaller marine organisms that are drawn into the plant with the seawater (entrainment). As a result of the pretreatment in SWRO plants, which involves chlorination at the intakes to control marine growth and the removal of suspended solids, it must be assumed that the survival rate of plankton and smaller organisms inside the plant is minimal.

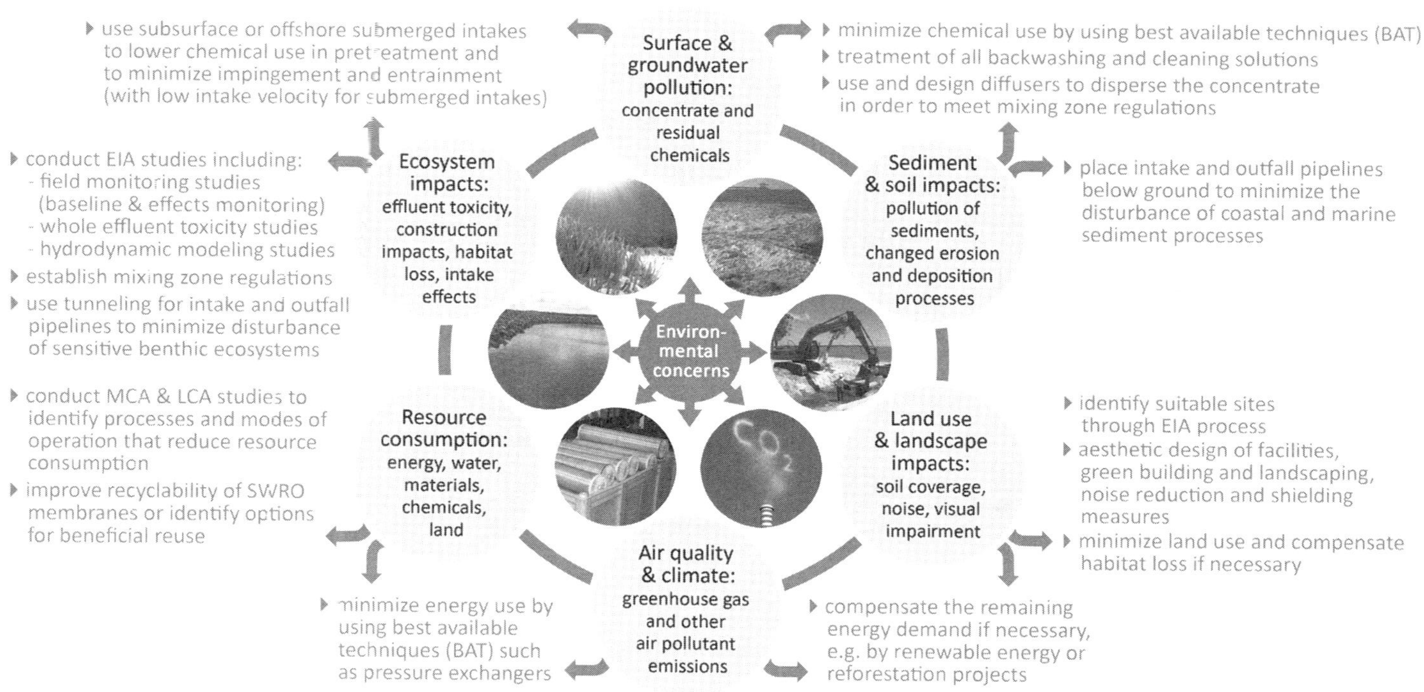

Figure 2.26 Key environmental concerns of seawater desalination projects and measures for impact mitigation.

Entrainment causes the loss of a large number of plankton organisms. The issue, however, is whether this represents a *significant, additional* source of mortality for the affected species, which negatively affects the ability of the species to sustain their populations, and that may affect the productivity of coastal ecosystems. These secondary ecosystem effects are difficult to quantify [22]. Plankton organisms are generally prevalent in coastal surface waters and have rapid reproductive cycles. Fish and invertebrate species produce large numbers of eggs and larvae to compensate for a high natural mortality rate as part of their reproduction strategy. The mortality caused by entrainment in a single facility therefore seems unlikely to have a substantial negative effect on population and ecosystem dynamics. The situation is different when endangered species, species of commercial interest, or marine protected areas are potentially affected by the intakes, and when cumulative sources of mortality (i.e., harmful conditions resulting from the proximity of other power or desalination plants) exist. While it is relatively straightforward to estimate the levels of entrainment for a single desalination project, it is difficult to evaluate the indirect impacts on the ecosystem, especially in places where cumulative sources of mortality are involved.

Furthermore, open intakes may result in the loss of larger marine organisms at the intake screens (impingement), if the intakes are not well designed. Impingement mortality can be caused by suffocation, starvation, or exhaustion of these organisms when they are pinned up against the intake screens or from the physical force of jets of water used to clear screens of debris (see Ref. 22, after Ref. 84). Similar to entrainment effects, the cumulative ecosystem effects are difficult to estimate. If the intake velocity of the feedwater were reduced to velocities of $\sim$0.1 m/s, which is comparable to background currents in the oceans, mobile organisms would be able to swim away from the intake area.

The California Coastal Commission concluded in 2004 that "the most significant potential direct adverse environmental impact of seawater desalination is likely to be on marine organisms: This impact is due primarily to the effects of the seawater intake and discharge on nearby marine life; however, these effects can be avoided or minimized through proper facility design, siting, and operation."

The US Clean Water Act, Section 316(b), requires that "the location, design, construction and capacity of *cooling* water intake structures reflect the best technology available for minimizing adverse environmental impact." The same standards may apply for seawater desalination plants in the future, although their intake volumes are smaller.

2.4.2 Concentrate and Chemical Discharges

Specific to all desalination processes is the discharge of a concentrate. This reject stream contains mainly the natural constituents of the intake seawater in a concentrated form. In addition, it may contain residual chemicals from pretreatment and cleaning operations and their reaction and degradation products. When evaluating the potential impacts of the concentrate and chemical discharges, one has to distinguish between the salt, which is a natural component, and the chemical additives. The discharge of potential pollutants depends mainly on the desalination process and the design of the facility with regard to pretreatment and existing backwash water treatment facilities.

2.4.2.1 Potential Impacts of Increased Salinity The salinity of the concentrate usually depends on the plant recovery rate. SWRO plants typically have a recovery rate of 40–45%, with an upper operational limit of 50% to control the risk of scaling. The concentrate may therefore have up to twice the salinity as the ambient seawater, and its discharge back into the sea causes elevated salinity levels in the mixing zone. The salinity increase can be mitigated by predilution with other waste streams such as cooling water from coastal power plants, or dissipation by a multiport diffuser, in combination with strong wave action and currents in the discharge site to effectively dissipate the salinity load (see Section 2.3.5).

Because of its high salt content, the RO concentrate has a higher density than ambient seawater. For example, a SWRO plant with a feedwater salinity[3] of 35 and operating at 50% recovery would produce a concentrate with a salinity of 70. At 20°C, the density[4] of the concentrate is 1051 kg/m^3, which is heavier (negatively buoyant) compared to an ambient density of 1025 kg/m^3. The plume would sink to the seafloor (unless adequately dissipated), forming a water mass of elevated salinity that spreads over the seafloor in the vicinity of the outfall pipe and that will possibly also diffuse into the sediment pore water. It is important to prevent such bottom-clinging plumes as they could potentially harm any sensitive benthic ecosystems.

It has been shown through several toxicity studies that locally elevated salinity levels caused by desalination plant discharges can be potentially harmful to marine organisms. In general, toxicity will depend on the sensitivity of a species to increased salt levels, its lifecycle stage, the exposure time, and the natural salinity variations of the habitat to which the species is adapted. For instance, salinity tolerance studies of the seagrass *Posidonia oceanica* in the western Mediterranean showed that a salinity of 43 may reduce growth rates by 50% and that a salinity of 45 may result in 50% mortality within 15 days [86]. These values compare to ambient salinity levels of 37–38 in the western Mediterranean. In contrast, two related seagrass species from western Australia, *P. australis* and *P. amphibolis*, seem to be adapted to naturally higher salinity levels, as densest covers are observed at a salinity of 40–50 (see Ref. 73, after Ref. 87). This indicates that even species within the same genera can have very different salinity tolerances, depending on their natural habitat salinity.

[3]The UNESCO definition of *practical salinity units* (psu) is used, which is the conductivity ratio of a seawater sample to a standard KCl solution and hence a *dimensionless* value. As salinity reflects the amount of total dissolved solids (TDS) in ocean water, it was traditionally expressed as parts per thousand (ppt). A salinity of 35 ppt equals 35 g of salt per 1000 g of seawater, or 35,000 ppm (mg/L), or in approximation 35 (psu).

[4]Density is a function of salinity and temperature (and pressure in greater water depths) and is typically calculated by software tools that implement the equation of state for seawater [e.g., 85].

Some macrofauna taxa such as echinoderms (e.g., sea urchins, starfish) are assumed to be generally more sensitive to salinity variations than other species, and young lifecycle stages of organisms, such as sea urchin embryos, are generally considered to be more sensitive than adults. Most marine organisms can adapt to minor deviations in salinity and might recover from extreme, short-term exposure to increased salinities. For example, *P. oceanica* seagrass that survived in a salinity of 43 over 15 days were able to recover when returned to normal conditions [86]. However, only a few species will be tolerant of high salt concentrations over prolonged periods of time, as caused by the continuous discharge of desalination concentrates.

Salt concentrations that continuously and significantly exceed the ambient levels to which the native species are adapted may result in osmotic stress. This will drive mobile animals away from the discharge site and can cause a die-off of the sessile fauna and flora. For example, salinity increases near the outfall of the Dhekelia SWRO plant in Cyprus were reported to be responsible for a decline of macroalgae forests, and echinoderm species were observed to have vanished from the discharge site [88]. Observations on the distribution of marine species from naturally hypersaline environments in the Arabian Gulf indicate that a salinity above 45 altered the benthic community considerably [89], which stresses the importance of salinity as a controlling factor. The Gulf is impoverished at the species level, mostly because of its harsh environmental conditions, and low species richness was found for most major benthic groups such as echinoderms and corals [14].

The concentration of salts, rather than the salt itself, and the exposure time prior to the discharge is the problem. Discharge into the sea is therefore an adequate means of concentrate disposal if dilution to ambient levels is achieved within a very short distance from the outfall and if sensitive ecosystems are not impacted by the dispersing plume.

An adequate approach to minimize the impacts of the concentrate is to establish a restricted mixing zone around the outfall in which dilution to a level close to ambient is achieved by a combination of artificial intervention (installation of multiport diffusers) and natural processes (strong currents). Mixing zone regulations combine parameters that define the spatial extent of the allowable mixing zone with a water quality standard for salinity and other pollutants that apply at the edge of this mixing zone. The following standards for salinity have been established for different SWRO projects in different parts of the world.

- The license for the Perth SWRO plant limits salinity to within 1.2 units of ambient levels within 50 m and within 0.8 units within 1000 m from the discharge point. This requires a dilution factor of 45–1 at a distance of 50 m in all directions of the diffuser. After 2 years of operation, the actually achieved dilution factors ranged from 50–1 to 120–1. Whole effluent toxicity (WET) tests were performed with five native species at commissioning and after 12 months of operation, showing that the required dilution factor to achieve a 99% species protection level is about 15–1, i.e., that 99% of the species in the marine ecosystem will be protected at this dilution level [90].

- For the Sydney SWRO project, WET tests were performed using effluents from a pilot plant and five indicator species. A dilution factor of 30–1 at the edge of the near field was calculated in order to achieve salinity levels within natural variation levels of 1 unit above ambient, and to achieve the desired species protection level of 95%. The studies showed that salinity was the key source of toxicity of the whole effluent [91]. Baseline studies have been conducted for a comprehensive field monitoring program to verify the results after startup [92].
- For the Gold Coast desalination plant, field measurements during startup confirmed an effective dilution factor of 60–1 at the edge of the near-field mixing zone in 60 m distance from the diffuser with no discernible difference to ambient salinity at that point. A minimum dilution ratio of 47–1 had been predicted by the hydrodynamic models. WET tests using effluent from the full-scale plant were performed on six marine species from more than three trophic levels representative of the local ecosystem. The results imply that a minimum dilution factor of 9–1, corresponding to a salinity of 37.6 or an increase of 2.3 units above ambient, should be achieved at the edge of the mixing zone to obtain a 95% species protection level [93].
- The most extensive WET testing was carried out for the Olympic Dam SWRO project. On the basis of WET tests with 15 species from four trophic levels, it was calculated that a dilution of 45–1 should protect 99% of the marine species in the area, corresponding to a salinity increase of 0.7 units above ambient. The hydrodynamic modeling studies predicted that this dilution would be achieved within 300 m from the outfall in 90% of all times. In 100 m distance from the outfall, the minimum dilution would be 8–1, corresponding to a maximum salinity increase of 3.7 units or 9% above ambient. The maximum extent of the 45–1 dilution contour would be $\leq$1.1 km for 99% of the time. A dilution of 85–1 or a salinity increase of 0.4 unit above ambient, which ensures 100% species protection at all times, would be achieved within 3.9 km of the outfall [94].
- On the basis of extensive field and laboratory studies of the seagrass *Posidonia oceanica* in Spain, it has been recommended that discharges of desalination concentrate nearby *Posidonia* meadows be avoided or that the discharge salinity be diluted to ensure that it neither exceeds a value of 38.5 in any point of the meadow for >25% of the observations (on an annual basis) or a value of 40 in $\leq$5% of the observations. These values compare to ambient salinities in the western Mediterranean of 37–38 [95].
- For the proposed Carlsbad SWRO project in California, long-term salinity tolerance and toxicity tests were carried out. In this study, 18 marine species were held in a tank containing a blend of concentrate and power plant cooling water at a salinity of 36 (expected to occur in the mixing zone in 95% of the time, compared to an ambient salinity of 33.5). All organisms remained healthy and showed normal activity and feeding behavior during the 5-month test. Three indicator species were also exposed to salinities of 37–40 over an extended period of time with 100% survival and normal behavior at the end

of the 19-day test. In 300 m distance from the point of discharge, the salinity near the bottom is expected to reach 34.4 on average and 40.1 under extreme conditions. An initial dilution factor of 15.5-1 has been assigned at the edge of the mixing zone [96].

- For a SWRO plant in Okinawa, Japan, a maximum salinity of 38 in the mixing zone and an increase of 1 unit where the plume meets the seafloor was established [97].

A common approach is to tailor regulations for local conditions, as it is generally difficult to develop a universal set of mixing zone and water quality standards that apply equally to the wide range of marine ecosystems. Moreover, single standards for each physical and chemical stressor (e.g., salinity and residual chlorine) do not take potentially synergetic effects into account. A single threshold or trigger value can be derived from a suite of bioassays that use the whole effluent to measure the acute and chronic toxicity to different local marine species representing different taxonomic and trophic levels.

The WET testing undertaken for the Perth, Sydney, Gold Coast, and Olympic Dam SWRO projects (Table 2.16) follows the Australian and New Zealand Guidelines for Fresh and Marine Water Quality [98]. In a first step, the NOEC[5] or EC_{10}[6] is established for each species. From this data set, a species protection trigger value (SPTV) is calculated, which is the safe dilution ratio for the concentrate that protects a certain percentage of the species from adverse impacts. A species protection level (SPL) of 95% is usually adopted for slightly to moderately disturbed ecosystems, and 99% for ecosystems of high conservation value. The mixing zone regulations are a combination of the dilution ratio (e.g., 1–50), which is required to protect a certain percentage of species (e.g., 95%) from experiencing no effect or a subchronic effect of 10%, and which has to be met at the edge of a spatially defined mixing zone (e.g., 100 m) in a given percentage of time (e.g., 90%) to avert negative effects. This approach reflects that toxicity is a function of species sensitivity, concentration, and exposure time.

A similar approach, but following a different order and different methodologies, was selected for testing the long-term salinity tolerance of marine species for two SWRO projects in California [99]. In a first step, the salinity level in the middle of the zone of initial dilution (ZID; defined as the area within 330 m from the point of discharge) in 95% of the time was predicted on the basis of hydrodynamic modeling studies. A long-term biometric test with 18 species in a single aquarium over a period of 5 months was then carried out to investigate chronic effects at this salinity. In addition, salinity tolerance tests were carried out over a range of salinities to investigate whether marine organisms will be able to survive periodic extreme (worst-case) salinity conditions. Three local species known to have the highest susceptibility to salinity stress were used (purple sea urchin *Strongylocentrotus purpuratus*, sand dollar *Dendraster excentricus*, and the red abalone

[5]No-observed-effect concentration.

[6]Statistically calculated concentration that causes a 10% effect.

Table 2.16 Whole Effluent Toxicity Testing Data[a]

Plant	SPL %	SPTV	Diffuser Details	WET Test Species	Reference
Perth	95 99	12.3–1 15.1–1	45–1	Tests at commissioning and after 12 months of operation: 72-h macroalgae germination (*Ecklonia radiata*), 72-h macroalgae growth test (*Isochrysis galbana*), 48-h mussel larval development (*Mytilus edulis*), 28-d copepod reproduction test (*Gladioferens imparipes*), 7-d larval fish growth test (*Pagrus auratus*)	95
Sydney	95	30–1	30–1 dilution at edge of near field (50–75 m) equal to salinity variations of 1 above ambient determined by modeling	Five target organisms: algae, crustaceae (prawn), mollusca (oysters), echinodermata (sea urchin fertilization & larval development), chordate (fish)	92
Gold Coast	95	9–1	47–1 minimum dilution in 60 m distance from the diffuser (edge of mixing zone) determined by modeling, validation during startup confirmed a dilution in excess of 9–1 at edge of the mixing zone	Six species from >3 trophic levels representative of the local ecosystem, targeting sensitive early lifecycle stages (fertilization, germination, larval development and growth): acute microtox (bacterium *Vibrio fischeri*), 72-h microalgae growth inhibition (*Nitzschia closterium*), 72-h macroalgae germination (*Ecklonia radiata*), 48-h rock oyster larval development (*Saccostrea commercialis*), 72-h sea urchin larval development (*Heliocidaris tuberculata*), 7-d larval fish imbalance (*Pagrus auratus*)	93

(*continued*)

Table 2.16 (***Continued***)

Plant	SPL %	SPTV	Diffuser Details	WET Test Species	Reference
Olympic Dam	99	45–1	45–1 dilution within 0.3 km (90% of time) 1.1 km (99% of time) 2.2 km (100% of time)	15 species from more than 4 trophic levels representative of the local ecosystem, including acute and chronic tests with early lifecycle stages, juveniles and adults:	94
	100	85–1	85–1 dilution within: 1.1 km (90% of time) 2.8 km (99% of time) 3.9 km (100% of time) 45–1 dilution would be achieved in 30% of the time at the edge of the near-field mixing zone (100 m); the salinity increases for dilution factors of 45–1 and 85–1 would be 0.7 and 0.4 units above ambient, respectively	72-h microalgae chronic growth rate inhibition test (*Nitzschia closterium* and *Isochrysis galbana*), 72-h macroalgae chronic germination success (*Ecklonia radiata* and *Hormosira banksii*), 48-h chronic copepod reproduction (*Gladioferens imparipes*), 96-h acute prawn postlarval toxicity test (*Penaeus monodon*), 21/28-d juvenile/adult prawn growth test (*Melicertus latisulcatus*), 7-d subchronic crab larval growth test (*Portunus pelagicus*), 48-h sub-chronic oyster larval development (*Crassostrea gigas* and *Saccostrea commercialis*), 72-h sea urchin subchronic fertilization success (*Heliocidaris tuberculata*), 96-h acute fish larval imbalance and mortality (*Seriola lalandi*), 7-d subchronic fish larval growth test (*Seriola lalandi, Pagrus auratus, Argyrosomus japonicus*), chronic developmental and hatching tests (*Sepia apama*)	

[a] The species protection trigger value (SPTV) is calculated from a range of test species and gives the minimum dilution ratio that should be achieved at the edge of the mixing zone for a given species protecting level (SPL). The SPTV is compared to the actual dilution as predicted for or achieved by the diffuser.

Haliotis rufescens). Methods for measuring the acute and chronic toxicity of effluents to marine organisms have also been established by the USEPA [100,101].

To conclude, stringent mixing zone regulations should be established site-specifically for desalination plants that specify (1) the spatial extent of the mixing zone in accordance with the local conditions (open coast, estuary, bay, etc.) and the sensitivity of the marine ecosystems in the vicinity of the outfall and (2) the dilution rate or the salinity standard that must be met at the edge of this mixing zone in a given percentage of time. The safe dilution ratio or salinity level that protects a given percentage of the local species can be calculated from a set of WET tests. The best location and design of the outfall to meet mixing zone regulations should be determined through field monitoring in combination with hydrodynamic modeling studies that investigate the mixing conditions under different ambient, including quiescent worst-case conditions.

The salt itself, however, is not the problem. Even in semienclosed sea areas such as the Gulf, the Mediterranean, or the Red Sea, natural evaporation exceeds the water abstraction rates by desalination plants by several orders of magnitude. The key to avoid impacts is to sufficiently dilute and disperse the salinity load to ambient concentrations. The same argument, however, does not necessarily hold true for the chemical additives.

2.4.2.2 Pretreatment and Cleaning Chemicals The pretreatment chemicals typically used in SWRO plants include chlorine or other disinfectants, sodium bisulfite (SBS) as a reducing agent for residual chlorine, ferric chloride, or ferrous sulfate as primary coagulant and organic polymers as coagulant aids, antiscalants, and acids. The chemicals are rejected along with different waste streams.

The *concentrate*, which is the largest reject stream by volume, contains the antiscalants and possible disinfection byproducts (the disinfectants themselves should have been reduced by sodium bisulfite). Another concern is reduced pH by the use of acids, and reduced oxygen levels. The latter may result from an overdosing of the reducing agent (SBS), the high salinity (water of higher salinity can dissolve less oxygen), and a possible density stratification effect caused by the dense discharge plume in the mixing zone that may counteract the reoxygenation of bottom water layers.

In conventional pretreatment, the *backwash waters* from media filters typically contain coagulants and coagulant aids. If concentrate is used for backwashing, the reject stream is also of elevated salinity. In UF pretreatment, backwashing is usually carried out with filtrate. The reject water may contain coagulants, where inline coagulation is practiced, and chlorine, where chemically enhanced backwashing (CEB) is practiced.

A chemical cleaning in place (CIP) of SWRO and UF membranes is typically performed in two stages to remove different types of foulants. Acidic solutions of pH 2–3 are effective against metal oxides and scales, while alkaline solutions of pH 11–12 are used to remove silt deposits and biofilms. CIP solutions for SWRO membranes may additionally contain detergents such as dodecylsulfates, oxidants such as sodium perborate, and organic or inorganic chelating agents such as

EDTA and tripolyphosphates. After cleaning, or prior to storage, SWRO membranes are typically disinfected. CIP of UF membranes often involves acidic or caustic solutions containing chlorine only.

Contrary to the concentrate, impact mitigation for pretreatment and cleaning chemicals should emphasize avoidance and minimization of chemical use through the implementation of best available techniques (BAT) rather than discharge and dilution of wastes into the sea. The concept of BAT typically involves a hierarchy of actions, which favors avoidance of wastes over treatment and disposal.

For SWRO projects, the identification of a site- and project-specific BAT solution should have the objective to lower the overall chemical use by:

- Reducing the occurrence of fouling through selection of an adequate intake design, thus minimizing cleaning and pretreatment requirements.
- Substitution of harmful substances with less harmful substances.
- Optimizing the application and dosage of pretreatment chemicals, where necessary, on the basis of pilot testing and/or monitoring of the feedwater quality.
- Treatment of waste streams before discharge into the sea, or disposal into the sewer.

The last point explicitly applies to all reject streams produced by intermittent filter backwashing and membrane cleaning operations. Untreated discharge of these wastes into the sea cannot be considered BAT, as collection and treatment are technically feasible and have been practiced on an industrial scale in several small and large SWRO projects. In essence, a BAT solution should give preference to no or low chemical use, respectively, no or low waste discharge designs.

If it were possible to choose freely between the different process options, leaving out technical, economical, and other site-specific environmental limitations and taking only environmental benefits into account, the most preferred design would be a SWRO plant with a subsurface intake and enhanced multiport diffuser design in a suitable oceanic site. A subsurface intake completely avoids impingement and entrainment of marine organisms and, as a biofiltration process, potentially can provide a consistently high feedwater quality with advantages for pretreatment, cleaning, and membrane life, hence reducing chemical consumption in various ways:

- As beachwells are biofilters that can reduce both organic and biofouling, further pretreatment after a beachwell is often minimal (the conventional steps of chlorination–dechlorination and coagulation–flocculation are seldom required). Moreover, energy use, land use, and landscape impacts are typically lower than for SWRO plants with an open intake followed by a conventional or UF pretreatment.
- As cleaning intervals often increase because of the lower fouling potential, chemical use for cleaning is reduced and less cleaning wastes are generated, which would otherwise require treatment onsite or in a municipal waste water treatment plant.

- Lower fouling potential and less frequent cleaning increases the membrane lifetime, thus reducing material and energy use in the manufacturing process.

The only pretreatment for a subsurface intake is often scale control. However, SWRO plants have been reported to work well without antiscalant additives and the need for antiscalants has to be questioned, because of more recent laboratory results indicating that scaling may rarely occur in SWRO systems [38]. More evidence is needed from laboratory studies, pilot testing, and operational plants. Until then, the two precautionary approaches to reducing the risk of scaling would be to (1) lower the recovery and thereby achieve completely chemical-free operation with a subsurface intake or (2) use antiscalants, preferably acid, to control calcium carbonate scaling in SWRO.

A low recovery rate increases the feedwater requirements of the plant, but since the water is taken from a subsurface intake, neither impingement, entrainment, nor increased use of other pretreatment chemicals is a concern. However, the specific energy demand of the desalination process and the energy needed for pumping increases with decreasing recovery. In essence, a tradeoff between chemical use on the one hand and a chemical-free but more energetically intensive process on the other hand must be made. If a need for antiscalants has been established through pilot testing, readily biodegradable, phosphor-free polymers or acid should generally be preferred. Acid has the advantage that surplus acidity is quickly neutralized by mixing with surrounding seawater, whereas antiscalants require some time until degradation is complete. If the recovery is reduced, the increase in energy demand can be compensated by climate change mitigation measures. Another benefit of a lower recovery is that the salinity of the concentrate is lower, and hence dilution to ambient levels can be more easily achieved.

The main concerns of subsurface intakes are the construction-related impacts, such as disturbance of soils and sediments, dispersing sediment plumes that increase turbidity and may affect water quality, habitat destruction, or the disturbance of sensitive wildlife, and possible adverse effects on groundwater processes and flows. As a BAT approach, trenchless techniques and best environmental practice, such as timing of construction activities, should be implemented to minimize the adverse effects on the coastal ecosystem. The hydrological conditions of the intake area should furthermore be investigated in order to avoid adverse changes in groundwater flows and conditions.

An acceptable alternative where a subsurface intake is not possible because of technical, geologic, or environmental constraints is a submerged intake in deeper water and in an offshore location fitted with a passive screen (large surface area with low flow velocities), velocity caps, and a combination of screens. Offshore submerged intakes can reduce chemical requirements by providing better feedwater quality than nearshore or surface intakes. Also, they reduce entrainment of planktonic organisms, which are usually less abundant in deeper water layers; and low intake flows, screens, and velocity caps help to reduce impingement of larger organisms. Nevertheless, chlorine use to control biofouling and coagulant use to remove solids and dissolved organics may be required. BAT would be low-level or pulse chlorination, and treatment of the filter backwash waters.

For seawater applications, there is also a growing interest in the application of chlorine dioxide, which is, for example, used in the Tampa Bay SWRO plant in Florida, followed by sodium bisulfite to remove residual oxidants ahead of the RO membranes [19,77]. Chlorine dioxide is—like chlorine—a powerful oxidant, but requires a shorter contact time and dosage. Unlike chlorine, however, it does not readily react with bromides to form bromine, with ammonia to form chloramines, and does not favor addition and substitution reactions that would produce chlorination-by-products such as halomethanes. A low impact is also predicted by the intermittent discharge of chlorine dioxide into marine waters that undergo rapid mixing on the basis of toxicity tests with species from three trophic levels. Effects were only observed at high chlorine dioxide doses of ≥25 mg/L. The study found chlorine dioxide to be markedly less toxic, with NOEC concentrations 1000 times higher than for total residual chlorine [102]. However, more studies are needed before a final conclusion can be drawn.

Co-location near a power plant with an once-through cooling (OTC) system, preferably with an offshore submerged intake, also provides certain environmental benefits over a stand-alone intake [80], but OTC systems are not considered BAT in some locations, especially in restricted water bodies, but also in open coast locations in California, mainly because of impingement and entrainment effects.

Also, UF with a low chemical approach may be a suitable alternative to conventional pretreatment for open intakes. UF pretreatment is still a relatively young technology. The potential for future improvements is therefore high, and knowledge can be gleaned from past shortfalls of conventional pretreatment systems; that is, chemical use needs to be minimized and backwash waters and spent cleaning solutions should be collected and treated rather than discharged. To increase the performance and acceptance of UF, most membrane suppliers and researchers have focused on operational and economic aspects in the past. It is clear that UF will not achieve a real breakthrough over conventional pretreatment if it does not perform equally well or better in terms of water quality. In case of similar performance, however, environmentally friendly designs may tip the balance in favor of UF. This may particularly be an advantage in emerging markets such as Australia, where UF pretreatment has been selected for two large SWRO projects.

2.4.3 Greenhouse Gases and Air Quality Pollutants

Emissions of greenhouse gases and air quality pollutants of SWRO projects are associated primarily with the use of electricity during plant operation.[7] Electrical energy can either be produced on site by a co-location power plant or purchased

[7]Energy is also used during the manufacturing and transportation of materials and the construction of project infrastructure. However, plant operation accounts for most of the energy use over the lifetime of a project. Lifecycle analyses of desalination processes found that material use and disposal has a low importance (10%) compared to plant operation (90%), due to the high energy demand of operation, irrespective of the desalination process used [103–105]. Assessments of the Sydney and Melbourne SWRO projects arrived at similar conclusions, i.e., greenhouse gas emissions associated with the materials and construction stages represented <5% of the total project-related emissions [71;72].

from the electricity grid. The total energy demand of a SWRO facility comprises the energy needed to drive the desalination process, and energy for pumping and pretreatment, for heating and air conditioning, for lighting and office supplies. The *specific* energy demand refers to the energy demand of the desalination process only. Modern SWRO plants can achieve a *specific* energy demand of <2.5 kWh/m^3 and a total energy demand <3.5 kWh/m^3 by using state-of-the-art equipment (see Section 2.3.4)

2.4.3.1 Energy Demand in Perspective As the treatment and distribution of water by conventional means also require energy, the *relative* increase in energy demand should be considered besides the total demand of the process. The electrical energy demand of treating local surface water is typically between 0.2 and 0.4 kWh/m^3, compared to a specific energy demand of modern SWRO plants of 3 kWh/m^3 (Table 2.13), resulting in the best case in a relative increase of 2.6–2.8 kWh/m^3 for seawater desalination. In locations where the water is transported over long distances, the relative increase of a local desalination plant may be much smaller.

In *California*, for instance, water is transferred in between water basins by an energy-intensive statewide conveyance system. The total *water*-related energy use in California represented one-fifth of the *total* energy use (viz., 48,012 GWh) in the state in 2001 [106]. If *additional* desalination projects with a capacity of 1.3–1.7 Mm^3/d were to be implemented by 2030 with an average energy use of 3 kWh/m^3, the water-related energy use might further increase by 3–4% over the 2001 levels [11,107]. Taking likely future energy savings in SWRO technology into account, which are estimated to be limited to 15% [19], the increase will be 2.5–3% for an average energy demand of 2.5 kWh/m^3.

Presently, the electricity needed to deliver water to San Diego County is 2.8 kWh/m^3 [108], and 3–3.2 kWh/m^3 if one assumes that the water still has to undergo treatment. San Diego County is the farthest point of delivery in the aqueduct systems [107] and 90% of the county's water supplies are imported [11]. The City of Carlsbad in San Diego County is planning to switch its entire water supply from imported water to desalinated seawater, with a 200,000 m^3/d SWRO project under development. The plant's electricity demand with present state-of-the-art technology is expected to be about 3.6 kWh/m^3 [108], which is 0.4–0.6 kWh/m^3 higher than for the imported water. Seawater desalination, in this case, is still the most energy intensive water supply, but the dependence on external resources, which have become increasingly limited, made desalination an attractive option, and it may well become competitive in terms of energy demand in the future.

In *Perth*, Western Australia, the energy demand for transporting water by a pipeline from the Kimberley river system in the north would have an estimated energy demand of 12 kWh/m^3 [52], compared to an electricity demand of the Perth SWRO plant between 3.3 and 3.5 kWh/m^3 for the whole plant [109], resulting in a metropolitan bulk water energy demand of 0.5 kWh/m^3 [52].

To conclude, desalination can be a significant energy consumer in some parts of the world that depend heavily on desalinated water. For example, desalination accounts for 14% of the *energy* demand on the Canary Islands [110]. On the mainland of Spain, however, desalination accounts for "only" 1.4% of the national *electricity* use [111], and this value will even be lower if compared to the total Spanish *energy* use taking other sectors such as transportation into account. In Sydney, the local desalination plant with a capacity of 250,000 m^3/d will result in as much as a 0.6% increase of New South Wales' *electricity* demand and will provide about 15% of the city's water [72,112]. The Perth SWRO plant with a capacity of 144,000 m^3/d is responsible for about 0.7% of the peak *electricity* demand in the Perth region in summer time and produces 17% of the city's water supply.

2.4.3.2 Greenhouse Gas Emissions A main environmental and public health concern of fossil energy use is the release of greenhouse gases (carbon dioxide, methane, etc.) and air pollutants (NO_x, SO_2, fine particulate matter <10 μM (PM_{10}) and <2.5 μM ($PM_{2.5}$) into the atmosphere. Greenhouse gas and air pollutant emissions generally depend on the fuel type, the efficiency of the power plant that produces the electricity for the SWRO plant as well as the installed exhaust purification equipment at that plant.

Carbon dioxide (CO_2) emissions can be estimated with a high degree of certainty, as they depend mainly on the carbon content of the fuel, and carbon sequestration techniques are not in use yet. Emission factors have been established as part of the international emission trading schemes. When electricity is taken from the grid, the energy mix of the respective grid must furthermore be taken into account.

Spain has the third largest (section 2.2.2.3) seawater desalination capacity on a global scale. ENDESA, the leading electric utility company, specifies a CO_2 emission factor of 0.51 kg/kWh for their plants on the Iberian Peninsula. According to the Spanish National Hydrological Plan, SWRO plants consume on average 4 kWh/m^3 in Spain [54]. This results in CO_2 emissions of about 2 kg per m^3 of desalinated water (Table 2.17). The increase in electricity consumption by 11 GWh/d caused by the Spanish AGUA program, which targets an estimated production of 2.7 Mm^3 of desalinated water in 2010, will result in additional CO_2 emissions of 5476 t/d, which represents a 0.6% increase in national CO_2 emissions compared to pre-2005 levels of 326 million tons [113].

In addition to CO_2 emissions, the total global warming potential should be calculated, taking *all* relevant climate change gases into account that arise from the combustion of fossil fuels, namely, CO_2, methane, nitrous oxide, perfluorocarbons, hydrofluorocarbons, and sulfur hexafluoride, as specified in the Kyoto protocol. The global warming potential is expressed as carbon dioxide equivalents (CO_2-e), which specifies the equivalent amount of CO_2 that would have the same global warming potential as all non-CO_2 emissions. Average emission factors for CO_2-e of the specific grid and energy mix should be available on request from utility companies or published by national authorities.

For example, a grid average value of 1.16 kg CO_2-e per kWh was used to calculate the emissions for the Gold Coast desalination project in *Queensland*, Australia. This value includes the direct emissions of CO_2, methane and nitrous

Table 2.17 Estimated CO_2 and CO_2-Equivalent (CO_2-e) Emissions of Selected Desalination Projects

	CO_2, kg/kWh	CO_2-e, kg/kWh	Electricity, kWh/m^3	Emissions, kg/m^3	Capacity, m^3/d	Emissions, t/d
Spain	0.51	No information	4.00	2.03	2,700,000	5476[a]
Ashkelon	0.20	No information	3.60	0.73	330,000	240[b]
Perth		0.98	4.00	3.92	—	—
		0.98	2.30	2.25	144,000	325[c]
		0.98	3.50	3.43	144,000	494[d]
Sydney		1.06	4.00	4.24	—	—
		1.06	3.60	3.82	250,000	954[e]
		1.06	4.00	4.24	250,000	1060[d]
Queensland		1.16	4.00	4.64	—	—
		1.16	4.10	4.76	125,000	595[e]
		1.16	4.70	5.43	125,000	679[f]
Melbourne		1.31	4.00	5.24	—	—
		1.31	4.20	5.50	434,783	2392[e]
		n/a	n/a	7.45	434,783	3239[g]
		n/a	n/a	7.80	434,788	3391[h]

[a] CO_2 emission factor for ENDESA plants.
[b] Gas-fired power plant onsite (without fuel lifecycle).
[c] Lowest specific demand.
[d] All operations.
[e] Desalination only.
[f] All operations and water transfer, 345 days of operation per year.
[g] All operations, water transfer, lifecycle analysis of operation, 345 days of operation, emissions of 1,117,950 CO_2-e t/y, capacity of 150 GL/y or 434,783 m^3/d.
[h] Desalination, water transfer, full lifecycle analysis including operation and construction (1,403,140 CO_2-e tons attributed to diesel generators, grid-connected power, construction equipment, transportation of workforce and materials, offsite waste decomposition, and embodied emissions of materials for the construction of the power and water grid connections, the desalination plant and the marine structures).

oxide from power generation as well as other factors such as grid transmission losses [74]. For the Victorian SWRO project in *Melbourne*, a grid average value of 1.31 kg CO_2-e per kWh was used, which is the highest in the whole of Australia [114], due to a high share of brown coal in the energy mix of the state of Victoria with a relatively high carbon content [71].

If one assumes an energy demand of 4 kWh/m^3 for both plants for reasons of comparability with the *Spanish* plants, CO_2-e emissions would be 4.6 kg/m^3 of desalinated water in the *Queensland* case and 5.2 kg/m^3 in the *Melbourne* case, compared to 2 kg CO_2/m^3 in the Spanish example, which refers to CO_2 emissions only. This illustrates that the global warming potential of desalination may be a factor of 2 higher if one takes other climate change gases in addition to CO_2 into account. It will be even higher if one furthermore includes the distribution of the water, construction activities, as well as CO_2 emissions associated with the use of materials and chemicals into the calculation.

The real energy demand of the *Queensland* plant, including all operations and pumping within the existing water storage network, is estimated to be 24.5 MW for a capacity of 125,000 m^3/d, which equates to 4.7 kWh/m^3 (4.1 kWh/m^3 for desalination alone). The indirect greenhouse gas emissions as a result of electricity use are estimated to be approximately 679 tons of CO_2-e per day, or 5.43 kg of CO_2-e per m^3 of desalinated water, which represents a 2% increase in emissions in the Gold Coast region [74].

The real energy demand arising from the *Melbourne* SWRO project was estimated to be 3,239 tons of CO_2-e per day which equates to 7.45 kg of CO_2-e per m^3 of desalinated water. This value covers electricity used to drive the process and transfer the water, as well as emissions from the transportation of workforce, wastes, and chemicals, from offsite waste decomposition, and embodied in the chemicals used during operation. If furthermore the energy used during construction of the project is added, amortized over the project life of 30 years, the energy demand amounts to 7.8 kg of CO_2-e per m^3. The construction process accounts for only 4% of the total project emissions. In total, 75% of the construction-related emissions stem from the desalination plant, 15% from the marine structures, and 10% from the power and water grid connections [71].

For *Sydney* and *Perth*, where other large desalination projects are to be located, emission factors for electricity from the grid are 1.06 and 0.98 kg CO_2-e per kWh, respectively [114] (Table 2.17). The electricity demand arising from the *Sydney* SWRO project with an initial capacity of 250,000 m^3/d may result in emissions of 4.24 kg of CO_2-e per m^3 or 1060 tons of CO_2-e per day. Similar to the Melbourne project, a LCA found that 5% of the total project emissions are associated with the materials and construction stages, and 95% with operation of the plant, which is mainly electricity use [72]. The electricity demand results in a 1.2% increase in NSW's electricity demand and compares with a predicted ongoing annual increase of around 3%. The Perth project has the lowest grid-specific emission factor and lowest reported energy demand of the Australian projects given in Table 2.17, resulting in the lowest CO_2-e emissions of 3.43 kg/m^3 for the whole plant. The desalination-specific emissions amount to 325 t/d.

The preceding examples all calculate the greenhouse gas emissions for electricity purchased from the grid applying the grid-specific emission factor. It would also be possible to co-locate a desalination plant to an existing power plant, or to build a power plant adjacent to the desalination plant, which would supply electricity "over the fence" to avoid transmission losses. In Ashkelon, the desalination process is driven by a gas-fired power plant on site, which produces 50 MW of electricity for the desalination of 330,000 m^3/d (3.6 kWh/m^3). Applying the CO_2 emission factor for natural gas (202 g CO_2/kWh [115]) results in a very low emission factor of 0.73 kg CO_2/m^3. Even if one adds a safety factor of 2 to factor in the full lifecycle of extracting and providing the natural gas and any other climate change gases that may be released during this process, the Ashkelon project still has the lowest global warming potential from the projects listed in Table 2.17. However, some of the Australian projects such as Perth or Sydney compensate their energy demand through newly erected windfarms, which feed electricity back into the grid.

2.4.3.3 Other Air Pollutants While CO_2 emissions can be estimated with a relatively high degree of certainty, emissions of other air pollutants such as NO_x, SO_2, and fine particulate matter (PM_{10} and $PM_{2.5}$) depend on the fuel type as well as on the technology that is used to minimize pollutant emissions at the source (if any), such as catalysts or scrubbers. Non-CO_2 emissions are therefore more difficult to quantify.

The daily direct and indirect emissions of SO_x, NO_x, particulate matter (PM_{10}), and other air pollutants (carbon monoxide, reactive organic compound—an ozone precursor substance) were estimated for a large SWRO plant in southern California with a projected capacity of 200,000 m^3/d [56]. The daily *direct* emissions are associated with landscaping, delivery trucks, and employee vehicles and amounted to 15 kg of SO_x, 27 kg of NO_x, and 29 kg of PM_{10}. The *indirect* daily emissions are caused by electricity production to provide the electrical energy for the facility and amounted to <0.1 kg of SO_x, 3 kg of NO_x and 0.1 kg of PM_{10}.

It is interesting to note that the environmental impact report (EIR) for this project concluded that *operation* activities including these direct and indirect emissions will not exceed any established air quality thresholds, but that *construction* activities may result in NO_x-emissions of 176 kg/d that could temporarily and locally exceed established emission thresholds. Estimates for SO_x and PM_{10} emissions during construction amounted to 15 kg and 14 kg per day, respectively.

Construction-related emissions include exhaust generated by construction equipment, trucks, and worker vehicles as well as fugitive dust generated by demolition of structures, site grading, and trenching. All air pollutant emissions are project-specific; however, the example above illustrates the order of magnitude of construction-related air emissions and indicates that construction causes a localized and temporal—but measurable—increase in air pollutants, which may violate air quality standards in the worst case. Project-specific emission estimates, based on the specific emission factors of construction vehicles and fuel type, existing background levels and other emission sources in the vicinity need to be accounted for when evaluating whether project-related construction activities may violate any existing air quality standards.

In conclusion, the reference values that we chose may influence whether we consider the energy use of desalination plants as a *significant* environmental concern. For example, the Perth SWRO plant provides 17% of Perth's water, enough for about 300,000 inhabitants, but accounts for "only" 0.7% of the peak *electricity* demand in the region in summer, compared to 30% as required for air conditioning [52]. However, the CO_2-e emissions of 325 t/d that result from operating the SWRO plant are the same as if additional 195,000 cars were put on the road, driving 13.3 km every day, which is the same as 2.59 million km or 65 times around the globe[8].

Compared to other activities and amenities of modern lifestyles, such as air conditioning or transportation, desalinated water does not seem to be an overly

[8]Assuming an economic car with a mileage of 5 L/100 km; a full fuel cycle emission factor of 2.5 t CO_2-e/m^3 of fuel, which comprises the fuel combustion at the point source (car) of 2.3 t CO_2-e/m^3 and an indirect/fuel extraction emission factor of 0.2 t CO_2-e/m^3 [114].

energy-expensive product. However, it is far more energy-intensive than the treatment and distribution of local ground- and surface-water sources by conventional processes, and is often still more energy-intensive than the import of water over long distances. The energy use of desalination is therefore a matter of controversial public debate. However, simply understating the problem by comparing the energy demand of desalination to even more energy-intensive forms of use is not instrumental in the discussion, nor is it to belabor the point if it is clear that conventional water supplies cannot meet the demand.

The wider implications in the debates about new desalination projects are usually the extent to which consumers need and value the additional water provided by the project. Desalination plants are normally not built because people lack water for drinking and sanitation, but to provide additional water as an amenity of modern lifestyles, which may include swimming pools and irrigation in home gardens. A second question is how much impact are consumers willing to accept in return for these amenities, and how much costs would they except on top of the costs of producing desalinated water to make the process more sustainable. The price tag includes not only energy use but also other environmental impacts. As the problem increases in complexity, double moral standards may be applied in the discussion. The sustainability of desalination projects is often questioned on the grounds of high energy use and potential marine impacts—under the tacit assumption that the status quo of the existing water supply or other alternatives are more sustainable, which is not necessarily the case.

Desalination is without question an energy-intensive option, but the status quo is that energy is often wasted in other sectors of use, e.g., by old and inefficient electrical appliances in households. For examples, it is estimated that the total saving potential of more energy-efficient household appliances in Germany is about 29% per household [116], without compromising comfort or living standards. It can be assumed that similar values apply for other industrial and developing nations. Energy saving in households could possibly make up for the additional energy demand of providing desalinated water to households [27].

The status quo of existing water supplies, which may involve the depletion of groundwater resources, or the damming, regulation or diversion of rivers, may also prove unsustainable. According to the World Commission on Dams, a considerable portion of the world's large dams is falling short of their physical and economic targets; that is, they deliver less water and electricity than promised while significantly overrunning costs, besides having extensive impacts on aquatic ecosystems, which, in many cases, have led to irreversible loss of species and ecosystems [117].

In the end, it depends on the perception and definition of significance and on local circumstances whether a community or individual considers energy use of desalination as a *significant* factor. In many parts of the world, energy use is generally recognized as a significant issue. This is reflected in policy initiatives and stricter standards in many countries to reduce energy consumption and increase energy use efficiency in all sectors. On the project level, the concept of environmental impact assessment stipulates that for all *significant* impacts of a new development project, impact mitigation measures have to be identified and implemented, involving a hierarchy of measures from prevention to minimization and compensation.

If energy use of desalination projects is considered a significant impact, then the logical consequence must be to reduce the energy demand as far as possible, and to compensate the remaining energy use. For example, CO_2 emissions are a central issue in all large Australian SWRO projects (e.g., Perth, Sydney, Melbourne, [59,72,73]), and project proponents are encouraged to provide for the use of energy from renewable sources, planting of plantations, or rehabilitation of vegetation to offset these emissions [73]. The electricity demand for two SWRO plants in Perth and the Sydney project, for instance, is compensated by newly erected windfarms, which compensate for the electricity taken from the grid [112,118,119].

2.5 SEAWATER DESALINATION AS A GREEN TECHNOLOGY

Earth Day 2009 on April 22 marked the beginning of a new campaign called the *green generation*, which seeks to foster the development of green technologies and solutions to urgent global issues such as climate change or the world water crisis [120]. Energy, water, and climate change are inseparable global problems. On one hand, energy is needed to deliver water, and water is needed to generate energy [121]. On the other hand, extensive energy consumption in all sectors of use has caused an increase in atmospheric greenhouse gases to 37% above the preindustrialization level, and research indicates stronger than expected forcing of climate change, which also affects the global water cycle. For instance, the eastern and southwestern parts of Australia have experienced substantial rainfall declines since 1950, which is assumed to be partly due to anthropogenic climate change, and models predict up to 20% more drought-months over most of Australia by 2030 [122]. The continent has recently been experiencing one of the harshest droughts in its history and turns to desalination in order to alleviate problems of water scarcity in most major cities.

Seawater desalination is a technology that can mitigate the problems of water shortage, and analysts agree that capacities will continue to grow rapidly in the coming years [123,124] (see also Section 2.2). However, the issue is whether desalination is also a green and sustainable solution. A critical examination and appraisal of the energy and chemical consumption and key environmental impacts of SWRO has been carried out in previous sections (Sections 2.3 and 2.4 and in Ref. 27) and shall only be briefly exemplified here.

To date, the world's largest operating SWRO plant produces 330,000 m^3/d, which is the equivalent of 132 olympic-size swimming pools. About the same amount of water is discharged as concentrate into the sea. The plant consists of 27,000 membrane elements with a total active surface area of about 99 ha (hectares) (or 200 football fields), which need to be replaced every 3–7 years. The energy demand attributed to the use of materials during construction and operation, however, is small (in the range of 5%) compared to the energy needed to operate the desalination plant over its lifetime. The energy demand of SWRO is about 3.5 kWh/m^3 using state-of-the-art equipment (Section 2.3.4), which is the equivalent of a moderately efficient laundry dryer, or about 330,000 laundry dryer loads every day in this example. Besides materials and energy, all desalination plants use chemicals. Their residues are discharged into the sea along with the concentrate (Section 2.4.2) and may amount to thousands of tons per day in some sea areas [27,125]. The discharges, and the intakes (Section 2.3.1), which may cause impingement and entrainment, are the main reasons why marine protection groups such as the Surfrider Foundation in California or the Clean Ocean Foundation in Australia campaign against proposed desalination projects. The other main public concern is pollution of the atmosphere (Section 2.4.3). For example, more than 75% of the formal submissions received in response to the environmental assessment of the proposed Sydney SWRO project expressed concerns about energy use and greenhouse gases [126].

A 2008 review in *Nature* described desalination as a water treatment technology that is often "chemically, energetically and operationally intensive, focused on large systems, and thus requiring considerable infusion of capital, engineering expertise and infrastructure." The costs as well as the environmental concerns are still an impediment to the widespread use of desalination technologies today [127]. Yet some project developers have more recently made headlines with buzzwords such as "green" [108,128] or "sustainability" [112,118]. The seemingly contradictory statements are indicative of the current debate on the extent to which SWRO plants will actually affect the environment. As desalination capacities continue to grow, it remains even more necessary to gain an objective understanding of the energy and chemical use and the environmental impacts of desalination technologies.

The objective of this publication is therefore to investigate whether present state-of-the art desalination technologies can be considered as green and sustainable solutions. "Green" or "clean" or "environmental" technology are all synonyms for the application of environmental science and technology to conserve the natural environment and its resources and to curb the negative impacts of human involvement. Sustainable development is the core of green technologies [129]. Consequently, "green desalination" should refer to desalination technologies that conserve the natural environment and its resources. This can be achieved by implementing technology standards in order to limit waste emissions and resource consumption, i.e., standards for best available techniques (BAT). Furthermore, if "green desalination" means the application of environmental science to curb the negative impacts of human involvement, this can be interpreted as a need for investigating, evaluating, and minimizing environmental impacts of each project individually, i.e., by conducting environmental impact assessment (EIA) and monitoring studies.

Both approaches, BAT and EIA, legitimately coexist because the first is a technology-based approach that refers to industry categories, whereas the other is applied at the project level in order to identify the best environmental option with regard to process design and project site from a set of alternatives. BAT standards can provide guidance in the selection of individual solutions, especially with regard to process design. The United Nations Environment Program released an EIA guidance manual for desalination projects in 2008 [111]. A comparable reference on BAT for desalination, which describes a standard for state-of-the-art desalination technologies that indicate the practical suitability for limiting resource consumption and waste products, is lacking so far.

2.5.1 Best Available Techniques (BAT)

The concept of best available techniques (BAT) has been established by different legislative systems, e.g., in Europe and the United States, and has been applied to similar industries and applications, such as power plants and seawater cooling water systems. In Europe and neighboring seas, the concept of BAT is introduced by the EC Directive on Integrated Pollution Prevention and Control (IPPC), the Conventions for the Protection of the Marine Environment of the North-East Atlantic (OSPAR) and of the Baltic Sea Area (HELCOM), and the Protocol for the Protection against Pollution from Land-Based Sources of the Mediterranean Action Plan (LBS protocol).

The IPPC directive imposes a requirement for certain industries with a high pollution potential to obtain a permit, which is issued if certain environmental conditions are met, such as the use of BAT. This applies to industries listed in Annex I of the directive, such as energy industries, but not water treatment installations. A reference document on BAT in industrial cooling systems has been provided under the IPPC directive [130], which could serve as a guideline for designing desalination plants.

The marine conventions, in contrast, clearly indicate the need for a BAT concept for seawater desalination technologies in European and neighboring sea areas. For example, the Mediterranean LBS protocol requires contracting parties to take BAT into account when adopting action plans, programmes and measures. In 2005, the countries of the Mediterranean Action Plan adopted the concept that desalination plants are "industrial facilities," which means that they need to be regulated and assessed through EIAs [131]. In addition, BAT standards for desalination plants should be developed and implemented through the Mediterranean Action Plan. In some countries, general BAT regulations exist already. In Israel, for example, a prerequisite for discharge into the sea is the use of BAT, which prohibits discharge directly at the coast with the exception of cooling water outfalls of power plants. Marine outfalls therefore have to be deep offshore outfalls, and the entire length of the pipeline has to be buried using BAT during construction to minimize damage to the coastal area [132]. While the first large SWRO plant in Israel (Ashkelon) is co-located to a power plant and therefore discharges directly at the coastline, tender documents for new projects set a request for an outfall to a depth of 20 m [133].

According to the IPPC directive and the marine conventions, BAT is defined as state-of-the-art processes, facilities, or methods of operation that indicate the practical suitability for limiting discharges, emissions and waste, and for reducing the impact on the environment as a whole. The term *technique* includes both the technology used and the way in which the installation is designed, built, maintained, operated, and dismantled. The techniques that are considered BAT should be economically and technically feasible, should be used or should have been tried out on an industrial scale, and should account for technological advances in scientific knowledge. Special consideration in the development of BAT is typically given to the consumption of raw materials, water, energy, less hazardous substances, and the possibility for recovery and recycling of any resources used or wastes generated. Applying the IPPC principle to an industry through implementation of BAT implies the need to take preventive measures when there is substantial evidence to suspect that an activity may cause harm to the environment even if there is no absolute proof (prevention principle), and to reduce the emissions into the atmosphere, water, and soil, as well as waste generation (control principle).

In the United States, BAT terminology is used in the Clean Air Act and Clean Water Act (CWA). Under the CWA, the Environmental Protection Agency (EPA) issues national standards for facilities discharging directly to surface waters. These so-called effluent guidelines apply to categories of dischargers and are technology-based, and not on the impacts on the receiving waters. The intent of technology-based standards is to establish a basic national standard for all facilities within a category using BAT, which becomes the minimum regulatory requirement in permits that are implemented through the National Pollutant Discharge Elimination System (NPDES) permit program [134].

Similar to the European IPPC directive, guidelines have been established for different industrial categories including power plants but not water treatment installations. CWA and NPDES regulations authorize the use of "best professional judgment" to derive technology-based effluent limitations on a case-by-case basis where standards are absent. Best professional judgment was for example used to derive the NPDES permit for the Carlsbad SWRO plant in southern California [96]. The EPA has now initiated a new rulemaking on drinking water treatment effluent guidelines to address the direct discharge of drinking water treatment residuals to surface water, such as suspended solids, aluminum or iron salts, organic matter, polymers, *desalination concentrates*, or other residuals [134]. Section 316(b) of the CWA furthermore requires that the location, design, construction, and capacity of cooling water intake structures reflect the BAT for minimizing adverse environmental impact, that is the impingement and entrainment of fish and other aquatic organisms. The regulations on cooling water intakes could serve as a basis for regulating seawater desalination plant intakes as well.

In order to develop BAT standards for SWRO plants, the main components of desalination plants (i.e., the intake, pretreatment, desalination process, cleaning, and concentrate disposal system) need to be compared with regard to environmental criteria such as energy, material and chemical use, and resulting emissions and likely environmental impacts. The identified BAT solution can be used as a reference in the determination of individual BAT solutions on a case-by-case basis, taking site- and project-specific considerations into account. It should be noted here that the environmental impacts of a desalination plant will depend on the *technology and mode of operation* on one hand, and the *environmental characteristics of the project site* on the other hand. A commonly used tool for comparing alternative technologies is the lifecycle assessment (LCA). A novel approach has been used in section 2.6., using multicriteria analysis (MCA) to compare different pretreatment alternatives for SWRO plants.

2.5.2 Environmental Impact Assessment

By definition, the primary goals of an EIA are to provide information on the environmental consequences of a project for decision-making, and to promote environmentally sound and sustainable development through the identification of appropriate alternatives and mitigation measures. EIAs are seldom limited to environmental aspects, but where appropriate also address public health and socioeconomic concerns. Public participation is therefore an integral element of EIAs in many legislative systems. As a result, EIAs are often complex multistage, multidisciplinary, and multiparticipatory processes.

To deal with this complexity and facilitate the implementation of EIAs for desalination projects on a wider scale, a 10-step process has been proposed by UNEP/ROWA, as shown in Figures 2.27 and 2.28 [111]. The initial EIA phase includes the steps of screening and scoping of the project, in which a decision is taken on whether an EIA is required for a particular project, and in which the scope, contents, and methodologies of the EIA and expert studies are specified in the terms of reference (ToR). A reference list for preparing the ToR has been provided in Reference 111, which may also serve as a blueprint for preparing the EIA report. During the main EIA phase, a detailed description of the preferred project configuration including site and process alternatives is provided, and other statutory permits applicable to the project are identified and obtained. The scientific studies and analyses are conducted during this phase, including baseline studies, the prediction and evaluation of impacts, and the identification of alternatives and impact mitigation measures. The final EIA phase involves decision-making and a review of the EIA process. An environmental management plan is often established at this stage, which specifies the monitoring requirements during the installation and operation of the plant.

In principle, EIAs for large desalination projects will not differ in terms of complexity and level of detail from those of other water supply infrastructure projects. Depending on the proposed project, it is generally the responsibility of the competent authorities to individually define the need, scope, and complexity requirements of each EIA study. When dealing with a larger number of desalination proposals, a collaborative effort between the main government agencies and participatory groups should be initiated to elaborate a national EIA guideline for desalination projects. It would facilitate the EIA process by establishing equal standards for the environmental studies to be undertaken and the information to be submitted as part of the EIA for each individual project in the future. Moreover, as a number of agencies usually have permitting authority over the project, a lead agency should be nominated to coordinate the process by involving other agencies and by informing the project proponent about permitting requirements.

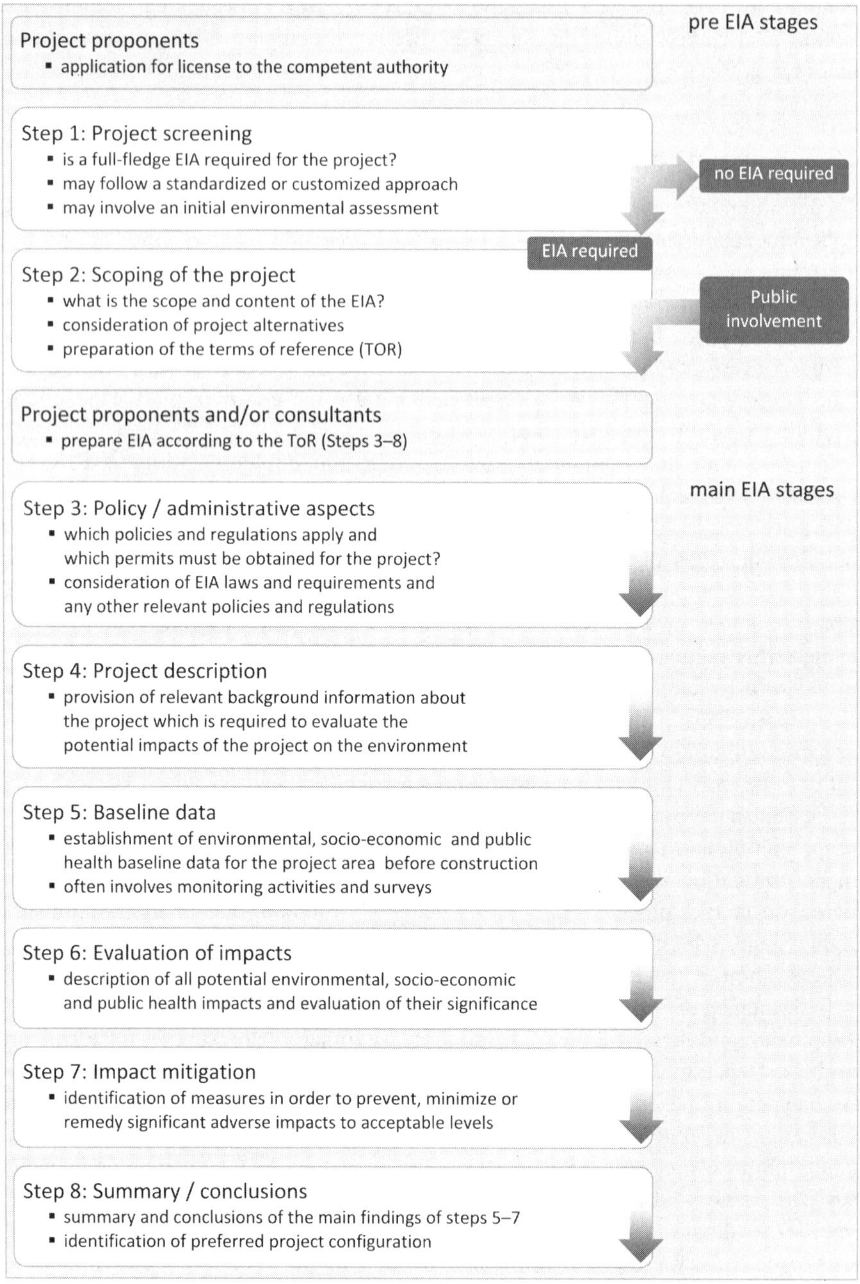

Figure 2.27 The 10-step EIA process: scoping, screening, and main EIA phase.

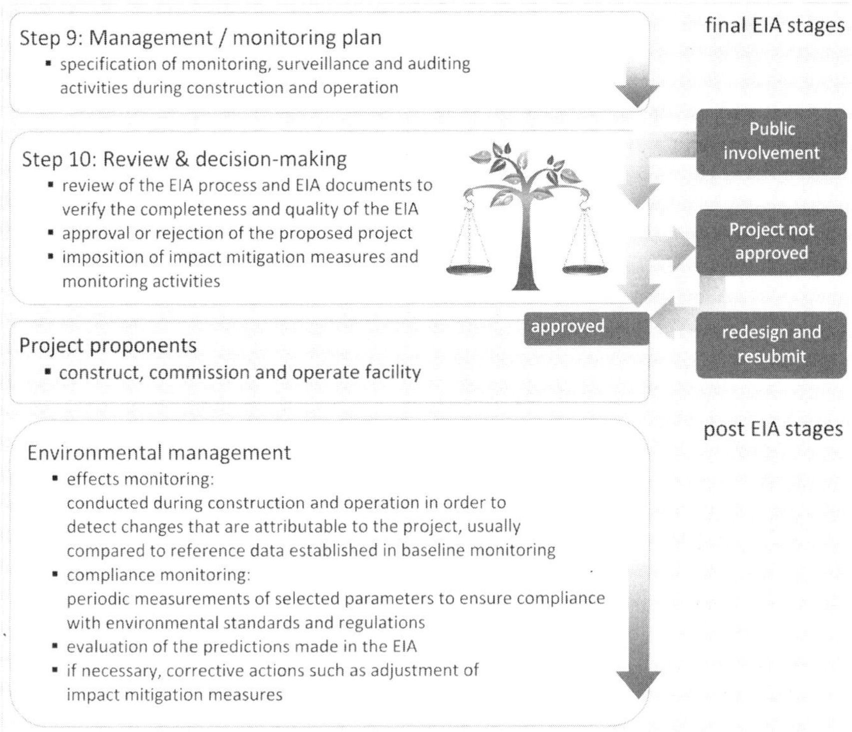

Figure 2.28 The 10-step EIA process: EIA decision phase and follow-up activities.

EIA studies are a widely recognized and accepted approach for identifying, evaluating, and mitigating potential impacts of major infrastructure projects on the environment. To this day, however, only a limited number of EIA studies for desalination projects have been carried out and published. Most of them are from Australia and the United States. In the EU, the EIA Directive[9] regulates which project categories have to be made subject to an EIA by member states. It lists major water supply projects such as groundwater abstraction schemes, dams, and works for the transfer of water resources between river basins. The list should be expanded to include desalination projects.

EIAs for desalination plants from other parts of the world, particularly the Middle East, are scarcely available. The reason may be that EIA studies are considered to be the intellectual property of the project proponent, even if projects may receive environmental assessments and may have state-of-the-art monitoring programs. Sheppard et al. criticize that too many environmental studies remain confidential for alleged commercial or security reasons, and that there is too little sharing of information, particularly on the Gulf [14]. This is in contradiction to the notions of

[9]Directive 85/337/EEC on the assessment of the effects of certain public and private projects on the environment, amended by Directive 97/11/EC.

transparency and public participation, and EIAs should generally be made available to a wider public audience, especially if the studies are conducted for municipal water supply projects.

Often, EIA investigations are carried out under immense time constraints. For instance, only 4 months were given for an EIA study for a large SWRO plant in Algeria [135]. This shows that environmental concerns can be of secondary importance when a ready supply of freshwater is urgently needed. The opposite is also true; comprehensive and time-consuming environmental studies and concerns have been the major hurdle in the permitting process of new projects in California. The first large SWRO project in California took more than 10 years to receive all necessary permits. In Spain, the government has announced plans to speed up the EIA process from the current average of 770 days for infrastructure projects to no more than 6 months. The reform could benefit vital water projects; however, doubts were raised from within the European Commission that it would be possible to condense the whole EIA procedure including public consultation into a six month period [136].

In 2004, Manuel Schiffler from The World Bank stated that an "internationally agreed environmental assessment methodology for desalination plants does not exist so far and its development would be desirable" [137]. The UNEP guidance document partially fills this gap. It offers guidance for designers of desalination projects, consultants, regulators, and decisionmakers on the methodology, scope, and contents of EIA studies and specifically for desalination projects. Still missing, however, are long-term monitoring studies that improve the basic understanding of the actual environmental impacts of desalination plants. Although an increasing number of EIA studies is being published, these are based mainly on conceptual models and laboratory experimental methods, including hydrodynamic modeling of the discharges and effluent toxicity testing, carried out *before project startup*. The results need to be verified for the majority of these projects in effects monitoring studies *during plant operation*.

2.5.2.1 *Environmental Monitoring* As EIA studies make predictions about the expected impacts of a desalination project on the environment, it is necessary to validate the accuracy of these predictions against observations during project implementation and operation. In 2008, the US National Research Council of the National Academies attested a surprising paucity of useful *experimental data*, either from laboratory tests or from field monitoring in 2008 [19]. Large projects and accompanying monitoring programs are only now being implemented. The longest effects monitoring programs for a few SWRO projects worldwide have accumulated only 2–3 years of cohesive monitoring data. Although this may allow for some conclusions regarding those particular projects, it is too early to use these results as conclusive evidence concerning the long-term impacts of desalination plants, or concerning the cumulative impacts of several desalination plants in a given sea area. A review of existing monitoring studies [111] revealed that most other monitoring studies for desalination projects were either limited in scope—addressing only one aspect such as salinity, short-term—without a continuous baseline and effects monitoring, and localized—without adequate spatial replication.

The importance of operational monitoring cannot be overemphasized, which is also illustrated by the following case study. Ambrose et al. [138] compared the actual impacts of the cooling water discharges from a nuclear power plant in southern California, established by a 15-year monitoring program, to predictions made in the EIA which had been generated in three different ways. The comparison showed that (1) almost all of the testimonies of scientists before the permitting agency, which were based on professional judgment with little scientific analyzes, were wrong; (2) the accuracy of the final environmental statement, based on standard assessment methods at that time, was mixed but generally not too high; and (3) the predictions of the marine review committee, based on a comprehensive baseline study over several years, were the most accurate but also showed inaccuracies. Although a clear correlation between effort and accuracy of the predictions seems to exist, the lesson here is that EIAs cannot predict with complete confidence what will happen in the environment, even if considerable resources are dedicated to monitoring [138]. EIAs, like other observational studies, are likely to be "messy" even after a conscientious effort to apply the appropriate techniques and mathematical statistics [139]. The impacts predicted in EIAs are not always the actual impacts, although they become the de facto impacts if there is a lack of follow-up studies [138].

The core of the problem is to design a monitoring program that provides sufficient statistical power to be able to distinguish the effects of the desalination project from natural processes. For this reason, sufficient spatiotemporal replication is needed in field monitoring studies. Field monitoring for desalination projects should include the project (impact) site and several control sites that adequately represent the habitats of the impact site. To capture the temporal variance at these sites, paired sampling should be carried out at several times before, during, and after project implementation (BACIPS approach). Baseline monitoring for major development projects is usually carried out over a 2-year period before project startup and effects monitoring for 2–3 years during commissioning and operation. A holistic monitoring framework for desalination projects should furthermore integrate field monitoring with hydrodynamic modeling and bioassay studies, preferably using whole effluent toxicity (WET) tests. These concepts have for example been implemented in the monitoring programmes for the Sydney and Gold Coast SWRO projects.

Regulatory agencies may still be reluctant to require rigorous operational monitoring studies, and project developers are understandably opposed to funding it [138]. However, there is an increasing tendency to regulate new developments worldwide under the requirement that predictions will be tested by measuring the real impacts by scientific means, and by imposing project modifications if impacts are found to differ from those predicted. Adequate monitoring could therefore also be understood as "insurance" against unwarranted claims [140]. In this context, it is also noteworthy that both effects monitoring and compliance monitoring only allow for reactive impact management. It should therefore be in the interest of all parties (and of the environment) that management responses are established in case unexpected or more severe impacts are detected during effects monitoring, or in case trigger values are exceeded during compliance monitoring.

Effects monitoring also serves to produce much relevant fundamental research, which is of particular value to industries that are not involved in one-off developments [141]. The desalination industry can thus learn from each experience to minimize impacts for the next development. As mentioned above, comprehensive environmental monitoring programs are underway for several large Australian SWRO projects, which will provide valuable results in the near future. An overview on how to design an environmental monitoring framework for desalination plants is given by Lattemann [27].

2.6 MULTICRITERIA ANALYSIS (MCA) OF INTAKE AND PRETREATMENT ALTERNATIVES FOR SWRO

Environmental impact studies are part of the permitting process of large infrastructure projects, including new desalination projects. As such, they are often complex multistage, multidisciplinary, and multiparticipatory processes, in which different site and process alternatives and a wide range of environmental impacts are being considered. A vast body of quantitative data and qualitative information is thus compiled, which results in lengthy reports with numerous appendices, and extensive evaluation tables, which are usually unsuitable for direct evaluation [142]. For example, the recently completed EIA of the Victorian desalination project in Australia covered 1600 pages, and the volumes stack nearly 1.5 m high. Moreover, different government agencies, stakeholders, and the wider public usually participate in the permitting process of a new desalination project. It is therefore necessary to communicate the results of the EIA to the decisionmakers, and conflicting references about the project need to be balanced in decision-making. This requires a structured and transparent approach. As a single and objectively best solution rarely not exists, the process of environmental decision-making has been described as a conflict analysis characterized by environmental, social, economic, and political value judgments, which is essentially a search for an acceptable compromise solution [143].

The increasing volume of complex and controversial information generated by EIAs and the limited capacities of individuals to integrate and process this information emphasizes the need for a formalized decision support tool, which aggregates the information in a transparent and consistent way. One such tool is multicriteria analysis (MCA), which can facilitate the EIA in different ways and in different stages (Table 2.18). In contrast to traditional evaluation tools such as cost-benefit analysis (CBA) often used in economics, or lifecycle analyses (LCA) used to quantify the resource consumption of a product, MCA allows for a comparison of alternatives by using nonmonetary and nonquantitative criteria. Such qualitative criteria are often more applicable in environmental contexts, and thus in EIAs.

Table 2.18 Requirements of EIA and Capabilities of MCA Studies

EIAs are . . .	MCA can . . .
Multistage processes: In a first selection round (typically scoping), all possible alternatives and their potential environmental concerns are identified, of which a limited number are selected for the second round. In the second design round, more detailed investigations are carried out for the selected alternatives. In the final step of decision-making, the preferred alternative is selected.	Be used in different stages (design rounds) of the EIA to eliminate, refine, or add alternatives in the light of new information, which is summarized in an evaluation table and aggregated by MCA in each stage/round.
Multidisciplinary processes that: Cover different natural and environmental science disciplines as well as human health and socio-economic aspects where appropriate. Involve various specialist studies.	Be used to integrate a large number of criteria, including noncommensurable, qualitative criteria, into decision-making processes.
Multiparticipatory processes: Involving different decision-makers, such as different government agencies, politicians, and stakeholders. Involving the wider public, particularly for projects that arouse public interest such as water supply.	Raise awareness of different value judgments of decision-makers and stakeholders, and highlight tradeoffs between alternatives.
Based on predictions: EIAs can only be as accurate as the information that is available at the time of project planning. Information gaps need to be clearly identified in EIAs. A precautionary approach should be applied in decision-making.	Include a sensitivity analysis to evaluate whether the ranking changes if variations occur in the input data, in case of uncertain data or unforeseen events.

MCA has been successfully used in a wide range of environmental planning and management contexts, including allocation of water resources, coastal development, or the management of coastal resources [143–145], such as fisheries [146]. Moreover, MCA has become part of the standard decision aid frameworks used in EIAs [142,147,148].

This section explores the usefulness of MCA as a decisions support tool for new desalination projects. MCA has been used before in other studies, e.g., to facilitate site selection for one large desalination plant in Australia and for two small plants in South Africa [149–151]. However, the present evaluation focuses on a comparison of technologies instead of different sites for a specific project. SWRO is the preferred process for most new desalination projects. The most important consideration in a SWRO system is the intake and pretreatment (Section 2.3) because a good and reliable water quality must be obtained with a low fouling potential to the SWRO membranes. The design of the outfall is another important consideration in terms of environmental impacts. However, concentrate disposal is more straightforward, i.e., it basically requires an effective diffuser in a suitable oceanographic site, and the alternatives are therefore rather limited. It was therefore decided to apply the MCA to the main intake and pretreatment options for SWRO plants. The information needed for this approach was obtained from literature sources and, to a limited extent, directly from plant operators. An overview on the methodology is given in Section 2.6.1, the input data are described in Section 2.6.2, and results and conclusions are presented in Sections 2.6.3 and 2.6.4.

2.6.1 Methodology

An MCA typically involves two phases (Fig. 2.29). In the first phase, the decision problem is defined, input data is generated, and the alternatives can be ranked according to the input data by means of a graphical evaluation. In the second phase, the alternatives are ranked using MCA, which involves the selection of an MCA model and standardization functions, applying weights to the criteria that reflect value judgments, and a sensitivity analysis of the ranking.

A prerequisite to MCA is problem definition. This consists in the establishment of an overarching, primary objective that clearly states what the decision seeks to achieve, and the identification of a complete set of *alternatives*. The decision-making objective has to be translated into a number of operational *criteria*. Structuring the criteria is an important prerequisite of MCA and results in the establishment of an *objective tree*. The final set of criteria should be complete, nonredundant, operational, and minimal (see Ref. 143, after Ref. 152). Although up to 100 criteria have been reported for complex infrastructure projects, a more manageable range would be 6–20, which is sufficient for well-founded decision-making in many cases [142]. The important point is to identify those criteria that are truly relevant to decision-making and to avoid redundancy.

The performance of every alternative against all criteria is measured by *scores*. The data are gathered in an *effects table*, which is an intermediate product that can be graphically evaluated without attaching weights to the criteria. This requires that the incompatible units of measurement of the various criteria be transformed into a compatible form of measurement, which is achieved by applying a suitable *standardization function* for each criterion. For example, scores can can be measured on quantitative scales (e.g., costs in euros or energy use in kWh), on qualitative scales (ranking on an ordinal scale from highest impact to lowest impact), or on

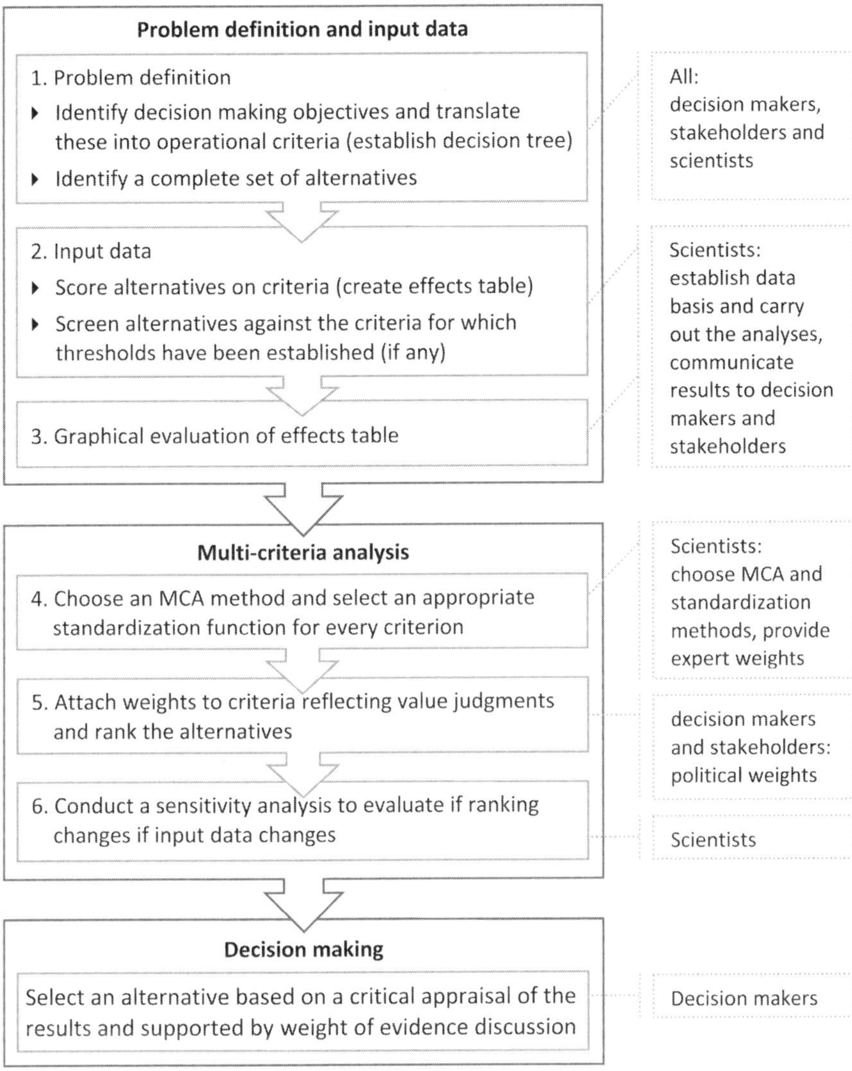

Figure 2.29 Roadmap to decision making using MCA.

binary scales (yes/no). By the step of standardization, these different scales are transformed into a dimensionless scale from zero to one that is used for comparing and aggregating the different criteria.

Furthermore, *weights* need to be generated and allocated to the different criteria. Weights represent value judgments of different people or groups of people. Typically, weights within a category of criteria are given by experts and weights between categories by decisionmakers. For example, an expert weight would be to say that a potential loss of seagrass beds is more severe than the loss of a sandy seafloor habitat inhabited by motile macrofauna species. Expert weights reflect the opinion of one or more experts, and do not create much controversy in the best case, although they also have to balance (sacrifice) one effect against another [142]. In contrast, political weights often create much controversy, as they reflect the tradeoffs between categories [142], such as economic versus ecologic aspects. How weights are chosen, i.e., the method used, can be as important as who chooses the weights. Weights can be established directly, or indirectly following a formal method, such as *pairwise comparison* or the *swing weight method* [153–155].

Subsequently, a suitable *MCA model* has to be selected in order to rank the alternatives. The MCA model should be well defined and easy to understand and should be capable of manageing the necessary number of alternatives and criteria, and support different decisionmakers, and the method should be able to handle inaccurate or uncertain criteria information, as uncertainty is inherent to many environmental decision contexts. There is usually no means for objectively identifying the best MCA model [143]. Weighted summation, one of the simplest MCA methods, performs well in most cases. It was found to be the most popular MCA model in Dutch EIAs, and because the model is methodologically sound, easy to explain, and transparent, it is also recommended by the Dutch Commission for EIA [142].

After ranking the results, one should conduct a *sensitivity analysis* to evaluate whether the ranking is robust, i.e., whether the ranking changes if a variation in the input values (scores and weights) occurs, in case that the input data are incorrect or variable by nature [156]. For example, the energy demand of a desalination plant will vary depending on the intake seawater temperature and salinity, which may vary between intake location sites (e.g., estuarine, surface, or offshore submerged intakes) and seasons. As energy demand will likely have a significant influence on the overall ranking of the alternatives, it is necessary to investigate how statistical variability in the input data will influence the ranking. Sensitivity analysis also investigates which adjustments of the scores or weights are necessary to bring about a change, particularly if two alternatives have only a small difference in their overall utility value. A ranking is considered as robust when it is not sensitive to minor variations in the scores or weights [156].

The final step is *decision-making*. MCA is a tool that can facilitate but not replace decision-making. The purpose of MCA is not to single out the "correct" or "best" decision, but to dynamically evaluate a set of alternatives in order to gain information about the effects of different courses of action [145]. It was noted that the attitude toward MCA often changes in the process of conducting an MCA [142]. In the beginning, it is often perceived as a "blackbox" that is easily manipulated, whereas, ironically, the confidence in the results is often too high in the end. To that end, not only should the MCA method be well documented and transparent, but its limitations should be kept in mind. No MCA technique can eliminate the need to rely heavily on sound knowledge, data, and judgments, or the need for a critical appraisal of the results [147]. The final selection of an alternative should therefore always be supported by a weight of evidence discussion and qualitative considerations.

2.6.2 Data Input

The objective of this study is to compare different intake and pretreatment systems for SWRO using MCA. SWRO intake and pretreatment systems are usually optimized to control fouling by bacteria, organic compounds, suspended and colloidal matter, and scaling inside the plant. Depending on local circumstances, solutions range from minimalist to operationally intensive; the latter consume considerable amounts of natural resources, including chemicals, materials, energy, land, and water. The overall environmental footprint of a desalination process can therefore be minimized by selecting the best pretreatment option in a given site. The methodology follows the MCA approach described in Section 2.6.1 and Figure 2.29, using the software tool DEFINITE 3.1 [154], which facilitates ranking and comparing a finite set of alternatives [153].

2.6.2.1 Alternatives The alternatives considered, as shown in Figure 2.30 included natural subsurface systems, i.e., beachwells (see Sections 2.3.1.2 and 2.3.2.1), and engineered pretreatment systems, distinguishing between conventional pretreatment (Section 2.3.2.2) and membrane pretreatment (Section 2.3.2.3).

It was assumed that the intake and pretreatment systems pretreat between 40,400 and 42,000 m^3 of seawater per day, depending on the water losses in the pretreatment system for backwashing (from 1% to 5%), and provide 40,000 m^3 of feedwater for a single-pass SWRO plant operated at 50% recovery and with a capacity of 20,000 m^3/d (Fig. 2.32). A nominal plant capacity of 20,000 m^3/d was selected because beachwells have a limited intake capacity and are therefore mostly used for smaller SWRO plants. Similarly, UF/MF pretreatment has so far mostly been applied to smaller SWRO plants, with the largest operational UF-SWRO plant having a capacity of 50,000 m^3/d [Table 6 in Ref. 27]. Most large SWRO plants still use a conventional pretreatment with one granular media filtration stage. Some plants have two media filtration stages, and a few plants have a third pretreatment

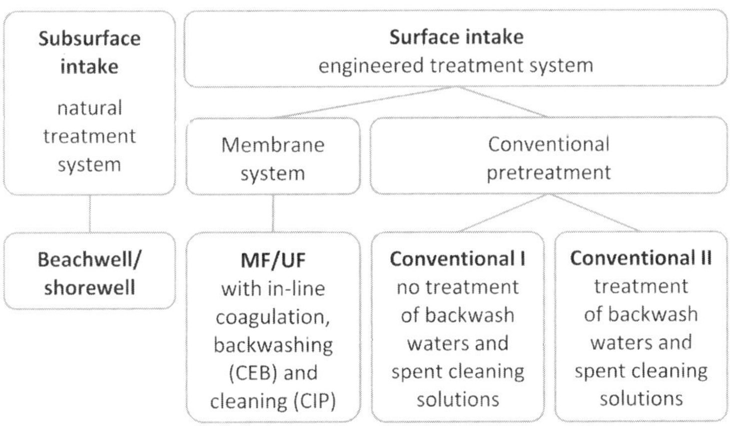

Figure 2.30 Pretreatment system alternatives considered in the MCA.

stage in addition to two stage media filtration (Section 2.3.2.2). The conventional pretreatment design considered in this study also consists of one stage, similar to most SWRO plants. As no large SWRO plant uses a beachwell intake or UF pretreatment to date, caution must be used when extrapolating the results of this study to large SWRO projects with capacities of $\geq 100,000$ m^3/d.

Neodren subseabed drains, which are a special configuration of natural subsurface systems, could provide sufficient flow rates for large facilities, depending on the number of drains installed (Section 2.3.1.2). Although the technology is used in a few SWRO plants and has been described as technologically sound and very environmentally friendly [157], it was decided to exclude it from the list of alternatives because of limited data availability.

We assumed that the beachwell intake is followed by an additional sand filter. Conventional pretreatment is assumed to consist of coagulation–flocculation followed by a single-stage pressurized DMF. Because cylindrical pressure vessels are limited in diameter to about 3 m, they are commonly used in smaller SWRO plants [18]. However, several *large* new plants in Algeria, Southern Europe and Australia [Table 5 in Ref. 27] also reported single- or dual-stage pressure filters. Single-stage gravity filters still seem to be more common in large SWRO plants though, mainly because they enable larger filtration areas, require less energy, and are therefore cheaper [18].

Two alternatives for conventional pretreatment were considered in the MCA: without treatment (conventional treatment I) and with treatment of backwash waters and spent cleaning solutions (conventional treatment II). Treatment will likely result in a higher land use, energy use, and cost, but lower marine environmental impact. Data on the costs and land use of a sludge and waste water treatment facility could not be obtained from the literature. To be accurate, one would also have to include the land use impacts for the sludge landfill, which depends on the sludge quantities and the disposal site, and the cost of operation and transportation [158]. Small sludge amounts may be dewatered in a simple and relatively inexpensive drying bed onsite. A worst-case scenario would require a clarifier/thickener followed by a sludge dewatering system, using a belt press or centrifuge, in a separate building. The energy use of sludge treatment can be estimated to range between 6% and 54% of the pretreatment energy costs [35] (Table 2.14). Because of the complexity and highly site-specific nature of land use, energy use, and costs, a general increase of 10% was assumed for the alternative conventional II in all three criteria.

For UF/MF pretreatment, it was assumed that the backwash and CEB wastes are discharged into the sea, as no UF-SWRO plant in the literature reported treatment. This standard practice has also been confirmed by a UF membrane supplier. Moreover, the use of a disinfectant, either continuously or intermittently, and the use of inline coagulation, is assumed to be common practice in pilot and full-scale UF plants to improve the process performance and filtrate quality of the pretreatment (see Section 2.3.2.3).

The study was limited to single-pass RO, as the need of a second RO pass depends on the product-water specifications. As pretreatment solutions are usually site-specific, the number of possible alternatives is theoretically infinite. To be practical, only the standard designs of the major pretreatment alternatives were considered, and it was assumed that the set of alternatives is thus complete, which is one prerequisite for MCA.

2.6.2.2 Criteria The primary objective of the MCA is to optimize the pretreatment of SWRO plants, which is subdivided into eight subobjectives and 15 criteria (Fig. 2.31). Objectives 1–6 reflect different aspects of resource consumption and environmental impacts, and objectives 7 and 8 refer to economic and operational aspects. For MCA, one should distinguish between *cost* criteria, which have a negative correlation between score and effect (the higher the score, the worse the effect), and *benefit* criteria, which have a positive correlation. The criteria of water

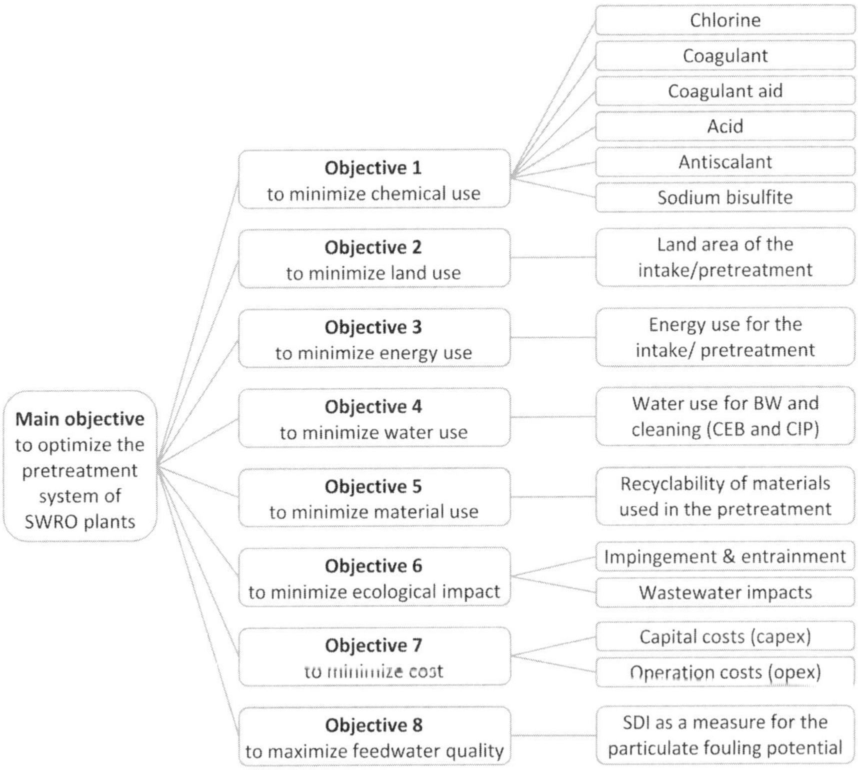

Figure 2.31 Objective tree (value tree) hierarchy of the MCA with main objective (left), subobjectives (middle), and evaluation criteria (right) (BW—backwashing, CEB—chemical-enhanced backwashing; CIP—cleaning in place). For material use, *recyclable* means the materials can be reused or reprocessed into new products in order to prevent waste. *Ecology* refers to the ecological impact due to seawater intakes, causing impingement and entrainment, and potential effects from the discharge of untreated backwash waters and cleaning solutions. SDI is a surrogate parameter that estimates the potential of SWRO fouling caused by fine suspended organic or inorganic colloids.

quality and material recyclability were defined as benefits, while the remaining criteria were all costs. Furthermore, the measurement scales of the criteria have to be specified. Chemicals, land, energy, and water use as well as costs are ratio scales; i.e., the importance of a criterion is proportional to its value; for instance, a twofold higher energy demand is 2 times as bad (or two times as good for benefits). Ecological impact, material recyclability, and SDI are binary scales, which only indicate whether an effect is fulfilled (yes) or not fulfilled (no). A "yes" means a benefit in the case of material recyclability and SDI, and a cost in case of ecologic impact.

2.6.2.3 Scores The four alternatives were scored on all 15 criteria, which can be categorized into quantitative and qualitative criteria. In a first step, a database was established for all quantitative criteria through a literature review of operational and pilot plant data and by personal communications with plant operators [Tables 3–7 in Ref. 27]. In a second step, a value was selected which is considered to be representative of the standard design of an alternative (Table 2.19). It should be mentioned that the compiled baseline data were quite variable, and that a single "true" value, which represents all operating conditions and feedwater qualities, does not exist. The selected values should therefore be understood as one scenario that can be refined and revised in the light of new or better data. To accommodate data variability in the input data, a sensitivity analysis was carried out (Section 2.6.3.1) and two revised scenarios were considered (Section 2.6.3.2).

The MCA scores (Table 2.19) were calculated using the reference value, assuming a reference plant size with a capacity of 20,000 m^3/d, and factoring in the respective flows and dosing points given in Figure 2.32. For example, the dosing concentration of 1 mg/L of chlorine is multiplied by the intake volumes of 41,200 m^3/d and 42,000 m^3/d, which results in a daily load of 41 and 42 kg/d for conventional and UF pretreatment.

The only qualitative criteria are water quality (SDI), ecological impact, and recyclability of materials. For SDI, a value of 3 is selected assuming that all pretreatment alternatives are running in good condition. An SDI range of 2–4 is typically required by membrane manufacturers for conventional pretreatment. UF/MF has a reported SDI range of 0.8–2.2 [46,62,159,161–163], and beachwells have reported values of 0.4–2.8 [Tables 3-7 in Ref. 27,31,160]. Concerning ecological impact, it is assumed that the open intakes of UF and conventional pretreatment systems have the potential to cause the impingement and entrainment of marine organisms ("yes"), which is not the case for beachwell intakes ("no"). The other criterion refers to the potential impact resulting from the discharge of untreated backwash waters from filters and UF/MF membranes and cleaning solutions in all three pretreatment systems ("yes"). The alternative conventional II assumes that the backwash waters are treated and that the sludge is deposited on land ("no" impact). For material use, only the UF/MF system was given a score of "no" assuming that the UF/MF membranes are currently not recycled, while it is being assumed that the filter media are recycled or have a beneficial reuse ("yes").

A linear interval standardization function was used in this study, both for graphical evaluation and for the following MCA, to transform the different units of

Table 2.19 Effects Table for the Multicriteria Analysis[a]

Criterion	⊖/⊕ Scale[b]	Beachwell Selected Reference Value	Calc. Score for 20,000 m^3/d	Conventional Pretreatment (Conv.) I[c] Selected Reference Value	Calc. Score for 20,000 m^3/d	UF Pretreatment Selected Reference Value	Calc. Score for 20,000 m^3/d
Chlorine (feedwater)	⊖ ratio	— (water from beachwells is not chlorinated)	—	1 mg/L If continuous chlorine use: 0.5–1 mg/L residual usually maintained throughout pretreatment line	41 kg/d	1 mg/L If continuous chlorine use: 0.5–1 mg/L residual usually maintained throughout pretreatment line	42 kg/d
Chlorine (backwash)	⊖ ratio	— (does not apply)	—	— (does not apply)	—	100 mg/L 1x/d varied within 5–20 mg/L 1×/15–60 min, up to ≤500 mg/L 1×/d	4 kg/d
Sodium bisulfite	⊖ ratio	— (no dechlorination required)	—	3 mg/L Dosing rate usually 3 times the residual chlorine concentration [6,18]	120 kg/d	3 mg/L same assumption as for Conv. I	120 kg/d
Coagulant	⊖ ratio	— (not used after beachwells)	—	1.2 mg/L Fe (≡3 mg/L $FeCl_3$, 40%) Values were 0.5–3 mg/L as Fe and <1–18 mg/L as $FeCl_3$ or $FeSO_4$	49 kg/d	0.4 ppm 0.1–0.6 mg/L Fe (≡0.3–1.5 mg/L $FeCl_3$, 40%) in full-scale plants, 0.2–4 mg/L as Fe in pilot plants	17 kg/d
Coagulant aid	⊖ ratio	— (not used after beachwells)	—	0.85 mg/L Values were 0.2–1.5 mg/L	35 kg/d	— (typically not used)	—

(*continued*)

Table 2.19 (*Continued*)

Criterion	⊖/⊕ Scale[b]	Beachwell Selected Reference Value	Calc. Score for 20,000 m^3/d	Conventional Pretreatment (Conv.) I[c] Selected Reference Value	Calc. Score for 20,000 m^3/d	UF Pretreatment Selected Reference Value	Calc. Score for 20,000 m^3/d
Acid (H_2SO_4)[d]	⊖ ratio	— [no pH adjustment required (no coagulation step)]	—	30 mg/L Values were 22–55 mg/L, 20–25 mg/L (membrane textbook, [18]) 30–100 mg/L (literature review, [6])	1236 kg/d	10 mg/L No doses available, 2 full-scale plants used acid in addition to Fe, which is assumed to be for coagulation and not for scale control, as both plants also used antiscalant, selected value is $\frac{1}{3}$ rd of H_2SO_4 dose in Conv. I	420 kg/d
Antiscalant[e]	⊖ ratio	2 mg/L Values were 0.9–1.2 mg/L for polymer antiscalants and 8 mg/L for phosphate antiscalant	80 kg/d	2 mg/L Values were 1–2 mg/L or 1–4 mg/L if one counts an earlier literature review [6] and company information	80 kg/d	2 mg/L Probably similar to Conv. I; 1 pilot plant reported 3 mg/L [159], a literature review reported 1–3 mg/L [62]	80 kg/d
Land use for pretreatment	⊖ ratio	0.8 m^2 per 1000 m^3/d inst. capacity 4.7 m^2 was given for a 6000 m^3/d plant with a vertical beachwell [160]	16 m^2	38 $m^2/1000\ m^3/d$ value selected from a range of 35–40 m^3/d [18]	Conv. I: 760 m^2 Conv. II: 836 m^2	27 $m^2/1000\ m^3/d$ ≡30% reduction from Conv. I, 25–33% reduction reported [18,39]	540 m^2

Energy use for pretreatment	⊖ ratio	0.1 kWh/m^3 of pretreated water for a vertical beach well including sand filter [160]	4000 kWh/d	0.15 kWh/m^3 of pretreated water for intake + 1 stage pressurized DMF, reported values were 0.035 kWh/m^3 RO permeate ≈0.018 kWh/m^3 pretreated water for single GF excl. intake [35] (Table 2.14) 0.35 kWh/m^3 RO permeate ≈0.16 kWh/m^3 pretreated water for 2 stage DMF including intake 0.12 kWh/m^3 pretreated water for single-stage pressure filter [65]	Conv. I: 6000 kWh/d Conv. II: 6600 kWh/d	0.15 kWh/m^3 of filtrate, values ranged between 0.05 and 0.2 kWh/m^3 filtrate	6000 kWh/d
Water use for BW, CEB, CIP	⊖ ratio	1% of pretreated water assuming a sand filter backwashed 1× every 3 days	400 m^3/d	3% of pretreated water assuming a 1-stage DMF backwashed 1× every day; 2–3% of the filtrate capacity per filtration stage used for backwashing [18]	1200 m^3/d	5% of filtrate assuming a BW/CEB 24–48×/d and CIP 1–2×/y	2000 m^3/d

(*continued*)

Table 2.19 (*Continued*)

Criterion	⊖/⊕ Scale[b]	Beachwell Selected Reference Value	Calc. Score for 20,000 m^3/d	Conventional Pretreatment (Conv.) I[c] Selected Reference Value	Calc. Score for 20,000 m^3/d	UF Pretreatment Selected Reference Value	Calc. Score for 20,000 m^3/d
Costs	⊖ ratio	No data, assumed to be 50% of the conventional pretreatment cost		Assumed costs for a single-stage pressure DMF [65]:	Conv. I / II in US$/d:	Assumed costs for a DOW SFP2880 operated at 75 LMH [65]:	
		Capex: 0.4 US¢/m^3 of pretreated water	160 US$/d	Capex: 0.7 US¢/m^3	280 / 308	Capex: 1.2 US¢/m^3 of filtrate	480 US$/d
		Opex: 1.1 US¢/m^3 of pretreated water	440 US$/d	Opex: 2.2 US¢/m^3	880 / 968	Opex: 1.9 US¢/m^3 of filtrate	760 US$/d
		Total: 1.5 US¢/m^3 of pretreated water	600 US$/d	Total: 2.9 US¢/m^3	1160 / 1276	Total: 3.2 US¢/m^3 of filtrate	1280 US$/d
Material use	⊕ binary	Material is mostly recyclable, and no large material wastes are produced?	Yes	Material is mostly recyclable, and no large material wastes are produced?	Yes	Material is mostly recyclable, and no large material wastes are produced?	No

Potential ecological impact	$\ominus$ binary	Impingement/entrainment of organisms?	No	Impingement/entrainment of organisms?	Yes	Impingement/entrainment of organisms?	Yes
		Discharge of backwash water/sludge?	Yes	Discharge of backwash water/sludge?	Yes for Conv. I; No for Conv. II	Discharge of backwash water/sludge?	Yes
Water quality	$\oplus$ binary	Is an SDI <3 achieved?	Yes	Is an SDI <3 achieved?	Yes	Is an SDI <3 achieved?	Yes

[a] The MCA reference values listed in this table were selected from Tables 3–5 in a report by Lattemann [27] and used to calculate the scores for a 20,000 m^3/d plant, assuming the flow rates and dosing points shown in Figure 2.32.
[b] This column indicates whether the criterion is a *cost* ($\ominus$) or *benefit* ($\oplus$) criterion, and the scale of the criterion (ratio or binary).
[c] For the alternative conventional pretreatment (Conv.) II (with treatment of backwash and waste waters), which is not included in this table, a 10% higher energy use, cost and land use was assumed.
[d] Acid can be added to adjust the pH for coagulation and/or to control scaling. Since the dosing of an antiscalant is already assumed ahead of the RO units, acid dosing is assumed to take place before the pretreatment and primarily for the purpose of coagulation. This is also the case in several large SWRO plants, e.g., Barcelona, Fujairah, Perth, Sydney, and Tugun/Gold Coast.
[e] Antiscalant dosing takes place after the pretreatment system and concentrations were assumed to be the same for all four alternatives.

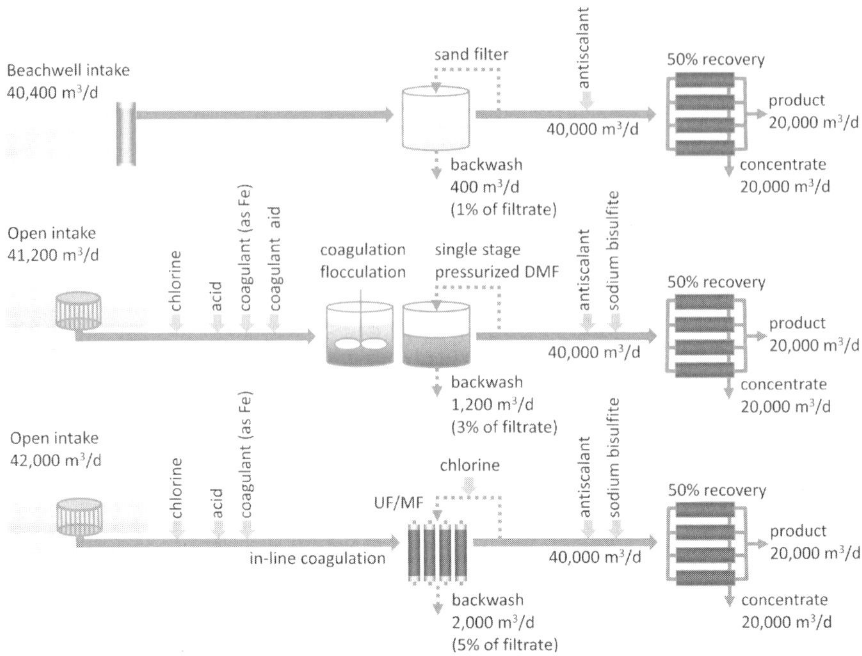

Figure 2.32 Process flows for a 20,000-m^3/d facility. In beachwells (top), the seawater is naturally filtered through the sand bed and then pumped into the SWRO plant for further pretreatment, which usually consists of a sand filter and acid and/or antiscalant addition. Either conventional (middle) or membrane pretreatment (bottom) is used where the feedwater is received from an open intake. Both usually involve chlorination followed by coagulation, filtration, acid and/or antiscalant addition, and dechlorination. For the purpose of this study, it is assumed that conventional pretreatment uses a single-stage pressurized dual-medium filter. In UF/MF pretreatment, coagulants are typically used inline and lower dose rates are possible, no coagulant aid is used, and no flocculation step is required, but UF/MF typically requires chemically enhanced backwashing with chlorine.

measurement into dimensionless values between zero and one. The best score receives the value of one and the worst score receives the value of zero. For benefit criteria, the best score is the highest absolute value. For cost criteria, the best score is the lowest absolute value. The other alternatives are linear interpolations between zero and one according to their relative position. The standardized values are not proportional to the original scores and differences are accentuated. Although in practice the relation between a criterion score and its value is usually more complex, a linear standardization is often an acceptable approximation if the range of the scores is not too large [153].

2.6.2.4 *Graphical Evaluation* The effects table (Table 2.19), which contains the pretreatment alternatives' scores on the various criteria, as well as the units of measurement or type of scales, and a statement whether a criterion is a cost or a benefit, is an intermediate working product, which can be graphically evaluated before the start of the MCA (see Fig. 2.29, MCA roadmap). Graphical evaluation has the objective to illustrate the relative performance of the alternatives without assigning weights to the criteria. The scores were transformed into a dimensionless value between 0 and 1 by a linear interval standardization function, which is the default setting of the DEFINITE software.

The graphical presentation in Figure 2.33 shows that a beachwell:

- Is the best alternative in terms of chemical use (chlorine, sodium bisulfite, coagulant, and acid) because these chemicals are typically not required.
- Has the lowest land use, water use, energy use, and costs, and does not cause ecological impacts from impingement and entrainment (intake effects).

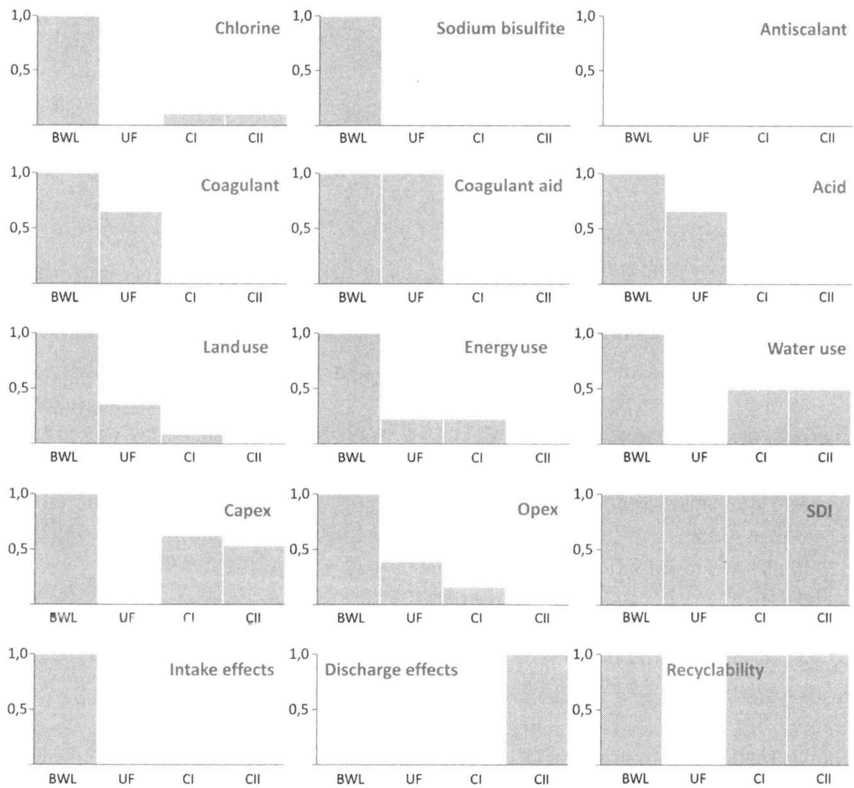

Figure 2.33 Graphical evaluation of scores using interval standardization, in which the best score receives a value of one (full bar), the worst a value of zero (empty bar), and all remaining alternatives are scaled in between. For *benefit* criteria (recyclability, SDI), a higher bar represents the better alternative. Note that for *cost* criteria (all remaining criteria, such as chemical or energy use), a higher bar indicates a lower negative effect and therefore also a better alternative.

- Is equal to the other pretreatment alternatives in terms of antiscalant use (all are assumed to use 2 mg/L), and with regard to SDI (all assumed to achieve an SDI<3).
- Is equal to UF pretreatment with regard to the use of coagulant aid (none).
- Is equal to UF and conventional pretreatment I in terms of ecological impact from filter backwash waters if these are assumed to be discharged into the sea.

Accordingly, UF pretreatment:

- Is the second best alternative in terms of coagulant and acid use, but worst in terms of chlorine and sodium bisulfite use similar to conventional pretreatments I and II.
- Is the second best in terms of land use and operating costs (opex), but worst in water use and investment cost (capex).
- Is equal to conventional pretreatment I in terms of energy use.
- Is the worst alternative in terms of recyclability of materials and ecological impact, if one assumes that the intake causes impingement and entrainment, and that the backwash waters from CEB and CIP are discharged without treatment.

Conventional pretreatments I and II:

- Are both equal to beachwells in terms of recyclability of materials.
- Are the second best alternatives in terms of water use and investment cost as well as chlorine use (or second worst in the latter case, depending on the perspective), but the worst with regard to all other chemicals, as well as land use and operating costs.
- Are the worst alternative in terms of ecological impact (the same as UF), if one assumes that the intake causes impingement and entrainment, and that the backwash waters from filters are discharged without treatment.

It is noteworthy that conventional pretreatment I is equal to UF in terms of energy use. Conventional II is the best alternative in terms of impacts caused by waste water discharges, because a waste water treatment step has been included in this alternative, but at the expense of additional energy and land requirements of this treatment step.

It can be concluded that beachwells are dominant over all other alternatives if one excludes the criterion of discharge effects. Dominance occurs when one alternative performs at least as well as another alternative on all criteria and strictly better than the other on at least one criterion [164]. In the considered scenario, it is assumed that the untreated discharge of backwash waters from sand filters is the norm for small SWRO plants. However, three out of four beachwell plants [Table 3 in Ref. 27] reported a sludge treatment and landfill [31,32], and one plant injected its waste water into a deep well [160]. Even if the backwash is discharged without treatment, sludge amounts are small because of a low solids content in the feedwater and no coagulant use, which results in a low backwash frequency of once every few days [32]. It is thus safe to assume that beachwells are dominant over the other alternatives. In practice, dominance of one alternative is rare, and the extent to which it can support real decisions is correspondingly limited [164].

2.6.2.5 Weights An advantage of MCA is that subjective references about a project are made explicit, which increases transparency of the decision-making process and allows for a systematic evaluation of different stakeholder perspectives. The subjective references are expressed in the form of weights, which are attached to the decision-making criteria. In order to gather different perspectives, a questionnaire was prepared and sent to four groups of participants with different perspectives in the area of seawater desalination:

- Environmental perspective (three responses): two persons with university background, one person working in an international company that develops desalination projects.
- Operators' perspective (four responses): three plant managers and one person working in a water authority overseeing desalination projects, all four representing plant capacities between 50,000 and 330,000 m^3/d (two from the Mediterranean region, two from Australia).
- Commercial perspective (five responses): two persons from an international company, three from commercial research institutes.
- Research perspective (six responses): six professors and researchers from universities with expertise in process operation and optimization.

The questionnaire was sent to an equal number of stakeholders within each group, but not all questionnaires were returned. The imbalance between the groups, with three responses for the environmental perspective and six responses for the research perspective, results in a different reliability of the weights derived for each group; i.e., more responses in the group of environmental experts may have resulted in a different set of weights.

The expected-value method was used to elicit the expert weights because it is easy to explain and use. The aim was to have a single page with questions that would take no more than 10 min to answer. The main group of criteria [chemical, energy, water, and land use; material recyclability; water quality (SDI); costs; and ecological impact (see value tree in Fig. 2.31)] were ranked from highest to lowest importance, based on the personal background and experience of each participant. In addition, the single chemicals within the chemicals group were ranked. It was possible to assign the same rank for two or more criteria if these were considered equally important. Weights within a category are often given by experts, as in this study. However, weights between categories that require tradeoffs between environmental, social, and economic aspects should be given by politicians with a mandate for decision-making.

The average rank was calculated for each criterion within each group, and the criteria were sorted accordingly in decreasing order of priority from rank 1 to the lowest rank. In this study, persons with an environmental and university background considered ecologic impacts as most important, followed by energy use, chemical use, and water quality as lower ranks. In contrast, persons who represent the operators' and commercial perspectives gave the highest priority to water quality (SDI) and costs, followed by energy use, chemical use, and ecology as ranks 3–4. The criteria of water use, land use and material use occupied the lowest ranks in all groups.

Concerning the different chemicals, all four groups considered coagulant use (rank 1 or 2) of high priority, which indicates that it is important for process operations as well as in terms of environmental impact. Chlorine was also considered very important from an environmental perspective (rank 1), but less important from all other perspectives (rank 3 or 4). Antiscalant and acid are considered moderately important (ranks 2–4) in all groups except for the university group, which considers antiscalant as most important (rank 1) and acid as least important (rank 6). The criteria of coagulant aid and sodium bisulfite (SBS) ranked lowest in all groups (ranks 3–5).

The expected-value method converts the rank order of the criteria ($c_1 \geq c_2 \geq c_3$) into a set of quantitative weights ($w_1 \geq w_2 \geq w_3$) by the following algorithm (see Ref. 165, after Ref. 166):

$$w_k = \frac{1}{K} \sum_{i=k}^{K} \frac{1}{i} \tag{2.1}$$

where K represents = the number of criteria. The sum of all quantitative weights is 1. When two or more criteria are considered equally important (i.e., have tied ranks), the weight given to each criterion is the average weight for the tied ranks. For example, the responses from the university group led to the following ranking and weights: ecology ($w_1 = 0.340$) $\geq$ chemicals use ($w_2 = 0.215$) $\geq$ energy use = water quality ($w_{3,4} = 0.131$) $\geq$ costs ($w_5 = 0.079$) $\geq$ land use ($w_6 = 0.054$) $\geq$ material use ($w_7 = 0.033$) $\geq$ water use ($w_8 = 0.016$). The higher the importance, the higher the weight of the criterion and the greater the difference between the next smaller criterion. The expected-value method therefore emphasizes the more important criteria, and gives only little weight to the less important criteria. For example, ecology accounts for 34% of the total weight, followed by chemical use with 21%, whereas the two least important criteria account for less than 4% and 2%.

The weights derived for the main group are first-level weights as shown in the pie diagrams in Figure 2.34. Second-level weights were established for the single chemicals by expert ranking and the expected-value method for each group (see Ref. 27 for details). The first-level weights are multiplied by the second-level weights to derive the overall weight of a criterion. The criteria capex and opex in the main group *costs* were considered equally important, as were the intake and discharge effects in the main group *ecologic impact*.

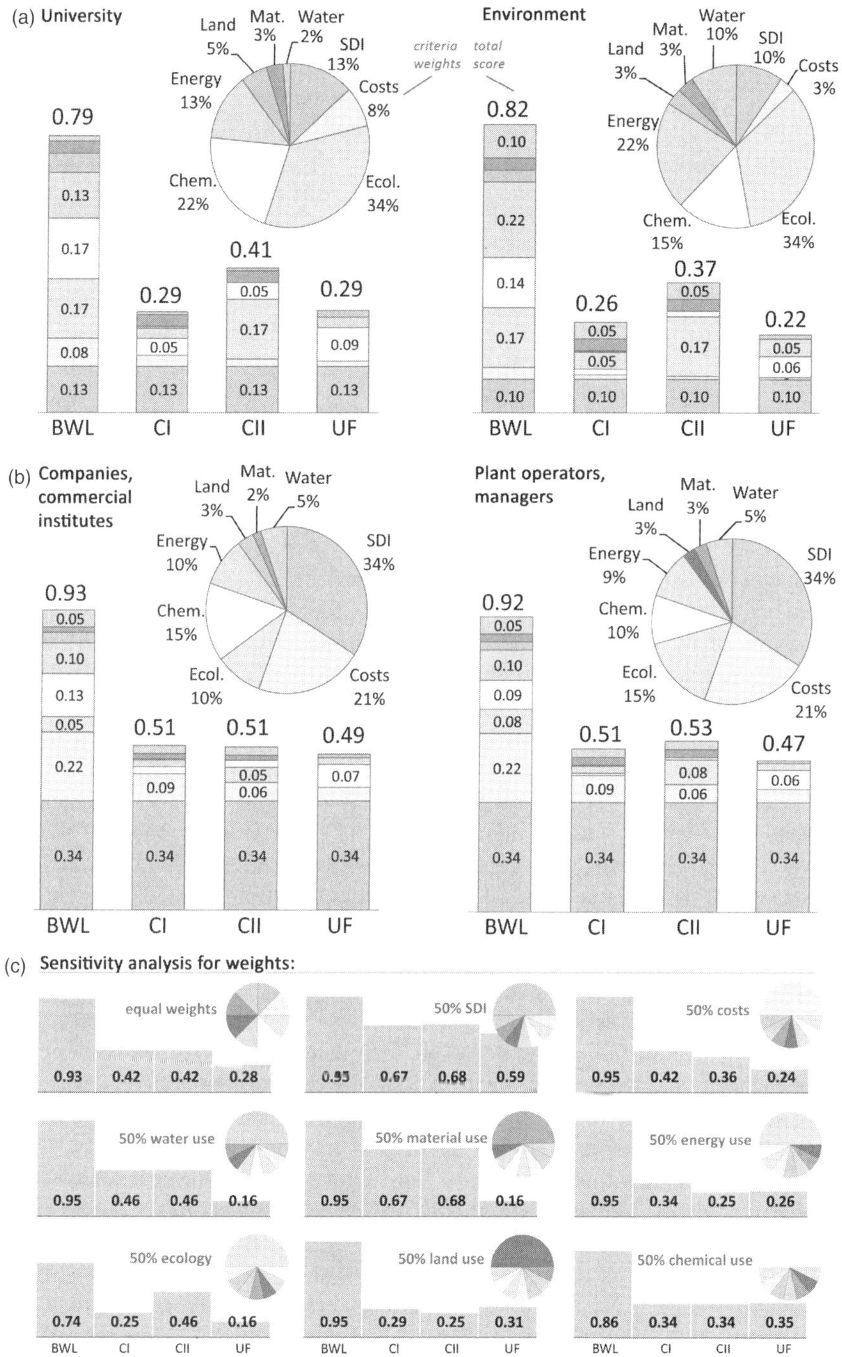

Figure 2.34 MCA ranking results of the four alternatives by four expert groups (a and b) and variation of weights to investigate different perspectives (c) [BWL—beachwell; UF—ultrafiltration; CI—conventional pretreatment I; CII—conventional pretreatment II with waste water (sludge) treatment]. A colored graph is published in Ref. 27

2.6.3 Ranking Results

The synthesis of the weights and scores into a ranking is a computational step that depends on the selected MCA model. Weighted summation, which was used in this study, is a compensatory aggregation method, in which poor performance in any one criterion can be compensated for by overall good performance in the other criteria. The overall performance of an alternative is the sum of the alternative's score for each criterion multiplied by the weight given to that criterion [153]:

$$\text{Total score } a_j = \sum_{i=1}^{N} w_i \cdot \hat{s}_{ij} \tag{2.2}$$

where A = set of alternatives with $a_j (j = 1, \ldots, M)$
C = set of criteria with $c_i (i = 1, \ldots, N)$
$\hat{s}_{ij}$ = standardized score of alternative a_j for effect c_i
w_i = weight of effect c_i

The ranking results for the four groups are presented in Figure 2.34. The outcome in which beachwells are also the favored option from different perspectives was to be expected, because the alternative was already found to be dominant by graphical evaluation of the effects table. The analysis, however, shows that the ranking is generally similar in the two groups that represent plant operators and companies, and in the two groups that represent university and environmental backgrounds. Beachwells obtained the best result in the company and operators groups (total scores of 0.93 and 0.92), followed by conventional pretreatment (0.51, 0.53) and UF pretreatment (0.49, 0.47) as ranks 2–4 with only marginal differences. Beachwells reached a lower total score in the university and environment groups (0.79, 0.82), followed in some distance by conventional pretreatment II with sludge treatment as the second best alternative (0.41, 0.37), followed by conventional pretreatment I and UF with small differences in total scores as ranks 3 and 4 (0.22–0.29).

2.6.3.1 *Sensitivity Analysis* The sensitivity of the ranking to changes in weights and scores was investigated. First, the *weights* of the main criteria group were systematically altered. Weights were distributed equally (12.5%) between criteria and then successively changed to 50% for each criterion to emphasize that perspective, while the remaining 50% were equally distributed among the remaining seven criteria (7% each). The subcriteria (in chemicals, costs, ecology) were given equal weight within their groups (Fig. 2.34c bottom). The same analysis was carried out for the expert weights while maintaining the original weights within the subgroups.

In the following list, only the changes in ranks 2–4 are summarized, as the ranking of beachwells is not sensitive to changes in weights. The sensitivity analysis of weights, with results shown in Figure 2.34c (bottom), shows that:

- Conventional pretreatment I and II rank second, with similar or equal scores, followed by UF as the last rank, if equal weights are allocated to the main criteria, and if 50% is allocated to the criteria water quality (SDI), water use, or material use.

- Conventional pretreatment I is the second best alternative if 50% weight is given to cost or energy use, and conventional pretreatment II is the second best alternative if 50% weight is given to ecological impact.
- UF ranks last and only becomes second if 50% weight is allocated to the criteria chemical and land use, or third if 50% weight is given to energy use. However, the differences are marginal between UF and the next lower alternative in all three cases.

Then the effects of *score* uncertainty on the ranking was analyzed under the assumptions that the scores could be 50% or 100% higher or lower than the selected scores listed in Table 2.19. The program calculates the probability that an alternative ranks at a certain position on the basis of the specified uncertainty values. The probability is calculated from 2000 repetitive outcomes, in which scores are drawn at random from the specified uncertainty limits, assuming a normal distribution function.

Figure 2.35 shows that beachwell has a high probability of 88–100% to rank at first position in all groups, even given a high uncertainty in the scores. In all groups, conventional pretreatment I or II is likely to rank second. In the university and environment groups, conventional pretreatment II is most likely (69–92%) at the second rank. UF has a high probability of ranking at third or fourth position.

2.6.3.2 Revised Scenarios The decision problem was revised in two consecutive scenarios, in which both the set of alternatives and the number of criteria were changed.

Scenario 1: Comparison with Existing Plants *Case studies* with different intake and pretreatment systems were used to investigate whether these produce the same ranking results as the selected reference values in the previous MCA. As the reference values had been chosen independently for each criterion from a wide range of literature values, it is possible that the overall result gives a distorted image of the operating conditions of existing plants. For example, chlorine use is still often reported as continuously in the literature, whereas it is, in fact, used intermittently in many plants. Moreover, beachwells are generally assumed to have a low energy demand in the literature; however, the reported energy demand of a Neodren beachwell intake was slightly higher than that of an identical plant in the same location with an open intake [31]. A third example is that many of the more recently commissioned conventional plants have a sludge–waste water treatment system.

The revised scenario has the objective to compare the "hypothetical textbook values" to real plant data. The intake and pretreatment alternatives included the twin plants at San Pedro del Pinatar, Spain [Table 4 in Ref. 27], one of which has a Neodren beachwell followed by a sand filter and the other, a conventional pretreatment with a two-stage dual-medium filter. A second conventional pretreatment system was included, assuming a single gravity filter as in the Tugun plant in

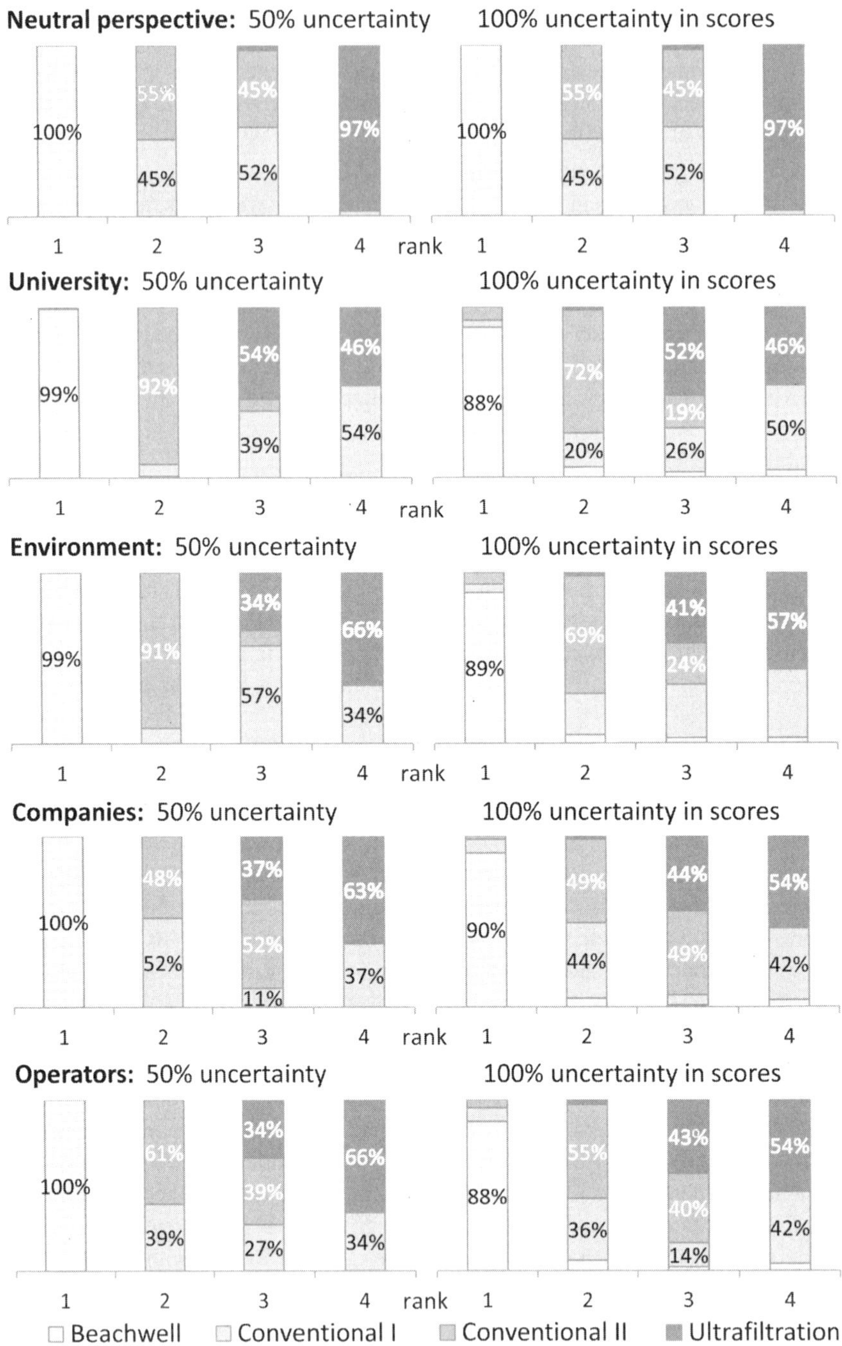

Figure 2.35 MCA sensitivity analysis of scores.

Australia [Table 5 in Ref. 27]. Both conventional systems had a sludge treatment, but were quite different in terms of chemical usage and dosage. Furthermore, a large SWRO plant with a UF pretreatment (Jumeirah Dubai; see Table 6 in Ref. 27) was included.

The value tree was reduced to four main environmental criteria: chemical use, energy use, water use, and ecological impact. The other two environmental criteria, i.e., land and material use, were no longer considered because their scores were already rather hypothetical without a good database in the first place. Moreover, all expert groups considered them of very low importance. As cost data for the plants could not be obtained, and all alternatives were assumed to achieve an SDI<3, these two criteria were also eliminated. The main objective of the revised scenario was therefore to identify the intake and pretreatment system with the lowest overall impact on the environment.

Although the case studies generally provided a good data basis, a few assumptions were still necessary, as outlined in Table 2.20. The case study data were used to calculate the scores for a reference plant of 60,000 m^3/d operated at 45% recovery, a design that bears more resemblance to the selected case studies than the assumptions of the original MCA. The scores were standardized using interval standardization. The criteria ranking order from the environmental expert group was used to calculate the weights and the total scores, using weighted summation. A perspectives analysis was carried out, consecutively assigning equal weights and then 50% weight to each main criterion in order to investigate how the ranking changes. The results were as follows:

- The Neodren beachwell turned out to have the lowest overall environmental impact given the selected criteria (total score 0.73), followed by the twin plant in the same location with an open intake and minimal chemical pretreatment (0.54), closely followed by the more extensive conventional pretreatment (0.52).
- This ranking applied to all perspectives except if 50% weight were placed on energy use, in which case UF pretreatment ranked first (0.59) but is almost equal to the two conventional pretreatments (CIIb—0.58, CIIa—0.57). UF pretreatment ranked at the lowest position except for this and the chemical use perspective, in which it was only marginally better (0.01) than the more extensive conventional pretreatment.

The results of the revised scenario using the operation and design specifications of real full-scale plants support the findings of the previous MCA, in which beachwells ranked first, followed by conventional pretreatment and UF pretreatment.

Scenario 2: Comparison of UF Pretreatments In order to implement a UF pretreatment successfully, the backwashing and CEB intervals, and the chemical doses added to the feed and CEB need to be optimized. For example, it would be possible to operate a UF plant with no or only little coagulant and chlorine addition to the feedwater, but with more frequent backwashing and chlorine enhanced

Table 2.20 MCA Reference Values and Calculated Scores: Revised Scenario 1

Criterion	Beachwell San Pedro del Pinatar I	Calc. Score for 60,000 m^3/d	Conventional IIa San Pedro del Pinatar II	Score 60,000 m^3/d	Conventional IIb Tugun, Australia	Calc. Score for 60,000 m^3/d	UF Jumeirah, Dubai	Calc. Score for 60,000 m^3/d
Details of case study plant	65,000 m^3/d capacity, Neodren wells, sand filters, sludge treated		65,000 m^3/d capacity, open intake, two stage DMF, sludge treated		125,000 m^3/d capacity, offshore submerged intake, gravity DMF, sludge treated		64,000 m^3/d, UF recovery of 90%, backwash every 45 min, backwash discharged	
Chlorine (feedwater)	— (none used)	—	— (none used)	—	10 mg/L shock dose, assumed to take place for 1 h/d	57 kg/d	15 mg/L for 20 min 1×/day	31 kg/d
Chlorine (backwash)	— (does not apply)	—	— (does not apply)	—	— (does not apply)	—	200 mg/L chlorine CEB 1×/day	83 kg/d
Sodium bisulfite	— (none used)	—	— (none used)	—	3 mg/L prior to RO, assumed to take place for 1 h/d	17 kg/d	— [normally not required (only if hypochlorite breakthrough is detected)]	—
Coagulant	— (none used)	—	1.6 mg/L Fe calc. from 4 mg/L $FeCl_3$ assumed 40% active ingredient	213 kg/d	7.2 mg/L Fe calc. from 18 mg/L $FeSO_4$ assumed 40% active ingredient	982 kg/d	0.3 mg/L as Fe	44 kg/d
Coagulant aid	— (none used)	—	— (none used)	—	0.2 mg/L 0–0.4 mg/L polyelectrolyte	27 kg/d	— (none used)	—

Acid (H_2SO_4)	— (occasionally)	—	50 mg/L acid to pH 6.5, assumed to require 50 mg/L H_2SO_4	6,667 kg/d	35 mg/L H_2SO_4	4,772 kg/d	275 mg/L for 20 min 1×/day	560 kg/d
Antiscalant	1 mg/L; range 0.9–1.2 mg/L	133 kg/d	1 mg/L	133 kg/d	2 mg/L	267 kg/d	2 mg/L	267 kg/d
Energy use	0.18 kWh/m^3; 0.40 kWh/m^3 permeate for intake/ pretreatment, value of 0.18 kWh/m^3 calculated assuming 45% RO recovery	24,000 kWh/d	0.16 kWh/m^3 0.35 kWh/m^3 permeate for intake/ pretreatment, value of 0.16 kWh/m^3 calculated assuming 45% RO recovery	21,000 kwh/d	0.14 kWh/m^3 assumed, value extrapolated from 0.16 kWh for intake + 2 stage DMF [31], 0.018 kWh for 1 gravity DMF [35]	18,667 kWh/d	0.12 kWh/m^3 filtrate 0.03 kWh/m^3 for UF pretreatment 0.08 kWh/m^3 for intake 0.01 kWh/m^3 for screening	16,000 kWh/d
Water use	— (backwash with concentrate)	—	— (backwash with concentrate)	—	2.25% of pretreated water assumed (1000 m^3 1×/8 h)	3000 m^3/d	10% of filtrate	13,333 m^3/d
Potential ecologic impact	Impingement/ entrainment?	No	Impingement/ entrainment?	Yes	Impingement/ entrainment?	Yes	Impingement/ entrainment?	Yes
	Backwash discharged?	No	Backwash discharged?	No	Backwash discharged?	No	Backwash discharged?	Yes

backwashing. Another factor in this optimization problem is energy use. The TMP increases when deposits build up on the UF membranes, and backwashing is usually carried out at 2–4 times the filtration flow. Both cause an increase in energy demand. Moreover, some UF manufacturers employ an air scrub, in which pressurized air is introduced with the backwash water. This improves the cleaning process, possibly reducing chemical but increasing energy demand.

The hypothesis was that different UF pretreatment systems optimize their chemical use, energy use, CEB, and backwashing intervals differently. The objective was to compare different membrane types in terms of chemical, energy, and water use. Four different UF membranes were considered as alternatives; two are pressurized inside-out UF membranes (Norit, Inge), and two are pressurized outside-in membranes (Dow, Pall). A reference value was selected, based on an inventory of full-scale plants, pilot plants, and general manufacturer information, and used to calculate the scores. Preference was given to data from full-scale plants, if available (Table 2.22).

The value tree was further reduced from scenario 1, eliminating the criteria coagulant aid, acid (none used), and antiscalant. Antiscalant is usually added after the filtration stage and does not provide meaningful information about the different UF systems. Acid can be used in SWRO to lower the risk of scaling on the RO membranes and is not related to the filtration step in that case. Acid can also be used to control the pH during coagulation before the filtration step (as assumed in the previous MCA). However, most UF pretreatment systems use inline coagulation, for which acid use is not anticipated here to reduce complexity. The *ecologic impact* criterion was eliminated as all UF systems alike are assumed to operate on open intakes and to discharge their backwash waters without treatment. The remaining criteria were chlorine, SBS, and coagulant addition to the feedwater, chlorine use in backwashing, energy, and water use.

As in scenario 1, interval standardization and weighted summation was used to calculate the scores for a reference plant size of 60,000 m^3/d (Table 2.21). A perspectives analysis of the weights was carried out assuming (1) the original ranking of the environmental group; (2) equal weights for all main criteria; and (3) 66% weight allocated to energy, chemicals, or water use. The results were as follows:

- The inside-out UF systems (Norit, Inge) ranked at first and second positions and the outside-in systems (Dow, Pall) at third and fourth positions in all perspectives, except when 50% weight was given to water use. In that case, Pall ranked first, followed by Inge, Dow, and Norit in final position. As PVDF in outside feed formats (Dow and Pall membranes) allows the use of air scour, which reduces water consumption for backwashing, one would have expected Dow to score better. However, the same water use (7%) has been reported for backwashing and CIP of Dow and Inge systems (Table 2.22).
- Norit ranked first for the environmental group's perspective and the energy use perspective, while Inge ranked first for the equal weight and chemicals perspectives.

Table 2.21 MCA Reference Values and Calculated Scores: Revised Scenario 2

Criterion	UFa (Dow) Selected Reference Value	Calc. Score for 60,000 m^3/d	UFb (Pall) Selected Reference Value	Calc. Score for 60,000 m^3/d	UFc (Inge) Selected Reference Value	Calc. Score for 60,000 m^3/d	UFd (Norit) Selected Reference Value	Calc. Score for 60,000 m^3/d
For further details on reference values, see Table 2.22	Partial data from 3 full-scale and 2 pilot plants and empirical values of the manufacturer		Partial data from 1 full-scale and 3 pilot plants		Partial data from 1 full-scale and 1 pilot plant		Partial data from 3 full-scale and 4 pilot plants	
Chlorine (feedwater)	2 mg/L resulting in 0.5 mg/L chlorine residual	285 kg/d	1 mg/L	140 kg/d	1.5 mg/L	214 kg/d	15 mg/L for 20 min 1×/day	31 kg/d
Chlorine (backwash)	15 mg/L every 60 min (every backwash)	140 kg/d	10 mg/L every 30 min 500 mg/L every day	67 kg/d 69 kg/d	20 mg/L every 24 h	6 kg/d	200 mg/L chlorine CEB 1×/day	83 kg/d
Sodium bisulfite	1.5 mg/L 3× residual chlorine level	200 kg/d	0.75 mg/L 1 mg/L chlorine dose assumed to result in 0.25 mg/L chlorine residual, extrapolated from Dow value	100 kg/d	1.2 mg/L 1.5 mg/L chlorine assumed to result in 0.4 mg/L chlorine residual, extrapolated from Dow value	150 kg/d	— [normally not required (only if hypochlorite breakthrough is detected)]	—
Coagulant	— (none used)	—	0.4 mg/L Fe	56 kg/d	— (none used)	—	0.3 mg/L as Fe	44 kg/d
Energy use (UF only, without intake/ screening)	0.095 kWh/m^3	12,000 kWh/d	0.1 kWh/m^3 assumed	13,333 kWh/d	0.05 kWh/m^3	6667 kWh/d	0.03 kWh/m^3 filtrate	4000 kWh/d
Water use	7%; 93% UF recovery	9,333 m^3/d	5%; 95% UF recovery	6667 m^3/d	7%; 93% UF recovery	9,333 m^3/d	10%; 90% UF recovery	13,333 m^3/d

Table 2.22 UF Membrane Types and Their Pretreatment and Energy Requirements

Membrane Supplier	Chlorine to Feedwater	Coagulant	CEB	CIP	Water	Energy
Inside Feed—Always in Housing, Always Pressurized, Mostly PES (Norit, Hydranautics, Inge Membranes)						
Norit: partial data from 3 full-scale and 4 pilot plants	1.5–25 mg/L ClO_2 1×/day for 15–30 min in 1 full-scale plant reported; 2 pilot plants did not use chlorine, the rest did not specify this information	0.3 mg/L as Fe in fullscale plant, 6 out of 7 pilot/fullscale plants used coagulant in doses of 1.3–1.5 mg/L as Fe and 5 mg/L as $FeCl_3$ (≡2 mg/L Fe, 40% active ingredient)	200 mg/L Cl_2 1 min/d (full-scale), NaOCl used in 4 pilot plants, 2 reported a dose of 200 mg/L 1×/6–18 h 3 pilot and 1 full-scale plant used acid (HCl, pH 2.5) 1×/3–24 h	CIP: 0.5% oxalic acid and 0.25% ascorbic acid 1×/9 months (full-scale)	94% recovery/backwash (BW) every 45 min (pilot plant), 90% recovery (full-scale plant)	Full-scale plant: 0.03 kWh/m^3 filtrate plus 0.08 kWh/m^3 for intake and 0.01 kWh/m^3 for screening 0.12 kWh/m^3 total; 0.1–0.2 kWh/m^3 reported range in papers
Selected reference values[a]	Shock chlorination 15 mg/L for 20 min 1×/day	0.3 mg/L as Fe	200 mg/L Cl_2 1×/day	0.5% oxalic acid and 0.25% ascorbic acid CIP 1×/9month	90% recovery; BW every 45 min	0.12 kWh/m^3
Hydranautics: partial data from 2 full-scale and 5 pilot plants and empirical values of the manufacturer	3 full-scale plants gave the information that CaOCl is used for shock treatment at the intake; 1–5 mg/L manufacturer's empirical value for intermittent chlorine use (optionally)	3 pilot plants specified a dose of 0.1-0.7 mg/L as Fe; 1 full-scale plant found the design dose of 1 mg/L as Fe not necessary; manufacturer's empirical value: 0.5–1 mg/L as Fe	Cl_2 in 1 full-scale and (20 mg/L Cl_2 1×/2-8 h) 1 pilot plant (20–50 mg/L, pH 12, 1×/6 h); acid CEBs with H_2SO_4 (full-scale) and pH 1.5–2 (pilot); manufacturer's empirical value: 50 mg/L Cl_2 1×/4 h	CIP 1× every 1–2 months	95% recovery; BW every 45–70 min (1 full-scale), 94% recovery/BW every 30 min (1 pilot), i.e. 6% for BW and rinse air enhanced BW 1×/day	Manufacturer's empirical value: 0.1 kWh/m^3 of filtrate

Selected reference values[a]	Shock chlorination 3 mg/L assumed for 20 min 1×/week	None (0.5 mg/L Fe optionally)	20 mg/L Cl_2 1×/12 h H_2SO_4 (pH 1.5–2)	No dosing information CIP 1 × per month	95.5% recovery BW every 45 min	0.1 kWh/m^3
Inge: partial data from 1 full-scale and 1 pilot plant	1–2 mg/L in pilot plant, not specified in full-scale plant	Full-scale plant: none; pilot: 0.25 mg/L as Fe found to be sufficient	20 mg/L Cl_2 1×/day in full-scale plant; 50 mg/L NaOCl 1×/2–3h in pilot plant	CIP with HCl 1×/3d in full-scale plant if necessary (pH 12); citric acid (1% or 10,000 mg/L) in pilot plant	93% recovery in full-scale plant, BW every 45 min	Pilot plant: TMP of 0.1–0.2 bar without $FeCl_3$, 0.2–0.25 bar with $FeCl_3$, 0.05 kWh/m^3 filtrate at 0.25 bar full-scale plant: 0.3 bar without $FeCl_3$
Selected reference values[a]	1–2 mg/L optionally	None (0.25 mg/L Fe optionally)	20 mg/L Cl_2 1×/24 h	Acid CIP 1×/3 days HCl at pH 12	93% recovery BW every 45 min	0.05 kWh/m^3
Outside Feed—in Housing and Pressurized, Mostly PVDF (Dow/Omexell, Pall/Asahi)						
Dow: partial data from 3 full-scale and 2 pilot plants and empirical values of the manufacturer	Full-scale plants: (a) 2 mg/L (0.5 mg/L residual), followed by SBS; (b+c) no chlorine; pilot plant: 0–6 mg/L; empirical value: 0.5 mg/L as residual	No coagulant in full-scale plants; 0–2.4 mg/L as Fe used in pilot plants	Full-scale plant (a) also uses 15 mg/L NaOCl with every BW (1×/h); full-scale plant; (b) no CEB pilot plants: 20 mg/L NaOCl 1×/30 min and 500 mg/L 1×/12–120 h, respectively	2 full-scale plants use CIPs: (a) 500 mg/L NaOH and 2000 mg/L NaOCl followed by 3600 mg/L HCl once per year; (b) 20,000 mg/L oxalic acid followed by 2000 mg/L NaOCl	Empirical value: 7% for BW, CEB, CIP, 93% recovery; BW every 60 min with air scrub	Empirical value: TMP 0.5 bar 0.09–0.1 kWh/m^3 filtrate

(*continued*)

Table 2.22 ***(Continued)***

Membrane Supplier	Chlorine to Feedwater	Coagulant	CEB	CIP	Water	Energy
Selected reference values[a]	2 mg/L (0.5 mg/L residual) optionally	none	15 mg/L NaOCl 1×/h optionally	500 mg/L NaOH and 2000 mg/L NaOCl followed by 3600 mg/L HCl, 1×/year	93% recovery BW every 60 min	0.095 kWh/m^3
Pall: partial data from 1 full-scale and 3 pilot plants	1 pilot plant disinfects the MF filtrate with UV and chlorine (1–2 mg/L for 30 min/d), no information besides this	0.3–1.5 mg/L $FeCl_3$ (≡ 0.1–0.6 mg/L Fe, 40% active ingredient) in full-scale plant; pilots used 0–3.5 mg/L $FeCl_3$ (≡0–1.4 as Fe)	10 mg/L Cl_2/NaOCl or 25 mg/L H_2O_2 every 17–40 min and/or 500 mg/L Cl_2/NaOCl 1×/day in the pilot plants, no information given for the full-scale plant	CIPs with caustic, chlorine and acid in 2 pilot plants: 10,000 mg/L NaOH and 1000 mg/L NaOCl, 10,000 mg/L citric acid 1×/month	Recovery in pilot plants was 95–96% with a BW every 20–40 min, simultaneous air scrub reverse flow	No information
Selected reference values[a]	1 mg/L assumed	0.4 mg/L as Fe	10 mg/L NaOCl 2×/h 500 mg/L 1×/day	10,000 mg/L NaOH and 1000 mg/L NaOCl and 10,000 mg/L citric acid 1×/month	95% recovery BW every 30 min	0.1 kWh/m^3 assumed

[a] Selected reference values: Information from full-scale plants was preferred over pilot plants, where available. If several values were available, a representative midpoint value was selected. If only data from pilot plants was available, the lowest reported value was selected to account for the fact that pilot studies may test higher doses than would be used under real operating conditions.

- Pall ranked last for the environmental group's perspective, the energy and chemicals perspective, while Dow ranked last when equal weights were assigned to all criteria.

It is noteworthy that the Dow reference values are based on the full-scale plant in Wang Tan, China (9600 m^3/day UF capacity, 3120 m^3/day RO capacity), which does not use any coagulant but high amounts of chlorine, due to both feed and backwash chlorination. If one assumes the Dow Magong plant in Taiwan as baseline instead, which uses neither coagulant nor chlorine, the ranking of Dow changes to first position in the *chemicals perspective* and to second position behind Norit in the *equal-weights perspective*. A main difference between the Wang Tan and Magong plants is the frequency of CIP employed for the UF membranes, which are given with once a year for Wang Tan and once a month for Magong [40,64].

CIP is another decisive factor in the design of an integrated membrane system, and refers to the CIP frequency of both the UF and SWRO membranes. CIP has not been included in the MCA because representative quantitative values are difficult to establish. The CIP intervals for UF plants generally vary between once a month to once a year, as do the intervals for the SWRO membranes. Most typically, strong alkaline solutions up to 10,000 mg/L NaOH are used to clean the UF membranes, in combination with chlorine (2,000 mg/L NaOCl), and/or followed by acid (up to 10,000 mg/L, Table 2.22). For example, the CIP in Magong involves a 2% oxalic acid ($H_2C_2O_4$) solution followed by a 0.2% sodium hypochlorite (NaOCl) solution [31]. The cleaning solutions are circulated through the UF modules in concentrations equal to 20,000 mg/L $H_2C_2O_4$ and 2000 mg/L NaOCl, which would be equivalent to about 0.4 and 4 mg/L if they were to be used continuously [65].

2.6.4 Summary and Conclusions

As outlined in Section 2.1, a comparison of different intake and pretreatment systems for SWRO plants was chosen as the case study for the MCA. The approach deliberately eclipsed other desalination technologies and other aspects of designing a desalination plant in order to limit the complexity of the decision problem. This can be justified because the decision between the two main technologies—reverse osmosis and distillation—is a fundamental one and usually depends on the availability of a cheap energy source (therefore, it could have taken place in a preliminary design round). Other design considerations, such as the number of RO stages or posttreatment requirements, depend on the product-water specifications and are also independent from the pretreatment.

Even though the decision problem had been narrowed down in complexity and level of detail, it proved difficult to establish a complete set of relevant intake and pretreatment alternatives and operational criteria, and to gather the necessary input data for the MCA. On one hand, a promising alternative with a presumably low impact on the environment (Neodren) had to be eliminated because data were mostly confidential or unavailable. On the other hand, certain criteria, such as those used to measure water quality, simply proved inadequate. Although SDI had been included in the main MCA, it contained no real information that allowed for a differentiation between the alternatives. For some criteria, such as land use or costs, information was hardly available, so that values had to be estimated and extrapolated. Even for an important aspect such as energy use, which is widely discussed in the literature, it was difficult to establish reference values. A value of 0.1–0.2 kWh/m^3 is usually given for intake and pretreatment in general, which needed to be broken down into the specific demands of different media filters for conventional and membrane types for UF pretreatment.

The problems that were encountered may be attributed to the hypothetical nature of the study. Data uncertainty is an intrinsic problem of EIAs, and the same holds true for the MCA. In both cases, the results can only be as good as the underlying data. In this case, the MCA had been deliberately narrowed down after it was realized that the given data would not support a more complex decision problem, as would possibly be the case in a real-life scenario. The main obstacle in this study was clearly to establish a complete set of representative reference values for all alternatives on the various criteria. The confidence in some of the scores, despite an extensive literature search and communications with consultants and plant managers, is therefore limited and could be improved by more accurate data from operational plants. A sensitivity analysis was therefore carried out to investigate the effects of data uncertainty on the ranking. Despite this data uncertainty, a few general conclusions can nevertheless be drawn from the MCA results:

- Beachwells were found to be dominant over the other alternatives, that is, beachwells scored better or at least as well as the other alternatives on all criteria. Beachwells therefore also ranked first in the MCA, irrespective of the weights that were attached to the criteria to represent different perspectives.
- The MCA showed that the value judgments were generally similar in the two groups that represented plant operators and companies on one hand, and in the two groups that represented university and environmental backgrounds on the other hand. While plant operators gave highest priority to the criteria water quality and costs, the university and environmental groups gave highest priority to ecological impact. Chemical and energy use varied in importance between ranks two to four in all groups. The most important chemical substance was considered to be coagulant.
- The MCA ranking showed that conventional pretreatment II (with sludge treatment) was the second best alternative according to the value judgments of the expert groups. The reference regarding conventional pretreatment I was more distinct in the university and environmental groups (>0.1 total score difference) than in the plant operators group, in which conventional II ranked second by a narrow margin (0.02 difference), and in the company group, in which both conventional pretreatments had equal reference.

- Ultrafiltration ranked at the last position in all groups, although the difference to the next best alternative, i.e., conventional I, was small in all four groups (0–0.4 difference).
- A sensitivity analysis of weights showed that conventional pretreatment I would rank before conventional II if the weights of the criteria cost or energy were to be increased to 50%. This can be attributed to the assumption in this study that sludge treatment causes a 10% increase in cost and energy use. UF would rank second if the weights of the criteria chemical or land use were to be increased to 50%.
- A sensitivity analysis of scores, assuming 50% and 100% score uncertainty, showed that beachwells had a high probability of ranking first. Either conventional pretreatment I or II was likely to rank second. UF pretreatment had a high probability of ranking on third or fourth position. In order to reverse a ranking, either discharge or intake effects had to be eliminated, or the scores for the various aspects of resource consumption had to be reduced by ~20–60%. A decisive criterion was energy use. If reduced by ≤11%, it brought about a change in the ranking between the alternatives conventional pretreatments I and II, which can again be attributed to the assumed higher energy use for conventional pretreatment II.

The MCA ranking and sensitivity analysis support the conclusion that a beachwell is the preferred intake and pretreatment alternative for SWRO plants. Where a beachwell is not feasible, for example, due to an impermeable geologic substratum or due to size restrictions, the preferred option would most likely be a conventional pretreatment, either with or without sludge treatment, followed by ultrafiltration. The results of the revised scenario 1, in which the specific operational conditions of selected full-scale plants were used as baseline for the MCA instead of selected literature values, also support these findings. Natural intake systems (i.e., Neodren) were also the preferred choice for larger SWRO plants, followed by conventional and UF pretreatment.

The results of the revised scenario 2, in which different UF membranes and modes of operation were compared, showed that the successful operation of an integrated membrane system is essentially an optimization problem, which has to be solved plant- and site-specifically, balancing energy demand, chemical use, filtration time, CEB and CIP intervals. Similarly to conventional pretreatment systems, which have diversified into various pretreatment options over the years ranging from minimalist to an extensive three-stage design, not all UF systems are alike, let alone their modes of operation. The range of possible operation modes of UF systems shows that a more sustainable approach with a low energy and chemical demand is feasible in principle, and that this approach could be altogether equal to or even better than a conventional pretreatment.

This MCA should be understood as an exercise, which can always be revised and refined in the light of better data and new information. The results are only valid for the given alternatives and criteria. The inclusion of new or modified alternatives and criteria may alter the ranking results. For example, the alternative

UF pretreatment will score better on the ecology criterion if one assumes that the backwash water is treated instead of discharged, i.e., by modifying the present assumptions for the alternative. Although discharge seems to be the common practice of the few operational UF-SWRO plants today, future projects in Australia or California may require a treatment of the backwash. Also, UF might perform better if better indicators for water quality were available. Pilot studies often found UF pretreatment superior to conventional pretreatment in difficult feed waters, which may have secondary beneficial effects on plant operations, such as lower cleaning frequencies of SWRO membranes, lower energy demand, or lower operating costs. However, quantifying these effects in an MCA would still be highly speculative at the moment, due to a lack of reliable indicators for both water quality and data.

Natural intake and pretreatment systems, such as beachwells or subseabed infiltration galleries, performed best in this MCA. One recommendation for MCAs is therefore to include a criterion which considers the feasibility of a natural system during site selection. However, a natural intake will not be feasible for all projects, for example, because of size limitations of the beachwell, or may not be the preferred alternative in all cases, for example, due to anoxic or anaerobic conditions in the ground water. Anoxic or anaerobic feedwater from a beachwell poses the risk that if oxygenated, iron or manganese flocs may form, which need to be retained by a filtration step. If the solids content of the filter backwash is too high to allow for discharge, treatment of the backwash water may be required. In that way, the advantages of a beachwell over an open intake with conventional pretreatment or ultrafiltration could be diminished.

An acceptable alternative where a subsurface intake is not possible is an open intake with conventional pretreatment, followed by an open intake with ultrafiltration. However, only conventional pretreatment systems have been implemented and successfully used as a pretreatment for large SWRO plants to date. Experiences with regard to subsurface intakes, which would probably have to be horizontal drains in the offshore sediments to provide a sufficient feed flow, and ultrafiltration for large SWRO projects is lacking.

The conclusion of this MCA that conventional pretreatment systems altogether outrank ultrafiltration is supported by a lifecycle analysis (LCA) carried out by Beery and Repke [167]. Input data for the two analyses were exchanged and discussed in several meetings. While the MCA allows for an objective analysis of the performance of alternatives as well as subjective value judgments with regard to multiple criteria, the LCA primarily evaluates the performance with regard to chemical and energy streams over the lifecycle of a project. The researchers at TU Berlin concluded that "the gravity media filter is currently still a more sustainable technological solution" when working with "nondifficult waters," and call for "further optimization of UF design and operation concerning the overall process sustainability" [167].

The potential for future improvements is high, given the fact that UF pretreatment is still a young technology in its learning stage. In this regard, it may also learn from past shortfalls of conventional pretreatment systems, such as that backwash waters and spent cleaning solutions should be collected and treated, rather than discharged. To increase the acceptance of UF, most membrane suppliers and researchers have focused on operational and economic aspects so far. According to the value judgment of plant operators and company representatives in this case study, these aspects have a higher priority than ecological aspects, energy demand, and chemical use. However, it may be shortsighted to neglect the latter because they are considered to be of secondary importance. Obviously, UF technology will never achieve a real breakthrough over conventional pretreatment if it does not perform equally well or better in terms of water quality. However, environmentally friendly designs may increase the acceptance and could tip the balance in favor of UF despite increased costs, especially in emerging markets such as Australia. In Australia, everything that was required has been done so far "to ensure desal works environmentally, with price being a secondary consideration" [168].

It can be concluded that MCA is a suitable tool for assessing both the site and process alternatives for desalination plants. As outlined in Section 2.1, the approach has previously been used and proved useful in selecting desalination plant sites. This study showed that MCA can also be used to compare and rank process alternatives, despite the limitations of a hypothetical study. It is anticipated that MCA will prove to be more powerful in real applications, in which uncertainty is limited to weights and scores, and where the decision maker does not have to worry about the additional uncertainty that is inherent in a study that is basically hypothetical in nature.

REFERENCES

1. IDA, *International Desalination Association Worldwide Desalting Plant Inventory*, No. 20, MS Excel (CD ROM), Media Analytics Ltd., Oxford, UK, 2007.
2. IDA, *International Desalination Association Worldwide Desalting Plant Inventory*, No. 22, MS Excel spreadsheet, Media Analytics Ltd., Oxford, UK, 2010.
3. Pearce GK, The Mayor of Londons challenge to the Beckton desalination plant, *Proc. IDA World Congress Desalination and Water Reuse*, The Palm, Dubai, 2009.
4. Moore BJ, Malfeito BJ, Zorrilla J, An overview of the 150,000 m^3/day Beckton desalination plant in London, *Proc. IDA World Congress Desalination and Water Reuse*, The Palm, Dubai, 2009.
5. UNESCO-WWAP, *Water—a Shared Responsibility, the United Nations World Water Development Report 2*, United Nations Educational, Scientific and Cultural Organization (UNESCO) and United Nations World Water Assessment Programme (WWAP) (www.unesco.org/water/wwap/wwdr/wwdr2/; last accessed 12/09/09), 2006.
6. Lattemann S, Höpner T, *Seawater Desalination—Impacts of Brine and Chemical Discharges on the Marine Environment*, Balaban Desalination Publications, L'Aquila, 2003.
7. Lattemann S, Guidelines for the environmental sound management of seawater desalination plants in the Mediterranean region, in *Seawater Desalination in the Mediterranean—Assessment and Guidelines*, *MAP Technical Reports Series No. 139*, UNEP/MAP/MEDPOL, Athens, 2003, p. 83.

8. GWI, www.desaldata.com, subscription-based internet platform incorporating the IDA desalination plants inventory, 2008.
9. Media Analytics Ltd., Israel unveils drought-busting water strategy, *Global Water Intell.*, **9**(6):18 (2008).
10. Media Analytics Ltd., Israel's brackish water challenge, *Global Water Intell.*, **9**(7):24 (2008).
11. Voutchkov N, Advances and challenges of seawater desalination in California, Proc. *IDA World Congress Desalination and Water Reuse*, Maspalomas, Gran Canaria, 2007.
12. Pankratz T, Melbourne desal shortlists two. (water desalination report by Media Analytics Ltd.), Global Water Intell., **44**(36):1 (2008).
13. Höpner T, Windelberg J, Elements of environmental impact studies on coastal desalination plants, *Desalination* **108**:11–18 (1996).
14. Sheppard C, Al-Husiani M, F, Al-Jamali, F, Al-Yamani F, Baldwin R, Bishop J, Benzoni F, Dutrieux E, Dulvy N, Durvasula S, Jones D, Loughland R, Medio D, Nithyanandan M, Pillingm G, Polikarpov I, Price A, Purkis S, Riegl B, Saburova M, Samimi Namin K, Taylor O, Wilson S, Zainal K. The Gulf: A young sea in decline, *Marine Pollution Bulletin* **60**:13–38 (2010).
15. UNEP, *Overview on Land-Based Sources and Activities Affecting the Marine Environment in the ROPME Sea Area*, UNEP Regional Seas Reports and Studies No. 168, UNEP/GPA Coordination Office, The Netherlands, and Regional Organization for the Protection of the Marine Environment (ROPME), Kuwait, 1999.
16. Regional Organization for the Protection of the Marine Environment (ROPME), *State of the Marine Environment Report* (CD ROM), 2003.
17. Cotruvo J, Voutchkov N, Fawell J, Payment P, Cunliffe D, Lattemann S, *Desalination Technology: Health and Environmental Impacts*, CRC Press, 2010.
18. Wilf M, Awerbuch L, Bartels C, Mickley M, Pearce G, Voutchkov N, *The Guidebook to Membrane Desalination Technology*, Balaban Desalination Publications, L'Aquila, 2007.
19. Committee on Advancing Desalination Technology, *Desalination: A National Perspective*, Water Science and Technology Board, National Research Council of the National Academies, National Academies Press, Washington, DC. (www.nap.edu/catalog/12184.html; last accessed 12/09/09), 2008.
20. Watson I, Morin O, and Henthorne L, *Desalination Handbook for Planners*, 3rd ed., Report 72, US Bureau of Reclamation, Desalination and Water Purification Research and Development Program, 2003.
21. Cartier G, Corsin P, Description of different water intakes for SWRO plants, *Proc. IDA World Congress Desalination and Water Reuse*, Maspalomas, Gran Canaria, 2007.
22. Damitz B, Furukawa D, Toal J, *Desalination Feasibility Study for the Monterey Bay Region*, final report prepared for the Association of Monterey Bay Area Governments (AMBAG) (www.ambag.org; last accessed 12/09/09), 2006.
23. Wang S, Leung E, Cheng R, Tseng T, Vuong D, Carlson D, Henson J, Veerapaneni S, Ocean floor seawater intake and discharge system, *Proc. IDA World Congress Desalination and Water Reuse*, Maspalomas, Gran Canaria, 2007.
24. Allen J, Cheng R, Tseng T, Wattier K, Update for the pilot and demonstration-scale research evaluation of under-ocean floor seawater intake and discharge, *Proc. AWWA Membrane Technology Conf.*, Tennesse (USA), Memphis, 2009.

25. Poseidon Resources Corporation, *Carlsbad Seawater Desalination Project, Alternative Project Intake Source Water Collection Systems. Beach Wells, Infiltration Galleries and Seabed Filtration systems*, 2004.

26. Bascompte D, Peters T, Seawater intake and pre-treatment using Neodren-technology based on sub-seabed drains, *Proc. IDA World Congress Desalination and Water Reuse*, Maspalomas, Gran Canaria, 2007.

27. Lattemann S, *Development of an Environmental Impact Assessment and Decision Support System for Seawater Desalination Plants*, dissertation, CRC Press/Balkema (http://repository.tudelft.nl), 2010.

28. Sanz M, Cremer G, Beltran F, Bonnélye V, Del Campo I, Europe's largest SWRO plant takes shape, *Int. Desal. Water Reuse Q*. **17**(4):12–18 (2008).

29. Pascaline P, *Algeria Seawater Reverse Osmosis Desalination Plants: A Study of the Intake and Pre-treatment Options*, Masters of Mechanical Engineering Final Year Project Report, 2007.

30. Voutchkov N, Thorough study is key to large beach-well intakes, *Int. Desal. Water Reuse Q*. **14**(1):16–20 (2002).

31. García Molina V, personal communication, Dow Water and Process Solution, Germany, 2009.

32. Visiers M, personal communication, Acciona Agua, Spain, 2008.

33. Bonnélye V, Guey L, Del Castillo J, UF/MF as RO pre-treatment: The real benefit, *Proc. EDS Conf. Desalination and the Environment*, Halkidiki, Greece, 2007.

34. Busch M, Chu R, Rosenberg S, Novel trends in dual media membrane systems for seawater desalination: Minimum primary and low environmental aspect treatment scheme, *Proc. IDA World Congress Desalination and Water Reuse*, The Palm, Dubai, 2009.

35. Ludwig H, Energy consumption of reverse osmosis seawater desalination—possibilities for its optimisation in design and operation of SWRO plants, *Proc. EDS Conf. Desalination for the Environment*, Baden Baden, Germany, 2009.

36. Wilf M, Schierach M, Improved performance and cost reduction of RO seawater systems using UF pretreatment, *Desalination* **135**:61–68 (2001).

37. Pankratz T, Red tide demo results are encouraging water Desalination Report by Media Analytics Ltd., *Global Water Intell*. **45**(22):1 (2009).

38. Waly T, Saleh S, Kennedy M, Witkamp G, Amy G, Schippers J, Will calcium carbonate really scale in seawater reverse osmosis? *Proc. Euromed 2008 (conf.), Desalination Cooperation among Mediterranean Countries of Europe and the MENA Region, Dead Sea, Jordan*, 2008.

39. Pearce G, The case for UF/MF pretreatment to RO in seawater applications, *Desalination* **203**:286–295 (2007).

40. Busch M, Chu R, Kolbe U, Meng Q, Li S, Integrated ultrafiltration and reverse osmosis membrane system for seawater desalination: 1000 days field experience with DOW^{TM} UF and FILMTEC^{TM} technology in the WangTan DaTang power plant, *Proc. Euromed 2008 (conf.), Desalination Cooperation among Mediterranean Countries of Europe and the MENA Region, Dead Sea, Jordan*, 2008.

41. Pearce G, Bartels C, Wilf M, Improving total water cost of desalination by membrane pre-treatment, *Proc. IDA World Congress Desalination and Water Reuse*, Maspalomas, Gran Canaria, 2007.

42. Gille D, Czolkoss W, Ultrafiltration with multi-bore membranes as seawater pre-treatment, *Desalination* 182:301–307 (2005).

43. Katsube M, Yagi T, Nishida M, Fujiwara N, Effective chlorine injection for maintaining stable operation in seawater desalination plant, *Proc. IDA World Congress Desalination and Water Reuse*, Maspalomas, Gran Canaria, 2007.

44. Knops F, te Linteli R, Long term operation experience of Seaguard UF pretreatment to SWRO in the Mediterranean region, *Proc. Euromed 2008 (conf.), Desalination Cooperation among Mediterranean Countries of Europe and the MENA region, Dead Sea, Jordan*, 2008.

45. Glueckstern P, Priel M, Wilf M, Field evaluation of capillary UF technology as a pretreatment for large seawater RO systems, *Desalination* **147**:55–62 (2002).

46. Wolf P, Siverns S, Monti S, UF membranes for RO desalination pretreatment, *Desalination* **182**:293–300 (2005).

47. Pankratz T, Membrane pretreatment for SWRO (Water Desalination Report by Media Analytics Ltd), *Global Water Intell.*, **45**(17):2 (2009).

48. Bu-Rashid K, Czolkoss W, Pilot tests of multibore UF membrane at Addur SWRO desalination plant, Bahrain, *Desalination* **203**:229–242 (2007).

49. Lorain O, Hersant B, Persin F, Grasmick A, Brunard N, Espenan J, Ultrafiltration membrane pre-treatment benefits for reverse osmosis process in seawater desalting. Quantification in terms of capital investment cost and operating cost reduction. *Desalination* **203**:277–285 (2007).

50. van Hoof S, Minnery J, Mack B, Dead-end ultrafiltration as alternative pre-treatment to reverse osmosis in seawater desalination: A case study. *Desalination* **139**:161–168 (2001).

51. Edzwald JK, Tobiason JE, Enhanced coagulation: US requirements and broader view. *Water Science and Technology* **40**:63–70 (1999).

52. Crisp G, Seawater desalination in Australia. The Perth experience—a sustainable solution, *Proc. Conf. Water, Finance and Sustainability. New Directions for a Thirsty Planet*, London, 2008.

53. Goebel O, Überblick über Entsalzungsverfahren und Einbindungsmöglichkeiten für erneuerbare Energien, *Clean Energy Power Fachtagung, Regenerative Meerwasser-entsalzung*, ICC Berlin, 2007.

54. von Medeazza G, "Direct" and socially-induced environmental impacts of desalination. *Desalination* **185**(1–3):57–70 (2005).

55. Tal A, Seeking sustainability: Israel's evolving water management strategy. *Science* **313**(5790):1081–1084 (2006).

56. City of Carlsbad and Poseidon Resources, *Environmental Impact Report for the Carlsbad Seawater Desalination Facility*, 2005.

57. Sanz M, Stover R, Low energy consumption in the Perth seawater desalination plant, *Proc. IDA World Congress Desalination and Water Reuse*, Maspalomas, Gran Canaria, 2007.

58. Faigon M, Koutsakos E, Van-Rooij F, Liberman B, Moxie D, Heferr D, Larnaca desalination plant 51,000 m^3/day—6 years of operation, *Proc. IDA World Congress Desalination and Water Reuse*, Maspalomas, Gran Canaria, 2007.

59. Melbourne Water and GHD, *Melbourne Augmentation Program Seawater Desalination Feasibility Study*, 2007.

60. MacHarg J, Seacord T, Sessions B, ADC baseline tests reveal trends in membrane performance, *Int. Desal. Water Reuse Q*. **18**(2):30–39 (2008).

61. Cardona E, Piacentino A, Marchese F, Energy saving in two-stage reverse osmosis systems coupled with ultrafiltration processes, *Desalination* **184**:125–137 (2005).

62. Fritzmann C, Niewersch C, Jansen A, Wintgens T, Melin T, Comparative life cycle assessment study of pretreatment alternatives for RO desalination, *Proc. IDA World Congress Desalination and Water Reuse*, Maspalomas, Gran Canaria, 2007.

63. Knops F, personal communication, X-Flow BV, The Netherlands, 2009.

64. García Molina V, Chang R, Busch M, First year performance review of Magong UF/RO seawater desalination plant, *Proc. EDS Conf. Desalination for the Environment*, Baden Baden, Germany, 2009.

65. Busch M, personal communication, Dow Water and Process Solution, Germany, 2009.

66. Rapenne S, Port C, Roddy S, Croué J, Pre-treatment prior to RO for seawater desalination: Sydney pilot-scale study, *Proc. IDA World Congress Desalination and Water Reuse*, Maspalomas, Gran Canaria, 2007.

67. Côté P, Cadera J, Coburna J, Munro A, A new immersed membrane for pretreatment to reverse osmosis, *Desalination* **139**:229–236 (2001).

68. Czolkoss W, personal communication, Taprogge GmbH, Germany, 2009.

69. Berg P, personal communication, Inge Watertechnologies AG, Germany, 2009.

70. Niepelt A, Bleninger T, Jirka G, Desalination brine discharge modelling. Coupling of hydrodynamic models for brine discharge analysis, *Proc. 5th Int. Conf. Marine Waste Water Discharges and Coastal Environment*, Croatia, 2008.

71. Victorian Government, Department of Sustainability and Environment, *Victorian Desalination Project Environment Effects Statement*, Vol 3, Chap. 2 (desalination plant project description), 2008.

72. Sydney Water, GHD, Fichtner, *Environmental Assessment for Sydney's Desalination Project*, 2005.

73. WEC, *Perth Metropolitan Desalination Proposal Environmental Protection Statement*, prepared by Welker Environmental Consultancy for the Water Cooperation, 2002.

74. GCD Alliance, *Material Change of Use Application, Gold Coast Desalination Project* (www.goldcoast.qld.gov.au; last accessed 12/09/09), 2006.

75. Mickley M, *Membrane Concentrate Disposal: Practices and Regulation*, 2nd ed., Desalination and Water Purification Research and Development Program Report 123, US Dept. Interior, Bureau of Reclamation, Water Treatment Engineering and Research Group, Denver, CO, 2006.

76. City of Huntington Beach (California, USA) and Poseidon Resources, *Environmental Impact Report for the Huntington Beach Seawater Desalination Facility*, 2005.

77. Tampa Bay Water (www.tampabaywater.org.).

78. Einav R, Lokiec F, Environmental aspects of a desalination plant in Ashkelon, *Desalination* 156(1–3):79–85 (2003).

79. Sinapsa Ltd, *Summary of the Environmental Document for the Desalination Plant in Hadera (Israel)*, 2007.

80. Alspach B, The evolution of SWRO co-location in the US: Analysis and alternatives, *Proc. IDA World Congress Desalination and Water Reuse*, Maspalomas, Gran Canaria, 2007.

81. Woodward Clyde Consultants, *Final Environmental Impact Report (FEIR), City of Santa Barbara and Ionics Incorporated's Temporary Emergency Desalination Project*, State Clearing House Report 9010859.
82. Cotruvo C, Voutchkov N, Fawell J, Payment P, Cunliffe D, Lattemann S, *Desalination Technology, Health and Environmental Impacts*, CRC Press Taylor and Francis Group and IWA Publishing, Boca Raton, London and New York, 2010.
83. WHO, *Guidelines for Drinking-Water Quality*, 3rd ed., World Health Organization, Geneva (www.who.int/water_sanitation_health/dwq/gdwq3rev/en/; last accessed 12/09/09), 2006.
84. California Water Desalination Task Force, *Draft Feedwater Intake Working Paper*, 2003.
85. Bleninger T, Jirka G, Al-Barwani H, Purnama A, Lattemann S, Höpner T, *Environmental Planning, Prediction and Management of Brine Discharges from Desalination Plants*, Middle East Desalination Research Center Project Report Series No. 07-AS-003 (www.ifh.uni-karlsruhe.de/science/envflu/research/brinedis/default.htm; last accessed 12/09/09), 2008.
86. Latorre M, Environmental impact of brine disposal on Posidonia seagrasses, *Desalination* **182**(1–3):517–524 (2005).
87. Walker D, Regional studies: Seagrass in Shark Bay, the foundations of an ecosystem, in Larkum A, McComb A, Sheperd S, ed., *Biology of Seagrasses. A Treatise on the Biology of Seagrasses with Special Reference to the Australian Region*, Elsevier, Amsterdam, 1989, pp. 182–210.
88. Argyrou M, *Impact of Desalination Plant on Marine Macrobenthos in the Coastal Waters of Dhekelia Bay, Cyprus*, Dept. Fisheries, Ministry of Agriculture, Natural Resources and Environment, Cyprus, 1999.
89. Coles S, McCain J, Environmental factors affecting benthic infaunal communities of the Arabian Gulf, *Marine Environ. Res*. **29**:289–315 (1990).
90. Christie S, Bonnélye V, Perth, Australia: Two-year feed back on operation and environmental impact, *Proc*. IDA World Congress Desalination and Water Reuse, The Palm, Dubai, submitted draft.
91. C, personal communication, Sydney Water Corporation, Australia, 2008.
92. Trousdale S, Henderson E, Sydney's desalination plant: Addressing environmental issues using innovative design, planning and monitoring, *Proc. IDA World Congress Desalination and Water Reuse*, The Palm, Dubai, 2009.
93. Cannesson N, Johnstone P, Mitchell M, Boerlage S, Community, environmental and marine impact minimisation at the Gold Coast desalination plant, *Proc. IDA World Congress on Desalination and Water Reuse*, The Palm, Dubai, 2009.
94. Olympic Dam EIS Project, *Olympic Dam Expansion Draft Environmental Impact Statement*, Chap. 16, Marine environment. (www.bhpbilliton.com/bb/odxEis/project Components/waterSupply.jsp; last accessed 12/09/09), 2009.
95. Sánchez-Lizaso J, Romero J, Ruiz J, Gacia E, Buceta J, Invers O, Fernández Torquemada Y, Mas J, Ruiz-Mateo A, Manzanera M, Salinity tolerance of the Mediterranean seagrass Posidonia oceanica: Recommendations to minimize the impact of brine discharges from desalination plants. *Desalination* **221**:602–607 (2008).
96. California Regional Water Quality Control Board, *Waste Discharge Requirements for the Poseidon Resources Corporation, Carlsbad Desalination Plant Project, Discharge*

to the Pacific Ocean via the Encina Power Station Discharge Channel, NPDES Report CA0109223, 2006.

97. Okinawa Bureau for Enterprises, *Environmental Impact Assessment Report for the Seawater Desalination Project in Okinawa, Japan, Executive Summary*, 1997.
98. ANZECC/ARMCANZ, *Australian and New Zealand Guidelines for Fresh and Marine Water Quality*, Environment and Conservation Council and Agriculture and Resource Management Council of Australia and New Zealand (www.mincos.gov.au; last accessed 12/09/09), 2000.
99. Voutchkov N, Salinity tolerance evaluation methodology for desalination plant discharge, *Desalination Water Treat*. 1:68–74 (2009).
100. USEPA, *Methods for Measuring the Acute Toxicity of Effluents and Receiving Waters to Freshwater and Marine Organisms*, US Environmental Protection Agency, Office of Water (www.epa.gov/waterscience/methods/wet/; last accessed 12/09/09), 2002.
101. USEPA, *Short-Term Methods for Estimating Chronic Toxicity to Marine and Estuarine Organisms*, US Environmental Protection Agency, Office of Water (www.epa.gov/waterscience/methods/wet/; last accessed 12/09/09), 2002.
102. Hose J, Fiore D, Parker H, Sciarrotta T, Toxicity of chlorine dioxide to early life cycle stages of marine organisms, *Bull. Environ. Contam. Toxicol*. **42**:315–319 (1989).
103. Raluy R, Serra L, Uche J, Valero A, Life-cycle assessment of desalination technologies integrated with energy production systems. *Desalination* **167**(1–3):445–458 (2004).
104. Raluy R, Serra L, Uche J, Life cycle assessment of water production technologies—Part 1: Life cycle assessment of different commercial desalination technologies (MSF, MED, RO), *Int. J. Life Cycle Assess*. **10**(4):285–293 (2005).
105. Raluy R, Serra L, Uche J, Life cycle assessment of MSF, MED and RO desalination technologies, *Energy* **31**(13):2025–2036 (2006).
106. Klein G, Krebs M, Hall V, O'Brien T, Blevins B, *California's Water Energy Relationship*, California Energy Commission, final staff report (www.energy.ca.gov/2005publications/CEC-700-2005-011/CEC-700-2005-011-SF.PDF; last accessed 12/09/09), 2005.
107. Cooley H, Gleick P, Wolff G, *Desalination with a Grain of Salt. A California Perspective*, Pacific Institute, California (www.pacinst.org/reports/desalination/desalination_report.pdf; last accessed 12/09/09), 2006.
108. Voutchkov N, Carlsbad project plan for green SWRO, *Int. Desal. Water Reuse Q*. **18**(3).30–36 (2008).
109. Stover R, Affordable desalination and osmotic power, *Proc. Osmosis Membrane Summit*, Amsterdam, The Netherlands, 2008.
110. von Medeazza G, Water desalination as a long-term sustainable solution to alleviate global freshwater scarcity? A north-south approach. *Desalination* **169**:287–301 (2004).
111. Lattemann S, Mancy K, Khordagui H, Damitz B, Leslie G, *Desalination—Resource and Guidance Manual for Environmental Impact Assessments*, United Nations Environment Programme (UNEP), Nairobi, Kenya (www.unep.org.bh/Publications/Type7.asp; last accessed 12/09/09), 2008.
112. Port C, Roddy S, Trousdale S, Sydney's SWRO project—working towards sustainability, *Int. Desal. Water Reuse Q*. **18**(4):21–33 (2009).

113. European Commission, *Spain. Energy Mix Fact Sheet*, Eurostat (ec.europa.eu/energy/energy_policy/doc/factsheets/mix/mix_es_en.pdf; last accessed 12/09/09), 2007.
114. Australian Government, Dept. Climate Change, *National Greenhouse Accounts (NGA) Factors* (www.climatechange.gov.au/workbook/index.html; last accessed 12/09/09), 2008.
115. German Environmental Protection Agency (Umweltbundesamt), *National Inventory Report for the German Greenhouse Gas Inventory 1990-2007 under the United Nations Framework Convention on Climate Change*. (www.umweltbundesamt.de/klimaschutz-e/index.htm; last accessed 12/09/09), 2009.
116. Esser A, Möst D, Rentz O, Energy demand of German households and saving potential, *Int. J. Environ. Technol. Manage.* **9**(4):378–389 (2008).
117. The World Commission on Dams, *Dams and Development, a New Framework for Decision Making*, Earthscan Publications Ltd., London and Sterling (www.dams.org; last accessed 12/09/09), 2000.
118. Crisp G, Rhodes M, Procter D, Perth seawater desalination plant. Blazing a sustainability trail, *Proc. IDA World Congress Desalination and Water Reuse*, Maspalomas, Gran Canaria, 2007.
119. URS Australia Pty Ltd., *Southern Seawater Desalination Project: Draft Commonwealth Public Environment Report*, 2009.
120. Earthday Network, ww2.earthday.net.
121. Webber M, Catch-22: Water vs. energy, *Sci. Am.* **Earth 3.0**:34–41 (2008).
122. Australian Bureau of Meteorology and CSIRO, *Climate Change in Australia. Observed Changes and Projections* (www.climatechangeinaustralia.gov.au/documents/resources/Summary_brochure.pdf; last accessed 12/09/09), 2009.
123. Bremere I, Kennedy M, Stikker A, Schippers J, How water scarcity will affect the growth in the desalination market in the coming 25 years, *Desalination* **138**(1–3):7–15 (2001).
124. GWI, *Desalination Markets 2007—a Global Industry Forecast* (CD ROM), Global Water Intelligence, Media Analytics Ltd., Oxford, UK (www.globalwaterintel.com), 2007.
125. Lattemann S, Protecting the marine environment, in Micale G, Cipollina A, Rizzuti L, eds., *Seawater Desalination—Green Energy and Technology*, Springer, 2009, pp. 271–297.
126. Sydney Water, GHD, Fichtner, *Preferred Project Report for Sydney's Desalination Project*, 2006.
127. Shannon M, Bohn P, Elimelech M, Georgiadis J, Marinas B, Mayes A, Science and technology for water purification in the coming decades. *Nature* **452**(20):301–310 (2008).
128. Crisp G, Australias first big plant thinks green, *Int. Desal. Water Reuse Q.* **16**(3):18–27 (2006).
129. Environmental Technology (en wikipedia.org/wiki/green_technology) *Wikipedia* (www.en.wikipedia.org).
130. European Commission, *Reference Document on Best Available Techniques in the Industrial Cooling Systems*, European Integrated Pollution Prevention and Control (IPPC) Bureau (eippcb.jrc.es/reference; last accessed 12/09/09), 2001.

131. Abousamra F, personal communication, MEDPOL Programme, UNEP, Greece, 2008.
132. Safrai I, Zask A, Reverse osmosis desalination plants—marine environmentalist regulator point of view, *Desalination* 220:72–84 (2008).
133. I, Amir R, personal communication, Ministry of the Environment, Israel, 2008, 2009.
134. USEPA, *Effluent Limitations Guidelines and Standards*, US Environmental Protection Agency, Office of Water (cfpub.epa.gov/npdes/techbasedpermitting/effguide.cfm; last accessed 12/09/09), 2009.
135. Mooij C, Hamma water desalination plant, *Desalination* **203**(1–3):107–118 (2007).
136. Media Analytics Ltd, Spain fast-tracks EIA process, *Global Water Intell*. **9**(9):20 (2008).
137. Schiffler M, Perspectives and challenges for desalination in the 21st century, *Desalination* **165**(1–3):1–9 (2004).
138. Ambrose R, Schmitt R, Osenberg C, Predicted and observed environmental impacts. Can we foretell ecological change?, in Schmitt R, Osenberg C, eds., *Detecting Ecological Impacts. Concepts and Applications in Coastal Habitats*, Academic Press, 1996, pp. 345–369.
139. Stewart-Oaten A, Problems in the analysis of environmental monitoring data, in Schmitt R, Osenberg C, eds., *Detecting Ecological Impacts. Concepts and Applications in Coastal Habitats*, Academic Press, 1996, pp. 109–131.
140. Underwood A, Beyond BACI: The detection of environmental impacts on populations in the real, but variable, world, *J. Exp. Marine Bio. Ecol*. **161**:145–178 (1992).
141. Bamber R, Environmental impact assessment: The example of marine biology and the UK power industry, *Marine Pollut. Bull*. **21**(6): 270–274 (1990).
142. Janssen R, On the use of multi-criteria analysis in environmental impact assessment in The Netherlands, *J. Multi-Criteria Decision Anal*. **10**(2):101–109 (2001).
143. Lahdelma R, Salminen P, Hokkanen J, Using multicriteria methods in environmental planning and management, *Environ. Manage*. **26**(6):595–605 (2000).
144. Beinat E, Multi-criteria analysis for environmental management, *J. Multi-Criteria Decision Anal*. **10**(51):51 (2001).
145. Linkov I, Satterstrom F, Kiker G, Batchelor C, Bridges T, Ferguson E, From comparative risk assessment to multi-criteria decision analysis and adaptive management: Recent developments and applications, *Environ. Int*. **32**(8):1072–1093 (2006).
146. Stewart T, Joubert A, Janssen R, MCDA framework for fishing rights allocation in South Africa, in *Group Decision and Negotiation, Amsterdam*, Springer, Amsterdam (online publication, DOI 10.1007/s10726-009-9159-9), 2009.
147. Kiker G, Bridges T, Varghese A, Seager T, Linkov I, Application of multi-criteria decision analysis in environmental decision making, *Integr. Environ. Assess. Manage*. **1**(2):95–108 (2005).
148. Balasubramaniam A. Voulvoulis N, The appropriateness of multicriteria analysis in environmental decision-making problems, *Environ. Technol*. **26**(9):951–962 (2005).
149. S, personal communication, South East Queensland Desalination Company Pty Ltd., Sure Smart Water, Australia, 2009.

150. Bornman T, Klages N, *Ecological Pre-feasibility Study of the Establishment of a Desalination Plant at Knysna*, Inst. Environmental and Coastal Management, Univ. Port Elizabeth, South Africa, 2004.
151. Bornman T, Klages N, *Ecological Pre-feasibility Study of the Establishment of a Desalination Plant at Plettenberg Bay*, Inst. Environmental and Coastal Management, Univ. Port Elizabeth, South Africa, 2004.
152. Keeney R, Raiffa H, *Decisions with Multiple Objectives: Preferences and Value Trade-offs*, Cambridge Univ. Press, 1993
153. Herwijnen M, Janssen R, Software support for multicriteria decision-making, in Giupponi C, Jakeman A, Karssenberg D, Hare M, (eds.,) *Sustainable Management of Water Resources*, The Fondazione eni Enrico Mattei (FEEM) Series on Economics, the Environment and Sustainable Development, Edward Elgar Publishing, 2006, pp. 131–150.
154. Janssen R, Herwijnen M, and Beinat E, *DEFINITE 3.0 Case Studies and User Manual*, Inst. Environmental Studies, Vrije Uni. Amsterdam (www.ivm.vu.nl/en/projects/Departments/spatial-analysis/DEFINITE/index.asp; last accessed 12/09/09), 2003.
155. Belton V, Stewart T, *Multiple Criteria Decision Analysis—an Integrated Approach*, Kluwer Academic, 2002.
156. van Drunen M, Janssen R, *Evaluation Methods. (Online user's manual for DEFINITEprogram)*. Inst. Environmental Studies, Vrije Uni. Amsterdam (www.ivm.vu.nl/en/projects/Departments/spatial-analysis/DEFINITE/index.asp; last accessed 12/09/09), 2009.
157. Peters T, Pinto D, and Pinto E. Improved seawater intake and pre-treatment system based on Neodren technology, *Desalination* **203**(1–3):134–140 (2007).
158. Pankratz T, personal communication, Water Desalination Report, USA, 2007.
159. Brehant A, Bonnélye V, Perez M, Comparison of MF/UF pretreatment with conventional filtration prior to RO membranes for surface seawater desalination, *Desalination*, **144**:353–360 (2002).
160. Crowley K, personal communication, Consolidated Water Company, Cayman Islands, 2009.
161. Pearce G, Talo S, Chida K, Basha A, Gulamhusein A, Pretreatment options for large scale SWRO plants: Case studies of UF trials at Kindasa, Saudi Arabia, and conventional pretreatment in Spain, *Desalination* **167**:175–189 (2004).
162. Remize P, Laroche J, Leparc J, Schrotter J, A pilot scale comparison between granular media filtration and low pressure membrane filtration for seawater pretreatment, *Proc. Euromed 2008 (conf.), Desalination Cooperation among Mediterranean Countries of Europe and the MENA Region, Dead Sea, Jordan*, 2008.
163. Vial D, Doussau G, The use of microfiltration membranes for seawater pre-treatment prior to reverse osmosis membranes, *Desalination* **153**:141–147 (2002).
164. Baltussen R, Niessen L, Priority setting of health interventions: The need for multi-criteria decision analysis, *Cost Effect. Resource Alloc*. **4**(14): (2006). www.biomedcentral.com/1458-7547/4/14
165. Edwards W, Barron F, SMARTS and SMARTER: Improved simple methods for multiattribute utility measurement, *Organiz. Behav. Human Decision Process*. **60**:306–325 (1994).
166. Jeffreys I, Lawrence P, Principles and methods for developing a MODSS process for farm forestry, in Harrison S, Herbohn J, ed., *Agroforestry and Farm Forestry:*

Support SA systems to assess the Viability of Whole-Farm and Regional Agroforestry Enterprises, RIRDC, Canberra (espace.library.uq.edu.au/eserv/UQ:8107/Ch_13_-_MODSS_pr.pdf; last accessed 12/09/09), 2008, pp. 166–179.

167. Beery M, Repke J, Sustainability analysis of different SWRO pre-treatment alternatives, *Desal. Water Treat*. **16**:218–228 (2010) (in press).

168. Pankratz T, California—who or what is to blame? (Water Desalination Report by Media Analytics Ltd.), *Global Water Intel*. **45**(28):1–3 (2009).

CHAPTER 3

Economic Aspects of Water Desalination

AMITZUR BARAK

Advances in Water Desalination, First Edition. Edited by Noam Lior.

3.1 INTRODUCTION

3.1.1 Objectives and Scope

The main role of the economics of any production—in the case discussed in this chapter, desalted water—is to assist in making decisions. These decisions may be managerial, technical, financial, and/or administrative in nature. They may refer to a near-future small project as well as to long-term broad programs.

This assistance in decision-making, obtained by using economics, stems from a reduction in the number and type of the criteria by which the production alternatives to be decided on are assessed. Instead of coping with many technical, physical, temporal, logistic, workforce, and many other factors, these production alternatives are measured quantitatively mainly according to the amount of monetary funding that is allocate to them. This monetary value serves as a common denominator for comparing these alternatives on a money-per-project, money-per-program, or money-per-unit product (where *money* denotes the amount of monetary funding allocated to or spent on that project). Effective and proper comparisons are performed for identical times and places and for equal product quantities and qualities.

The economic criteria used as the bottom line in economic analyses are costs, prices, and values of the products; investments; percentage of local currency in the product cost and the investment; cashflow; payback period; and rate of return.

All these criteria, as well as related aspects, are discussed and described in this chapter. The related aspects are financing, economy of scale, technoeconomic optimization and scheduling. Particular attention is given to dual-purpose plants, nuclear desalination, and solar desalination. Some aspects are treated quantitatively; others, only qualitatively. The economics of hybrid desalination plants, energy supply alternatives, trasnport of desalted water to users, and optimal plant location are also discussed.

Specifically, economic aspects relating uniquely to desalination, rather than similar industrially produced materials, are described and discussed. These aspects include the challenge to ensure extremely low cost per unit product, the effect of feedwater quality on the cost and the initial investment, changes in cost and investment with respect to the quality of the desalted water (salinity, boron content, etc.), and the possible effect of hydrology–desalination integration on the cost of desalted water and other aspects.

Finally, a tentative forecast of desalted water costs is presented and explained.

Two examples of cost calculations are appended. One is for a single-purpose seawater reverse-osmosis (SWRO) desalination plant and the other, for a cogeneration plant employing electric power and desalted seawater with multieffect distillation (MED) evaporation. Another appendix discusses the principles of basic optimization of the distillation process.

3.1.2 Economics of Production in General

Desalted water is often produced by a single industrial process and in quite a few cases by a combination of several processes. As such, the economical aspects of water desalination are, in principle, identical to those of any other industrially produced material such as cement, sulfuric acid, or table salt (NaCl). Thus, both as a reminder and for readers not acquainted with the details of the subject, the economics of the process industry are briefly presented in this section.

3.1.3 Specific Features of Desalination

However, water desalination has several unique technoeconomic features in addition to those of all other industrially produced materials:

1. *Large Quantities*. The amounts produced, at what is considered a typical large plant, are much larger than any other industrially produced material. [*Note*: Desalination processes are defined as industrial ones. A distinction exists between *industrial production* of materials and alternative means of materials provision (supplying), procurement, or preparation (by mining, harvesting, or treatment such as filtering or chlorination). These alternative means are simpler, less sophisticated, and therefore less expensive than industrial production.]
2. *Very Low Acceptable Price*. The price of desalted water that is acceptable to the common water user is at least one order of magnitude lower than any other industrially produced material. Thus the cost of production per unit mass should not exceed a few US dollars per metric ton, preferably even less than US$1 per ton even for small production rates, while for any other industrially produced material the production cost is at least several dozen US dollars per ton. Moreover, very large desalted water production rates are acceptable only for ***considerably*** less than US$1 per ton (1 ton of potable water is equal to one cubic mater, in the context discussed here).
3. *Sensitivity of Cost to Economic Environment (Particularly Energy Prices)*. The economics of a desalination project is strongly dependent on local factors, the prevailing market prices of electricity and water, and a series of institutional factors that may influence the economic assessment of the project. Energy consumption and cost are considered crucial for desalination, partly because of issues mentioned here and in list item 2, above. Throughout the more recent decades, energy current prices have undergone large fluctuations, sometimes drastic; thus, issue of forecasting costs and determining prices, both short- and long-term, becomes more precarious than for other materials.
4. *National system of long term production plants*. For many desalination projects the need for desalted water exists due to a local water shortage, thus defining the desalination plants production and installation timing with very little flexibility. However, in some states and countries water shortage is forecasted for long term and wide national or regional scope. In the former

type of local enterprises the economic considerations are similar to those of the common process industry while for a national desalination program the economic considerations are different for various reasons which will be discussed in section 3.15.

5. *Public, Political, and Media Intensive Involvement*. Since water is a common utility commodity, the amount of public interest in the availability and quality of water in general and in desalted water in particular is much greater than for other materials. Any difficulty—financial, managerial, technological, administrative, constructional or operational (schedule delays, accidents, etc.)—will impose additional economical penalties and create a poor image of the project and the parties involved, of the type of desalination process, and even of the entire field of desalination. This poor image may negatively affect the market demand for desalination systems and desalted water (in this sense it is somewhat similar to nuclear power).
6. *Need for Good, Effective, and Highly Reliable Peripheral Infrastructure*. Water is almost always transported and distributed by pipelines, unlike most other chemicals, which are transported by trains, trucks, and ships. Thus, poor infrastructure, namely, leaky or clogged pipelines and inadequate pumps, between the desalination plants and the desalted water users will sabotage the water supply. The reliability of this supply will deteriorate, the cost of transportation will increase as a result of excessive energy losses and high maintenance expenses, and—above all—expensive high-quality desalted water will disappear through the leaks. All these factors—although external to the desalination system itself—will raise the actual net cost of each unit (measured in m^3) of desalted water used, in addition to the resulting poor image that will be attributed by the public and media to the entire branch of desalination and the impacts described above.

This aspect of the need for high quality infrastructure obviously holds for all water sources, not only for desalination. However, the sensitivity of the public and the water suppliers in this regard is considerably higher in the case of desalted water.

3.1.4 Competition of Desalination with Alternative Means of Water Supply

The relatively high cost of desalination is one of the major reasons for the reluctance to implement it more often. Paradoxically, broader use of desalination could have decreased its overall cost, at least slightly. This point will be discussed further in Section 3.18. However, the cost difference between water desalination and other methods of reducing or preventing water shortage is one of the major reasons to reject, avoid, or at least delay desalination and turning to one or more of the following alternatives:

- Transporting water from distant sites
- Water reuse such as sewage reclamation
- Lowering water consumption by education and propaganda, enforcing drastic water saving, water rationing, and/or raising water prices
- Discovering new aquifers
- Preventing water loss such as leakage or evaporation from piping and reservoirs
- Decreasing and improving agricultural activities (e.g., reducing irrigation)
- Cloud seeding to increase or create artificial rain

Despite the tough competition from these other methods, desalination quite often becomes the preferred solution because of economic considerations, and/or flexibility and ease of control, availability (where sufficient amounts of saline water exist), and the possibility of improving not only the *quantity* but also the *qualtity* of the water. This may be achieved by dilution: mixing the poor-quality existing water with the added desalted water, while ensuring that this water has been desalted to a very high quality (at minimal additional cost). Such dilution yields an accumulative mixture of desired or at least acceptable quality, thus upgrading the economic value of the water above that of the existing natural water. For certain sites or applications, such as where the existing natural water is of relatively poor quality as well as being in short supply, such upgrading may be quite meaningful.

In certain cases desalination is performed *after* the site-specific relevant alternatives have been applied but are unable to economically satisfy the entire need for water. This may occur if these relevant alternatives have limited, but not sufficient, potential and their resources have been fully exploited, or if the marginal water from the alternative source is more expensive to obtain (or more complicated to process) than water that has been desalted.

Thus, the existence of potential alternatives obliges us to consider the economic aspects of these alternatives as well as the economic aspects of desalination in our decision as to the best way to supply potable water. All the relevant economic criteria (described below and in Section 3.2) must be evaluated and compared.

3.1.5 The Roles of Economics at the Planning Stage

The main role of economics of any production—in this case, desalted water—is to facilitate the decision-making process. These decisions may be *fundamental*, such as whether to solve the water shortage by desalination or take one or several of the above mentioned alternative solutions; *technical*, such as which process is best for the specific site in consideration; *economical* and *financial*, such as the price at which the water should be sold or which are the preferred sources of money for the initial investment; *managerial*, such as whether to bid for desalination plants or for desalted water; and *administrative*, such as whether to hire the maintenance staff as employees or utilize the services of a contractor for periodic maintenance work. The scope and timescale of these decisions may vary from a near-future small project of a single plant with one unit to long-term programs involving multiple large plants.

The assistance in decision-making obtained by applying economic tools, stems from the reduction in the number and type of the criteria by which the production alternatives, to be decided, are measured. Since the required decisions involve choices between alternatives, comparisons of the considered alternatives have to be accomplished (even if the decision is only between "do" and "do not"). The comparison process is simplified by the said economic tools as instead of dealing with many technical, physical, economical, temporal, workforce *I still prefer* manpower *or* personnel and other criteria while preparing these comparisons—all alternatives are measured mainly by money which serves as a common denominator. More precisely, the assessment of the alternatives is done mainly by comparing either money per unit product or money per program (consisting of one or several projects).

Note: an effective and proper comparison has to be based on equal terms for all the alternative scenarios: identical times, similar sites and equal products' quantities and qualities.

3.2 ECONOMIC CRITERIA

The following economic criteria are used as the basis for economic analyses:

1. *Cost of Product*. This is often the most important, "popular," and useful factor, per unit product—in this case a metric ton of desalted water—for a given plant or for completion of a given project on a fixed date. Product cost is often considered as the *sole* criterion for deciding on whether to produce the needed material or to obtain it by other means.
2. *Investment*. This criterion is in most cases a major economical factor, which determines the *capital cost*, which, in turn, is the largest (or second largest) product cost component. (As will be explained in detail in the following section, the investment is usually proportional to the capital cost.) In principle, if the only impact of the investment were its contribution to the overall product cost, it should not be presented as a separate economic criterion. However, it has another parallel impact—on the financing problem, and is tightly associated with the risk policy of the desalination plant owner as well as with the product pricing. (Product price is another economic criterion, the seventh in this list). For example, if an evaluation of two alternatives yields practically the same product cost, the one with the lower investment will probably be preferred, because of lower risk and easier financing (which is another economic and management issue). A further possible recommendation for a possible scenario may be to prefer a meaningfully lower investment even if the predicted product cost is ***slightly*** higher.

3. *Local Currency Percentage*. A specific double criterion, related to the two previous ones, is the *percentage of local currency* in the product cost and in the investment. When two alternatives show practically (i.e., almost) the same product cost but the differences in local currencies are meaningful, the alternative with the higher percentage of local currency will be preferred.
4. *Cashflow*.
5. *Payback Period*.
6. *Rate of Return*. This is usually of concern to private investors more than to large public authorities (such as a government) as owners. Cashflow, payback period, and rate of return depend on the investment and the annual expenses. They depend also on the next criterion.
7. *Price of the Product*. *Price* differs from *cost* according to the profit (or subsidy); whereas the cost constitutes the expenses for production, i.e., the producer's outgoing money, the price is what the customer pays back to the producer, i.e. incoming money. The price is determined by (among other factors) the next economic criterion.
8. *Value of the Product to the Customer*. In the case of desalted water, there is an added value, as compared with many other water resources, due to the high reliability of quantities, qualities, and availability, which are controllable to a high degree.

The considerations described in list items 1–3 above leed to criteria 4–8.

Cost, price, and value are measured and expressed in the same units (currency per product unit). They usually differ in magnitude. Under normal conditions the cost is lower than the price, which, in turn, is lower than the value to the customer. In many cases the differences between the cost, the price, and the value are quite lower than the cost. These three criteria are of interest to the customers (i.e., desalted water consumers) as well as to the investors and plant owners involved in the desalination systems. The other criteria are of interest mainly to the investors, plant owners, and sometimes also economists and politicians.

Criteria 7 and 8, price and product value, differ from the other criteria as they do not result solely from the decisions made by project owners, designers, and managers. Rather, they result mainly from the existing economic conditions. They have to be quantitatively estimated, (or determined, if objective estimation is not feasible).

All these criteria, as well as related aspects, are discussed and described in this chapter. (Another set of economic criteria, similar to these, is presented in Section 3.13 regarding economical evaluation of bids). The related aspects are financing, economy of scale, technoeconomic optimization, project scheduling, economics of hybrid desalination plants, energy supply alternatives, desalted water transport to users, and optimal plant location.

3.3 BASIC EVALUATION OF DESALTED WATER COST

3.3.1 Definition of Cost

The true product cost is the *total* expenditure needed to supply the product to the customer per unit product. In practice, two product costs exist. One fits the definition fully and refers to the cost at the location and time of transition from the supplier to the end user and is termed *delivered* or *gross* product cost. The other cost refers to the product leaving the plant/factory and may be termed *plant* or *net* product cost. The first includes the expenses that accrue between the plant exit and the point of delivery to the customer, covering the costs of water storage, transportation, possible losses, and distribution. These factors have to be added to the net cost, where the gross cost is more relevant. However, reference to the net product cost is more often found in literature and in feasibility studies as the "cost of desalted water." This is so because the expenses occurring between the plant exit and the point of delivery are highly site-dependent and can be analyzed only on a case-by-case basis and are not process-dependent. Thus, in pricing the desalted water for the customer, the *gross* product cost is the relevant one whereas for comparing desalination methods or plants the *net* product cost is more convenient.

3.3.2 Actual Cost

The basic principle of determining the actual (real) unit product cost, net or gross C, of a given desalting plant is quite simple: C is the sum of the payments of all expenses of the desalination plant throughout the lifetime of the plant divided by the number of the entire amount of desalted water units during the same period, namely, the average mix of expenditures per unit product. However, this simple costing method does not take the economic time factor, namely, interest, into consideration.

In order to properly introduce the time factor, the unit cost definition is slightly modified: C is equal to the unit price of the capitalized incoming payments for all sales of the water desalted by the plant throughout its lifetime that balances the sum of the capitalized payments of all the expenses of the desalination plant throughout the plant's lifetime. Where this balance exists, the stated price equals the cost. This price of the water is virtual, being fixed at the cost value; the real price is naturally higher because of the expected profit added to the cost (under normal conditions, at least). The capitalization is done by summing the discounted values of the sales and expenses to any specified date (D). Usually the date on which the plant commercial operation starts is taken for the discounting as follows:

Mathematically, the balance is expressed as follows:

$$\sum(\text{Expense})_i = \sum(\text{Sale})_j \tag{3.1}$$

where "Expense" is an expense payment, paid by the desalination plant, discounted to a given date as will be explained below; "Sale" is a water sale payment, received by the desalination plant, also discounted to the same given date; indices i and j are the serial numbers assigned to the individual payments of expenses and sales, respectively; and $\sum$ is the summation notation.

For a sale of W_j product units [W_j for desalted water is measured in tons, m^3 (=0.997 ton), acre-feet (=1233.5 m^3), or kilogallons (=3.785 m^3)], under payment j at (virtual) price C per unit, the *actual* payment j would have been $C \cdot W_j$. The *discounted* value, however, is

$$\text{Sale}_j = C \cdot W_j \cdot F_{dj} \tag{3.2}$$

where F_{dj} is the discounting factor

$$F_{dj} = (1+d)^{D-D_j} \tag{3.3}$$

where d is the annual discount or interest rate (see Section 3.3.2.1 next page), D is the reference date for capitalization, and D_j is the date of payment j (the time difference $D - D_j$ is expressed in years).

Thus, summing the plant life income payments for product sales yields

$$\sum \text{Sale}_j = \sum (C \cdot W_j (1+d)^{D-D_j}) = C \cdot \sum W_j (1+d)^{D-D_j} \tag{3.4}$$

and hence

$$C = \frac{\sum \text{Sale}_j}{\sum \text{W}_j (1+d)^{D-D_j}} \tag{3.5}$$

but since the cost-determining principle states that $\sum(\text{expense})_{\text{i}} = \sum(\text{Sale})_j$, the last equation can also be written as

$$C = \frac{\sum \text{Expense}_i}{\sum W_j (1+d)^{D-D_j}} \tag{3.6}$$

Discounting the expense payment Ex_i is similar to the sales payments

$$\text{Expense}_i = \text{Ex}_i (1+d)^{D-D_i} \tag{3.7}$$

So the final result is

$$C = \frac{\sum_{(i)} \text{Ex}_i (1+d)^{D-D_i}}{\sum_{(j)} \text{W}_j (1+d)^{D-D_j}} \tag{3.8}$$

where D_i is the date of payment i and the time difference $D - D_i$ is expressed in years.

3.3.2.1 *Interest and Discount Rates*

The *interest rate* is the annual *cost* of money required for the entire work or part of it. Similarly, the *discount rate* is the annual *value* of this money for the owner of the plant. In principle, both interest and discount rates may be quantitatively equal, especially if the total investment is financed by loans that have the same interest rates and a payback period identical to the lifetime of the project. If the project is financed by loan(s) having shorter (or longer) payback periods and/or when the project is funded by several sources, with different interest rates, the *discount rate d* may differ from the *interest rate* of the single loan or from the mean value of interest rates of the funding (financing) sources. In such cases d is used in the preceding formula [Eq. (3.8)] to calculate the capitalized values of the payback sums of these loans. Frequently, for feasibility or preproject studies, there is no reference to loans, and thus the interest rate is taken equal to the discount rate, and a value identical to d is used to calculate F_{d_j}, Expense_i, and C.

The exact value of the discount rate is often determined by the government (sometimes even arbitrarily to a certain extent) according to high-level economic and financial considerations. For example, a government that has incurred large debts may decide on a discount rate value that is higher than the interest rate value, even if the total investment is financed by a single loan that does have a payback period identical to the lifetime of the project.

Usually the value of the discount rate per year is determined within the range of 5–10%, relative to *constant currency* at a given date, thus bypassing inflation. More representative annual values are 7%, 8%, or 9%. Extreme known values are 4% and 12% per year. The values in most cases are round figures since they are not calculated but determined arbitrarily, albeit within narrow range limits.

3.3.2.2 *Formulas Used to Determine Actual Cost*

A few remarks should be added regarding the last formula. [Eq. (3.8)]:

1. $(\text{Expense})_i$ and Ex_i refer to
   a. Payments for
      (1) Equipment, including instrumentation and controls, feedwater supply, concentrate removal, desalted water posttreatment and transport to users, in addition to the obvious main process equipment
      (2) Spare parts
      (3) materials for the preliminary stock [such as chemicals, oils, paints and possibly fuels],
      (4) services and labor during plant construction, planning and design,
      (5) Civil engineering [buildings, storages, excavation, fencing, pavements and roads, piping, drainage etc.],
      (6) Electrical engineering [cables, transformers, contactors and switches, meters, cathodic protection etc.]
      (7) Other infrastructural items, including (a) payments for rented or purchased land; (b) preliminary and construction management; (c) offices,

documentation, communication, and transportation (vehicles used), and handling of the equipment and materials (cranes, etc.); (d) workshop, warehouse, and laboratory; (e) preliminary insurance policies; (f) legal expenses and fees; (g) environmental surveys, permits, and licensing; (h) employee salaries; (i) contractors, consultants, and services; (j) recruiting and training desalination equipment operators; and (k) preliminary taxes.

All these payments occur mainly before the commercial commissioning of the plant. This initial part of the expenses payments is defined as the *investment* (Inv or Capex).

b. Payments occurring after commercial commissioning of the plant, including those for

(1) Salaries and fees of equipment operators, plant management and other service staff, maintenance workers, employees, and various service contractors

(2) Electricity and steam

(3) Materials for current use (chemicals, oils, paints, and possibly fuels)

(4) Replaced items of equipment

(5) Rentals, insurance, and annual taxes

These latter expenses payments (items 1b1–1b5) are collectively defined as the *running cost* (Rn or Opex).

c. Payments occurring during and after the decommissioning of the plant (Dec). The latter is very small relative to Inv and Rn, sometimes even negative if the remains of the plant are sold for a good price. The *discounted* value of Dec is even lower than, almost negligible, relative to the discounted values of Inv and Rn since it occurs at a very late stage in the plant history.

d. The following formula results from a combination of points 1a–1c above:

$$\sum_{(i)} \mathrm{Ex}_i = \mathrm{Inv} + \mathrm{Rn} + \mathrm{Dec} \approx \mathrm{Inv} + \mathrm{Rn} \tag{3.9}$$

e. If part or all of the financing is implemented by loans, the back payments should also be considered as plant/project expenses and included in Expense. In this case the incoming sums loaned to the project should be considered as negative expenses. Both the incoming loan sums and the loan back payments should be discounted to the capitalization reference date D (preferably the plant commissioning day).

2. Some incoming payments may be allocated for sales other than ordinary desalted water. Typical examples are payments

a. Used equipment

b. Buildings and/or land at the end of the plant life

c. Excess energy where the desalination plant has a built-in or self-supplied energy system

d. Concentrated brine to salt ponds

e. High-purity (e.g., double-distilled) water for specific applications such as feedwater for boilers

These payments should be considered for the cost calculation as negative expenses rather than sales and treated as such for calculating the desalted water cost.

3. Introducing real and accurate values into the last formula—namely, the values of all the payments $(\text{Expense})_i$, the water quantities W_j and the dates D_i, D_j, and D—is possible only at the end of the plant life so that the accurate value of the product cost can be obtained only at this very late time in the plant history.
4. Even this accurate value of C depends on one factor that is not measured or calculated but determined rather arbitrarily—the discount rate d, discussed in Section 3.3.2.1. Thus, for given data of all the payments $(\text{Expense})_i$, the water quantities W_j and the dates D_i, D_j and D the desalted water cost C is either calculated for a certain fixed value of the discount rate d, or is presented as a one-variable function by simple tabulation of several values of d or is plotted in a graph for a specific range of d values.
5. When the formula for C [Eq. (3.8)] is used prior to the end of the plant's lifetime (i.e., at plant startup, before or during plant operation, or even earlier, before plant construction, during the bidding period, when the possible erection of a desalting plant is only in the decision or planning stage), this C formula serves only as an *estimate*—rather than a firm *calculation*—of the actual cost of the desalted water. This can be attributed to the fact that some or all of the other input values—including expenses $(\text{Expense})_i$, water quantities W_j, and their dates D_i, D_j, and D—are not known but are only assessed or assumed. In other words, the earlier in the plant history that this formula [Eq. (3.8)] is applied, the lower will be the degree of accuracy.
6. In practice, using the C formula is very laborious as it requires numerous calculations. Therefore, a shorter procedure is quite often used, even though it may introduce additional inaccuracies. This shorter procedure, and two other alternative methods, is discussed in Section 3.3.3. Nevertheless, the present method is strongly recommended as it can be used to calculate the real cost after the plant is finally closed and decommissioned. The data obtained using this method can be extrapolated for comparisons with previous product cost estimates and to identify any deviations of previous assumptions from the actual cost values.

3.3.2.3 *A Sample Cost Calculation* The per unit price of desalted water C can be calculated according to the last two formulas above [Eqs. (3.8) and (3.9)] as follows. Suppose that the following summarized and processed data exist in the plant's records:

- No loans are used for financing.
- Plant life production is $\sum_{(j)}(W)_j = 385$ Mt (million metric tons).

- Plant production life is 32 years
- Plant erection time, including design and planning, is 3.5 years.
- Discount rate $d = 0.075$ per year (7.5%/y).
- Total investment is Inv = \$31.5M.
- Capitalized/discounted investment covering all the payments before commissioning:

$$\sum(\text{Expense})_i = \sum_{(i)} \{\text{Ex}_i (1+d)^{D-D_i}\} = \$35.2\text{M} \tag{3.10}$$

where D is the date of plant commissioning $0 \leq D - D_i \leq 3.5$.
- Running cost during plant life for an average year is = 2.65M\$/y electricity + 0.95M\$/y salaries + 0.6M\$/y materials + 0.7M\$/y equipment + 0.3M\$/y taxes & insurance = \$5.2M /year,
so that Rn = \$5.2M/y* 32 years = \$166.4M.
- Capitalized/discounted value of running cost is $\sum(\text{Expense})_i$ = \$52.3M for payments after commissioning.
- Decommissioning expenses are negligible.

The *capitalized/discounted value* of incoming payments for (virtual) water sales at cost price means:

$$\begin{aligned}\sum \text{Sale}_j &= \sum (C \cdot W_j (1+d)^{D-D_j}) \\ &= C \cdot \sum_{(j)} \{(W)_j (1+d)^{D-D_j}\} = \$(120 \cdot C)^* 10^6\end{aligned} \tag{3.11}$$

as $\sum_{(j)} \{(W)_j (1+d)^{D-D_j}\} = 120 \cdot 10^6$ tons.

Thus $\$(120 \cdot C)^* 10^6$ = \$35.2M+ \$52.3M = \$87.5M,
so that C = \$87.5M/120M tons ≈ \$0.73 *per ton*.

Note: Capitalization of sum of many items $\sum_{(i)} \{\text{Ex}_i (1+d)^{D-D_i}\}$ can be expressed as being equal to: $\{\sum_{(i)} \text{Ex}_i\}(1+d)^{D-D_{i\,\text{av}}}$, where $D_{i\,\text{av}}$ is defined as the average date of all the payments of expenses i. Similarly

$$\sum_{(j)} \{(W)_j (1+d)^{D-D_j}\} = \left\{\sum_{(j)} (W)_j\right\} (1+d)^{D-D_{j\,\text{av}}} \tag{3.12}$$

where $D_{j\,\text{av}}$ is defined as the average date of all the incoming payments of desalted water sales j.

Using the values of this sample to find these average dates yields the following:

- For the investment: $D - D_{i\,\text{av}} = 1.54$ year; that is, $D_{i\,\text{av}}$ occurs about 1.5 years before commissioning, close to the midpoint of the 3.5-year construction period.
- For the running expenses: $D - D_{i\,\text{av}} = -16.0$ years; that is, $D_{i\,\text{av}}$ occurs 16 years after commissioning, at the midpoint of the 32-year production period.

- For the water sales incoming payments: $D - D_{j\,\mathrm{av}} = -16.12$ years; that is, $D_{j\,\mathrm{av}}$ occurs 6 weeks after $D_{i\,\mathrm{av}}$ of the running expenses, also almost midway through the production period.
- A lesson from this analysis is that approximate discounting can be easily done by assuming that all the investment payments are made equivalently on one representative date at the midpoint of the construction period, all running expenses similarly occur equivalently on one representative date midway through the production period, and all the water sales incoming payments are received equivalently on the same representative date midway through the production period (or a short time later). Thus a very useful and convenient approximate formula evolves:

$$C \approx \frac{\mathrm{Inv}(1+d)^{t_{\mathrm{con}}/2} + \mathrm{Rn}(1+d)^{-t_{\mathrm{prod}}/2}}{W(1+d)^{-t_{\mathrm{prod}}/2}} \tag{3.13}$$

$$C \approx \frac{\mathrm{Inv}}{W}(1+d)^{(t_{\mathrm{con}}+t_{\mathrm{prod}})/2} + \frac{\mathrm{Rn}}{W} \tag{3.14}$$

where W is the total plant water production, t_{con} is the construction period in years, t_{prod} is the production period in years, and t_{proj} is the entire project period in years.

$$t_{\mathrm{proj}} = t_{\mathrm{con}} + t_{\mathrm{prod}} \tag{3.15}$$

$$C \approx \frac{\mathrm{Inv}}{W}(1+d)^{t_{\mathrm{proj}}/2} + \frac{\mathrm{Rn}}{W} \tag{3.16}$$

This formula, however, is not in common use. For $d > 5\%$ per year and for $t_{\mathrm{prod}} > 20$ years, the deviations from the accurate results are not negligible.

3.3.3 Practical Methods of Product Cost Evaluation

There are several ways to evaluate the cost of product:

1. Summation of all the cost components
2. Using the cost practically experienced in other plants
3. Various combinations of methods 1 and 2

Obviously each method has specific disadvantages due to the level of inaccuracies and the limits of use. In principle, the most accurate evaluation is done by summing up the cost components, provided the input data are complete and accurate. Such an evaluation process may be quite laborious and time-consuming (as indicated above p. 209). If a high degree of accuracy is not necessary, as explained in the following paragraph, and/or if the evaluation results are needed within a short time period, then one of the other evaluation methods may be used.

Product costs are evaluated for the following reasons:

1. To compare alternatives for production methods, including the "reference" alternative which is to avoid production altogether and seek other solutions; specifically, in the case of water shortage, if the cost of the desalted water

is higher than that obtained using an the alternative means of water supply, then one of these alternative means must be used. Optimization of design, achieved by comparison of several different designs of the same desalting process, is a special case of such comparison.

2. To prepare for financing by forecasting the cashflow using the product cost.
3. To price the product.
4. To obtain information for future projects (data extrapolation).
5. To provide an outlook on the economy of the country on the domestic, intranational level for large scale and long-term desalination (in particular for a desalination *program* rather than a specific *project*). This objective is relevant only for very large-scale production items such as water and energy at national scales.

In the following text the general term *product* refers to desalted water, although many aspects discussed in the text are relevant to most industrial products.

The specific reason or objective for which the product cost is evaluated will determine the degree of accuracy.

3.3.3.1 Summation of All Cost Components This approximate method differs from the more accurate one described in Section 3.3.2 with respect to the following features:

- Calculation of the total cost by summing its three major components, described later Sections 3.3.3.2, 3.3.3.3 and 3.3.3.4.
- Capitalization of large *sums* of expenses rather than individual expenses.
- Levelization of both the expenses and the water production for the plant lifetime.
- Assumption that the annual water production and running expenses remain the same throughout the plant's commercial operation life.
- Assumption that no loans are needed for financing, and that the plant owner's self-financing, estimated at "money value" by using the time factor of the discount rate, will provide sufficient funding for the project.
  This is equivalent to using a loan for every expense payment prior to the plant commercial commissioning date under the conditions of: (1) interest rate equal to the discount rate and (2) equal periodical back-payments throughout the economical lifetime period of the plant.
- Assumption that the plant's running expenses are paid at the same time that equivalent sums of money are incoming from water sales—or, more accurately but less practically stated, that equal amounts of money are expended and collected at the end or the midpoint of each "payment period."

The calculation according to these principles is as follows. There are two principal product cost components: (1) the *capital cost* (Cc), which is closely associated with the total investment (Capex), and (2) the overall operation, maintenance, and other current expenses, namely, the *running cost* (Opex). In the case of the desalination plant, the latter is divided into *energy cost* (Ec) as a separate item, and all other operation and maintenance costs are designated as *O&M cost* (Omc).

The contribution of individual cost components to the overall cost of water depends on many factors, such as production capacity, type of desalination process, site characteristics, and types of energy sources. However, experience shows that for most seawater desalination plants[1] capital cost makes up 30–50% of the total desalted water production cost; energy cost, 30–50%; and O&M balances, 15–25%. The cost percentage of O&M decreases as the capacity of the plant increases.

3.3.3.2 Capital Cost The capital cost (C_c) is quite often the largest component in the cost of desalted water.

This cost component differs from the other ones because, in addition to all the actual expenses mentioned above, namely, the initial payments before commercial water production defined above as the investment (Inv), the value of interest during construction (Idc) is added by capitalization to the convenient date D of commercial commissioning:

$$I_{\text{tot}} = \text{Inv} + \text{Idc} \tag{3.17}$$

where "I_{tot}" is the total investment capitalized to date D. Thus I_{tot}

$$I_{\text{tot}} = \sum (\text{Expense})_i = \sum\nolimits_{(i)} \{\text{Ex}_i (1+d)^{D-D_i}\} \tag{3.18}$$

for payments before commissioning, where D is the date of plant commissioning.

In more practical terms, I_{tot} is often calculated not by summing all the multiplications

$\text{Ex}_i (1+d)^{D-D_i}$, but by a quite accurate approximation if the construction period t_{con} is shorter than 3 years:

$$I_{\text{tot}} = \text{Inv}(1+d)^{t_{\text{con}}/2} \tag{3.19}$$

Thus,

$$\text{Idc} = I_{\text{tot}} - \text{Inv} \approx \text{Inv}\{(1+d)^{t_{\text{con}}/2} - 1\} \tag{3.20}$$

If the period t_{con} is ≥ 3 years, or if the investment payments are not evenly distributed between the two halves of t_{con}, then t_{con} can be divided into two equal periods:

$$I_{\text{tot}} = \text{Inv1}(1+d)^{3/4 t_{\text{con}}} + \text{Inv2}(1+d)^{t_{\text{con}}/4} \tag{3.21}$$

[1] Seawater desalination is more commonly practiced than is saline-water desalination, involving desalting of brackish-water, reclaimed wastewater, marginal water, and other types of non-sea-origin water.

Here Inv1 and Inv2 are, respectively, the accumulating investment payments during the first and second halves of the construction period t_{con}.

If the investment payments are far from being evenly distributed along the 3–4-year construction period, t_{con} can be divided into annual periods:

$$I_{tot} = \text{Inv1}(1+d)^{0.5} + \text{Inv2}(1+d)^{1.5} + \text{Inv3}(1+d)^{2.5} + \text{Inv4}(1+d)^{3.5} + \cdots \quad (3.22)$$

Here Inv1, Inv2, Inv3, and Inv4 are, respectively, the sums of the investment payments during the last and the previous years of the construction period t_{con}.

Once I_{tot} is known, C_c is usually calculated, per unit product, as

$$C_c = \frac{I_{tot} \cdot (\text{CRF})}{P_A} \quad (3.23)$$

$$P_A = \frac{W}{t_{prod}} \quad (3.24)$$

where: I_{tot} is the total capitalized investment, as explained in detail above;
CRF is capital recovery factor (at year end); and
P_A is the average annual production (in unit product). Ideally, it equals the nominal design capacity P_{NA}, but in practice it may be quite lower or in rare cases even higher.

The average annual production P_A is factually known only at the end of the plant life time, as are both W and t_{prod}. At the earlier stages of a project, P_A can be only estimated. This is done by multiplying the nominal production rate P_N [usually given in units per day (or per hour)] by 365 days (or 8760 h) and by the anticipated load factor

$$P_A = 365\, P_N \, {}^* \, \text{LF} \quad (3.25)$$

where P_N is the nominal production rate and LF is the load factor.

The load factor LF (also referred to in the literature as *capacity factor* or *plant factor*) is associated with the plant's availability. It is a quantitative representation of the fact that the plant or the unit produces less than its theoretical annual potential at full-time operation and full nominal capacity. Desalination plants—like most process plants—do not operate at full design capacity along the entire year and even less so throughout the plant lifetime. There are planned and unplanned outages, expressed quantitatively by a factor termed *time-availability*. Also, even during operation there are usually periods of time, hopefully short and few, during which production is only partial because of poor equipment performance, reduced power supply, or reduced demand for the water. The quantitative combination of partial load and planned and forced outages of the plant is known as the *load factor* (LF).

Optimistic LF values of are 0.9–0.93 for reverse osmosis (RO) and almost the same for electro-dialysis (ED) and mechanical vapor compression (MVC), as these processes are energized by electricity, which usually has very high time availability. Multieffect distillation and multistage flash distillation (MED and MSF) processes operating with steam and electricity have lower load factors because the steam source time availability is often lower. LF is approximately the product of the time availabilities of the energy sources and the desalination system LF. In case of MED and MSF, there are at least two energy sources; heat (steam—quite often, hot water or hot gas) and mechanical work (electricity, most often; higher-pressure steam, seldom), each having its time availability. Optimistic of the MED and MSF load factor values are 0.8–0.9.

The second factor (CRF, *Capital Recovery Factor*), in the formula $C_c = I_{tot} \cdot (CRF)/P_A$ [Eq. (3.23)], is usually calculated (on the basis of end-of-year-recovery) by the formula used for calculating unit loan end-of-year back payments, namely

$$\mathrm{CRF} = \frac{d}{1-(1+d)^{-n}} \tag{3.26}$$

where d is the annual discount or interest rate as explained previously and n represents years of economical lifetime of plant. It may differ from the above mentioned t_{prod}, as will be explained later.

3.3.3.2.1 Practical Capital Recovery Factor and Other Values Assuming that $d = 7.5\%$ per year = 0.075/y and $n = 30$ years, inserting these values into Eq. (3.26) yields CRF = $0.075/(1-1/1.075^{30}) = 0.0847$/y = *8.47% per year*. Although this expression for CRF is very commonly used, it is not accurate because it is based on the assumption that all payments for water sales—part of which funds the total investment—is accepted at the end of each year. In most cases this is not correct. Where these payments are more frequent, namely when time periods Δt between these payments are much shorter than a year, say, every 2 months, and the number of payments per year is $f = 12/\Delta t$, (where Δt is expressed in months), the formula for CRF becomes

$$\mathrm{CRF} = f \cdot \{(1+d)^{1/f} - 1\}/\{1-(1+d)^{-n}\} \tag{3.27}$$

This gives somewhat lower values for CRF. For the values assumed above, $\Delta t = 2$ months and $f = 12/\Delta t = 6$ periods per year, we obtain CRF $= 6\cdot(1.075^{1/6} - 1)/(1-1/1.075^{30})$ = 0.0822/yr = *8.22% per year* rather than *8.47% per year, as given above.*

The Specific Investment: The formula for the capital cost [Eq. (3.23)] C_c can be modified and presented as

$$C_c = \frac{I_{tot}}{P_A} \cdot (\mathrm{CRF}) \tag{3.28}$$

The first factor on the right side, i.e. the ratio (I_{tot}/P_A), is the *actual specific investment*, namely, the investment per unit capacity (in m^3/year). The specific investment is a very important criterion of desalting systems and processes.

This actual specific investment per factual annual production P_A is ideally equal and practically rather close to I_{SA}, the investment per unit design capacity P_{NA}:

$$I_{SA} = \frac{I_{tot}}{P_{NA}} \tag{3.29}$$

which will be referred to simply as the *annual specific investment*. Thus, for cost estimates before the end of the plant production we obtain

$$C_c = \frac{I_{tot} \cdot (CRF)}{P_{NA}} = I_{SA} \cdot CRF = \frac{I_{SA} \cdot d}{1 - (1 + d)^{-n}} \tag{3.30}$$

Alternatively

$$C_c = \frac{I_{SA} \cdot f \cdot \{(1 + d)^{1/f} - 1\}}{1 - (1 + d)^{-n}} \tag{3.31}$$

where I_{SA} is commonly expressed in US$ per annual m^3 [$/($m^3$/year) = ($ year)/m^3]. However, since the actual annual load factor of a plant may fluctuate, the technical and engineering design of a desalination plant is done for a specified nominal *daily* (sometimes hourly) rather than *annual* production rate. Thus, it is much more common and reasonable to express the specific investment by money invested per daily capacity I_S (expressed by $/($m^3$/day) = $d/t) rather than by I_{SA} per annual capacity. The ratio between the values of these two definitions of I_S is

$$I_S \text{ (per daily output)} = 365^{*}\text{LF}^{*}I_{SA}\text{(per annual output)} \tag{3.32}$$

Thus if, for example, I_{SA} = $2.4/($m^3$/year) and LF = 92% = 0.92, then I_S per daily capacity is

$$I_S = 365^{*}0.92^{*}2.4 = \$806/(\text{m}^3/\text{day})$$

The actual load factor LF is obtained only at the end of the plant lifetime. Preliminary values of LF can be only estimated; deviations from this preliminary estimated LF may occur during the plant lifetime. In view of its previous definition (per daily capacity), I_S does not depend on LF, while I_{SA} does. Hence it should be presented by I_{SA} as function of the actual LF:

$$I_{SA} = \frac{I_S}{365\text{LF}} \tag{3.33}$$

Practical Values: The practical, experience-based values for existing plants range from ~$2 to >$6 (US dollars) per cubic meter per annum; the lower values apply for very large modern plants.

For example, a 100M (=10^8) m^3/y plant requires an investment of at least $0.2billion.

Using these assumed values and assuming $I_S = \$2.4/(m^3/year)$, a numerical result is obtained for the capital cost component:

$C_c = \$2.4/(m^3/year)*0.0822/year = \$0.197/m^3 = ¢19.7$ per $m^3 \approx ¢20/m^3$.

These figures refer to seawater; the values for brackish-water may be lower.

More specifically, typical values of I_S for efficient single-purpose plants in year 2012 are:

- \$700–\$1000 per unit design capacity (m^3/day) for large SWRO and for very large low-temperature MED plants
- Higher values, \$900–\$1500/(m^3/year), for MSF and high-temperature MED,
- Lower values, \$500–\$1500/(m^3/year), for brackish-water RO and ED (electrodialysis); the latter is used only for brackish-water.
- Highest values, \$1000–\$2000/(m^3/year), for vapor compression distillation.

3.3.3.2.2 Factors influencing the Specific Investment Various factors affect the specific investment of each process:

1. Plant capacity.
2. Quality of feedwater.
    a. Salinity of both seawater and brackish-water
    b. For brackish-water: hardness and concentration of problematic materials
3. Required quality of product water.
4. Energy supply—the lower values of specific investment given here refer to plants with external energy supply such as electricity and steam.
5. Site-specific costs of investment in the systems of feedwater supply and product supply to users.
6. Site-specific costs of energy prices, salaries, transportation, etc. paid during plant erection.
7. Land cost—the lower values given here do not include payments for the purchased land, which are very site-specific, nor other special expenses, such as special brackish-water feed system (deep wells) and complex concentrate removal means. This item is not included since it is not always relevant; for some projects the use of the land is granted free of charge or the land is leased to the plant owners or the land belongs to the plant owners.
8. The discount rate d (or the interest rate if the investment expenses are financed by loans) and the construction period t_{con}, as indicated in the formulas above, such as

$$I_{tot} = \text{Inv}(1 + d)^{t_{con}/2} \qquad (3.34)$$

The effects of factors 1–4 are investigated and discussed below.

The Plant Capacity The investment increases less than proportionally with the capacity according to the principle of the *economy of scale*; for example, the investment in a SWRO plant of 100 mgd (millions of gallons per day; 1million gallons = 3785 m^3) is roughly only 20–25 times as much as in a plant of 3 mgd ($\approx$11,350 m^3/day), reducing the capital cost component of m^3 desalted water by 30–40%.

The relationship between the investment I_{tot} (in US dollars) in a SWRO dual-purpose plant and the nominal production rate P_N (m^3 per year) within the range of capacities given above can be demonstrated very roughly by

$$I_{tot} \approx A^* P_{NA}{}^{m} \tag{3.35}$$

with A ranging between 25 and 30 and m, between 0.8 and 0.9. The values of the experienced-based factor A and perhaps even the power factor m have to be updated following changes in technology and the market place. A more focused example for the last formula [Eq. (3.35)], relevant around year 2000 is

$$I_{tot} \approx 28^* P_{NA}{}^{0.86} \tag{3.36}$$

where the constant factor A in the economy-of-scale formula for the specific investment per *annual* output unit is $A = 28$ and the power factor is $m = 0.86$.

This relationship stems from the convenient and popular (although approximate, simplified, and limited to a specified range of plant capacities) *power law*, connecting the ratio of investments and the ratio of production rates of any two similar desalination plants as well as of other industrial products as

$$I_{tot1}/I_{tot2} \approx (P_{N1}/P_{N2})^{m} \tag{3.37}$$

where in most cases: $0.5 < m < 0.9$. (*Note:* More accurate and more logically theoretical relationships exist, but they are more complex and less convenient to use).

Following the preceding definition [Eq. (3.37)] of the specific investment $I_S = I_{tot}/P_N$, the result is

$$I_{SA}(P_N) \approx \frac{A}{P_{NA}{}^{(1-m)}} \left\{ \approx \frac{28}{P_{NA}{}^{0.14}} \right\} \tag{3.38}$$

for P_{NA} in m^3 per year. For P_N in m^3 per day, the result is

$$I_S(P_N) \approx \frac{A_1}{P_N{}^{(1-m)}} \approx \frac{4000}{P_N{}^{0.14}} \tag{3.39}$$

The constant factor A_1 replaces A in the economy-of-scale formula for the specific investment per *daily* output.

Thus, for example, if the daily nominal production rate is 100,000 m^3, the expected specific investment is approximately $4000/(10^5)^{0.14} = 4000/10^{0.7}$ = \$798/(m^3/day) $\approx$ \$800/(m^3/day). It should be emphasized that these latest approximate formulas are useful in the very early phases of a project only for preliminary cost estimates and comparisons of alternatives.

The economy of scale leads in general to lower specific investment. Nevertheless, at least three components of I_S may increase with the plant capacity:

- *The Interest during Construction (Idc/P).* The reason is that the construction time of a larger plant is naturally longer, which obviously increases one factor of Idc/P. This cost increase may, however, be avoided if the plant is built and operated modularly with several independent units.
- *The Desalted-Water Distribution System*. Usually the greater the plant capacity and the amounts of product water, the more users to whom the water is delivered. With larger desalting plants, many users are located farther away from the plant than are users of smaller more local plants. The distribution system size and cost increase with the average distance of delivery.
- *Possible Surplus Capacity during the Initial Years of Plant Operation*. Since the demand for water increases more or less gradually where large plants are considered, the plant size and capacity correspond to the period of water shortage time. For example, if the predicted average demand in a certain region increases by 20 million m^3 annually, the shortage can be reversed by incrementally installing one 20Mm^3/y plant each year or, say, a 50Mm^3/y plant every $2\,^1\!/_2$ years or 100Mm^3/y plants every 5 years. The largest plants are least expensive per unit product having the best I_{tot}, but might not be fully productive in the first 4 years, and their inferior load factor LF will increase the capital cost C_c, despite their better specific investment, as expressed by the following formula, which combines several of the relationships presented above:

$$C_c = I_{SA} \cdot f \cdot \frac{(1+d)^{1/f} - 1}{1-(1+d)^{-n}} \tag{3.40}$$

$$= \frac{I_S/365\text{LF} \cdot f \cdot (1+d)^{1/f} - 1}{1-(1+d)^{-n}}$$

The following three factors, which depend negatively on plant size, limit the tendency for increasing plant capacity.

Quality of Feedwater. It is obvious that the higher the salinity of the feedwater, the higher is the effort and hence the cost to produce potable water from this feed. This stems from the basic need to separate more dissolved materials from the feedwater to obtain desalted water. This higher effort and cost are reflected either in energy increase or in purchase of more expensive equipment; in practical terms, both cost components are higher under optimal design. Since the maximum concentrate (brine) salinity is limited to about 7% or slightly higher to avoid scale deposition, and the process design has to maintain this value, it influences the design with respect to two parameters:

1. The flow rate of the needed feedwater increases. For example, as the salinity content of ocean water is ~3.5%, 2 tons of feedwater is needed per ton of

desalted water, whereas with Red Sea water, which contains ~4.4% salinity, almost 2.7 tons of feedwater is needed. The feedwater supply is larger by ~34%: pipes and their supporting structures, pumps and motors, control and ordinary valves, filters, chemical and physical feed pretreatment, and in evaporating processes, the vacuum system. The concentrate removal system (brine outfall) is 68% larger. In some cases the latter is only a short pipe and the added investment is marginal, but in other plants this system is rather expensive because of environmental requirements. These expenses may add a few percent to the investment and the capital cost. In the example mentioned above, C_c may be higher by ¢1–2/m^3. (Additional energy for pumping and deaerating the higher feed flow rate and additional chemicals for feed pretreatment are part of the added running costs, to be discussed later.) In the case of inland brackish-water, the effect of larger feedwater supply and concentrate removal systems due to higher feed salinity is even more pronounced, since the inland systems are more expensive per ton feedwater and per ton concentrate, respectively.

2. The average salinity of a salty stream that is desalted is higher. In example I above (for ocean feedwater) the average process salinity is (3.5% + 7%)/2 = 5.25%, whereas with the Red Sea feedwater it is (4.4% + 7%)/2 = 5.7%. The practical effect of the higher average process salinity is higher average osmotic pressure in RO plants or higher average boiling-point elevation in distillation units. These forms of thermodynamic resistance either increase the energy consumption, using more expensive motors and pumps for RO and larger steam lines for distillation, or require more RO membranes and larger heat transfer area for evaporation–condensation. Another effect of the higher mean process salinity, mainly with RO, is the higher product salinity, which often may require extra equipment and energy to correct.

In RO, the combination of parameters 1 and 2 results in larger high-pressure (main) pumps, which are quite expensive, as well as much larger pressure recovery devices such as water turbines. In the systems compared above for ocean versus Red Sea feedwaters, the pumps are almost 42% larger; 34%, due to higher flow rate and ~8%, to higher pressure difference. The pressure recovery device is 77% larger; 68%, due to higher flow rate and ~9%, to higher pressure difference.

Required Quality of Product Water: Similar to the salinity of high feedwater, the salinity of product water is also indicative of the *desalination effort*—the lower the required salinity, the greater the effort to ensure salt separation, in terms of energy and equipment. It is impractical to expect zero product salinity, because in the distillation processes tiny saline droplets drift with the vapor and in membrane processes, either the salt rejection is not 100% (RO) or the salt separation is asymptotic (ED); thus it is necessary to control the migration of salt into the product stream to acceptable levels that are allowed or determined by the authorities, the customers, or international standards. Previously the acceptable salinity was 500 ppm TDS (parts per million of total dissolved solids), specifically, 0.5 g/L, since sea and brackish feedwaters do not contain specific problematic solutes. More recently, further restrictions have been imposed, mainly by the product users, regarding chlorides, boron, and nitrates at low levels.

The reduction in product-water salinity is quite simple and inexpensive for *distillation* processes, employing droplet separators such as louvers or sieves; the addition of such separators can reduce the salinity down to very few ppms. In *membrane* processes, however, 500 ppm or even 300 ppm can be attained with very little effort, but lower values require the use of additional equipment and energy. Quite often the 300–600-ppm product (or at least part of it) is passed through a secondary RO unit as the optimal way to lower the salinity meaningfully as required. This is an expensive additional investment.

In certain situations the problematic aspect of boron should be stated where product-water boron contents should be kept very low. In SWRO the boron is poorly separated by the RO existing membranes and an additional subsystem is often needed; This is an expensive part of the posttreatment production.

Energy Supply System–Electricity: The lower specific investment values given above pertain to plants with external energy supplies such as electricity and steam. However, some plants may not have these external energy sources. Also, for large plants operated by electricity such as RO, MVC [in principle, and perhaps in the future also freeze-desalination (desalination obtained at freezing temperatures)] and ED, even if externally supplied electricity is fully available at a reasonable price, plant self-supplied electricity, with grid connection for backup, is still preferable. This type of arrangement has several advantages:

1. Expenses per kWh are lower since only the *cost* is relevant, while for power obtained from the grid (in kWh), the *price* is relevant. This price is much higher—50 to 200%—than the cost of the kWh power generated by the efficient and large grid-connected units because with the grid:
   a. The average generation cost of the mix of all the power units—some of which may be old and less efficient, while other operate only at peak power demand—is very expensive per kWh, even with efficient power units operating at lower load factor than a power unit dedicated to the desalination plant. (The average load factor of the entire power generation system, including the efficient units, is relatively low, ~60%, whereas the load factor of a power generation system at the desalination plant is about 90%.)
   b. the price of 1 kWh from the grid covers the capital cost of the grid, the grid energy losses of 5–15%, and the grid maintenance.
   c. the price of 1 kWh from the grid further includes the overhead of the utility company: management, planning and design, research and administration, as well as profit.
2. The reliability of power supply is higher with two energy sources: one in situ, independent of grid failures, employee strikes, forced outage of power plants, or excess power demand that results in reducing the power supply, with the second energy source—the grid—as backup.

3. The investment in a self-generating system is relatively low since the infrastructure for the power system is mostly shared by the desalination system, and there is another saving; cooling-water is not required, as the desalation system easily absorbs all the rejected heat. Moreover, the heat rejected from the power unit can be absorbed by the desalination process to improve it in many cases, since both heat transfer and mass transfer increase with the temperature.
4. If the power unit of the desalination plant is generating more than needed for the desalination, the excess electricity may be sold to the utility at some profit (since any marginal product usually has low marginal production cost).

On the other hand, the full-load calculated generation cost of power by the larger and more efficient units generating many hundreds of megawatts in most grids—roughly 4(±20%) US¢/kWh at the power plant–is about 10–40% lower than that of the much smaller efficient units needed for desalination plants because of the economy of scale of the power units. Even for very large plants of, say, 150 Mm^3/year SWRO, the electric power consumption is about 60–80 MW, an order of magnitude smaller than power units in conventional grids. However, considering the advantages of self-supplied electricity listed above, the latter option is preferred in many cases, particularly for large desalination plants. The above mentioned advantages of self-supply versus the disadvantages, which relate mostly to the economy of scale of the in situ generating system, balance at a certain desalination plant with a consumption of roughly 20 MW. However, in some cases self-supplied power is preferable even for very few megawatts, whereas in other cases self-supplied power is economically justified only above 50 MW.

In those cases where this self-supplied power option is preferred, there are two differences with respect to the cost of desalted water as discussed above:

- For the structure of the main cost components, the term *fuel cost* is used instead of the more expensive option, *energy cost*. The latter item shrinks to cover the back-up grid power.
- The capital cost increases, as the total investment includes the additional expenditure on the installed power system. Quantitatively, this is roughly \$600–\$1500 per installed kW_e electrical kilowatt for fossil fuel and \$1000–\$3000 per installed kW_e for nuclear power. These values result from experience and feasibility studies. The exact value of the investment depends on the plant capacity, site economic and physical conditions, specific technology, and other factors. As for solar and wind power, the investment is even higher and very site-specific.

Typical I_S values of for plants with self-supplied power per m^3/day of desalted water capacity (as well as for dual-purpose plants with the power plant part allocated to desalination) will be, therefore, at least \$880–\$900 per m^3/per day, as follows. With \$750 per m^3/day for SWRO consuming 4 kWh_e per m^3, plus

\$800 per installed kW_e for fossil fuel (which are quite low desalination and power investments), the sum is $\$750/(m^3/d)+\$800/kW_e * (4kWh_e/m^3)/(24\ h/d) = \$883/(m^3/day)$.

This increase in investment is significant not only because of the additional capital cost component C_c, which is more than compensated for by the decrease resulting from the differences between the *fuel cost* versus the more expensive *energy cost*, but also in terms of financing.

It should be noted that even if power from the grid is used, and the desalting plant owner's investment is low, the national or regional development of the water sector still bears the needed investment for the energy required for the desalination, with the impact on financing mentioned above. This impact is sometimes problematic, sometimes marginal, and sometimes negligible.

Energy Supply System–Heat: A somewhat different situation may exist with thermal distillation plants such as MED and MSF, which usually required ~60–150 kg of low-pressure steam per ton of desalted water (in the range of 0.25–3 bar absolute) and 1–4 kWh_e electricity (plus very few kg of higher-pressure steam for the vacuum system).

If these energies are supplied from external sources, the specific investment is \$700–\$1500/m^3 per day, as given previously. However, if such external energy sources are not available, low-pressure boilers have to be installed, the specific investment may increase by about \$130–\$200/(m^3 per day), without backup and still with external electricity, provided this external electricity is available, namely, with partial energy independence only. Full energy independence for MED or MSF plants may increase the specific investment by roughly \$200–\$300 m^3/day, still without backup.

This increase in specific investment by full or partial self-energy supply of MED or MSF is the smaller part of extra cost of energy independence. The main penalty is the high fuel cost, which is discussed in the next section with respect to the energy cost component. Because of this penalty, MED and MSF are preferably installed coupled with industrial plants that have to reject large amounts of heat, usually steam power stations and occasionally gas turbines. Such coupling results in an inherent reduction of the generating capacity of the power units, which is significant in steam plants and combined cycle systems but almost negligible in gas turbines. This reduction in generating capacity can be regarded as if part of the investment in the power plant were dedicated to evaporative (MED or MSF) desalination.

For a virtual example, a high-temperature MED with $I_S = \$900/m^3$ per day is connected to a nuclear power unit from which it consumes 6 kWh_e/m^3 desalted water; that is, each m^3 reduces the electricity generation and sales of the power plant by 6 kWh_e. The specific investment of the nuclear unit is \$2400 per installed nuclear kW_e. The total specific investment allocated to desalination will be quite high: $\$900/m^3$ per day + $\$2400/kW_e*(6\ kWh_e/m^3)/(24\ h/d) = \$1500/(m^3$ per day).

3.3.3.2.3 Direct and Indirect Capital Costs Capital costs are often divided into direct and indirect costs, usually during an early phase of the preliminary evaluation of the project.

The *direct costs* are payments for equipment and buildings, civil and electrical engineering services, and stocks of materials, fuels, and spare parts. *Indirect costs* include

- Interest incurred during construction, assumed at 10% of the direct cost
- Provisional sums, representing 10% of the direct cost
- Contingencies and architect-engineer charges, representing 15% of the direct cost
- Working capital, making up 5% of the direct cost

Another aspect of capital costs involves the definition of—in addition to the direct costs, mentioned above—three "higher-encompassing" terms, expressed in Eqs. (3.41)– (3.43):

$$\text{Base cost} = \text{direct cost} + \text{indirect cost} \tag{3.41}$$

Indirect cost includes construction management, site services, home office engineering and services, and field office engineering and services.

$$\text{Fore cost} = \text{base cost} + \text{owner's costs} + \text{spare parts} + \text{contingency} \tag{3.42}$$

This term is sometimes called *overnight cost*.

$$\begin{aligned}\text{Total (capital) cost} = {} & \text{fore cost} + \text{escalation} \\ & + \text{interest incurred during construction}\end{aligned} \tag{3.43}$$

3.3.3.2.4 Infrastructure Development Cost These costs constitute a special indirect investment component, which should be considered capable of producing benefits by promoting the country's overall development. It consists of many items, some of which relate to specific plants, while others relate to the entire national desalination–energy program. They include the cost incurred in planning, studies, R&D activities in support of various energy and desalination alternatives, workforce teaching and training, national infrastructure development, local (state) participation promotion, technology transfer, and regulatory and licensing costs. These costs are difficult to evaluate per plant. Theoretically they should have been included in the specific investment category I_s as a separate, identified item; practically, this is complicated. However, as they represent a relatively very small part of I_s, they can be included as "miscellaneous" I_s items. Some items of this component are referred to in the following text.

3.3.3.2.5 Capital Cost Reduction For a given plant, once the site, the timing, the production rate, the quality of the desalted water, and the desalination-process have been determined, the capital cost component is minimized by the following six-step procedure (in addition to sound engineering, efficient organization, and diligent management):

1. *Determining the Daily Nominal Production Rate*. Unless constrained, the plant nominal capacity should be maximized to the *optimal limit* of demand—considering not only the immediate and very near-term water shortage but also potential future needs, as long as this strategy does not significantly decrease the plant's load factor or increase construction time and distribution distances.
2. *Reducing the Interest and Discount Rates*. Successful negotiations to reduce these rates can result in the following:
   a. The total investment I_{tot} decreases, as well as the specific investment I_{SA}, according to Eq. (3.19): $I_{tot} \approx \text{Inv}(1+d)^{t_{con}/2}$
   b. The levelized value of the investment I_{tot} attains lower value according to Eq. (3.31): $C_c = I_{SA} \cdot f \cdot \{(1+d)^{1/f} - 1\}/\{1-(1+d)^{-n}\}$.

   Note: Since the discount rate is very often determined by the government, a considerable reduction can be achieved because it is customary for governments to approve fairly low or even very low discount rates for "national projects." Whereas the current discount rates might be 8–10% per year, national projects are sometimes entitled to rates as low as 3%. If, for example, this value of 3% per year is assumed and compared to 7.5% while the construction time T_{con} is 2 years and the economic plant lifetime n is 30 years, then the accumulating effect of steps 2a and 2b will be $C_c = 0.0883 \cdot \text{Inv}/\text{P}_\text{N}$ for 7.5% annual discount rate versus $C_c = 0.0519 \cdot \text{Inv}/\text{P}_\text{N}$ for 3% annual discount rate. Lowering the interest rate might be achieved by governmental or international (e.g., the World Bank, UN) guarantee backing the financing of large desalination programs or projects. Such support reduces the risk and thus may lower the interest rate. This can sometimes also be achieved by negotiations with the financing entities.
3. *Reducing Construction Time*. With diligent preparations, The construction time t_{con} can be shortened by applying the same formula as in step 2a: $I_{tot} \approx \text{Inv}(1+d)^{t_{con}/2}$ [Eq. (3.19)].
4. *Bidding and Negotiations for a Purchased Plant*. The strategies outlined above to reduce the total investment I_{tot} by reducing the interest and discount rates, the construction time, and other means can be bypassed by the plant owners by bidding for a complete plant ready to begin commercial operation often called the *turnkey bid*. The competition is apparently on I_{tot}-net, which is cost to the owner and the price for the bid-winner contractor, as it includes the contractor's profit-and-cost, namely, the contractor's direct and indirect expenses. The overall owner's investment I_{tot}-gross must include another item, the *owner's cost*, which is equal to a few percentage point of I_{tot}-net (typically 10%). In some cases the land is directly purchased or given by the owner. This has to be added to I_{tot}-net and the owner's cost or included

in the latter to obtain the true I_{tot} (i.e., I_{tot}-gross). An experienced, efficient, and diligent contractor might be motivated by the competitive bid to offer I_{tot}-net that is low enough to ensure an overall owner's investment I_{tot}-gross that is better than the next-best alternative despite the gap between I_{tot}-net and I_{tot}-gross. The contractor can allow the low-price offer without losing money by lowering Inv, d, and/or t_{con}. Similarly, with or without bidding, reduction of the total investment or parts of it may be achieved by smart negotiations.

5. *Increasing the Frequency of Investment Recovery*. Recovering the investment several times a year rather than once at year end, by increasing the frequency of payments for water, if possible monthly or bimonthly, will reduce the value of the factor $f \cdot \{(1+d)^{1/f} - 1\}$ in the formulas given above.
6. *Extending the Economic Plant Lifetime n*. This will reduce the value of the factor $1/\{1-(1+d)^{-n}\}$ in the preceding formulas.

The contribution to C_c in steps 3 and 4 is a few percentage points each. The contribution of step 5 is ~1.3% per year extension beyond $n = 25$ years and ~0.9% per year extension beyond $n = 30$ years.

3.3.3.2.6 Plant Lifetime In general, the expected lifetime of a desalination *unit*[2] today ranges from 20 (which is too low) to ideally 30 years after commissioning. The factors t_{con} (years of construction time) and n (years of economic lifetime) used in the formula for C_c have a definite meaning for a single-unit plant or a plant that consists of several units that are installed, commissioned, and decommissioned simultaneously—which seems very unlikely; then the plant lifetime may be identical to the unit life. However, if the units that operate in parallel are installed one after the other (consecutively)—which is often preferable economically—and/or if the units do not have the same lifetime each, the overall plant lifetime can be quite a few years longer than the average unit life. The cost calculations become somewhat more complicated and less accurate. One can calculate the capital cost component for each separate unit and obtain somewhat different results, but this might be difficult with part of the investment in the infrastructure of the plant shared by all the units while the more important result is the cost of water from the entire plant even though the product cost from each unit is also of interest. In this case averaging the cost of the units and assuming some scheme of sharing the common infrastructure can resolve this issue.

The effect of the unit life on the water cost is moderate, in particular where the discount rate is high. This is demonstrated in Table 3.1 for a typical plant of

- Specific investment $I_s = \$800/(m^3/day)$

[2]So far the simplified description of the economical aspects did not really differentiate a desalination *plant* from a desalination *unit*. This differentiation, however, is necessary for the discussion here; a *plant* may often consist of several units, mainly in parallel, but sometimes these units are connected in series via the product stream or the saline-water stream.

Table 3.1 Effect of Plant Economic Lifetime on Water Cost

Case	Plant Economic Lifetime n or t_{prod}, years	Capital Recovery Factor (CRF), %/year	Capital Cost C_c, ¢/m^3	Relative Change in Capital Cost ΔC_c, %	Product Cost C, ¢/m^3	Change in Water Cost ΔC, ¢/m^3	Relative Change in Product Cost ΔC, %
Reference	25	0.0897	21.85	Reference	55.0	Reference	Reference
+20%	30	0.0847	20.63	−5.6	53.8	−1.22	−2.2
+40%	35	0.0815	19.85	−9.1	53.0	−2.00	−3.6

- Discount rate d = 7.5%/year
- Load factor LF = 90%
- A reference case of n = 25 years for which the assumed product cost is 55¢/m^3

Table 3.1 shows that the relative change in product cost is an order of magnitude lower than the change in the economic lifetime of the plant.

Also, a distinction should be made between three plant lifetime periods:

1. Years of *expected* economic lifetime of the plant, denoted as n; this is assumed or assessed at the beginning of the plant life and used to calculate the early estimated water cost.
2. The period t_{prod}, years of the *actual* production period that is known at the end of plant lifetime and used to calculate the real factual water cost. It may obviously differ from n.
3. The *potential physical* lifetime of the plant. In many cases it may be the same as t_{prod}. In many other cases it may be longer than the economic lifetime or production period. The commercial operation of the plant may be terminated after t_{prod} years, before the end of the potential physical lifetime because of emerging new and improved technologies (expected or factual), higher maintenance expenses in later in the plant lifetime, and for other reasons.

3.3.3.3 Energy and Fuel Cost Component This cost component depends on the types and amounts of the needed energies and the price or cost of each kind of energy.

3.3.3.3.1 Energy Consumption for Desalination *Reverse Osmosis* requires (RO) mainly mechanical energy, which is conveniently supplied by electric motors. The plant energy quantities are theoretically proportional to the amount of desalted water, so it is justified to refer to the specific energy consumption, usually given in kWh/m^3. Typical values for large, efficient SWRO units of approximately $\geq 10^4$ m^3/day or higher are (year 2012)~3–5 kWh$_e$/m^3 desalted water, not including pumping of desalted water from the plant. Much larger modern units, with capacities of $\geq 10^5$ m^3/day or so, are more sophisticated and consume ~2.8–4 kWh_e/m^3. Brackish-water consumes 1–3 kWh_e/m^3. The deviations are due to:

1. Technology versions, such as type of pressure recovery device.
2. Plant size and capacity; the larger the equipment, the lower the friction flow losses and the electric motors losses.
3. Plant location and topography, which determine the pumping power required to and from the plant.
4. Salinity of feedwater; the higher the salinity, the higher are the feed needed, the osmotic pressure, and the permeate salinity (the latter may need correction by secondary desalination) as discussed previously. Each one of these process features bears its share of energy. The difference between Atlantic and Indian Ocean feedwater (3.5% and 4.5% salinity, respectively) is at least 1 kWh_e/m^3.
5. Required quality of product water. Reducing product salinity requires an auxiliary system that consumes additional energy; the lower the product salinity, the higher is the additional energy.

The cost of 1 kWh depends on the source of electricity. Several possibilities exist:

- *The Grid*. In most countries the prices of electricity supplied to the industrial sector are significantly lower than that supplied to the domestic sector. In 2004 it was 5.27¢/kWh in the United States for the industry vs. 8.97 for domestic use, 16.15 vs. 19.13 in Italy (highest), 3.28 vs. 3.76 in Austria (lowest; subsidized??), and 9.39 vs. 11.89 in Israel. A typical energy cost component for power grid supply for very large units would be 3.8kWh/ton* 5.5¢/kWh = 20.9¢/m^3
- *Load Shedding*. It is possible to negotiate with the power utility on a certain reduction of the price, say, 20%, as is the case in some grids, for the option of "load shedding" under the condition of extremely high peak power demand.
- *Self-supply*. The possible means are diesel generators; steam turbines; gas turbines; combined cycles (of gas with steam turbines)' after"steam turbines; and nuclear, solar, wind, tidal, (in principle; unlikely in practice) or geothermal power. The investment in a self-generating system is relatively low since the infrastructure for the power system is mostly shared by the desalination system and there may be no need for cooling-water as the desalination system can absorb all the rejected heat. The grid is used as backup, as mentioned. The kWh_e costs can range from 4.5¢ to 15¢ (US), depending mainly on the type of fuel and its price. A typical result for very large units would be about the same as from the grid, may be even slightly lower than 20¢/m^3.
- *Directly from Adjacent Power Plant*. Electricity could be supplied from a neighbouring power plant or an industrial plant that has its own power system. Thus very little expenses and losses would be involved with transformation and transmission. Typical results depend on the price of purchased kWh.

- *Being Part of a Dual-Purpose Plant*. This possibility is a very popular and attractive arrangement, as discussed in Section 3.4.

Mechanical vapor compression (MVC), like RO, is also energized mostly by electricity, the sources of which are the same as mentioned above for RO. The amounts of energy, however, are about twice those for RO, at least 6–8 kWh/m^3. Because of the economy of scale, the cost of 1 kWh is somewhat lower than for RO if the power is self-supplied *when compared with equal production*. This is seldon relevant because of the higher energy consumption, but a reasonable market for MVC still exists. The specific energy of MVC also depends on technology versions, but to a lesser extent on plant and unit capacities, site, feedwater salinity, and product quality requirements. A typical result for a very large and sophisticated battery of units would be 6.8 kWh/ton*5.5¢/kWh = 37.4¢/m^3.

Electrodialysis (ED) is energized only by electricity, the sources of which are the same as mentioned above for RO. The amounts of energy, however, range widely since ED is useful mainly with brackish-water. This application requires smaller amounts of electricity per unit product, and the plant production rates are lower than those of large seawater desalination plants, so there is inadequate justification for self-supplied power. Thus, electricity is preferably purchased from the grid or from a nearby industrial or power plant, unless the local conditions do not allow purchase of electricity at "acceptable prices" of below ~10–12¢/kWh.

There is perhaps one potential advantage for self-supply of electricity to ED: the fact that ED requires mostly DC (direct current). External power sources hardly supply this kind of electricity. Thus, where DC is needed while the naturally available power supply is AC, a conversion system is needed, but this can be expensive and consume a few percent of the energy, due to unavoidable inefficiency. A self-supply source can bypass these extra expenses if a DC generator or a photo-voltaic solar system is used. A typical result for very large units is difficult to define because of the variability of site-specific conditions, as the feedwater is brackish. The following formula might be taken as somewhat typical: 2.0kWh/ton*5.5¢/kWh = 11¢/m^3.

Multieffect and Multistage Flash (MED and MSF) distillation, unlike RO, ED, and MVC, require mainly thermal energy, which is convenient to supply by low-pressure or medium-pressure steam (subatmospheric or <3 bar, respectively). Alternative heat sources are hot streams of liquids or gases. Both processes also require secondary mechanical energy for pumping, which is conveniently to supplied by electricity, and tertiary energy (mechanical work, electricity, medium-pressure or high-pressure steam) for starting vacuum and maintaining it (i.e., with gases pumping from the process to the atmosphere).

The costs of these energies depend on their sources (in addition to their amounts as mentioned above).

Primary Thermal Energy The most common source for heat in a single-purpose desalination plant is a low or medium-pressure steam boiler. This source is expensive, relative to other alternatives. The specific investment in a fully installed medium-pressure boiler is in the order of magnitude of a few dozen US dollars per thermal kilowatt. Assuming

- \$60/kW$_t$, which is equivalent to \$36,000/(t/h) of dry saturated steam at 3 bar absolute. (Since the latent heat of such steam is 516 cal/g and 1Wh$_t$ = 860 cal, it follows that 516/860 = 0.6 kWh$_t$/kg steam = 600 kWh$_t$ per ton of steam at 3 bar, i.e., \$60/kWt*600 kWh/t steam = \$36,000/(t/h) of boiler capacity.)
- A very high gain output ratio (GOR) of 12.5 m^3 desalted water per ton of steam in an MSF plant
- A very high load factor (LF) of 89%; each ton of steam thus will produce 12.5 m^3/t steam*8760 h/y*0.89 = 97,500 (m^3/y)/(t steam/h).

The annual CRF for a discount rate of 7.5% and 30-years boiler + plant lifetime is ~0.085, as shown in Table 3.1, so that the boiler part of the capital cost component of the desalted water is \$36,000*0.085/97,500 = \$0.0314/m^3 = *3.14¢/m^3*.

The main part of the steam cost is the fuel. Assuming low-cost coal of \$50/ton having a lower heating value (LHV) of 6200 cal/g, with a boiler efficiency of 85%, we obtain \$50/t-coal*516 Mcal/t steam/(6200 Mcal/t coal*0.85) = \$4.9/t steam; thus \$4.9/t-steam/12.5 m^3/t-steam = \$0.3916/m^3 = *39.16¢/m^3*. Thus the capital cost and fuel cost components of the steam contribution to the water cost add up to 3.14¢/m^3 + 39.16¢/m^3 = 42.3¢/m^3.

With other boiler expenses such as electricity and maintenance, the heat cost for a very efficient MSF plant is at least *43¢/m^3*.

With MED higher values of GOR are attainable, up to 20 m^3 desalted water per ton of steam, so the heat cost is almost proportional to the reciprocal of GOR. In this case, due to the economy of scale of the boiler, it is at least 27¢/m^3, which is somewhat higher than the net calculation result: 43¢/m^3*12.5/20 = 26.8¢/m^3.

Secondary (Pumping) and Tertiary (Lights, Services, Workshop, etc.) Mechanical and Other Energy Forms These are effected mostly by electricity. The possible sources and resulting costs for electricity are about the same as those above given for RO and for MVC. The amounts of electricity needed for MSF are 2–4 kWh$_e$ per m^3 desalted water and for MED, 1–3 kWh$_e$/m^3. Assuming the average values in the range 3kWh$_e$/m^3 for MSF and 2kWh$_e$/m^3 for MED, and taking the low price of electricity from the grid or from self-generation as 5.5¢/kWh$_e$, we see that almost 17¢/m^3 and 11¢/m^3 are added to the heat cost of MSF and MED, respectively.

This power cost can be reduced if the steam from the boiler to the evaporators is directed to expand in a steam turbine to drive a generator that will supply the secondary and tertiary mechanical or electrical energies. This will circumvent the need to purchase power from the grid (except as backup).

However, this strategy involves increasing the boiler size and the fuel consumption by 5–10% as well as installing a higher-pressure boiler, which increase the cost, plus a steam turbine and generator that are relatively expensive per unit product considering their economies of scale. The added turbine and generator increase the system complexity, and require more labor to operate and maintain and more spare parts, oil, etc. The increased cost due to the boiler and its fuel consumption is about 2–2.5¢/m^3 for MED and 2.3–4¢/m^3 for MSF. The capital cost of the

Table 3.2 Tentative Energy Cost Components with Self-Supplied Steam, Large Plants, 3.5% Seawater Salinity, and High-Temperature (~120°C) Seawater

	Purchased Electricity for Pumping		Self-Generation of Electricity (for Pumping) by Plant's Own Steam	
Component	HT-MED	MSF	HT-MED	MSF
Boiler	2.3¢/m^3	3.2¢/m^3	3¢/m^3	4¢/m^3
Coal, 50$/ton	24.7¢/m^3	39.5¢/m^3	27¢/m^3	42.5¢/m^3
Power, 5.5¢/kWh	11¢/m^3	17¢/m^3	Backup 0.3¢/m^3	Backup 0.5¢/m^3
Turbogenerator	—	—	0.7¢/m^3	1¢/m^3
Total	~38¢/m^3	~60¢/m^3	~31¢/m^3	~48¢/m^3
% more than RO at 5.5¢/kWh	~75%	~170%	~40%	~120%

added turbine generator is about 0.5–1¢/m^3 for MED and 1–2¢/m^3 for MSF (see Table 3.2).

The total energy cost, where the heat is self-supplied, typically averages 17 + 43 = 60¢/m^3 for MSF and 11 + 27 = 38¢/m^3 for MED. These values are very high compared to RO—almost 80% more for MED and 170% for MSF. Moreover, the optimum design for self-steam generation calls for high-temperature evaporators, which are expensive and increase the capital cost compared to the alternative possibilities of buying much cheaper, "used" steam, from nearby power plants or factories, if available.

Combing MED and MSF with TVC (thermal vapor compression) is a design that reduces the amount of heat consumption per unit product by a few percents but does not allow for self-power generation, unless an expensive high-pressure boiler is used. This alternative of both TVC and self power generation might be economically justified only with very large plants. Because the thermal vapor compressor, namely, the steam ejector, has very low mechanical (compression) efficiency, it is inherently inferior and is useful in only a narrow niche of circumstances where it might be the preferred economically as an operational compromise between purchasing the needed electricity or generating it in situ.

3.3.3.3.2 Costs of Energy Units from Various Primary Sources

Fossil Fuels The preceding discussion of the costs of thermal energy, in particular self-generated heat, referred quantitatively to coal as the primary heat source as being relatively inexpensive with moderate price changes. Obviously, all kinds of fuel oils as well as hydrocarbon gases can also be used. They were widely used before 1974, when they became much more expensive and much less stable. They are used today, where available, and are inexpensive in the same way that oil is in oil-producing countries. However, globally there is a strong tendency to shift to coal and occasionally gas for steam generation in the power industry. Steam generation for desalination follows the same principles and economic motivations as does steam generation for electricity, at least for large plants; thus coal is often preferred as the cheapest fossil fuel. For much smaller MED and MSF single-purpose desalination plants, the boilers are smaller and it might sometimes be more convenient to use liquid or gaseous fossil fuels, even if they cost more than coal.

Nuclear Desalination. Several nuclear possibilities have been offered for use with single-purpose desalination, ranging from very small units of 7–10 MW_t (SAKHA-92, SES-10, NHR-10) for MED or MSF that can desalt water at rates of 3000–4000 m^3/day, up to 500 MW_t (AST-500), at a desalting rate of $\leq$400,000 m^3/day. Nevertheless, no single-purpose nuclear desalination plant has been installed yet. One of the main reasons is the high cost. The main advantage, in principle, of nuclear energy is the lower *nuclear fuel* cycle cost component as compared with *fossil fuel* cost—roughly 0.15–0.3¢/kWh_t (thermal kWh) nuclear versus 0.7–1.4 ¢/kWh_t from coal. The cost of fossil fuel, such as coal, includes the cost of transportation from the pit to the plant site, thus depending on the location of the site and increasing with the distance. It is usually 3–5 times higher than nuclear fuel cost per kWh_t.

However, this key factor in the economics of nuclear energy is partly eroded by the much higher *capital* cost of nuclear energy and even more so, considerably, with evaporative desalination. MED and MSF require an additional investment in the complex safety thermal coupling equipment that transfer the heat from the nuclear source to the MED or MSF evaporators. This coupling equipment, which is designed to isolate the heating stream in the nuclear system from the heated stream in the desalination system, is expensive. It also consumes electric-power directly, while indirectly lowering the thermodynamic efficiency of the plant. (directly—for pumping the circulating the heat carrier inside coupling equipment; indirectly—by reducing the useful process exergy).

The nuclear heating system for single-purpose desalination also requires

- Larger investment, which might increase financing difficulties
- More land area due to safety requirements and public acceptance aspects
- More workforce for O&M
- Higher skill of these workers, with higher salaries
- Long periods of refueling, i.e., water supply problems that are solved for extra cost.

Nuclear desalination may also be performed using mechanicoelectrical energy for RO, MVC, and ED. The *fossil fuel* cycle cost component per electric kWh as compared with *nuclear fuel* cost is usually 2–4 times higher—about 0.5–1.0¢ per nuclear kWh_e versus 1.7–3¢/kWh_e from coal. The above mentioned thermal kWh fossil/nuclear ratio (3–5), is more favorable to nuclear than electrical kWh ratio (2–4), because the thermodynamic efficiency of fossil systems is higher than for most nuclear systems. On the other hand, the extra costs for specific nuclear safety requirements are lower where nuclear electricity rather than nuclear heat is used.

However, as mentioned above regarding evaporative desalination systems, most—and sometimes all—of this relative advantage of nuclear fuel cost is eroded by the higher capital cost of nuclear installations. Thus, the costs of electricity (at a given capacity and site) differ only a little (a fraction of a cent per kWh_e) regardless of whether nuclear or fossil fuel is burned to generate this electricity. For large-scale SWRO, this small difference in electricity cost adds up to about $\pm 2¢/m^3$, sometimes in favor of fossil fuel. With the above mentioned additional expenses for larger investment with potential financing difficulties, more land area due to safety requirements, problematic public acceptance aspects, more laborforce for O&M, higher skills and salaries of these workers, and long periods of refueling, the scope of using nuclear energy for *single purpose* desalination is limited for a narrow combination of site-specific conditions. On the other hand, the scope for using nuclear energy for *dual-purpose* power generation with desalination is much wider.

Solar Desalination There are three methods for using solar energy for desalination, which differ economically:

1. In the most popular method, using "solar stills," the solar radiation vaporizes the water directly at a moderate temperature slightly above ambient temperature. This method requires very little purchased (mechanical or electrical) energy, for low-pressure and low-volume pumping, and can be fully automated at a reasonable cost, significantly reducing O&M expenditure.
2. Solar radiation can be used to generate *steam* or higher-temperature heat by employing solar collectors or solar ponds at higher temperatures. This heat is transferred to MED or MSF evaporators. The expenses for secondary and tertiary energy as well as labor are the same as for self-generation of heat by fossil fuels.
3. Solar radiation can be used to generate *electricity* by steam or by photovoltaic cells, and this electricity can be used for RO, MVC, or ED. No secondary and tertiary energy is purchased. In this case, the economic potential of solar power for desalination is about the same as for ordinary power use.

All three methods suffer from high investment in equipment and large land area for capturing the solar radiation, with impact on potential financing difficulties. Another problematic feature is the load factor, which fluctuates even during the 48% = 11.5 theoretical hours of solar radiation per clear average day, as it is dependent on weather and climate (clouds, fog, rains etc.), latitude, and sun position and movement. Solar radiation methods may thus require additional investment either in storage of desalted water and/or energy, or in a backup energy system.

Therefore, the economic potential of solar desalination is appropriate to only limited combinations of site-specific conditions, such as secluded locations where fuels are significantly more expensive, justifying the high investment.

Geothermal Desalination Like nuclear or solar energy, thermal energy sources require very little electricity but involve high investment. Geothermal desalination may be useful only if the following three conditions are met: (1) the thermal energy is available near sites where desalted water is needed and in sufficient quantity, (2) the cost of the geothermal heat (geothermal well, pumping, etc.) plus the system that transmits the heat to the MED or MSF evaporators is less than that for alternative heat sources (this may be possible if the wells are not very deep), and (3) the geothermal well has proved reliable by operation for a sufficiently long time to justify the investment.

Wind Energy This source supplies primary mechanical energy suitable for RO, ED, and MVC. It involves quite high investment, except if the winds are high-velocity and sustained, which is quite rare. Large land areas might be required for installation of many wind turbine units. Occasionally such areas may be found where desalted water is needed, good winds are blowing, sufficient low-cost area is available, and—if an inland site—there is enough saline water to feed the plant.

The cheapest heat for MSF and MED can be obtained from the following sources:

1. Low-cost "used" steam, or "waste heat" from nearby power plants or factories, if available.
2. Low-cost heat from solar collectors or geothermal wells. So far these heat sources have very seldom supplied the cheapest heat where desalted water is needed.
3. A high-pressure steam boiler that supplies energy for a large hybrid desalination plant that includes RO as well as MED or MSF. This is discussed in greater detail in Section 3.4.
4. Heat produced by a desalination plant that is part of a dual-purpose plant that uses heat to produce other products. (see Section 3.4). The most useful and frequent product is electricity, so the discussion here focuses on dual-purpose plants that cogenerate power and water. The economic considerations of dual-purpose plants that produce other industrial products are similar to those for power generation and of buying heat purchased from nearby factories (list item 1 above).

3.3.3.3.3 External Heat for MSF and MED Buying low-cost used steam or waste heat from nearby power plants or factories is economically worthwhile if the price of the heat is below a certain value. In general, the price is determined by negotiations. Each side may have alternatives that set the value; the desalination plant may or may not use RO—which for high production rates ensures very low allowable heat prices—or other heat sources, such as heat self-generation. The potential heat supplier may or may not use the heat to improve the economy of the own process and plant—for a certain cost—or sell it to a competitive heat customer. In some extremely rare cases the heat supplier has to dispose of excess heat and might be willing to provide the heat free of charge.

Beyond negotiations, by which the price is determined possibly subjectively, there are the objective aspects:

1. The heat has a cost, which is complicated to calculate since it is delivered by used steam and since it depends on questionable and debated assumptions.
2. The heat has a thermodynamic value, expressed in energy units. It can be translated into economic units by practically using the thermodynamic value in a technical system such as power generation or heating applications such as for industrial processes, buildings, water streams, etc. The *thermodynamic value* depends strongly on the temperature at which the heat is used. It increases roughly proportionally with the temperature difference above the ambient temperature $\Delta T_{th} = T_{heat} - T_{amb}$. Similarly, the practical *economic equivalent value* is roughly proportional to the practical, smaller temperature difference

$$\Delta T_{pr} = \Delta T_{th} - T_{\Delta} \tag{3.44}$$

where T_{Δ} ($\approx 10^{\circ}$C) is the common economic temperature difference needed to reject heat from steam to the ambience.

This economic value of the consumed heat is slightly ($\sim$10–40%) less than the cost of the net electricity that could have been generated by this heat. This value depends—sometimes in a complex way—on the amounts of the purchased heat, on the temperature at which this heat is supplied (if by superheated steam, then the pressure and temperature of the steam), and obviously on the distance between the heat source and the desalination plant. In efficient MED plants this nongenerated electricity is $\sim$4–6 kWh_e/m^3; for MSF it is roughly 70% higher.

Thus, for a typical optimistic example, at an efficient MED plant purchasing heat according to its technological value, which, is assumed, say, at 80% of the cost of the net nongenerated electricity, which, in turn, is assumed at 4.5¢/kWh_e then, at 5kWh_e nongenerated power per m^3 desalted water, the cost of heat per m^3 is approximately 5 kWh_e/m^3*4.5¢/kWh_e*0.8 = 18¢/m^3.

As conveying the heat costs more, some profit for the heat supplier should also be added so that well over 20¢/m^3 is charged for the heat cost component. The total energy cost—considering the additional secondary and tertiary mechanical and electrical energies (1–3 kWh_e/m^3 for MED and 2–4 for MSF), assuming 1.5 kWh_e/m^3 in this example for power from the grid at 5.5¢/kWh_e price—accumulates to more than 30¢/m^3.

Another aspect is the time availability of the heat supply. If it is relatively low, the load factor of the desalination plant also becomes low, which increases the water cost mainly because the capital cost rises.

3.3.3.3.4 The Effect of Discount Rate on Energy Cost This effect is quite small. If the electricity and the heat are procured from external sources, their payments occur generally at the same time as do payments for the water. A time gap of a few days is negligible. If the time gap is a few weeks, the effect is about $d/52 = d$ divided by 52 weeks per year per week, i.e. with d = 8%/year and, say, 3-week lag in water payments after energy purchase, the latter becomes almost 0.5% more expensive.

The effect is greater where the energy is self-generated. The fuel is often purchased a few months before use, and the effect may be an increase of a few percentage points.

3.3.3.4 *Operation and Maintenance (O&M) Cost Component* The numerous items covered by O&M costs can be categorized in three groups:

1. The annual *variable-cost items*, which are theoretically proportional to the annual amount of desalted water P_A; specifically, they have constant value per unit product. These items include chemicals for pretreatment and post-treatment; at least part of the membrane and spares replacement; and some maintenance activities, such as filter cleaning and possibly part of the taxes. In practice, the proportionality is only approximate as a small cost reduction per unit product is expected for larger quantities of chemicals, membranes, and other materials due to the economy of scale.
2. The annual *fixed-cost items*, which are theoretically proportional to the designed production rate of the plant, i.e., the nominal capacity P_N, but do not depend on the number of annual operating hours and load factor. These include insurance and possibly another part of the taxes, part of the staff salaries, Fee for special workers hired for intermittent short periods of maintenance activities, most of the periodic maintenance activities, the other part of membrane and spares replacement, and part of the administrative and management activities.
3. The annual *independent-cost items*, which do not depend on plant size, in principle, at least within a wide range of plant capacities. These include the other portion of the salaries such as those of the minimum shift operators (at least nine people working without overtime pay), one plant manager or supervisor plus minimum maintenance and administration staff, and basic office expenses.

The larger items covered by O&M costs are

1. *Wages and Salaries*. If the plant is fully automated, which applies to small plants, this item is not expensive. Belonging partly to category 2 and partly to category 3 (above), with large plants, this may cost between \$0.5M and \$2M per year.
2. *Consumables, Mainly Chemicals*. Belonging to category 1, this usually costs 4–5¢/m^3 desalted water.
3. *RO or ED Membrane Replacement*. Belonging partly to category 2 and partly to category 1, this ranges within the extreme values of 2–6¢/m^3. Common values with large plants are 3 ± 0.5¢/m^3.
4. *Equipment Replacement and General Maintenance (ER&M)*. Belonging partly to category 2 and partly to category 1, this is preliminarily often

assumed to cost within the extreme range of 2% per year of the direct investment and 3% per year of I_{tot}, the total capitalized investment. Common values with large plants are 2% per year of I_{tot}. Assuming US$720/m^3 per day and a load factor of 92%, this item is

$$2\% \text{ per year}*\$720\ (m^3d^{-1})/(365*0.92) = \$0.0429/m^3 \approx 4.3\,¢/m^3.$$

5. *Insurance*. This represents 0.5–1% per year of the direct investment. Assuming the latter to be \$500 (m^3d^{-1}), with a load factor of 92%, insurance costs:

$$0.75\% \text{ per year}*\$500/(\ m^3d^{-1})/(365*0.92) = \$0.0112/m^3 \approx 1.1\,¢/m^3.$$

These Five items cover most of the O&M costs. In the calculation demonstrated above, for large-scale SWRO at $10^8 m^3$/year per plant, these costs total

Salaries [\$2M per year/($10^8 m^3$/year)]	2¢/m^3
Chemicals	4¢/m^3
Membrane replacement	2.5¢/m^3
Equipment replacement and maintenance (ER&M)	4.3¢/m^3
Insurance	1.1¢/m^3
Total	14¢/m^3 (approximate)

3.3.3.5 Cost Comparison of Single-Purpose Desalination Processes

The following comparison is obviously tentative, approximate, and virtual. It refers to the following [dispersed and normalized] economic data of year 2004 for:

- 100,000 m^3/day plant capacity
- Seawater of 3.5% salinity and
- Maximum product salinity of 500 ppm TDS
- 7.5% discount rate, 30-year plant lifetime, 6 payments per year (thus, annual amortization is 8.214%)
- 5.5¢/kWh$_e$ electricity cost
- With \$1.5M/year salaries and wages
- Equipment replacement, general maintenance and insurance (ERMI) at 2.5% per year of I_{tot}.

The following processes, which are at present the most popular for *large*-scale *seawater* desalination, are compared (see also Table 3.3):

- MSF (high temperature: 120°C $T_{max\ brine}$, 125°C T_{steam})
- HT-MED (high temperature: 120°C $T_{max\ brine}$, 125°C T_{steam})
- LT-MED (low temperature: 70°C $T_{max\ brine}$, 125°C T_{steam})
- SWRO

Table 3.3 Comparison of Desalting Processes[a]

Cost Parameter	MSF	HTMED	LTMED with Heat Pump	SWRO	Unit
GOR	12	20	16	—	Product/steam
Load factor	0.89	0.90	0.90	0.92	
Annual production	32.5	32.9	32.9	33.6	Mm^3/year
Electricity per ton	3.0	2.0	1.3	3.7	kWh/m^3
Total investment	$120M	$130M	$117M	$80M	
Steam	347	208	260	—	tons steam/h
Coal	34.7	20.8	28.5	—	tons coal/h
Equivalent hours/year	7802	7890	7890	8065	h/year
Annual capital cost, 8.214%/year	9.86	10.68	9.61	6.57	$M/year
Annual electricity cost	5.36	3.62	2.35	6.84	$M/year
Annual fuel cost	13.54	8.21	11.24	—	$M/year
Salaries	0.8	0.8	0.8	0.8	$M/year
Annual ERMI cost	3.0	3.25	2.92	2.0	$M/year
Chemicals	1.82	1.84	1.31	1.35	$M/year
Annual membrane replacement cost	—	—	—	0.84	$M/year
Total annual cost	34.38	28.40	28.23	18.40	$M/year
Water cost	100.6	86.3	85.8	54.8	¢/m^3
Water cost with[b] self-supply of power	~89[b]	~80[b]	~82[b]	~64[c]	¢/m^3

[a]Tentative Desalted Water Cost with Self-Supplied steam (via coal), 3.5% seawater salinity, large plants: 100,000 m^3/day plant capacity.

[b]As described in Section 3.3.3.1, the steam for MSF distillation and even MED, if self-generated at ~10 bar, can expand in a turbine before being fed to the evaporator, to generate the electricity for the process, saving about 70% of the purchased electricity.

[c]For SWRO product salinity of 50 ppm TDS, to ensure adequate comparison of equal product quantity (daily) and quality. The cost of 54.8¢/m^3 is for the required product salinity at 500 ppm TDS, easily obtained by SWRO, whereas with MSF and MED the product salinity is almost "automatically" 5–50 ppm TDS, much better than required. Since the latter product has a higher value, while the "adequate comparison" principle calls for equal product quality, the SWRO cost is modified to meet that higher quality. Reducing the product salinity by 100 ppm TDS costs roughly 2¢/m^3, so the extra cost for reduction from 500 to 30 ppm TDS equals ~9.5¢/m^3 to the tentative cost of 54.8¢/m^3 (estimated above), bringing the cost to about 64¢/m^3.

Conclusions for large-scale single-purpose desalination are as follows:

1. Reverse osmosis is the significantly cheapest process, while multistage flash evaporation is more expensive, and multieffect distillation falls in between these extremes.

2. Multieffect distillation at high temperatures and low temperatures, each having relative advantages, differ negligibly in cost.
3. Despite potential complications of the equipment and O&M, self-generation of electric power significantly reduces the product cost.

3.3.3.6 Socioenvironmental Costs A major concern nowadays is to increase the sustainability of desalination plants, preferably by accounting economically for the "externalities," which are miscellaneous costs of environmental and social aspects that were not considered in the past. This will include the (1) costs of improved brine discharge that does not allow for discharge of chemicals or damaging levels of salt concentration and temperature into the sea (or wherever possible), (2) costs of reducing CO_2 emissions when fossil fuels are used, or (3) cost of proliferation prevention when nuclear power is used. There is rapidly growing criticism of the unsustainable design of desalination plants, and it is unlikely that desalination plants will be designed in the future without compensating for these problems in some way or another. Authorities such as municipalities, governments, and international organizations will impose stricter regulations. These costs may be for preventive purposes, thus expended prior to commercial operation of the desalination plant, or as compensation by corrective steps and even as fines for violating the stated regulations and expended during or after the commercial operation.

Quantifying these costs is still problematic and to some extent premature. The regulations are often site-specific, depending on local physical conditions and possibly on the personal requirements of the executives and consultants serving within the local authorities. Also, in many plants the regulations are not yet at final stage. Moreover, the technical and administrative solutions to the environmental problems are still in rather early phases of development. New solutions are expected to emerge, and the costs are thus expected to change.

Effluent discharge costs for seawater desalination may depend on the topography and other local conditions of the seabed. As for brackish-water desalination, the effluent discharge costs may depend on groundwater local conditions and geography such as distance from the nearest sea and transportation options.

The improved brine discharge may consist of acceptable dilution of the outgoing brine returned to the sea and/or longer outfall piping. It may alternatively or additionally consist of additional chemical treatment of the brine to deal with flocculants and iron. The detailed design of such candidate solutions may change drastically as a result of numerous factors—physical, environmental, technical, economical, institutional, administrative, and even political (public relations, electoral, popularity, etc.). The cost of these candidate solutions obviously follows the design of each solution.

As for costs of CO_2, solid particles (fly ash), NO_x and SO_x emissions, and nuclear proliferation prevention, where fossil fuels and nuclear power, respectively are used, the terms *technical* and *administrative solutions* do not refer specifically to desalination but rather to the much broader scope of power and heat generation, thus becoming a component of the pertinent power or heat cost, automatically included in the energy price paid by the desalination facility. The cost of desalted water will increase equally to the costs of cement, steel, beer, sugar, and general household supplies. Although desalted water is still a small component of the overall energy cost, it should not be ignored. Since the energy cost component usually amounts to only 30–40% of the total water cost, the additional costs of these factors is—although certainly not negligible—fairly marginal. In any case, it is calculated and considered, locally, as mentioned previously, by the power generation sector that supplies the energy *to*, not *by*, the desalination plants.

This convenient bypassing of these energy and environment issues holds in most cases except where the desalination plants have self-supplied power and heat. Then these aspects—mainly CO_2, NO_x and SO_x emissions, which hardly represent proliferation prevention—become more difficult to manage, with wider scope relative to externally energized desalination and possibly higher cost due to the negative economy-of-scale effect (because energy generation for desalination is usually an order of magnitude smaller than for common power stations). Roughly speaking, the added cost may range between a few percent and very few percent of the total water cost.

One cannot ignore the possibility that in some cases where desalted water is critically essential, the environmental issues and problems may be partly offset if the cost of their solutions is considered too high for the users or the community. The possibility of providing desalted water of a quality higher than that of the "predesalination" status has a positive environmental impact with an equivalent economic value.

The social aspects are even more site-dependent than the environmental impacts. One positive socioeconomical aspect is employment in various phases of the desalination plants—mainly construction and operation. Further long-term aspects associated with national and multinational desalination programs are the establishment of desalination research and training institutes and the emerging desalination courses in universities, technology institutes, etc. Another aspect is the development of the area near the plant, which is meaningful mainly for large single-purpose desalination plants (for small plants it has a negligible effect, and for dual-purpose plants it is credited to power plant as being the more dominant component of the complex). These social aspects do not reflect, however, on the direct cost of the desalted water.

3.3.3.7 Cost of Desalted Water Conveyance This cost component is not included in water production. However, it concerns the nuclear–conventional desalination controversy in two possible ways:

1. If the energy needed for pumping the desalted water through the distribution system differs in cost, due to differences between power sources
2. If the conveyance distance is longer than in the alternatives methods, due to specific requirements imposed on nuclear plant siting.

For large amounts of desalted water, the costs of conveyance through flat and horizontal topography range between approximately 0.1 and 0.3¢/m^3 per kilometer.

Economy of scale plays a significant role in water conveyance; a double flow rate reduces the cost of transporting per unit water by roughly 7%. The cost may be an order of magnitude higher for small amounts, for seismically problematic distances, or for meaningful elevation differences.

3.4 COST EVALUATION OF COMPLEX DESALINATION SYSTEMS

These systems include dual-purpose plants, hybrid desalination plants, and two-qualities product water streams.

3.4.1 Dual-Purpose Plants

By definition, desalination plants that are part of an industrial complex that produces other commodities may be considered as multipurpose plants. In principle, the economic rationale for desalting water in *multi*-purpose plants is to reduce the cost by sharing facilities, resources, equipment, and services. In practice so far, mostly cogeneration plants have been considered for this sort of sharing. A cogeneration plant, which produces both electricity and good-quality water, is often called a *dual-purpose plant*. The combination is logical for power plants that are situated on coasts using large amounts of screened and filtered seawater for cooling.

The potential power-generating-components of the dual-purpose plant are as follows:

1. Fossil fuel boiler and steam turbine.
2. Nuclear reactor and steam turbine
3. Fossil fuel gas turbine
4. Nuclear reactor and gas turbine
5. Fossil fuel combined cycle (gas turbine, boiler, and steam turbine)
6. Diesel engines
7. Geothermal heat from brines or steam and steam turbine
8. Garbage incineration and steam turbine
9. Solar collectors or ponds and steam turbine
10. Solar photovoltaic panels or cells
11. Wind turbines

All these power generation systems are suitable for the mechanically or electrically driven processes (RO, MVC, or ED) as obviously part of the generated power is diverted to the desalting sector rather than to the grid or more distant users/customers. Generation system types 9 an 10—wind and photovoltaic—which do not have thermal elements, are suitable only for RO, MVC, or ED. Photovoltaic panels or cells in particular are suitable for ED, as the primary electricity generated by the photovoltaic system is lower-voltage DC while the basic electricity for use by the ED system is of the same type, saving costs for transformation to AC at the photovoltaic section and rectification at the ED section, thus saving investment in equipment and unavoidable energy losses. However, wind and photovoltaic systems are relatively small power sources and not very common, in particular as candidates for dual-purpose plants, so they have a narrow potential niche in the desalination market and only for small desalination plants.

Regarding the nine power systems that convert heat as primary energy to electric power, they are naturally suitable for evaporative desalination by low-temperature MED and low-temperature MSF (the latter is very rare). Most of these systems are also suitable for high-temperature MED and MSF: fossil fuel boilers with steam or gas-turbines, nuclear reactors with steam or gas turbines, fossil fuel combined cycle systems, garbage incineration with steam turbines. However, and solar concentrating collectors with steam turbines; diesel engines, geothermal heat, and solar ponds are suitable only for low-temperature distillation [$<170^{\circ}$F (77°C)].

Diesel engines and gas turbines, both fossil fuel and nuclear, are very attractive for *distillation* because they may supply low-grade heat from exhaust gases and cooling-water. This heat is not useful in single-purpose power plants. Using it for desalination does not involve burning fuel, and the only costs are those for the necessary heat exchange equipment; thus this type of low-grade heat is available virtually, free of charge. On the other hand, diesel engines and gas turbines have a narrow niche in the desalination market because

1. Diesel engines and regular fossil fuel gas turbines are considerably less common than steam turbines, either fossil fuel or nuclear.
2. They do not require cooling-water and are often located inland rather than at the seashore, and thus are not suitable for seawater desalination or for brackish-water either, since the latter is desalted by membranes rather than evaporation.
3. They are smaller than steam power plants, so the amount of desalting by evaporation is of narrower scope, losing the economy-of-scale advantage even for the few plants that are located at seashore.
4. They use mostly fuel gas or fuel oil, which is often more expensive than coal per unit heat.
5. Gas turbines have lower thermodynamic efficiency than steam turbines.
6. Because of features 1–5 (above), they are not used as base-load units when connected to the grid, but as peak-load units with a low load factor, which renders the desalination costs quite high.

In principle, when one of the listed power technologies is preferred solely for power generation in view of site-specific conditions while a need for both power and desalted water exists, the dual-purpose plant is a good potential solution, which may sometimes offer significant economic advantages, as with diesel engines and gas turbines. But supplying very low-cost heat from diesel engines and gas turbines with LT-MED, which yields very low-cost desalination, seldom justifies use of a dual-purpose plant with gas turbines.

Each such power source has specific advantages and disadvantages for combinations with desalination in addition to their common economic advantages over single-purpose plants. These common advantages are detailed below:

3.4.1.1 Common Advantages

There are four basic advantages:

1. *Lower Financial Investment Due to Sharing of Facilities*. Seawater supply and brine outfall, land area, site preparation, fences and roads costs are less, and some ancillary equipment such as pumps, compressors, lighting, cranes, transformers as well as administrative buildings can be used in common. For evaporative desalination there is also a large saving in boilers or steam generators and final condensers. Solar photovoltaic and Wind turbines share only part of these items.
2. *Less Fuel and Energy Consumption*. The total combined fuel consumption is less, since some main components are larger in dual-purpose plants, thus having higher efficiencies—in some cases much higher. Also, energy (in the form of electricity, steam, or hot water) is saved in shared facilities and common subsystems as well as by short-distance energy transport. Where large MSF and MED are installed using steam from the turbine exhaust (backpressure turbine), the turbine efficiency is somewhat higher than in a regular steam turbine (condensing turbine) because in the latter low-pressure stages the expanding steam is wet and its moisture decreases the turbine efficiency, as the moisture increases the energy losses due to higher friction.
3. *Lower Charge per Unit energy*. This is a *potential* (not actual) advantage because the energy consumed by the desalination system may be rated less than the price at which it would have been sold to a nearby single-purpose plant. It may even be charged as low as the cost of generating this energy since it is in essence "selling to itself"; however, this possibility is not clearcut, as it might be complicated and difficult to evaluate the unit cost of energy, which is generated at a dual-purpose plant sharing resources with the water section. This difficulty can be overcome by assuming (to a reasonable extent or arbitrarily) the share of the power and the water. An extremely low charge may be obtained if the energy is charged according to the cost of the *marginal* amount of the energy that the desalination sector of the dual-purpose plant consumes, rather than the *average* cost of the generated energy.
4. *Lower Staffing Requirement*. A dual-purpose plant requires fewer staff than the two single-purpose plants (one for power generation and the other for desalination) because of joint operation and maintenance of common facilities.

However, a dual-purpose plant also has disadvantages, compared to two single-purpose plants, as described below.

3.4.1.2 Common Disadvantages These are

1. *Less Overall Flexibility*. The operation of a dual-purpose plant is not as flexible as that of a single-purpose plant. There is economic pressure to maximize the combined production of water and power by increasing the load factor and thus reducing the cost of both water and power. The load factor is maximized by operation at full load for as many hours during its life time as possible. However, such design and operation may reduce the plant flexibility and cause indirect penalties; for instance, the demand for either water or power may occasionally be low—rarely for water, which can be stored, and much more often for power. Where heat from gas turbines or where backpressure steam is used to operate MED or MSF, any reduction in the power generation rate due to low demand results in less steam or hot gas to the turbine and thus from the turbine via exhaust, followed by reduced water production. In the reverse circumstances, when the evaporative desalination—which functions as a *heat sink* for the power generation section—operates at partial load while the power generation is fully loaded, the matching is problematic. Solutions for both situations exist, for an extra cost and risk of operation procedures complications. Certain designs do, indeed, provide for variation in the water–electricity ratio, but at the cost of efficiency or extra investment.

   Operation flexibility is inversely proportional to the degree of coupling; the higher the degree of coupling, the lower the flexibility. The degree of coupling is

   a. Maximum in the classical dual-purpose plant where backpressure steam from the steam turbine exhaust is used to operate MED or MSF, as mentioned above; the flexibility is lowest compared to alternative designs 1b and 1c (below).
   b. Slightly less where the steam for MED or MSF, comes from the *cross-over point* at the steam entrance to the low-pressure section of the steam turbine or from a steam extraction point between turbine stages, or if such extracted steam drives RO pumps or MVC compressors by steam engines or, more likely, steam turbines.
   c. Much lower with RO and MVC that require electricity and get it directly from the power generation section of the dual-purpose plant.

   The higher the degree of coupling, the lower is the flexibility of the combined system, and consequently the higher is the cost.

2. *Lower Availability Factor and Load Factor*. Any incident interrupting the output of one of the two commodities may lead to a disturbance or stoppage in the production of the other, thus increasing the cost per unit product. It is possible to improve the plant availability as a whole and reduce this cost increase by connecting SWRO to the grid or adding devices such as bypass steam lines (in steam power units) so that steam can be directed to the evaporative desalination plant if the turbo-generator is out of operation. Likewise, addition of an auxiliary condenser enables the power plant to be operated if the desalination plant is shut down. However, these devices involve extra investment.

a. As described above regarding the flexibility, power stations are operated to meet the fluctuations in demand due to daily and seasonal variations. Load fluctuations are not favorable for plant economics, in particular for nuclear and desalination systems, which have a significant fixed cost due to their relative high capital cost component. There are various means to minimize the effect of these fluctuations on the load factor. One is to nominate dual-purpose plants for base-load operation. Also, depending on the type of desalination process, the coupling of a desalination system to a power plant can improve the utilization of the latter and thereby decrease power and water costs. On the other hand, it may impose some penalty on the other units where all generation plants are connected to the grid. It is easy to maintain high load factors for RO systems, even if they are coupled to a power plant that is not at base-load. In the case of power outage of the latter, the RO system can be energized from the grid, which is technically easy to install as backup. (Moreover, the continuous and uniform load due to RO improves the load factor of the coupled power plant.) Thus, the expected LF value for RO is ~0.9–0.92.

b. A different situation takes place regarding the operation of a *thermal* desalination system coupled to a power plant. As stated above, the coupling is definitely not flexible. The desalination plant can be operated economically only when the power plant is producing electricity, as any backup arrangement is quite expensive to install and maintain. Moreover, any negative fluctuation in power generation, which reduces the power unit load factor, has a similar effect on the desalination plant load factor. The expected value of LF for MED or MSF is about 90–97% that of the power system's LF; specifically, it is ~0.65–0.8.

3. *Off-Optimum Site and Timing*. The optimal location and/or commissioning date of the desalination plant, in terms of water conveyance, distribution, and supply, may not coincide with those of the power plant, bearing another penalty. As the power section is dominant in most cases, the water section bears the penalties.

Economic comparison of dual- and single-purpose plants involves more than a balance of their respective investment and operating costs. Indirect economic effects may exist, and the environment in which the plants are located should also be considered. It is important to note that while the economic weight of the above mentioned *advantages* of dual-purpose plants is fully expressed in the quantitative cost calculations, it is difficult to do so with the *disadvantages*, due to their narrower scope, complex nature, and indirect role; thus, they are only partly accounted for in the cost calculation.

3.4.1.3 Comparison of Power Generation Systems The most popular candidates for power generation are

- *Fossil fuel*—used in either boilers–steam-turbines combination, or in Combined Cycles (gas turbine–boiler–steam-turbine).

- *Nuclear Fuel*—used in nuclear reactors in combination with either steam-turbines, or with gas-turbines.

It is therefore important to identify and weigh the basic economic differences between dual-purpose plants using fossil fuel and nuclear fuel. These are described below.

Nuclear Fuel is favored for the following reasons:

1. *Energy Charge*. Obviously the source that supplies lower-cost energy is economically favored, whether it is nuclear or fossil, according to the site conditions. Also, in general, the cheaper the thermal energy is, the lower is the cost of the generated electricity. For generic, large-scale models, nuclear energy seems to cost less than the power or heat from fossil fuel (by ~10–20%, but updated data may show different values); nevertheless, fossil fuel is sometimes preferred for non-cost-related reasons such as use of a smaller power grid, negative public acceptance of nuclear energy, political reasons such as governmental refusal to supply nuclear technology or objections to adopt it, licensing problems (e.g., plant location in a seismically unstable area, dense population), or lack of appropriate infrastructure. However, fossil fuel may occasionally be preferred to nuclear fuel for the following *economic* reasons:
   a. The plant in question, either existing or planned, may not be large enough to accommodate a nuclear system to be economically competitive with fossil fuel, or the local fossil fuel prices may be low enough to have lower energy cost than nuclear.
   b. Heat that is usable for MED or MSF is, indeed, quite cheap, but the cost of the power needed to produce desalted water using nuclear fuel is higher than that using alternative energy sources (e.g., from gas turbines vs. steam cycles, as mentioned above).
2. *Maximum Production Rate*. For RO or MVC there is no difference between fossil or nuclear dual-purpose plants with respect to maximum production rate. However, for large-scale MED and MSF desalination, the available *nuclear* reactors enable much larger desalination plants than the existing large fossil fuel steam boilers. This is because for large-scale plants the available heat sources are of fixed dimensions rather than being "tailor-made" for particular cases. With given dimensions of large fossil fuel boilers (~2 GW thermal) and nuclear reactors (~3–4 GW thermal), the economy-of-scale principle can be applied for nuclear thermal desalination. Moreover, the amounts of usable heat for evaporative desalination from nuclear reactors are much greater than from fossil plants since the latter lose 10–15% of the heat generated by the fuel in flue-gases and also since its thermodynamic features allow for less usable heat to be exhausted to the heat sink. Thus, even with equal heat generation capacity of nuclear and fossil sources—the most common water-cooled nuclear power reactors release 60–70% more heat, usable for MED or MSF, compared to fossil fuel steam power units.

This added heat is proportional to the production rate and further increases the economy-of-scale effect.

3. *Load Factor and Time Availability*. As nuclear power involves considerably higher investment than fossil power and much lower output energy cost than fossil power, it is highly preferable economically to operate nuclear plants connected to the grid as base-load units with the fossil plants second in line. Thus the load factor of nuclear power plants is higher, as is the load factor of the linked desalination units. In particular, it is effective with regard to MED and MSF that have virtually no thermal back-up.
4. *Bypass Line*. In case of turbine generator outage, with the steam generator or boiler functioning, working with a bypass line is less expensive in the case of evaporative desalination at a nuclear plant because of the lower initial quality of the steam; in other words, the exergy lost with single-purpose instead of dual-purpose steam is lower under nuclear steam supply system (NSSS) thermodynamic conditions than with the use of a fossil fuel boiler system.
5. *Cost Fluctuations*. Nuclear fuel operation is much less vulnerable to cost fluctuations than is fossil fuel operation.
6. *Geographic Location*. Nuclear plant economics do not vary to any great extent with geographic location. This factor is important for the regions with low availability of fossil fuel. Nuclear desalination offers an additional degree of freedom in securing local energy requirements.
7. *CO_2*. The carbon footprint (i.e., due to CO_2 emissions) is much smaller for nuclear operation, and fewer potential carbon penalties or costs are incurred.

Fossil Fuel is favored for the following reasons:

1. *Safety*. The requirement to ensure no radioactive penetration whatsoever from the nuclear section of the dual-purpose plant to the water section imposes an additional expense on nuclear desalination. This expense is quite small with RO, where such penetration can occur only if the feedwater of the desalination plant comes from the cooling-water of the power section heat rejection (steam condenser for steam turbines or gas cooler for gas turbines), and only if there is leakage in the condenser or gas cooler. This low-probability, low-damage hazard can be eliminated at low cost by various means. However, MSF and even more so MED require extensive additional investment in the thermal coupling equipment that transfers the heat from the nuclear source to the MED or MSF evaporators, as explained above for the potential use of single-purpose nuclear desalination. This high-temperature, corrosion-resistant coupling equipment, which isolates the heating stream in the nuclear system from the heated stream in the desalination system, is expensive not only in terms of the large amount metal employed; it also, includes additional instrumentation, control and alarm equipment, and consumes pumping energy and thermodynamic exergy, thus reducing the process efficiency significantly,

at least with MED. The isolation loop increases the cost of the MED evaporator by roughly 10–15%, the specific thermal energy by ~5–12%, and the pumping energy by ~10–20%.

2. *Licensing*. Licensing procedures are longer and require more effort and expense.
3. *Security and Proliferation Considerations*. In addition to safety and licensing considerations, there is major concern about security and proliferation potential problems with the use of nuclear energy. There is an additional cost for the extra design efforts, security devices, special personnel training, and so on, reflected not only on the power section—which is included in the energy cost component of the desalination section—but also possibly on other cost components of the desalination. In particular, the need for security imposes an economic burden for the potential single-purpose nuclear desalination plants with relatively small nuclear reactors.
4. *Conveyance Distance*. Nuclear plants are located at greater distances from the population centers that they serve than are fossil plants; thus transportation costs for water delivery are higher.
5. *Availability of Suitable Steam Turbines*. The selection of large-scale nuclear steam turbines, which are designed to operate under steam conditions that are different from those for other steam turbines, is rather narrow. This drastically limits the flexibility of adapting MED or MSF to nuclear steam supply systems. Alternatively, appropriate turbines can be designed and developed, but the economic efforts involved are too high: roughly over $50 million range—and the prospective market does not seem to justify such development investment.

It is interesting to note that the first issue—energy charge—is traditionally considered as the main reason to prefer nuclear over fossil fuel as the primary energy source. However, in practice, and *on average*, nuclear energy is preferred for desalination only slightly.

3.4.1.4 Cost Calculation With dual-purpose plants there is a problem of true cost calculation for each product, especially in view of the advantages and disadvantages listed above. The total production cost of the combined two commodities (desalted water and electricity) within the boundaries of a cogeneration plant can be obtained by evaluating the fixed and variable costs of the cogeneration plant. However, when seeking to establish a separate cost for each product, a difficulty arises due to sharing of common cost items such as seawater intake and outfall systems, labor, land use, site development, offices, cranes, and other equipment, transportation vehicles, laboratories, and other common auxiliaries. The main difficulty involves placing an economic value on the *energy* delivered to the desalination plant.

There are several techniques for allocating costs to the two final products of a dual-purpose plant—water and electricity—or to an intermediate product such as steam delivered to the desalination plant.

3.4.1.4.1 The Power Credit Method This is the most applicable method. It is based on comparison between the dual-purpose plant studied and an imaginary (hypothetical) reference single-purpose power plant with both plants using identical power generation methods and major systems (or an alternative primary heat source). For the common cases of fossil fuels and nuclear power generation, the same maximum temperatures and pressures of steam and gases and the same amounts of the input primary heat are attributed to both the specific dual-purpose plant and the hypothetical reference single-purpose power plant.

The annual amount of the net electric energy generated by the hypothetical reference single-purpose plant (E) and the total expenses, including the stabilized capital costs, incurred per average year (C_{E}), are calculated first, from which the cost per salable kWh (C_{kWh}) is derived:

$$C_{\mathrm{kWh}} = \frac{C_{\mathrm{E}}}{E} \tag{3.45}$$

Then the annual amounts of both the desalted water (W) and the lesser amount of the net salable power (E_2) produced by the existing dual-purpose plant (let's call this plant A), as well as its total expenses per average year (C_2), are calculated. E_2 is less than E if the load factors in both cases are equal because of the energy needed for the desalination in the dual-purpose plant. Obviously C_2 is higher than C because of the additional desalination expenses. The desalted water is then charged by these expenses and afterwards, credited by the net salable power cost: $C_2 - E_2{}^{*}C_{\mathrm{kWh}}$, to yield the net expenses attributed to the desalted water.

Assuming that the cost of salable power (C_{kWh}) is the same for plants A and B (the hypothetical single-purpose power plant) [see Eq. (3.45)], the cost of the desalted water can be derived as follows:

$$C = \frac{C_2 - E_2{}^{*}C_{\mathrm{kWh}}}{W} = \frac{(C_2 - E_2{}^{*}C_{\mathrm{E}}/E)}{W} \tag{3.46}$$

The unit price of desalted water C is obtained in \$/m^3 when the average annual quantities of the following parameters are given in m^3, \$, and kWh as appropriate:

- The desalted water W
- The levelized power expenses C_{E}
- The net electric energy gencrated by plant B (the hypothetical single-purpose plant) E
- The total expenses C_2
- The net salable power E_2

For example, a coal power unit of 1580 MW$_{\mathrm{t}}$ is commissioned to desalt 300,000 m^3/day using SWRO while it is capable to generate 600 MW$_{\mathrm{e}}$ net salable power as a single-purpose plant (plant B). Additional data for the dual-purpose plant (plant A) are

Total investment: \$950M.

Load factor for power generation: 0.86; i.e., 7534 full-capacity hours per year

Load factor for desalination: 0.91; i.e., 7972 full-capacity hours per year

Coal consumption: 220 t/h, \$50/t coal = \$11,000/h

O&M expenses: \$54M per year

Electricity from grid: purchased at 5.53¢/kWh

Specific energy for desalted water during power generation: 3.40 kWh/m^3 (3.44 kWh/m^3 during power outage)

Additional data for plant B are

Total investment: \$730M

Load factor for power generation: 0.86 (same as for plant A)

Coal consumption: 220 t/h (same as for plant A)

O&M expenses: \$25M per year

Both plants are amortized at 7.5% discount rate, with 30-year life and bimonthly payments as given above; annual capital cost is 8.214% per year for each plant.

Calculation proceeds as follows:

1. Plant B annual expenses:
   a. Amortization: 8.214%/y*\$730M = \$60M
   b. Fuel: \$11,000/h* 7534 full-capacity hours per year = \$82.9M
   c. O&M expenses: \$25M per year
   d. Total: C_E = \$60M+\$82.9M+\$25M = \$168M
2. Net salable power generation: E = 600 MW*7534 h/y = 4518 GWh/y
3. Cost of kWh: $C_{kWh} = C_E/E$ = (\$168M/y/(4518 GWh/y) = 3.72¢/kWh
4. Plant A annual expenses:
   a. Amortization: 8.214%/y*\$950M = \$78M
   b. Fuel: \$11,000/h* 7534 full-capacity hours per year = \$82.9M
   c. O&M expenses: \$37M per year
   d. Electricity for desalination at power outage: 365(0.91 – 0.86)d/y*300,000 m^3/d*3.44 kWh/m^3 = 18.83 GWh/yr
   e. Grid electricity annual cost: 18.83 GWh/y*\$0.055/kWh = \$1.04M
   f. Total: C_2 = \$78M+\$83M+\$37M+\$1M = \$199M
5. Annual amount of desalted water: $\mathrm{W} = 300,000\mathrm{m}^3/\mathrm{d}^*(365^*0.91)\mathrm{d/y} = 99.64\mathrm{M}\ \mathrm{m}^3/\mathrm{y}$
6. Electricity for desalination during power generation: E_w = (365*0.86)d/y* 300,000 m^3/d*3.4 kWh/m^3 = 32 GWh/y
7. Gross power generation: E_1 = 600 MW*7534h/y = 4518 GWh/y

8. Net salable power generation:

$$E_2 = E_1 - E_w = (4518 - 320)\text{GWh/y} = 4398\ \text{GWh/y} \tag{3.47}$$

9. Cost of the desalted water:

$$\begin{aligned} C &= \frac{(C_2 - E_2{}^{*}C_{\text{kWh}})}{W} \\ &= \frac{(\$199\text{M} - 4198\ \text{GWh}^{*}3.72¢/\text{kWh})/\text{y}}{99.64\text{Mm}^3/\text{y}} \\ &= \$(199 - 156.1)\text{M}/99.64\ \text{Mm}^3 = \$0.43/\text{m}^3 = 43¢/\text{m}^3 \end{aligned} \tag{3.48}$$

Further analysis of this example shows that the three major cost components for desalted water are the capital, O&M, and energy costs.

1. Capital cost:
   a. Direct investment: \$950M–\$730M = \$220M
   b. Specific investment: \$220M/300,000 m^3 = \$733/(m^3 per day)
   c. Power plant indirect investment in electricity diverted to desalination: \$730M*320/4518 = \$51.7M
   d. Gross investment in water: 220 + 51.7 = \$272M
   e. Gross specific investment: \$272M/300,000 m^3 = \$906/(m^3 per day)
   f. Capital cost \$906/(m^3 per day)*0.08214 per year/(365*.91) = \$0.2243/m^3 = 22.43¢/m^3
2. Operation & Maintenance cost:
   a. Direct O&M cost per annum: (\$37M-\$25M)/99.64 Mm^3 = 12.04¢/m^3
   b. Indirect O&M cost of electricity diverted to desalination per annum: \$25M/year*(320 GWh/4518 GWh)/99.64 Mm^3 per year = 1.78¢/m^3
   c. Total O&M: 12.04 + 1.78 = 13.82¢/m^3
3. Energy cost:
   a. Annual fuel attributed to the water:
   b. \$82.9M*(320 GWh per year/4518 GWh per year) = \$5.87M
   c. Annual electricity during power outages: 3.44 kWh/m^3*5.5¢/kWh = 18.92¢/m^3
   d. 18.92¢/m^3*300,000m^3/day*[365(0.91–0.86)day/y] = \$1.04M
   e. Total: 5.87 + 1.04 = \$6.91M per year
   f. Annual energy cost per m^3: \$6.91M/99.64 Mm^3 = 6.94¢/m^3

Thus the *total water cost*: 22.43 + 13.82 + 6.94 = *43.19¢*/m^3 ≈ *43.2¢*/m^3.

The result in this example may seem to be quite low, but it is quite realistic. Four factors contribute to this result: the size of the unit, the low, "standard" feed salinity (only 3.5%; in many cases it is ~4%, requiring larger equipment and more energy), the mild "standard" product salinity (as high as 500 ppm TDS; in many cases it is considerably lower, requiring more equipment and energy), and the status of being part of a dual-purpose plant.

3.4.1.4.2 Modified Power Credit Methods The power credit method benefits the desalted-water industry with economic advantages such as sharing the various resources of the dual-purpose plant (plant A in the example above). Other cost calculation methods allow the cogenerated power to benefit economically (viz., reduced cost) from the shared facilities; these methods will be presented later. However, there are two power credit method modifications that enable even lower cost results:

1. *The Optimal Load Factor Cost Method*. Let us assume that the load factor of plant A (above) is increased, as compared to the LF of plant B (above), enabling plant A to generate the same salable electricity if possible. This can be done by operating more hours per year and/or by increasing the generation load to the fullest permissible and possible extent. If it is not possible to generate and sell the same amounts of electricity from plants A and B, it may still be possible to raise the load factor of plant B to somewhat higher values. The possibility for both situations exists only where the optimal or near-optimal operation of the entire grid system allows it.
2. *The Marginal Water Cost Method*. The difference between this costing method and the LF method (method 1, above) becomes obvious on comparison between the dual- and single-purpose plants (plants A and B above). According to the unmodified power credit method, plants A and B have identical primary heat sources, however, the marginal water cost method assumes identical net salable power and energy as well as the same load factor; thus $E_2 = E$. Accordingly, the primary heat source of plant A is larger than that for plant B, as well as the turbine, the generator, and, in the case of RO, also the heat rejection unit (steam condenser or gas cooler). Consequently, lower water costs (C_{water}) are derived using the economy-of-scale principle, with all the advantages of enlarged size and resource sharing allocated to the desalted water. This costing method has been used only a few times in previous feasibility studies.

The accuracy of the power credit methods might be questionable if the required quantities of power and water were to lead to single- and dual-purpose plants with two different primary heat sources.

In principle, an opposite "symmetric" costing method can be applied by using the water credit method where the single-purpose plant (B) is a desalination system, rather than a power plant. This method apparently has not been used, since reduction of the desalted water cost is the crucial economic issue. Another reason may be the fact that (so far, at least) the main product of common dual-purpose plants is electricity, which is more amenable to use of the water credit method—either because electricity is more basic than desalted water or because it is the main product.

Various different costing methods are discussed below.

3.4.1.4.3 Cost Allocation Methods Based on Exergy, Value, or Caloric Prorating The exergy of a system is a measure of the value of energy. Whereas energy can be neither produced nor destroyed, and therefore it is nondepletable, exergy is variable, and it is the upper limit of the share of energy that is transferable to mechanical work or electric energy in the a process of bringing a system from its present thermodynamic state to stable equilibrium with the environment. The exergy of mechanical and electric energy is higher than the exergy of heat, and the exergy of heat coming from high-temperature steam is higher than the exergy of heat coming from low-temperature steam.

Although thermodynamic analysis has been traditionally based on energy and the first law of thermodynamics, a system's performance can be more appropriately evaluated using the exergy approach [where the term *exergy* means "the potential of a system, or a given amount of heat at a certain temperature or temperature range, to produce mechanical work"]. An exergy analysis describes (1) how this heat-to-work potential is used, (2) how it is divided between several products (final or intermediate)], and (3) wherein the system the losses of this potential take place.

In the exergetic cost allocation method, the overall annual expenditures (C_2) of the cogeneration plant are divided into the following cost components:

1. Direct electricity generation expenditures C_{Ee} allocated exclusively to the generation of electricity supplied to both the grid δC_{Ee} and the desalination plant $(1-\delta)C_{\mathrm{Ee}}$
2. Direct steam production expenditures for providing heat to the desalination plant C_{Se} allocated exclusively to the production of potable water
3. Common electricity and steam production expenditures C_{com}
4. Remaining water production expenditures $C_{\mathrm{w}*}$

Thus

$$C_2 = C_{\mathrm{Ee}} + C_{\mathrm{Se}} + C_{\mathrm{com}} + C_{\mathrm{w}*} \qquad \text{(in US \$/year)} \tag{3.49}$$

The method for calculating each component (1–4, listed above) according to the plant exergy flows is described in Reference 15. The principle is that C_{com}, which is a large component of C_2, be divided between the power βC_{com} and the water $(1-\beta)C_{\mathrm{com}}$ in proportion to the exergy that each of them consumes:

$$\beta = \frac{(\text{power}) - (\text{exergy})}{(\text{total}) - (\text{exergy})} \tag{3.50}$$

The water unit cost is therefore

$$C = \frac{C_2 - \delta C_{\mathrm{Ee}} - \beta C_{\mathrm{com}}}{W} \tag{3.51}$$

It seems that according to this exergy prorating method all the advantages of increased size and resource sharing are allocated decently (and "equivalently") between the power and the desalted water, while with the *power credit method* the water derives relatively *more* benefit; and with the *marginal water method—most* of the benefit.

As discussed above, the power credit method has two additional versions based on the same principle: the optimal load factor cost method and the marginal water cost method. Similarly, the product exergy method has a few more versions: the basic exergic cost theory, the functional approach, and the split-factor method.

There are a few other prorating methods:

- The *caloric method*, also called the *product energy method*, where the sharing is proportional to the enthalpies consumed by the power and water production
- The *proportional value method*, where the sharing is proportional to either (1) the market value of the power and desalted water or (2) their production costs in single-purpose plants
- The *reference cycle method*
- The *work loss method*
- The *enthalpy drop method*

Those methods that are not described here, are described in detail in the References list (at end of the chapter)

The exergetic and caloric methods should include consideration of expenses shared parallel to process energy such as land area, buildings, roads, fences, feedwater, cooling-water, workforce, and equipment (cranes, etc.).

3.4.1.5 Discussion The dozen or so cost calculation methods mentioned above for desalted water versus generated power in dual-purpose plants, each having its own justification, might seem confusing. Several questions may a rise:

1. Should all the possible methods be used to calculate the stated costs in order to provide the plant owners and designers with a comprehensive view of the plant economics?
2. If one were to calculate the costs of water and power using all the methods mentioned above, what should one do with all the results? Which technical and economic decisions regarding the project (e.g., process selection or pricing the products) are taken or changed in view of the results?
3. Which method(s) should be used or mostly recommended?
4. Are any of these methods most or more appropriate for a given plant or preferred with another plant? For example, the power credit method might be most suitable under one or more of the following circumstances:
   a. Where there is a high power–water market price revenues ratio? (i.e., where the amount of power is high and the amount of water is low, measured by the expected annual gross income for each product sale). In this situation a typical low-temperature MED may be heated by backpressure steam from a 1500-MW_t steam boiler, producing 500

$MW_e + 10,000$-t/h salable products. Typical prices might be 10¢/kWh and 70¢/t desalted water; the power sale might be \$50,000/h, while the water sale might be only \$7000/h; thus the main product would be electricity, with water as the secondary product.

b. Where there is a high water : power market price revenues ratio (i.e., where the amount of water is high and the amount of power is low, measured by the expected annual gross income for each product sale). In this situation a typical self-power-supply SWRO plant might employ a 250-$\dot{MW}_t$ steam boiler; producing 35 $MW_e + 13,000$-t/h salable products. With typical prices of 10¢/kWh and 70¢/t desalted water (the same as in example 4a, above), the power sale might be \$3500/h and the water sale \$9100/h; thus the main product would be water while the electricity would be a secondary byproduct.

c. Where the water : power market price revenues ratios are about the same (i.e., where the amounts of water and power, measured by the expected annual gross income for each product sale, are close). In this scenario there would be no secondary product.

Determining which of these scenarios is most amenable to the power credit method is not simple. However, one should be aware of the many existing split-factor methods and their problematic applications.

In practice, the power credit method seems to be the more popular method. This may be explained as follows:

1. A major problem is the compound decision on (a) whether to desalt water and, if the answer is positive, (b) which desalination technology is best for the given project. For this twofold question, and where a dual-purpose plant is considered, it is preferable to use the energy–water splitting (split-factor) methodology that yields the low water cost. This is the power credit. Using another allocation method will yield more expensive water (although lower-cost power), which may, in turn, yield a negative decision—to abandon the option of desalination by using the heat from the power plant and resort to an alternative solution: either to desalt water by, say, RO, or even provide fresh-water conveyance and/or water curtailment. In such a case the power section would be operating at a loss economically.
2. In addition, if all the desalination processes are compared on an economic basis, the power credit method seems to be preferred; otherwise, the comparison is considerably more complicated. The extra complication stems from the need to include different energy cost comparisons in addition to the desalted-water costs comparisons with a questionable weighing method, namely, comparing somewhat complex economic *vectors* rather than the simpler scalar values. The greater complexity of vector comes from frequent cases where the amounts of the water as well as those of the power might differ for the processes being compared, leaving the economic vectors in question not fully defined.

It seems that further work has to be done to try and find the the most suitable yet the most lucrative allocation method for each of the three possible *secondary*-byproduct dual-purpose arrangements: water, power, or neither, thus reducing the number of the relevant cost allocation methods to two: the most suitable (for the circumstances) versus the most lucrative.

3.4.2 Hybrid Desalination Plants

A special qualitative plant design is the possible concept of a *hybrid complex desalination plant*, which, by definition, consists of several desalination units of **different** technologies operating in parallel or connected in series. In most cases two technologies are involved, often RO with MED or MSF. Serieal connection can be effected via the concentrate to feed-water, namely, where the waste stream of one unit (e.g., the RO unit) becomes the feed stream of the second unit (the MED unit), or with product going to feedwater (rarely), or with exhaust heat from MED or MSF turned to the RO or MVC preheater.

The concept of hybrid desalination plants is mentioned frequently. Naturally it has advantages and disadvantages as compared to single-technology plants. In the hybrid case, the disadvantages are indicated first because the idea is in principle quite negative. The philosophy of this concept is to combine the specific advantages of each method. However, in practice this concept seldom shows any economic benefit, except under a few rather unique circumstances.

3.4.2.1 Disadvantages

1. *Using Inferior Technology*. The probability that the different methods are economically equal is very low; under the same conditions one method is slightly or considerably better than the other one. Naturally, if one desalination method is economically superior over the other under the prevailing conditions, adding less competitive systems is economically wrong.
2. *Negative Economy of Scale*. Moreover, for a given total plant capacity, a single system is larger than any of the two or more units of the hybrid plant; thus each system in the hybrid plant has a negative economy of scale compared to a single-process plant.
3. *Complications*. Finally, it is easier to install, operate, control, and maintain a single desalting technology, rather than several ones that operate in parallel, and even more so if they are connected in series.

3.4.2.2 Advantages

1. *Mixing Product Waters*. The lower quality of low-cost product water from the RO unit can be improved, if desired or required, either by product treatment within the RO plant or by mixing with very low-salinity product water from the distillation unit. Sometimes the latter solution may work better, in particular if strict boron content limit is imposed on the desalted water, which is expensive with RO and free with distillation.

2. *Insufficient Low-Cost Heat*. Where very low-cost heat is available, low-temperature MED is the most economic process, but the amount of heat is not sufficient to desalt the needed quantity of water; thus the addition of an RO system may be the best solution.
3. *Correcting Power Load Fluctuations in Dual-Purpose Plants*. Although MED may show better economic promise than RO, large power load fluctuations are expected with the power generated by the dual-purpose plant; hence the practical performance and economics of the coupled MED is damaged. Coupling additional RO units—provided the additional desalination is needed—will improve the power load factor as well as the relative fluctuation in water supply and render the MED more economical.
4. *Raising RO Operation Temperature*. This option might be preferred if MED is less expensive than RO—which might occur at low temperatures (which in turn are favorable for MED but not for RO), and where the economics of RO using the cooling-water from the MED as a higher-temperature feed is still better than both "MED-alone" and "RO-alone" possibilities. Such cases are seldom encountered, however.
5. *Late Addition of Modern Units to Older Ones*. Where the economy of older units falls behind that of new technologies, or where their performance deteriorates but they are still worth operating since the investment has already taken place, or where a local need for more desalted water develops, a good solution might be to enhance production by combining additional units, often using different desalting technology.

It should be noted that the disadvantages of hybrid plants are essentially general principles, whereas the advantages are of a more specific nature that identify niches where the hybrid plants have nevertheless positive potential. In practice, these niches are quite wide.

3.4.2.3 Cost Calculation Where the product waters from the various components of the hybrid plant are mixed, it is easy to calculate the cost of the common product, according to the procedures described above for single-purpose desalting processes.

Calculating the specific cost of each product is somewhat more problematic. For the major cost components, most of the investment and energies is specific to each unit. The questionable issue is how to allocate expenses on shared systems and resources such as intake and outfall, labor, and some chemicals for pre- and posttreatment. Several prorating formulas can be offered—according to production, investments, market values of specific products of different qualities or other criteria—or crediting any of the units, similarly to the costing methods used for dual-purpose plants.

3.4.3 Two-Quality Product Water Streams

Some processes may produce two streams of desalted water, each having different qualities, such as double-distilled water in MED or MSF—the condensate resulting from flashing of the product in high-temperature effects or stages. Similarly, permeates may be collected from the front end and the back end of RO pressure vessels. In most of these cases it is difficult to isolate the cost of each such physically separate stream; however, the main issue is their different *values*, which depend on their different qualities as well as on market demands, which are eventually expressed in their *prices*.

3.4.4 Linear Correlation Common Principle

In all three complex plants discussed here—dual-purpose, hybrid, and two-quality—the arithmetic sum of the two multiplications of the average annual quantity of a product by its cost is equal to the sum of all the levelized expenses for the year. For a dual-purpose plant, it is

$$W^*C + E_2^* C_{\mathrm{kWh}} = C_2 \tag{3.52}$$

Thus, no matter what prorating method is used for charging the desalination unit within the dual-purpose plant for the energy received from the power generating section, and/or for the shared facilities such as seawater intake, this linear correlation of the resulting costs of the salable power and the desalted water according to Equation (3.52) must be satisfied.

Similarly, for hybrid plants and for two-quality systems the linear correlation is

$$W_1{}^*C_{\mathrm{w1}} + W_2{}^*C_{\mathrm{w2}} = \Sigma\text{expenses} \tag{3.53}$$

where W_1 and W_2 are the annual water production values from each of the two sources, C_{w1} and C_{w2} are the costs attributed, respectively, to these water products via any of the allocation methods, and Σexpenses is the sum of all net levelized annual expenses.

In a complex that produces more than two products (e.g., electric power, desalted water, salt and high-pressure steam for a nearby factory), the correlation becomes multilinear:

$$\sum P_k{}^*C_k = \Sigma\text{expenses} \tag{3.54}$$

where P_k, C_k, and k are, respectively, the amounts, the allocated costs, and the given serial number of the product indicated by the index k.

3.5 RELIABILITY OF COST ESTIMATES

In principle, the reliability and accuracy of product cost calculation depends on the method of calculation, as well as its goal and its timing.

3.5.1 Methods

As stated in Section 3.3, there are several ways to evaluate the cost of a product:

1. *Summation of all the cost components*. This method has been described above in detail. The reliability has been discussed in Section 3.3 and is summarized and elaborated on in Section 3.5.2. This method is relevant for any stage of the plant history—late as well as very early. The next method (listed item 2, below) is relevant mainly to the preliminary stages, quite some time before commercial operation, while the last method (3, below) is applicable to preliminary and early stages, up to very few years of commercial operation experience.
2. *Using the cost practically experienced in other plants*. Obviously these reference plants should be appropriate—use the same process, be of the same nature (single-purpose or dual-purpose), yield the same product quantities and qualities, have the same feedwater, be charged according to equal discount rates, expect the same lifetime, and pay the same prices for fuels, electricity, chemicals, and workforce. This method is quite simple to apply. However, the required fitness of data may be difficult to find except where a second or third unit is added to an existing plant, or where a plant is about to be constructed in an environment that is similar economically to that of the reference plant. Otherwise, if such fitness is not found or if it exists only partially, the differences must be identified, assessed, and modified quantitatively. Then, of course, this method is quite less simple to apply.
3. *Various combinations of (methods 1 and 2 (above))*. In view of the limitations of the above mentioned methods, several parts (or sometimes only one part) of the cost calculation input data—either the more reliable or the more complicated part—can be taken from the experience of previous plants while the other parts are calculated as in method 1. Thus, for example, the investment in the process equipment, the workers to be employed, and the amount of energy can be estimated from previous plant data while the site-specific investment and the operation & maintenance (minus labor) may be calculated, particularly for the plant under consideration. The best combination might be the full implementation of method 1 while comparing each step with similar projects, according to method 2 as assisting guidance.

The advantages of method 1 are the lower level of inaccuracies and the wide versatility of uses. The specific disadvantages of this method compared to method 2 are the effort and the time needed to apply it and the risk that certain minor cost components have been forgotten or gave the wrong values—erroneously assumed, calculated, or estimated. If high accuracy is not necessary, as explained in the following paragraph, and/or if the evaluation results are needed within a short time period, the other methods may be used.

To obtain maximum cost accuracy, the input data must be reliable, complete, and precise. Introducing real and accurate values, from the very beginning to the end of a desalination project, of all the payments and ensuring that they are made on time, as well as ensuring that all the water quantities are supplied each month is effective only at the end of the plant lifetime. As mentioned above, the accurate value of the product cost can be obtained only at this very late time in the plant history. The abovementioned input data require accurate listing and well-preserved, accessible documentation.

3.5.2 Levels of Accuracy

Although the best evaluation in principle is done by summing up all the discounted cost components, there are many potential pitfalls to be carefully avoided to attain high accuracy. Fortunately, those pitfalls are of rather minor magnitude.

First, even this accurate value of the product cost depends on the discount rate, which is taken rather arbitrarily, although more or less rationally. (*Hint*: This value is a round or semiround number of percents.)

Thus, for given data of all the payments, the water quantities, and the dates of incoming and outgoing payments, the desalted-water cost is either calculated for a certain stated value of the discount rate or is displayed as a sensitivity test as a one-variable function by simple tabulation of several values of the discount rate—often 5%, 8%, and 10% per year—or as a graph for a certain range of d values.

The post factum true cost is very difficult to obtain also because it should include positive and negative indirect effects such as environmental changes (e.g., warm concentrated brine stream, long-term possible damages), employment during construction and operation, experience, and possible impact on the electric grid.

In most cases the product cost is needed much earlier than at the end of the plant lifetime—namely, either during the period of plant operation, at plant startup, quite often before the beginning of operation, or even before plant construction: during the bidding period, when a decision is made whether or not to erect a desalting plant. Most of the input data are assumed by the person who performs the calculation with a certain degree of subjectivity. The plant supplier tends to assume long plant lifetime, low prices for purchased electricity, high load factor even late in the plant lifetime (ignoring the prediction that failures occur more frequently), and obviously no equipment malfunctioning. The tendency is to ignore the probability for emerging new competing technologies that might render the plant obsolete on one hand and on the other hand, the probability that novelties and improvements may be applied to the plant itself, such as better and/or cheaper membranes. Moreover, there is a tendency to adopt optimistic methodologies that favor desalination such as the marginal water cost method. Finally, positive and mainly negative escalation of prices of project components, such as salaries, fuels, and electricity are rarely considered and are difficult to predict quantitatively.

As the project progresses, more factual data accumulate and replace previously assessed or assumed input values. Thus the degree of the cost accuracy improves during the later stages of the plant history.

3.5.3 Accuracy in Terms of Cost Calculation Objectives

In this respect, distinctions should be made between cost estimates intended for economic comparisons of alternatives, and cost estimates needed for cashflow studies, pricing, or future project evaluations. More specifically, there are several different objectives for which product costs are evaluated:

1. Comparing economically relevant alternative methods for water production, within and out of desalination, including the basic alternative to avoid desalination and find other solutions; for example, in the case of water shortage, if the cost of desalted water is too high, an alternative to desalination may be preferable.
2. Optimization of design by economic comparison of several designs of the same desalting process is a special case of such evaluation.
3. Preparing for financing by forecasting the cashflow using the product cost.
4. Pricing the product.
5. Information for future projects.
6. Outlook on the economy of the country for large-scale, long-term desalination (in particular for desalination *program* rather than a specific *project*). This objective is relevant only to very large scale production items such as water and energy at national scales.

The specific objective for which the product cost is evaluated will determine the degrees of accuracy, detail, and caution at which this task of cost evaluation is carried out. Also, the timing at which the cost evaluation takes place is determined by the specific objective. For example, if the objective is pricing, the time should be very close to the plant commissioning, whereas for comparing alternative desalination processes, the water costs of the alternative candidate processes are evaluated a few years earlier, at the time of bidding or decisions making and well before the plant design stage.

A special duality exists regarding cost estimates for comparisons. Such estimates are prepared very often at the preliminary phase of a national program or at the time of decision regarding installation of a plant. At such early stages, the input data include many values that are assumed:

- On one hand, erroneous input data can be tolerated in cases of certain needed comparisons more than in cases of needed absolute results, since all the compared alternatives tend to deviate from real values to quite close degree. For example, if the size of a plant is the compared variable for which the cost estimates are calculated, then assuming an erroneous (too low) input data power price of 5.5¢/kWh instead of a correct price of 8¢/kWh contributes a very small cost difference (<0.5¢/m^3) between the erroneous and correct comparisons. This is because there are very small power consumption differences between the compared variables or parameters. The conclusions from the comparisons do not change, regardless of whether the power cost estimate is correct, although the total cost difference is more than 5¢/m^3 for double size and the difference in absolute cost is almost 10¢/m^3 for a specific energy of ~3.8 kWh/m^3, as shown in Table 3.4

- On the other hand, deviations of, say, ±5% caused by erroneous input data at early cost estimates are quite acceptable for overall project evaluation, but cannot be tolerated in other cases of needed comparisons. For example, if the decision on the process is the goal of the comparison for which the

Table 3.4 An Example of Low Sensitivity to Erroneous Input

Variable	"Correct" Production Comparison		"Erroneous" Production Comparison		Unit
Production	16.8	33.6	16.8	33.6	Mm^3/ year
Investment, $M	44	80	44	80	—
Annual capital cost, 8.214% per year	3.61	6.57	3.61	6.57	$M/year
Capital cost component	21.54	19.55	21.54	19.55	¢/m^3
Electricity per ton	3.85	3.7	3.85	3.7	kWh/m^3
Power price	8.0	8.0	5.5	5.5	¢/kWh$_e$
Energy cost component	30.8	29.6	21.18	20.35	¢/m^3
Salaries	0.7	0.8	0.7	0.8	$M/year
Annual equipment replacement ERMI	1.1	2.0	1.1	2.0	$M/year
Chemicals	0.68	1.35	0.68	1.35	$M/year
Annual membrane replacement cost	0.42	0.84	0.42	0.84	$M/year
Total annual O&M cost	2.9	5.0	2.9	5.0	$M/year
Water O&M cost component	17.26	14.88	17.26	14.88	¢/m^3
Water cost	69.6	64.03	59.98	54.78	¢/m^3
Absolute error due to erroneous ¢/kWh$_e$			69.6 – 59.98 = 9.62	64.03 – 54.78 = 9.25	¢//m^3
Comparison results: large-scale advantage	69.6 – 64.03 = 5.57		59.98 – 54.78 = 5.20		¢/m^3
Deviation of comparisons results due to erroneous input data			5.57 – 5.20 = 0.37		¢/m^3

cost estimates are calculated, then assuming the same low power price of 5.5¢ instead of 8¢ per kWh for 1.3 kWh/m^3 in MED contributes a very significant erroneous cost difference, about 6.5¢/m^3, to the total true cost difference that may be only 5¢/m^3 for the competing RO with a specific energy of 3.9 kWh/m^3.

Thus, each case has to be examined individually for accuracy:

Cost Estimates at Different Times. As the degree of sophistication and complexity increases, in particular with dual-purpose power-and-desalting plants, several economic estimates may be needed for evaluating the economics of these plants. Different estimations may be employed for the same project, each with a unique purpose. Two examples are listed next.

Costs at Time of Estimate. This is the most common procedure used in the water desalination literature for reporting capital and operating costs of a desalination plant. It has value for feasibility studies, for project screening, and for evaluation of the bid. Its value is diminished in later project stages, when considering real, rather than presumed, capital and operating costs over the plant life time.

Costs at Later Times, toward Project Completion. During the history of a desalination plant, cost components change inevitably, which affects the plant economics compared to preliminary cost data and predictions. Even more so, this occurs during the history of a large-scale desalination program of a water utility or a long-term national desalination program.

This upgrades the cost estimating procedure to the next higher levels of accuracy by the input of more factual data regarding updated changes from the original estimated expenditures on equipment, energy, salaries, and chemicals. This escalation of the cost estimation, as well as any other deviations from the originally estimated or planned production rates, and the lengths of—and expenses incurred on—any unplanned outages, must be accounted and adjusted for. Future inflation rate that will be faced by the project through to its completion must be estimated. This costing procedure is of particular importance when the project requires several years from the time of bid to startup, and/or when the procurement documents allow for progress payments and an extra charge for inflation.

It is relevant to include inflationary effects on payment schedules when estimating the future financial and pricing needs.

3.5.4 Inflation and Deflation

3.5.4.1 Levelized Costs in Terms of "Constant" Currency Inflation is the weighted result of all the price changes (equipment, energy, salaries, chemicals, services, etc. not only within the desalination project but within the entire market) with the time, and thus expresses the change in terms of the purchasing power of the relevant currency; if the changes decrease, the constant currency cost (i.e., the inflation rate) is negative, this is called *deflation*.

Inflation and deflation complicate these estimates and contribute to fuzzy results. Using "constant" currency with a known fixed value and purchasing power at the time of the estimate, or at any other convenient reference date, is a common method for bypassing this problem.

3.5.4.2 *Levelized Costs in Terms of "Current" Money* The levelized unit cost may be determined—in addition to "constant" currency, where the costs are expressed in terms of the purchasing power of the relevant currency—in terms of "current money" or as "spent money." The constant-currency cost is replaced by the actual increased or decreased values, usually by the rate of inflation or deflation of a given number of current dollars (or any other relevant currency).

The transition from actual expenses in current money of year j payments to constant money (constant of the plant commissioning date D) is effected by using the factor $(1+d)^{D-D_j}/\Pi_{(j)}(1+\text{inf}_y)$ instead of $(1+d)^{D-D_j}$ in the formulas presented in Section 3.3, where "inf" is the inflation rate and inf_y is the inflation rate during year y. (Π is the multiplication notation). Then y is counted from the date of commissioning between 1 and j. If j is also counted from the date of plant commissioning, then

$$D - Dj = j \tag{3.55}$$

Often in literature the factor $(1+d)^{D-Dj}/\Pi_{(j)}(1+\text{inf}_y)$ is presented as Z^j

$$Z^j = (1+d)^{D-Dj}/\Pi_{(j)}(1+\text{inf}_y) \tag{3.56}$$

where

$$Z = \frac{1+d}{1+e_\text{a}} \tag{3.57}$$

and e_a is the average inflation rate during the referred period of j years:

$$\begin{aligned} e_\text{a} = [\Pi_{(j)}(1+\text{inf}_y)]^{1/j} - 1 = [(1+\text{inf}_1)(1+\text{inf}_2) \\ (1+\text{inf}_3)\cdots(1+\text{inf}_y)\cdots(1+\text{inf}_j)]^{1/j} - 1 \end{aligned} \tag{3.58}$$

Regardless of which methodology is used, only the total *cost* of the product water is thus determined, not the selling *price*.

3.6 DESIGN AND OPTIMIZATION OF PLANT AND PROCESS PARAMETERS

There are two kinds of a desalination system design parameters—those that influence the economics, mainly the external *plant* parameters, which constitute input data, and those that are influenced by the economics, namely, *process* parameters. The latter are determined *in principle* by economical considerations via optimization, unless the optimal value is beyond a limiting technological constraint that replaces the stated optimal value, and *in practice* also by experience and professional tradition.

The first type includes *in principle* the quality and the production rate of the desalted water, timing and location of the project (the latter determines the feedwater properties—salinity, analysis, and temperatures—as well as other environmental, geographic, and infrastructural characteristics), the type of plant—contiguous (multipurpose) or standalone (single-purpose)—and the desalination method. These plant parameters may constitute user requirements or input data, expressing the plant owner's preference and possible constraints. Initially they seem to have a one-way effect on the economics. When methodically examined, however, almost all of them have mutual feedback effects to some extent with economic considerations. Thus, for example, the user may compromise on higher product-water salinity (e.g., 400 ppm TDS instead of the 250 ppm initially required level) to reduce the water cost by a few percentage points.

Parameters of the second type are internal plant specifications. They include selection of the process itself and the specific energies, the number and magnitude of effects or stages, and the top brine temperature—in thermal desalination plants—as well as the membrane configuration and type, flux, recovery ratio and rejection rate, operating temperature, and pressure in RO plants. These process parameters obviously influence the desalted-water production cost. However, the economic feedback is very effective and should be used, with sound engineering, to determine these parameters. This is done by the use of applicable optimization techniques.

The usual goal of optimization is to identify the parameters with which optimal results, in most cases lowest product cost, are obtained. The output of the optimization is a set of qualitative choices and quantitative values according to which a desired project is implemented. In the optimization procedure the relevant options are examined by weighing (to the fullest extent possible) their advantages and disadvantages. Eventually the product cost is calculated for all the promising candidate options, and the best one is found by discrete comparison or by the minimum-derivative technique.

One problem with optimization is the sensitivity of the results to variations in the input values; small changes in these data may yield large quantitative and qualitative differences with the output results. Thus, input data should be cautiously and accurately determined. This might be difficult as some of the input data are not known accurately during the optimization stage of the project.

Another problem may occur in case of a dual-purpose plant. If the lowest water cost is found, through optimization, for an option (i.e., one of a few possible sets of parameters) in which the electrical energy is generated at a higher cost than with one or more other options, a method should be selected to identify the best option. This situation is seldom expected, but has been encountered in a previous study citation reference No. 15, page 62 and may be encountered in future comparisons.

Four outstanding input parameters are the most characteristic of all desalination processes and have significant economical effects: the production rate, feedwater quality (hardness, turbidity, and total salinity), desalted-water quality (mainly total salinity), and the recovery ratio (their roles have been briefly described in Sections 3.3.2 and 3.3.3.1):

- The *production rate*, namely, plant capacity, is based on the economy-of-scale principle. The economic contribution of large capacities manifests itself in four aspects. In percentage cost reduction, the first one is the *labor* (workforce) expense; e.g., a plant of with a production rate of 100 mgd (million gallon per day) employs roughly only 4 times as many the workers as a 3-mgd plant. Second is the *investment*, as discussed in detail in Section 3.3.2. The third aspect of product cost reduction based on economy of scale is the *energy*, as the energy efficiencies of equipment items usually increase with equipment size. The fourth and final aspect is the potential benefit of *secondary improvements*, which also increases with plant size. This input parameter is apparently determined by the need for desalted water rather than by economic considerations, but for the reasons mentioned above, decisions regarding erection of a larger plant or expansion of an existing one can be made within the framework of long-term desalination program.
- The *feedwater quality* is also discussed in detail in Section 3.3.2, explaining the contributions of low feed salinity to the reduction of desalination cost. Decisions regarding this parameter cannot be based on economic considerations except where high-salinity feedwater can be mixed with other nonpotable water to reduce this salinity. Mixing may take place with available lower-salinity streams such as high-hardness brackish-water, concentrate from a brackish-water desalting plant, or even treated sewage water.
- The *quality of the product water* is also mentioned above and determined on the basis of health and agricultural criteria and standards. Economically, in the 600–50 ppm range it may add roughly 2¢/m^3 per 100 ppm product-water salinity reduction. Boron removal from the SWRO permeate, ~1.5–2 ppm, to the required level of 0.2–0.5 ppm today costs 2–10¢/m^3. Although the product quality is an input parameter, economic consideration may lead to compromises within a rather narrow range of salinity and boron levels.

A few notes regarding the economic aspects of the product-water quality are appropriate here. Assigning milder restrictions and requirements for the desalted-water quality where possible, will consequently reduce the product cost. At some desalination plants the required salinity of the desalted water is too low, driving the water cost up. This requirement may be too severe, since the average domestic water consumption rate is ~100 m^3 per year per capita, only $\leq$0.5–1 m^3 per year per capita of which is being used for drinking. Also, the issue of whether the quality of >99% of domestic water should be as high as that of the "best" drinking water should be studied. Furthermore, people obtain their beverage water from juices, milk, bottled mineral water, and solid food and only part from the faucet; thus, another issue to consider is whether faucet (tap) water has to be as good as the "best" drinking water. Finally, adding another economic criterion for water desalination—the cost of a unit of salt removed as a function of, inter alia, water salinity—might help in the optimal design of membrane desalination processes.

The three parameters listed above—production rate, feedwater salinity, and desalted-water quality—are plant characteristics. The fourth one is a process parameter:

- A high *recovery ratio*, namely, the amount of desalted water per unit feedwater, decreases the feed requirement for a given plant output and hence the pretreatment equipment and chemical consumption. For single-purpose plants of electrically driven processes (RO, ED, and MVC), a high recovery ratio also reduces the pumping energy, the intake, and outfall equipment size, including feed pumps. [In most dual-purpose plants the feedwater is taken from the cooling-water stream, so the intake and outfall systems are not affected by the recovery ratio; similarly, in seawater MFS and MED, the amount of seawater (feedwater plus cooling-water) does not depend significantly on the recovery ratio.] In the case of RO, a high recovery ratio also reduces the main process pumps and the pressure recovery device; smaller pumps and pressure recovery devices reduce the investment and conserve energy. These economic savings, however, bear some penalties. A higher recovery ratio means higher salt concentrations in the main desalination system and, in particular, toward the brine outlet. Consequently, it increases the specific energy in the main desalination system as well as the probability of scaling in the brine outlet zone of the main desalination system; and, especially in the case of RO, it adversely affects the product-water quality. This design parameter is apparently not an input factor but should result from design optimization. However, in many cases the recovery ratio is determined by a chemical constraint of a maximum brine salinity limit of 7–8.5% to avoid sulfate scaling in the evaporators and on the membranes. Thus, for seawater this constraint imposes an upper limit for a recovery ratio of 44–60%, depending on the seawater salinity, even if the optimization results in higher recovery ratio value.

Other major process parameters determined by a combination of technological considerations and economical optimization are

1. The *maximum process temperature* in distillation plants, which depends on several conditions and requirements, such as single- or dual-purpose plant, ratio of product to available heat, temperature, and price at which the heat is available. For example, if the amount of available low-cost heat per unit

desalted water is above 60 kWh_t per needed ton product, this temperature is about ~70°C. In some cases this parameter may lead, in view of the prevailing conditions, to an optimal arrangement where a heat pump is introduced.

2. The *average temperature drop per effect of MED*—ΔT_{eff}, which, in turn, determines partly (but to a large extent) the energy of the plant as well as the number, size, and cost of effects of the evaporator, which is the main component of the plant [when considering the maximum process temperature (see list item 1, above) and the feedwater temperature]. Thus, the main economic characteristics of the process are largely determined—the specific energy and the specific investment. This temperature drop has a significant thermodynamic meaning and depends on many variables, including prices of electric power and heat transfer equipment, heat transfer coefficients, power generation system efficiency, and load factor. This optimization step is demonstrated and discussed in detail in Section 3.A.2 (in the Appendix). Basically, it is the best compromise between the cost of the process energy, per unit product (ton desalted water) which increases linearly with ΔT_{eff}, and the cost per unit product of the evaporator, which decreases with ΔT_{eff}.
3. The *temperature increase in the brine heater of MSF*— ΔT_{BH}, which, in turn (similarly to MED), determines partly (but to large extent) the energy of the plant as well as the size and cost of the evaporator, which is the main component of the plant. Thus, the main economic characteristics of the process are largely determined—the specific energy and the specific investment. This temperature increase has a significant thermodynamic meaning and depends on many variables (similarly to MED), such as prices of electric power and heat transfer equipment, heat transfer coefficients, power generation system efficiency, and load factor. The optimal value is the result of a compromise between the cost of the process energy, which increases linearly with ΔT_{BH}, and the cost of the evaporator, which decreases with ΔT_{BH}.
4. The *number of MSF stages*, which is the best compromise between the cost of the envelope of the stage, i.e., the vessel or partitions, which increases linearly with the number of MSF stages, and the cost of the internal components of the evaporator, mainly the heat transfer surfaces and droplet separators, which decreases with the number of stages.
5. The *pressure of RO process*, which, in turn, determines the average water flux through the membranes. This is only theoretically true for most cases with the first RO pass, as the practical flux is lower than optimum, due to the fouling limit and concentration polarization phenomena, but it is practical for the second or third pass. This pressure is the best compromise between the costs of the energy and the pump, which increase with RO pressure and the cost of the membranes, pressure vessels, and associated equipment and construction, which decrease with RO pressure and the flux.

3.7 PROCEDURES AND PROGRAMS

To provide a common basis for comparisons and ensure a reliable cost calculation that includes all the relevant items and factors, it has been recognized that a formal procedure should be established for these purposes. Several procedures have been offered since the mid-1950s.

3.7.1 US Methods

In 1956 the US federal Office of Saline Water, Department of the Interior (OSW) published a 44-page detailed report titled *A Standardized Procedure for Estimating Costs of Saline Water Conversion* [1], which refers only to thermal desalination since membrane processes were in the earliest stages of development. The report is very informative as it supplies both the calculating procedure and input data on prices of energy and equipment. Almost all input numerical values are outdated, but most of the procedures, published as guidelines and cost component checklists, are still valid.

This was followed 16 years later by an OSW report that included a cost format in the 1972 edition of the *Desalting Handbook for Planners* [6]. It includes a tabulation of all direct capital costs and indirect costs (interest incurred during construction, working capital, and contingency expenditure). It also includes a tabulation of annual operating costs, including an annual cost for depreciating capital. The total cost for production of desalted water is then worked out as the total annual costs divided by the annual production.

3.7.2 International Methods

An adaptation of the 1972 OSW format was used in the cost updates prepared for the US Office of Water Research and Technology (OSRT) and is recognized by the International Desalination Association (IDA). A software package for calculation of costs for seawater desalination, following this format, is now available through IDA. It is designed to calculate cost for three commercial processes—MSF, MED, and RO—for seawater desalination. The desalted-water cost does not include product-water storage, distribution, or administrative costs. The sample cost figures for RO, MED, and MSF based on the IDA seawater desalting cost program are presented in Table 3.5. (*Note:* The values in this table are extremely pessimistic for present day's desalination. Currently, plants of this capacity can supply water at a cost ~30% less!!).

3.7.3 UN Methods

Fossil-fuel-based and nuclear methods of water desalination can be compared on an economic basis. The preferred source of primary energy for seawater desalination goes a long way in the economics of desalination system. Nuclear energy becomes

Table 3.5 IDA Seawater Desalting Cost Sample Data

Item	RO	MED	MSF
Plant capacity, mgd	5	5	5
Plant capacity, m^3/ day	19,000	19,000	19,000
Seawater TDS, ppm	37,000	37,000	37,000
Plant load factor, %	90	90	90
Interest rate, %	8	8	8
Electricity rate, $/kWh	0.08	0.08	0.08
Total investment, $\$/10^6$	25.6	44.7	44.3
Cost of water, $\$/m^3$	1.26	1.37	1.71

increasingly competitive in large installations compared to fossil-fuel-based desalination plants because fuel costs are considerably lower for nuclear power plants than for fossil fuel plants. The ratio of fuel cost to total cost in a fossil fuel plant is significantly higher than in a nuclear plant. This is likely to further increase as the inventory of fossil fuel is gradually depleting and the oil prices are very unstable and fluctuate widely. A third competitor is grid electricity for RO, ED, and MVC.

In the early 1960s the United Nations published a report [5] similar to the first OSW procedure. About 30 years later, the United Nations' International Atomic Energy Agency (IAEA) prepared the Desalination Economic Evaluation Program (DEEP). The DEEP program covers single- as well as dual-purpose plants. The latter is evaluated using the principles of the power credit method. Its main objective is comparison of possibilities.

During 1991/92, a generic investigation was conducted by IAEA on the technical approach and the comparative cost of utilizing nuclear energy with various state-of-the-art desalination technologies. A key outcome of the investigation was the development of a convenient methodology for rapidly calculating performance and costs of power and water production for various power and desalination plant couplings. DEEP is the IAEA's software package for the economic comparison of seawater desalination plants, including nuclear options. It has been validated with reference cases. A user-friendly version based on the validated version has been prepared along with a manual with information on how to use DEEP [16]. The methodology is incorporated in an EXCEL spreadsheet routine, which is available from the Nuclear Power Technology Development Section of the IAEA.

The methodology is suitable for economic evaluations and screening analyses of various desalination and energy source options. The methodology is imbedded in a spreadsheet routine containing simplified sizing and cost algorithms that are easy to implement and are generally applicable to a wide variety of equipment and representative state-of-the-art technologies. The spreadsheet methodology was substantially improved to include the capability of modeling many types of nuclear and fossil electric power and heat sources of varying magnitudes depending on site-specific demands. It embodies the basic technical and economic principles of power and desalination plant performance and can be adapted to any site conditions. Current cost and performance data are incorporated so that the spreadsheet can be quickly adapted to analyze a large variety of options with very little new input data required.

The output includes the levelized cost of water and power, breakdowns of cost components, energy consumption, and net salable power for each selected option. Specific power plants are modeled by adjustment of input data, including design power, power cycle parameters, and costs. The desalination systems are modeled for meeting the World Health Organization (WHO) drinking-water standards. However, modifications and changes can easily be incorporated on the basis of description in the manual.

The spreadsheet serves three important objectives:

1. It enables side-by-side comparison of a large number of design alternatives on a consistent basis with common assumptions.
2. It enables quick identification of the lowest cost options for providing a specified quantity of desalted water and/or power at a given location.
3. It gives an approximate cost of desalted water and power as a function of quantity and site-specific parameters, including temperatures and salinity.

For planning an actual project, final assessment of project costs should be assessed more accurately on the basis of more substantive information, including project design and vendor-specific data.

The use of the spreadsheet methodology is limited to the types of the power or heating plants, desalination processes, and coupling models described in the manual. The spreadsheet models the common commercial processes for large-scale seawater desalination as given in Table 3.6. The MED process is assumed to be either a low-temperature, horizontal-tube system or a high-temperature, vertical-tube evaporator. (Modification to a high temperature horizontal-tube evaporator is possible.) Heat is supplied in the form of low-grade dry saturated steam, and the maximum brine temperature is assumed to be limited to 70°C and 125°C for low- and high-temperature MED, respectively. MSF is modeled as a once-through process. Heat is supplied by hot water or low-grade saturated steam, and the maximum brine temperature is limited to 135°C.

Stand-alone (the IAEA term for single-purpose) construction assumes that the RO plant is only electrically coupled to the power plant. Contiguous RO (the IAEA

Table 3.6 Desalination Processes Contained in the Spreadsheet

Process	Abbreviation	Description
Distillation	MED	Multieffect distillation
(thermal desalination)	MSF	Multistage flash distillation
Membrane	SA- RO	Stand-alone reverse osmosis
	C-RO	Contiguous reverse osmosis
Hybrid	MED-RO	Multieffect distillation with reverse osmosis
	MSF-RO	Multistage flash distillation with RO

term for dual-purpose construction) assumes that the RO plant shares a common seawater intake and outfall with the power plant cooling system and may utilize the power plant reject heat for the feedwater preheating. In addition, there are two membrane options available: hollow-fiber and spiral-wound membranes.

According to the hybrid concept, the thermal desalination and membrane desalination plants together provide the desired water quality and demand. Feedwater to RO is taken from the condenser reject water of the thermal desalination plant or power plant condenser.

The spreadsheet includes different types of nuclear reactors (Table 3.7). Some of them can provide both electricity and heat and can be coupled with any of the desalination plant concepts such as thermal desalination, the membrane process, or both. Some of the energy sources are *heat-only* systems and are coupled only with thermal desalination systems using hot water or low-pressure steam. The other energy sources are considered as *power-only* systems and are coupled only with membrane processes using electricity. The input data in the spreadsheet can be adjusted to enable the model to approximate any type of nuclear steam power plant, including liquid metal systems.

In case of water-cooled nuclear reactors, the spreadsheet considers the specification of an intermediate isolation loop for either steam or hot-water supply to the desalination system to prevent the possibility of radioactive contamination of the product water. The model also includes a backup boiler for ensuring energy supply to the thermal desalination plant if the main energy source is not operating. This is a highly expensive addition, in terms of both investment and fuel.

The procedure for water and power cost calculations used by DEEP begins by selecting the site location, type of energy source, and type of desalination process at the site. Then the desalted-water demand at a certain timepoint at that site is defined. The size of the power plant should be large enough to meet the requirement of a desalination plant. Various data on site, performance, and cost input are then specified. Power plant cost input data of an imaginary reference single-purpose power plant (like plant B in Section 3.4.1.4) of identical heat source enables calculation of the base (uncoupled) power cost that is used to determine the energy cost for desalination. Desalination input data consists of information to determine output, energy consumption, equipment size, and quantity. For the case of thermal

Table 3.7 Energy Sources Contained in the Spreadsheet

Energy Source	Abbreviation	Description	Plant Type
Nuclear	PWR	Pressurized (light-) water reactor	Cogeneration (dual-purpose)
Nuclear	PHWR	Pressurized heavy-water reactor	Cogeneration
Nuclear	SPWR	Small pressurized water reactor	Cogeneration
Nuclear	HR	Heat reactor (steam or hot water)	Heat only

desalination, the maximum brine temperature needs to be specified to calculate the gained output ratio (GOR), heat requirement from the power plant, and modified power plant performance. Desalination cost input (excluding energy) consists of information relating to the equipment cost and O&M cost of thermal desalination and RO systems.

For each cogeneration (dual-purpose) plant, the spreadsheet calculates power and water production performance, including costs for coupling. It first calculates the base power plant performance. Then, it calculates the modified system performance for coupling with a thermal desalination system where relevant. A key output of the modified plant performance is the "lost electric power" resulting from elevation of the turbine backpressure. Next, the desalination plant calculations are performed. The size of the thermal desalination unit is set by the spreadsheet to a typical size, depending on the total desalination plant capacity. Electrical energy consumption and maintenance conditions are also calculated as input for cost calculation.

Calculations are performed for the stand-alone (single-purpose) and the contiguous (dual-purpose) RO plants where relevant. The unit size is determined manually. The spreadsheet calculates the number of units required, electricity consumption, and maintenance conditions for the cost analysis. System performance is adjusted for input values of the temperature and salinity of the feedwater. The recovery ratio may be reduced for cases of higher feedwater salinity to avoid scaling and to achieve required product quality. The spreadsheet also calculates the performance and cost of the hybrid desalination systems if the energy source produces both power and heat.

From the performance calculations and the cost input, the spreadsheet calculates the base power cost and the desalted-water cost by summing the annual capital, fuel (or energy), and O&M costs allocated to desalination and dividing this sum by the annual product output. The desalination cost includes the post-treatment of the distillate to produce drinking water as per WHO standards.

The DEEP spreadsheet can be used for comparative calculations, in order to determine under which conditions either nuclear or fossil fuel desalination is economically competitive. The water cost for a sample case is presented in Table 3.8.

Table 3.8 IAEA Desalination Economic Evaluation Program (DEEP): Sample Case

Parameter	RO	MED	Hybrid (MED-RO)
Plant capacity, m^3/d	500,000	500,000	500,000
Energy source	Nuclear	Nuclear	Nuclear
Seawater TDS, ppm	38,500	38,500	38,500
Interest rate, %	8	8	8
Stabilized power cost ($/kWh)	0.047	0.047	0.047
Water cost, $/m^3$	0.72	1.06	0.87

Source: Data from Reference 21.

3.8 DOMESTIC VERSUS FOREIGN CURRENCIES

In many countries the decision-makers may prefer to pay a little more for desalted water or power if the local currency percentage of the cost is higher.

In comparing the alternatives of desalination, nuclear, grid electricity, and fossil fuel energy supply sources for desalination, it is important to know the local (domestic)–foreign currency ratio in the product cost. This seems to be meaningful for developing (and possibly other) countries that are short in foreign currency. In countries that import fossil fuel, the foreign currency component in the energy cost becomes relatively high. Therefore the nuclear option might be preferred even if the total product cost for this option is not lower than those of the alternatives. Also, in many cases the civil engineering and to some extent even the electrical engineering cost components in the nuclear installation are relatively higher. This expense is usually local currency, giving additional motivation to prefer the nuclear option.

Another cost component is labor (workforce): employment and training. Many countries prefer to increase or maximize the local participation in infrastructural programs and projects. This may be manifested in recruiting most of the employed workforce from local sources for power generation and for water supply because of (1) local currency considerations, (2) better communication, and (3) increased employment. Large desalination programs as well as nuclear power plants require a meaningful amount of staffing at the high professional expertise level. For desalination in general and nuclear desalination in particular, the term *specialized expertise* refers not only to O&M tasks but also to design and construction activities. Selection of adequate staff, and preparing and training them, requires some preliminary organization such as establishing special academic desalination studies classes within the departments of chemical or mechanical engineering at some universities or at least courses—studies and tests—outside the academia.

3.9 VALUE OF DESALTED WATER

The *value* is the worth of the desalted water to the consumer. It depends on many factors such as water scarcity, specific applications, desalted-water quantity and salinity, and qualities and costs of additional and alternative water sources. Thus, for example, the value of water for luxurious hotels or military bases may be quite high, whereas for large-scale agricultural irrigation it is usually expected to be low.

In many cases it is difficult to assess the value of the desalted water. However, it is reasonable to assume that, because of the relatively high availability and reliability of the desalted-water controlled quantity and quality, the real value of desalted water may be higher than that of alternative-source water, in particular where the latter has lower quality, such as high salinity, boron content, or hardness.

In principle, it is justified to use desalted water if the real value of such marginal desalted water (considering the availability, quality, and quantity) is higher than its cost, provided, of course, that there is no better water source. The potential consumer of desalted water—in many cases the national or regional water utility or a large municipality—should consider this possibility while making decisions regarding payment for desalted water. Similarly, a government should consider this aspect while determining its desalination policy and desalination program.

3.10 PRICES OF WATER AND POWER

Price is the *money paid to the supplier* by the consumer. It differs from the *cost*, which the supplier expended to produce the water or the energy, by the net profit. Under normal conditions the price level is between the cost and the abovementioned value of the product to the consumer. Thus, both the supplier and the consumer benefit. If the consumer produces for her/himself, the price equals the cost. On the other hand, if the authorities intervene by taxing or by subsidizing (having economic, social, or political motives)—the price received by the supplier is different from that paid by the consumer.

The situation may become more complex for dual-purpose plants. Even under normal market conditions one product may be sold at a price that leaves a relatively modest profit while the other product yields a high profit. It is also possible to have an overall positive net profit due to a "good" price of one product, such as electricity, even if the other—say, the desalted water—is sold for a price lower than its cost. In such case, the latter is practically subsidized by the former. This might sometimes give a misleading impression (as some people mix the notions of cost and price) that the cost of desalination is quite low, with a risk that erroreous decisions may follow.

One issue that concerns power price affecting desalted-water cost refers to dual-purpose plants. In the preferred power credit method, the water is charged by the cost of the differential electrical energy that would have been sold if the same power plant were to operate on a single-purpose basis. One may question whether it is not justified to charge the water by the price of this "unsold" amount of electricity, which is the commercial loss of the entire plant.

Quantitatively, for example, if the water consumes 4 kWh/m^3 that cost 4.5¢/kWh to generate and the water cost is therefore 60¢/m^3, and if the net price of electricity is, say, 5.4¢/kWh and the net price of the water is 72¢/m^3 (i.e., 20% profit for each), then the plant loses [= 4 kWh/m^3 $*$ (5.4 − 4.5)¢/kWh] that have to be recovered by either increasing the price of the water or charging the water cost for the lost profit. The real profit of the water—if the price cannot be raised—is then only 8.4¢/m^3 rather than 12¢/m^3. Thus, the gross water cost is 72¢/m^3 − 8.4¢/m^3 = 63.6¢/m^3, rather than 60¢/m^3 given above.

Raising the price by the authorities or by the water supplier—probably with the approval of the authorities—may serve two additional goals: (1) to reduce the demand for water and encourage saving and (2) to finance current and future desalination projects or research and development (R&D) activities.

3.11 CASHFLOW, PAYBACK PERIOD, AND RATE OF RETURN

These three criteria are strongly interconnected. They concern mainly the investors and the owners, but to a lesser extent they may be of interest to the water customers and to the government. Each of them constitutes a special criterion regarding the financial aspects of the investment:

1. *Cashflow* provides two-dimensional information about the financial balance and timetable according to which investment components and other expenses—not only fixed payments—are paid, and which payments for the products are received.
2. The *payback period* is the time difference between the date of the cashflow turning zero from previously negative and the (representative) date of the investment.
3. The *rate of return* is the equivalent compounded interest at which the investment is recovered, by the revenues from product sales, considering the price.

These criteria, as well as others (such as the ratio of net profit: investment) depend strongly on product price, investment, plant load factor, and plant lifetime. They are significant in determining price and mainly in financing the investments.

It may be interesting to indicate that these criteria can be bypassed from the plant owners perspective, if the equipment installed in the plant, the price of which is the major part of the plant investment, will be rented by the contractor who installed the plant to the owners or the operator. Thus, almost all the plant operator's expenses will be continuous and annual, linked to the sales revenues rather than a preliminary large investment. Such an arrangement has not yet taken place in the desalination market.

It may also be interesting and important to note that when a long-term, multiplant program is initiated, investment may take place over a multiyear period. A single investor in such a program will have to endure very long payback periods, especially if the production and/or generation capabilities are expanding, because the cashflow of the program in later years consists of revenues from water/power sales (incomes) on one hand and *progressing investment* (expenditures) on the other hand.

3.12 FINANCING

As mentioned above, the overall gross investments in large seawater desalination plants are in the range of \$1000 per unit capacity of cubic meter per day. Similarly, it is around \$1000 per installed power unit capacity (kW_e) for fossil fuel plants and roughly 50–200% more for nuclear power plants.

The problems of financing nuclear desalination projects are quite similar to those for nuclear power.

The major issues of financing high-capacity plants are raising large capital funds under the optimal combination of the following conditions: low interest rate, long payback periods, and maximum local currency. One difficulty is the long period of time between the investment date and the start of the payback period. Experience in past projects proved that poor financing led to major difficulties.

As a general guideline, the financing entity gains profit, which adds to the price (and possibly to the cost) of the water and/or power unless this entity is the consumer or the government.

More aspects of financing are discussed in the following sections, describing the contractual approaches of turnkey, build–operate–transfer (BOT) build–own–operate–transfer (BOOT), and build–own–operate (BOO) within which financing is included. Another possible contractual approach is a version of turnkey with the plant being rented, rather than sold, to the water/power supplier, as mentioned above (section 3.11).

3.13 BIDS AND OWNERSHIP

3.13.1 The Entities Involved

There are several possibilities to establish a desalination plant, possess it, and operate it. The following legal distinct positions with this respect are

1. The water manufacture
2. The water supplier
3. The water user
4. The water owner (not to be confused with the desalination plant owner)
5. The authorities

Often the authorities are also the water owners, who, in turn, grant the right of using the water to the users, and the permission to produce—to the water manufacturer and the license to distribute the desalted water—to the water supplier. This mutual status of authorities, owner, manufacturer, supplier, and user imposes certain mutual responsibilities on each party.

In principle, the water supplier (listed item 2, above), who is responsible to the supply of potable water, buys the water from the manufacturers (1) and sells it to the water users (3). These positions 1–3, can be manifested either by three different entities, or by two—where the water supplier is also the water manufacturer such as in the case of some water utilities—or, relatively seldom, even by a single entity where the water user is also the water manufacturer, as described in Section 3.13.2, with no need and no room for the function of a water supplier.

The initiative of establishing desalination plants can originate with either one of the five entities listed and described above.

3.13.2 Full and Direct Implementation of Desalted Water Production

The organization that requires desalted water or is responsible to the supply of potable water initiates the desalination project. This initiating organization—which may be either a private entity such as an industrial factory or a resort, or a public entity such as the army for a military camp, municipal or state authorities for a town, or a region or an entire country—may independently decide to build, own, and operate the plant. In such cases, the cost (to these owners) of the desalted water is the lowest, as compared to the following alternatives, provided the owners and their teams (plant managers, operators, and maintenance workers) carry out their professional tasks diligently. Thus, two principal advantages are gained: (1) from the *performance* aspect—maximum control of the owners on the plant and its operation is achieved, and (2) *economically*, the net profits to contractors in the alternative arrangements (discussed below) are saved.

Nevertheless, although the desalted-water preliminary cost under such an arrangement is the lowest, in quite a few cases the abovementioned authorities or organizations feel that they lack the experience or are not equipped to handle efficiently all the numerous activities needed to accomplish the supply of desalted water, or have other priorities and limitations. Thus, they do not want to enter such obligation and prefer to pass considerable responsibilities to other entities according to the arrangements described below.

3.13.3 Turnkey Construction

The authorities or organizations that require the desalted water own the plant as in the abovementioned arrangement, but the entire plant construction is given to contractors under an arrangement called a *turnkey construction project*. Most of the preliminary investment is paid to the contractor, plus very roughly ~10% of the investment that is expended by the plant owners on miscellaneous items such as land, possibly employing an advisory and planning architect-engineering company, legal and accounting services, supervision, selection of contractors, and recruiting and training the operation and maintenance staff. Operation and maintenance are often carried out by the owners' direct employees, but may also be given to special contractors. These turnkey construction project contractors (and possibly the O&M contractors) are often determined by bids for the lowest price for the plant and for the O&M services. Other criteria, beside low cost, are also required or considered concerning the bid winner such as experience in similar projects, early and brief startup or implementation, good reputation, and having proven sufficient economic resources that will enable successful accomplishment of the implementation.

The bid concept has the advantages of granting the work to the winner on the basis of competence and impartiality as well as potentially lowering the price (and perhaps even the cost) of the work through the motivation created by competition. However, bidding procedures consume time and money for both the owners and candidate contractors, increasing the desalted water cost by a certain amount. Thus, the owner may alternatively contact contractors without bids and select them according to other criteria such as previous activities and acquaintance, recommendations, and other considerations. Avoiding bidding may save time and money but introduces the economic risk of not acquiring the lowest price plant or services. With or without bids, using contractors may raise the water cost for the plant owner as the net profits of the contractors are added to the plant owner expenses.

3.13.4 Build–Operate–Transfer (BOT) Approach

A less direct degree of responsibility of the entity that initiated the desalination project exists with the BOT option mentioned in Section 3.12. The initial financing and early O&M expenses are the direct responsibility of the company (or often a consortium of several companies) that erect and run the plant, rather than of the organizations that need desalted water. The latter are thus free from the risky stages of the project. Only after sufficient time for the project to prove successful is the desalination plant handed to or purchased by the entity that initiated the desalination project. Part or all of the investment is recovered by the contractor through the sale of the desalted water so that the price at which the plant is transferred to the entity that initiated the desalination project—often the government or the municipal authorities—is the relatively small remaining part of the entire initial investment. The longer the time at which the contractor operates the plant, the lower is the payment for transferring the plant.

The BOT option sometimes leads to the next possibility, BOOT, as detailed below.

3.13.5 Build–Own–Operate–Transfer (BOOT) Approach

With the build–own–operate–transfer (BOOT) approach, the responsibility for O&M and other expenses of the party who initiated the desalination project is even less direct than that with the BOT option (described in Section 3.13.4). The financing and current O&M expenses are the direct responsibility of the companies that erect and run the plant for a 20–30-year period, rather than those organizations that require desalted water, who, in turn, pay only for the water rather than for the plant, exempting them from involvement in issues related to the project, in particular, the financing. Only after a long period of time during which all the investment has already been recovered through payments for the consumed water is the plant transferred to the entity that initiated the desalination project.

(*Note:* The distinction between BOT and BOOT is small and is sometimes ignored. According to those who do distinguish between the two possibilities, the main difference is obviously the ownership. Whereas with BOT the plant is continuously owned by the initiating entity that grants the right to build and operate to the contractor, with the land possessed by the initiating entity, with BOOT the plant and the land belong to the contractor until the transfer. Another difference *may* be the length of the operation period before transferring the plant from the contractor to the initiating entity, which may be longer with the BOOT arrangement.)

In the broad sense and also in more detail, one information source refers to the BOT and BOOT option: as follows:

> Build-Operate-Transfer (BOT) is a form of project financing, wherein a private entity receives a franchise from the private or public sector to finance, design, construct, and operate a facility for a specified period, after which ownership is transferred back to the funding entity. During the time that the project proponent operates the facility, it is allowed to charge facility users appropriate tolls, fees, rentals, and charges stated in

their contract to enable the project proponent to recover its investment, and operating and maintenance expenses in the project. However, in some countries, the term used is *Build-Own-Operate-Transfer* (BOOT).

In a BOT arrangement, the private sector designs and builds the infrastructure, finances its construction and owns, operates and maintains it over a period, often as long as 20 or 30 years. This period is sometimes referred to as the "concession" period.

Traditionally, such projects provide for the infrastructure to be transferred to the government at the end of the concession period. BOT is a type of project financing. The hallmarks of project financing are:

The lenders to the project look primarily at the earnings of the project as the source from which loan repayments will be made. Their credit assessment is based on the project, not on the credit worthiness of the borrowing entity.

The security taken by the lenders is largely confined to the project assets. As such, project financing is often referred to as "limited recourse" financing because lenders are given only a limited recourse against the borrower.

Most project finance structures are complex. The risks in the project are spread between the various parties; each risk is usually assumed by the party which can most efficiently and cost-effectively control or handle it.

Once the project's risks are identified, the likelihood of their occurrence assessed and their impact on the project determined, the sponsor must allocate those risks. Briefly, its options are to absorb the risk, lay off the risk with third parties, such as insurers, or allocate the risk among contractors and lenders. The sponsor will be acting, more often than not, on behalf of a sponsor at a time when the equity participants are unknown. Nevertheless, each of the participants in the project must be satisfied with the risk allocation, the creditworthiness of the risk taker and the reward that flows to the party taking the risk. In this respect, each party takes a quasi equity risk in the project.

3.13.6 Build–Own–Operate (BOO) Approach

The entity that initiated the desalination project using the build–own–operate (BOO) approach, has the least direct responsibility for the project compared to the previous approaches. The financing and current operation and maintenance expenses are under the direct responsibility of the companies that erect and run the plant along its lifetime, rather than those organizations that need desalted water. The latter, in turn, pay for the water rather than for the plant, absolving them involvement any potential or actual issues regarding the project. On the other hand, they pay the full price of the water—the gross cost of desalination, possibly the coverage for risk elements of the project, and the net profit of the contractors.

3.13.7 Rented Plant

Another possible arrangement of plant ownership for the desalination project initiators, not in use so far, is aimed at bypassing financing issues by holding and operating the plant while it is being rented from those investors that finance the erection of the plant to the entity that operates the plant and sells the water. The desalination project initiators may operate and maintain the plant directly or hire contractors to operate and maintain it.

3.14 CONTRACTUAL PRICING DUE TO CHANGES AND DEVIATIONS FROM THE ORIGINAL PLAN

These modifications include linkages, bonuses, penalties, compensation for "excess water," and high-quality product water streams. The economic changes and risks are as follows. At the stage of decision-making regarding construction of a desalination plant (and even more so regarding a multiplant desalination plan), changes and risks are expected, even in the construction phase of the plant, which is near-future and rather short in duration. Such modifications are most likely to occur during the operation phase, which usually extends a few years into the future and is a rather long period, especially where construction of several plants is planned to take place at specified time intervals.

Special clauses regarding contractors' and water users' risks reduction are usually included in the contracts of BOT, BOOT, and BOO arrangements. The escalations risk is diminished by linking the water prices through an agreeable formula to the main operation cost components, namely, energy (fuels and electricity) and salaries and possibly the pertinent chemicals and spare parts.

Another contractor risk of the project economy calls for a compensation clause protecting the contractor against the possibility of refusal of water customers to accept the desalted water—due either to an early end of the project life, emergence of cheaper alternative water sources, or inclement wheather, to be included in the contract. The compensation for rejecting product water covers the water cost components that are not proportional to the amount of the water produced: capital cost and most of the staff salaries, but excluding energy, chemicals, and worker overtime.

On the other hand, there are risks on the side of the water customers or the authorities who granted the contract to the plant contractor—mainly negative deviations from the original terms of the contract:

- Late plant commissioning (which is also relevant in turnkey contracts)
- Low water production rate
- High TDS contents of and/or chlorides, boron, nitrates, etc.
- High energy or chemical consumption
- Low time availability
- Short life of main equipment components such as membranes, pumps, or heat transfer surfaces
- Short life of the entire plant or a unit

If these deviations are not too large, or if they can be reasonably corrected, they are usually compensated for by penalties paid by the contractors to the water users according to agreed-on formulas. Otherwise they can cause termination of the project.

Positively, the opposite deviations from the original terms of the contract may also occur, due to high safety factors taken in the design, high-diligence operation, excellent management, or for other good reasons. If such positive deviations are desirable to the water users, they may be compensated by bonuses paid to the contractors according to established formulas.

3.15 NATIONAL DESALINATION PROGRAMS

The complexity of national desalination programs is larger than that of specific projects in the following aspects:

1. *Different Decisions*. Since a national desalination program usually comprises several desalination plants, there is a certain flexibility regarding the size, location, and timing of each individual plant. In some cases even the detailed required quality of the desalted water may differ according to the location from one plant to another. These principal decisions—water quantity and quality, locations, and time schedule—are strongly dependent on economic considerations that may drastically change the final results (whereas for local water shortage problems the location is simply given, size is more or less known according to the needed amounts of water, and the timing flexibility is usually quite narrow so that the main decisions for a local water problem determine which desalination process will be used and who will erect and operate the plant).
2. *Broader Scope*. The scope of a national desalination programs is much broader than that of specific projects in the water quantities, length of time period, number of involved entities (suppliers, advisors, contractors, and subcontractors), and number of sites. Because of this broad scope, even small cost and price differences, which might be of marginal significance in a single plant for an acute water shortage, accumulate to large amounts of money and become important.
3. *Land Area*. To enable the smooth implementation of a national or regional desalination program, the authorities have to allocate and reserve appropriate sites of sufficient area for all the planned desalination projects. This might influence the economy both directly and indirectly: directly, by having to pay for the areas very early or compensate other entities that may have interest in the said areas, in particular sites near the sea that are more limited and have high demand; indirectly, by causing escalation in land prices due to the changes incurred in the demand versus supply within the real estate market. In view of large-scale desalination financing difficulties, a reasonable preliminary cost, cashflow and financial evaluation and planning should include, in some cases, these ground area impacts for dedicated plants or an additional area in dual-purpose projects.

4. *Distribution and Conveyance of the Desalted Water*. The large amounts of desalted water via the national program are, at least partly, delivered to relatively distant water users. This may require additional investments in pipelines and pumps and possibly also storage facilities, as well as in areas and/or passage permissions for these installations. Energy is also consumed by this water conveyance, in particular for delivery to water users located at elevated sites. Moreover, old pipes—either galvanized steel or concrete which rather often suffers from leakage might need repairs or replacement by plastic pipes to avoid loss of expensive desalted water, and even more so because the chemistry of the desalted water may enhance the leakage. All these activities, associated with the desalted water outside the desalination plants, impose an additional cost burden on the national program.
5. *Energy*. Similarly, another activity strongly associated with the desalted water outside the desalination plants is the energy generation for the national desalination program; the quantity becomes a factor of significant economic weight. For example, a very large local SWRO 50,000-m^3/day plant that can supply water to between quarter and half million people requires a ~8–9 MW_e energy source, which can be easily supplied from the grid; but a typical national program that will produce 500 million m^3/year needs, including distribution pumping, ~250–300 MW_e, which requires a significant addition to the nation's power generation capacity. With a specific investment of very roughly \$1000 per installed kW_e, it requires financing of ~\$0.25 billion (US) investment in addition to financing of the desalination plants and the financing of the pipelines and pumping equipment for the delivery.
6. *Applications*. Within the national program, the practical value of the desalted water can become higher because of the versatility of its applications, which result from the high degree of reliable supply, and especially from the controlled quality and quantity. This practical value has a reflected economic value as well. In addition to the obvious basic aspects of enhancing the amount of available water of desired solute concentrations, the applications include the following:
    a. The possibility of diluting of the existing waters with the low-salinity desalted water to improve the quality of the former. This application has impacts on health; scale deposits in pipes, kettles, boilers, washing machines, and other devices; and reduction of detergent consumption. Where a low-salinity mixture of desalted water and regular water is delivered for domestic use, the resulting sewage, which is more salty by ~100–150 mg/liter than the incoming water, is still suitable for agricultural irrigation. Consequently, sometimes there is a higher potential to reuse up to 70% or even 80% of the incoming water.
    b. Redemption of old water sources that have been abandoned because of exceedingly poor quality, thus adding more water—between roughly 20% and 200% of the desalted water used for such dilution—to the national water balance.

c. Similar potential addition to the water balance may be achieved by dilution of originally inappropriate brackish-water, if available. Thus, compared to the alternative of desalting the brackish-water, the entire amount of brackish-water is utilized, rather than only 70–90% of it; thus the cost of desalting the brackish-water is avoided, as well as the difficult problem of inland concentrate removal.

7. *Impacts on National Economy: Index of Living*. Since the production cost of desalted water is in most cases considerably higher than that of other waters and the prices of desalted water are obviously even higher, the large amounts of desalted water produced in a national program may affect the index of living. Introducing a new costly elementary commodity to the list of items that constitute the basis for calculating the index of living will raise the level of this index when this new commodity is widely consumed, which in many cases is eventually inevitable. This effect may, in turn, contribute to the salary increases of those employees whose wages depend on or are linked to the index of living—possibly the majority of employees throughout the country. Thus, part or all the government may object to such an effect and may try to reduce it by reducing the scope and pace of the national desalination program.
8. *Employment*. On the other hand, large-scale building and operation of desalination systems involves employment of many hundreds of local workers, which may somewhat motivate the government to support and encourage a broad desalination program.
9. *Impacts on National Economy: Subsidies*. Some water users cannot afford to pay the full price of desalted water, in particular the agriculture sector. The state may cover the difference between the price paid to the water suppliers and the *lower* price paid by these water users. In the case of desalted water, this subsidy is usually quite high per cubic meter and when multiplied by large quantities, may consume large amounts of money. Governments are reluctant to spend this money, and may consequently try to reduce the scope of national desalination programs.
10. *Taxes*. Taxing policy—the opposite of subsidies—relating to desalted water as part of national program may differ from that related to local desalination projects.
11. *Financing*. Similarly, financing policy relating to desalted water as part of a national program may differ from that related to local desalination projects in terms of both or either allocating and/or investing money and determining the discount rate. The latter can be reduced—in favor of cost reduction—directly by government decisions or legislation or indirectly by state guarantees to other financing entities that may decrease the discount or interest rate in view of the reduced financial risk resulting from this state guarantee.

3.16 DESALINATION AND HYDROLOGY

In many cases, in particular large desalination projects, the desalted water is one of several water sources. Integration of all water sources throughout the entire water system is obviously necessary. The desalted water can be integrated in several ways with the hydrology of the country, the region and/or the vicinity of an existing (or a candidate) desalination plant:

1. *Parallel and Complementary*. The potential amounts of water pumped from the available aquifers do not reach the existing or expected demand. The resulting gap, if no other (or better) sources exist, is covered by desalination.
2. *Parallel and Mixing*. In addition to closing the quantitative gap described above (1), mixing the desalted water with the groundwater may improve the quality of the latter and in some cases of salty brackish-water may reclaim inappropriate groundwater and redeem contaminated wells.
3. *Storage*. In years of high rainfall the demand for potable water may decrease, while during dry seasons, which may last several years, water shortage might develop even with desalination systems in place. Reducing the output of the latter during rainy years due to the low demand might prove to be unwise if dry spells or droughts were to develop later. An optimal strategy might be to operate the desalination plants in full capacity (or almost full) and store the excess water during rainy seasons. The storage can be performed directly, by infiltrating the desalted water into the ground and thus recharging the aquifers, or—perhaps even more effectively—to use all the desalted water while reducing the pumping from the aquifer and allow the water table to rise with the surplus of incoming rain water seepage above the lowered pumping rate, namely, to ensure a definite positive groundwater balance.
4. *Delaying Erection of Desalination Plants*. Postponing the installation of desalination plants by one or more years has several advantages:
   a. Reducing interest fees on the investments and O&M.
   b. Preventing or reducing risk of equipment deterioration.
   c. Using more advanced technology, as progress develops during the period of postponement (improved RO membranes etc.).
   d. Cost reduction due to application of economy-of-scale principle as the postponed plants need to be of larger capacity than the original or alternative planned plants.
   e. Accumulating more experience elsewhere.
   f. Since many desalination plants are and will be integrated with electric power generation, reducing the power plant capital costs or grid capital costs resulting from power unit operation may be achieved by postponing the project, if the larger desalination units are commissioned simultaneously with the power plant. (This strategy is effective particularly for

nuclear desalination.) The reason is that when a large power unit is added to the grid, there is a surplus of power supply for a few years, until the gradually increasing power demand consumes this surplus. Obviously this surplus increases the amortization of the extended generating system. This amortization can be reduced by connecting a large (step function) power consumer, namely, the large desalination system to the power plant, or even to the grid simultaneously with commissioning of the large power unit or very short time later, thus reducing the power surplus.

Postponement of the installation of dual-purpose or large-scale desalination plants, in particular nuclear plants, by several years may be preferred, because of the six advantages mentioned above, in those cases where the need for additional power is less crucial than the need for water, even if desalination is the best or the only solution for water shortage. More specifically, apparent hydrological constraints forbid additional pumping from the aquifers. Despite such hydrological constraints, it is still possible to use groundwater provided it is done temporarily and followed by recuperation as described here.

The obvious disadvantage of water shortage can be reversed by carefully integrating groundwater utilization and national or regional desalination activities in such a way that may allow project postponement while avoiding water shortage as described below.

One way to postpone desalination because of actual or potential water shortage is to deplete certain aquifers to a specified limit as follows. Those aquifers that can withstand a few years of overpumping have to be identified, located, and utilized. Tolerance for overpumping is determined by the following hydrological and management conditions:

- A temporary drop of the water table of the aquifer below the minimum level must not cause irreversible damage to the potable groundwater. This is a pure hydrological condition.
- The period of overpumping from those aquifers shall be immediately followed by a period of low-pumping, no-pumping, or even recharge of freshwater or desalted water to accelerate recuperation. This condition is of operational in nature and is related to hydrology management.

Another management condition and requirement is that larger desalination capacity (compared to the original or alternative plan) be available immediately after the overpumping period to cover the previous original need for desalted water, replaced by the overpumping, plus the added increasing demand for freshwater during the postponement period, plus the "overdraft" accumulated by the overpumping to be recuperated.

For example, if an aquifer is found where the water table in a 10-miles-radius can tolerate, for a few years, an average drop of 2 m, and assuming 30% effective soil porosity, the potential amount of water "loaned" from this aquifer is almost 500 Mm^3. This potential "overdraft" amount may, indeed, allow delay in installing desalination systems by a few years, even if the aquifer is not fully exploited.

3.17 ECONOMICS OF DESALINATION RESEARCH AND DEVELOPMENT (DESALINATION R&D)

Parallel to the vast of economic amount resources and efforts invested in desalination plants, another channel of activity exists in which additional funds are allocated for exploring new possibilities for desalination, investigating partly familiar phenomena, developing new processes, and improving existing systems and components. These desalination R&D activities take place at various levels, are financed by several sources, and are carried out by many entities. The financing comes from governments, desalination system manufacturers, desalination plant operators, water and power utilities, and other sources. Relatively small amounts of money are invested by the academia.

The R&D activities are categorized in the following groups according to cost of a typical activity:

- *Research*; specifically, investigation of phenomena. This is done by the academia, in research institutes, and by industrial entities, namely, various desalination equipment manufacturers and desalination systems users. Expenses for research are relatively low per year, but may accumulate over the years because research activities usually are long term. Expenses consist mainly of salaries, since the equipment is of small size. Funding sources for these activities may include the government of the state where the research takes place, sometimes other governments under multinational cooperative agreements, possibly other public sources, manufacturers and/or users of desalination equipment, and—usually only if the expenditure is relatively small—the institute's own internal funds. It is important to note that a significant part of the expenses involved are not recognized as desalted-water cost components and are not recovered by sale of the desalted water, at least not directly.
  - *Development Small-scale development* at bench scale or unit operation scale—in early phase of testing concepts or formulas underlying equipment items or subsystems. The expenses are relatively low per year, and most activities are carried out during relatively short periods, such as a few weeks of operation and very few months for preparation.
  - *Medium-scale development* at pilot plant scale—at advanced phase of testing ideas of subsystems or complete systems of rather smaller size relative to commercial units. Sometimes the tested system operates in closed-loop fashion to save part of the infrastructure. (A *closed-loop* desalination pilot plant is a plant at which the saline feedwater stream is artificially created by mixing the product desalted water with the stream of the concentrated brine, thus obviating the expenses of seawater or brackish-water intake and pretreatment and brine outfall system, and enabling the testing activity to be carried out at convenient sites, not necessarily near the sea or saline-water wells.) The capacity is usually about one order of magnitude smaller than a small commercial plant or even smaller. The expenses may very roughly

in the range of \$100,000 (US) per installed pilot plant plus at least a similar amount of money for operation, analyzing the results, reporting, and decommissioning.

- *Large-scale development* at demonstration plant scale—last phase of testing and validating semiproven ideas and concepts of complete systems of rather full size and open loop as (or almost as large as) commercial units, often with considerably more instrumentation and windows than in commercial units, to enable better follow-up and detailed measurements of the process and equipment performance. Because of the size of the equipment, the needed infrastructure, the instrumentation, and the relatively long periods of the tests, this phase is quite expensive. The expenses are relatively high but take place only once in several years and hopefully are lucrative because the demonstration plant yields salable potable water in addition to information, data, know-how, and experience. Since there it is highly probable that a demonstration plant will be installed at the site where desalted water is needed, selling the desalted water may compensate for most or all of the demonstration expenses. The expenses of such large-scale R&D may total between a few hundred thousands and a few tens of millions US dollars. Financing these expenses comes either from desalination system manufacturers or other investors who want to sell plants or desalted water or from governments of states that require desalted water or are interested in developing a desalination industry for export of desalination plants.
- *R&D priorities*—in principle two possible scenarios can exist with regard to financing R&D activities. In scenario 1, the initiative for such activities originates with the financing entities who allot a certain amount of money for the general purpose of advancing the technologies and offer this money to the technology entities, such as academia, research institutes, industry, and water and power utilities. In scenario 2, the initiative originates with the technology entities that want to carry out the R&D activities and request the money from external financing entities or raise it from their own resources.

In scenario 1 the specific subjects of the activities are not defined and the overall amount of money is offered for desalination in general or for the type of activity that the financing organization prefers, such as inland desalination, desalination environmental aspects, novel processes, or new RO membranes. Occasionally, very few research activities actually develop, due to a lack of definite ideas or lack of enthusiasm from the R&D community to participate because they are either reluctant to accept some of the conditions that usually accompany the granting of money, or otherwise too occupied. In such cases the few proposed activities are very likely to be fully financed. In the opposite case, where several definite activities in need of relatively large amounts of money compete on a limited budget that is less than the total requested sum, the money should be granted according to priority in which economic potential is the main consideration. Two other factors are the expected probability of success of the R&D efforts in the subject activity and the amount of resources—money and time—requested. For example, in a competition between two well defined development steps where one is improving the corrosion resistance of RO pumps to absolute durability versus raising the salt rejection of SWRO membranes to 99.9%, the priority will most probably be assigned to improving membranes, (unless the needed money for finding a way to raise the membrane rejection is orders of magnitude greater than that needed for the pump durability).

In scenario 2, where the initiative for R&D activities originates within the technology community, there might also be some competition between candidate activities, but more seldom, since the financing is requested for well-defined activities, normally after some self-criticism. On the other hand, and for the same reason, the lack of definite ideas or enthusiasm from the R&D community to participate hardly exists. Then the main two criteria for financing the R&D activities are the time period for full commercial recovery of the investment in the activity (or, more accurately, its probabilistic equivalent) and the expected cashflow resulting from this activity.

3.18 WORLD MARKET AND FUTURE PROSPECTS FOR DESALINATION

The global need for potable water is expected to grow quite rapidly, due to increase of world population, the striving to improve the quality of life in developing countries, and deterioration of the quality of part of the freshwater sources. A considerable part of the additional water will come from desalination.

The demand for desalted water and desalination plants will rise and may consequently bring the prices up according to the market forces. Another potential motive for future price escalation is the possibility of the authorities requiring more strict environmental limits from the desalination plants. A third risk is the rising energy prices.

On the other hand, there are several factors that may contribute to *price* reduction, most of them via *cost* reduction:

1. The economy of scale of the desalination systems' manufacturing enterprises.
2. The continuous development of better desalination equipment, such as (for RO) new membranes having lower price and/or improved performance (longer life, higher water flux per unit driving pressure, and/or better salt rejection) or better pumps, in particular, pressure energy recovery devices, or developing low-cost heat transfer surface elements with high-heat-transfer coefficients for MED process.
3. Improvement of desalination system manufacturing procedures and techniques.
4. Finding better solutions to existing operation problems, stemming from accumulating (and hopefully shared) experience.
5. Competition between desalination system suppliers (and operators where the plants are not operated by the owner or the desalination system supplier).

6. Accumulating experience with erection, operation, and management of desalination plants will reduce the level of technological and managerial risk associated with desalination projects and might bring down the interest rates and lower the capital costs. The same effect may occur with market client pressures in view of the vitality of water to the citizens. This might be additionally achieved by governmental or international (e.g., the World Bank, UN) guarantee backing the financing of large desalination programs or projects. Such backing further reduces the financial risk and thus may lower the interest rate.

APPENDIX

3.A.1 Sample Cost Calculations for Desalted Water at the Plant

The equipment prices given below do not represent any actual information. They are presented only for demonstrating a typical cost calculation procedure, so the reader should not use them as proven professional data. They constitute only rough approximations at some sites, and even at those sites they are subjected to negotiations and changes with time and technological and commercial developments.

Cases 1: Single-Purpose SWRO Plant of 100,000 m^3/d Capacity

Parameters: $P_{\mathrm{N}} = 100,000$ t/d $= 4167$ t/h; at load/plant factor of 92%, the annual production is $P_{\mathrm{NA}} = 33.6$ Mm^3/year.

Main Process Design

1. For *given* seawater salinity of 3.5%: sf $= 35$ kg/t.
2. *Required* permeate (product) salinity of 500 ppm TDS: sp $= 0.5$ kg/t.
3. *Allowed* maximum concentrate (brine) salinity of 7.0%: sb $= 70$ kg/t.
   a. The feed flow rate can consequently be calculated as the mass balances of the water and the salt yield: $P_{\mathrm{N}}(\mathrm{sb} - \mathrm{sp})/(\mathrm{sb} - \mathrm{sf}) = 100,000(70 - 0.5)/(70 - 35) = 198,570$ t/d $= 8274$ t/h.
   b. The *average* permeate flux is 8 gfd (US gallons per square foot per day) $= 8 * 1.698 = 13.58$ L/m^2 per hour ≈ 0.326 m/d ≈ 0.325 t/m^2 per day. [1 gfd $= 1.698$ Liter/m^2 per hour ≈ 1.7 mm/hour].
   c. At this point the net area of the membranes, which is the main equipment component of the process, can be calculated: Membrane area $= 100,000/0.325 = 306770$ m^2 $= 3.3 * 10^6$ ft^2.
   d. The number of elements, each of which has about 400 ft^2 of active membrane area per element, is approximately $3.3 * 10^6$ ft^2/400(ft^2 per element) $= 8250$ elements.
   e. However, according to the membrane manufacturer's data and computer program, which yields the plan with the numbers of vessels and elements, 1020 vessels with 8 elements per vessel $= 8160$ elements, are needed. (Either the exact membrane area of the elements is slightly larger than 400 ft^2 per element, or the flux taken by the computer program is 1.1% higher.)

f. Cost of installed elements: ~\$600 (US monetary units, throughout) per element including labor: $8160 * 600 =$ \$4,896,000, plus ~2% spare membranes, result in \$5M = \$50 per daily ton.

g. Cost of installed pressure vessels: ~\$2500 per vessel including labor: $1020 * 2500 =$ \$2.05M = \$21 per daily ton of desalted water.

h. Membrane racks and piping: ~15% of membranes and pressure vessels: 0.15*(4.9M\$+ 2.05M\$) = \$1.04M = \$10 per daily ton of desalted water.

i. Power of installed high-pressure pumps (pressure is given by the computer program Hydranautics, the membrane manufacturer): 8274 t/h/(1.03 t/m^3) $*$ 64 bar/0.87/(36 bar*m^3/h) = 16.41 MW = 20,417 hp (horsepower), installed power ~15% more −24,000 hp.

j. Cost of installed high-pressure pumps: ~\$200/hp, including labor: 24,000 hp*\$200 = \$4.8M = \$48 per daily ton of desalted water.

k. Power of installed high-pressure recovery device: (8274 − 4167) t/h/(1.06 t/m^3) $*$ 62 bar $*$ 0.87/(36 bar $*$ m^3/h) = 5.8 MW = 7900 hp, installed – 9000 hp.

l. Cost of installed high-pressure recovery device: ~\$200/hp, including labor: 9000 hp*\$200 = \$1.8M = \$18 per daily ton of desalted water. [*Note*: The recent use of the modern pressure exchanger as an energy recovery device (rather than the water turbine) reduces the size, cost, and energy consumption of high-pressure pumps by roughly 50%. The net energy and investive savings are, however, about 15% (0.4 kwh/m and 0.5M\$, respectively.]

m. Installed low-pressure pumps:
   (1) Feed: 8274 t/h/(1.03 t/m^3)*5 bar/0.86 = 1.3 MW = 1770 hp.
   (2) Product: 4167 t/h/(1.00 t/m^3)*5 bar/0.86 = 0.67 MW = 910 hp.

   Total: $1.3 + 0.67 \approx 2.0$ MW, installed −3000 hp.

n. Cost of installed low-pressure pumps: ~\$200/hp including labor: 3000 hp*\$200 = \$0.6M = \$6 per daily ton of desalted water.

o. Cost of pretreatment system: ~\$12M = \$120 per daily ton of desalted water.

p. Cost of seawater intake system and concentrate outfall: ~\$4M = \$40 per daily ton of desalted water.

q. Cost of instrumentation: ~\$1.5M = \$15 per daily ton of desalted water.

r. Cost of control system: ~\$1.5M = \$15 per daily ton of desalted water.

s. Cost of electrical work: ~\$4M = \$40 per daily ton of desalted water.

t. Cost of membrane cleaning system: ~\$1M = \$10 per daily ton of desalted water.

u. Cost of other piping: ~\$2M = \$20 per daily ton of desalted water.

v. Cost of buildings: ~\$4M = \$40 per daily ton of desalted water.

w. Cost of civil engineering: ~\$10M = \$100 per daily ton of desalted water.

x. Cost of product posttreatment: ~\$4M = \$40 per daily ton of desalted water.

y. Cost of furniture, laboratory, communication equipment, computers, transportation means, cranes, and miscellaneous: ~\$3M = \$30 per daily ton of desalted water.

z. Total: \$62.25M = \$623/t per day.

aa Cost of management, planning, and design: ~15% of the total (item z, above) = 0.15*62.3 = \$9.4M = \$94 (US)/t per day.

bb Owner's cost: ~10% = 0.1*(62.3+9.4) = \$7.2M = \$72/t per day.

cc Total, assuming 7.5%/py as interest during construction, effective period of construction 2.5 annual years: $1.075^{2.5/2}(62.3+9.4+7.2) = 1.0946 * 78.9 \approx$ \$86.4M = \$864/per day; hence the total percentage of interest during construction is $1.0946 - 1 = 0.0946 \approx 9.5\%$.

dd Energy:

(1) Pumps total: 16.4 MW + 1.3 MW + 0.67 MW = 18.4 MW mechanical work.

(2) Pressure recovery: 5.8 MW mechanical works.

(3) Net mechanical works: 18.4 − 5.8 = 12.6 MW.

(4) At 92% electrical system efficiency (motors, transformers, etc.), average electric power consumption: $12.6/0.92 = 13.7\ MW_e$.

(5) An additional 200 kW_e, is consumed by plant outside desalination process for facilities such as lights, workshop, kitchen, air conditioning, offices, and showers: $13.7\ MW_e + 0.2 \approx 14\ MW_e$.

(6) Specific energy of the desalted water at the plant is $14\ MW_e/4167$ t/h = 3.36 kWh_e/ton.

(7) Allowing ~2% additional losses for off-optimum operation, and another 2% for gradual decrease in the efficiencies of the pumps and pressure recovery devices, the estimated specific energy is 3.36/0.96 = 3.5 kWh_e/ton

(*Note*: Comparing the specific investment and energy calculated here with the values presented in Table 3.3 shows 8% higher investment but 5.4% lower energy. The higher investment is explained partly because of the rounded values assumed for the items of equipment and mostly due to conservative redundancies in equipment size and specification. The lower energy is explained partly because the value of 3.7kWh$_e$/t, shown in Table 3.3, is experienced in plants with very low product salinity that require secondary desalination of part of the product, namely, multipass systems.)

Table 3.A.1 lists typical capital expenses for a hypothetical SWRO plant.

Cost of Desalted Water
It is assumed here that the annual expenses are as follows:

- Salaries are the same as in Table 3.3: \$800,000 per year.
- Capital cost is 8.214% of the total investment: (as in 3.4.1.4.1, page 250) \$86.4M $*$ 0.08214 = \$7.1M/per year.
- Price of electric power is 5.4¢/kWh, as in Table 3.3. Thus the annual expense is 3.5 kWh/t*\$0.054/kWh*10^5t/d*365*0.92 = \$6.35M/year.

Table 3.A.1 Example: Capital Cost Summary for a SWRO Plant[a]

Item or Parameter	Cost, \$/t per day
Membrane elements	50
Pressure vessels	21
Membranes racks and their piping	10
High-pressure pumps	48
Low-pressure pumps	6
Pressure recovery	18
Pretreatment system	120
Seawater intake	40
Instrumentation	15
Control system	15
Electrical works	40
Membrane cleaning	10
Other piping	20
Buildings	40
Civil engineering	100
Posttreatment	40
Miscellaneous	30
Subtotal	623
~15% planning and management	94
Subtotal	717
~10% owner's cost	72
Subtotal	789
~9.5% interest during construction	75
Total specific investment	864

[a]Production rate: 10^5 tons per day.

- Chemicals are proportional to the feedwater flow rate. In this case with 3.9% feedwater salinity, the flow rate is calculated to be 206,950 t/d, while for the case shown Table 3.3, with 3.5% feedwater salinity, the flow rate is 186,250 t/d. Thus, using the assumption of Table 3.3 of 4¢ per ton product, the annual expense is 4¢/ton*206,950/186, 250 = 4.44¢ per ton product. The annual expense is \$0.0444/t*33.6 Mt/y ≈ \$1.5M per year.
- The average annual membrane replacement is 18% of the initial investment in the installed membrane elements (this is equivalent to replacing all of them every 4.8 years, although the life is guaranteed by the manufacturers to be at least 5 years). The annual expense is 0.18/ year *\$4.9M ≈ \$0.88M per year.
- Annual replacement, maintenance, insurance (ERMI) at 2.5% of total investment: \$86.4M * 0.025 = \$2.16M per year.
- Land rental assumed to be \$100,000 per year.
- The annual expenses sum is \$18.89M ≈ \$18.9M per year.
- The cost of the desalted water is \$18.9M/33.6Mt = \$0.562/t ≈ 56¢/m^3.

These data are summarized in Table 3.A.2 and compared with those in Tables 3.A.3 and 3.3 (Page 238).

Table 3.A.2 Water Cost Calculation

Parameter	Unit	SWRO Data Presented Above	SWRO Data Presented In Table 3.A.3	LT-MED, Dual-Purpose
GOR	Product/steam	—	—	11.5
Load factor	—	0.92	0.92	0.90
Annual production	Mm^3/ year	33.6	33.6	32.9
Electricity per ton	kWh/m^3	3.5	3.7	1.3
Total investment	\$M	86.4	80	\$105.7M
Steam	steam/h	—	—	362
Equivalent hours/per year	h/yr	8065	8065	7890
Annual capital cost, 8.214%	\$M/year	7.10	6.57	8.68
Annual electricity cost	\$M/year	6.35	6.84	2.6
Annual steam cost	\$M/year	—	—	8.3
Salaries	\$M/year	0.8	0.8	0.8
Annual ERMI cost	\$M/year	2.16	2.0	2.64
Chemicals	\$M/year	1.5	1.35	1.31
Annual membrane replacement cost	\$M/year	0.88	0.84	—
Land rental		0.1		0.1
Total annual cost	\$M/year	18.90	18.40	24.43
Water cost	¢/m^3	56.0	54.8	74.3

Table 3.A.3 Example: Capital Cost Summary for a MED Plant[a]

Item or Parameter	Cost, \$/t per day
Heat transfer surfaces	190
Vessels	150
Internal parts	100
Pumps	30
Vacuum pumps and system	10
Connection to power generating unit	30
Pretreatment system	30
Seawater intake	20
Instrumentation	15
Control system	15
Electrical works	20
Buildings	40
Civil engineering	60
Posttreatment	10
Miscellaneous	30
Subtotal	750
~15% planning and management	112
Subtotal	862
~10% owner's cost	86
Subtotal	948
~11.5% interest during construction	109
Total specific investment	1057

[a]Production rate: 10^5 tons per day.

Cases 2: Dual-Purpose Low-Temperature MED Plant of 100,000 m^3/day Capacity

Parameters: $P = 100,000$ t/d $= 4167$ t/h; at load/plant factor of 90%, the annual production is $P = 32.9$ Mm^3/year.

Main Process Design

1. For Mediterranean seawater salinity of 3.9%: sf = 39 kg/ton, temperature of motive steam supply 75°C, condenser vapor-side equilibrium temperature is 34°C. Thus the representative average latent heat in this temperature range is 0.66 kWh/kg = 2376 MJ/t.
2. Required product salinity: 500 ppm TDS (sp = 0.5 kg/t); expected, <50 ppm.
3. Concentrate (brine) salinity: 7.0%, sb = 70 kg/t-.
4. The feed water flow rate can consequently be calculated as the mass balances of the water and the salt yield: Feed $= P_N(\text{sb} - \text{sp})/(\text{sb} - \text{sf}) =$ $100{,}000(70 - 0.05)/(70 - 39) = 225{,}650$ t/d ≈ 9440 tons/h.

5. The average optimal heat flux is 5 kW_t/m^2. (See end of Section 3.A.2.) With the average latent heat of water at the temperatures range given above at 0.66 kWh/kg, the mass flux is thus 5/0.66 = 7.6 kg/h/m^2.
6. At this point the net heat transfer area of the evaporators, which is the main equipment component of the process, can be calculated: Area = 4,167,000(kg/h)/(7.6 kg $h^{-1}m^{-2}$) = 550,000 m^2. Because of inefficiencies, the needed area is ~8% larger, in addition to the condenser area, which is ~6% more; thus the total heat transfer area is $550,000\ m^2 * 1.14 = 630,000\ m^2$.
7. Cost of installed heat transfer area (aluminum alloy): ~\$30/m^2, including labor: 630,000 m^2*30 = \$19M = \$190 per daily ton of desalted water.
8. Cost of installed vessels including foundations: ~\$15M including labor = \$150 per daily ton of desalted water.
9. Internal items in the vessel (tube sheets, spray nozzles, partitions, droplet separators, tube supporting equipment, etc.) ~\$10M = \$100 per daily ton of desalted water.
10. Piping: ~\$3M = \$30 per daily ton of desalted water.
11. Connection to power generating unit: ~\$3M = \$30 per daily ton of desalted water.
12. Vacuum system: ~\$1M = \$10 per daily ton of desalted water.
13. Cost of installed pumps: ~8000 hp*\$200/hp including labor: \$16M = \$16 per daily ton of desalted water.
14. Cost of pretreatment system: ~\$2M = \$20 per daily ton of desalted water.
15. Cost of seawater intake system and concentrate outfall: ~\$2M = \$20 per daily ton of desalted water. This cost is low as the cooling-water system of the power plant is utilized.
16. Cost of instrumentation: ~\$1.5M = \$15 per daily ton of desalted water.
17. Cost of control system: ~\$1.5M = \$15 per daily ton of desalted water.
18. Cost of electrical work: ~\$3M = \$30 per daily ton of desalted water.
19. Cost of buildings: ~4M = \$40 per daily ton of desalted water.
20. Cost of civil engineering: ~\$6M = \$60 per daily ton of desalted water.
21. Cost of product post treatment: \$1M = \$10 per daily ton of desalted water.
22. Cost of furniture, laboratory, communication equipment, computers, transportation means, cranes and miscellaneous items: ~\$3M = \$30 per daily ton of desalted water.
23. Total: \$75M = \$750/t per day.
24. Cost of management, planning, and design: ~15% = 0.15*75 = \$11.2M = \$112/t per day.
25. Owner's cost: ~10% = 0.1*(75 + 11.2) = \$8.62M = \$86t per day.

26. Total, assuming 7.5% interest during construction, effective period of construction 3 years: $1.075^{3/2}$ (75 + 11.2 + 8.6) = 1.1146*94.8≈ \$105.7M = \$1057/t per day.
27. Energy:
    a. Loss of electricity generation: due to 362 ton steam/hour extraction at 75°C, 0.385 bar is 21 MWe generation, 19.5 MW_e loss of power sales.
    b. Pumps total, including vacuum: 5.5 MW mechanical work.
    c. At 92% electrical system efficiency (motors, transformers, etc.), average electric power consumption: 5.5/0.92 = 6 MW_e.
    d. An additional 100 kW_e, is consumed by plant outside desalination process for facilities such as lights, workshop, kitchen, air conditioning, and showers: 6 MW_e + 0.1 = 6.1 MW_e.
    e. Specific energy of desalted water of plant: (19.5 + 6.1) MW_e/ 4167t/h = 25.6/4167 = 6.14 kWh_e/ton.

Cost of Desalted Water
It is assumed here that the annual expenses are as follows:

- Salaries are the same as in Table 3.3: \$800,000 per year.
- Capital cost is 8.214% (as in 3.4.1.4.1, page 250) of the total investment: \$105.7M*0.08214 = \$8.68M/year.
- Price of electric power is 5.4¢/kWh = \$54/MWh, as in Table 3.3. Thus the annual expense is 6.1MW*\$54/MWh*8760 h/y*0.9 = \$2.6M per year.
- The price of steam for reduced electric power compensation is also 54\$/MWh, as in Table 3.3. Thus this annual expense is 19.5MW*\$54/MWh*8760 h/y*0.9 = \$8.3M/year.
- Chemicals are as in Table 3.3; \$1.31M/year.
- Annual ERMI at 2.5% of total investment: \$105.7M*0.025 = \$2.64M/year.
- Land rental assumed at \$100,000 per year.
- Annual expenses sum: \$18.89M ≈ \$24.43M/year.
- Cost of desalted water: Thus \$24.43M/32.9M ton = \$0.743/t ≈ 74.3/m^3.

These data are summarized in Table 3.A.2 and compared to those in Table 3.A.3.

3.A.2 Optimization of the Main Process Component: An Example

In this example, we will consider the magnitude and number effects in a low-temperature MED plant. Assume, as an example, that an effect of MED is taken. Vapor is introduced at a certain temperature and pressure to be condensed to potable water while heat is being supplied to evaporate an equivalent amount of pure water from a stream of saline water. This heat is ejected from the said condensing vapor and is transferred to the said saline water.

The parameter to be optimized is the temperature difference that is the driving force for heat transfer θ. In turn, by knowing θ, we can easily calculate the optimal heat and mass fluxes, as well as the number of effects. With increasing θ, the heat transfer surface is reduced and the investment in the heat transfer area and associated equipment (IHT) decreases. On the other hand, the practical useful thermodynamic exergy consumed for heat transfer by the thermal driving force (EXR) increases:

$$\mathrm{IHT} = \mathrm{Area}^{*} C_{\mathrm{a}} \tag{3.A.1}$$

where

$$\mathrm{Area} = \frac{q}{U\theta} \tag{3.A.2}$$

where q is the transferred heat (in kW), U is the heat transfer coefficient (in kW $\mathrm{m}^{-2}\ \mathrm{K}^{-1}$), and C_{a} is the cost of unit area (in \$/m^2):

$$\mathrm{EXR} = \frac{q\eta\theta}{T} \tag{3.A.3}$$

Here, T is the absolute temperature (in kelvins) at which a representative effect operates (i.e., average of condensing and evaporating sides: $\sim 54^{\circ}\mathrm{C} = 327$ K) and η is the efficiency involved in converting gross mechanical energy to salable electrical power.

The economic value of IHT per year is

$$C_{\mathrm{area}} = (\mathrm{CRF} + \mathrm{ERMI}) * \mathrm{IHT} \tag{3.A.4}$$

where $\mathrm{CRF} = d/\{1 - (1 + d)^{-n}\}$ is the capital recovery factor presented in Section 3.3.2 (where d is the annual discount rate and n is the lifetime of the equipment in years; for $d = 7.5\%$ per year and $n = 30$ years, $\mathrm{CRF} = 8.46\%$ per year) and ERMI (equipment replacement, maintenance, and insurance) is the annual fraction of plant maintenance, assumed to be 2.5% of the investment per year in the cost estimates presented above.

The economic value of EXR per year is

$$C_{\mathrm{XRG}\theta} = C_{\mathrm{kWh}} * \mathrm{EXR} * 8760 * \tau \tag{3.A.5}$$

where C_{kWh} is the cost of kWh (\$/kWh), which is not generated because of the exergy consumed by heat transfer, (8760 h per year; 365*24), and τ is the time fraction of heat transfer occurrence; thus 8760τ is the life average number of hours of heat transfer performance per year (7884 for the assumed $\tau = 90\%$).

Optimum θ is when

$$\frac{\partial(C_{\mathrm{area}} + C_{\mathrm{XRG}\theta})}{\partial\theta} = 0, \tag{3.A.6}$$

namely

$$\frac{\partial}{\partial\theta}[(\mathrm{CRF}+\mathrm{ERMI})*\mathrm{IHT}+C_{\mathrm{kWh}}*\mathrm{EXR}*8760*\tau]=0 \tag{3.A.7}$$

Substituting these formulas for IHT and EXR, we can perform the following derivatives:

$$\frac{\partial}{\partial\theta}\left[(\mathrm{CRF}+\mathrm{ERMI})*C_{\mathrm{a}}*\frac{q}{U\theta}\right]+\frac{\partial}{\partial\theta}\left(C_{\mathrm{kWh}}*\frac{q\eta\theta}{\mathrm{T}}*8760*\tau\right)=0 \tag{3.A.8}$$

Thus

$$\frac{\partial}{\partial\theta}\left[(\mathrm{CRF}+\mathrm{ERMI})*C_{\mathrm{a}}*\frac{q}{U\theta}\right]=-(\mathrm{CRF}+\mathrm{ERMI})*C_{\mathrm{a}}*q/U/\theta^2 \tag{3.A.9}$$

$$\frac{\partial}{\partial\theta}\left(C_{\mathrm{kWh}}*\frac{q*\eta*\theta}{T}*8760*\tau\right)=C_{\mathrm{kWh}}*\frac{q*\eta}{T}*8760*\tau \tag{3.A.10}$$

From which:

$$C_{\mathrm{kWh}}*\frac{q*\eta}{T}*8760*\tau+\left\{-(\mathrm{CRF}+\mathrm{ERMI})*C_{\mathrm{a}}*\frac{q}{U\theta^2}\right\}=0 \tag{3.A.11}$$

$$\begin{aligned}\theta^2&=\frac{(\mathrm{CRF}*C_{\mathrm{a}}/U)}{(\mathrm{CRF}+\mathrm{ERMI})*C_{\mathrm{a}}/U}\\&=(\mathrm{CRF}+\mathrm{ERMI})*C_{\mathrm{a}}*\frac{T}{C_{\mathrm{kWh}}*\eta*U*8760*\tau}\end{aligned} \tag{3.A.12}$$

$$\text{Optimum }\theta=\sqrt{\frac{\{(\mathrm{CRF}+\mathrm{ERMI})*C_{\mathrm{a}}*T}{C_{\mathrm{kWh}}*\eta*U*8760*\tau\}}} \tag{3.A.13}$$

The numerical sample calculation for MED is

$\mathrm{CRF}=0.0847/\mathrm{year}$

$\mathrm{ERMI}=0.025/\mathrm{year}$

$(\mathrm{CRF}+\mathrm{ERMI})=0.0847/\mathrm{year}+0.025/\mathrm{year}=0.1097/\mathrm{year}$

$C_{\mathrm{a}}=\{\mathrm{net}-30\$/\mathrm{m}^2\};\ \mathrm{gross}-70\$/\mathrm{m}^2$

$T=54+273=327\mathrm{K}$

$C_{\mathrm{kWh}}=0.055\$/\mathrm{kWh}$

$\eta=0.8$

$U=3.5\ \mathrm{kW/Km}^2$

$\tau=0.9$

Thus $\theta_{op} = \surd\{0.1097 * 70 * 327/[0.055 * 0.8 * 3.5 * 8760 * 0.9]\} = 1.44$ K, optimum heat flux $= \mathrm{U}\theta_{op} = 3.5 * 1.44 = 5.05\mathrm{kW/m^2}$, and optimum water mass flux $= (5.05\mathrm{kW/m^2})/0.66(\mathrm{kWh/kg}) = 7.65\ \mathrm{kg/(m^2h)}$.

It is interesting to note the following:

1. According to the design of the evaporators, based on the optimal heat transfer driving force θ_{op}, we obtain

$$C_{\mathrm{area}} = C_{\mathrm{XRG}\theta} \tag{3.A.14}$$

In other words, the gross capital cost component of the desalted water due to that part of the evaporator that is proportional to the heat transfer area is equal to the energy cost component of the desalted water due to that part of the energy charge for reducing the power generation by the exergy consumed for heat transfer instead of for generating power. This equality is obtained by substituting the expression for θ_{op} in the expression for C_{area}:

$$\begin{aligned}\frac{(\mathrm{CRF}+\mathrm{ERMI}) * C_a * q}{U\theta_{op}} &= \frac{(\mathrm{CRF}+\mathrm{ERMI}) * C_a * q * \surd[C_{\mathrm{kWh}} * \eta * U * 8760 * \tau]}{U\surd[(\mathrm{CRF}+\mathrm{ERMI}) * C_a * T]}\\ &= \frac{q\surd\{(\mathrm{CRF}+\mathrm{ERMI}) * C_a * C_{\mathrm{kWh}} * \eta * 8760 * \tau\}}{\surd(T * U)}\end{aligned} \tag{3.A.15}$$

The same result is obtained by substituting the expression for θ_{op} in the expression for $C_{\mathrm{XRG}\theta}$:

$$\begin{aligned}C_{\mathrm{kWh}} * \mathrm{EXR} * 8760 * \tau &= \frac{8760 C_{\mathrm{kWh}}\tau q\eta\theta_{op}}{T}\\ &= 8760 C_{\mathrm{kWh}}\tau q\eta\ \ T\sqrt{\frac{(\mathrm{CRF}+\mathrm{ERMI}) * C_a * T}{C_{\mathrm{kWh}} * \eta * U * 8760 * \tau}}\\ &= \mathrm{q}\sqrt{\frac{\{(\mathrm{CRF}+\mathrm{ERMI}) * \mathrm{Ca} * \mathrm{C_{kWh}} * \eta * 8760 * \tau}{T * U\}}}\end{aligned} \tag{3.A.16}$$

The quantitative data given in this example yield $C_{\mathrm{XRG}\theta} = q\surd\{0.11*70*0.055*0.8*8760*0.9/327/3.5\} = 1.528q$ in \$/year $= 8760 C_{\mathrm{kWh}}\tau q\eta\theta_{op}/T = 8760 * 0.055 * 0.9 * 0.8 * 1.44q/327 = 1.528q$. Thus

$$q = \frac{P_A(1+\varepsilon)\lambda}{8760 * 3600\tau} \tag{3.A.17}$$

where P_A is the annual water production, and ε is the additional fraction of the net heat transfer area needed to transfer parasitic heat (whereas the net heat transfer area transfer the latent heat of the condensing product). In

Section 3.A.1 ε was estimated as 14%, and λ is the average latent heat of the condensing product, 2374Jl/g $= 2.374 * 10^6$ kJ/t; 8760*3600τ is the active heat transfer time per average year in seconds. Thus $q = P_A 1.14 * 2.374 * 10^6 \text{kJ/t}/(8760 * 3600 * 0.9) = 0.0954 P_A$ (in kW), and $C_{XRG\theta} = 1.528q$ in \$/year = 1.528*0.0954 $P_A = 0.1457P_A$. This cost component per unit product is $C_{XRG\theta}/P_A$ = \$0.1457/t, and since $C_{area} = C_{XRG}$, that part of the annual charge that depends on the thermal driving force θ totals 2*0.1457 = \$0.292/m^3. This sum represents ~40% of the total water cost, estimated in Section 3.A.1 to be ~74¢/m^3.

2. The full annual charge C_{XRGT} for the heat consumed by the MED desalination process is almost twice that of $C_{XRG\theta}$. Theoretically, it is calculated according to the same thermodynamic principles as applied for $C_{XRG\theta}$ and has almost the same formula, with one difference—θ is replaced by the average temperature drop per effect of MED, ΔT_{eff}:

$$C_{XRGT} = 8760 C_{kWh} \tau q \eta \frac{\Delta T_{eff}}{T} = C_{XRG\theta} * \frac{\Delta T_{eff}}{\theta} \qquad (3.A.18)$$

where ΔT_{eff} is essentially the sum of four components. The largest is θ. The second quantitatively is BPE (the boiling-point elevation), which is proportional to the thermodynamic free energy needed to separate pure water from saline water. For the range of the low-temperature MED process feedwater and brine salinities and temperatures, it is about ~0.7 K; at other distillation processes or in high-temperature MED it can be as high as 1.2 K.

The third and fourth components are pressures. One pressure value is the sum of all local and longitudinal friction pressure drops Δp of the flowing vapors; the other is the partial pressure of noncondensable gases p_{ncg} in the vapor. Vapor pressure differences are equivalent to saturation temperature differences according to the thermodynamic principle known as the *Clausius–Clapeyron relation* (for small differences):

$$\Delta T = \frac{RT^2 \Delta p}{p * \lambda}. \qquad (3.A.19)$$

The equivalent equilibrium temperature differences are (1) approximately ~0.2–0.3 K for the friction pressure drop Δp of the flowing vapors and (2) <0.1 K for p_{ncg}. Thus

$$\Delta T_{eff} = \theta + \text{BPE} + \Delta p + p_{ncg} = 1.44 + 0.7 + 0.25 + 0.1 \approx 2.5\text{K} \qquad (3.A.20)$$

The charge of heat per unit product can now be estimated approximately:

$$C_{XRGT} = C_{XRG\theta} * \frac{\Delta T_{eff}}{\theta} = 0.1457 * \frac{2.5}{1.44} = \$0.253/\text{m}^3 \qquad (3.A.21)$$

This represents ~34% of the total water cost, estimated in Section 3.A.1 to be ~74¢/m^3. The economically optimal number of effects N_{eff} can be calculated on the basis of the optimal known ΔT_{eff}. We assume the maximum process temperature to be 72°C due to scaling deposits and aluminum corrosion considerations and the lowest condensing temperature to be 34°C, as in steam power plants. The range between these temperatures is $\Delta T_{\text{pr}} = 72 - 34 = 38$°C. Then

$$N_{\text{eff}} = \frac{\Delta T_{\text{pr}}}{\Delta T_{\text{eff}}} \approx \frac{38}{2.5} \approx 15 \text{ effects} \tag{3.A.22}$$

[*Note*: The small deviation of $0.5 = \Delta T_{\text{pr}} - N_{\text{eff}} * \Delta T_{\text{eff}}$ means that either the heat will be supplied at 0.5 K lower temperature, or the average ΔT_{eff} and θ will be higher by ~0.03 K. (The first alternative is slightly better.)] Thus, the average production rate is $P_N/N_{\text{eff}} = P_N/15$ t/d per effect or $2.778P_N$ effect. Since it was found above that the optimum water mass flux = 7.65 kg/h per m^2, the net heat transfer surface per effect is a $2.77P_N/7.65 = 0.363P_N$ (in m^2/effect). The gross area is about 8–10% higher, ~$0.4P_N$ m^2 per effect. In a 100,000-m^3/d MED plant consisting of two units $P_N = 50{,}000$ m^3/d per unit, the economically optimal heat transfer area per effect is therefore $0.4*50000 = 20{,}000$ m^2 per effect.

3. Modifying the evaporator design by assigning a different value for θ that is not the optimal one will result in a very small relative increase in the water cost if the deviation $\theta - \theta_{\text{op}}$ is within ± 15% of θ_{op}. The result of such change is investment reduction and capital cost decrease with slightly higher increase of the energy charge for $\theta > \theta_{\text{op}}$, or vice versa for $\theta < \theta_{\text{op}}$. To demonstrate, if, instead of the optimal value of $\theta_{\text{op}} = 1.44$ K calculated above, the design value of the heat transfer driving force is taken as ~15% higher, namely, $\theta = 1.65$ K and with 14 smaller effects instead of 15 optimal ones, then $C_{\text{XRG}\theta} = 0.1457*1.65/1.44 = \$0.167$/t, while $C_{\text{area}} = 0.1457*1.44/1.65 = \$0.1271$/t, and the off-optimum sum will be $C_{\text{XRG}\theta} + C_{\text{area}} = 0.167 + 0.1271 = \$0.2941$/t, compared to the optimum design sum of $2*0.1457 = \$0.2914$/t. The difference is $0.2941-0.2914 = \$0.0027$/t = 0.27¢ per ton, which is only 0.36% of the total water cost.

   On the other hand, at a production rate of 100M/t per year, this 0.27¢/t cost increase accumulates to over \$8 million during 30 years of the plant lifetime!! (*Note*: The last example is a plausible strategy of "pay less now and more later" taken by decision-makers who prefer to lower the initial investment and thus reduce risk and financing scope—even if the expected higher energy expenses will push the product cost to values a bit above the minimum.)

NOMENCLATURE

A	an experience-based factor for the relation $I_{tot} \approx A * P_{NA}^{m}$ between the annual production and the total plant investmenty
A_1	constant factor that replaces A in the economy-of-scale formula $I_{tot} \approx A_1 * P_{N}^{m}$ for the specific investment per daily output
Area	heat transfer area
C	price of desalted water per unit
C_2	total expenses per average year at specific dual-purpose plant studied
C_a	cost of unit heat transfer area
C_{area}	economic value of IHT (see "Abbreviations" list below) per year
C_c	capital cost
C_{com}	common electricity and steam production expenditures, according to cost allocation method based on exergy
C_E	total (imaginary) expenses, including levelized capital costs, incurred per average year of this imaginary reference single-purpose power plant
C_{Ee}	direct electricity generation expenditures, allocated exclusively to electricity generation, according to cost allocation method based on exergy
C_k	allocated cost of the product indicated by index k in a complex that produces several products (e.g., electric power, desalted water, salt, and high-pressure steam for a nearby factory)
C_{kWh}	cost per kilowatt hour, either actual, or vrtual, generated at an imaginary reference single-purpose power plant: $C_{kWh} = C_E/E$
CRF	capital recovery factor: Water payments at year end $$\text{CRF} = \frac{d}{1-(1+d)^{-n}}$$ Water payments per year at end of f short periods (annually) $$\text{CRF} = \frac{f \cdot [(1+d)^{1/f} - 1]}{1-(1+d)^{-n}}$$
C_{Se}	direct steam production expenditures for providing heat to the desalination plant, allocated exclusively to electricity generation, according to cost allocation method based on exergy
C_{w*}	remaining water production expenditures in a dual-purpose plant, according to cost allocation method
C_{w1}, C_{w2}	costs attributed respectively to the desalted water quantities W_1 and W_2 (see below, this list)
C_{XRGT}	economic value per year of the exergy consumed by ΔT_{eff} (see below, this list)

$C_{XRG\theta}$	economic value of EXR (see "Abbreviations" below) per year
d	annual discount or interest rate
D	reference date for capitalization
Dec	payments occurring during and after decommissioning of the plant
D_i	date of incoming payment i with time difference $D - D_i$ expressed in years
$D_{i\,av}$	average date of all payments of expenses i
D_j	date of incoming payment j with time difference $D - D_j$ expressed in years
$D_{j\,av}$	average date of all incoming payments of desalted water sales j
e_a	average annual inflation rate during a referred period plant history counted from the commissioning date of the plant (the reference date)
E	net salable electricity annual generation of an imaginary reference single-purpose power plant that uses the same heat source as the actual dual-purpose plant
E_1	gross power generated by specific dual-purpose plant studied
E_2	net actual salable power generated by specific dual-purpose plant studied
Ec	energy cost
E_w	gross annual electricity consumed for desalination at specific dual-purpose plant studied
Ex_i	actual expense payment
Expense	an expense payment, paid by the desalination plant, discounted to a given date
f	number of incoming payments for water sales per year ($f = 12/\Delta t$)
F_{dj}	discounting factor: $F_{dj} = (1 + d)^{D-D_j}$
i, j	serial number indices assigned to individual payments of expenses and sales, respectively
inf	inflation rate
inf_y	inflation rate in year y, namely, y years after plant commission
Idc	value of interest during construction
Inv	investment; Inv1, Inv2, Inv3, Inv4—sums of the annual investment payments during the final and the previous years of the construction period T_{con}, respectively.
I_S	specific investment per daily unit design capacity
I_{SA}	investment per annual unit design capacity $P_{NA} : I_{SA} = I_{tot}/P_{NA}$
I_{tot}	total investment capitalized to date D
k	serial number of product indicated by index i ($i \leq k$) in a complex that produces several products (e.g., electric power, desalted water, salt, and high-pressure steam for a nearby factory)
kWh_e	electrical kilowatthour

kWh_t	thermal kilowatthour
LF	load factor; ratio between actual plant life production and theoretical (i.e., full time at full design capacity) plant life production $\{P_A = 365P_N * LF\}$
m	economy-of-scale power factor for relation $I_{tot} \approx A * P_{NA}^m$ between the annual production and the total plant investment.
n	years of economic lifetime of plant (may differ from t_{prod}, defined below, this list)
Omc	all other operation and maintenance costs, designated as O&M cost
p	absolute pressure of the vapor
p_{ncg}	partial pressure of noncondensable gases in vapor
P_A	average *annual* production (in unit product); ideally equal to P_{NA}
P_k	amount of product indicated by index k in an industrial complex that produces several products (e.g., electric power, desalted water, salt, and high-pressure steam for a nearby factory)
P_N	nominal *daily* design capacity usually given in units per day
P_{NA}	nominal *annual* design capacity
q	transferred heat (in kW)
Rn	running cost (i.e., later part of the expenses payments)
Sale	a water sale payment, received by the desalination plant, discounted to a given date
sb	brine salinity
sf	seawater salinity
sp	product salinity
t_{con}	construction period in years
t_{prod}	production period in years
t_{proj}	entire project period in years
T	absolute temperature [in degrees Kelvin (K)] at which a representative effect operates
T_{amb}	ambient absolute temperature
T_{heat}	active heat temperature, specifically, the absolute temperature at which the heatsource is used in distillation processes
T_Δ	common economical temperature difference needed to reject heat from exhaust steam to ambience or to heatsink ($\approx 10 \pm 3^\circ C$)
U	heat transfer coefficient kW m^{-2} K^{-1}
W	plant total water production
W_j	product units measured in tons: cubicmeters (m^3) or kilogallons
W_1, W_2	annual water productions from each of the two sources (hybrid plant or two-quality plant)
y	ordinal number representing one year in a desalination plant history y years after plant commission (y may have a negative value for years of construction and other preparations)

Z	annual discount factor corrected by average inflation rate, $Z = (1 + d)/(1 + e_a)$
Δp	average sum of all drops in vapor pressure due to local and longitudinal vapor friction as a result of the vapor moving inside of an average effect of a multieffect distillation plant
Δt	time periods, expressed in months, between water sales payments (usually very few months)
ΔT_{BH}	temperature increase in the brine heater of multistage flash distillation
$\Delta T_{eff\_}$	average temperature drop per effect of multieffect distillation
ΔT_{pr}	practical, thermodynamic temperature difference, which is roughly proportional to the practical *economical equivalent value* of the heat: $$\Delta T_{pr} = \Delta T_{th} - T_{\Delta}$$
ΔT_{th}	temperature difference above the ambient temperature: $\Delta T_{th} = T_{heat} - T_{amb}$
ε	fraction of net heat transfer area needed to transfer parasitic heat
η	efficiency of conversion of theoretical gross mechanical energy to electrical power
θ	average heat transfer driving force
θ_{op}	optimal heat transfer driving force
λ	average latent heat of condensation
τ	average annual time fraction of heat transfer occurrence
Π	multiplication notation
Σ	summation notation
Σexp	expenses sum of all net levelized annual expenses of hybrid plant or two-quality plant.

Abbreviations

AC	alternating (electrical) current
BPE	boiling-point elevation
DEEP	Desalination Economic Evaluation Program: IAEA program for economic comparisons between power generation and desalination plants
ED	electrodialysis
ERMI	equipment replacement, maintenance, and insurance percentage of total investment per year
EXR	practical thermodynamic exergy consumed for heat transfer by thermal driving force
gfd	gallons per square foot per day (US)
GOR	gained output ratio
IAEA	International Atomic Energy Agency

IHT	investment in heat transfer area and associated equipment
MED	multiple-effect (multieffect) Distillation
mgd	million gallons per day
MSF	Multistage flash distillation
MVC	mechanical vapor compression
O&M	operation and maintenance
ppm	parts per million (g/ton or mg/kg)
RO	reverse osmosis
SA	stand-alone
SW	seawater
TVC	thermal vapor compression
VC	vapor compression
WHO	World Health Organization

REFERENCES

1. US Office of Saline Water (OSW), *A Standardized Procedure for Estimating Costs of Saline Water Conversion*, US Dept. Interior, 1956.
2. MW Kellogg Company, Saline Water Conversion Engineering Data Book for the OSW, 1965 (Sec. 3. 500).
3. IAEA, *Costing Methods for Nuclear Desalination*, Technical Report 69, Vienna, 1966.
4. IAEA, *Guide to the Costing of Water from Nuclear Desalination Plants*, Technical Report 80, Vienna, 1967.
5. Barnea J, Costing of Desalted Water?, United Nations, early 1960s—(a report similar to the first OSW procedure [1]).
6. OSW, *Desalting Handbook for Planners*, 1972, a cost format, [an adaptation of which was used in cost updates prepared for the US Office of Water Research and Technology and is recognized by the International Desalination Association (IDA)].
7. IAEA, *Guide to the Costing of Water from Nuclear Desalination Plants*, 1973 Edition, Technical Report 151, Vienna, 1973.
8. IAEA, *Guidebook on the Introduction of Nuclear Power*, Technical Report 217, Vienna, 1982, pp. 75–94, 147–154, 235–243, 286–288, 306.
9. Awerbuch L, Rogers AN, Siebentthal CD, Battey RF, *Desalination Technology—Report on the State of the Art*, Bechtel Group Inc. Job 15669 (agreement 1462), San Francisco, Feb. 1983.
10. IAEA, *Expansion Planning for Electrical Generating Systems—a Guidebook*, Technical Report 241, Vienna, 1984, pp. 151–241.
11. IAEA, *Use of Nuclear Reactors for Seawater Desalination*, International Atomic Energy Agency Technical Document IAEA-TECDOC 574, Vienna, Sept. 1990.
12. IAEA, *Technical and Economical Evaluation of Potable Water Production through Desalination ... Using Nuclear Energy and Other Means*, International Atomic Energy Agency Technical Document IAEA-TECDOC 666, Vienna, 1992.
13. IAEA, *Options Identification Program for Demonstration of Nuclear Desalination*, IAEA-TECDOC 898, Vienna, 1996.

14. IAEA, *Potential for Nuclear Desalination as a Source of Low Cost Potable Water in North Africa*, IAEA-TECDOC 917, Vienna, 1996.
15. IAEA, *Thermodynamic and Economic Evaluation of Co-production Plants for Electricity and Potable Water*, IAEA-TECDOC 942, Vienna, 1997.
16. IAEA, *Methodology for the Economical Evaluation of Cogeneration/Desalination Options: A User's Manual*, IAEA Computer Manual Series Vol. 12, Vienna, 1997.
17. Leitner GF, Breaking the cost barrier for seawater desalination, *Int. Desal. Water Reuse*. **8**(1):14–20 1998.
18. Barak A, Economic aspects of desalination in Israel, *Water Irrig*. (Israeli journal, in Hebrew), **380**:28–32 (June 1998).
19. IAEA, *Desalination Economic Evaluation Program (DEEP): User's Manual*, IAEA Computer Manual Series Vol. 14, Vienna, 2000, pp. 133–150.
20. IAEA, *Examining the Economics of Seawater Desalination Using the DEEP Code*, IAEA-TECDOC 1186, Vienna, 2000.
21. Barak A, Steps of planning, development, research and additional technological activities for seawater desalination cost reduction, *Water Irrig*. **406**:4–11 (Aug. 2000).
22. IAEA, *Introduction of Nuclear Desalination*, Technical Report 400, Vienna, 2000, pp. 135–160, 188–189, 203, 221, 230–235, 240–245, 282.
23. Glueckstern P, Priel M, Potential cost reduction in large SWRO systems, *Proc. 5th IWA Conf. Membranes in Drinking Water and Industrial Water Production*, Muelheim an der Ruhr, Germany, Sept. 2002.
24. IAEA, *Optimization of the Coupling of Nuclear Reactors and Desalination Systems*, IAEA-TECDOC 1444, Vienna, 2005.
25. Middle East Desalination Research Center (MEDRC), *Proc. Int. Conf. Desalination Costing*, Lemesos/Limassol, Cyprus, Dec. 6–8, 2004.
26. Glueckstern P, History of desalination cost estimations, *Proc. MEDRC Int. Conf. Desalination Costing*, Lemesos/Limassol, Cyprus, 6–8, Dec. 2004, pp. 1–17, this paper cites 45 references on costing of desalted water during 1966–2004.
27. Dore MHI, Forecasting the economic costs of desalination technology, *Desalination* **172**:207–214 (2005).
28. Blank JE, Tusel GF, Nisan S, The real cost of desalted water and how to reduce it further, *Desalination* **205**:298–311 (2007).
29. Methnani M, The influence of fuel costs on seawater desalination options, *Desalination* **205**:332–339 (2007).
30. Reddy KV, Ghaffour N, Overview of the cost of desalinated water and costing methodologies, *Desalination* **205**:340–353 (2007).
31. Abdulnasser AM, Nafey AS, Fath HES, Thermoeconomics analysis of some existing desalination processes, *Desalination* **205**:354–373 (2007).
32. Yongqing W, Lior N, Fuel allocation in a combined steam-injected gas turbine and thermal seawater desalination system, *Desalination* **214**:306–326 (2007).

CHAPTER 4

Advances in Hollow-Fiber Reverse-Osmosis Membrane Modules in Seawater Desalination

ATSUO KUMANO

Advances in Water Desalination, First Edition. Edited by Noam Lior.

4.1 INTRODUCTION

The reverse-osmosis (RO) method has become widely accepted for reliable and long-term seawater desalination. In particular, the RO method based on the use of hollow-fiber RO modules has earned recognition for excellent performance in high-salinity waters such as occur in the Mediterranean and the Middle East. A cellulose triacetate (CTA) hollow-fiber RO module is recognized for providing high-quality permeate water and reliable performance for long-term operation in many practical seawater applications.

Reverse-osmosis membranes were first developed from cellulose acetate (CA) and cellulose triacetate CTA. These CA and CTA membranes have been in commercial use for many years. The CTA membrane, in particular, is still in widespread use throughout the world. This global use proves that the CTA membrane is excellent as an RO membrane. Its performance is enhanced by a combination of high permeability and high selectivity, due to a balance of excellent hydrophilic and hydrophobic properties. Moreover, the CTA membrane has high tolerance to oxidizers such as chlorine. This characteristic makes it useful as a RO membrane. The CTA membrane is the only RO membrane among many other RO membranes currently marketed that offers this high degree of chlorine resistance. On the basis of the prediction model of performance change of RO module due to hydrolysis and oxidation by chlorination, the intermittent chlorine injection (ICI) method has been developed and is effectively used with the CTA membrane to control biofouling growth. It also enables the stable operation of seawater desalination plants, especially in the Middle East, where there is increased biofouling potential due to high temperatures.

Following development of aromatic polyamide hollow-fiber RO membranes, CTA with a hollow-fiber membrane structure was developed to control biofouling. Because of its superior resistance to chlorine in comparison with polyamide hollow-fiber RO membranes, the CTA membrane is relatively impervious to contamination by the biofouling material to which it is exposed. Furthermore, the CTA membrane is capable of significantly reducing flux per unit membrane area, because its hollow-fiber configuration provides an element in a module that has a membrane area that is approximately 10 times larger per unit volume than that of a spiral-wound membrane element. Therefore, with the CTA membrane, it is possible to decrease the membrane load, which makes it more difficult for fouling to occur. These key attributes lead to minimum chemical cleaning and long membrane life. The hollow-fiber RO module design and the optimization of flow pattern in the module are based on an analytical model of hollow-fiber RO modules. The analytical model has been presented as a friction–concentration–polarization (FCP) model, in which the concentration polarization and the pressure drop in the fiber bore were considered. The validity of this FCP model has been verified by actual performance data on hollow-fiber modules.

A more cost-efficient and high-efficiency RO desalination module has been developed that further reduces product-water cost. This is achieved by a higher recovery capability due to a higher pressure-resistant membrane and module capability. An innovative module has been developed by Toyobo for the next generation of improvements. This newest module can provide more than 2 times the product-water capacity compared with previous conventional modules. The success in developing this newest module will lead to even further reduction of product-water costs. Toyobo seawater RO modules provide high reliability and low cost capability to produce freshwater. The ability to ensure low-cost, fresh drinking water by RO with this latest innovative technology undoubtedly will lead to future statements such as "the 21st century is [was] an age of water" (i.e., reclamation of water as a natural resource).

4.2 BACKGROUND

The US Saline Water Act was enacted on July 3, 1952. At that time, the US Office of Saline Water (OSW) was commissioned by the Department of the Interior to develop a method for obtaining freshwater from seawater and brackish-water economically. This effort accelerated the development of reverse-osmosis (RO) membranes. The RO process was proposed as a desalination methods by Dr. C. E. Reid of the University of Florida in early 1953 [1]. Workers at that university studied various commercial polymer films in 1957, and announced the development of a cellulose diacetate film with a semipermeable membrane that provided a salt rejection rate of $\geq$ 96%. However, the water permeability was negligible. In 1960, Loeb and Sourirajan of the University of California succeeded in developing a method of manufacturing a new asymmetric membrane [2]. The membrane had high permeability and consisted of heat-treated cellulose diacetate. These enhancements in membrane performance lead to a practical application of the RO membrane module [1–3].

In more recent years, seawater RO technology has been applied for desalination in large power plant projects, increasingly replacing the older thermal distillation technology. In the case of single-purpose (with no excess power production) seawater desalination plants, RO technology has displaced MSF (multistage flash) technology. Key examples in the Middle East include Jeddah phases 1 and 2 [each producing 15 million gallons per day (mgpd)], Yanbu (33 mgpd), and Jubail (24 mgpd). No other technology offers lower water production costs and the other advantages of RO over thermal processes in single-purpose desalination plants. The advantages of RO over thermal methods include greater mitigation of the effects of environmental pollution, wider availability of suitable site, for desalination plant construction, less complex plant design, lower maintenance costs, and lower operating costs.

Since seawater RO technology was first commercialized in the mid to late 1970s, it has become the most practical and cost-effective method for water desalination. The only exceptions to the automatic selection of RO technology for desalination are those large projects where energy costs are low (such as in the Middle East) and where both power and water needs are required. However, even in these cases, which might be typical scenarios in the Middle East, seawater RO technology outperforms MSF technology with its additional hybrid approach. In a hybrid approach, desalinatied waters produced by the MSF plant and a separate RO plant are blended to meet final drinking-water requirements. An example of a large scale hybrid approach is found in the Fujairah, United Arab Emirates (UAE) plant, which can blend 62.5 migd (million Imperial gallons per day) MSF water with 37.5 migpd RO water.

Further application of seawater RO technology into the realm of dual-purpose operations will be achieved by continued successful operation of large desalination plants (and the attendant confidence that such operation brings to potential end users), improved plant optimization via biofouling elimination, and broader understanding of the cost savings and additional flexibility of RO technology versus thermal processes in dual-purpose desalination plants.

Desalination plant capacity using RO processes has increased more rapidly in comparison with evaporation methods such as multistage (MSF) flash distillation. In particular, the successful operation of the Jeddah 2 seawater RO plant since startup provides a key example of optimization of the large-scale plant performance. High availability has been demonstrated over 5 years of operation following development of the intermittent chlorine injection (ICI) method as a means to effectively eliminate biofouling. This technique is unique for application with the CTA hollow-fiber RO module. Hollow-fiber membranes offer greater versatility and flexibility in plant operations as well as excellent long-term membrane durability.

4.3 SEPARATION MEMBRANES AND PRODUCTION METHODS

4.3.1 Separation Characteristics

In general, semipermeable membranes used for salt–water separation are divided into the following four categories: microfiltration (MF), ultrafiltration (UF), nanofiltration (NF), and reverse-osmosis (RO) membranes. (see Table 4.1). MF membranes are used for rejecting particles and UF membranes, for rejecting polymer materials, including those soluble in water. The criteria for MF and UF membranes are roughly similar; in terms of non-selectivity for salt the main difference is pore size. In most situations, a membrane with the pore size of more than several tens of nanometers (e.g., >10–20 nm) is classified as MF, while membranes with pore size <10–20 nm, for example, are classified as UF. Both NF and RO membranes can reject inorganic ions such as sodium ion, chloride ion, calcium ion, and sulfate ion, while MF and UF membranes cannot. The criteria for NF and RO membranes are roughly similar; however, the NF membrane can reject multivalent ions better than a an RO membrane can; thus, NF membranes are used to soften hard water and are classified as low-rejection-type RO membranes. In many circumstances, membranes with low rejection (e.g., <90%) of sodium chloride are designated as NF membranes and membrenes with a higher rejection rate (e.g., >90%), as RO membranes. Both MF and UF membranes reject objects mainly by a sieving mechanism because the distance between the object and the inside wall of the pore is

Table 4.1 Water–Salt Separation Process Using Four Types of Membrane

Membrane Type	Abbreviation	Object for Separation
Microfiltration	MF	Particles, size 0.01–10 μm
Ultrafiltration	UF	Polymers, MWCO[a] 1000–300,000
Nanofiltration	NF	Polymers, MWCO[a] several hundreds to several thousands
Reverse osmosis	RO	Polymers, MWCO[a] 60–350

[a]Molecular weight cutoff.

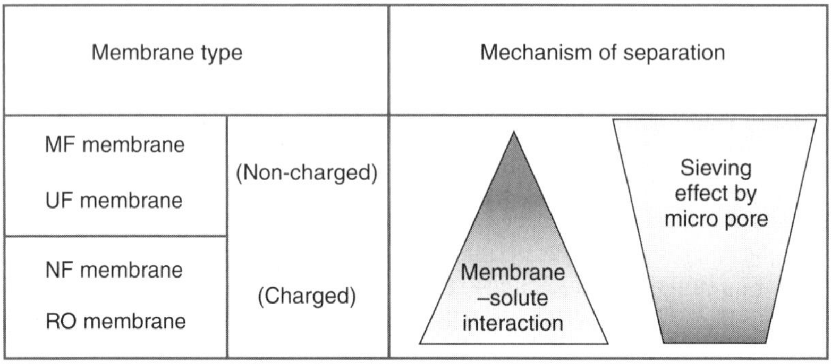

Figure 4.1 Membrane and mechanism of separation (membrane–solute interaction = solution–diffusion, adsorption–diffusion, electrostatic interaction).

greater than that of NF and RO membranes. On the other hand, NF and RO membranes reject objects membrane–solute interaction as will as by the because the distance between the object and the inner pore wall is less than that with the UF membrane. A schematic representation of these separation mechanisms is shown in Figure 4.1.

4.3.2 Membrane Structure

The separation membranes are also divided into two types with respect to structural symmetry: symmetric and asymmetric. microfiltration membranes are mainly symmetric; RO membranes, asymmetric. A typical structure cross section is shown in the Figure 4.2.

As shown in Figure 4.2, porosity in a symmetric membrane is uniform and the porosity distribution does not depend on the direction or orientation of the thickness in the membrane. On the contrary, porosity in an asymmetric membrane is not uniform and the porosity distribution depends on thickness orientation in the membrane; this is especially evident in the high-density surface layer, which is called the *dense layer*, as it has a selectivity function; the other layer, composed of porous material, is called the *support layer*.

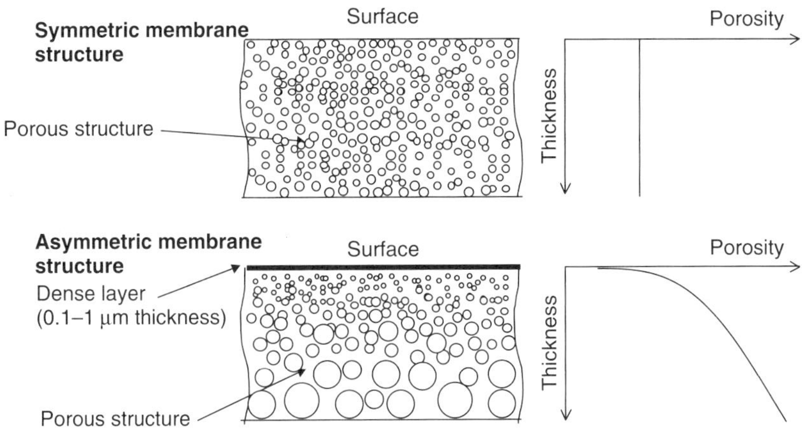

Figure 4.2 Cross sectional diagrams of symmetric and asymmetric membranes.

In the asymmetric membrane, the dense and support layers are usually composed of the same material; if the material composition differs between these two layers, the asymmetric membrane is called a *composite membrane*.

In addition to the asymmetric membrane, a broad class of thin-film composite membranes is offered by a number of desalination equipment manufacturers. The thin-film composite membranes are formed by an interfacial polymerization process. An example of the thin-film membrane is shown in Figure 4.3. The thin membrane layers are generally of different polymeric types. The support films are generally of polysulfone chemistry, and the fabric support materials are made of polyester. Similar to the asymmetric hollow fiber, the membrane film determines the extent and of the quality rejection, and the more porous support material usually controls flow. Thin-film composite membranes, because of their multilayer construction, can be difficult to manufacture. The multiple layers must be constructed together to yield an ultrathin film, uniform throughout the sheet and free of imperfections. Manufacturers of thin-film composites strive to automate

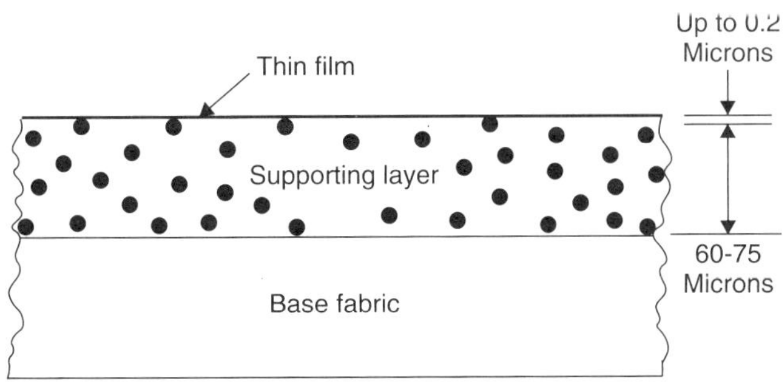

Figure 4.3 Cross section of a thin-film composite membrane.

manufacturing, as much as possible, to minimize the opportunity for embrittling or scratching the films.

4.3.3 Membrane Shape and Module Configuration

Reverse-osmosis modules are available in four configurations, flat-sheet (plate-and-frame), spiral-wound, hollow-fiber, and tubular. For water treatment such as seawater desalination and water purification applications, hollow-fiber and spiral-wound modules are mostly used. Flat-sheet (plate-and-frame) and tubular modules are rarely used, especially in typical applications using liquids with high concentrations of suspended solids, or highly viscous liquids such as those in beverages, food products, pharmaceuticals, and wastewater. Flat-sheet (plate-and-frame) modules have been used more recently for treatment of leachate found near refuse incinerator plants.

4.3.3.1 Flat-Sheet (Plate-and-Frame) Type One module configuration used for flat-sheet RO membranes is the plate-and-frame type. In this configuration, circular sheets of membrane are sealed to molded plates. These plates provide protective support against operating pressure and in flow channels against permeate water. The plate-and-frame configuration is conceptually similar to the conventional plate-and-frame format used in gas separation and low-pressure applications, for example, except that a typical higher fluid operation pressure is applied, especially in RO applications. As shown in Figure 4.4, the membrane package is installed in a pressure vessel designed and fabricated to withstand

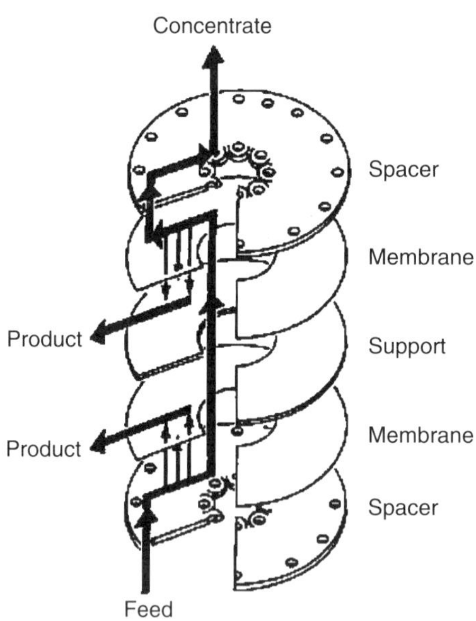

Figure 4.4 Plate-and-frame configuration.

operating pressure ranging from 30 to 200 bar. A stack of parallel porous plates is used to support the membrane on each side of the porous plate. Pressurized feedwater enters the top of the pressure vessel and flows between the parallel stacks of membranes and porous plates. Permeate water passes through the membrane and into the porous plate to be routed to the product water collection system and then out of pressure vessel. As the feed/reject water stream passes across the membrane, it becomes concentrated and eventually leaves the pressure vessel as a concentrate or reject.

4.3.3.2 Spiral-Wound Type The spiral-wound configuration is shown Figure 4.5. Essentially, the spiral-wound element consists of two sheets of membrane separated by fabric material. This fabric material both supports the membrane against the operating pressure and provides a flow path for withdrawal of the permeate water. The membrane envelope is sealed with an adhesive on three sides to prevent contamination of the permeate water. The fourth side is attached to a permeate-water tube that has holes within the edge seal to allow for removal of permeate water from the porous product-water spacer.

The membrane envelope is rolled up around the central permeate-water tube, with a plastic mesh spacer positioned between the facing membrane surfaces, in

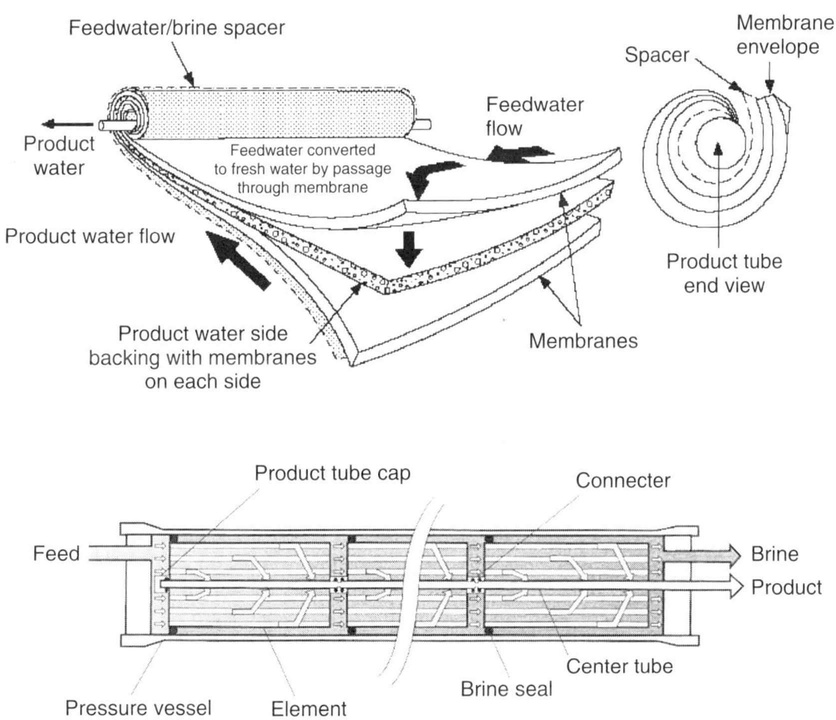

Figure 4.5 Spiral-wound element and modules.

a spiral pattern. The mesh spacer not only separates the membrane surfaces but also provides a flow path for and turbulence in the feed/reject water flow of each element. The elements are equipped with an outer wrap to contain the feed/reject water flow in the mesh passageway and brine seal to ensure that the feed/reject water flow passes through, around, the element and not. Spiral-wound elements are typically available in lengths of 40 in. and diameters of 4 or 8 in. More recently, 16-in. diameter spiral-wound elements have been used for seawater and brackish-water desalination.

Spiral-wound elements are installed in a pressure vessel that is usually fabricated from fiberglass-reinforced plastic. The pressure vessel inside diameter is sized to match the outside diameter of the element's brine seal. Pressure vessels are designed and fabricated to accommodate one to eight elements and operating pressures ranging from 60 to 90 bar in, for example, seawater desalination applications. Figure 4.5 shows a pressure vessel with several elements installed. Feedwater enters on one side of the pressure vessel and flows through the first element and then through the permeate-water spacer into the permeate-water tube. The reject water from the first element flows to the second element, and the reject from this second element becomes the feed to the next element. The reject water from the last element flows out from the pressure vessel. The element permeate-water tubes in the pressure vessel are interconnected by filaments that contain O-ring seals to prevent permeate-water contamination. The permeate water can flow out the pressure vessel from either end of the pressure vessel. The minimum brine (reject) water flow rate in an element is limited and therefore is used in low-recovery elements. In high-recovery systems, a arrangement of pressure vessels are arranged in a multistage configuration.

This spiral-wound type is one of the most widely used module configurations; however, a moderate amount of pretreatment is required for some feedwater to prevent fouling of feedwater–brine-side spacers.

4.3.3.3 Hollow-Fiber Type The hollow-fiber configuration is most commonly used in microfiltration and ultrafiltration UF membranes. Hollow-fiber filaments have been produced for the textile industry for a long time. Initially, this technology was applied to polymers to induce reverse osmosis into hollow fine fibers. The available polymers in hollow-fiber elements for RO are cellulose triacetate, cellulose diacetate, and aromatic polyamide. The hollow fibers are very fine (e.g., ~0.1–0.3 mm) in outer diameter, similar to a human hair. The inner diameter is about ~0.04–0.12 mm, and the outer: inner diameter ratio is >2 because the fiber requires a thick-walled cylinder to provide the compressive strength necessary to withstand the operating pressure. The hollow-fiber membrane is so fine that the membrane area is larger than that in flat-sheet and spiral-wound membrane elements. Also, because of their high packing density, hollow-fiber membranes also provide considerably lower flux than do flat-sheet and spiral-wound membranes.

A schematic representation of a typical hollow-fiber membrane module is shown in Figure 4.6. A continuous hollow fiber is wound into a bundle around a core tube as a feedwater distributor. The core tube pipe is sealed at the product-water end

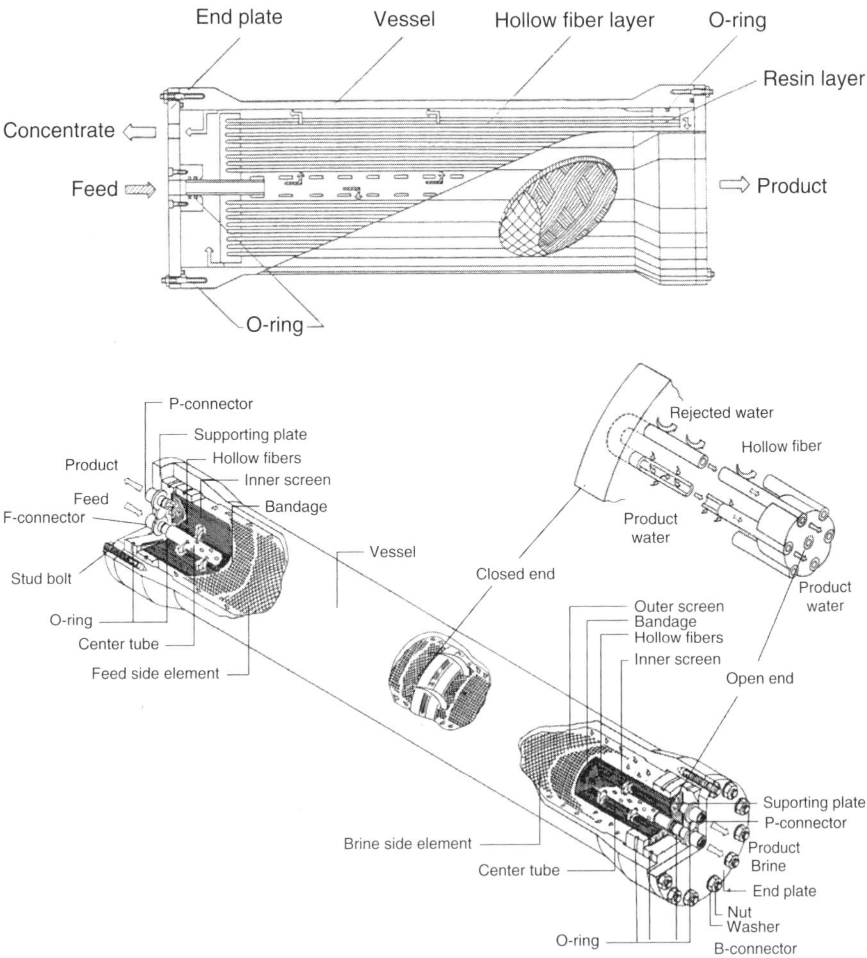

Figure 4.6 Hollow-fiber membrane modules.

and is positioned in the center of the hollow-fiber bundle. Both ends of the fiber bundle are embedded in an adhesive material as a resin layer. The resin layer is cut to open the fibers at each end.

The hollow-fiber membrane module consists of a pressure vessel and an element containing the hollow-fiber bundle. The hollow-fiber bundle is then installed in a pressure vessel and an O-ring seal is placed in the product-water end of the tube sheet, against the pressure vessel wall, to prevent the high-pressure brine water flow from entering the pressure vessel and mixing with the product-water flow. The feedwater flow enters the hollow-fiber element from a perforated feed tube placed in the center of the membrane bundle. The feedwater stream then flows radially outward from the center core tube to the brine collection channel at the outside of the element.

When the feedwater stream attains maximum velocity, it enters into the module; at minimum velocity, it approaches the outer edge of the membrane bundle. The rate of recovery from a hollow-fiber element ranges from 40% to 75% for feedwater and brackish-water, respectively, and between 25% and 60% for seawater. The recovery rate is typically higher in hollow-fiber than in spiral-elements wound elements.

In general, pressure vessels are consist of fiberglass-reinforced plastic pipe and are designed to apply operating pressures of 30–82 bar. One or two hollow-fiber membrane elements are installed in the pressure vessel. Hollow-fiber membrane elements are available with nominal diameters of 3, 5, 8, 9, and 10 in.

The high packing density and the elimination of membrane support materials are main the advantages of the hollow-fiber element.

4.3.3.4 Tubular Type The tubular membrane module consists of a tube-type membrane that is similar in structure to the hollow-fiber membrane. The tubular RO membrane module is shown Figure 4.7. The permselective (i.e., permeate-water-selective) membrane is installed inside the tube, which serves as the pressure vessel. Its inner diameter is ~0.5–1 in. Feedwater enters through one end of the tube, where the membrane is sealed to the tube to prevent contamination of the product water. The brine water flows through the other end of tube and is routed to additional tubes in series or to wastewater. The main advantages of the tubular RO membrane module are (1) tolerance to high concentrations of turbidity and suspended solids in feedwater and (2) contamination resistance mechanical cleaning. The main disadvantage is the design complexity, with a large number of components (tubes, O-ring sealing devices, etc., in the membrane area of the pressure vessel.

4.3.4 Materials of RO Membranes

In the use of particular polymers for membranes it is convenient to classify the membranes as either porous, such as microfiltration and ultrafiltration membranes, or nonporous or dense, as in reverse-osmosis membranes.

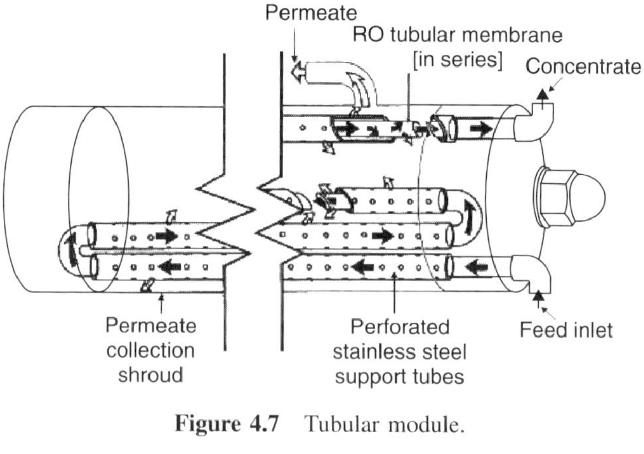

Figure 4.7 Tubular module.

For porous membranes, the important factors in selecting a material are formability (ductility) property and stability. For nonporous membranes, the flux and selectivity characteristics of the material are important factors. In addition to its formability property (e.g., it can be shaped into a thin film without a pinhole), the RO membrane is hydrophilic, and although it is not water-soluble, this is another important criterion for selecting the material. Therefore, material selected for RO membranes must be both hydrophilic and hydrophobic.

Cellulose triacetate and aromatic polyamide materials are widely used in RO membranes.

4.3.4.1 Cellulose Derivative Material Cellulose derivative material such as cellulose diacetate and cellulose triacetate are hydrophilic but not water-soluble. This material has long been thought regarded as one of the most suitable materials for use in RO membranes.

As mentioned in Section 4.2, development of RO membrane technology began in earnest in the 1950s and 1960s, following passage of the Saline Water Act in 1952, creation of the Office of Saline Water (OSW) by the US Department of the Interior, and experimentation by Reid, Breton, and others [1–3] to find a method for extracting salt from seawater and brackish-water to provide freshwater on a wide scale but at reasonable cost. Reid and his collegues studied the properties of several commercial cellulose-based polymer films with respect to their ability to reject salt from water. They found that the reverse-osmosis mechanism in these materials rejected the salt at a very high rate (>96%) and proposed that RO membranes be developed for commercial water desalination applications.

Gulf General Atomic, a company funded by OSW, conducted studies on the spiral-wound module development. The company also applied for a patent of the basic structure in 1968 [4]. The patent, used an example of the cellulose diacetate membrane developed by Loeb and Sourirajan earlier [2]. Following publication of these patents, other companies (e.g., UOP, Hydranautics, Envirogenics, Toray Industries, Daicel) began using the spiral-wound configuration in a cellulose acetate membrane, which they put on the market.

Work on various models of tubular-type modules also began in the 1960s. Many companies, including the plant where Loeb and others had developed the RO process, developed and marketed these tubular models [5].

A hollow-fiber module was developed and a fundamental patent was filed by Dow Chemicals in 1960. The patent was based on RO module using a cellulose triacetate hollow fiber [6]. A significant portion of the research and development carried out at Dow Chemicals, was based on their research contract with OSW. The development results on use of the RO module for brackish-water were published in 1970 and for seawater, all in 1974. The RO module using a cellulose triacetate hollow fiber for brackish-water was marketed in 1974 [7,8].

Research and development (R & D) on the hollow-fiber module using a cellulose acetate membrane was conducted by Monsanto, Toyobo, and other companies in addition to Dow Chemicals. In 1979 Toyobo announced a RO module for one-pass desalination of seawater that used a cellulose triacetate hollow-fiber membrane

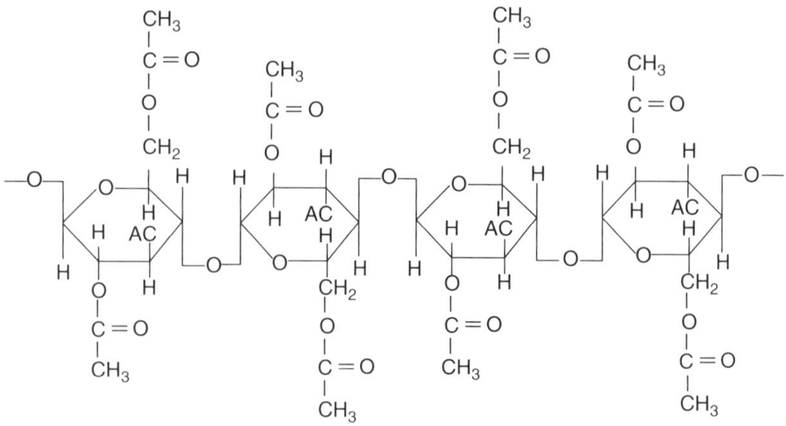

Figure 4.8 Structure of typical RO membrane material (cellulose triacetate).

module [9,10]. The typical structure of cellulose triacetate used in RO membranes is shown in Figure 4.8.

Although they are limited in terms of chemical and thermal stability, cellulose diacetate and cellulose triacetate are used in RO membranes because these compounds have higher chlorine resistance than does polyamide.

4.3.4.2 Polyamide Material A RO membrane with linear aromatic polyamide was developed and commercialized by Dupont Company in 1969 for water desalination after a long period of research. It has been used for hollow-fiber membranes that can desalinate seawater. Initially only brackish-waters were desalted. Later, in 1973, seawater was converted for potable use The polyamide materials are described as aromatic polymers with substantially linear condensation. The hollow fibers may be spun by use of a suitable solvent and additives. Some groups bearing pendant ionic groups may be used to provide additional hydrophilic properties. The typical structure of the aromatic polyamide is shown in the Figure 4.9. The fibers exhibit satisfactory rejection of monovalent ions posttreatment, but the rejection rate is sometimes lower while the fibers are being spun. The polyamide hollow-fiber membranes have several characteristics.

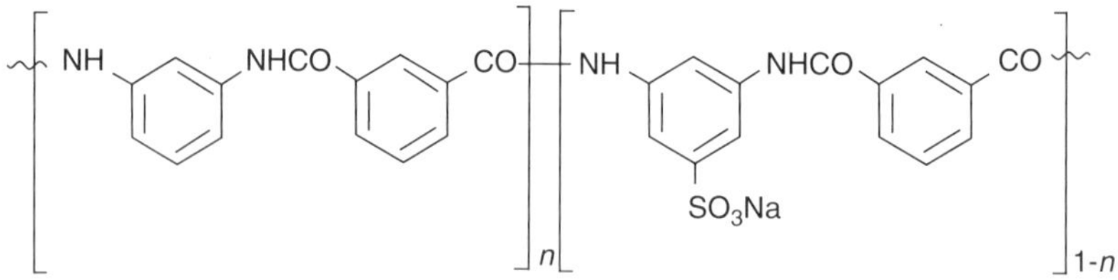

Figure 4.9 Structure of typical RO membrane material (linear aromatic polyamide).

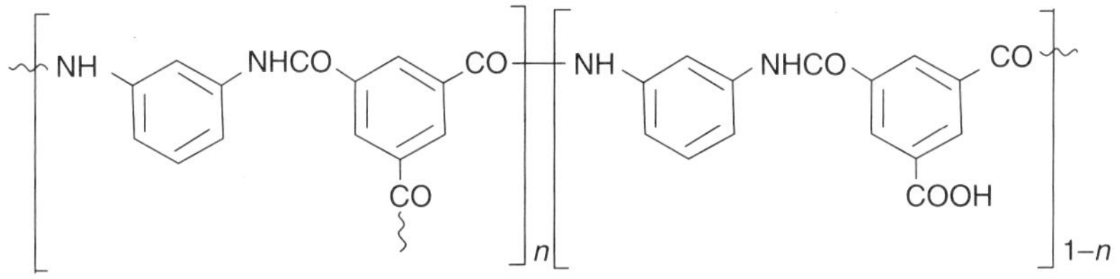

Figure 4.10 Structure of typical RO membrane material (crosslinked aromatic polyamide).

Microorganisms are unable to evolve enzymes to react with the synthetic polymer, Membranes are capable of operating at higher temperatures and in a pH range of 4–11 without hydrolysis. On the contrary, membranes are extremely sensitive to oxidants such as chlorine. The outer diameter of the polyamide hollow fiber membrane averages 85 μm and the inner diameter, 42 μm. However, the polyamide RO membrane has not be manufactured since 2000 because the manufacturer withdrew it from the market.

Aromatic polyamides, rather than aliphatic polyamides, are used for RO membranes. Aromatic polyamide, especially crosslinked polyamide, is applied to a thin-film composite of the RO membrane.

The structure of typical crosslinked polyamide is shown in Figure 4.10.

4.3.4.3 Other Materials Someother RO membrane materials are shown in Figure 4.11. Polypiperazineamide is used for low-pressure RO membranes and also for nanofiltration membranes at present. A composite membrane using polyether was developed in 1982. Development of material on RO membranes is shown in Figure 4.12.

4.3.5 Membrane Production Method

4.3.5.1 Outlines of Membrane Production Several methods of producing polymeric membranes have been developed; the most common are the phase separation, stretching, and composite methods. The phase separation method is used for producing both symmetric and asymmetric structures, with a wide range of

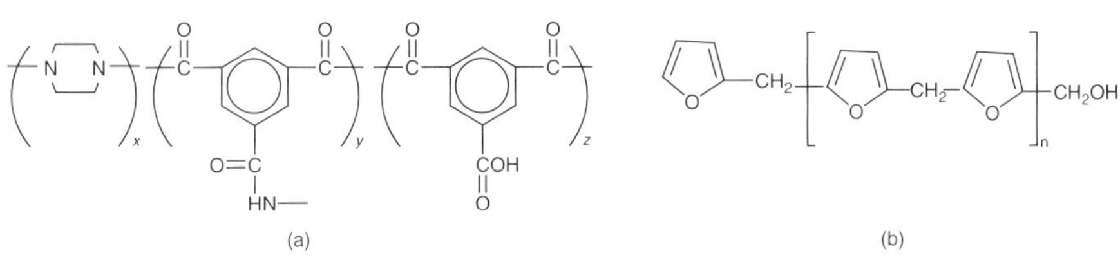

Figure 4.11 Structures of typical RO membrane material: (a) polypiperazineamide; (b) crosslinked polyether.

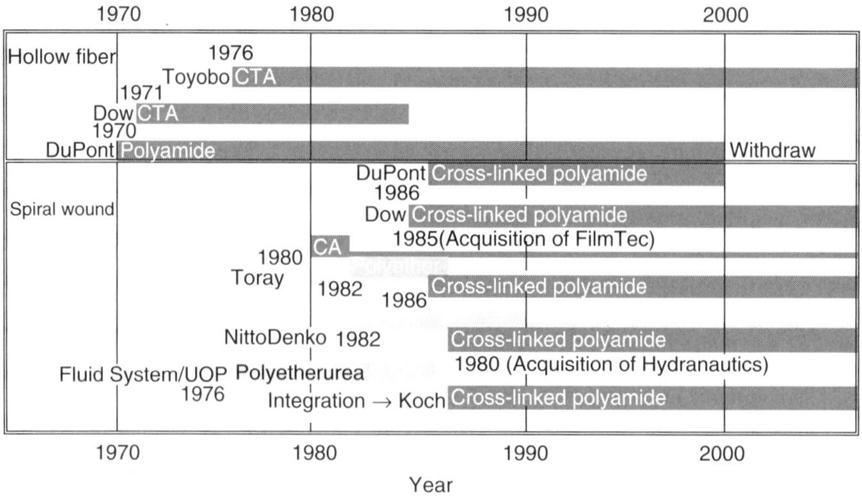

Figure 4.12 Timeline for development of RO membrane material.

Table 4.2 Outline of Polymeric Membrane Preparation Method

Method	Membrane Structure	Membrane MF	UF	NF	RO
Phase separation					
Solution					
Dry	Symmetric	○	○	○	○
Dry and wet	Asymmetric	○	○	○	○
Wet	Asymmetric	○	○	○	○
Melting	Symmetric and asymmetric	○	○	—	—
Stretching	Symmetric and asymmetric	○	—	—	—
Electron irradiation/etching	Symmetric	○	—	—	—
Composite					
Polymer coating	Composite	—	○	○	○
Monomer polymerization	Composite	—	—	○	○
Interfacial polymerization	Composite	—	—	○	○

characteristics, such as those found in MF, UF, NF, and RO membranes. Composite methods are used for preparing NF and RO membranes. An outline of the polymeric membrane preparation method is shown in Table 4.2.

4.3.5.2 Phase Separation Method The vast majority of polymeric membranes can be produced by a method known as *phase separation*. This method is classified into two broad categories: (1) the thermally induced phase separation method (TIPS) and (2) the non-solvent-induced phase separation method (NIPS). TIPS is almost always used for microfiltration membrane and Ultrafiltration membranes. The RO membrane is mainly prepared by NIPS. In the NIPS phase separation process, a polymer solution inverts to a distended three-dimensional macromolecular complex. The membranes with porous structure are produced from a >2-component dope mixture containing polymer solvent and nonsolvent, and in some cases with additive agents.

The procedure is to first dissolve the polymer in the solvent solution with some additive agents. For flat-sheet membranes, the solution is usually spread onto a suitable support using a casting slit. The support consists of nonwoven polyester. The cast film is transferred to the nonsolvent solution bath, where the exchange between solvent and nonsolvent occurs, and it leads to polymer precipitation.

The performance characteristics of a membrane produced by the phase separation method depend on many parameters, such as polymer concentration, evaporation time before immersion, temperature, humidity, composition of casting solution, and coagulation bath composition and conditions.

The procedure for producing hollow-fiber membranes is different from that for flat-sheet membranes. The hollow-fiber membrane is produced by wet spinning, dry–wet spinning, dry spinning, and melt spinning. The preparation method takes the viscous polymer solution and pumps it through a nozzle. The nozzle ordinarily has an annular structure, and the annular section enables the polymer solution to be produced in a cylindrical form with an inner coagulation fluid that can be used in case of wet spinning and dry–wet spinning. After exposure to the atmosphere for some time the fiber is immersed in a coagulation bath that is filled with nonsolvent solution.

A schematic diagram of the wet-spinning hollow-fiber production process is shown in Figure 4.13 [11], and a preparation diagram is shown in Figure 4.14 [12].

A typical manufacturing process for cellulose acetate RO and UF membranes consists in dissolving the polymer in a solvent mixture such as acetone and formamide. The casting solution is spread as a thin film or extruded as a hollow fiber while exposed to the air and then dropped into the coagulation bath. Generally the formation of a casting or extruded polymer solution affects membrane structure in the subsequent stages of the membrane producing process.

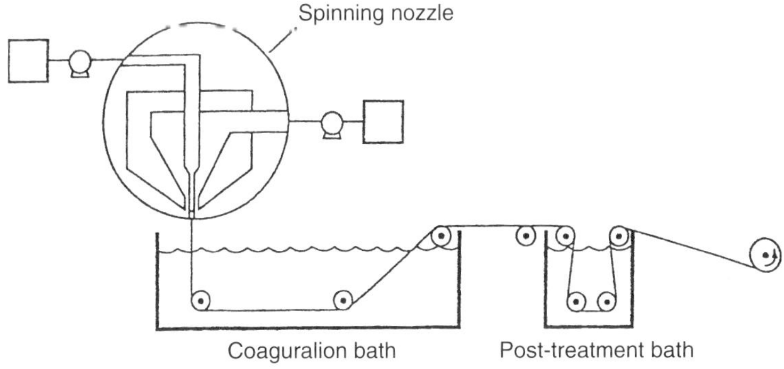

Figure 4.13 Outline of hollow fiber preparation [11].

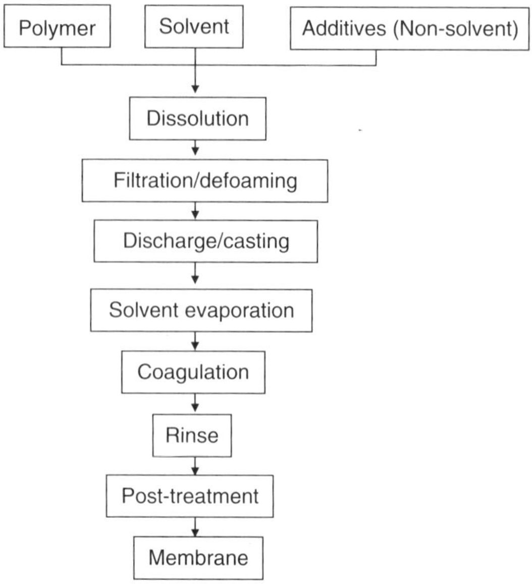

Figure 4.14 Flowchart showing typical preparation of an asymmetric membrane [12].

In principle, the Loeb–Sourirajan cellulose acetate membrane is made in the same method as described above. The Loeb–Sourirajan technique has been used to prepare thin-skin membranes in flat-sheet, tubular, and hollow-fiber configurations. The Loeb–Sourirajan membrane is extremely thin, with very high water flux. This minimum effective thickness is exceedingly important in RO membranes, since this was easily the thinnest membrane prepeared without imperfections using any known techniques at the time of discovery.

One of the key points in asymmetric membrane formation is the thin skin that forms on the air-dried surface. Figure 4.15 shows a view of the a frame format used in formation of an asymmetric hollow-fiber RO membrane [13].

4.3.5.3 Stretching Method Thermoplastic polymers can be melted and extruded through a nozzle to produce microfiltration membranes with a pore structure induced by stretching the material. The partially crystalline polymeric material is stretched perpendicular to the direction of extrusion so that crystalline regions will be located parallel to the direction of extrusion. Under mechanical stress, small slitlike ruptures occur in the membrane; these small crevices are generally 0.2 μm in length and 0.02 μm in width. The porosity of the membranes produced is high, approaching 90% is some cases. The most widely used materials are polyethylene and polypropylene. The method is not used for RO membranes.

4.3.5.4 Interfacial Polymerization Method The composite membrane configuration was developed in order to increase water permeability of the membrane by creating a thinner top active layer. Fabrication of a thin-film composite

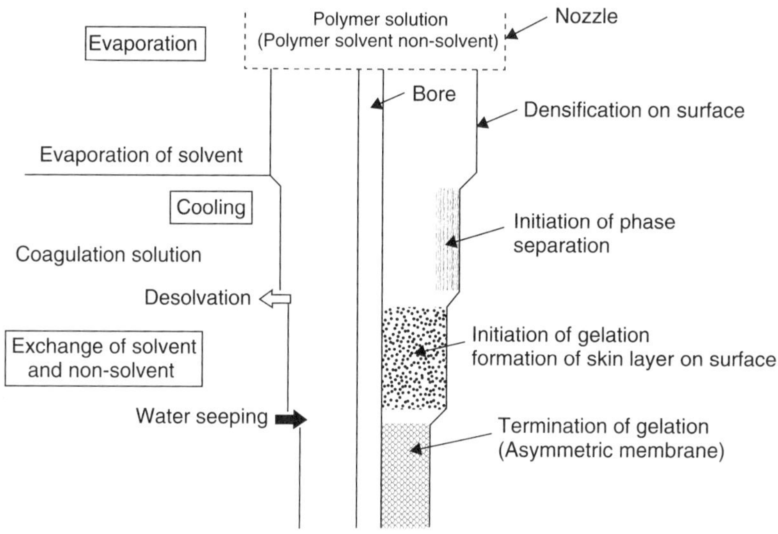

Figure 4.15 Typical diagram showing preparation of an asymmetric membrane [13].

membrane is typically more expensive than that for RO membranes since it involves a two-step procedure, unlike the one-step phase separation procedure. However, it offers the possibility of each individual layer being tailor-made for maximum performance. The composite membranes are formed on a porous support layer, typically a polysulfone UF membrane, which possesses most of the desired properties of a support layer. The skin layer, which serves as the top desalting layer of the composite membrane, must be very thin (typically $<0.2\ \mu m$), and is therefore very fragile. Several methods are used to prepare composite membranes, including the dip coating method, plasma polymerization, and interfacial polymerization. The most common preparation method for commercial composite reverse-osmosis membranes (thin-film composite membranes) is interfacial polymerization.

Figure 4.16 illustrates a typical preparation of a thin-film composite membrane using interfacial polymerization [12]. A microporous polysulfone sheet is saturated

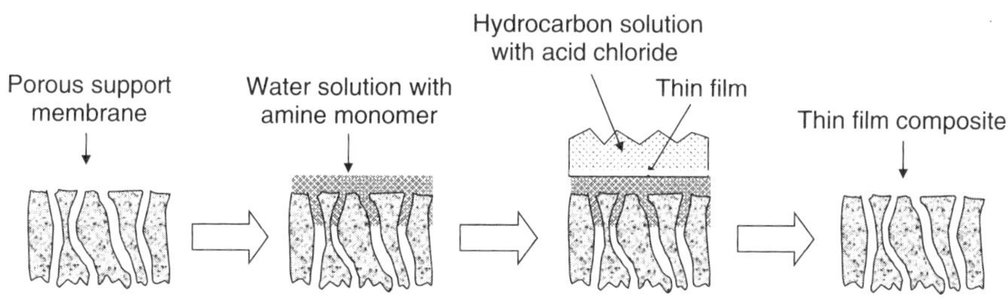

Figure 4.16 Steps in typical preparation of a thin-film composite membrane by the interfacial polymerization method.

with a water solution of amine monomer. After excess solution is drained off, the surface of sheet is immersed in organic solution such as hydrocarbon with acid chloride. This forms a very thin crosslinked polyamide layer on the surface of sheet.

4.4 REVERSE-OSMOSIS DESALINATION PROCESS

4.4.1 Fouling Control

Successful performance of reverse-osmosis (RO) membrane modules in long-term depends on several factors, such as pretreatment, system design, and operation and maintenance (O & M). All three of these factors are important. However, pretreatment and system design are fundamentally interrelated because proper pretreatment depends on proper design. All fouling involves trapping materials within the RO membrane itself or precipitation of soluble compounds on the membrane surface. The causative factors are different; most commonly those are colloids, metal oxide precipitation, membrane scaling, suspended organics, and biological matter. Membrane fouling is a complex phenomenon that involves several effects on RO performance. The effects are different but interrelated. The influence of fouling on membrane performance and membrane flux is shown schematically in Figure 4.17. The upper (solid-line) curve in the figure indicates normal membrane flux loss due to membrane compaction. The dashed-line and sawtooth-shape dotted-line curves in the figure indicate the effect of fouling and periodic cleaning. Other parameters such as salt rejection and pressure drop, which are also affected by fouling, can be represented by similar curves.

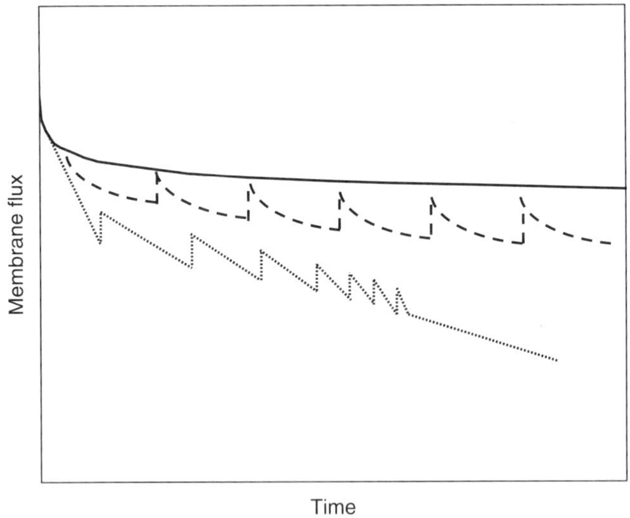

Figure 4.17 Relationship between membrane flux and time.

4.4.1.1 Biological Fouling Performance of the RO module may be affected by increased amounts of biological matter such as the growth of microorganisms, which can sometimes damage cortain types of membranes and modules. However, proper treatment for preventing growth of microorganisms is effective in preventing biological fouling. A typical treatment is proper disinfection of the membrane. If the feedwater contains sufficient nutrient source to sustain rapid growth of microorganisms in the module, growth of biological matter can occur. This growth can lead to the formation of biofilms on the membrane, and biofilms and their metabolites are known to compromise module performance.

In actual operation, with rapid pressure loss in the module, biological growth is normally first indicated. This pressure loss is the pressure difference between the feed side and the brine side of the module. In the hollow-fiber module, plugging across the fiber bundles with matter leads to increased pressure loss. In case biological fouling occurs, biological cleaning treatment is recommended. There are several cleaning methods, such as acid/alkali shock treatment, sodium bisulfite (SBS) shock treatment, and chlorine injection. The acid or alkali shock treatment requires a large quantity of chemicals to effectively decrease the pressure loss in modules. Meanwhile, SBS shock treatment is seldom effective in decreaseing pressure loss, especially after numerous operations, because SBS is an antibacterial material, not a disinfection agent. In contrast, chlorination by injection of chlorine is very effective for preventing the growth of biologicalmaterial that causes fouling in RO modules.

A typical desalination plant using a polyamide RO membrane module uses chlorine as a disinfectant in the pretreatment process. Because the polyamide RO membrane module is affected by chlorine, SBS is injected to dechlorinate the feedwater before it enters the RO module. Since chlorination treatment does not destroy all the microorganisms, some of them will survive and continue to grow, resulting in biological fouling, also called *aftergrowth*. However, some membranes have no resistance to chlorine as an oxidant; for instance, chlorine cannot be used for biological cleaning of polyamide RO membranes, which are the most common type of RO membranes used in the world. Meanwhile, a cellulose triacetate RO membrane does have resistance to chlorine and therefore be disinfected directly to prevent biological fouling. Monitoring biological activity is also effective in preventing biological fouling of RO membrane modules, especially in surface seawater applications.

4.4.1.2 Physical Fouling Physical fouling normally consists of fouling by metal oxides; scaling is the most common form of fouling. Scaling is caused by the precipitation of relatively insoluble salts ($CaCO_3$, $CaSO_4$, $BaSO_4$, $SrSO_4$, SiO_2 etc.) in the module. The scaling potential of any feedwater is determined from chemical analyses of that feedwater, recovery of the RO system, and the solubility limits of various salts. $CaCO_3$ is usually the only salt in seawater for which pretreatment is required. In normal operation using commercial RO modules, recovery is 30–60% in seawater, and the solubility limits of $CaCO_3$, $BaSO_4$, $SrSO_4$, and SiO_2 are rarely exceeded in this recovery.

The second type of fouling that can occur in RO modules is precipitation of metal oxides. This is caused by the formation of the oxides and hydroxide of iron and manganese. For seawater containing iron, the iron will probably be present in the form of soluble ferrous bicarbonate. If the iron remains in the soluble ferrous form, the RO modules will reject it quite effectively. However, if the soluble (ferrous) iron is oxidized to the ferric state in the RO module, iron fouling will occur with the formation of $Fe(OH)_3$ precipitates.

4.4.2 Disinfection Method for Pretreatment

Implementation of an adequate disinfection procedure is critical for reliable operation of RO modules, especially when source feedwater is obtained from surface water. The rationale for disinfection is to prevent growth of microorganisms in the RO module that can adversely affect the performance (permeate water flow capacity and salt rejection). In addition to prevention of microorganism growth, the proper disinfection procedure must also ensure that membrane damage does not occur because of the presence of microorganisms or the disinfectant chemical itself. The two key factors to consider in discussing biological matter include (1) degradation of the membrane and (2) biological fouling of the membrane. These two factors are

In this section, the following described with respect to a disinfection method used to eradicate biological fouling in cellulose triacetate RO modules that have chlorine resistance:

1. *Membrane Degradation*. The cellulose triacetate membrane used is compatible with chlorine disinfectant and hence can be easily protected from biological attack.
2. *Membrane Biological Fouling*. Biological matter can grow inside RO modules when the feedwater contains sufficient nutrients to sustain rapid growth. In more recent years this rapid biological growth has been recognized as '*aftergrowth*', even in situations where chlorination of feedwater was initially employed. This possibility is greater for polyamide membranes than cellulose acetate membranes since a dechlorination step must be performed before the treated feedwater enters the membrane element. Elimination of chlorine in the feedwater has been observed to allow potentially rapid aftergrowth since the initial chlorination step does not completely kill all microorganism. In fact, it has been speculated that the chlorination step reacts with organic matter such as humic acids to render them more assimilable as food to support microorganism growth in the RO modules. Ultimately this biological organism growth can lead to formation of biofilms on the membranes, leading, in turn, to poor RO module performance.

In the case of cellulose triacetate membranes, compatibility with chlorine disinfection prevents the potential rapid aftergrowth of microorganisms in the module. Generally, at plant startup the chlorine dosing amount is optimized to minimize the cost of chlorine chemicals, maximize membrane life, and minimize generation of disinfectant byproducts (DBPs).

4.4.2.1 Disinfection by Intermittent Chlorine Injection (ICI) In place of the conventional continuous chlorine injection (CCI) method, disinfection procedures successfully adopted since the year 2000 at large seawater RO plants on the Red Sea coast have used conventional Toyobo RO membranes. These membranes have a 6.8 MPa pressure capability rating, and more recent long-term pilot tests on seawater using the newer Toyobo RO membranes have revealed even higher operation pressure capability ratings (at 8.2 MPa), demonstrating the advantages of intermittent chlorine injection (ICI) in feedwater. Figure 4.18 shows the newer ICI method with the conventional continuous chlorine dosing regimen. In the ICI process, RO modules are directly disinfected by chlorine containing water. These tests clearly show the advantages of the ICI method over continuous chlorine injection, including

- Successfull elimination of biological fouling of CTA membranes
- Reduction of chemical consumption compared to the continuous chlorine dosing method
- Provision of a more environmentally friendly (i.e., "greener") process since disinfectant byproducts (DBPs) are not produced continuously.

Optimum chlorine dosing for the ICI method will depend upon the biological activity in the feedwater, which varies from site to site. It has been demonstrated for the case of high biologically active surface waters, such as those near the Red Sea coast, that ICI dosing levels of 0.2 mg/L for 3 hours each day (1 h for each 8-h period) successfully eliminated biological fouling. In the case of more normal surface seawaters and less biological activity, ICI dosing levels of 0.2 mg/L for 1 h of each day was adequate to prevent biological fouling. The ICI method has been

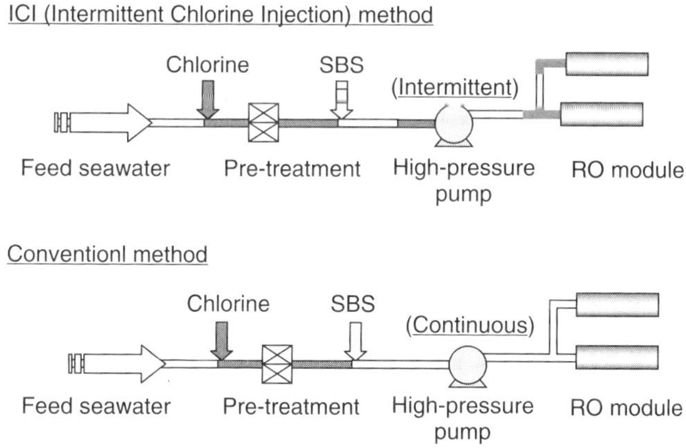

Figure 4.18 Intermittent chlorine injection method.

successfully demonstrated in plant operations with recovery ranges of 30–55%, and offers an opportunity to maximize plant availability by minimizing time and cost of frequent cleaning.

4.4.2.2 Control by Shock Treatment In cases where chlorination and dechlorination do not provide adequate control of microbiological fouling condition and rapid growth of microorganisms resumes in the modules, chemical shock treatments have been employed using sodium bisulfite ($NaHSO_3$, SBS) and acid such as sulfuric acid.

The intermittent injection of SBS at concentrations of ~500–1000 mg/L in the RO modules may be effective. As mentioned above, SBS is an antibacterial material, not a disinfection agent. The effectiveness of this method using SBS is limited, although SBS does not react with the membrane, such as a polyamide RO membrane.

The acid shock treatment needs a large quantity of chemicals to effectively stabilize the microbiological condition. In general, microorganisms may develop higher thresholds to the effects of the chemicals after many, repeated shock treatments using acids.

4.4.2.3 Bacteria Content of RO Product Water While RO membranes will reject almost 100% of bacteria, all RO devices have mechanical seals, but it is possible for minute, trace quantities of feedwater to bypass or breach these seals and enter the product-water stream. In addition, any broken fibers or microcracks in spiral-wound membranes, although very small in quantity, can also permit feedwater to enter the product-water stream. For these reasons, RO modules are not guaranteed to produce sterile water. If sterile water is desired, ultraviolet (UV) radiation can be used. This technique is commonly used in the electronics industry.

4.4.3 Prediction of RO Performance for Long-Term Operation

4.4.3.1 Membrane Compaction Some degree of membrane compaction usually occurs during the operating period, which results in reduced permeate water flow rate. This phenomenon is known as *membrane flux decline*. The effect of pressure, temperature, and time on membrane flux is expressed as a *membrane flux retention coefficient* (MFRC) as shown in the following equation

$$\mathrm{MFRC} = \frac{Q_{\mathrm{pt}}}{Q_{\mathrm{pt0}}} \tag{4.1}$$

where Q_{pt} is the permeate flow rate at time t and Q_{p0} is the permeate flow rate at initial time t_0. Both Q values have the same operating parameters such as pressure, temperature, and water recovery. The rate of decline depends on the operating parameters such as temperature and pressure. The greater the temperature and/or pressure, the greater the rate of decline of membrane flux.

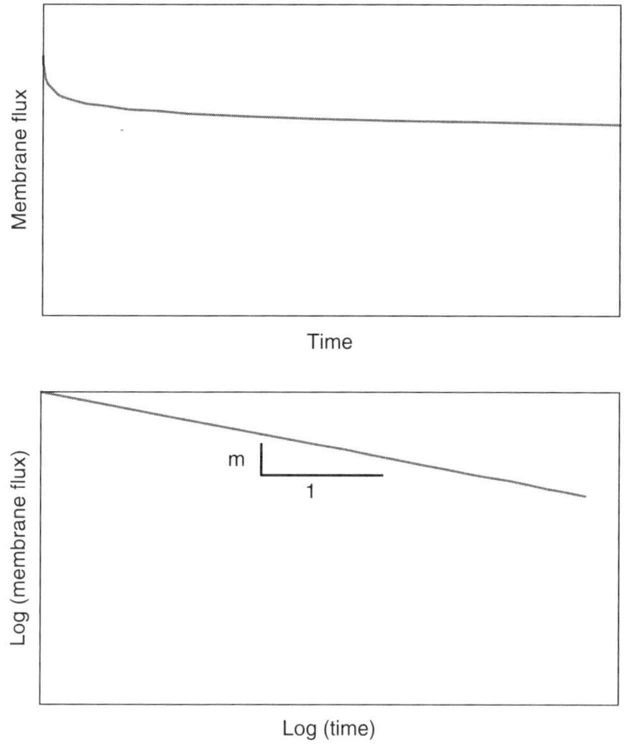

Figure 4.19 Relationship between membrane flux and time.

MFRC is a logarithm function of time as shown in the following equation. (See curves in Fig. 4.19; as shown, the time factor is not so significant after the initial decline of membrane flux)

$$\mathrm{MFRC} = \left(\frac{t}{t_0}\right)^{-m} \tag{4.2}$$

where t is time in hours and m is the slope of the curve, called the *m value*.

The following equation was also suggested to explain the relationship between membrane flux at the initial time and operation time [14].

The equation also derived by experimental results.

$$\frac{1}{A_t} = \frac{1}{A_0} + a \cdot t - b \cdot e^{-t/t_0} \tag{4.3}$$

where A_t is the pure-water permeation coefficient at time t and A_0 is the pure-water permeation coefficient at initial time t_0; a and b are constant determined by fitting the experimental values.

4.4.3.2 Performance Change Rate of Cellulose Triacetate RO Membrane in Case of Degradation [15] Generally, cellulose triacetate RO (CTA-RO) membranes undergo hydrolysis and oxidation. Hydrolysis is accelerated at very high or low pH levels, and oxidation is accelerated at high concentration of an oxidizing agent. Both hydrolysis and oxidation result in loss of acetyl groups and decrease in molecular weight, leading to a decline in salt rejection in RO membranes. It is important to be able to predict any potential variation in RO performance with time for both the design and the operation of RO plants.

To evaluate the potential effects of hydrolysis and oxidation on RO performance, B-value increases in B- and K-value, rate constants were measured at various pH, temperature, and residual chlorine concentration levels were measured, and a formula that could predict K-value changes in the CTA-RO hollow fiber was derived. This formula analyzes the potential temporal effects of hydrolysis and oxidation on performance of the CTA-RO hollow-fiber membrane [see Eqs. (4.4)–(4.8)].

4.4.3.2.1 Increasing Rate Constant of B-Value (K-Value) It is known that the salt permeability coefficient (B value) of cellulose acetate RO membranes increases exponentially with time when hydrolysis is present. Analysis of changes in performance of the experimental data on that the CTA-RO membrane in presence of an oxidizing agent such as chlorine revealed B-values of CTA-RO in the oxidation also increase exponentially with time, as shown in Figure 4.20. Therefore, we assumed the following.

- The B-value changes in CTA-RO membrane performance with time in presence of oxidation can be expressed using a formula similar to that for determining hydrolysis-related changes.

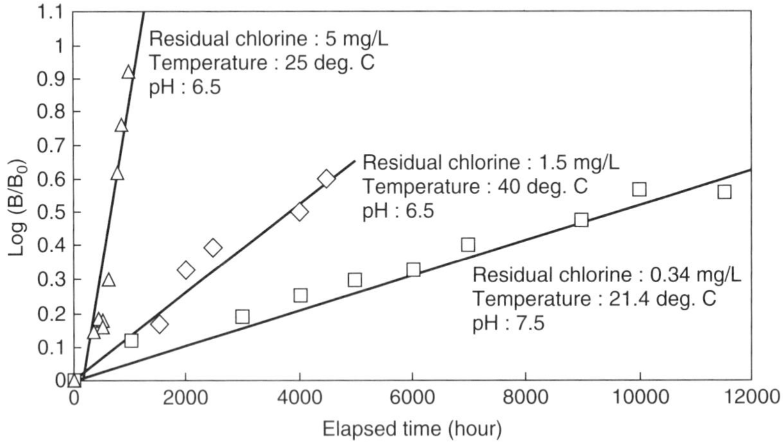

Figure 4.20 Variation in permeability (B) of CTA-RO membrane with time in presence of oxidation.

- The B-value increasing rate constant (K value) is equal to the sum of K_H and K_O, where K_H and K_O are B-value increasing rate constants in hydrolysis and oxidation, respectively. In short, an additive applies to K_H and K_O.

Therefore, we defined the B-value increasing rate constant for both hydrolysis and oxidation (K value) by the formulas

$$\log \frac{B}{B_0} = Kt \tag{4.4}$$

$$K = K_H + K_O \tag{4.5}$$

where B_0 = initial B value, cm/s
t = elapsed time, h
K = B-value increasing rate constant, h^{-1}
K_H = B-value increasing rate constant in hydrolysis, h^{-1}
K_O = B-value increasing rate constant in oxidation, h^{-1}

4.4.3.2.2 Derivation of Prediction Formula Using the B-Value Increasing Rate Constant (K Value)

Derivation of Prediction Formula for K_H (in Hydrolysis). Data on K_H-value changes recorded on continuous operation of CTA-RO hollow-fiber modules were measured at various temperature and pH levels. The following formula for predicting changes in K_H values with respect to temperature and pH was derived:

$$\log(K_H) = \frac{-6400}{273.2 + T_{fw}} + 0.80\text{pH} + 11.0 \tag{4.6}$$

where T_{fw} = feedwater temperature, °C
pH = feedwater pH (−)

Derivation of Prediction Formula on K_O (in Oxidation). Since K_O could not be measured directly, it was obtained as the difference between K and K_H, which was calculated using Equation (4.4). The K value, that is, $K_H + K_O$, as shown before, was measured by experimental results in the presence of an oxidizing agent such as chlorine. The following prediction formula for K_O was derived by a multiregression analysis:

$$\log(K_O) = 1.072 \log(C_{\text{Cl}_2\ \text{res}}) \frac{-1850}{273.2 + T_{fw}} + 0.89\text{pH} - 4.246 \tag{4.7}$$

where $C_{\text{Cl}_2\ \text{res}}$ is the residual chlorine concentration in feedwater (in mg/L).

Prediction Formula for K-Value. By rearranging Equations (4.5)–(4.7), we obtain the following formula:

$$K = 1 \times 10^{11} \times \exp\left(\frac{-14,740}{273.2 + T_{\mathrm{fw}}} + 1.842\mathrm{pH}\right) + 5.675 \times 10^{-5}(C_{\mathrm{Cl_2\,res}})^{1.072} \times \exp\left(\frac{-4260}{273.2 + T_{\mathrm{fw}}} + 2.049\mathrm{pH}\right) \quad (4.8)$$

The calculated K values are in fair agreement with experimental values. Since the B-value increasing rate constants for both hydrolysis and oxidation can be calculated independently, This formula presumably applies to the prediction of performance changes with time, including the period of suspension in RO plants.

4.4.4 High-Recovery Process

Reverse-osmosis technology is becoming the key method for obtaining freshwater from the sea, especially in the Middle East. Membrane manufacturers are working to develop membranes offering higher water recovery and lower energy and installation costs in order to establish world wide recognition of the RO process as the most popular method for supplying freshwater.

In a seawater RO process, the water is subjected to disinfection, coagulation–filtration, and acidification processes in the pretreatment section and forwarded as feedwater to the RO section. When the recovery fraction, the ratio of product flow rate to feed flow rate, is high, the amount of feedwater required for desired production volume is lower; hence, cost of chemicals, equipment sizing, and energy cost are for the pretreatment system significantly reduced.

For high-recovery operation, in comparison with conventional low-recovery operation, higher feed pressure (i.e., driving force) is required. In order to enable the high-recovery operation, the RO module must be designed and manufactured to withstand the higher pressure.

4.4.4.1 *Higher-Efficiency RO Single-Stage High-Recovery System*

Figure 4.21 shows a higher-efficiency RO single-stage high-recovery system at 60% recovery level in comparison to a conventional 40% recovery process. Application of the higher-pressure capability module compared to the conventional module can reduce the feed seawater flow rate by ~40% compared with the conventional process. This higher recovery makes it possible to reduce the construction and running costs of the pretreatment process. In addition, with the lower amount of seawater that must be pressurized, energy can be recovered efficiently from the high-pressure brine by use of energy recovery devices. Therefore, energy consumption per cubic meter of water produced would presumably also be less than that in the conventional process.

This is a simple and a reliable system with a reliable track record in many operations, and facilitates control of the operation.

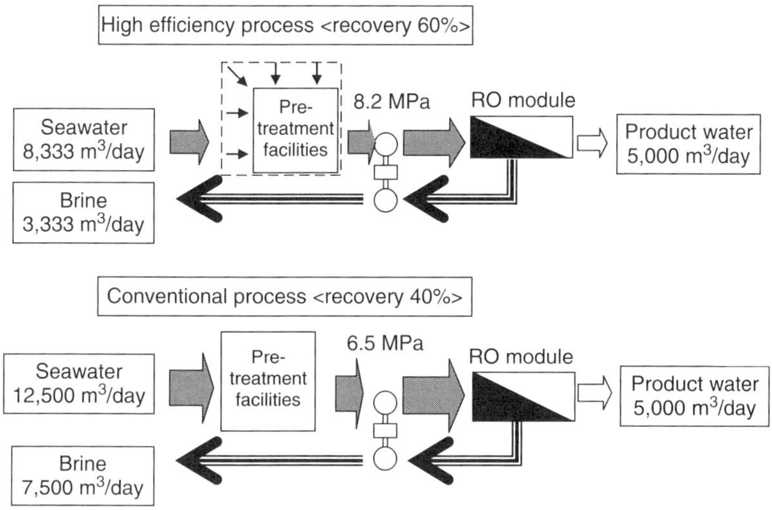

Figure 4.21 High-efficiency RO single-stage high-recovery system.

4.4.4.2 Brine Conversion Two-Stage RO Seawater Desalination System The newly developed brine conversion two-stage RO seawater desalination system uses less energy than the conventional system to produce the same amount of freshwater. This system produces more freshwater from the concentrated water (e.g., containing 5.8% salt), which was exhausted to the sea in the conventional system. In the system, membranes are subjected to of high-pressure (e.g., 9 MPa) and high-salt-concentration (e.g., 5.8%) operating conditions. The system is shown in Figure 4.22.

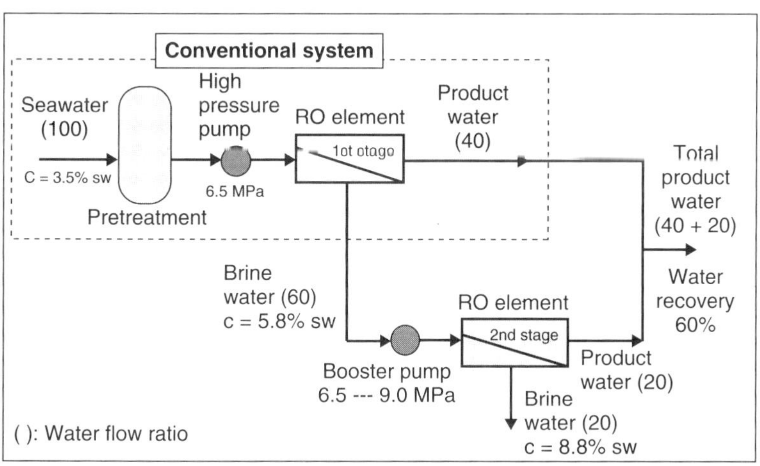

Figure 4.22 Two-stage RO seawater desalination system for brine conversion.

4.5 DEVELOPMENT OF HOLLOW-FIBER RO MEMBRANE MODULE FOR SEAWATER DESALINATION

4.5.1 Hollow-Fiber Bundle Configuration

Hollow-fiber membrane modules are classified into several types according to hollow-fiber bundle configuration and flow pattern. Hollow-fiber membrane modules are classified broadly into cross-wound configuration type and parallel configuration type. The parallel configuration is necessary to enable the spacer to bundle the hollow fibers together, while the cross-wound configuration does not reguire a spacer because the hollow fibers automatically pack themselves into a bundle. Therefore, the cross-wound type is preferred for fouling resistance.

The flow patterns in the hollow fiber bundle are classified into axial and radial. In general, feedwater velocity in the module in axial flow is higher than that in radial flow. Therefore, pressure loss in the axial-flow module is higher than that in the radial-flow module. The greater the module length is, the greater the difference between velocity and pressure values.

In RO module applications, feedwater is usually pressurized outside the hollow-fiber membrane and permeated water flows from outside to inside of the hollow fiber, especially in high-pressure operation such as seawater desalination applications. In some RO membrane modules designed for low-pressure operation, feedwater is fed into the hollow-fiber bore. It is than pressurized inside the hollow fiber, and permeated water flows from inside to outside of the fiber. The "both open-ended" (BOE) module configuration is reguired for this internally pressurized type module.

4.5.2 Features of Hollow-Fiber RO Membrane Module for Seawater Desalination

The cellulose triacetate RO membrane module typically has a hollow-fiber configuration.

4.5.2.1 Module Structure (Hollow-Fiber Configuration) Hollow-fiber modules can offer greater surface membrane area in the module than can spiral-wound modules. Because of the greater compactness of the hollow-Fiber bundle, the production rate is higher in these modules and their footprints in the soil or sand banks near the desalination plant are smaller. Figure 4.23 shows a cellulose triacetate hollow-fiber array fixed in epoxy resin at both ends. This arrangement provides mechanical stability to the fiber array. The fibers at one end of the element are cut precisely so that product water can be discharged from the fibers bore.

The cellulose triacetate RO module used for seawater desalination has a double-element configuration, as shown in Figure 4.23. Because permeate water of each element in a pressure vessel can be obtained directly, the quality of permeate water throughout the vessel can be measured directly, which Facilitates RO plant maintenance [16].

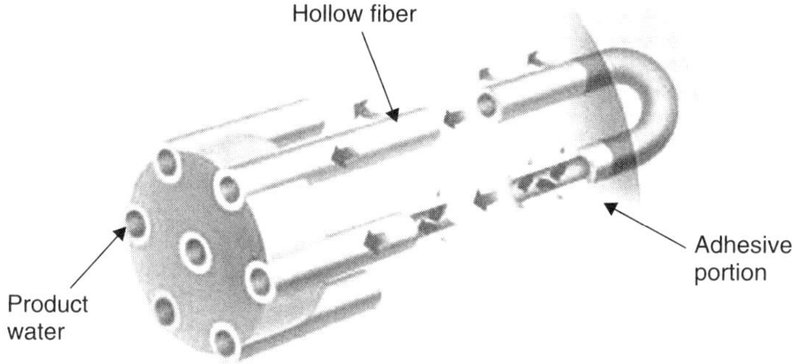

Figure 4.23 Hollow-fiber module.

The conventional hollow-fiber RO module consisted of a single element inserted in one pressure vessel. However, economic and logistic considerations demanded modules with a number of elements inserted in one pressure vessel, and in 1979, Toyobo was the first company in the world to successfully produce double-element modules on a commercial basis, following experimentation with various types of double-element modules designed to accommodate flow path.

Toyobo improved the double-element module design with further advances developed in 1981. This module structure and flow pattern is shown in Figure 4.24.

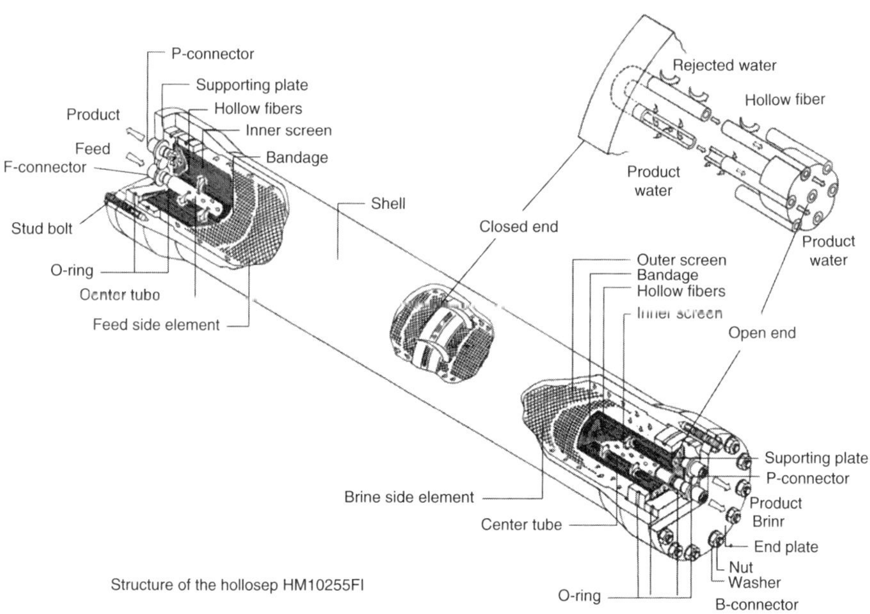

Figure 4.24 Hollow-fiber RO module (Hollosep HM series type).

Feedwater enters the feed center pipe of the feed-side element, flows radially from the center pipe, past the hollow fibers in the bundle, and away from the element as concentrated brine flow of the feed-side element. Feed flow is uniformly distributed in the RO module throughout the cross-wound hollow-fiber bundle. The brine water of the feed-side element flows toward the periphery of the element, passes through the narrow space between the pressure vessel and the element, and flows toward the periphery of the brine-side element, and enters and the hollow fibers in the bundle radially from outside to inside. Brine water flows to and passes through the center tube of the brine-side element toward the brine port and away from the module.

Permeate water is collected with a supporting plate and passes through the permeate pipe on each side of the element.

4.5.2.2 Stable RO Performance The physical structure of the cellulose triacetate hollow-fiber membrane RO module is optimized to provide mechanical strength resistance to collapse in a hydrostatic pressure. This is achieved by spinning the hollow-fiber geometry at dimensions consistent with the mechanics of a tiny hollow tube or pipe and the material properties of a strong cellulose triacetate polymer.

The cellulose triacetate hollow-fiber RO membrane itself has outstanding pressure resistance retention, and the hollow-fiber design and selection of suitable dimensions provide high-pressure resistance. The outer diameter is ~165 μm and the inner diameter, ~70 μm. The resistance of this hollow fiber membrane against high pressure is at a practical level.

A microscopic view of a hollow-fiber membrane used for seawater desalination is shown in Figure 4.25.

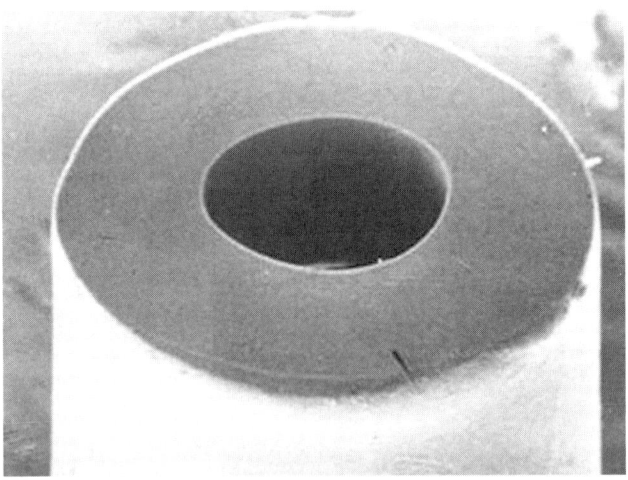

Figure 4.25 A microscopic view of a hollow-fiber membrane used for seawater desalination.

4.5.2.3 Superior Fouling Resistance

4.5.2.3.1 Chlorine Resistance Membranes made from cellulose triacetate, which provides improved membrane performance, are used widely today because of their high performance and long-term reliability. The cellulose triacetate hollow fiber features superior chlorine tolerance compared with that of the polyamide membrane, as shown Figure 4.26. In biologically active seawater sterilization by chlorine is considered a very effective solution to prevent biological fouling during the RO process. The material properties of cellulose triacetate help protect the hollow-fiber membrane against biological fouling in the RO process simply by chlorine injection. The sterilization operation mode can be optimized by the continuous or intermittent chlorine injection (ICI) disinfectant method.

4.5.2.3.2 Optimum Permeability Continuous flux (permeability or flow per membrane surface area) across membranes can complicate membrane performance in a desalination process, due to deposition of fouling materials on the RO membrane surface. Generally, the higher is the flux, the sooner will the membrane surface become coated with fouling material. Spiral-wound membrane elements offer relatively low membrane surface area, and hence maximum flow must be restricted to prevent fouling.

Hollow-fiber membrane elements offer about 10 times greater membrane surface area than do spiral-wound membrane elements. Figure 4.27 compares the fouling vulnerabilities of CTA and polyamide membranes. The larger membrane surface area advantage of the hollow-fiber element allows the same quantity of permeate water to be produced at a permeability rate ~10% that for spiral-wound elements.

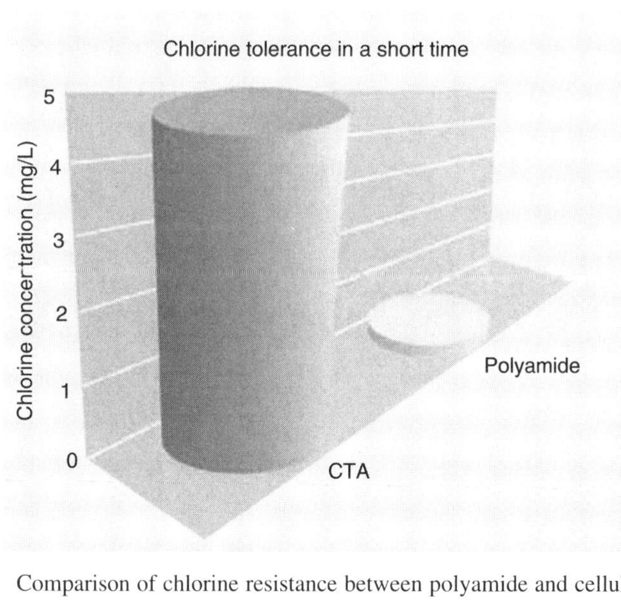

Figure 4.26 Comparison of chlorine resistance between polyamide and cellulose triacetate in a hollow-fiber RO membrane.

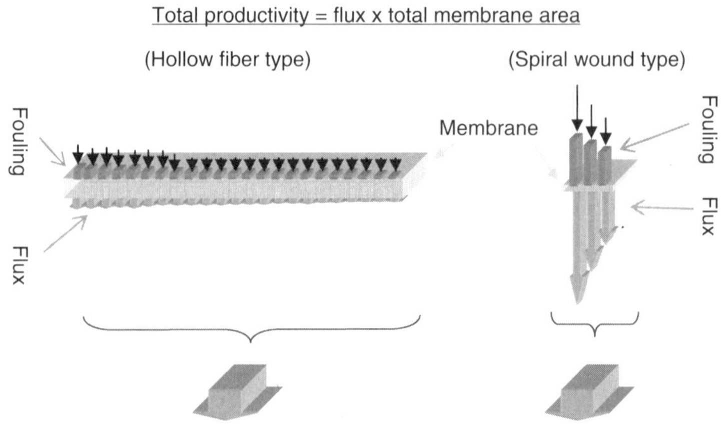

Figure 4.27 Comparison of fouling tendency between hollow-fiber and spiral-wound membranes.

This results in much fewer restrictions in operation and less frequent need for cleaning.

4.5.2.3.3 Cross-Winding Style A hollow-fiber membrane manufactured via a multifilament spinning process allows element construction with parallel fiber arrays or filament winding, cross-layered arrays. A bundle of several thousands of hollow fibers is fabricated into an element and assembled in a module. Hollow fibers in the element are arranged in a mutually cross-wound configuration without any type of supporting materials between the hollow-fiber layers.

The hollow-fiber membrane is wound into a bundle in a layered, cross-wound arrangement as shown in Figure 4.28, and moderate-size, regular intervals are left

Figure 4.28 Cross-wound arrangement of hollow fiber.

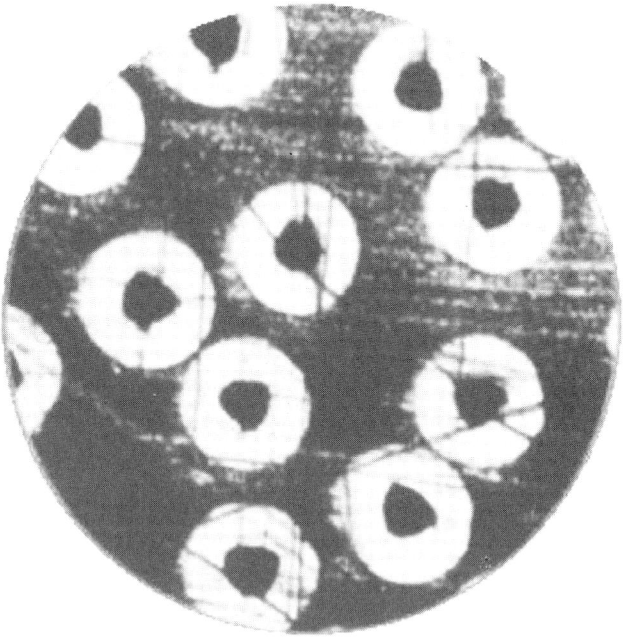

Figure 4.29 Cross-wound arrangement of hollow fiber.

between the follow fibers to minimize pressure loss and allow uniform flow and small pressure drop, thus minimizing concentration polarization and extending the allowance of the fouling index of feedwater up to SDI = 4. This structure is also less prone to block age by fouling matte. As seen in the microscopic view in Figure 4.29, the space of Toyobo's hollow fiber is remarkably large, and accordingly rapid increase of deferential pressure is not observed in the RO process. This characteristic allows a RO plant to be operated with easy maintenance.

4.5.3 Analytical Model of Hollow-Fiber RO Modules

Hollow fibers have been selectively employed in the desalination industry because of their economic efficiency among several membrane types. Several analyses of hollow-fiber RO modules in which permeate water flows from outer side to inner side of the hollow-fiber bore have been attempted since the 1980s. Gill and Bansal [17] proposed a mathematical model that considers the pressure drop in the fiber bore and a constant salt rejection, assuming no concentration polarization they later improved this model by introducing the solution–diffusion model and the solution–diffusion–imperfection model [18,19]. The membrane transport was improved from a constant rejection model to a more realistic model. Ohya proposed a simple and practical model based on the Dupont B9 permeator by ignoring the pressure drop in the fiber bore and considered the concentration polarization, which other researchers had assumed to be negligible [20]. Therefore, the transmembrane pressure was in fact overestimated, lowering the pure-water permeability constant in the calculation. The analytical model mentioned above, namely, the (friction–concentration–polarization (FCP) model, in which the concentration polarization and the pressure drop in the fiber bore were considered, has already been presented as an analytical model of a hollow-fiber RO module, and the validity has been proved from verification by a wide range of actual performance data on a hollow-fiber RO module by Sekino and coworkers [21,22].

4.5.3.1 Model Formulation of Friction Concentration Polarization (FCP) Model The governing equations in the FCP model are membrane transport, fluid pressure drop, and material balance equations. More recently considered in this model are the introduction of concentration polarization and pressure drop in the fiber bore. This model is applicable to the radial expansion of feed flow in Toyobo's hollow-fiber module.

4.5.3.1.1 Membrane Transport Equations The solution–diffusion model is practical for membrane permeation equations, such as the following:

$$J_{\mathrm{w}} = A(\Delta P - \Delta\pi) \tag{4.9}$$

$$J_{\mathrm{s}} = B\Delta C \tag{4.10}$$

Impermeable solutes accumulate on the membrane surface, gradually developing a concentration polarization layer. The material balance in the layer and concentration polarization coefficient ϕ is expressed as:

$$\Phi = \exp\left(\frac{J_{\mathrm{v}}}{k}\right) = \frac{C_{\mathrm{M}} - C_{\mathrm{P}}}{C_{\mathrm{B}} - C_{\mathrm{P}}} \tag{4.11}$$

where J_{v} is solution flux and the mass transfer coefficient k [from Eq. (4.4)] is expressed by

$$\mathrm{Sh} = \frac{kd_{\mathrm{O}}}{D} = \mathrm{Sh(Re,Sc)} \tag{4.12}$$

In this analysis, k is calculated by the following correlation equation, derived by [21] Sekino and Fujiwara for Toyobo's modules:

$$k = 0.048\left(\frac{D_{\mathrm{L}}}{d_{\mathrm{O}}}\right)\mathrm{Re}^{0.6}\mathrm{Sc}^{1/3} \tag{4.13}$$

4.5.3.1.2 Fluid Pressure-Drop Equations The pressure drop in the fiber bore is expressed by the Hagen–Poiseuille equation:

$$\frac{dP}{dz} = \frac{128\,\mu Q_{\mathrm{P}}}{\pi d_{\mathrm{I}}^{4}} \tag{4.14}$$

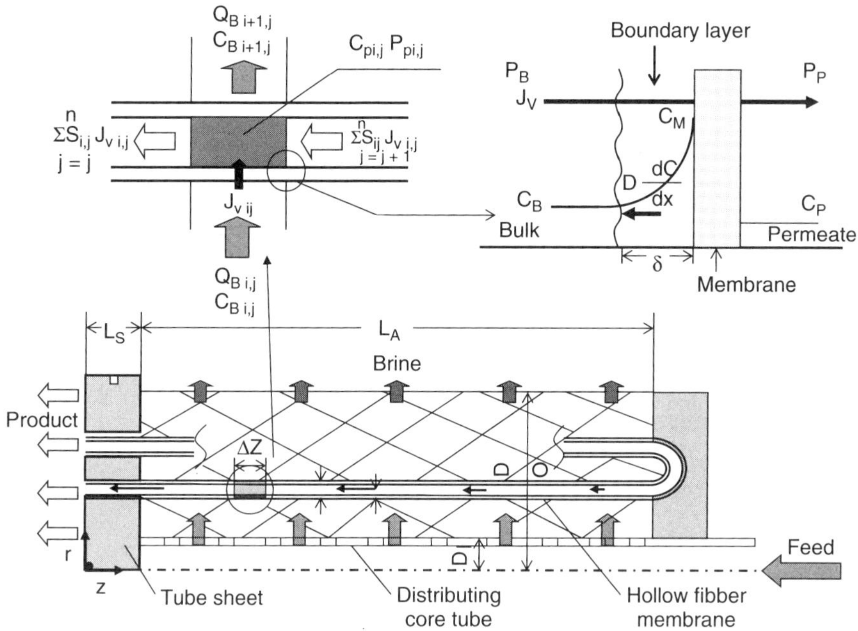

Figure 4.30 A schematic diagram of the hollow-fiber system.

4.5.3.1.3 Numerical Analysis The equations presented above are solved simultaneously, and a finite-differential method is applied to solve them numerically. A schematic diagram of the hollow-fiber system is shown in Figure 4.30.

Measured and calculated values are compared in Figures 4.31–4.33 under the conditions of various recovery ratios, feed concentrations, and applied pressures. This model is described in the figures as the friction–concentration–polarization (FCP) model. These values are compared with respect to results calculated by two other models, the nonfriction (NF) model and non–concentration–polarization (NC) model, which neglect pressure drop in the fiber bore and concentration polarization, respectively.

This analytical model, considering concentration polarization and pressure drop, shows a reasonable agreement with Toyobo's hollow-fiber RO membrane module in brackish-water and seawater desalination [21].

4.5.3.2 Optimum Design of Double-Element Hollow-Fiber RO Module

Experimentation to determine the optimum design for a double-element RO membrane module, which contains two RO elements in a cylindrical pressure vessel, included trials with the module structure design, the fluid stream design, and analytical model simulation.

The double-element hollow-fiber module is classified into several types in terms of flow path in the module as shown in Figure 4.34, where F, B, and P represent the feed port, the concentration port, and permeate port, respectively. Calculation

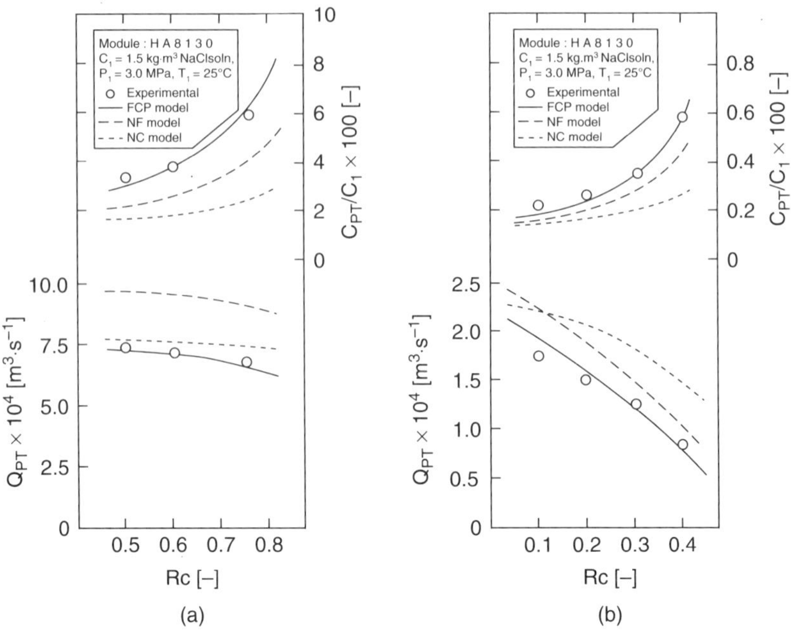

Figure 4.31 Dependence of module productivity and salt passage on recovery ratio for brackish-water (a) and seawater desalination (b) [(◯) experimental; (—) FCP model, considering the concentration polarization phenomenon and the pressure drop in the fiber bore; (– –) NF model, neglecting the pressure drop in fiber bore; (---) NC model, neglecting the concentration polarization].

results using the FCP model are shown in Figures 4.35 and 4.36. Figure 4.35 shows the radial profiles of bulk-side concentration and velocity of bulk water in the module. Figure 4.36 shows comparison of module characteristic.

In terms of feed flow velocity and differential pressure between two elements, the expanding–contracting series flow type was found to be the best among several double-element RO membrane modules. This type of module also reflected the scaleup effect.

4.5.4 Development of RO Module for Higher Recovery

The RO seawater desalination process has many advantages in terms of energy conservation, lower capital cost, short startup and shutdown times, short construction period, less installation space, and lower total water cost. In a seawater RO process, the seawater is subjected to disinfection, coagulation–filtration, and acidification processes in the pretreatment section and forwarded as feedwater to the RO section. When the recovery fraction (the ratio of product flow rate to feed flow

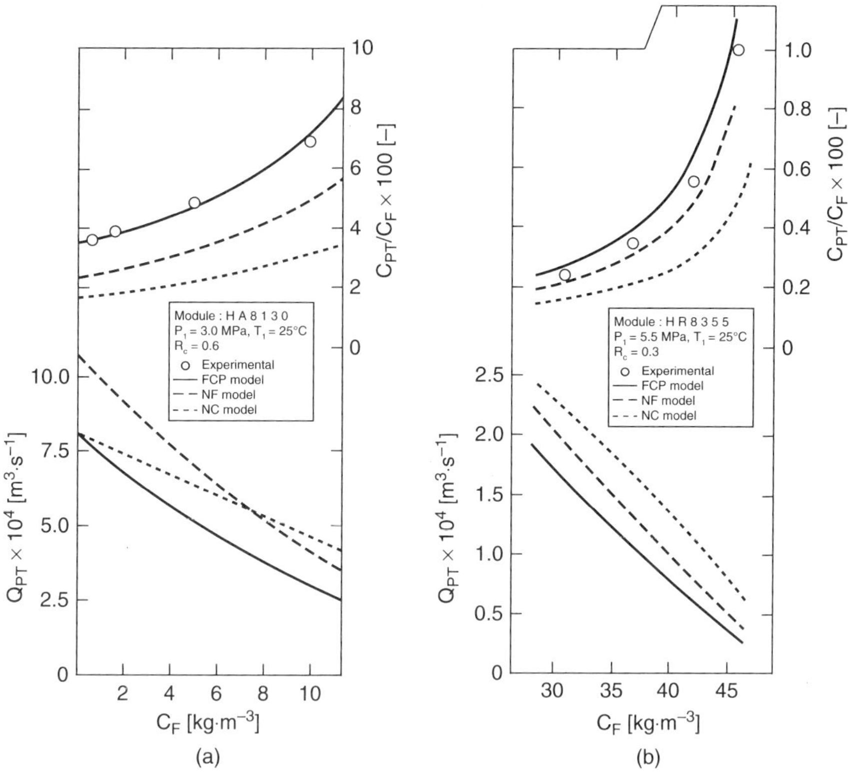

Figure 4.32 Dependence of module productivity and salt passage on solute concentration of feedwater for brackish-water (a) and seawater desalination (b) [(○) experimental; (—) FCP model, considering the concentration polarization phenomenon and the pressure drop in the fiber bore; (-----) NF model, neglecting the pressure drop in fiber bore; (---) NC model, neglecting the concentration polarization].

rate) is high, the amount of feedwater required for a desired production is lower, and hence the pretreatment system, chemical cost, equipment sizing, and energy costs are significantly reduced [23,24].

Membrane manufacturers are working to develop membranes offering higher water recovery, lower energy, and lower installation cost in order to ensure the RO process recognition of as the most popular method for supplying freshwater around the world.

In areas such as the Middle East, where seawater has high salinity, commercial seawater RO desalination plants were normally designed to operate at approximately 35% recovery. This relatively low recovery was due to the very high osmotic pressure of the seawater, and most commercially available RO membranes did not allow operating pressures above 7.0 MPa. Osmotic pressure of seawater near the Red Sea coast is approximately 3.2 MPa. For operation with recovery at 35%, the osmotic pressure of brine is about 4.8 MPa in the RO module, as shown in Figure 4.37.

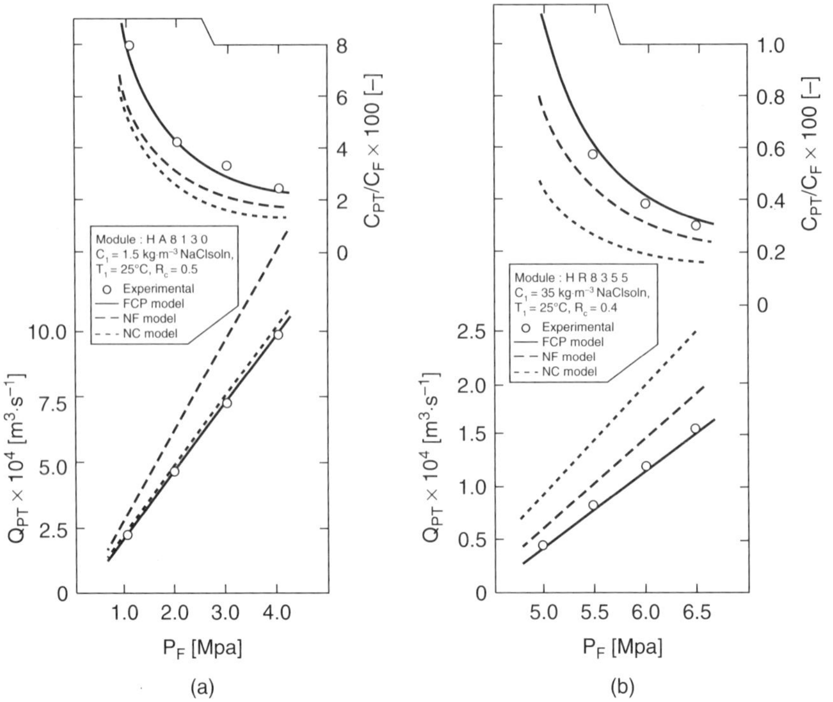

Figure 4.33 Dependence of module productivity and salt passage on applied pressure for brackish-water (a) and seawater desalination (b) [(◯) experimental; (—) FCP model, considering the concentration polarization phenomenon and the pressure drop in the fiber bore; (-·-·-) NF model, neglecting the pressure drop in fiber bore; (---) NC model, neglecting the concentration polarization].

If the recovery increases to ~50%, the osmotic pressure of the brine increases to ~6.1 MPa. Therefore, higher feed pressure (i.e., driving force) is required for high-recovery operation in comparison with conventional low recovery operation. In order to enable the high-recovery operation, the RO module must be designed and manufactured to withstand the higher pressure.

Toyobo has developed a new type of RO module to achieve higher product-water recovery in order to further reduce the cost of RO desalination. Toyobo's hollow-fiber RO modules are widely used a in RO desalination plants in many countries. On the basis of data accumulated from the long operating experience and more recent research efforts, Toyobo developed the high-pressure high-flux HB series modules. The HB series is an improved version of the conventional HM series type of module and contain the same materials [25–27]. The hollow-fiber membrane in the HB series module is wound in a cross arrangement, designed to minimize pressure loss and allow uniform water flow in the module. The hollow fiber incorporated in the HB series has higher pressure resistance based on change in the hollow-fiber outer/inner-diameter (od/id) dimension and optimization of manufacturing conditions. The specification of HB series modules are shown in Table 4.3.

No.	Type	Module structure	ΔP between two elements
I	Parallel (expanding)	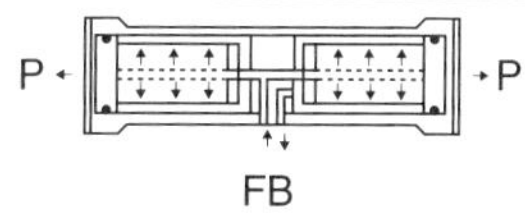	0
IIa	Series (expanding-expanding)	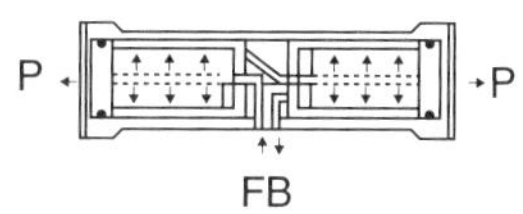	ΔP_2
IIb	Series (contracting-expanding)	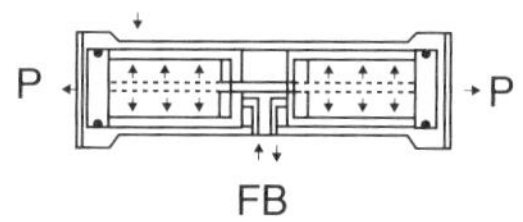	$\Delta P_1 + \Delta P_1$
IIc	Series (expanding-contracting)	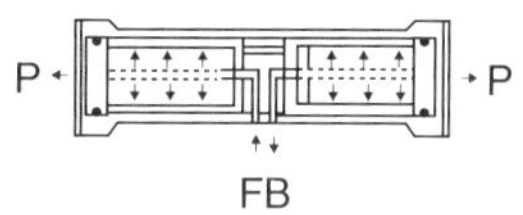	0
III	Series (expanding-contracting)	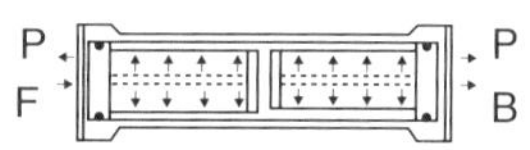	0

Figure 4.34 Double-element module structure.

The product flow rate of the new model improved by ~1.4 times compared with the conventional type.

A high-pressure, single-pass desalination process using the new HB series modules with a high recovery rate was successfully conducted for the first time at an RO test plant on the Red Sea coast in conditions of > 52% recovery [28].

4.5.5 Development of Both Open-Ended (B.O.E.) RO Module

Toyobo's newest innovative RO module builds on the proven reliability of the HM-type module. The new technology is based on both open-ended (BOE) hollow-fiber membrane structures and single open-ended hollow-fiber membrane structures. The BOE hollow-fiber membrane structure allows reduction in the pressure drop along the hollow-fiber bore, which leads to both greater salt rejection due to greater dilution effects and greater permeate-water flow.

The effective fiber length L_e is reduced by opening both ends of the fiber to allow flow from each end, and less so when only one end is opened, as illustrated in Figure 4.38. The Figure also compares the average bore pressures for the BOE type versus the single open-ended type.

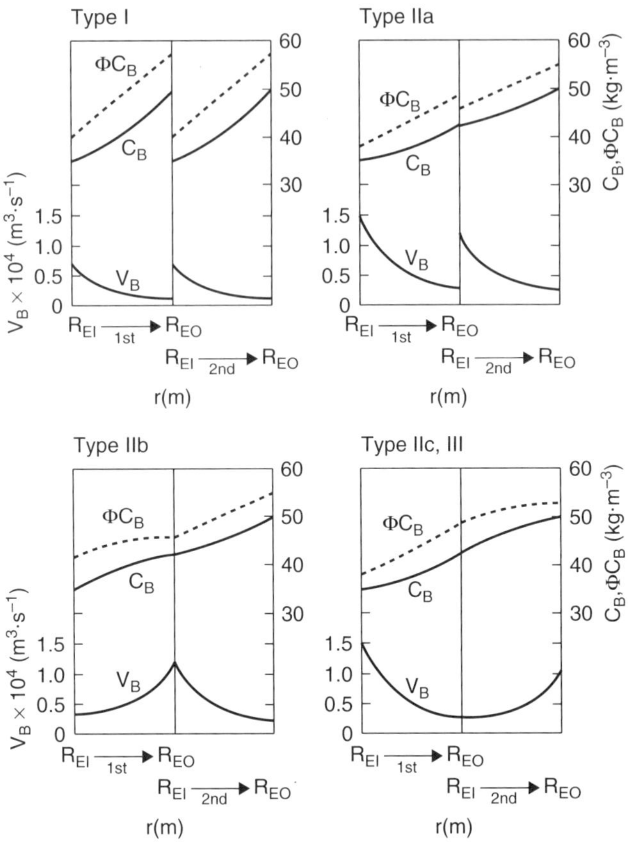

Figure 4.35 Radial profiles of C_B, φC_B, and V_B for each module type.

The maximum bore pressure in both cases is the pressure in the fiber at the furthest distance from the open end of the fiber. In the BOE type, the maximum bore pressure is transferred from the end of a longer fiber of length L to the midpoint of fiber length $L/2$, and the average pressure is reduced. The net effect of this change increases the amount of water flow since less bore pressure drop occurs for the same applied pressure.

The BOE RO element structure and flow pattern are illustrated in Figure 4.39.

The module structure also consists of two elements and is shown in Figure 4.40.

The center pipe of the RO module is of a concentric configuration. Permeate water flows in to the inner tube of the concentric pipe and feedwater flows between the outer and inner tubes of the concentric pipe. Feedwater enters the center pipe of the feed-side element through the feed-side distribution connector. The distribution connector divides the intersecting flow of the feed and permeate waters. Feedwater flows radially outward from the center pipe, past the hollow fibers in the bundle, and away from the element as concentrated brine flow. Feed flow is

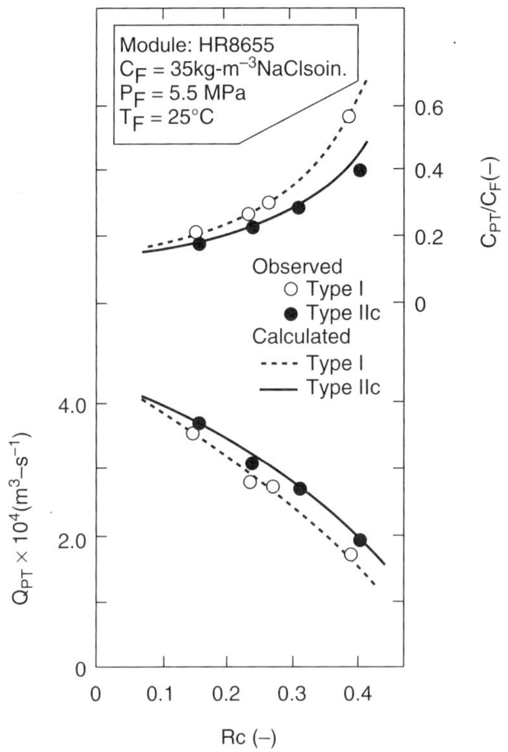

Figure 4.36 Comparison of module characteristics between types I and IIc.

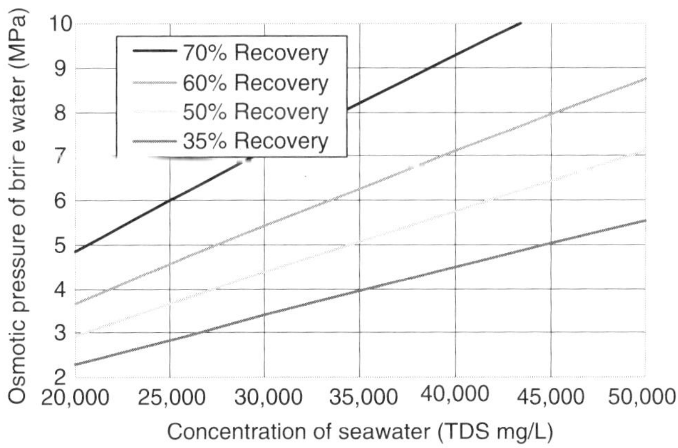

Figure 4.37 Osmotic pressure of brine water.

Table 4.3 Specifications for HB Series Modules

	Conventional Model	New Model HB Series	
Model Parameter	HM10255	HB10255	HB9155
Element diameter, mm	280	280	216
Product flow rate, m^3/day	45	62	15
Product salt rejection,[a] %	99.6	99.6	99.6
Number of elements	2	2	1
Test conditions			
NaCl Concentration, mg/L	35,000	35,000	35,000
Pressure, MPa	5.4	5.4	5.4
Temperature, °C	25	25	25
Recovery, %	30	30	30
Operation conditions (maximum)			
Pressure, MPa	6.9	8.2	8.2
SDI	4	4	4
Temperature, °C	40	40	40
Residual chlorine,[b] mg/L	1.0	1.0	1.0

[a] Salt rejecton = (1 − salt concentration in product water/salt concentration in feedwater) ×100.
[b] Residual chlorine concentration is limited by feedwater quality.

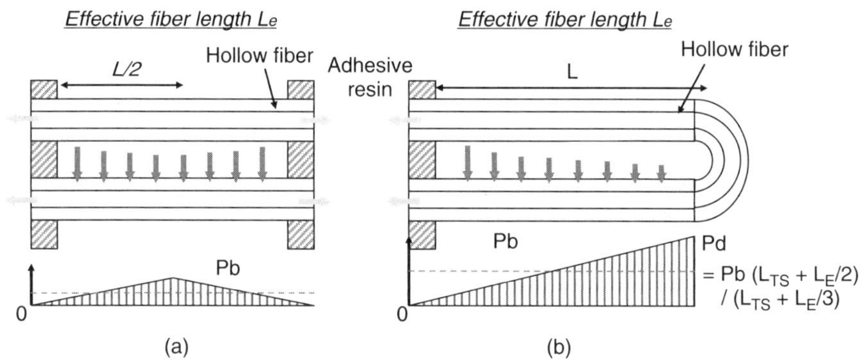

Figure 4.38 Comparison of both open ended (BOE) versus single open ended types of hollow-fiber membrane.

uniformly distributed in the RO module throughout the cross-wound hollow-fiber bundle.

Permeate water is recovered from both ends of the BOE RO module, and is collected with each supporting plate, respectively. At the opposite side of the feedwater inlet (brine side), permeate water is collected through the supporting plate and passes through the permeate pipe inside the concentric pipe. This permeate combines with the feed-side permeate water and flows to the permeate port of the pressure vessel. The supporting plate is held in position using compression snaps and an O-ring seal is placed on the face of the open-end, so that concentrate water cannot penetrate into the permeate-water area. The compression snap

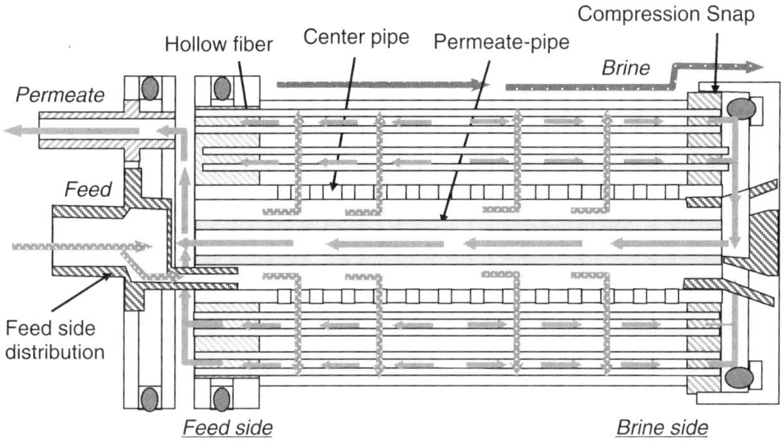

Figure 4.39 Cross section and flow pattern of both open ended (BOE) hollow-fiber element.

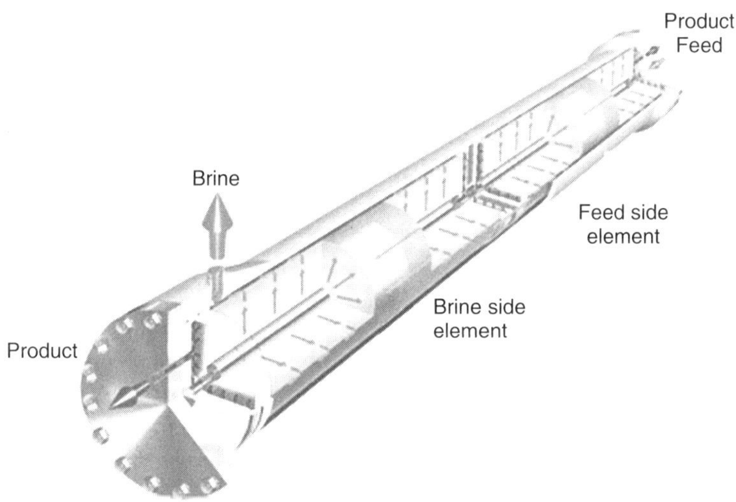

Figure 4.40 Both open-ended (BOE) hollow fiber module.

has another function, which is to center the element in the pressure vessel and form a narrow space between the pressure vessel and the element for the brine water flow.

Brine water flows on the periphery of the RO element, passes through the narrow space between the pressure vessel and the supporting plate and flows to the brine port of the pressure vessel.

The specifications for the BDE modules are shown in Table 4.4.

Table 4.4 Specifications for BOE Module

Model Parameter	New Model HB10255FI (HD10155) Both Open-Ended (BOE)
Element diameter, mm	280
Product flow rate, m^3/day	67
Product salt rejection,[a] %	99.6
Number of elements	2
Test conditions	
NaCl concentration, mg/L	35,000
Pressure, MPa	5.4
Temperature, °C)	25
Recovery, %	30
Operation conditions (maximum)	
Pressure, MPa	8.2
SDI	4
Temperature	40
Residual chlorine,[b] mg/L	1.0

[a] Salt rejecton = (1-salt concentration in product water/salt concentration in feedwater) × 100.
[b] Residual chlorine concentration is limited by feedwater quality.

The product flow rate of the BOE module improved by ~1.5 times compared with that of conventional (SOE) type [28].

4.5.6 Advanced Larger RO Module

Toyobo developed an advanced, large-sized new-style module, utilizing the technology of the BOE module. Since the bore pressure loss in a hollow fiber is smaller for the BOE module, even if the fiber length is increased, greater productivity performance is obtained. This new, more advanced, larger RO module has the same basic structure of the BOE module mentioned above and has a length ~1.5 times greater. The specifications for this module are shown in Table 4.5.

The amount of product water of one module has a capacity of more than double or a ~100 m^3/day. The membrane material is the same as CTA, has the same cross-winding hollow-fiber arrangement and the same excellent chlorine resistance, and maintains stable operation [30].

Figure 4.41 shows trends in the development of a ~11-in-diameter hollow-fiber RO module.

4.6 ACTUAL PERFORMANCE OF HOLLOW-FIBER RO MEMBRANE MODULE FOR SEAWATER DESALINATION

4.6.1 Actual Global Performance

The main features of a cellulose triacetate (CTA) membrane are summarized as follows:

Table 4.5 Specifications for Advanced Large BOE Module

Model Parameter	Conventional Model HM10255 Single Open-Ended (SOE)	Advanced Large-Szed HL10255 Both Open-Ended (BOE)
Element diameter, mm	280	280
Element length, m	1.35	2.0
Product flow rate, m^3/day	45	100
Product salt rejection,[a] %	99.6	99.6
Number of elements	2	2
Test conditions		
NaCl Concentration, mg/L	35,000	35,000
Pressure, MPa	5.4	5.4
Temperature, °C	25	25
Recovery, %	30	30
Operation Conditions (maximum)		
Pressure, MPa	6.9	6.9
SDI	4	4
Temperature, °C	40	40
Residual chlorine,[b] mg/L	1.0	1.0

[a] Salt rejecton = (1 - salt concentration in product water/salt concentration in feedwater) × 100.
[b] Residual chlorine concentration is limited by feedwater quality.

1. tolerance to chlorine, which is an effective disinfectant and allows direct sterilization of the osmosis RO membrane module to be directly sterilized, and
2. hydrophilicity, which reduces the adhesion of biofouling material to the membrane surface on exposure to organic matter.

These characteristics are most important for a seawater desalination plant because seawater has a high potential for microbial growth in the module since there are many sources of nutrients. If the seawater RO module cannot be sterilized by chlorine, high microbe multiplication occurs and propagates on the membrane surface. Microbe propagation within a RO module will result in sharp decrease in product-water quantity and deterioration of water quality. Moreover, frequent membrane cleaning is required, plant downtime increases, and the amount of chemicals increases because of the to cleaning. However, such performance deterioration and other problems do not occur in CTA-RO modules which can be sterilize directly by chlorine as a disinfectant. Many CTA-RO membranes have been developed the Middle East, where the possibility of microbe multiplication is high because of the open-surface intake and high temperatures. The main large-scale seawater desalination plants that have adopted CTA-RO module are listed in Table 4.6.

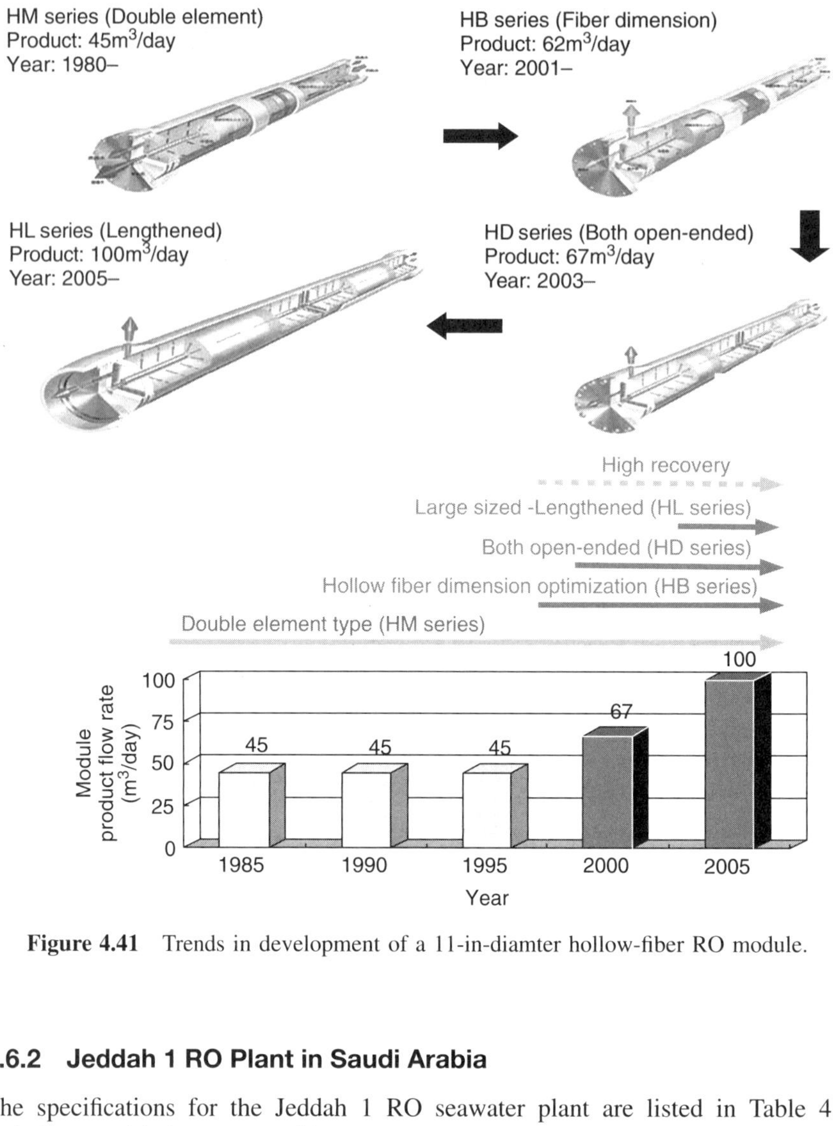

Figure 4.41 Trends in development of a 11-in-diamter hollow-fiber RO module.

4.6.2 Jeddah 1 RO Plant in Saudi Arabia

The specifications for the Jeddah 1 RO seawater plant are listed in Table 4.7 and discussed below to provide an example of a Red Sea coastal plant in the Arabian Peninsula. The Jeddah 1 RO phase 1 plant, which has a capacity of 15 mgpd (56,800 m^3/day), went into operation in 1989. The phase 2 plant, of the same size, came onstream in 1994, giving the plant a total capacity of 30 mgpd (113,600 m^3/day) [31]. The specifications for the Jeddah Phase I and Phase II RO plants are shown in Table 4.7.

The two plants are of almost the same construction, except that the separation membrane at the phase 2 plant was guaranteed for 5 years. Product water is blended with that from the MSF plant and distributed to Jeddah. The tight water supply in Jeddah just meets demand, and so product water from the plant is of crucial

Table 4.6 Main Supply Record of Toyobo Hollow-Fiber RO Module

Plant	Country	Capacity, m^3/day	Startup
Ras Alkhair	Saudi Arabia	345,000	2014
Jeddah 3	Saudi Arabia	260,000	2013
Shuqaiq	Saudi Arabia	240,000	2010
Rabigh	Saudi Arabia	218,000	2008
Medina-Yanbu	Saudi Arabia	128,000	1998
Al-Jubail	Saudi Arabia	85,000 (90,900)	2007
Jeddah 1 phase 1	Saudi Arabia	56,800	1989
Jeddah 1 phase 2	Saudi Arabia	56,800	1994
Marafiq-Yanbu	Saudi Arabia	50,400	2004
Fukuoka	Japan	50,000	2005
Ad Dur	Bahrain	45,500	2005
Tanjun-Jati B	Indonesia	10,800	2005
Florida	USA	11,400	2005
Tanajib	Saudi Arabia	6,000	2001
Duba	Saudi Arabia	4,400	1989
Haql	Saudi Arabia	4,400	1989

Table 4.7 Plant Specifications of Jeddah 1 RO Plants

Specification	Phase 1	Phase 2
Number of trains	10	10
Capacity, m^3/day	1.5 mgd × 10	1.5 mgd × 10
Permeate quality Cl^-, mg/L	<625	<625

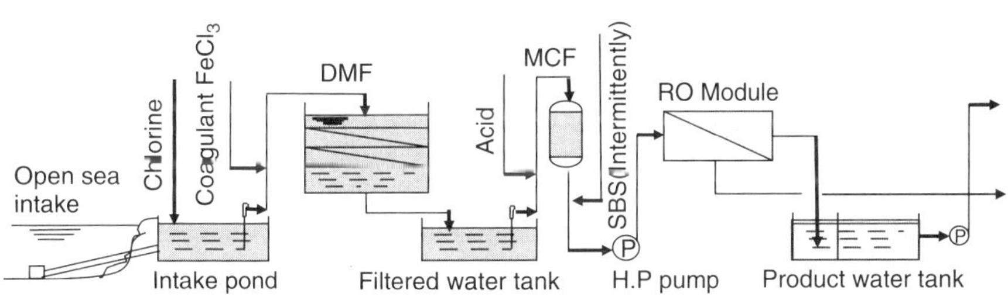

Figure 4.42 Schematic flow of Jeddah Phase II plant.

importance. The maintenance period must be kept to a minimum as the plant continuously operates at full capacity.

Toyobo double-element hollow-fiber RO modules (HM10255) are used for Jeddah phase 1 and phase 2 plants. Figure 4.42 illustrates process flow in the Jeddah 1 phase 2 plant.

Raw seawater is taken from beneath the Red Sea surface, and then disinfected by sodium hypochlorite produced by chlorine generation plant using filtered seawater as feed. Then ferric chloride as a coagulant is added to seawater feed ahead of the dual-medium filter (DMF) to help reduce the SDI (fouling index) to the membrane manufacturer's specification. The filtered feedwater is then collected in a clear well, and a cartridge filter is set in place to prevent particles measuring >10 μm from clogging the membrane pores. Sulfuric acid is added to the feedwater to adjust the pH to approximately neutral (~6.5). Sodium bisulfite (SBS) is injected for 7 h per shift (8 h) ahead of the high-pressure pump to nullify residual chlorine and protect the membrane from oxidation by residual chlorine in presence of heavy metals. To avoid biological fouling, 0.2 mg/L residual chlorine is allowed to pass through the membrane for 1 h every shift (8 h) intermittently using the (intermittent chlorine injection (ICI) method instead of the conventional continuous chlorine injection (CCI) method.

The disinfecting capacity of chlorine was tested by cultivating bacteria taken from the Jeddah seawater and sent to Japan. The results were as shown in Figure 4.43 [32].

It was confirmed that even when the bacteria count was high, disinfection using a chlorine concentration of 0.2 mg/L reduced it almost to zero in 30 mins.

To determine whether the disinfection was sufficient, the lead-time concept was considered [33]. *Lead time* is the time from (1) when injection of SBS is halted and chlorine is fed into the module to (2) when a certain chlorine concentration is detected in filtered water in the module. Except for certain times when the SDI of feed seawater exceeded 4.5 and the seawater was polluted, lead time remained stable at 10 mins. This signifies that a 10-min injection of chlorine at a concentration of 0.2 mg/L was sufficient to disinfect the module. In fact, it was confirmed that differential pressure in the module remained stable and no biofouling occurred during the test period.

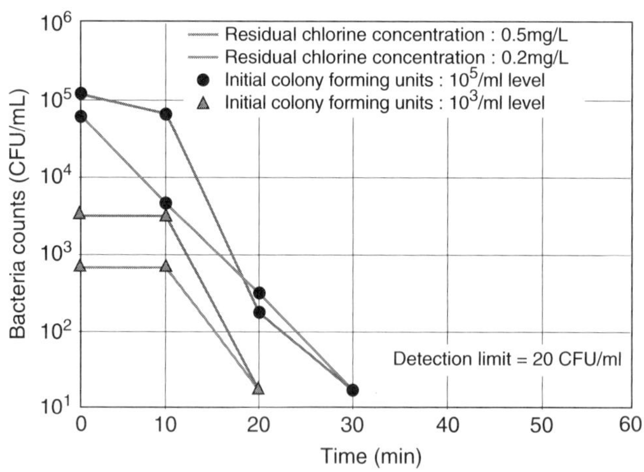

Figure 4.43 Disinfecting power of chlorine [30].

4.6.2.1 Operation Results of Phase II Plant Due to the successful site testing of ICI method at Jeddah, the operating conditions were kept the same for the phase 2 plant as shown in Table 4.8.

This plant is of crucial importance to Jeddah, which is prone to water shortages, and barring times when the feed seawater is polluted and SDI exceeds 4.5, it operates continuously at 99% capacity. It is important to note that the membrane has not been replaced at all in the 5 years since the plant went into operation.

4.6.2.2 Product Flow Rate and Quality Changes in the permeate flow rate are shown in Figure 4.44.

The permeate flow rate remained stable at the rated 56,800 m^3/day throughout the entire operation period. As is evident from the diagram, all 10 units at the plant were in virtually continuous operation. Permeate TDS (shown in Fig. 4.44) remained at <500 mg/L without membrane being replacement at all in 5 years, which is highly satisfactory [32].

4.6.2.3 Differential Pressure One point borne in mind about ICI was the importance of preventing biofouling. While biofouling should not occur when the module is disinfected by CCI, intermittent disinfecting of the module necessitates

Table 4.8 Operating Conditions of Jeddah Phase 2 Plant

Feed TDS	Concentration at 43,300 mg/L (Open-Sea Intake)
Feed temperatures, °C	24–34
Feed pressure, bar	57–64
Recovery, %	35

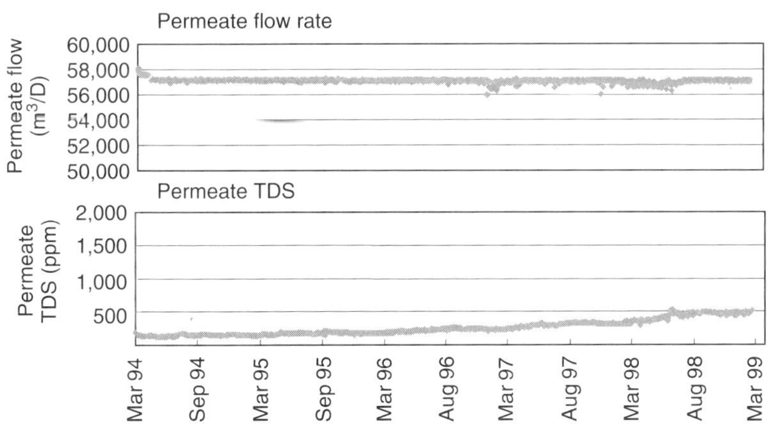

Figure 4.44 Permeate flow rate and permeate TDS of Jeddah phase 2 plant.

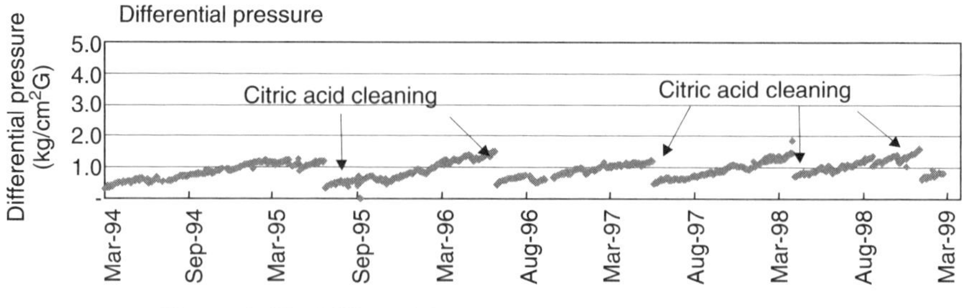

Figure 4.45 Differential pressure of Jeddah phase 2 plant.

caution. Disinfection tests were therefore conducted by cultivating bacteria taken from Jeddah seawater as described earlier, and extended testing was carried out using site test units to confirm that bio-fouling did not occur. As a result, differential pressure was stable at the phase 2 plant, as shown in Figure 4.45, and it was found that periodic cleaning with citric acid once a year reduced differential pressure to almost initial level [32].

At the Jeddah phase 2 plant, as a result of ICI cleansing, product flow and product quality were excellent, even without membrane replacement for 5 years.

The significant improvement in the performance of an existing plant following exchange of a non-CTA-RO membrane for a CTA-RO membrane is discussed below. The operation test results at the AdDur plant in Bahrain are introduced as an example of a typical desalination plant on the Arabian Gulf coast in the Middle East.

4.6.3 AdDur Plant in Bahrain

The AdDur RO desalination plant was designed to produce 10 migd (45,500 m^3/day) with polyamide RO hollow-fiber membrane modules from seawater containing $\leq 45,000$ mg/L as TDS. This plant was commissioned in 1990. The AdDur plant process consisted of a dual-medium filter (DMF) with coagulants, dechlorination by sodium bisulfate, high-pressure pumps, RO modules, and posttreatment by lime injection. Specifications for the AdDur plant are shown in Table 4.9 [31].

In 2000, rehabilitation work was conducted to obtain improved performance of the pretreatment using ultrafiltration (UF) membranes, which proved to be the most effective among the tests conducted. The work included modification of the existing dual-medium filter to a single-medium filter and additional installation of spiral-wound type UF membranes with a capacity of 130,000 m^3/day to reduce SDI. Chemical dosage was stopped completely with modification of the medium filter. The filtered seawater from the UF process was fed to the existing eight first-pass RO trains of polyamide membrane module. However, even after installation of the UF system, the RO membranes required frequent cleaning. It was concluded that additional work was needed to achieve more stable operation. The UF membrane was suspected to have failed remove to the natural organic materials from the raw

Table 4.9 Plant Specifications for AdDur RO Plant

Parameter	Specification
Operation Conditions of Design	
Feed temperature	16–36°C
Feed pressure	$\leq$68 bar
Feedwater TDS	45,000 mg/L
Recovery	35%
Total number of trains	8 trains, each divided into 4 banks
Specifications of Design	
Plant capacity	10 migd (45,500 m^3/day)
Product TDS	<500 mg/L

Source: Al-Badawi et al. [29].

seawater, thus allowing there materials to act as nutrients for biological matter in the RO membranes.

4.6.3.1 Schematic Flow Diagram of CTA Membrane Test Unit The RO feedwater used in the test plant is taken from acidified UF filtrate and is exactly the same in composition as the actual RO feedwater to the plant. SBS is dosed the in feedwater to remove residual chlorine contained in pretreated water. The SBS dose is interrupted 3 times a day for 1 h to introduce residual chlorine into the RO membrane to sterilize it purpose. Three Toyobo RO modules, HB9155 model, were used in this test. A schematic flow diagram of the test unit is shown in Figure 4.46.

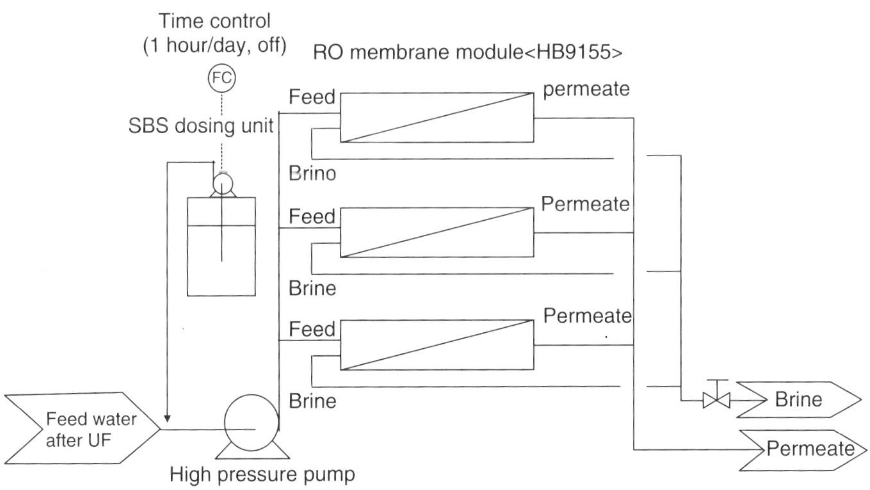

Figure 4.46 Schematic flow diagram of test unit.

Table 4.10 Test Conditions and Test Requirements

Test Conditions	Test Requirements
Feed pressure: ≤6.9 MPa	Permeate quality: <500 mg/L as TDS
Feed temperature: 16–37°C	Permeate flow rate per module: 10.77 m^3/day
Feed TDS: 45,000 mg/L	Differential pressure: normal increase or no change
Recovery: 35%	
Chlorine injection : ICI[a]	

[a] Intermittent chlorine injection method.

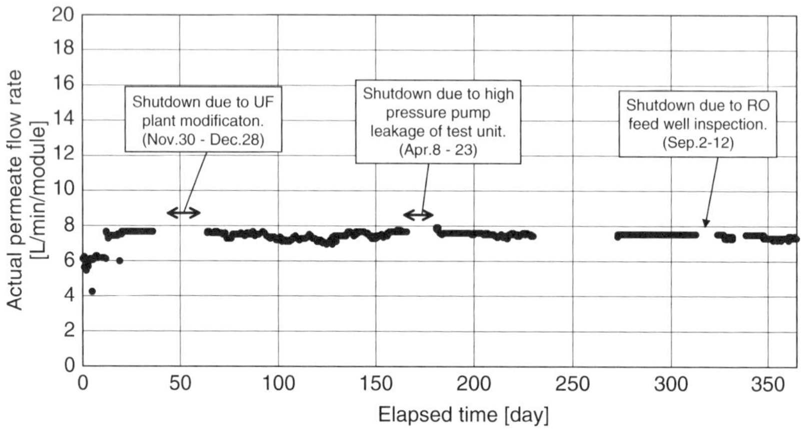

Figure 4.47 Permeate flow rate of test unit in AdDur plant.

4.6.3.2 Test Conditions

Test conditions and test requirements are listed in Table 4.10.

The objective of the test was to satisfy the plant specifications. The ICI method was found to effectively prevent biological fouling.

4.6.3.3 RO Performance at Site Test

4.6.3.3.1 Permeate Flow Rate and Permeate Quality

The actual permeate flow rate at the AdDur plant is shown in Figure 4.47.

The permeate flow rate was gradually increased to ~7.8 L/min per module, following adjustment of operation conditions during ~10 days from startup. Then the permeate flow was set at ~7.5 L/min per module (10.77 m^3/day per module at design conditions). During winter the permeate flow rate was maintained by increasing feed pressure. The flow rate was stable and satisfied the plant requirement for a 12-month test period. The actual permeate quality is shown in Figure 4.48.

The permeate quality was 764 μS/cm at operation startup and then gradually decreased to ~200 μS/cm, due to natural membrane compaction. The permeate quality was stable at a level of 200 μS/cm (95 mg/L as TDS at design conditions)

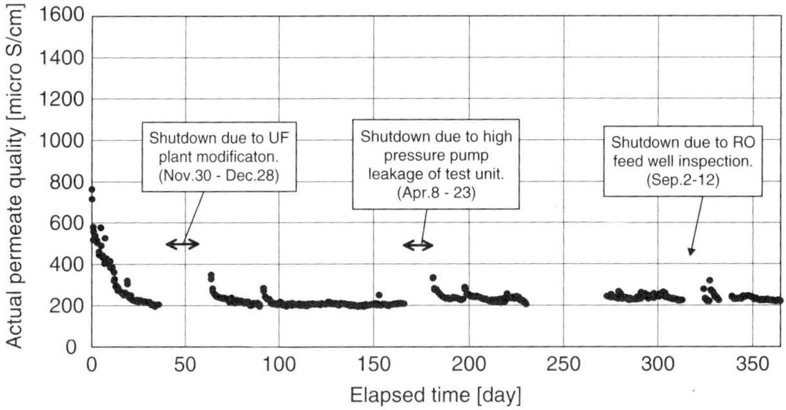

Figure 4.48 Permeate quality of test unit in AdDur plant.

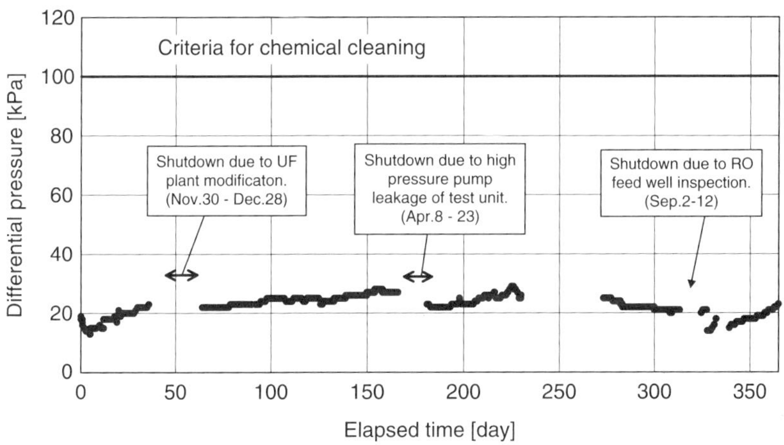

Figure 4.49 Differential pressure of test unit in AdDur plant.

throughout the test period. This performance was much better than the plant requirement of 500 mg/L as TDS. RO performance (permeate flow rate and permeate quality) was very stable and satisfactory during the entire test period [34].

4.6.3.4 Differential Pressure Differential pressure can be used as an indicator for biolfouling growth in the RO module (see Fig. 4.49).

The differential pressure was stable at the low level of 20 kPa without chemical cleaning. Therefore, it was concluded that biological fouling had not occurred in RO modules during the 12-month test period, and also that the intermittent chlorine injection ICI (method) worked effectively to prevent biological fouling.

Throughout the 12-month test period, RO performance of HB9155 membranes (see Table 4.3, rightmost column, for parameters) based on CTA material remained very stable and satisfied the test requirement. It was further demonstrated that

RO plant performance could be recovered by replacement of existing polyamide membranes with CTA membranes.

In general, the Arabian Gulf region is known as an area of physical hardship for RO plant installation because of the high temperatures and density of biological matter in the surrounding seawater. Control of biological activity in the RO membrane is key to ensuring stable desalination performance in the region. The results obtained in the 12-month test described above should shed some light on RO plants in this region.

4.6.4 Ikata Power Station in Japan

The Ikata power station has two types of desalination plant: one using the MSFD process; and the other, the RO process. The Ikata desalination RO plant consists of seawater intake pumps, pretreatment facilities for the seawater, and RO membranes. The seawater is drawn up by intake pumps, and some of it is transported to the pretreatment facilities. The remainder is transported to the discharge pit to dilute the brine from the RO plant. The system is designed to obtain treated seawater with SDI <4. After the pretreated seawater has passed through the safety filter, it is transported to the RO modules, where it is separated into desalted water and brine. The RO modules have 36 vessels in parallel. The desalted water is transported to the purification system through another tank, and then the brine is drained to the discharge pit. The maximum recovery is designed at 45% in the RO system; it is 40% in normal operations. Specifications for of the seawater RO desalination systems in the Ikata power station are shown in Table 4.11. The RO desalination system has been operating well for at least 10 years. The operating results are for years 1992–2005 shown in Figure 4.50 [35].

4.6.5 Fukuoka Seawater RO Desalination Plant in Japan

The Fukuoka seawater desalination plant has a capacity of 50,000 m^3/day, the largest plant in Japan. The 60% recovery of the RO desalination system is the highest seawater recovery level in the world. To achieve this high recovery

Table 4.11 Specifications for Seawater RO Desalination Systems in Ikata Power Station

Specification	Ikata
Membrane module type	CTA hollow fiber (HM9255)
Production capacity, m^3/day	2000
Number of modules	72 (36 × 2 lines)
Feed pressure, MPa	6.8
Recovery, %	40 (at 17°C) 33.3 (at 10°C)
Inlet seawater TDS, mg/L	35,000
Product water TDS, mg/L	350

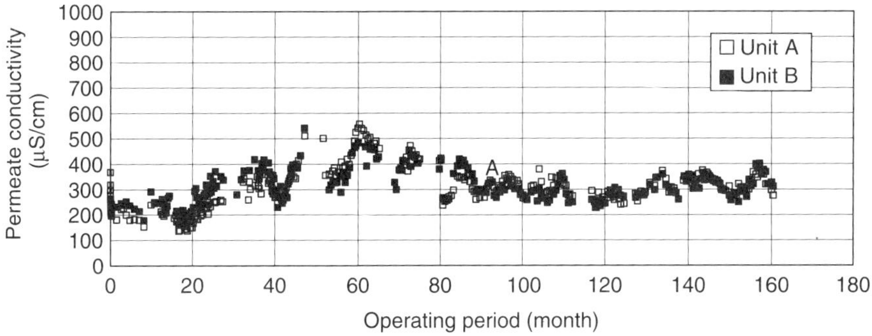

Figure 4.50 The operating results D in IKATA power station.

Table 4.12 Specification of Fukuoka Plant

Parameter	Specification
Production capacity	50,000 m^3/day
Product TDS	<200 mg/L
Plant system	
Intake	Infiltration intake
Pretreatment	Ultrafiltration
Desalination	High-pressure RO
Posttreatment	Low-pressure RO (Partial 2 pass)
Operating conditions	
Feed seawater TDS	35,000 mg/L
Seawater temperature	10–30°C
Recovery	60%
Feed pressure	
High-pressure RO	≤8.2 MPa
Low-pressure RO	≤1.5 MPa

operation, Toyobo's advanced RO modules were adopted. The specifications for the plant are shown in Table 4.12.

The production capacity is 50,000 m^3/day and the permeate TDS is 200 mg/L. The process flow diagram and membrane arrangement for this plant are shown in Figure 4.51.

This plant has adopted an infiltration intake system, an ultrafiltration system, and a high-pressure RO system that provides 60% recovery. A low-pressure RO membrane partial second pass is used in the posttreatment system and helps improve the quality of product water as needed.

Specifications for the high-pressure RO system are listed in Table 4.13, and views of the high-pressure RO system are shown in Figures 4.52 and 4.53.

The high-pressure RO system consists of five units. Each RO unit has a capacity of 11,988 m^3/day. Recovery is maintained within 57.5%–62.5% in response to water variation temperature [36,37].

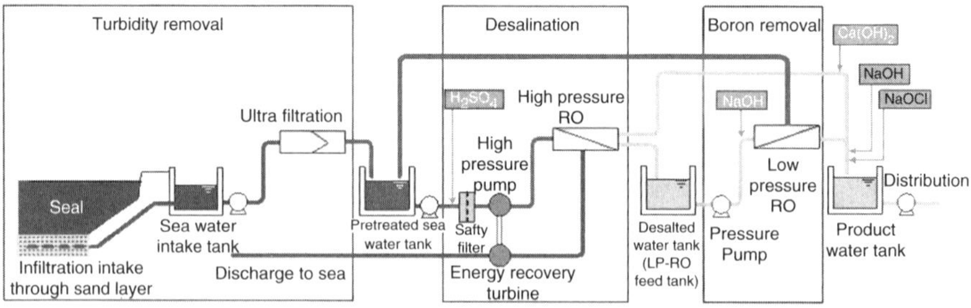

Figure 4.51 Process flow diagram and membrane arrangement.

Table 4.13 Specifications for High-Pressure RO system

Item	Specification
Plant System	
Number of units	5
Production capacity	11,988 m^3/day × 5 units
Product TDS	<350 mg/L
RO membrane	
Model	Toyobo, Hollosep
Type of membrane	Hollow-fiber
Material of membrane	Cellulose triacetate
Number of modules	200 pieces × 5 units

Figure 4.52 High-pressure RO module unit (partial, closeup view).

Figure 4.53 High-pressure RO module unit (whole view).

Plant operation began on June 1st, 2005. Depending on the water demand in Fukuoka City, the plant has produced up to the designed maximum capacity of 50,000 m^3/day, as needed. Since January 2006, operation of the plant at full capacity of 50,000 m^3/day has continued for at least several months.

Figure 4.54 shows mean performance of each high-pressure RO train.

The feed pressure was maintained constant for the first several months, and later tended to increase. This was attributed to increase in water recovery rate due to higher temperatures. After the mid-of August, low temperatures affected the increase in feed pressure. The recovery rate was estimated according to the feed temperature. It was confirmed that this estimated value agreed with the real value. Performance of the high-pressure RO desalination was stable under the severe conditions of 60% average recovery at a wide range of temperatures (12–30°C) in the feed [38].

4.7 ECONOMICS OF SEAWATER DESALINATION WITH RO MEMBRANE MODULES

Since the 1990s, the total water cost (TWC) index has been widely accepted as the preferred criterion for evaluation of seawater desalination processes. The TWC criterion replaces the initial cost or capital cost criterion for equipment selection, and its use is desalination increasing as a result of proliferation of seawater megaprojects and the need to better match expected lifetime operating costs with financing requirements of BOOT project development schemes. A standardized methodology for developing TWC is necessary in order to compare reverse osmosis (RO) and thermal desalination processes since the technologies greatly differ. A key conclusion almost always reached is that the production of potable water by RO processes is less expensive than producing freshwater from any of the various thermal distillation processes.

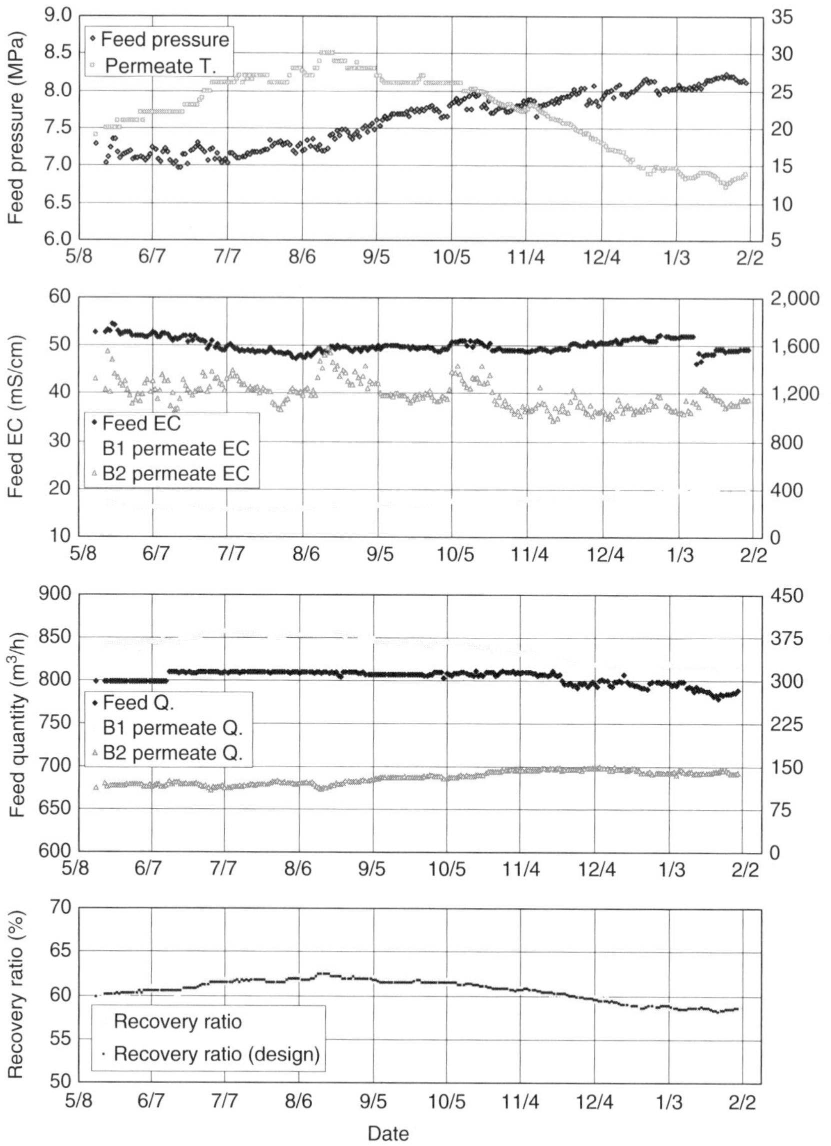

Figure 4.54 Performance of high-pressure RO.

There are several reasons for the significant RO economic advantage:

1. RO requires only 25–40% of the energy needed for thermal processes. With improved energy recovery devices and more recently, commercially introduced work exchangers for energy recovery, further energy savings over the more mature thermal distillation processes are expected.

2. Continuous technical improvements have been made in RO membranes. The Hollosep module model HB10255, for example, introduced over the last few years permits operation of ≤8.2 MPa, allowing RO plant operation at the 60% recovery level for moderate-salinity waters such as 35,000 mg/L as TDS oceanic seawater. Equally important is the HB10255 module capability to produce potable water in a single pass from higher-salinity water (i.e., 40,000 mg/L as TDS) such as that found in Arabian Gulf seawater.
3. Additional significant improvements associated with the Hollosep HB10255 module are the fact that this dual-element module results in lowering system capital cost and operating costs. These improvements are associated with (a) greater water production capacity of the larger module, allowing fewer pressure fittings, smaller manifolds, etc: (b) greater availability of the plant due to lower risk for fouling with use of continuous or intermittent chlorine injection (ICI): and (c) minimization of chemical usage and easy cleaning techniques.
4. Generally the construction materials for an RO plant are less costly than those for thermal desalting plants. Since RO plants operate at ambient temperatures, plastic or nonmetallic components can be used throughout, except for in high-pressure pumps and piping. The lower operating temperatures and easy replacement of components leads to lower maintenance costs for an RO plant compared those for thermal systems.
5. Extensive operator training is not necessary, due to the simple startup/shutdown procedures, modular plant design, and operation.
6. RO plant construction based on the modular design approach eliminates the need for shutdown of the complete plant for routine or unscheduled maintenance. Quality water production is achieved in minutes after startup.
7. RO plants can be delivered and put into operation in 1–2 years, depending on size. Thermal plants require about 50% greater delivery times.
8. RO plants need only about one-third the amount of feed seawater that is required for MSF and MED systems (including cooling-water). This allows for construction of smaller intake structures and less volume of feedwater to be pumped and pretreated. Therefore, the environmental impact of RO is more favorable than that of thermal desalting processes.

The estimated price of water for seawater desalination is shown in Figure 4.55 [39]. As shown in this figure, even in large-capacity seawater desalination plants such as a 200,000-m^3/day plant, the price of water for seawater desalination using the RO process is less than that for evaporation process nowadays. Naturally, the RO process is less harmful for to the environment than evaporation processes because of the lower energy consumption.

Figure 4.56 shows a timeline for the development of seawater desalination water cost using the RO process [40]. The cost of water for seawater desalination has decreased with each passing year, due to progress in membrane module performance, RO operation technology, decrease in RO module price, and so on. The

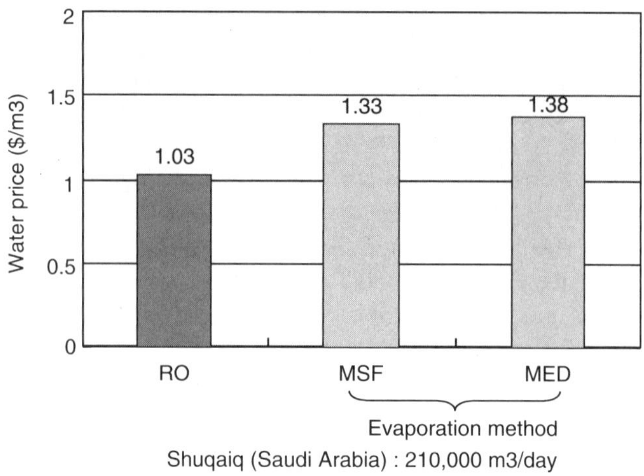

Figure 4.55 Water price by seawater desalination.

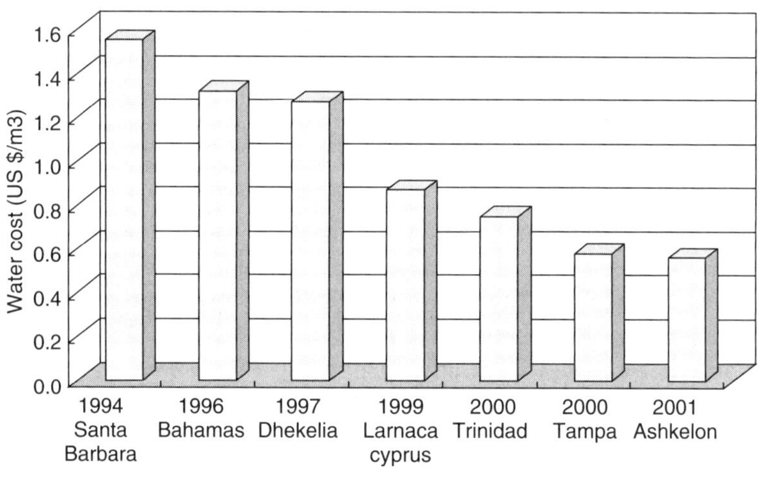

Figure 4.56 Water cost by seawater desalination.

cost of water for seawater desalination depends on prerequisites such as conditions of location, seawater condition, specifications for product water (capacity and quality), and contractual/financial conditions or requirements (insurance, apportionment of depreciation, etc.), as is well known. Therefore, because there are so many variables involved, it would be simplistic or inappropriate to evaluate the cost or price of water using a simple metric such as $\$0.5/m^3$ alone.

4.8 CONCLUSIONS

The reverse-osmosis (RO) seawater desalination process has many advantages, in terms of energy saving, lower capital cost, short startup and shutdown times, short construction period, less installation space needed, and less total water cost. The RO process is becoming the key technology for obtaining freshwater from the sea, especially in the Middle East. Desalination equipment manufacturers are working to develop membranes offering higher product-water recovery, lower energy consumption, and lower installation costs, in the hopes that the RO process will be adopted worldwide as the most reliable, efficient method for supplying freshwater.

Reverse-osmosis membranes were first developed from cellulose acetate (CA) and cellulose triacetate (CTA). These CA and CTA membranes have been in commercial use for many years. The CTA membrane, in particular is still used throughout the world. This wide spread use demonstrates the superior properties of CTA in RO membranes. As for performance, the coexistence of high permeability and high selectivity is enabled by an outstanding balance of hydrophilic and hydrophobic properties in the CTA RO membrane. Moreover, the CTA membrane is the only RO membrane currently marketed that offers such a high degree of chlorine resistance. The intermittent chlorine injection (ICI) method is effective with the CTA membrane to control biofouling growth and allows stable operation in seawater desalination plants, especially in the Middle East, where there is very high biofouling potential due to high seawater temperatures. RO membranes that not are chlorine-resistant are biofouling, more susceptible to which prevents the RO plants, from operating in stable mode because of the need for frequent cleaning. In those cases where CTA membranes have replaced non-chlorine-tolerant RO membranes because of the risk of biofouling, stable operation has been attained.

Moreover, the CTA material, when used in hollow-fiber bundles, offers excellent resistance to biofouling because it is generally hydrophilic, reducing the potential for adhesion of organic material to the membrane surface. Also, a hollow-fiber CTA module element provides a membrane area that is ~10 times larger per unit volume than that provided by a spiral wound membrane element, and this property significantly reduces the flux per unit membrane area. Therefore, with CTA it is possible to decrease membrane load, which helps prevent biofouling, to occur. These key attributes lead to minimal chemical cleaning and long membrane life.

At the time of writing, CTA-RO hollow-fiber membrane technology continues to expand throughout the world, especially in seawater desalination applications. This technology is expected to contribute significantly to the increasing global demand for freshwater. With high-volume commercial operation of large-scale, recently constructed desalination plants in the Middle East and North Africa [e.g., the Hamma desalination plant, Algeria, with a capacity of 200,000 m^3/day (73 million m^3/year)] now well underway, with plans for future plant installations in thos areas, CTA-based hollow-fiber RO membrane technology—developed primarily by Toyobo Co., Ltd. of Japan—seems to be the premier method for seawater desalination.

NOMENCLATURE

Symbols

A	pure-water permeability constant
B	solute transport constant
C	solute concentration
D	diffusion coefficient
d_I	inner diameter of hollow fiber
d_O	outer diameter of hollow fiber
J_S	solute flux through membrane
J_V	solution flux through membrane
J_W	water flux through membrane
k	mass transfer coefficient
P	pressure
Q	flow rate
Rc	recovery ratio
Re	Reynolds number, do $V\rho/\mu$
Sc	Schmidt number, $\mu/\rho D$
Sh	Sherwood number, kd_O/D
T_{FW}	Feedwater temperature
V	fluid velocity
z	axial coordinate
Δ	difference
μ	viscosity
ρ	density
π	osmotic pressure
ϕ	concentration polarization

Subscripts

B	bulk
F	feed
P	permeate

REFERENCES

1. Reid CE, Breton EJ, Water and ion flow across cellulosic membranes, *J. Appl. Polym. Sci*. **1**:133 (1959).
2. Loeb S, Sourirajan S, *High Flow Porous Membranes for Separating Water from Saline Solutions*, US Patent 3,133,132 (1964).

3. Breton EJ, Water and ionic flow through osmotic membranes OSWRDP Report 16, New York, 1957.
4. Westmoreland JC, *Spirally Wrapped Reverse Osmosis Membrane Cell*, US Patent 3,367,504, (1968).
5. McCutchan JW, Johnson JS, Reverse osmosis at Coalinga, California, European Federation of Chemical Engineering, *J. Am Water Works Assoc.*, **62**:346 (1970).
6. Mahon HI, *Permeability Separatory Apparatus, Permeability Separatory Membrane Element, Method of Making the Same and Process Utilizing the Same*, US Patent 3,228,876 (1966).
7. Dance EL, Davis TE, Mahon HI, McLain EA, Skiens WE, Spano JO, *Development of Cellulose Triacetate Hollow Fiber Reverse Osmosis Modules for Brackish-Water Desalination*, OSWRDP Report 763, New York, 1971.
8. Ammons RD, Mahon HI, *Development of a One-Pass Hollow-Fiber Seawater Desalination Module Having a Capacity of 2500–3000 gpd*, Report 924, US Office of Saline Water Research and Development Program (OSWRDP), US Government Printing Office, Washington, DC, 1974.
9. Orofino TA, *Development of Hollow Filament Technology for Reverse Osmosis Desalination Systems*, OSWRDP Report 549, 1970.
10. Ukai T, Nimura Y, Hamada K, Matsui H, Development of one pass sea water reverse osmosis module,"HOLLOSEP", *Desalination* **32**:169 (1980).
11. Aptel P, Buckley CA, Categories of membrane operations, in Mallevialle J, Odendaal PE, Wiesne MR, eds., *Water Treatment Membrane Processes*, McGraw-Hill, Johannesburg, 1996, p. 2.16.
12. Kumano A, Reverse osmosis membranes and the method of formation, *Seikei-Kakou* **14**(12):766 (2002).
13. Nita K, The aromatic polyamide hollow fiber reverse osmosis membrane with chlorine resistance, *Plant Process*, (Japan) **34**(2):63 (1992).
14. Ohya H, An expression method of compaction effects on reverse osmosis membranes at high pressure operation, *Desalination* **26**:163 (1978).
15. Kumano A, Matsui Y, Numata K, Fujiwara N, Iwahashi H, Nagai M, Performance change formula of cellulose triacetate hollow fiber RO membranes due to oxidation and hydrolysis *Desalination*, **96**:451 (1994).
16. Kannari T, Module performance monitoring system, *Proc. IDA World Congress Desalination and Water Reuse*, IDA, Abu Dhabi, 1995, Vol. 3, pp. 3–11.
17. Gill WN, Bansal B, Hollow fiber reverse osmosis systems analysis and design, *Am. Inst. Chem Eng. (AIChE) J.* **19**:823 (1973).
18. Dandavati MS, Doshi MR, Gill WN, Hollow fiber reverse osmosis experiments and analysis of radial flow systems, *Chem. Eng. Sci.* **30**:877 (1975).
19. Kabadi VN, Doshi MR, Gill WN, Radial flow hollow fiber reverse osmosis experiments and theory, *Chem. Eng. Commun.* **3**:339 (1979).
20. Ohya H, Nakajima H, Takagi K, Kagawa S, Negishi Y, An analysis of reverse osmotic characteristics of B-9 hollow fiber module, *Desalination* **21**:257 (1977).
21. Sekino M, Fujiwara N, Analysis of hollow- fiber reverse- osmosis module, *Kagaku Kogaku Ronbunshu* **17**:1088 (1991).

22. Sekino M, Precise analytical model of hollow fiber reverse osmosis modules, *J. Membrane Sci*. **85**:241 (1993).
23. Ohya H, Suzuki T, Nakao S, Kato S, Tsuji M, Sugi J, Proposal and technological breakthrough of an integrated system for the complete usage of sea water, *Bulle. Soc. Sea Water Sci*. (Japan), **50**(6):389 (1996).
24. Nakao S, Sea water desalination process for high recovery of fresh water by reverse osmosis, *Bull. Soc. Sea Water Sci. (Japan)*, **50**(6):406 (1996).
25. Ohnishi J, Nita K, Sekino M, Development of high permeate flow rate and high pressure resistance RO module for seawater desalination, *Proc. IDA World Congress Desalination and Water Reuse*, IDA, Madrid, Oct. 6–9, 1997, pp. 479–489.
26. Sekino M, Important conditions and present state for seawater desalination RO module, *Bull. Soc. Sea Water Sci*. (Japan) **52**(2):107 (1998).
27. Fujiwara N, Tanaka T, Kumano A, Sekino M, New economical RO desalination system by a recovery ratio of 60%, *Proc. IDA World Congress Desalination and Water Reuse*, IDA, San Diego, CA (USA), Aug. 29–Sept. 3, 1999, pp. 101–109.
28. Kumano A, Fujiwara N, Nishida M, Sekino M, A single-stage RO seawater desalination process at higher recovery with improved hollow fiber membrane modules, *Proc. IDA World Congress Desalination and Water Reuse*, IDA, Paradise Island, Bahamas, Sept. 28–Oct. 3, 2003.
29. Kumano A, Focus on the high pressure and high recovery operation system and single stage RO module in Fukuoka plant, *Bull. Soc. Sea Water Sci*. (Japan) **58**(3):248 (2004).
30. Marui K, Kumano A, Kotera H, Fujiwara N, Tanaka T, Sekino M, Higher recovery with improved hollow fiber membrane modules, *Proc. IDA World Congress Desalination and Water Reuse*, IDA, Paradise Island, Bahamas, Sept. 28–Oct. 3, 2003.
31. Al-Badawi AR, Al-Harthi SS, Imai H, Iwahashi H, Katsube M, Fujiwara N, Operation and analysis of Jeddah 1—phase 2 Plant, *Proc. IDA World Congress Desalination and Water Reuse*, IDA, Abu Dhabi, UAE, Nov. 18–24, 1995, Vol. 3, pp. 41–53.
32. Fujiwara N, Tanaka T, Katsube M, Farhan MA, Al-Harthi ASS, Imai H, Iwahashi H, Five-year operational performance of membrane at 15 MGPD Jeddah RO phase 2 plant, *Proc. IDA World Congress Desalination and Water Reuse*, IDA, San Diego, Aug. 29–Sept. 3, 1999, pp. 89–98.
33. Nada N, Iwahashi H, Umemori F, Test result of the intermittent chlorine injection method in Jeddah 1 plant, *Desalination* **96**:283 (1994).
34. Alawadhi A, Kannari T, Katsube M, Umemori F, Fujiwara N, Elimination of biological fouling in seawater desalination plant in Bahrain, *Proc. IDA World Congress Desalination and Water Reuse*, IDA, Singapore, Sept. 11–16, 2005).
35. Kumano A, Application of environmental reverse osmosis membrane of hollow fibers, *Nuclear Viewponits* **53**(2): 25–28 (2007).
36. Kotera H, Kumano A, Marui K, Fujiwara N, Tanaka T, Sekino M, Advanced RO module in the largest seawater desalination plant in Japan, *Proc. IDA World Congress Desalination and Water Reu*se, IDA, Singapore, Sept. 11–16, 2005.
37. Matsumoto Y, Kajiwara T, Funayama K, Sekino M, Tanaka T, Iwahori H, 50,000 m^3/day Fukuoka sea water RO desalination plant by a recovery ratio of 60%, *Proc. IDA World*

Congress Desalination and Water Reuse—Water for a Better Future, IDA, Manama, Bahrain, March 8–13, 2002.

38. Hamano T, Tsuge H, Goto T, Innovations perform well in first year of operation, *Desalination Water Reuse* **16**(1):31 (2006).
39. RO comes out top of fight for Shuqaiq. *Global Water Intelli*. **7**(8):9 (Aug. 2006).
40. Henthorne L, Introduction to the desalination market and pretreatment use, *Proc*. Int. Desalination Assoc. RO Pretreatment Workshops 2004, Tampa, FL. and San Diego, CA, 2004.

CHAPTER 5

Adsorption–Desalination Cycle

ANUTOSH CHAKRABORTY, KYAW THU, BIDYUT BARAN SAHA*, and KIM CHOON NG*

*Corresponding authors.

Advances in Water Desalination, First Edition. Edited by Noam Lior.

5.1 INTRODUCTION

Freshwater and energy are essential for sustaining human life on earth. The supply of clean, potable water requires energy expenditure and unfortunately, many countries of the world are deficient in energy sources, such as fossil fuel, natural gas, and coal. So, the combination of a renewable energy source, such as waste heat, wind, and solar and geothermal energy, with desalination systems holds immense promise for improving potable-water supplies that does not produce air pollution or contribute to the global crisis of climate change. In recent years, researchers in Singapore and Japan have experimented on the adsorption cycle for desalination, as it is both cost-effective and environmentally friendly and requires only low-temperature waste heat to operate.

The search for fresh or potable water remains a pressing concern throughout many regions of the world. The World Health Organization (WHO) reported that about 41% of the earth's population lives in water-stressed areas, and the number of people in the water-scarce regions may climb to 3.5 billion by the year 2025. At least one billion people do not have access to clean and potable water, and over one billion people live where water is economically scare, or places where water is available in rivers and aquifers, but the lack of infrastructures renders this water unavailable to people [1,2]. Figure 5.1 shows a projection of global water scarcity in 2025 [3]. According to the International Water Management Institute (IWMI), water scarcity is not a factor of absolute quantity; rather, it is a relative concept comparing the availability of water to actual use. Ultimately, it is necessary to limit the effects of water shortages by improving the efficiency of water use, implementing desalination technologies and policies to encourage water conservation and reuse, slowing population growth, and tapping nontraditional sources of freshwater such as seawater, fog water, and atmospheric water vapor. [2,4,5]. It is essential to discuss the water cycle, as water in the form of solid, liquid, or gaseous phase moves perpetually through the water cycle of evaporation and transpiration, precipitation, and runoff usually reaching the sea.

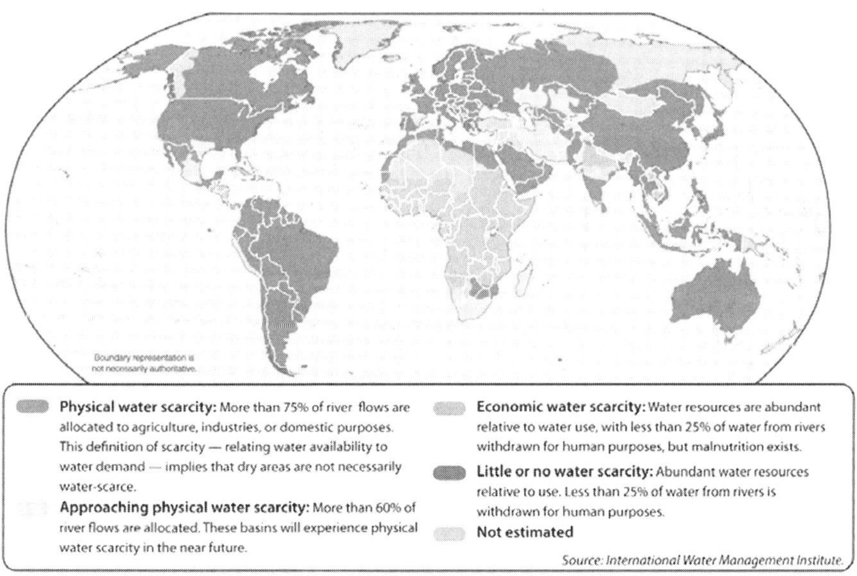

Figure 5.1 Global water scarcity in 2025: a prediction.

5.1.1 Water Cycle

Water is essential to all known forms of life, and the term *water* refers not only to its liquid state but also to its solid state, such as ice, and a gaseous state or water vapor. Approximately 1460 teratonnes (metric tons) (Tt) of water cover 71% of the earth's surface, mostly in oceans and other large water bodies, with 1.6% of water below ground in aquifers and 0.001% in the air as vapor, clouds (formed of solid and liquid water particles suspended in air), and precipitation. Saltwater oceans hold 97.0% of surface water, 2.4% of glaciers and polar icecaps, and 0.6% of other land surface water such as rivers and lakes. Distribution of the earth's water is shown in Figure 5.2. Water moves continuously in the form of evaporation and transpiration, precipitation, and runoff through the water cycle, which is plotted in Figure 5.3. The sun, which drives the water cycle, heats the water in oceans and seas. Water evaporates into the air. *Evapotranspiration* refers to water transpired from plants and evaporated from the soil. Rising air currents take the vapor up into the atmosphere, where cooler temperatures cause it to condense into clouds. Air moves water vapor around the globe; cloud particles collide, grow, and fall out of the sky as precipitation. Some precipitation falls as snow or hail, or sleet, and can accumulate as icecaps and glaciers, which can store frozen water for thousands of years. Snowpack can thaw and melt, with the melted water flows over land as snowmelt. Most water falls back into the oceans or onto land as rain, wherein water flows over the ground as surface runoff.

Actually, thermal desalination is a result of evaporation and condensation, and the next section discusses various desalination technologies.

5.1.2 Current Desalination Technologies

Desalination is the process by which fresh or potable water is produced from the sea or brackish-water of highly dissolved and suspended salts or solids. In general,

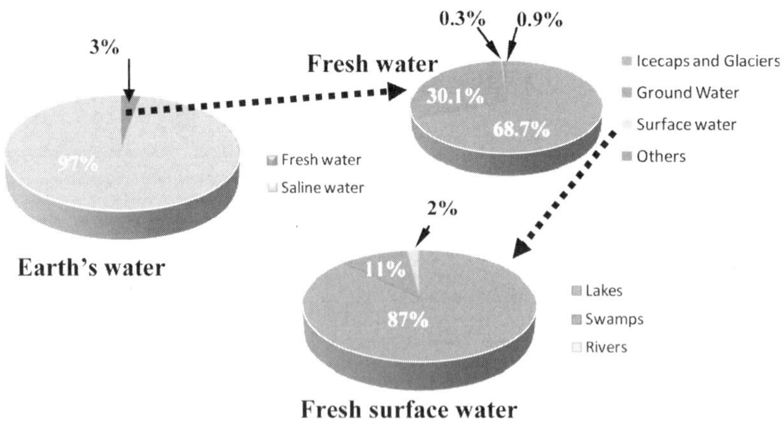

Figure 5.2 Distribution of earth's water (*source:* US Geological Survey [6]).

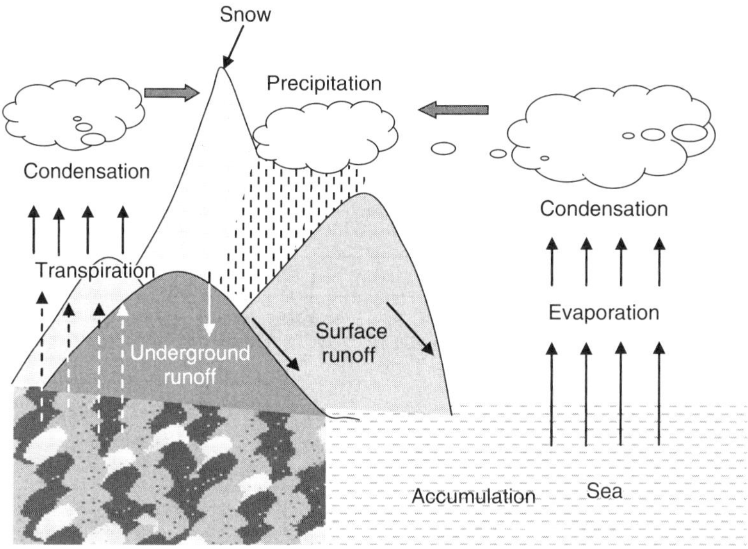

Figure 5.3 Conceptualization of the water cycle.

desalination methods can be categorized into three major groups: (1) thermally activated systems that utilize thermal energy to split freshwater by evaporation and then condense the vapor; (2) pressure-activated systems that split potable water from the saltwater by applying a certain pressure higher than osmotic pressure through a semipermeable membrane; and (3) chemical-activated desalination, which includes ion exchange desalination, liquid–liquid extraction, and gas hydrate or other precipitation schemes. Categories of the desalination process are shown in Figure 5.4. Thermally activated desalination systems include the well-known multistage flash distillation (MSFD), multieffect desalination (MED), and solar distillation [7–15]. These methods employ thermal energy to evaporate water vapor from salt solution. MED is sometimes referred to as *multieffect evaporation* (MEE). The membrane systems utilize semipermeable membranes or nanotubes to separate the water molecules from the salt solution. These are known as *reverse-osmosis* (RO) and *electro-dialysis* (ED) techniques and consume electrical energy [16]. Several researchers [11,14] concluded that the MEE process has more attractive attributes as compared to the MSFD method. The main benefits of MEE over the MSFD are lower energy consumption, lower sensitivity to corrosion and scaling, and greater development potential. In addition, in contrast to MSFD, the MEE process can efficiently operate with low-temperature heat sources (<100°C). In a sun-driven reverse-osmosis (RO) process, the efficiency of the collectors (e.g., of the photovoltaic cells) is independent of the RO efficiency [17–19].

5.1.2.1 Thermal Desalination Membrane distillation (MD) is an emerging technology for desalination, and the driving force for MD is the difference in vapor pressure of water across the membrane, rather than total pressure. The membranes

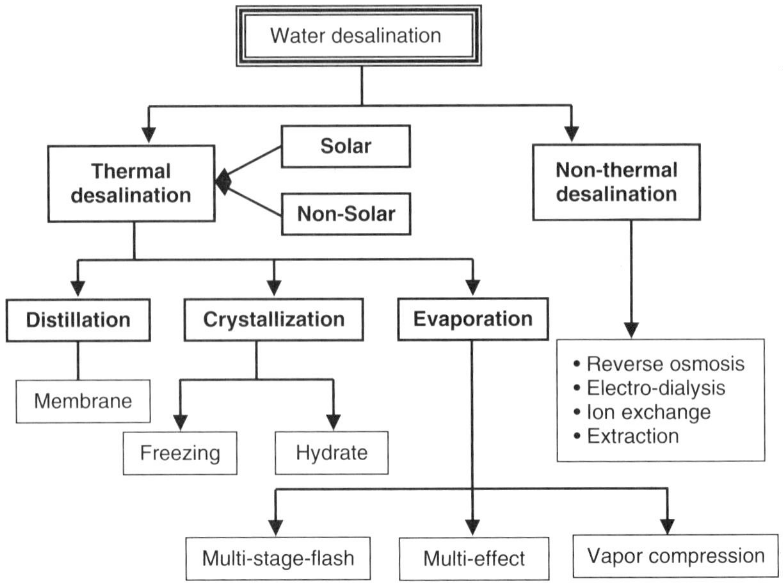

Figure 5.4 Categories of desalination processes.

for MD are hydrophobic, allowing water vapor to pass. The vapor pressure gradient is created by heating up the source water, employing low-grade waste heat or solar energy, thereby elevating its vapor pressure. The main advantages of membrane distillation are

- Production of high-quality distillate
- Distillation of water at relatively low temperatures (0–100°C)
- Use of low-grade heat (solar, industrial waste heat or desalination waste heat)
- No extensive pretreatment required for the water

The following methods are used for heat-based desalination:

Crystallization. This method depends on the fact that the freezing point of water is higher than that of the brine. The temperature of saltwater is reduced until ice crystals are formed; then, the formed crystals are separated from the mixture where they can be melted to form freshwater.

Multistage Flash Distillation. Multistage flash (MSF) distillation (MSFD) as shown in Figure 5.5 is the most common form of thermal desalination in use today. The MSF process uses a series of chambers that operate at progressively lower pressures. Each chamber can be divided into three sections. The top section contains a bundle of tube heat exchangers, which carry the seawater. A distillate collection chamber is positioned below these heat exchangers. The distillate condenses on the outer surface of the tubes, collects

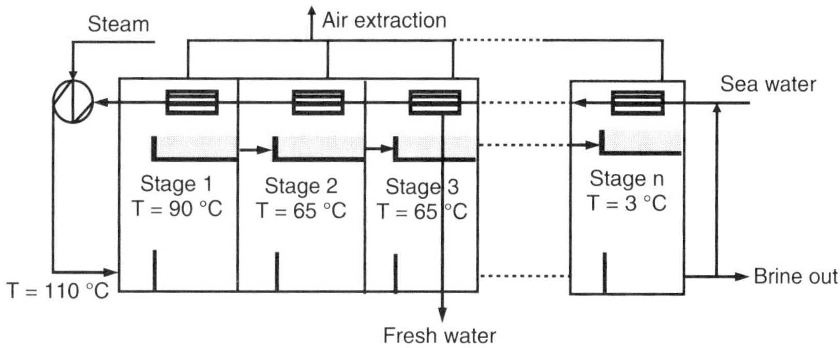

Figure 5.5 Multistage flash (MSF) desalination.

in the channel, and flows into the next stage in the direction opposite to the movement of seawater through the heat exchanger tubes. One of the main advantages of MSF plants, especially large ones, is the production of large quantities of waste heat, which can often be paired with cogeneration. However, the operating costs for MSFD are high when waste heat is not available, and MSFD also provides a higher rate of corrosion.

Multieffect Distillation. The multieffect (or multiple-effect) distillation (MED) process employs a series of chambers that operate at progressively lower pressures. The purpose of employing multiple chambers at lower pressures is to maximize the recovery of waste heats and exploit these heats for vaporization at lower pressure. In MED, each chamber is fitted with a bundle of tube heat exchangers. A schematic diagram depicting the operational flows of a MED unit is illustrated in Figure 5.6. In the MED method, saltwater is fed to the *first effect* or chamber where vapor is generated initially by an external heat source in a heat exchanger. The remaining saltwater is either

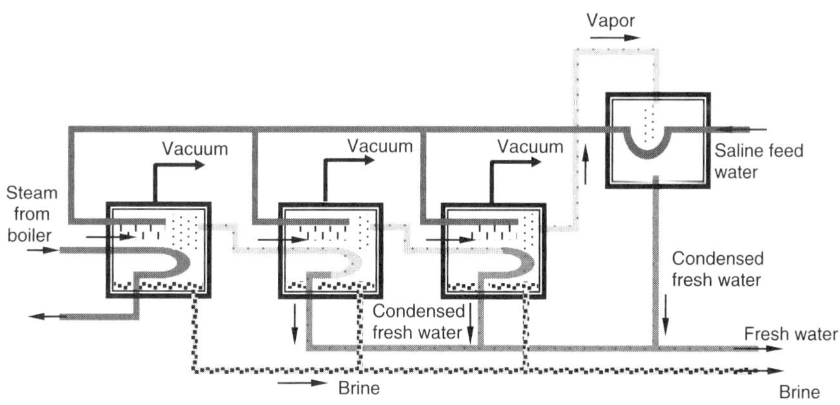

Figure 5.6 A schematic diagram of multieffect-desalination (MED).

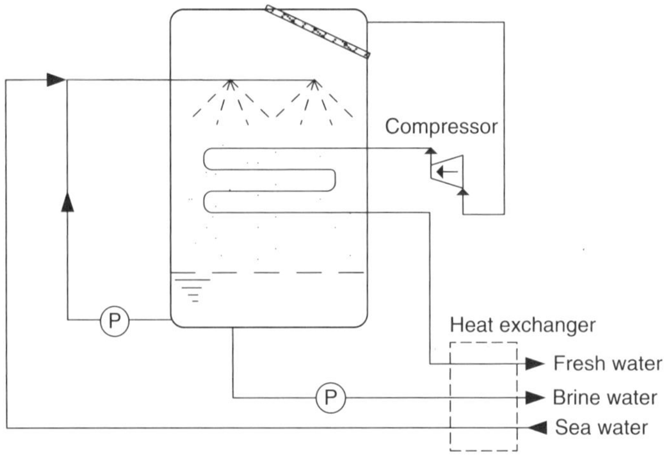

Figure 5.7 Schematic diagram of vapor compression unit for freshwater production.

pumped or refluxed (by gravity) to the next effect or chamber, which is at lower pressure than the former. The solution from a stage of higher pressure is sprayed on the outside surface of tubes on the subsequent chamber. The inner core of tubes carries the generated vapor from the previous effect, condensing on the cooler surfaces as it expands through. The latent heat of condensation produces vapor from the water films falling over the external surfaces of tubes. Condensate from the inner cores of tubes is collected and pumped to the ambient.

Mechanical Vapor Compression. Mechanical vapor compression (MVC), as shown in Figure 5.7, uses mechanical energy rather than steam as a source of thermal energy. MVC processes use a series of chambers that contain a set of heat exchanger tubes. A mechanical compressor draws the water vapor from the chambers, thus maintaining the vacuum in it. The discharged vapor from the compressor is at elevated temperatures before it condenses inside the tube banks where heat is used to vaporize the sprayed brine feed. The condensate from the tube banks is discharged as freshwater. The heat of condensation is used to evaporate a thin film of seawater that is recompressed on the inside of the tubes within the evaporation chambers. The smallest MVC systems are typically single-effect units that operate slightly above atmospheric pressure at a temperature of 102°C.

5.1.2.2 *Nonthermal Desalination* The following methods are used for non-heat-based desalination:

Electrodialysis. This method is effective for brackish-water or low water salinity ($\leq$10,000 ppm). It is a cost-effective method for desalting brackish-water. One of the principles on which electrodialysis operates is that most salts dissolved in water are ionic. These ions are attracted to electrodes with different charges. In order to desalinate water, membranes through which ions (either positive or negative but not both) are allowed to pass are placed between a pair of electrodes. These membranes are arranged alternately with opposite selective membranes followed by a negative selective membrane. A spacer sheet that permits water to flow along the face of the membrane is placed between each pair of membranes.

Reverse Osmosis (RO). Membrane desalination processes depend on the ability of membranes to differentiate and selectively separate water and salts. The most common application is RO. Osmosis uses a semipermeable membrane to separate solutions of different concentration. The solvent flows at a higher rate than do the dissolved solids from low concentration to the higher concentration. RO relies on a difference in chemical potential between the solutions on either side of the membrane. The chemical potential is a function of concentration, pressure, and temperature, and the solvent flows across the membrane. The basic concepts of osmosis and reverse osmosis are shown in Figure 5.8. In a system of finite volume, the liquid level on the low-concentration side of the membrane decreases and results in a hydrostatic pressure difference between the two sides. Once the hydrostatic pressure difference is equal to the driving force of flow, the system has reached equilibrium, and the net flow of solvent ceases. The equilibrium hydrostatic pressure level is known as *osmotic pressure*. RO operates by pressurizing the saline feed solution to a pressure greater than osmotic pressure. This causes the chemical potential of the solution to fall below that of the pure solvent and drives solvent flow from the solution side to the pure solvent side of the membrane. The pressurization process is the single greatest energy consumption process in the entire operation. In RO, the pretreatment of saline water is required to prevent fouling, scaling, and membrane degradation so as to increase the efficiency and operating life of the membrane being used for separation. The membrane-based desalination process is limited by several factors: (1) high osmotic pressure,

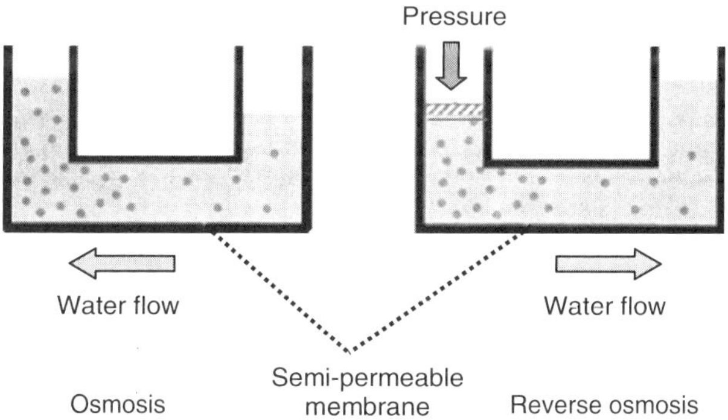

Figure 5.8 Schematic representation of the basic concepts of osmosis and reverse osmosis.

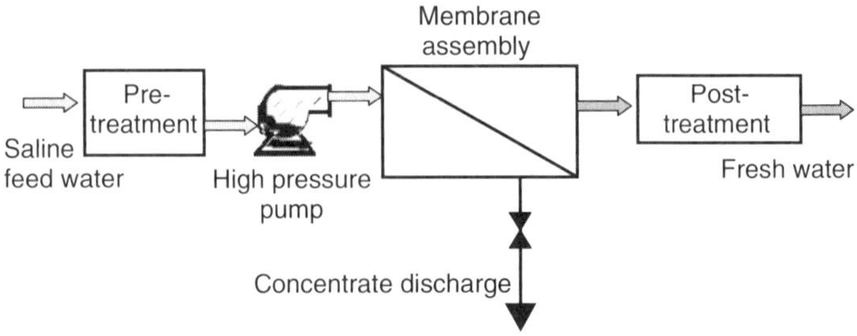

Figure 5.9 Flowchart of a simple reverse-osmosis desalination process.

(2) chemical composition of the feed, and (3) feed temperature. The basic operating principle of a RO system is shown in Figure 5.9.

5.1.2.3 *Discussion*

However, all existing desalination methods have shortcomings such as energy intensiveness and high maintenance arising from the cost of membrane replacement and corrosion. So some promising new approaches employing cheaper materials and more efficient equipment could be applied to ensure large-scale extraction of clean water in the battle against global thirst. In 2005, adsorption desalination was patented as a novel desalination technology [20]. This technology is believed to provide lowest specific energy input, and the desalting process is environmentally-friendly, as only adsorbent materials such as silica gel are regenerated with low-temperature waste heat. The advantages and limitations of conventional desalination technologies are compared presented in Table 5.1.

A central challenge of desalination science today is the development of a system that is (1) scalable, (2) free of moving parts, (3) driven by a low-temperature heat source or renewable energy, (3) environmentally benign (low or negligible carbon footprint), (4) able to produce chilled and drinkable water or conjugated with a cogeneration plant, and (5) economically viable. It should be noted here that adsorption desalination fulfills all the abovementioned requirements. In this chapter, we have proposed a novel adsorption desalination system.

5.1.3 Objectives of this Chapter

The objectives of this chapter are (1) to present a theoretical insight into the adsorption desalination (AD) system from the rigors of thermodynamic property surfaces of silica gel–water systems, adsorption isotherms, and kinetics; (2) to show numerical investigation for evaluating AD cycle performance in terms of daily water production rate, and to optimize the operation of both two-bed and four-bed AD configurations; (3) to investigate experimental tests on performance of the AD plant for a wide range of heat sources supplied to the AD cycle; (4) to discuss the concept of an advanced adsorption desalination cycle; and (5) to evaluate the economic analysis of the AD plant at both laboratory and large-scale plants. Finally, the specific water production cost of the AD plant is to be compared with the reported

Table 5.1 Advantages and Limitations of Desalination Techniques

Desalination Type	Usage	Advantages	Limitations
Multistage flash distillation (MSFD): distills seawater by flashing a portion of the water into steam in multiple stages of heat exchangers	Accounts for 85% of all desalinated water; in use since early 1950s	MSFD plants produce large quantities of waste heat, and can then be paired with cogeneration	High operating costs when waste heat is not available for distillation; high rates of corrosion
Multieffect evaporation (MED/MEE)	Using the heat from steam to evaporate water; widely employed since 1845	Highly efficient, relatively inexpensive	A large heating area is required
Vapor compression (VC) evaporation	Used mainly for wastewater recovery	Copes well with high salt content in water	Not environmentally friendly
Evaporation/condensation: freshwater is produced following consecutive condensation of the generated humid air at ambient pressure	Widely used; recommended for seawater or brackish-water	Easiest method of distillation; simple in design, environmentally benign	Time-consuming and inefficient in comparison to other techniques
Electrodialysis reversal (EDR): electrochemical separation process that removes ions and other charged species from water and other fluids	Widely used since early 1960s	Long membrane lifetime and high efficiency ($\leq$94% water recovery, usually ~80%)	High capital and operating costs
Reverse osmosis (RO)	Widely used; first plant installed in 1979	In water purification, effectively removes all types of contaminants to some extent	Requires more pretreatment of the seawater and more maintenance than MSFD plants

Table 5.1 (*Continued*)

Desalination Type	Usage	Advantages	Limitations
Nanofiltration (NF)	Emerging technology	Very high efficiency	High capital cost; unknown lifetime of membrane; no large-scale plant built yet
Forward osmosis (FO)	Emerging technology	Low or no hydraulic pressures; no energy needed for seperation; effect separation of water from dissolved solutes	Cannot produce pure water; only concentrated solutions
Membrane distillation (MD)	Widely used	Low energy consumption; low fouling	Cooling is needed on the permeate side; sensitivity to pressure and surface tension of liquids low rejection of volatile compounds

cost data found in the literature. For understanding adsorption desalination, the next section provides the general idea of physical adsorption and desorption.

5.2 ADSORPTION AND DESORPTION PHENOMENA

Adsorption phenomena have been known to humankind for a very long time, and they are increasingly utilized for separation, purification, gas storage, and cooling applications. The adsorption process occurs in a porous solid medium. The fundamentals of adsorption distinguish the relations between physical adsorption and chemisorption. Physical adsorption is attributed to van der Waals forces and the electrostatic force between adsorbate molecules and the surface atoms. On the other hand, chemisorption involves the formation of a chemical bond between the sorbate molecule and the surface of the adsorbent. The heat of adsorption accompanying adsorption can be higher than the heat of vaporization (condensation) of the adsorbate by as much as 30–100%. When an adsorbent is in contact with the surrounding fluid of a certain composition (adsorbate), adsorption takes place in the Henry region, and after a sufficiently long time, the adsorbent and the adsorbate reach equilibrium [21], as shown in Figure 5.10. The amount of adsorbate

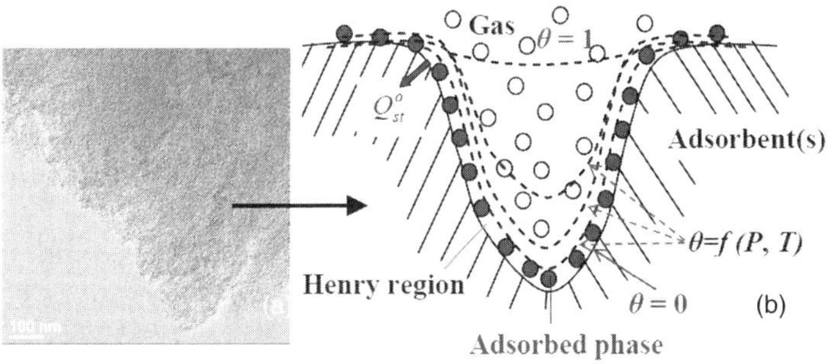

Figure 5.10 (a) Transmission electron micrographs (TEM) photos of activated carbon (type Maxsorb III) at nanoscale magnifications, (b) how adsorbate molecules huddle in adsorbent pores for any pressure P, temperature T, and surface coverage θ.

adsorbed by an adsorbent at a given pressure and temperature under equilibrium is a function of the nature of the adsorbate and adsorbent. The maximum capacity of adsorbent cannot be fully utilized because of mass transfer effects involved in actual fluid–solid contacting processes [22]. Regeneration or reactivation of the adsorbent is needed for recycling so as to produce a useful effect. Basically, the regeneration of adsorbents, which plays a key role in adsorption cooling as well as in the adsorption separation process, has two objectives: (1) to restore the adsorption capacity of exhausted adsorbent and (2) to recover valuable components present in the adsorbed phase.

To be feasible, the regeneration process must be as cost-effective as possible. If a waste heat source is obtainable, an adsorption system might be feasible for use. There are several alternative processes for the regeneration of an adsorbent [23]:

1. Desorption by thermal swing
2. Desorption accomplished by pressure swing
3. Purge gas stripping
4. Displacement desorption

Methods 1 and 2 are commonly used for regeneration of adsorbents used for gaseous phase adsorption. Method 3 is applicable when the adsorbed species are weakly held. Finally, displacement desorption, which avoids thermal aging of adsorbent, is effective for both gas and liquid systems. The choice between the possible modes of regeneration, for any particular system, depends on economic factors as well as technical consideration [23]. Availability of an inexpensive source of steam or waste heat tends to favor thermal swing adsorption over the others. Thermal swing is the most common system because it is effective for the strongly adsorbed species, where a small change in temperature gives rise to a large change in amount adsorbed (c^*).

In the next section we give the brief view of adsorbents that can be used in adsorption–desalination processes. Surface characteristics and pore structures of adsorbents are the main properties in determining adsorption equilibrium and the rate properties that are needed for the adsorption–desalination plant design.

5.3 ADSORBENTS SUITABLE FOR ADSORPTION–DESALINATION

It is well known that the adsorbent of an adsorption process is a critical variable, and that system performance depends on how the solid adsorbents perform in both equilibrium and kinetics. For instance, a material with high sorption capacity but slow kinetics is not a good choice as it takes adsorbate or water molecules a long a time to reach the equilibrium. On the other hand, the adsorbent material with rapid kinetics and low adsorbate capacity is also not a good choice, due to the requirements of more solid adsorbents for the system. A good adsorbent material is one that provides good adsorptive capacity as well as good kinetics. To satisfy these two requirements, the material must have (1) reasonably high surface area or micropore volume and (2) a relatively large pore network for the transport of molecules to the interior of solid adsorbents. To satisfy the first requirement, the porous solid should have small pore size with a reasonable porosity. This suggests that an adsorbent must have a combination of two pore types, ranging from micropores (pore width <2 nm) to mesopores (pore width varying within 2–50 nm according to IUPAC [24]). The efficacy of the adsorption–desalination (AD) cycle depends on the sorption characteristics of adsorbent and adsorbate (hence water vapor) for the production of potable water. For desalination purposes, the adsorbent material should be hydrophilic in nature with lower regeneration temperatures. The suitable adsorbents for desalination are described in the following subsections.

5.3.1 Silica Gels

It is well known that pure silica (SiO_2) is a generally chemically inactive nonpolar material. However, the surface becomes polar and hydrophilic when it contains a hydroxyl functional group (silanol group) [25]. Silica gel is prepared by coagulation of a colloidal solution of silicic acid with controlled dehydration. At higher regeneration temperature, silica gel rejects almost all water vapor and causes hydrophobic phenomena. Because silica gel is both hydrophilic and hydrophobic, it has been considered as the most suitable adsorbent for water-refrigerant-based cooling applications. It should be noted here that the regeneration temperature of silica gel is low compared with other adsorbents, as shown in Table 5.2. The literature shows that silica gel is used in most industries for water removal because of its strong hydrophilicity toward water. Some applications of silica gel are [26]: (1) water removal from air (dehumidification), (2) drying of nonreactive gases, (3) drying of reactive gases, (4) adsorption of hydrogen sulfide, (5) oil vapor adsorption, and (6) adsorption of alcohols.

The properties of silica gel are analyzed using the N_2 adsorption method. The amount of gas adsorbed or desorbed from a porous surface at a predetermined equilibrium (vapor) pressure is measured by the static volumetric method with

Table 5.2 Types of Adsorbent and Their Regeneration Temperatures

Type of Adsorbent	Regeneration Temperature, °C
Silica gel	55–140
Activated alumina	120–260
Zeolite molecular sieve	175–370

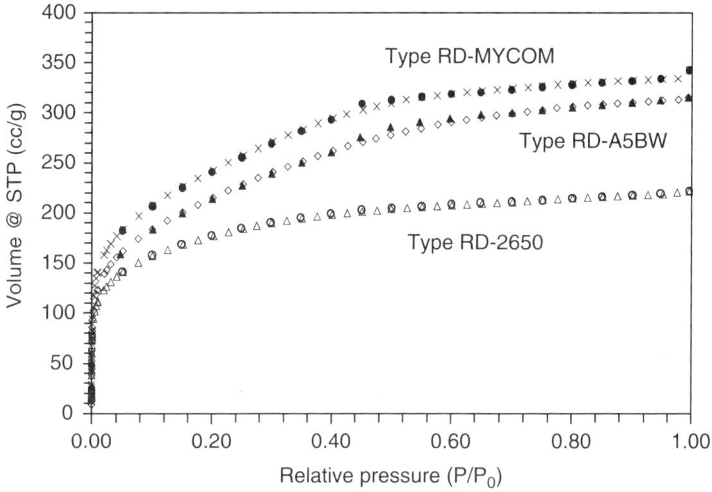

Figure 5.11 Nitrogen (N_2) uptake of different types of silica.

liquid nitrogen at 77 K as the filling fluid. The adsorption and desorption of N_2 by different types of silica gels are shown in Figure 5.11, where it is observed that MYCOM-manufactured silica gel has the highest uptake.

The surface area of the adsorbent (silica gel), as determined by using the Brunauer–Emmett–Teller (BET) method and the BET equation, is expressed as

$$\frac{1}{W\left(\dfrac{P_0}{P}-1\right)} = \frac{1}{W_m C} + \frac{C-1}{W_m C}\left(\frac{P}{P_0}\right)$$

where W is the weight of the gas adsorbed at a relative pressure P/P_0, W_m is the weight of adsorbate at a monolayer coverage, and C is the BET constant. This is related to the adsorption energy of the first adsorbed layer, indicating the magnitude of the adsorbent–adsorbate interactions. Three types of silica gel from different manufacturers were investigated: type RD-MYCOM, type RD-2650, and type A5BW. Figure 5.12 gives the BET plot for these silica gels. The

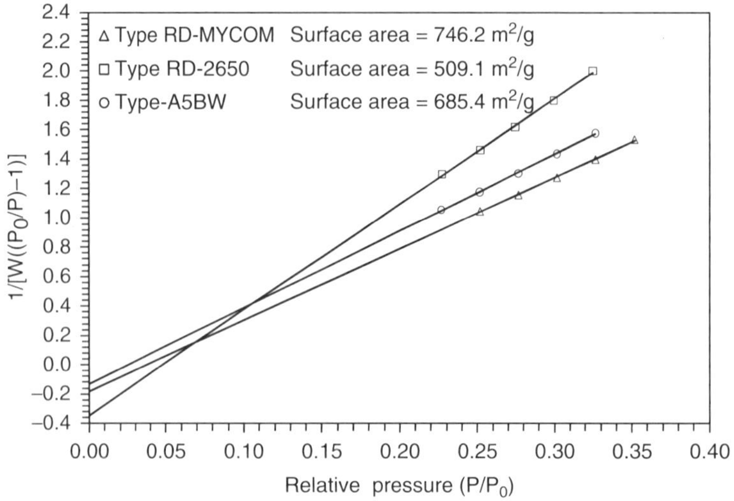

Figure 5.12 BET plot of different types of silica gel.

Table 5.3 Summary of BET Analysis

Parameter	Type RD-MYCOM	Type RD-2650	Type A5-BW
Slope	4.849	7.186	5.215
BET constant, C	−25.667	−19.81	−37.953
Y-intercept, i	−0.1818	−0.3453	−0.1339
Correlation coefficient, r	0.999373	0.999248	0.999622
Surface area, m^2/g	746.17	509.07	685.38

BET results show that type RD-MYCOM silica gel has the highest surface area, at ~746.2 m^2/g [27]; the surface area analysis BET method is summarized in Table 5.3.

Pore size distribution analysis of these silica gels is conducted using the density functional theory (DFT) method with the application of the provided software package of the AutoSorb (see Fig. 5.13). The results showed that all three parent silica gels have a two-maximum distribution or bimodal type, with pore diameters ranging between 12 and 17 Å.

In an adsorption–desalination cycle, the adsorbate is water vapor, which is an important factor for the selection of silica gel. Such an analysis can be performed using volumetric–gravimetric analysis. Figure 5.14 compares the water vapor uptake of silica gel at 25°C, which keeps the adsorbent at isothermal condition. The results indicated that MYCOM-type silica gel possesses the highest equilibrium uptake, at 537 cm^3/g. The inset diagram in Figure 5.14 shows the adsorption and desorption of the water vapor by MYCOM-type silica gel.

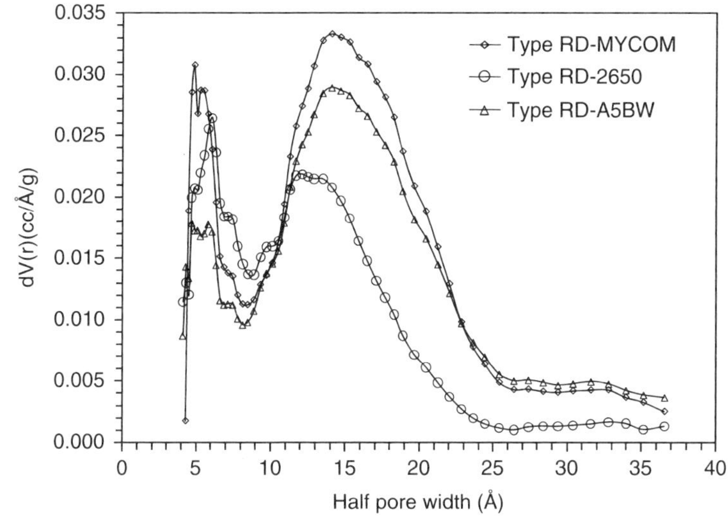

Figure 5.13 Pore size distribution of three different parent silica gels.

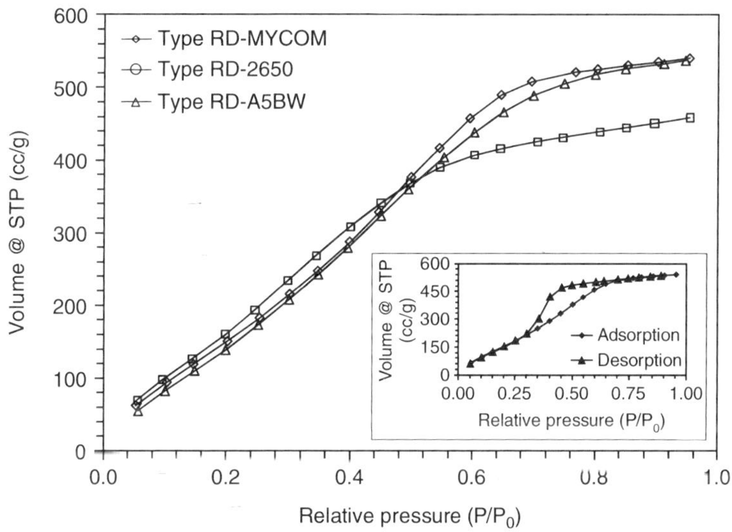

Figure 5.14 Different types of water vapor uptake.

5.3.2 Zeolites

Zeolites are porous crystalline aluminosilicates. The zeolite framework consists of an assemblage of SiO_4 and AlO_4 tetrahedra joined together in various regular arrangements to form an open crystal lattice containing pores through which the guest molecules can penetrate during adsorption [28]. Zeolites can be found

Table 5.4 Typical Characteristics of Zeolite 5A

Property	Value
Density	1.57 g/cm^3
Particle density	1.1 g/cm^3
Macropore porosity	0.31
Macropore volume	0.28 cm^3/g
Micropore volume	0.31 cm^3/g
Exterior surface area	1–20 m^2/g
Mean macropore radius	30–1000 nm
Mean micropore radius	0.5 nm

Source: Simonot-George et al. [29].

naturally or formed synthetically. The applications of natural zeolite are not as widespread as those of synthetic zeolite because of the higher specificity of synthetic zeolite. There are many types of synthetic zeolite, including types A, X, and Y; mordenite; and ZSM. The typical characteristics of type 5A zeolite are listed in Table 5.4 [29].

From Table 5.4, we observe that zeolite is a good candidate for adsorption and desorption, due to high micro- and macropore volume and the sizes of micro- and mesopores, which are suitable for cooling and desalination applications. However, the desorption temperature is very high (generally 120–200°C). It would be better to use zeolite–water and silica gel–water systems together for higher performance and full utilization of waste heat if higher-temperature sources are available.

5.3.3 $CaCl_2$-in-Silica Gel

$CaCl_2$-in-silica gel is a new generation of composite materials named *selective water sorbents* (SWSs) [30–33]. An SWS material comprises a porous host matrix and a hygroscopic substance impregnated into its pores. This adsorbent material has good adsorption properties and is considered to be very promising for cooling and desalination applications. Here the properties of $CaCl_2$-in-silica gel are based on the mesoporous commercial product KSKG, with silica gel as a host matrix and calcium chloride as a hygroscopic salt. Calcium chloride has the ability to absorb water. The porous silica gel contains a BET specific surface area of 350 m^2/g, pore volume 1.0 cm^3/g, and average pore diameter of 15 nm. The pore volume–pore size distribution is shown in Figure 5.15. The local maximum of $CaCl_2$-in-silica gel is observed at 15 nm. This indicates that $CaCl_2$-in-silica gel is mainly mesoporous The confined calcium chloride in the porous host silics gel matrix is prepared by filling pores of the silica gel with a 40 wt% aqueous solution of $CaCl_2$. Then the samples were dried at 200°C until their weight remain constant. The calcium chloride content in the anhydrous samples is measured at 33.7 wt%.

The amount of water vapor uptake was measured experimentally for temperatures ranging from 303 to 358 K and pressures varying up to 10 kPa by a thermogravimetric analyzer (TGA). The adsorption isotherms results are provided in

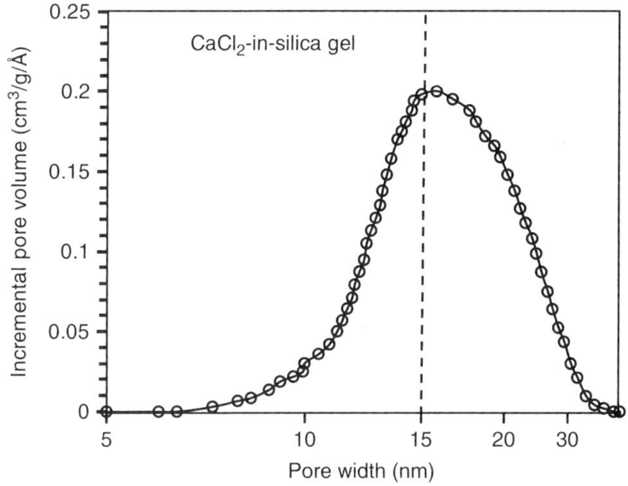

Figure 5.15 Pore size distribution of $CaCl_2$-in-silica gel.

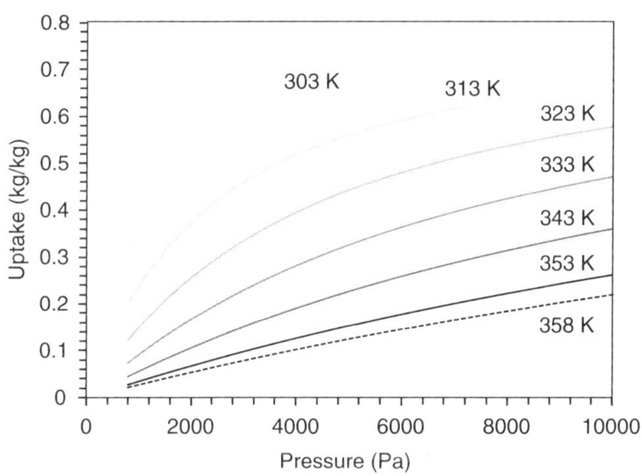

Figure 5.16 Adsorption isotherms of $CaCl_2$-in-silica gel water system.

Figure 5.16. From the isotherm graph, it is observed that the amount of water vapor uptake is very high and $CaCl_2$-in-silica gel is suitable for adsorption–desalination.

5.3.4 Metallorganic Frameworks

Metallorganic frameworks (MOFs) are crystalline compounds comprising metal ions or clusters coordinated to rigid organic molecules to form one-, two-, or three-dimensional structures that can be porous. The stable pores capture the guest molecules and can be used for storage of gases such as hydrogen, methane, and carbon dioxide. Other possible applications of MOFs are found in gas purification, gas separation, and catalysis. In more recent years, there has been significant research on the development of micro- and mesoporous adsorbent materials for closed-system heat transformation applications, such as thermally driven adsorption chillers. For adsorption–desalination application, the adsorbent material should be hydrophilic; it behaves hydrophobically at relatively high temperatures (e.g., 80°C), and the choice of the adsorbent–water working pair is determined by the amount of heat that can be extracted from the evaporator and is rejected from the condenser per adsorption cycle [34]. The heat extracted as a result of cooling load from the evaporator is proportional to (1) the amount of water vapor adsorbed and (2) the evaporation enthalpy of water vapor. Therefore, the loading spread, or the difference between the water vapor uptakes and offtakes over a given adsorption–desorption cycle, provides the best criterion for selection of a suitable adsorbent. The present adsorption chillers and desalination plants use silica gels mostly as adsorbents. The metal metallorganic framework (MOF) is a good adsorbent for heat transformations applications. The amount of water vapor uptake is quite high as compared with other adsorbents such as zeolites and silica gels [35]. It should be noted here that the metallorganic framework is less hydrophilic than silica gel or zeolites; thus, it releases water at lower temperature and lower partial pressure, which also indicates the potentiality of MOF materials for heat transformation processes or desalination. The dehydratable–hydratable water-stable MOF material with a large loading spread of 0.25 kg/kg will be the best candidate in heat transformation cycles for refrigeration, heat pumping, and heat storage.

5.4 FUNDAMENTAL STUDY OF ADSORPTION–DESALINATION

The physical adsorption process occurs mainly within the pores of adsorbent and at the external adsorbent surface, and is determined by its adsorption isotherms, heats of adsorption, and adsorption kinetics. In a physicsorption, the adsorbed phase is held near the pores of the adsorbent by the existence of van der Waals forces. Development of the adsorption–desalination (AD) system is based on the thermodynamic property surfaces of the silica gel–water system and the thermal compression of water, and lies in its ability to operate with motive energy derived from fairly low-temperature waste heat sources or solar energy, which indicates the adsorption process as an avenue for avoiding the use of ozone depleting substances [36–40]. The major components of the typical AD system are the evaporator, the condenser, and two or more adsorbent beds in which the adsorbents are packed in tube-fin heat exchangers. Operation of the AD cycle is the union of adsorption-triggered evaporation and the desorption-resulting condensation. This means that the pre-treated seawater is evaporated and adsorbed onto silica gel at low pressure and temperature. When the vapors are adsorbed on the adsorbent, the heat of adsorption is developed, and this heat is rejected into the cooling-water circuit running across the adsorbent bed. This adsorption process continues until onset of the switching time. For the regeneration of saturated silica gels, low-temperature waste heat is supplied to the bed at the onset of the desorption mode. The sorption bed is connected to the condenser, where the water vapor migrates to the cooler tube surfaces of the condenser and the potable water is collected in the collection tank.

5.4.1 Description of Adsorption–Desalination (AD) Cycle

The adsorption–desalination (AD) cycle is a novel desalination method for producing potable water from seawater or brackish-water, utilizing only low-temperature hot water from renewable energy such as process exhaust and solar energy. The AD cycle is based on the patented cycle by Ng et al. [20] that produces both desalted water and cooling by using low-temperature waste heat. A typical AD plant consists of three major components: (1) the evaporator, (2) single or multireactor beds where the adsorbent is placed, and (3) the condenser. A schematic diagram of the AD cycle is given in Figure 5.17 showing the major components. The subsystems involved in AD cycle are

1. *Pretreatment System*. Seawater is first pretreated where unwanted physical particles such as debris in the seawater are screened using a conventional filter (mesh size 20–30), and the dissolved air in the seawater is removed by deaeration.
2. *Adsorption-Initiated Evaporation System*. The deaerated seawater is pumped intermittently into the evaporator, where evaporation is achieved by water vapor uptake by the adsorbent (type RD silica gel). The evaporation process is enhanced by a spray system using full-cone type nozzles. The evaporation process is maintained by an external water circuit that provides the heating capacity to sustain the evaporation. The energy required for the evaporation is

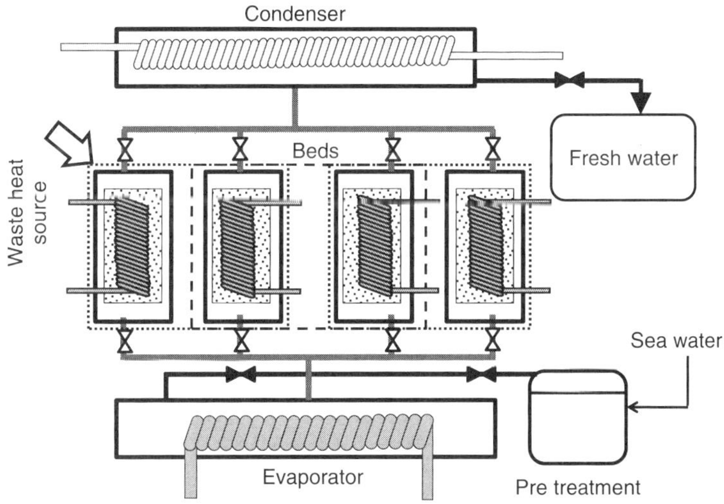

Figure 5.17 Schematic diagram of the adsorption–desalination system.

either supplied from the ambient or recovered internally within the condenser or adsorption process. The concentrated brine is occasionally discharged from the evaporator. Water vapor from the evaporator is adsorbed by the unsaturated silica gel, which is packed into a tube-fin heat exchanger housed inside a reactor bed. When the vapor valve is opened, a continuous vapor uptake is achieved. Owing to the exothermic nature of the adsorption process, heat is released during vapor uptake, and the adsorption process continues at a preset cycle time. In general, process cycle time is one of the input parameters for controlling the AD process, and its time interval of operation is a direct function of hot-water inlet temperatures.

3. *Heat-Activated Desorption–Condensation System*. The saturated silica gel (from adsorption of previous half-cycle) can be regenerated by introducing low-temperature hot water (typically $<85^\circ$C); this procedure is known as the *desorption process*. With the connecting vapor valves set to open position, the regenerated water vapor flows to the external surfaces of the water-cooled tubes and condenses to yield the distilled water. The condensation process is accompanied by rejection of the latent heat of condensation into the cooling-water passing through the condenser, and the condensate is collected as potable water.
4. *Potable-Water Collection System*. The condensate inside the condenser is removed using by a freshwater pump or a 10-m U-tube arrangement to overcome the differential pressure.

5.4.2 Theoretical Insight into AD

The AD method offers merits of scale independence since adsorption and desorption are primarily surface, rather than bulk, processes. The performance of AD depends on (1) the adsorption isotherms and kinetics of silica gel–water pair and (2) structural features such as pore width d, pore volume v_p, and specific surface area of silica gels. The heat of adsorption is of secondary importance since it is simply rejected to the environment. The thermodynamic behavior of the AD cycle consists of two isosters and two isobars, characterized by three temperatures: evaporator (T_{evap}), condenser (T_{cond}), and the driving heat source (T_{des}) temperatures. In this section, we have presented a thermodynamic framework, which has been derived from the rigor of classical thermodynamics, the Boltzmann energy distribution function, the condensation approximation approach, and the Lennard–Jones potential model for calculation of the energetic performances of AD in terms of specific water production and performance ratio for various pore sizes and volumes of silica gels. Employing the proposed formulations, we report a simulation study here for calculation of the maximum isosteric heats and characteristics energy for the adsorption of water vapor on silica gels as a function of pore width.

5.4.3 Thermodynamics of an AD Device

The energy flow for each sorption component of the AD system is derived and developed from the rigor of thermodynamic property surfaces of the silica gel–water system [41–43]. The enthalpy of adsorbed phase (h_a) is expressed by

$$h_a = h_g(P,T) + \int_{T_0}^{T} \left[Q_{st} \left\{ \frac{1}{T} - \frac{1}{v_g}\left(\frac{\partial v_g}{\partial T}\right)_p \right\} - \left(\frac{\partial Q_{st}}{\partial T}\right)_c \right] dT$$

$$+ \int_{T_0}^{T} (c_{p,s} + c_{p,g})\, dT + \int_{0}^{c} (h_g - Q_{st})\, dc \quad (5.1)$$

where T_0 indicates the reference temperature and c_0 defines the uptake of water vapor at initial state, c_p is the specific-heat capacity, v refers to the specific volume, and Q_{st} is the isosteric heat of adsorption. Here the subscripts "s" and "g" indicate the adsorbent and the gaseous phase, respectively. Employing Equation (5.1), the thermodynamic property surfaces of silica gel–water system [44] are plotted in a pressure–enthalpy–uptake (P–h–c) coordinate system as represented in Figure 5.18.

From Figure 5.18, one can easily calculate the energetic performances of adsorption cycle in terms of cooling capacity and water production. Due to the cooling load (Q_{evap}), the enthalpy of evaporation $h_{fg}(= h_g - h_f)$ evolves at the evaporator, and the evaporated water vapor is adsorbed in silica gel surfaces of the sorption bed (H–A). During regeneration phase (lines A–B–C), the pressure in the adsorber rises from evaporator pressure (P_{evap}) to condenser pressure (P_{cond}) and the desorption of water vapor from silica gels occurs when the adsorption bed is connected to the condenser. The amount of uptake falls from c_{ads} to c_{des}, and the enthalpy of adsorbed phase changes from h_A to h_C. At the condenser, the water vapor is condensed and heat is released to the environment. The amount of condensed water

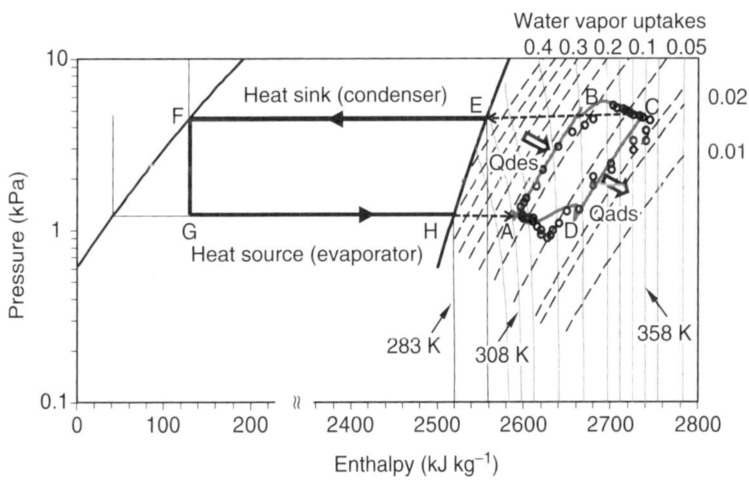

Figure 5.18 Pressure–enthalpy (P–h) diagram of AD cycle.

is calculated by $M_{sg}(c_{ads} - c_{des})$, where M_{sg} is the mass of silica gel. During that phase, the enthalpy changes from h_C to h_F. Finally the condensed water goes to the freshwater tank and completes the cycle. During adsorption phase (lines C–D–A), the sorption bed is cooled and the pressure falls from P_{cond} to P_{evap}. The refrigerant vapor is adsorbed on the adsorbents and the amount of water vapor increases up to c_{ads}. The overall performance ratio (OPR) is given by $(Q_{evap} + Q_{cond})/Q_{des}$, where Q_{des} is the amount of heats during regeneration periods and Q_{cond} is the heat removal for condensation. From Figure 5.18, we find that during the preheating stage ($A \rightarrow B$), the sorption bed is isolated from the evaporator and the condenser, and the pressure is increased from evaporator pressure to a pressure depending on c_{ads} and regeneration temperature. During desorption phase (line A–B–C of Fig. 5.18), the quantity of heat applied to the sorption bed is given by

$$Q_{des} = \int_{T_{ads}}^{T_{reg}} \left[c_{p,s} + c_{ads} \left\{ c_{p,l} + Q_{st} \left(\frac{1}{T} - \frac{1}{v_g} \frac{dv_g}{dT} \right) \right\} \right] dT$$

$$+ \int_{c_{des}}^{c_{ads}} \{ h_a(P, T, c) - h_g(T) \} \, dc + \int_{P_{evap}}^{P_{cond}} Q_{st} \left. \frac{\partial c}{\partial P} \right|_T dP$$

$$+ \int_{T_{reg}}^{T_{des}} \left[c_{p,s} + c_{ads} \left\{ c_{p,l} + Q_{st} \left(\frac{1}{T} - \frac{1}{v_g} \frac{dv_g}{dT} \right) \right\} + Q_{st} \left. \frac{\partial c}{\partial T} \right|_P \right] dT \qquad (5.2)$$

The cooling capacity (Q_{evap}) and the amount of condensation heat are calculated as

$$Q_{evap} = \int_{c_{des}}^{c_{ads}} h_{fg}(T_{evap}) \, dc - \int_{c_{des}}^{c_{ads}} \{ h_g(P, T) - h_g(T) \} \, dc \quad \text{and}$$

$$Q_{cond} = \int_{c_{des}}^{c_{ads}} h_{fg}(T_{cond}) \, dc + \int_{c_{des}}^{c_{ads}} \{ h_g(P, T) - h_g(T) \} \, dc$$

respectively.

5.4.4 Effects of Adsorbent Pore Size on AD

The experimentally measured adsorption isotherm data of silica gel–water systems are fitted well with Dubinin–Astakhov (DA) adsorption isotherm equation. Parameters of the D–A model are presented in Table 5.5. The D–A isotherm equation is given by

$$c = c^{\circ} \exp \left[\left\{ \frac{RT \ln(P_s/P)}{E} \right\}^n \right]$$

Table 5.5 Adsorption Characteristics of Silica Gel–Water Systems

Type of Silica Gel	Characteristics Energy E, kJ/mol	n	Pore Volume, cm^3/g	Pore Width, nm
RD	4.2	1.3	0.4	0.8–2
A	3.88	1.3	0.36	0.8–2
Grace 125	3.9	1.3	0.4	—
$CaCl_2$-in-silica gel	5.2	1.5	0.8	—

where R defines the gas constant, c° is the limiting uptake, and P_s is the saturation pressure. The heterogeneity coefficient n lies in the range 1–3 depending on the types of adsorbents; E represents the characteristics energy of adsorbent–adsorbate system and is linked to pore width.

Stoeckli and Morel [45] calculated E for a wide range of adsorbates and a variety of activated carbons, given by the formula $Q_{st}^* - Q_{st} = (0.85 \pm 0.07)E$, where Q_{st}^* is the isosteric heat of adsorption at zero surface coverage. When any adsorbate comes into contact with an adsorbent surface, the number of adsorbent surface atoms (N_s) with binding energy (Q_s) is related to the number of adsorbate surface atoms (ranging from Henry region, N^*, to condensation phase, N°) with energy varying from Q_{st}^* to Q°, where Q° is the least adsorption energy [43]. Using the Boltzmann distribution law, we can write $N_s/N^* = \exp\{-(Q_s - Q_{st}^*)/N_A kT\}$ for the Henry region and $N_s/N^\circ = \exp\{-(Q_s - Q^\circ)/N_A kT\}$ for the condensed phase, where k is the Boltzmann constant and N_A is the Avogadro number. By employing these two equations, we could obtain $Q_{st}^* - Q^\circ = N_A kT \ln(N^\circ/N^*) \approx E$. This means that E approaches $(Q_{st}^* - Q^\circ)$ [46]. The isosteric heat of adsorption at zero surface coverage is $Q_{st}^* = RT - N_A V$, where V defines the external wall potential as a function of pore width [47]. In the present analysis, we use the D–A model to establish the nature of water vapor on micro-, meso-, and macropores as a function of pore width and pore volume; we can then express the D–A equation as

$$c = v_p \rho_l \exp\left[-\left\{\frac{\ln(P_s/P)}{1 - (N_A V/RT) - (Q^\circ/RT)}\right\}^n\right]$$

with $V = \int_0^\infty w(y)\exp\{-w(y)/kT\}dy / \int_0^\infty \exp\{-w(y)/kT\}dy$, where v_p is the pore volume, ρ_l defines the density of water at liquid phase, y indicates the direction of pore width, and w represents the external wall potential. The adsorption characteristics energy is given by

$$E = RT - N_A \frac{\int_0^\infty w(y)\exp\{-w(y)/kT\}\,dy}{\int_0^\infty \exp\{-w(y)/kT\}\,dy} - Q^\circ \tag{5.3}$$

The adsorbent wall potential is described by $w(y) = U_{sf} + U_{sf}(H - y)$, where U_{sf} defines the adsorbent–adsorbate interaction potential; H is the distance between the

Table 5.6 Parameters for Silica Gel-Water System

Parameters	Values
Adsorbent–adsorbate collision diameter (σ_{sf})	3.43 Å
Well depth potential (ε_{sf})	8.1189×10^{-21} J/mol
Separation between adsorbent planes (Δ)	3.35 Å
Density of solid adsorbent (ρ_s)	0.055

Source: Chakraborty et al. [46].

nuclei of the outer adsorbent atoms on opposite walls. The adsorption interaction potential of water vapor along the pore width direction y of silica gel pore is written as [48,49]

$$U_{sf} = 2\pi \varepsilon_{sf}\rho_s\sigma_{sf}^2\Delta \left[\frac{5}{2}\left(\frac{\sigma_{sf}}{y}\right)^{10} - \left(\frac{\sigma_{sf}}{y}\right)^{4} - \frac{\sigma_{sf}^4}{3\Delta(0.61\Delta + y)^3}\right] \tag{5.4}$$

where σ_{sf} and ε_{sf} are the solid–fluid collision diameter and the solid–fluid well depth potential, respectively. The delta Δ defines the separation between the adsorbent planes, and ρ_s is the density of solid adsorbent (silica gel). These values are tabulated in Table 5.6. Employing Equation (5.2), we can determine the characteristics energy (E) with respect to the physical characteristics of the adsorbent (v_p and pore width). The amount of water vapor uptake (c_{ads}) and offtake (c_{des}) during adsorption and desorption phases are given by

$$c_{ads} = v_p\rho_l \exp\left[-\left\{\frac{\ln(P_s(T_{ads})/P_{evap})}{1 - (N_A V_{ads}/RT_{ads}) - (Q^\circ/RT_{ads})}\right\}^n\right] \tag{5.5}$$

$$c_{des} = v_p\rho_l \exp\left[-\left\{\frac{\ln(P_s(T_{des})/P_{cond})}{1 - (N_A V_{des}/RT_{des}) - (Q^\circ/RT_{des})}\right\}^n\right] \tag{5.6}$$

5.4.5 Results and Discussion

Employing the proposed formulations, we calculate and plot E and Q_{st}^* as a function of pore width for various types of silica gel–water systems, as shown in Figure 5.19. The parameters used to describe the proposed model are given in Tables 5.5 and 5.6, respectively. From Figure 5.19, we observe that a rapid change in E and Q_{st}^* occurs for pore sizes <3 Å. For larger pores, there is a rapid decrease in Q_{st}^* and E. The relationship between the E and pore size was correlated on the basis of experimental data resulting from X-ray scattering in micropores of carbonaceous materials. These results showed that the potential energy in the micropore decreases when the pore width increases, which indicates the validity of the proposed formulation as shown in Equation (5.3).

The experimentally measured characteristics energy data for type A, RD silica gels are also shown in Figure 5.19, where the simulated results match the experimental data. The effects of pore width on the specific water production (SWP; kg

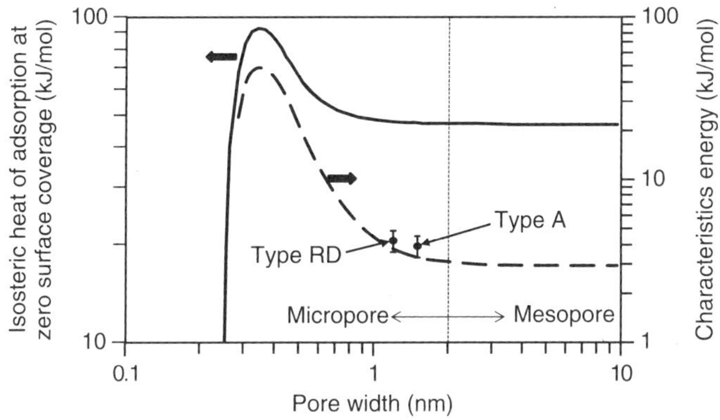

Figure 5.19 The isosteric heat of adsorption at zero surface coverage and the characteristics energy as a function of pore width.

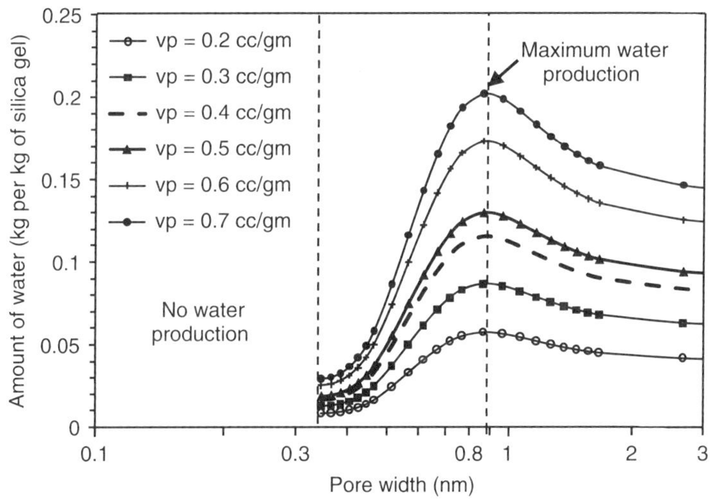

Figure 5.20 Effects of pore width on water production at evaporator temperature 280 K.

water/kg silica gel) for various pore volumes are shown in Figure 5.20. We find that the maximum water is obtained for pore width 0.9 nm; this is valid for all types of silica gel.

Specific water production (SWP) for various heat source temperatures is plotted in Figure 5.21. For evaporator temperatures ranging from 274 to 288 K, the maximum SWP is obtained for pore width 0.9 nm. These variations are due to movement of the isosters as a function of characteristics energy (E). The SWP increases with higher evaporator temperatures. At heat source temperatures of 288–303 K, the production of water is not influenced by the micro- and mesopores of silica gels. The overall performance ratio of the adsorption cycle, defined as the ratio of net useful effects to the useful heat input, is shown in Figure 5.22.

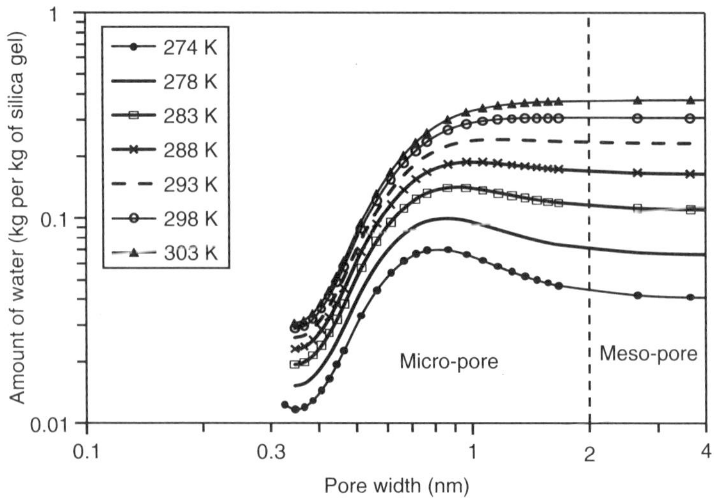

Figure 5.21 Plot of SWP against pore width for various evaporator temperatures; here the regeneration temperature is 353 K.

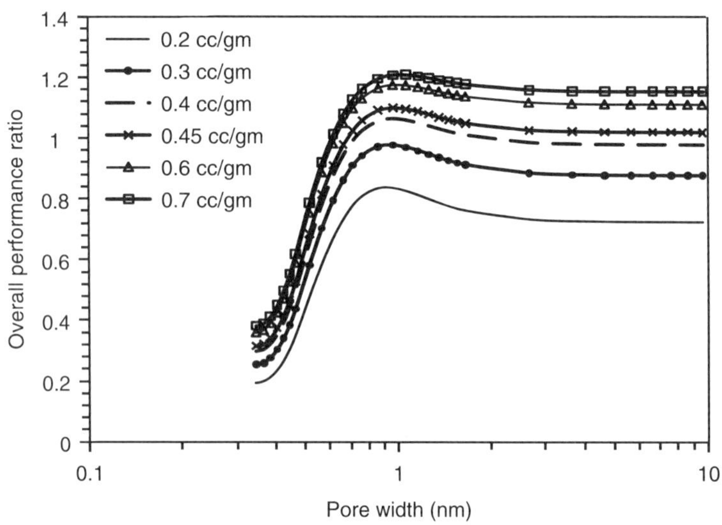

Figure 5.22 Overall performance of the AD cycle.

As there are two useful effects—cooling-water and desalinated water—high performance is expected. It could have a value of 1.1 at evaporator temperature 280 K for the pore width and volume of 0.9 nm and 0.45 cm^3/g.

5.4.6 Summary

From the present analysis, we notice that silica gels optimized with respect to pore size and volume for a given evaporating temperature can entail an increase in overall performances in the order of 20–40%, which cannot be neglected. If the evaporator temperature varies, especially when it decreases, there is a decreasing tendency of specific water production. Another conclusion of this study is that an increase in pore volume does not entail systematically an improvement of performance. The best adsorption–desalination performance is obtained from a compromise between the greatest pore volume and the optimal pore size. From the present analysis, it is observed that there is water production for pore widths varying from 0 to 0.3 nm. The best performance of silica gel–water–based AD is obtained at pore width 0.9 nm with higher pore volume. The characteristics energy (E) is found to be higher in micropore ranges. This indicates that the adsorption and desorption rates are higher in micropores. Adsorbent materials such as silica gel should be designed with micropores ranging from 0.8 to 1 nm and higher micropore volume for AD applications.

5.5 ADSORPTION–DESALINATION SYSTEM MODELING

The AD system utilizes the silica gel–water characteristics and produces drinkable water at the condenser by the amalgamation of *adsorption-triggered evaporation* and *desorption-resulting condensation*. These phenomena are expressed mathematically using the mass and energy balances between major components of the adsorption chiller system [50].

5.5.1 Evaporator

For continuous water production operation, initially, sea- or brackish-water is evaporated in the evaporator. Seawater is fed into the evaporator, and potable water is extracted from the condenser of the AD cycle, and the overall mass balance of the cycle is thus given by

$$\frac{dM_{\mathrm{s,evap}}}{dt} = \dot{m}_{\mathrm{s,in}} - \dot{m}_{\mathrm{d,cond}} - \dot{m}_{\mathrm{b}} \tag{5.7}$$

where $M_{\mathrm{s,evap}}$ is the amount of seawater in the evaporator, $\dot{m}_{\mathrm{s,in}}$ is the rate of feed seawater, $\dot{m}_{\mathrm{d,cond}}$ is the mass of potable water extracted from the condenser, and $\dot{m}_{\mathrm{b}}$ is the mass of concentrated brine rejected from the evaporator. The feed seawater is intermittently pumped into the evaporator depending on the amount and level of seawater, while brine is discharged once the concentration in the evaporator reaches the predetermined limit. The mass–salt balance for the evaporator of the

AD cycle is given as

$$\frac{dM_{\text{s,evap}}}{dt} = \overbrace{\theta\, \dot{m}_{\text{s,in}}}^{\text{feed}} - \overbrace{\gamma\, \dot{m}_{\text{brine}}}^{\text{brine discharge}} - \overbrace{n \cdot \frac{dc_{\text{ads}}}{dt} M_{\text{sg}}}^{\text{vapor uptake by adsorption processes}} \tag{5.8}$$

$$M_{\text{s,evap}} \frac{dX_{\text{s,evap}}}{dt} = \theta\, X_{\text{s,in}}\, \dot{m}_{\text{s,in}} - \gamma\, X_{\text{s,evap}}\, \dot{m}_{\text{brine}} - nX_{\text{D}} \frac{dc_{\text{ads}}}{dt} M_{\text{sg}} \tag{5.9}$$

where $X_{\text{s,in}}$ and $X_{\text{s,evap}}$ are concentrations of feedwater and seawater in the evaporator, respectively and, X_{D} is concentration of the vapor.

Seawater evaporation is attributed to uptake of the water vapor by the adsorbent, and external cooling load is used to maintain the evaporation process at a certain temperature level. The energy balance of the evaporator becomes

$$\begin{aligned}
&[M_{\text{s,evap}} c_{p,\text{s}}(T_{\text{evap}}, X_{\text{s,evap}}) + M_{\text{HX,evap}}\, c_{p,\text{HX}}] \frac{dT_{\text{evap}}}{dt} = \theta \times h_{\text{f}}(T_{\text{evap}}, X_{\text{s,evap}})\, \dot{m}_{\text{s,in}} \\
&\quad - n \times h_{\text{fg}}(T_{\text{evap}}) \frac{dc_{\text{ads}}}{dt} M_{\text{sg}} + \dot{m}_{\text{chilled}} c_p(T_{\text{evap}})(T_{\text{chilled,in}} - T_{\text{chilled,out}}) \\
&\quad - \gamma \times h_{\text{f}}(T_{\text{evap}}, X_{\text{s,evap}})\, \dot{m}_{\text{brine}}
\end{aligned} \tag{5.10}$$

where $M_{\text{HX,evap}}$ is the total mass of the evaporator, M_{sg} is the mass of silica gel, $\dot{m}_{\text{brine}}$ is the mass flow rate of brine discharge, and c_{ads} denotes water vapor uptake by the adsorption processes. The first term in the right-hand side (RHS) of Equation (5.10) represents the sensible heat of the supplied feed seawater; the second term, the heat removal by uptake of adsorbent; the third and fourth terms, energy supplied by the chilled water and the sensible-heat removal by brine discharge. The specific heat ($c_{p,\text{s}}$) and enthalpy (h_{f}) of seawater are calculated as functions of temperature, pressure, and salinity. The vapor pressure depression due to the concentrated saltwater is factored in with respect to the boiling-point elevation (BPE), whereas the vapor pressure depression affects the uptake of water vapor by the adsorbent that varies with temperature and pressure.

5.5.2 Adsorption Beds

The adsorption process results in the liberation of heat of adsorption at the adsorber, providing a useful heat energy output and a cooling effect in the condenser–evaporator heat exchanger. Also, the adsorbed bed is heated by an external heat source and the adsorbed water vapor is desorbed from the adsorbent and then condensed in the water-cooled condenser through the cooling tower. As the evaporated refrigerant is associated onto the solid adsorbent by the flow of cooling fluid at ambient conditions during adsorption period, and the desorbed water vapor is dissociated from the solid adsorbent by the flow of heating fluid during desorption period, heat energy is exchanged between cooling/heating fluid and the adsorption bed. The energy balance equation of the adsorber bed connected with the evaporator as well as that for the desorber bed communicating with the condenser are given by

$$[M_{\text{sg}}c_{p,\text{sg}} + M_{HX}c_{p,HX} + M_{\text{abe}}c_{p,\text{a}}]\frac{dT_{\text{ads/des}}}{dt} = \pm nQ_{\text{st}}\, M_{\text{sg}}\frac{dc_{\text{ads/des}}}{dt} \pm \dot{m}_{\text{cw/hw}}\; c_p\; (T_{\text{cw/hw,in}} - T''_{\text{cw/hw,out}}) \tag{5.11}$$

where $T''_{\text{cw/hw,out}}$ is the outlet water temperature of the master adsorber–desorber bed and Q_{st} and $c_{p,\text{a}}$ are the isosteric heat of adsorption and the specific heat of the adsorbed phase, respectively [51]. Using the heat transfer equation, the outlet temperature of the water from each heat exchanger is estimated using log mean temperature difference method, expressed as

$$T_{\text{out}} = T_0 + (T_{\text{in}} - T_0)\ \exp\left(\frac{-UA}{\dot{m}c_{\text{p}}(T_0)}\right), \tag{5.12}$$

where $T_0 (= T_{\text{ads}}/T_{\text{des}})$ is the temperature of the heat exchanger.

The adsorption–desorption rate is calculated from knowledge of adsorption equilibrium and kinetics and is given by the conventional linear driving force (LDF) equation [49]

$$\frac{dc}{dt} = \frac{15D_{s0}e^{-E_{\text{a}}/RT}}{R_{\text{p}}^2}\{c^* - c\} \tag{5.13}$$

where D_{s0} defines a pre-exponential factor of the efficient water diffusivity in the adsorbent, E_{a} represents the activation energy, R is the universal gas constant, and R_{p} is the average radius of the adsorbent grains. Kinetic data were taken from the literature [52,53]. Hence the adsorption uptake at equilibrium condition is expressed as a function of pressure (P) and temperature (T). The authors have measured the isotherms of water adsorption on silica gel Fuji–Davison type RD, type A [54,55], and SWS-1L [56,57]. These experimentally measured data are fitted using the Dubinin–Astakhov equation [58]

$$c^* = c^\circ \exp\left\{-\left(\frac{RT}{E}\ln\frac{P_{\text{s}}}{P}\right)^n\right\} \tag{5.14}$$

where c^* is the adsorbed adsorbate at equilibrium conditions, and c° denotes the limiting amount of adsorbate uptake.

5.5.3 Condenser

After desorption, the desorbed refrigerant is delivered to the condenser as latent heat, and this amount of heat is pumped to the environment by the flow of external cooling fluid. In the modeling, we assume that the condenser tube bank surface is able to hold a certain maximum amount of condensate. Beyond this the condensate would flow into the water collection tank via a U-tube. The energy balance of the condenser that is in communication with the desorber bed is given by

$$[M_{\text{cond}}c_p(T_{\text{cond}}) + M_{HX,\text{cond}}\ c_{p,\text{HX}}]\frac{dT_{\text{cond}}}{dt} = -h_{\text{f}}\frac{dM_{\text{d}}}{dt} + nh_{\text{fg}}(T_{\text{cond}})\frac{dc_{\text{des}}}{dt}\ M_{\text{sg}} + \dot{m}_{\text{cond}}\ c_p(T_{\text{cond}})(T_{\text{cond,in}} - T_{\text{cond,out}}), \tag{5.15}$$

where $M_{\text{HX,cond}}$ is the total mass of the condenser and M_{d} is the mass of distillate extracted from the condenser.

5.5.4 Energy Balance of AD System

The energy required to remove water vapor from the silica gels [here adsorption or desorption (Q_{des})] can be calculated using the inlet and outlet temperatures supplied to the reactors:

$$Q_{\text{ads/des}} = \dot{m}_{\text{cw/hw}}c_p(T_{\text{Mads/Mdes}})\ (T_{\text{cw/hw,in}} - T''_{\text{cw/hw,out}}) \tag{5.16}$$

The heat of evaporation (Q_{evap}) and the condensation energy (Q_{cond}) rejected at the condenser are given by

$$Q_{\text{evap}} = \dot{m}_{\text{chilled}}\ c_p(T_{\text{chilled}})\ (T_{\text{chilled,in}} - T_{\text{chilled,out}}) \tag{5.17}$$

$$Q_{\text{cond}} = \dot{m}_{\text{cond}}\ c_p(T_{\text{cond}})\ (T_{\text{cond,out}} - T_{\text{cond,in}}) \tag{5.18}$$

Finally, performance of the AD cycle is assessed in terms of specific daily water production (SDWP), and the performance ratio (PR), which can be calculated as follows:

$$\text{SDWP} = \int_0^{t_{\text{cycle}}} \frac{Q_{\text{cond}}}{h_{\text{fg}}M_{\text{sg}}}\ dt \tag{5.19}$$

$$\text{PR} = \frac{1}{t_{\text{cycle}}}\int_0^{t_{\text{cycle}}} \frac{\dot{m}_{\text{d}}h_{\text{fg}}}{Q_{\text{des}}}\ dt \tag{5.20}$$

The model predicts the performance of different operation modes of the AD cycle with and without heat and mass recovery schemes by changing the coefficients θ, ε, and γ, the values of which are given for different operation modes in Table 5.7 [50].

The mathematical modeling equations of the AD cycle are solved using the Gear backward differentiation formula method from the IMSL library linked by the simulation code written in FORTRAN PowerStation, and the solver employs double-precision arithmetic with a tolerance value of 1×10^{-6} using the input parameters summarized in Table 5.8.

Table 5.7 Values of Indicators for Changing Operation Mode of AD Cycle

Mode	Parameter	Two-Bed Mode	Four-Bed Mode with Master–Slave Arrangement
Operation	n	2	1
	θ	1 (charging seawater)	1 (charging seawater)
		0 (otherwise)	0 (otherwise)
	γ	1 (brine discharge)	1 (brine discharge)
		0 (otherwise)	0 (otherwise)
Switching	n	2	1
	θ	1 (charging seawater)	1 (charging seawater)
		0 (otherwise)	0 (otherwise)
	γ	1 (brine discharge)	1 (brine discharge)
		0 (otherwise)	0 (otherwise)

Source: Thu [50].

5.5.5 Simulation Results

The predicted performances of the AD cycle are discussed in this section. The transient temperature profiles of the major components such as adsorber, desorber, evaporator, and condenser of the AD cycle operating as a two-bed mode are given in Figure 5.23.

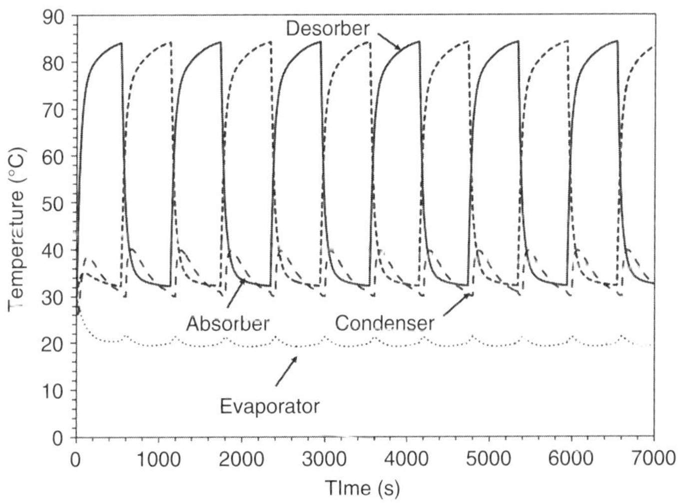

Figure 5.23 Temperature profiles of major components of AD cycle $T_{hw} = 85°C$, $T_{cw} = 29.8°C$, $T_{chilled} = 30°C$, $\dot{m}_{hw/cw} = 0.8$ kg/s, $\dot{m}_{chilled} = 0.8$ kg/s, half-cycle time = 600 s, switching time = 40 s.

Table 5.8 Parameters Used in Simulation

Parameters or Material Properties	Values	Unit	Reference
Sorption Thermodynamic Properties			
Kinetic constant D_{s0}	2.54×10^{-4}	m^2/s	59
Activation energy E_a	4.2×10^{-4}	J/mol	59
Average radius of silica gel particle R_p	1.7×10^{-4}	m	Manufacturer
Specific heat of silica gel c_{psg}	921	J/(kg·K)	59
Adsorber Bed			
Mass of silica gel per bed	36	kg	Authors' experimental data
Adsorber/desorber bed heat transfer area (A_{bed})	41.7	m^2	
Tube length	0.576	m	
Tubes/cake	15	—	
Passes/distribution	30	—	
Thermal mass of adsorber bed including fins and support	284.6	kJ/K	
Overall heat transfer coefficient of adsorber U_{ads}	250	$W/(m^2 \cdot K)$	
Overall heat transfer coefficient of desorber U_{des}	180–330	$W/(m^2K)$	
Condenser			
Condenser heat transfer area A_{cond}	5.08	m^2	Authors' experimental data
Thermal mass of condenser including fins and support	18.61	kJ/K	
Condenser heat transfer coefficient	2657.5	$W/(m^2 \cdot K)$	
Mass of condensate in condenser	10	kg	
Evaporator			
Evaporator heat transfer area A_{evap}	3.5	m^2	
Evaporator heat transfer coefficient	1715.2	$W/(m^2K)$	
Thermal mass of evaporator including fins and support	25.44	kJ/K	
Mass of refrigerant in evaporator	250	kg	
Concentration of feed	35,000	ppm	
Concentration limit to discharge brine	110,000	ppm	

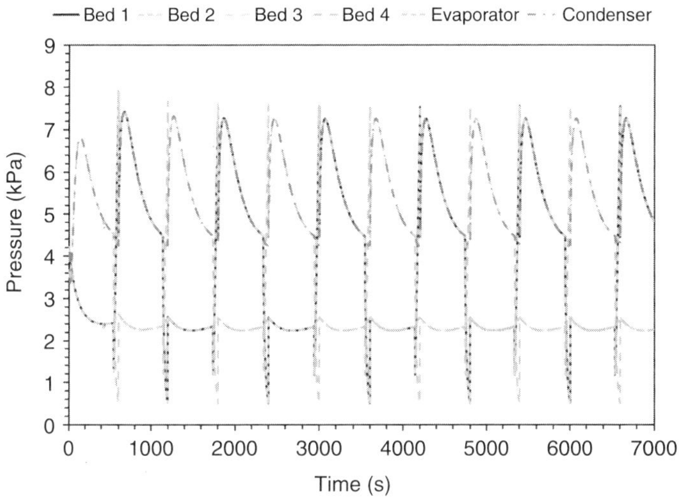

Figure 5.24 Pressure profiles of AD cycle operating in two-bed mode: $T_{hw} = 85°C$, $T_{cw} = 29.8°C$, $T_{chilled} = 30°C$, $\dot{m}_{hw/cw} = 0.8$ kg/s, $\dot{m}_{chilled} = 0.8$ kg/s, half-cycle time = 600 s, switching time = 40 s.

The predictions show that cyclic steady state is achieved after three half-cycles. The corresponding pressure profiles in the predictions are shown in Figure 5.24, and it is found that pressure drop is insignificant and can be neglected.

The transient water production rates of the cycle are captured in Figure 5.25. The equivalent specific daily water production (SDWP) is estimated to be ~7.4 m^3/t

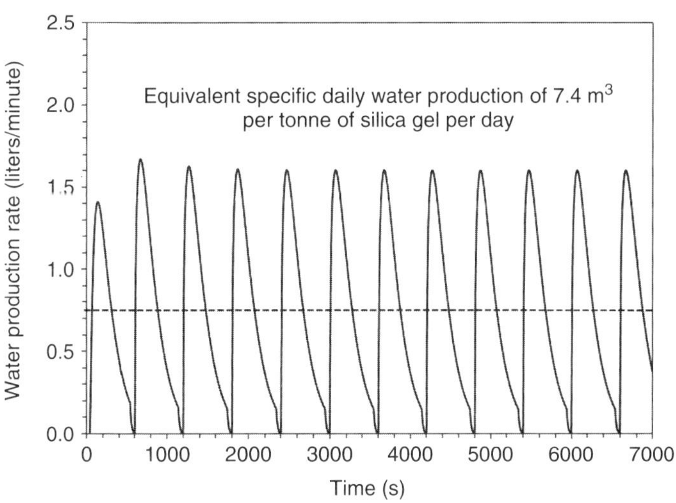

Figure 5.25 Water production rate of a two-bed AD cycle.

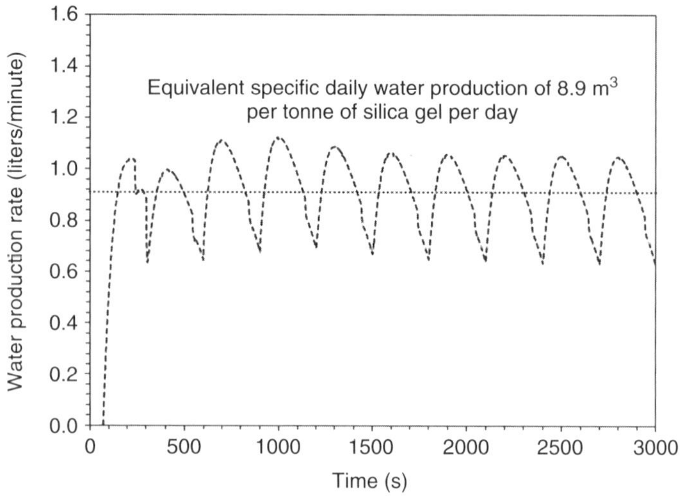

Figure 5.26 Water production rate of a four-bed AD cycle.

silica gel per day, while the production ratio, defined here as the ratio of the latent energy of condensate to the heat input to produce the condensate or PR, is ~0.64.

Figure 5.26 illustrates the water production rates of the four-bed mode, which gives a more continuous water production rate because of the master–slave arrangement. The minimum water production rate at switching period is ~0.7 Ls/min.

5.6 EXPERIMENTAL INVESTIGATION OF ADSORPTION–DESALINATION PLANT

5.6.1 Experiments

This section describes the experimental investigation of AD cycles. Figure 5.27 shows pictorial views of the adsorption desalination plant. The plant is designed with a conditioning facility for the heat source, cooling-water, and chilled-water temperatures, which enable operation under different operation modes such as two- or four-bed and with or without heat and mass recovery schemes [60]. Temperature measurements are achieved using 5-kΩ thermistors with a 3-s time constant (±0.2°C, YSI). The flow rates of the heat transfer fluids are monitored using electromagnetic flowmeters (±0.5% of reading). The absolute pressure sensors that are used have an accuracy ±0.125 kPa (Yokogawa). It is estimated that the accuracy of cycle-averaged cooling capacity measurements is ±3.5%, COP (coefficient of performance) measurements is ±3.8%, and heat input measurements is ±1.7%. A low-fin-tube bundle with internal corrugation is used for the evaporator, while the load-water flow rate is 0.8 kg/s. The condenser cooling-water flow rate is set at 2.0 kg/s. The plate-type heat exchanger manufactured by Mayekawa Manufacturing Co. Ltd. is employed for the reactor bed, where 36 kg of silica gel is packed

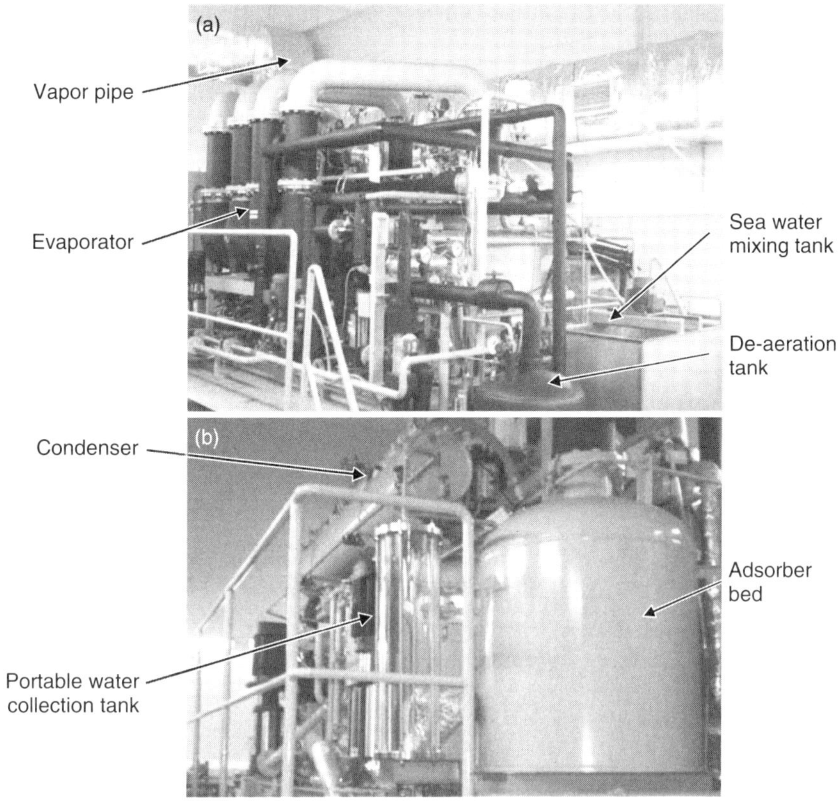

Figure 5.27 Pictorial views of the AD chiller showing (a) feedwater tank and evaporator and (b) bed, condenser, and potable-water collection tank.

in-between the fins. Cooling and hot water with rated flow rate of 0.8 kg/s are used during adsorption and desorption processes. Four temperature sensors of the same type are inserted in the reactor bed to measure the adsorbent temperature.

The cycle time is set at 600 s, while the switching period for both modes of operation is fixed at 40 s. The electrically fired rating facility is able to supply heat source water, cooling-water, and hot water at a practically constant flow rate and a temperature process control accuracy of $\pm 0.38^{\circ}C$ during cyclic steady state with an occasional drift to $\pm 0.5^{\circ}C$.

5.6.2 Results and Discussion

The quality of product water of the AD cycle was measured, and Table 5.9 gives the range of water qualities for different feedwater concentrations. It is found that the product water from the AD cycle has high purity (TDS $<$ 15 ppm) even with highly concentrated feedwater (recovery rate 70%). The TDS and conductivity of product water tend to increase slightly with the salinity of seawater feed, while the pH remains constant.

Table 5.9 Potable-Water Quality at Different Feedwater Concentrations

Feedwater Quality, ppt	Recovery, %	TDS, ppm	Conductivity	pH
10	—	7.12	13.1	8.35
35	—	7.54	13.5	8.41
67	48	10.5	13.63	8.42
110	70	12.7	13.88	8.37

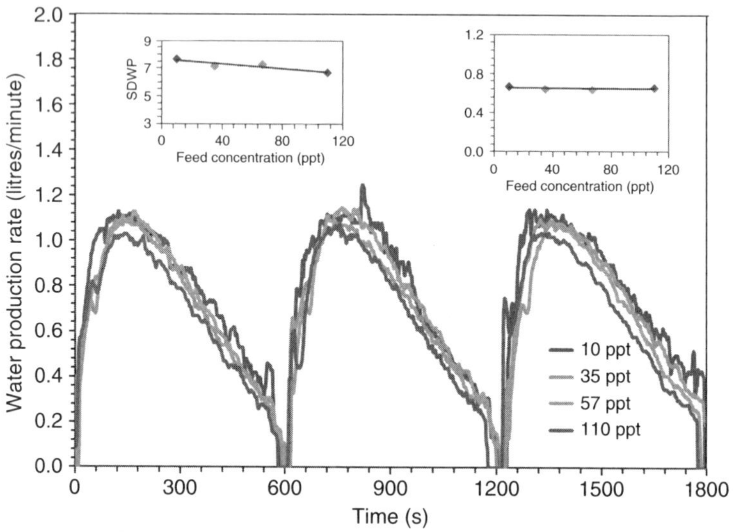

Figure 5.28 Water production rate of a four-bed AD cycle.

Figure 5.28 shows the specific daily water production (SDWP) of the AD cycle, while the insets in the figure show the SDWP and the PR at different saline feedwater concentrations. It is noted that the performance of the AD cycle is not significantly affected by feedwater concentration. The AD plant is still capable of producing 6.7 m^3 of potable water per metric ton of silica gel per day at feedwater concentration 110 ppt, which is equivalent to 70% recovery, while the cycle's performance ratio remains unchanged. This is because evaporation of the feedwater in the evaporator of the AD plant is initiated by the affinity of the adsorbent (silica gel) to the adsorbate (water vapors) in contrast to the existing desalination technologies such as RO, MSFD, and MED, where the evaporation or separation is achieved by an external energy input such as thermal energy or pressure input. In fact, the potable-water production in the AD plant is essentially governed by the nature of the isotherm properties of the adsorbent–adsorbate pair, which is dependent only on the temperature and pressure of the adsorption and desorption conditions.

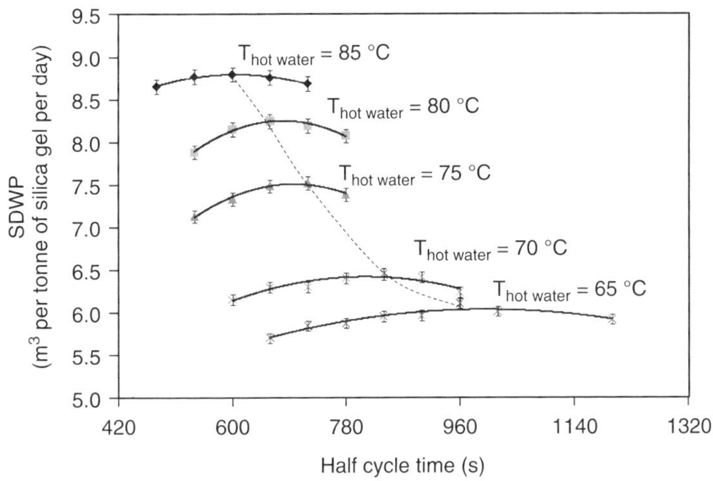

Figure 5.29 SDWP at different cycle times for two-bed mode showing optimum cycle times.

The SDWP of the AD plant slightly decreases at 70% extraction, which is at 110 ppt feedwater (Fig. 5.29). This phenomenon is attributed to the boiling-point elevation (BPE) of the feedwater at higher salt concentration. As the feedwater concentration increases, the BPE increases proportionally, resulting in the drop of the vapor pressure inside the evaporator. Consequently, the pressure of the adsorber bed that is in contact with the evaporator also decreases. According to the isotherm properties of the silica gel–water pair, the decrease in adsorption pressure results in decrease in vapor uptake. Hence, there is a slight degradation in the SDWP of the AD plant at higher percent (%) of potable water. However, it is experimentally confirmed that the potable-water extraction of the AD plant can be obtained as high as 70% without major increment in the power consumption of the plant. It is also remarkable that the recovery rate of the AD cycle is much higher than that of the RO plant, whose maximum recovery rate is 40% [61]. The high recovery rate of the AD plant may translate to a decrease in the operation cost of the plant and improvement in the reliability of the plant

Investigations on the optimum operation cycle time of the AD cycles (two- and four-bed modes) are discussed in this section. In practice, it is obvious that the temperature of the available hot water for operation of the AD cycle might not be constant, but it will fluctuate depending on the process from which the waste heat is extracted. The desorption rate is faster for high temperatures, and thus a shorter cycle time may be required to regenerate the water vapor. The AD cycle was investigated using different hot-water temperatures ranging from 65°C to 85°C to evaluate the optimal cycle time at which the cycle gives the highest water production rate. Figures 5.29 and 5.30 show SDWP versus cycle times for two- and four-bed operation modes of the AD cycle at different hot-water inlet temperatures. These results denote the existence of optimal cycle times at the specific hot-water

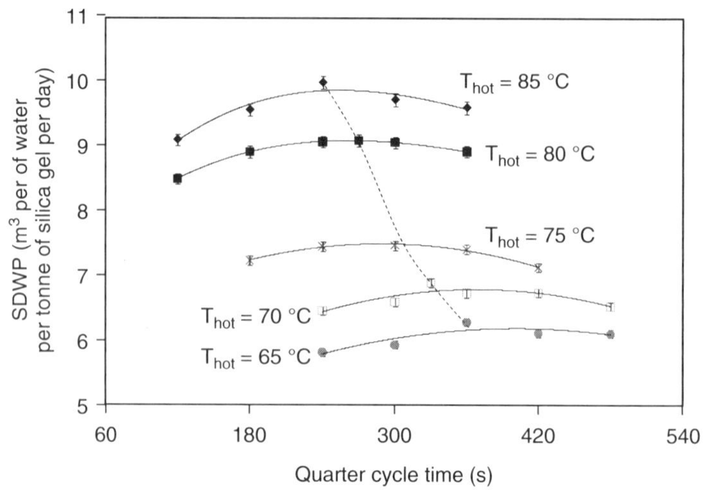

Figure 5.30 SDWP at different cycle times for four-bed mode showing optimum cycle times.

inlet temperature as the SDWP varies with cycle time. Lower SDWP is obtained at relatively shorter cycle times, because the regeneration process during desorption is not completed and the water vapors from silica gels are not fully emitted for condensation and water production in the condenser. This means that the affinity of the adsorbent (silica gel) for the uptake of water vapor in the next cycle will be damped as a result of the incomplete regeneration.

On the other hand, the lower SDWP at longer cycle time is due to waste of thermal energy resulting from the excessively long supply of hot water to the sorption elements during the desorption process. As a result, the additional energy supplied by the hot water has no effect on desorption other than simply heating the adsorbents and the heat exchanging components. In addition, the total numbers of operation cycles per day is also significantly reduced. It should be noted that the AD cycle is performed under heat and mass recovery schemes, namely, bed cooling and pressure equalization.

Figure 5.31 shows the effects of cycle time on heating fluid inlet temperature, and it is found that the optimum cycle time varies linearly with the hot-water inlet temperature in both operation modes. The longer cycle time is observed at lower hot-water inlet temperatures as a result of slower regeneration process in the desorption bed(s).

Figure 5.32 shows the comparison of the specific daily water production and performance ratio of AD cycles operating with four- and two-bed operation modes at respective optimal cycle times with assorted hot-water inlet temperatures. It is noticed that the AD cycle in four-bed operation mode produces higher SDWP compared with two-bed operation mode, and significant improvement can be realized at higher hot-water inlet temperatures typically above 70°C. This is because of the master–slave configuration in the four-bed mode, which results in the better

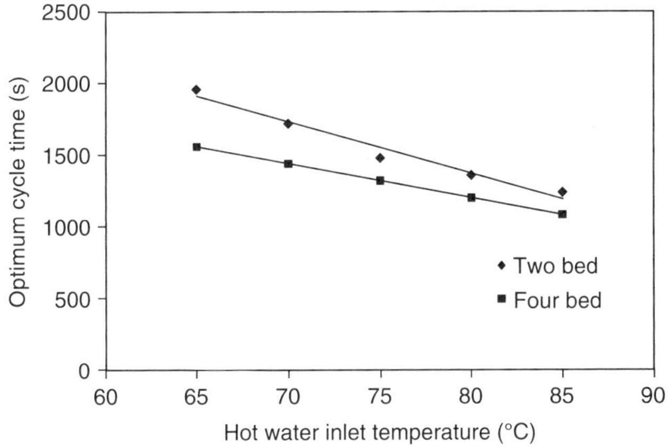

Figure 5.31 Optimum cycle time at different hot-water inlet temperatures.

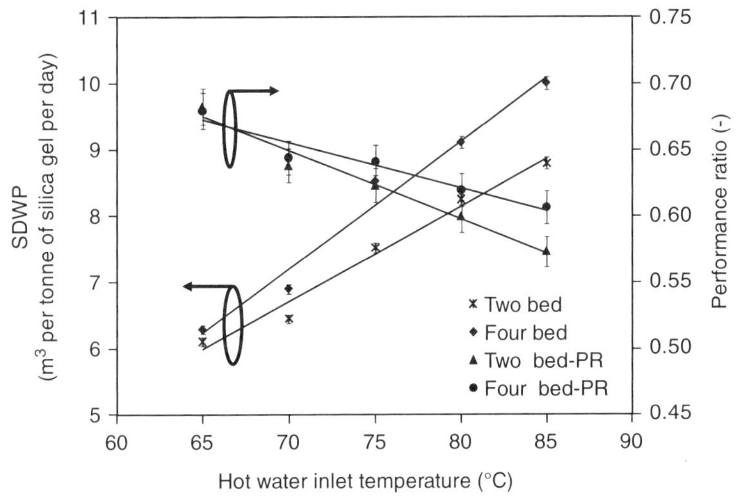

Figure 5.32 SDWP at different hot-water temperatures for two- and four-bed modes.

energy utilization of the hot water. For the lower hot-water inlet temperatures, the master–slave configuration becomes less effective because the outlet hot-water temperature from the master bed is not high enough to perform further desorption in the slave desorber. Thus, the advantage of four-bed operation mode diminishes at lower heat source temperatures. The PR of the AD cycle for both modes is decreased with increase in hot-water inlet temperature.

5.6.3 Bed Cooling Scheme

The energy for the evaporation of saline water can be recovered from the ambient or internally. In the bed cooling scheme, the ambient energy is extracted for

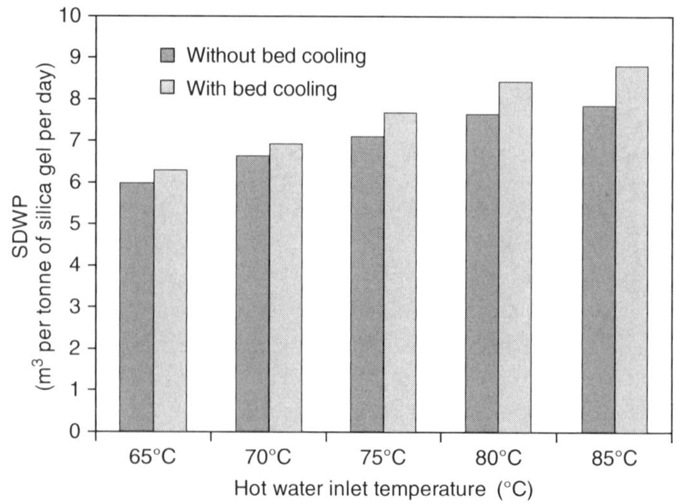

Figure 5.33 Comparison of SDWP values for a two-bed AD cycle without and with bed cooling scheme.

evaporation while simultaneously cooling the adsorber bed to boost the performance of the cycle. Figures 5.33 and 5.34 compare the SDWP values of two- and four-bed AD cycles with and without the bed cooling scheme for assorted hot-water inlet temperatures. It is found that SDWP of the AD cycle is 8.8 for two-bed mode and 9.7 for four-bed mode, while the percentage increment in SDWP is 12.5% and 16.0% for two- and four-bed operating modes, respectively.

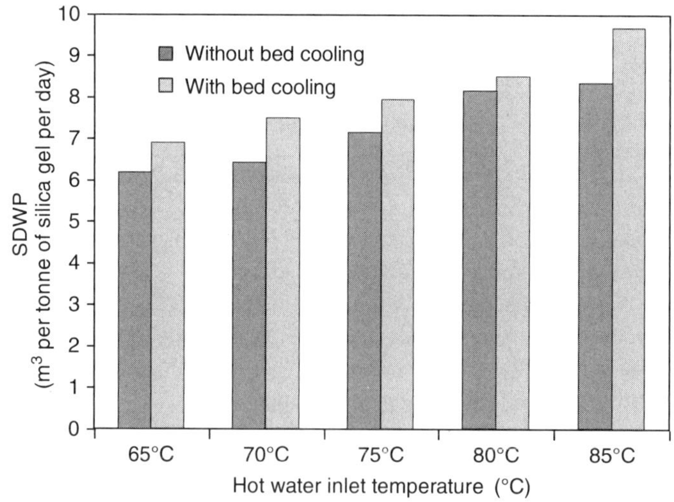

Figure 5.34 Comparison of SDWP values for a four-bed AD cycle without and with bed cooling scheme.

5.6.4 Summary

Performance levels of AD cycles have been investigated experimentally. AD cycle performance is evaluated for different operating conditions such as hot-water inlet temperatures, cycle times, load water temperature, and flow rates. Evaluation of the optimal cycle time for different hot-water inlet temperatures showed that the longer cycle time is required for lower hot-water temperatures. The effects of heat and mass recovery schemes such as water recirculation, pressure equalization (PE), and bed cooling schemes were also investigated experimentally, and the results showed the optimum PE time to be 10 s, enhancing SDWP and PR of the cycle by 12% and 21%, respectively. Similarly, the bed cooling scheme improves the SDWP by 16%. The specific unit cost for desalination of the advanced AD cycle is deemed to be the lowest.

5.7 ADVANCED ADSORPTION-DESALINATION CYCLE

In this section, an advanced adsorption desalination cycle is investigated using the heat of condensation process in the condenser to effect the desalting and cooling processes in the evaporator. The details of the advanced cycle are discussed along with performance analyses of the cycle.

5.7.1 Basic Concept of Advanced AD Cycle

Using the principles of internal heat recovery, it is proposed that the latent heat of condensation of water vapor, which is rejected during condensation at vapor pressures of 5–7 kPa, is recovered to the evaporative process of the evaporator where the system pressure is much lower (2–3 kPa). Heat recovery of this magnitude can be achieved by using either (1) *an integrated condenser–evaporator* layout within a single housing for direct heat and mass transfer or (2) a *recirculating coolant medium* that flows between the evaporator and the condenser, transferring the latent heat from the condenser to the evaporator. The second method is suitable for retrofitting an existing AD cycle and will be discussed in Section 5.7.3. Figure 5.35 shows the schematic diagram of an advanced adsorption–desalination (AD) cycle comprising three major components: (1) an adsorber bed, (2) a desorber bed, and (3) an evaporator–condenser device. Both methods of heat recovery, as outlined above, will be simulated in this chapter.

It is noted here that by employing the integrated condenser–evaporator layout, both the condenser and the evaporator water circuits of the conventional cycle are eliminated, resulting in a significant saving in the pumping cost. The heat is transferred directly across the walls of the condenser tubes, which are placed within the evaporator shell. This arrangement reduces the heat transfer resistance, leading to an improvement in the evaporation rates of water vapor from the seawater solution. Another consequence of integrating the condenser and evaporator units is the higher vapor pressure in the evaporator, and this has a direct effect of increasing the vapor uptake by the silica gel during the adsorption process. Thus,

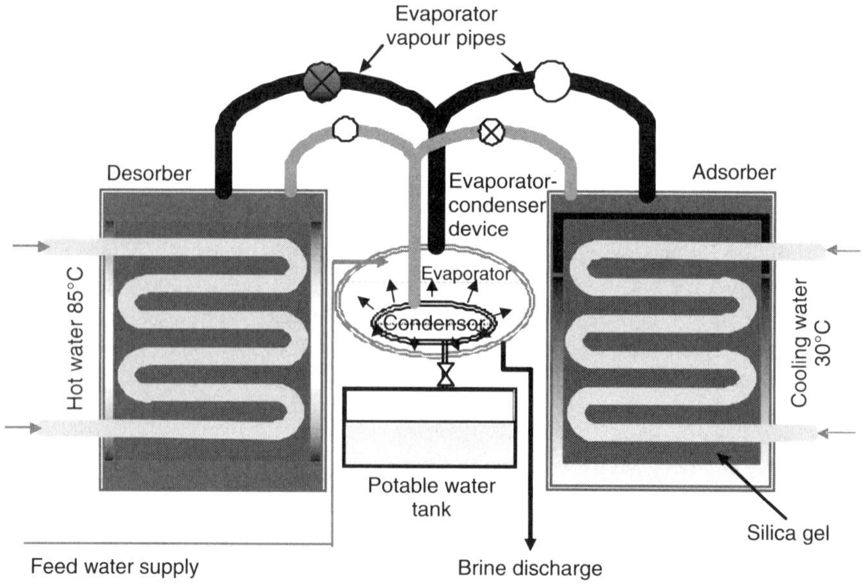

Figure 5.35 Advanced AD cycle with integrated evaporator–condenser device.

the advantages of the advanced AD cycle (with integrated condenser-evaporator) are summarized as follows:

1. A reduction in the parasitic electrical power owing to the elimination of pumps for the chilled- and cooling-water circuits
2. An improvement in the adsorption capacity of silica gel due to the pressurization effect
3. A lowering of the effective condensation temperature in the condenser as this tends to desorb more vapor during the desorption process
4. The ability to operate with a low temperature heat source in the range between 50 and 90°C.

   For performance verification, a mathematical model, based on the mass, energy balances, and the thermodynamic properties (isotherms and kinetics) of the adsorbent/adsorbate system, has been developed. In the following sections we describe the development of the numerical model to predict the performance of the advanced AD cycle.

5.7.2 Simulation of Advanced AD Cycle

This section describes the predictions from numerical analysis of the advanced AD cycle on the basis of the model discussed in the previous section. Analyses of the cycle under various operation modes such as different cycle times, hot/cooling-water temperatures, and different hot/cooling-water flow rates will be discussed.

Figure 5.36 shows the temperature–time history of the adsorber, desorber, evaporator, and condenser of the advanced AD cycle at cyclic steady-state

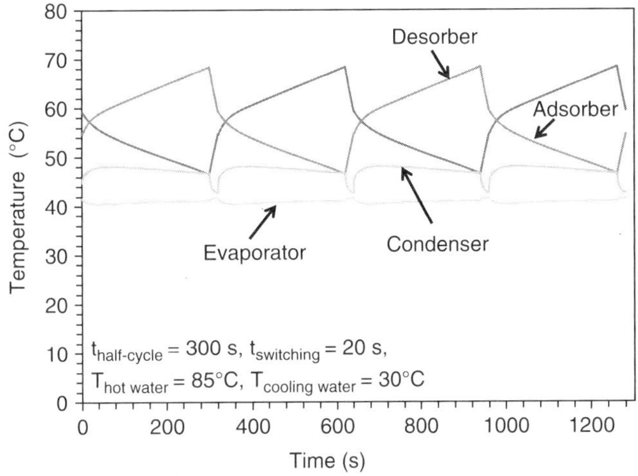

Figure 5.36 Predicted temperature–time histories of the adsorber, desorber, evaporator, and condenser of the advanced AD cycle.

conditions. The horizontal axis (abscissa) represents the time in seconds, and the vertical axis (ordinate) gives the temperature distributions. Owing to the integral evaporator–condenser design, it can be seen that the vapor pressure increased during the adsorption process. This increase is caused by an increase in the evaporator chamber temperature, which varies from 35°C to 40°C. The higher adsorption pressure enhances the adsorption capacity of the adsorbent.

The predicted water production rate of the advanced AD cycle (in L/min) is given in Figure 5.37, where the equivalent specific daily water production (SDWP)

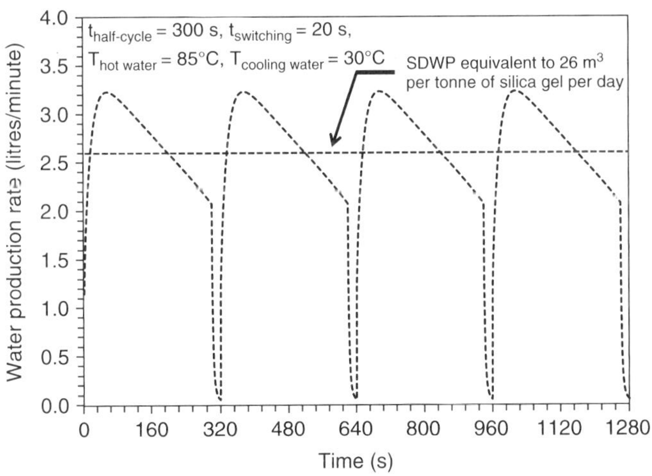

Figure 5.37 SDWP of advanced AD cycle compared with conventional AD cycle at normalized cycle time.

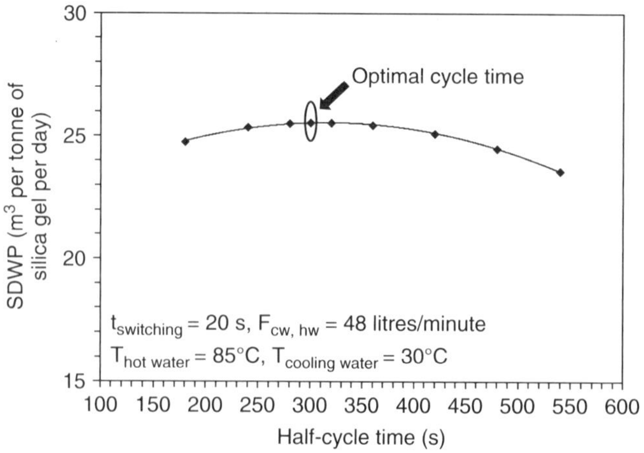

Figure 5.38 Optimal cycle time for SDWP of the advanced AD cycle.

of the cycle is ~26 m^3/t silica gel per day. It is almost 3 times higher than the yield from a conventional AD cycle. The high SDWP in the advanced AD cycle is attributed to two factors: (1) an improvement in the adsorption capacity of the silica gel (≤40% of the dry mass) owing to the pressurization effect during the adsorption process and (2) decrease in the effective condensation temperature in the condenser due to lesser resistance in heat transfer; this tends to desorb more vapor out of the desorber during the desorption process.

A parametric analysis on the advanced AD cycle has been conducted to study the performance of the cycle under various operating conditions, such as different cycle times and different hot/cooling-water inlet temperatures to the adsorber beds. The effects of the operation cycle time on the specific daily water production (SDWP) of the cycle at a fixed hot-water temperature (85°C) are shown in Figure 5.38. It is noted that the optimal half-cycle time for the advanced AD cycle is 300 s, which is 50% shorter than that for the conventional AD cycle at the same heat source temperature. A shorter cycle time can lead to higher water production yield, as seen below.

Figure 5.39 presents the specific daily water production (SDWP) and overall heat transfer coefficient (U) of the advanced AD cycle for regeneration temperatures ranging from 50°C to 85°C. The results show that the water production rate of the advanced cycle varies linearly with the hot-water temperature. This is due to the improved regeneration process for higher hot-water temperatures. The results also show that the overall heat transfer coefficient increases with the increase in hot-water temperature.

Another significant advantage of the advanced cycle is that it can be operated at a hot-water temperature as low as 50°C, which a conventional cycle is unable to do. This temperature of activation is easily available from the solar thermal sources, and the specific daily water production (SDWP) is found to be about 8.1 m^3. This production rate is comparable with that of the conventional cycle, even at this low heat source temperature.

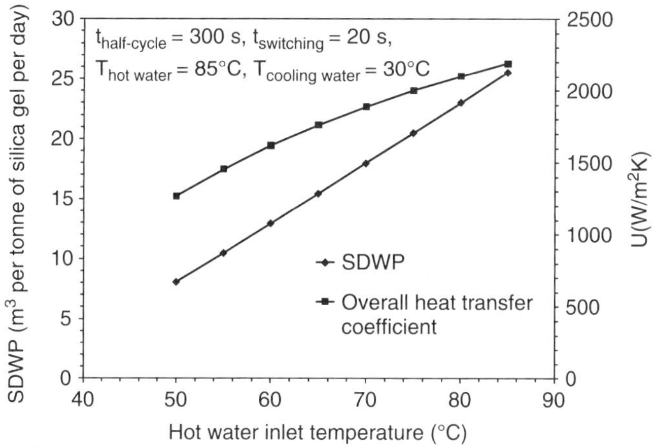

Figure 5.39 Predicted SDWP values and overall heat transfer coefficients of advanced AD cycle at assorted hot-water inlet temperatures.

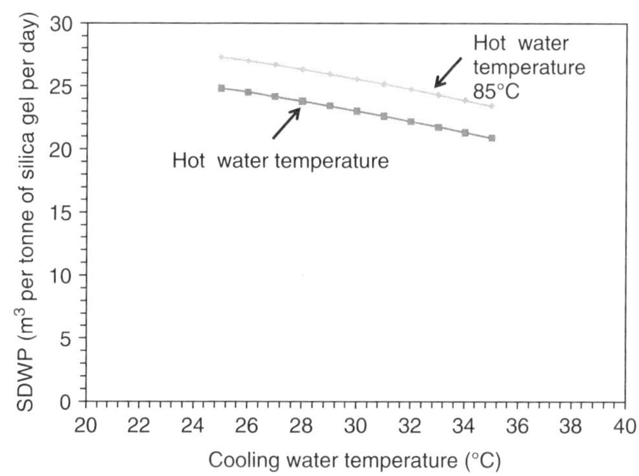

Figure 5.40 The predicted potable-water production of advanced AD cycle for different cooling-water inlet temperatures.

The effect of the cooling-water temperature to the adsorber on the performance of the advanced AD cycle is also investigated. The temperature ranges between 25°C and 35°C, which is a practical range for cooling-water. Effects of the cooling-water temperature on the cycle are investigated for two hot-water inlet temperatures: 85°C and 80°C. The performance results are presented in Figure 5.40, showing that the cycle performs more efficiently at lower cooling-water temperature, giving higher SDWP.

As the cooling-water temperature increases, the SDWP of the cycle decreases. This phenomenon can be attributed to the isotherm behavior of the silica gel; the higher saturation temperature (hence pressure in condenser) retards the desorption of water vapor when the desorber beds undergo heating.

In summary, the advanced AD cycle has several distinctive advantages: (1) its SDWP rate is 3 times higher than as that of the conventional AD cycle, and (2) the advanced cycle is found operational even at a hot-water temperature as low as 50°C, and yet produces 8.1 m^3/t silica gel per day. All these improvements can be achieved with only one heat source input.

5.7.3 AD Cycle with a Coolant Circuit between Evaporator and Condenser

From the heat transfer perspective, the AD cycle retrofitted with a coolant flowing circuit between the condenser and evaporator is inferior to the cycle described in Section 5.7.3, specifically, the cycle comprising an integrated evaporator–condenser unit. Nevertheless, the coolant flowing through the condenser to the evaporator has a similar heat recovery but incurs higher heat transfer resistance due to flow resistance and other factors. The key merits of the retrofitted coolant circuit are (1) the simplicity in retrofitting the additional coolant circuit outside the high-pressure vacuum environment and (2) the ease in switching the AD cycle from single-mode to dual-mode operation (cooling cum desalting) by simply closing the coolant flow. A numerical simulation is presented in this section, and the predictions can be compared with the experimental tests.

5.7.3.1 Description of Evaporator–Condenser Heat Recovery Circuit

In this arrangement, there is no change to the heat exchange units of the conventional AD cycle, except that the coolant flowing in the condenser to the cooling tower is now diverted to the evaporator. Condensation heat is now recovered and dumped into the evaporator by a simple runaround coolant between them. A schematic diagram of the retrofitted AD cycle showing the internal heat recovery is presented in Figure 5.41. The coolant lines are shown in red, indicating warmer fluids after receiving the condensation heat; the blue lines indicate the cooler coolant after heat is rejected in the evaporative process. The saturation temperatures of the evaporative and condensation processes in the evaporator and the condenser would maintain equilibrium level according to the rate of heat transfer in these exchangers during batch operation of adsorbers and desorbers.

5.7.3.2 Simulation Results and Discussion The transient temperature profiles of the major components such as the adsorber, the desorber, the evaporator, and the condenser of the proposed AD cycle are shown in Figure 5.42. In this analysis, the hot-water temperature at inlet is maintained at 85°C, while the cooling-water inlet temperature is kept at 30°C. The pressurization effect on the adsorption process can be seen from the higher equilibrium evaporator temperature of 30°C, which is achieved solely by the recovered condensation energy. Thus, the amount of vapor uptake by the silica gel is expected to increase and hence, the water production capacity should improve as well.

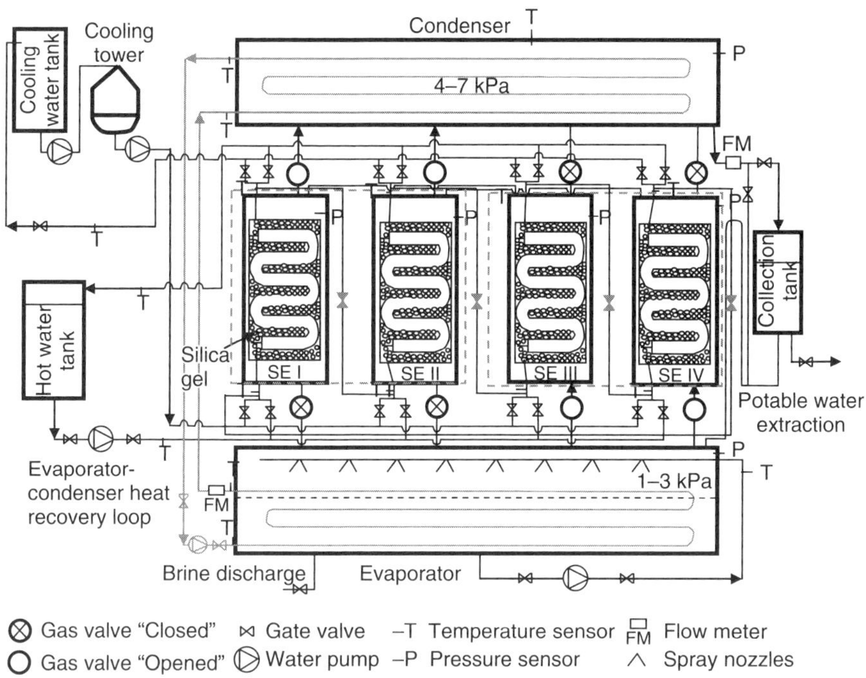

Figure 5.41 Schematic diagram of advanced AD cycle, showing internal heat recovery with an evaporator–condenser water circulating circuit.

Figure 5.43 gives the transient water production rate by the advanced AD cycle with the evaporator–condenser heat recovery circuit. The equivalent specific daily water production (SDWP) of the cycle is ~13.5 m^3 of potable water per metric ton of silica gel per day. This production rate is inferior to that of the integrated condenser–evaporator design, but it is about 2 times that of a conventional AD cycle. This is because of the additional energy losses incurred from the multiple heat transfer processes through a medium (water) involved in the evaporator–condenser circuit design.

5.7.4 Experimental Investigation of Advanced AD Cycle

For verification of any improvement in the heat recovery cycles, the existing prototype AD plant is retrofitted with a coolant circuit positioned in between the condenser and the evaporator. Extensive experiments have been conducted using different hot-water inlet temperatures ranging from 50°C to 70°C and different hot/cooling-water flow rates (ranging within 50–125 L/min) to investigate the performance of the cycle. The experimentally measured transient temperature profile of the major components of the advanced AD cycle that incorporates the heat recovery scheme between the condenser and the evaporator of the cycle is shown in Figure 5.44. In this experiment, the hot-water inlet temperature was maintained

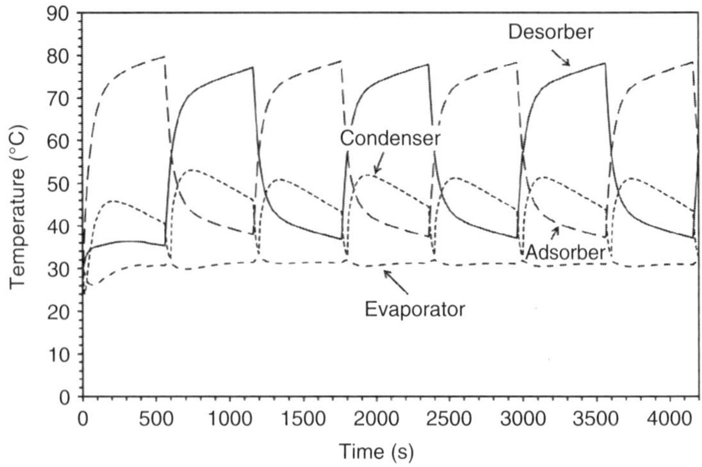

Figure 5.42 Temporal temperatures of the advanced AD chiller with the evaporator–condenser heat recovery circuit.

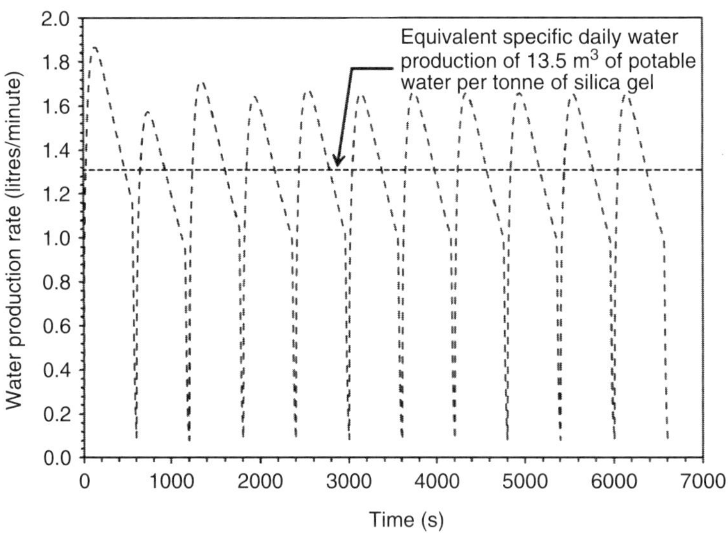

Figure 5.43 Transient water production rate of advanced AD cycle with evaporator–condenser heat recovery circuit.

at 55°C and the half-cycle time was 600 s, with a switching time of 40 s. The inlet and outlet water temperature profiles of the heating and cooling fluids are shown in Figure 5.45. It can be seen that the fluctuations of the hot and cooling-waters occurs during the switching period; this is due to preheating of the adsorber and precooling of the desorber reactors during the switching period.

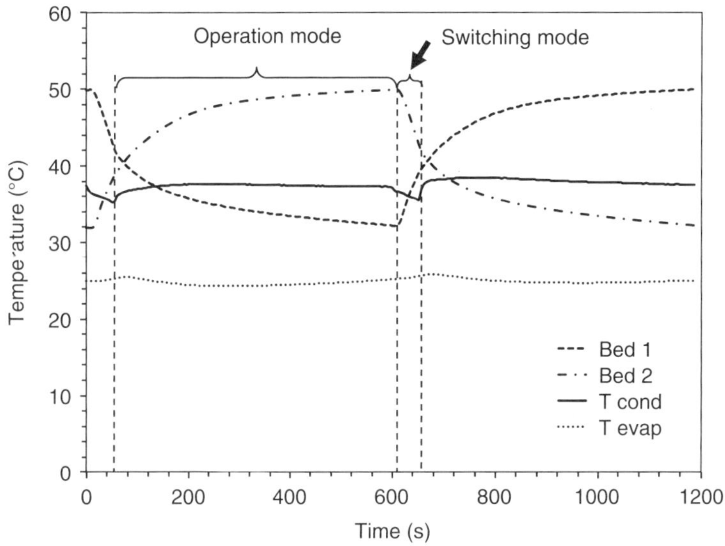

Figure 5.44 Experimentally measured temperature profiles of the major components of advanced AD cycle.

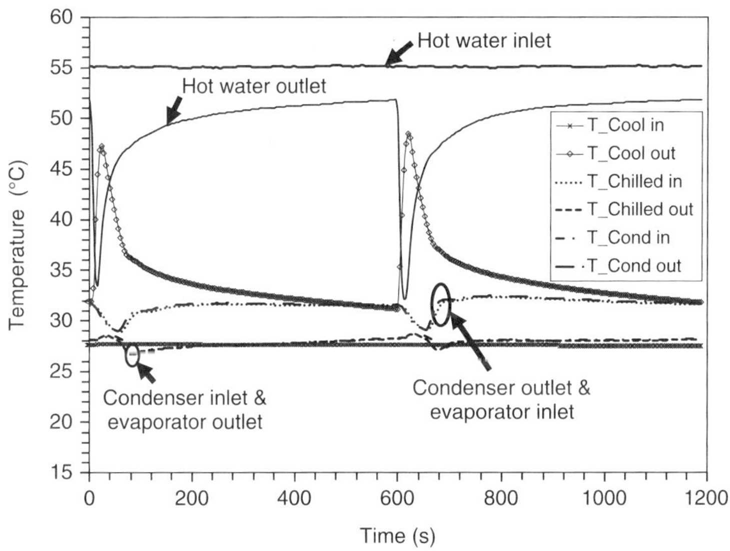

Figure 5.45 Temperature profiles of inlet and outlet of heat transfer fluids measured experimentally.

In the advanced AD cycle with the evaporator–condenser heat recovery circuit, the outlet water from the evaporator is channeled to the condenser and vice versa. The evaporator outlet water is properly insulated to prevent heat leak as much as possible, and it can be seen in Figure 5.45 that the temperature differences between the evaporator outlet and the condenser inlet, and that between the condenser outlet and the evaporator inlet, are negligibly small. This slight temperature change along the circuit results from the effective thermal insulation of the pipes involved in the heat recovery circuit between the condenser and the evaporator.

5.7.5 Validation of Experimental and Simulation Results

Experimental data from the prototype advanced AD plant are used for validation of the mathematical model. The comparison is performed on two 2-bed AD cycle operating over a half-cycle period. Figure 5.46 compares the simulated and experimentally measured temperatures of the major components of the adsorption cooling–desalination cycle. In this analysis, the hot-water inlet temperature to the desorber was maintained at 70°C while the half-cycle and the switching times are kept at 600 and 40 s, respectively. It can be seen that the experimental temperature measurements are subject to the time constant of the sensors. The results showed that the present formulation of the advanced AD cycle gives reasonable results, as validated by the experimental data. The temperature profiles of the simulation

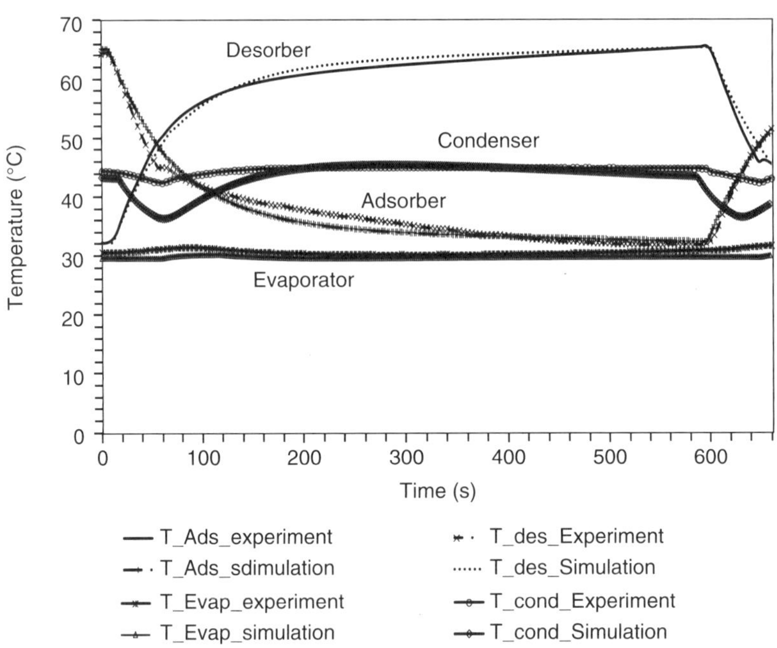

Figure 5.46 Temperature profiles of major components of a two-bed advanced AD cycle comparing the experimentally-measured and simulated values.

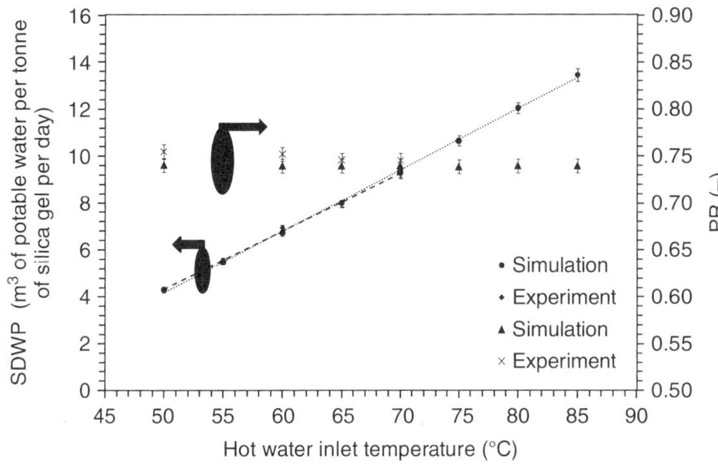

Figure 5.47 SDWP and PR of the advanced AD cycle with hot-water inlet temperatures of 50°–70°C.

and the experiments are found to be in good agreement within experimental error, except for that of the condenser. The condenser temperature measured in the experiments seems to be higher than temperatures from numerical simulation during the switching periods. This could be due to the location of the condenser temperature sensor. As the sensor is placed near the desorber vapor pipes, it may be influenced by the high temperature of the condenser shell as well as the vapor pipes and fail to capture the temperature drop by the cold front from the evaporator at the beginning of the cycle.

A parametric study of the advanced AD cycle has been conducted to investigate the performance of the cycle under various hot-water inlet temperatures selectively between 50°C and 80°C, which reflects the typical temperatures that could be derived from a solar energy collectors or the waste heat sources. Figure 5.47 gives the specific daily water production (SDWP) and the performance ratio (PR) of the advanced AD cycle during operation at different hot-water inlet temperatures. Both simulation and experimental results are presented and show that the SDWP varies linearly with the hot-water inlet temperature, while the SDWP is 9.24 at 70°C hot-water inlet temperature.

The improvement in SDWP with increase in the hot-water temperature is attributed to the improved desorption–condensation process at higher desorbing temperatures, resulting in the improved evaporation–adsorption process and thus giving higher SDWP. It is remarkable that the retrofitted advanced AD cycle is able to operate at a 50°C hot-water inlet temperature—a temperature range usually beyond the capability of a conventional AD cycle. At this low inlet hot-water temperature, the SDWP produced is about ~5.2 m^3 potable water/t silica gel per day. On the other hand, at the rated 85°C hot-water inlet temperature, the SDWP of the advanced AD cycle is found to be 13.5 m^3, which is almost 90% higher than that of the conventional AD cycle.

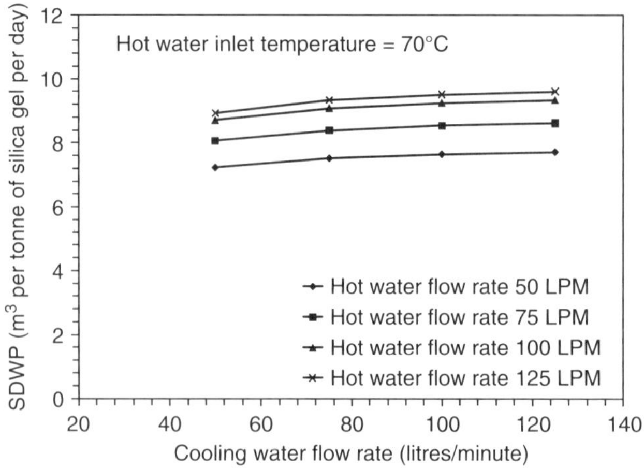

Figure 5.48 SDWP of advanced AD cycle for different cooling-water flow rates.

The *performance ratio*—defined here as the ratio of the useful effects to the heat input—of the advanced cycle is found to remain relatively constant at 0.73–0.75 for all hot-water inlet temperatures. This is a characteristic of the internal heat recovery.

Figure 5.48 provides the SDWP of the advanced AD cycle for various cooling-water flow rates at specific hot-water flow rate and a fixed temperature of 70°C. The results showed significant increase in the water production rate when the cooling-water flow rate was changed from 50 to 100 L/min. This is due to the improvement in the heat transfer coefficient at higher flow rates. It is also noted that the increase in SDWP becomes less significant at flow rates of >100 L/min, due to finite amount of adsorption and desorption capacities of the adsorbent.

Effects of the hot-water flow rate on performance of the advanced AD cycle are presented in Figure 5.49. These results show an increase in SDWP by ≤20% as the flow rates increases, but this effect decreases when the flow rate is >100 L/min.

Performance of the advanced AD cycle under different flow rates of heating and cooling fluids is presented in Figure 5.50. It is found that the SDWP of the cycle increases with increase in flow rate, while the heat source and cooling temperatures are held at 70°C and 29°C, respectively.

Effects of the cooling-water inlet temperature on the performance of the advanced AD cycle are shown in Figure 5.51. As can be seen, the SDWP of the cycle for two hot-water inlet temperatures (i.e., 70°C and 50°C) varied with coolant temperature in the range 26–32°C. The water production rate is inversely proportional to the coolant temperature.

As discussed in Section 5.6, the cycle time at which the AD cycle gives the optimal potable-water production varies with the available hot-water inlet temperatures. The optimal cycle times for the advanced AD cycle have been experimentally evaluated for two different temperature levels of hot water and are shown in Figure 5.52.

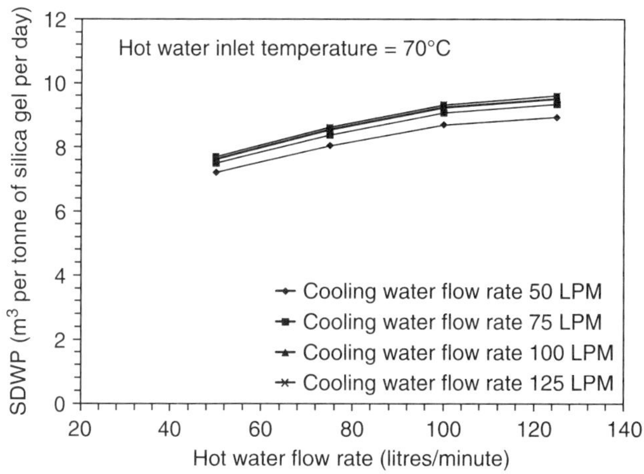

Figure 5.49 Specific daily water production of advanced AD cycle for assorted hot-water flow rates.

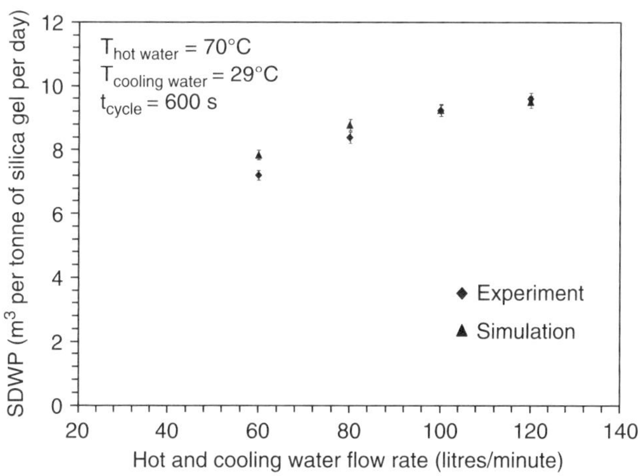

Figure 5.50 Performance of advanced AD cycle using different flow rates of heating and cooling fluids at constant source temperatures.

The results show that the optimal cycle times exist at the half-cycle times 570 and 600 s for 70°C and 65°C, respectively, while the corresponding SDWPs are 9.39 and 8.3.

The optimal cycle time for the advanced AD cycle is shorter than that of the conventional AD cycle that has been reported. This is because the evaporation–adsorption and desorption–condensation processes in the advanced AD cycle are greatly accelerated, owing to the pressurization of the adsorption, and the lower condensation temperature was a result of the heat recovery process between the condenser and the evaporator. The pressurization of the adsorption process is achieved since the higher-temperature water from the condenser is utilized

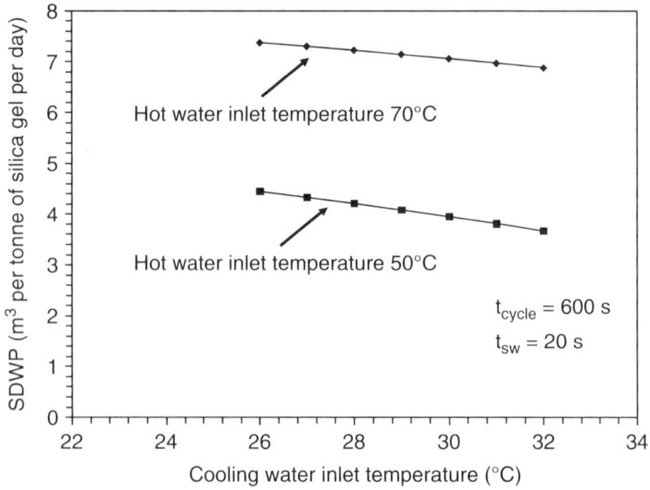

Figure 5.51 SDWP of advanced AD cycle at different cooling-water inlet temperatures.

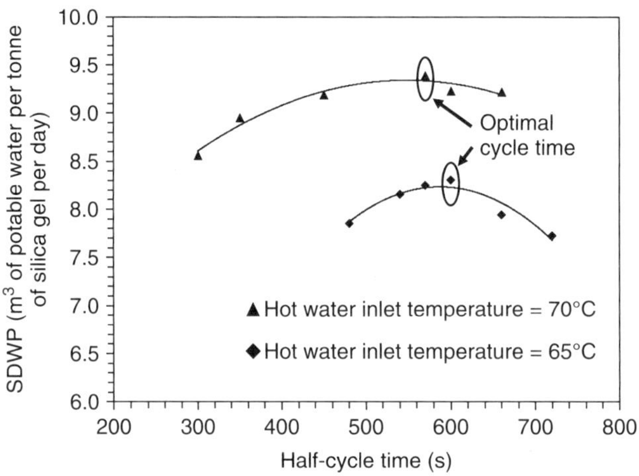

Figure 5.52 Performance of advanced AD cycle showing the optimal cycle times at hot-water inlet temperatures 70°C and 65°C.

for the evaporation. On the other hand, the temperature of the outlet water from the evaporator that is channeled to the condenser is ~25–27°C, and this cold-front lower condensing temperature ensures the better desorption process. These effects are more significant at the beginning of the operation mode after the switching process since the adsorption and desorption rates were at maximum at that period.

Finally, the water production rates of the AD cycles are compared in Figure 5.53, where the horizontal axis (abscissa) shows the normalized half-cycle time and the vertical axis (ordinate) gives the water production rate.

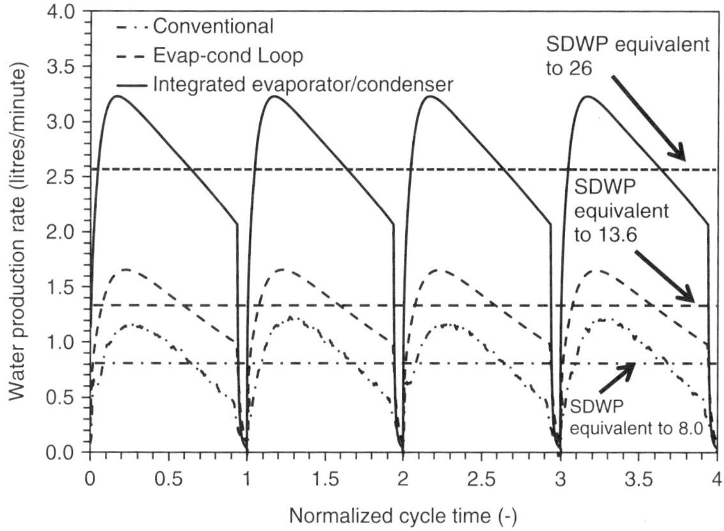

Figure 5.53 Comparison of water production rates of different AD cycles.

5.7.6 Summary

The innovative advanced adsorption–desalination (AD) cycle has been successfully developed and modeled. It achieves a superior improvement in the specific daily water production rates, which could potentially increase to 3 times those of the conventional cycle. This is achieved by the use of internal heat recovery of the processes in the cycle: (1) the integrated evaporator–condenser design, which fully recovers the evaporation–condensation energy, thus reducing the heat transfer resistance substantially, and (2) improvements in both adsorption and desorption offtake or uptake, due to favorable pressures of evaporator and condenser, respectively. In addition, the advanced AD cycle saves much of the parasitic pumping power of the condenser and evaporator circuits without compromising the salient features of the conventional AD cycle. An important outcome from tests conducted on the retrofitted AD cycle is that it has proved its ability to operate safely at a heat source temperature of only 50°C—a level not achieved, hitherto, by any other heat-activated equipment. With these advantages, the specific unit cost for desalination of the advanced AD cycle is deemed to be the lowest.

5.8 LIFECYCLE ANALYSIS OF AD SYSTEM

The specific water production cost is probably the most important indicator for assessing a desalination technology in terms of economic merits and commercialization. Desalination is a procedure in which a number of processes, such as the pretreatment process, main desalting process, and the posttreatment process, are included to produce potable water. All these processes require significant input of energy and chemicals, and thus the related cost must be considered in the cost analysis. A lifecycle analysis is adopted as a platform for the evaluation of unit production cost of potable water from desalination. This method enables estimation of the unit cost over a specific lifespan of the plant considering both capital and operational costs with inflation and interest rates [62,63]. Several factors influence the desalting cost and an overview of the factors affecting the desalination costs is presented below.

5.8.1 Factors Affecting Cost of AD Plant

Cost factors considered for selection of a desalination technology are generic to most desalination plants, but some of them are more site-specific, weighing in on factors such as location of the plant and access for feedwater and brine discharge. However, it is possible to categorize most common factors that are subject to most desalting plants. The general factors contributing to the unit cost calculation of desalination plants are listed in Table 5.10. The less apparent or indirect factors, such as the plant load factor, availability of skilled labor, storage and distribution of freshwater, and plant capacity and location, could also affect the unit cost, although to a lesser extent. For these indirect factors, a simple cost estimation methodology has been applied, and in the following section we describe these parameters in detail.

5.8.2 Total Cost of AD Plant

The total costs of a desalination plant comprise the capital, operational, and replacement costs of key components such as heat exchangers and membranes. Depending on the desalination methodology and the water production capacity, the plant lifetime could affect the annualized capital cost via the amortization period (n) and the interest rate (i) through a capital recovery factor (CRF), namely, the product between the initial investment and the CRF [64]. The operational cost comprises contributions from fuel and electricity rates, maintenance and replacement, pumping requirements, and chemical treatment of feedwater and output water. In reality, all operational costs can be subjected to the inflation effect or rates (j), arising from primary fuel, electricity price fluctuations, and other factors. Such increases over a period of time in the future could be incorporated using the inflation-weighted factor, $\mathrm{IWF} = (\mathrm{CFR}(i,n))/(\mathrm{CRF}(i^i,n))$. A lifecycle approach is adopted here for calculation of the unit cost of desalination, i.e.,

$$A(\$/\mathrm{m}^3) = \left(\sum_{k=1}^{m} C_{\mathrm{capital}}(\mathrm{CRF}(i,n))_k\right) + \left(\sum_{l=1}^{p} C_{\mathrm{operational}}\left(\frac{(\mathrm{CRF}(i,n))}{\mathrm{CRF}(i^i,n)}\right)_l\right) \tag{5.21}$$

where A is the unit cost of desalination on a volumetric basis, m is the number of items of capital investment, and p is the number of items related to operational cost. With Equation (5.21), the unit production cost of AD plants of varying

Table 5.10 Factors to Consider in Estimation of Desalting Cost

Cost Factor	Description
Performance ratio	Ratio of freshwater output to energy input to desalination cycle or alternatively, the ratio of condensing energy to heat input
Plant lifetime	Affects capital costs of plant throughout amortization period
Plant costs	The most important factor for decisions on desalination technology; may vary with plant size and capacity
Interest rates	Influence capital cost and total investment for plant
Inflation rates	Have considerable effect on unit production cost
Site costs	Relate to land cost, which varies with location and footprint area of plant
Seawater intake and brine discharge	Important role for the desalination process, since the desalination plant should have easy access to sea for seawater intake, as well for discharging brine; short distance to sea may reduce this cost
Feedwater quality	A direct function of the desalination cost for most of desalination processes, especially membrane-type processes; feedwater quality determines the type of pretreatment and thus also influences pretreatment cost
Output-water quality	Quality of desalted water may be a deciding factor for membrane desalination process where membrane life and replacement costs vary with output-water quality; however, thermally activated desalination process yields high-quality water with TDS <15 ppm, and output-water quality may not be a controllable parameter for such systems
Energy sources	All desalination processes require energy to separate salts from seawater; types and quality of energy source contribute to selection of desalination technology; renewable energy becomes a possible energy source for some desalination technologies where the adsorption desalination (AD) process utilizes low-temperature hot water from waste heat or solar energy
Pretreatment cost	Pretreatment is a necessary step in desalination processes to prevent or reduce performance degradation of separation unit such as evaporators of heat-driven systems and membranes of pressure-activated desalination systems
Chemical costs	Contribution significantly to desalination cost through operation cost and may vary with type of feedwater

Table 5.11 Equations Used to Calculate Capital and Operational Costs of an AD Plant

Item	Equation
V_L, volume of water produced by plant over lifespan (years)	$V_L = (\text{SDWP})(M_{sg})(365)(N)$, where SDWP is the specific daily water production rate in m^3/t of adsorbent, M_{sg} = mass of adsorbent, N = lifespan in years
Capital recovery factor [CRF(i,n)]	$\text{CRF}(i,n) = \dfrac{i(i+1)^n}{(i+1)^n - 1}$
Inflation-weighted factor with inflation rate j (IWF)	$\dfrac{\text{CRF}(i,n)}{\text{CRF}(i',n)}$ where $i' = (i-j)/(1+j)$
Capital cost per unit volume	$C_{\text{capital}} = \dfrac{D_{\text{capital}}}{V_L}$
Pumping power	$W_{\text{pumps}}(\text{kW}) = \sum_j \dfrac{\Delta P_j(\text{kPa}) \times V_{\text{pump},j}\,\text{m}^3/\text{s}}{\eta_j}$ where V_{pump} is volumetric flow rate η, is typical pump efficiency, j is number of pumps
Pumping cost per unit volume	$C_{\text{element}} = \dfrac{E_{\text{pumps}} \times \text{yearly operating hour}}{V_L}$ where E_{pump} (\$/h) $= W_{\text{pumps}}(kW) \times$ electricity rate (\$/kWh)
Maintenance cost per unit volume	$C_{\text{maintenance}} = (D_{\text{capital}} \times \beta)/V_L$ where $\beta = 4.89\%$ is percentage of the unit capital cost
Labor cost per unit volume	$C_{\text{labor}} = (D_{\text{capital}} \times \alpha)/V_L$ where $\alpha = 8.57\%$ is percentage of unit capital cost

water production capacity are realistically estimated, and Table 5.11 depicts all the equations used for the capital and operational costs of an AD plant over the plant lifetime.

For a fair comparison, we have selected a reasonably large desalination of 1000 m^3/day where the assumed plant lifetime is 30 years, interest rate is 5%, and inflation rate is 2%.

Using a lifecycle cost approach, the unit cost of the AD cycle was evaluated on the basis of performance data of the pilot AD plant [65,66]. The parameters used in the calculation of the unit production cost are summarized in Table 5.12. Owing to the low maintenance of the AD plant, the maintenance cost factor (β) of the AD plant is assumed to represent 4.63% of the direct capital cost. The feed seawater is pretreated only with microsize air bubbles in the prototype plant to remove the suspended solids by flocculation, and almost no chemicals are used, thus reducing the pretreatment cost significantly to only 0.35% of the direct capital cost. The labor (workforce) cost consists of the salary of operators in the plant and varies

Table 5.12 Key Parameters Used in the Unit Cost Estimation of AD Cycle

Plant lifetime, years	30
Interest rate, %	5
Electricity rate, US$/kWh	0.133
$(SDWP)_{advanced\ cycle}$	25
$(SDWP)_{conventional\ cycle}$	8

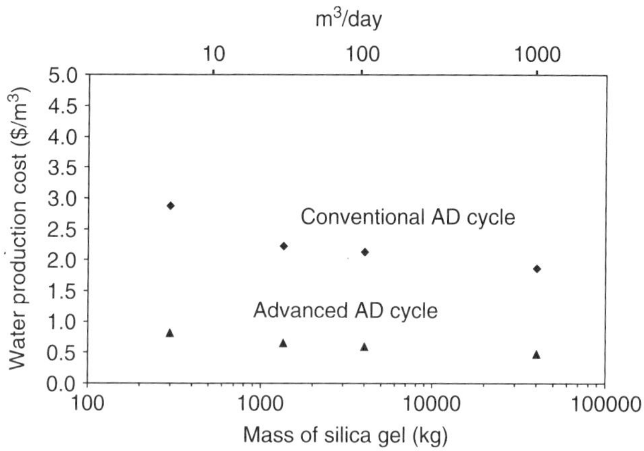

Figure 5.54 Lifecycle cost of conventional and advanced AD cycles.

with plant location. According to survey data, the labor cost represents 8.11% of the direct capital cost. As the thermal energy input to the AD plant is deemed free of charge, there is no cost for the consumption of waste heat. However, there is a cost assigned to equipment used for thermal energy extraction, such as heat exchangers, and these costs have been included. For fair comparison with the RO plants, the electricity cost ($/kWh) is selected as the same rate.

Figure 5.54 shows the unit potable-water production cost of the different capacities of conventional and advanced AD plants. It can be noted that the asymptotic unit production cost of the advanced AD process is ~$0.457/m^3, while that of the conventional AD cycle is ~4 times higher, at $1.91/m^3. The lower unit cost of the advanced AD cycle is attributed to (1) improvements in design, (2) reduction in pumping power, and (3) improvement in overall heat transfer coefficient of the condenser-evaporator integrated design.

The lowest unit production cost of the AD plant, i.e., the capital and operation costs, which sums to $0.457/m^3 for a plant capacity of 1000 m^3/day, is shown in Figure 5.55. The operation cost reduces asymptotically with respect to the output capacity, while the capital cost decreases more rapidly because of the upscaling of the plants. At 1000 m^3/day, the relative contributions

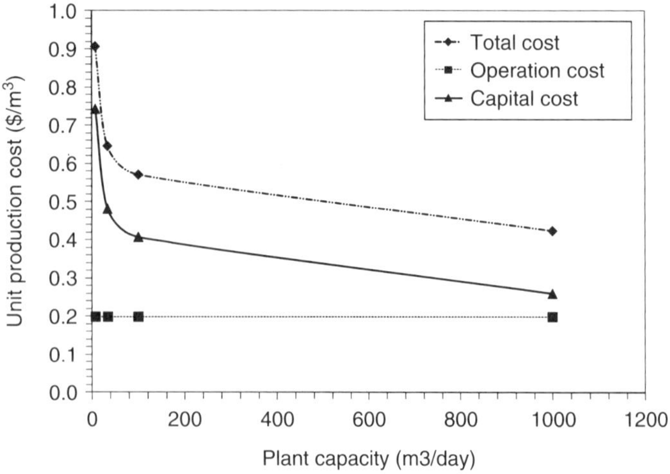

Figure 5.55 Potable-water production cost by AD cycle for different capacities.

from the capital and operation costs are roughly equal but at lower capacities, the capital cost contribution increases with decreasing water production per day.

5.8.3 Production Cost Comparison between AD and RO Plants

For a fair comparison with the desalination methods, the electricity rate is kept the same for the study. Among the available desalination systems such as multistage flash (MSF) distillation, multieffect (ME) distillation, and reverse osmosis (RO), the RO gives the lowest unit production cost. By adopting the parameters of RO plants and designing the AD cycle to the same production capacity of 1000 m^3/day, the common parameters are (1) plant lifetime is 30 years and (2) the nominal interest rate is 5%. Here, the inflation rate effect is omitted to give a lower-bound value for the unit production costs; Table 5.13 summarizes the parameters used. The electricity rate (in US dollars) is adjusted for the published rates of the utility companies

Table 5.13 Cost Parameters of AD Plant and Reference RO Plant with Adjusted Electricity Rate and Interest Rate

Parameter	AD Plant	Reference RO Plant 1 [67]	Reference RO Plant 2 [68]
Water production capacity, (m^3/day)	1000	1000	1000
Plant lifetime n, years	30	30	30
Interest rate i, %	5	5	5
Electricity rate, US$/kWh	0.133	0.133	0.133

Table 5.14 Contributions to Total Costs for AD and RO Plants at 1000 m³/ day

Unit Cost Factor[a]	AD Plant $/m³	AD Plant % of Total		RO Plant 1 [67] $/m³	RO Plant 1 [67] % of Total		RO Plant 2 [68] $/m³	RO Plant 2 [68] % of Total	
$C_{capital}$	0.259	56.7		0.215	22.8		0.127	15.9	
$C_{electrical}$	0.164	35.9	43.3	0.260	27.5	77.2	0.260	32.6	84.1
C_{labor}	0.021	4.6	—	0.050	5.3	—	0.021	2.6	—
$C_{pretreatment}$	0.001	0.2	—	0.035	3.7	—	0.0035	0.4	—
$C_{chemical}$	—	—	—	0.035	3.7	—	0.1	12.5	—
$C_{maintenance}$	0.012	2.6	—	0.061	6.5	—	0.039	4.9	—
$C_{membrane}$	—	—	—	0.060	6.4	—	0.02	2.5	—
C_{others}	—	—	—	0.228	24.1	—	0.228	28.6	—
Total cost = $\sum_i C(\$/m^3)$	0.457	100		0.944	100		0.799	100	

[a]The parameters $C_{pretreatment}$, $C_{chemical}$, $C_{membrane}$, and C_{others} represent the pretreatment cost, chemical cost, membrane replacement cost, and cost per unit volume of the potable water, respectively.

in Singapore since the 1990s. Table 5.14 shows the results of the calculations with contributions from items such as the capital, operational, maintenance, personnel, and other costs.

From Table 5.14, the unit cost of an RO plant is found to be twice that of the AD unit cost at \$0.944/m³. The higher cost contributions of RO are from the operational costs such as those for electricity, chemicals, maintenance, separation and membranes. Most reports of RO plants in the literature omitted or usually left vague the membrane replacement cost. The effective lifetime of a separation membrane is known to vary from 1 to 5 years, depending on the quality of water feeding through it. In both RO plants, an item denoted as C_{others} was defined as "costs attributed to factors not discussed here." This is likely the membrane replacement cost amounting to ~\$0.228/m³. This factor is depicted as "others" in row 8 of Table 5.14, and it contributes ≤20% of the unit production cost of water.

Using the cost data available in the literature, Figure 5.56 compares the unit potable-water production costs using conventional desalination methods such as MSFD, MED, membrane distillation (MD), brackish-water and seawater reverse osmosis (BWRO and SWRO), with the AD cycles. Despite the higher capital cost, the AD cycle still offers the lowest production costs for the seawater desalination process for the following reasons: (1) the AD cycle is operated by waste or renewable heat, which is available free of charge, and the parasitic electricity consumption in the plant is deemed the lowest; (2) the AD plant has almost no major moving parts, and as desalting occurrs at low temperatures, the maintenance cost is reduced to the lowest possible level for a plant (with only water pump maintenance); and (3) most importantly, the AD cycle requires no chemicals for cleaning during pretreatment of seawater and posttreatment of freshwater. With such key advantages and robust cycle, the AD plant is believed to be the most efficient desalination process thus far.

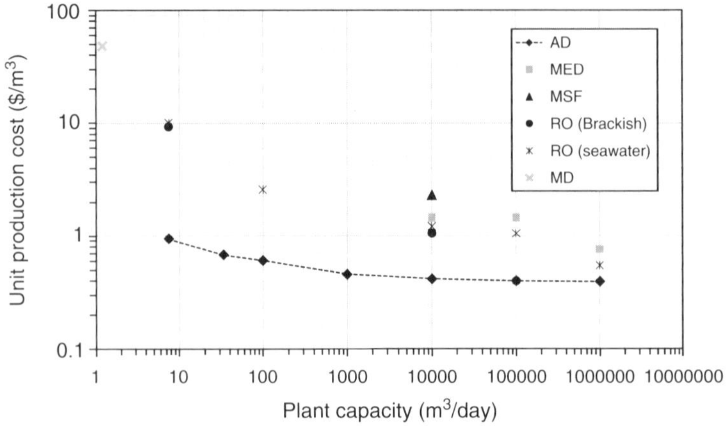

Figure 5.56 Comparison of unit production costs by different desalination methods and plant capacities.

5.8.4 Energy Sources of Desalination

The theoretical minimum energy for desalination of seawater, with an incremental recovery of freshwater, is ~0.706 kWh/m^3 as reported by Spiegler [69]. In practice, desalination processes are not ideal with perfect efficiency. Table 5.15 gives the thermoelectrical energy consumption of some of the commercially available desalination methods. However, it should be noted that the energy consumption figures of desalination are dependent on several factors such as design configuration, and site location and conditions. etc. The minimum-energy requirement for

Table 5.15 Energy Consumption of Different Desalination Methods

Desalination Method	Thermal Energy Consumed, kWh/m^3	Electrical Energy Consumed, kWh/m^3
Multistage flash (MSF) distillation	19.4	5.2
Multieffect distillation (MED)	16.4	3.8
Vapor compression (VC)	—	11.1
Reverse osmosis (RO)—single pass	—	8.2
Reverse osmosis (RO)—double pass	—	9.0

Source: All data are extracted from *Seawater Desalination in California*, California Coastal Commission, Chap. 1, Energy Use Section (available at http://www.coastal.ca.gov/index.html) [54].

desalting using the Gibbs free-energy approach is discussed in Section 5.A.1 of the Appendix.

It is reasonable to state that the MSFD process consumes the most energy, despite its relative maturity (≥50 years). It is followed by MED (or hybrid MED) systems and then vapor compression systems. The energy consumption of MSF distillation, by far the most widely used thermal process, is still at least 20–30 times the theoretical minimum. RO is a newer technology (30 years) that, with more recent improvements in energy recovery, is remarkably efficient, consuming only 3–10 times the theoretical minimum. It is important to consider, however, that RO consumes energy in the form of electricity. On the other hand, MSFD uses heat (or fuel) more directly. The conversion of thermal energy to electrical energy is only ~35% efficient. Therefore, in terms of fuel, RO consumes 9–30 times the theoretical energy requirement.

Overall, it is reasonable to conclude that today's desalination costs are ~\$5/m^3 and the energy consumption is at least 5 times that of the theoretical thermodynamic limit. It has been found that the adsorption–desalination process mitigates shortcomings such as high maintenance cost and high energy consumption that the current desalination technologies are facing. Adsorption–desalination is a novel desalination technology that has the potential of becoming the most energy-efficient method for production of potable water typically less than the benchmark energy consumption of 1.5 kWh/m^3.

5.9 CO_2 EMISSION SAVINGS

A CO_2 emission study was conducted that included a direct comparison to the existing desalination methods with respect to the thermoelectrical energy consumed. If the AD cycle is utilized for freshwater production, some CO_2 emission can be avoided because the AD cycle consumes only low-temperature heat source, thus, reducing global warming directly. The environmentally friendly aspect of the AD plant is demonstrated by comparing the amounts of CO_2 emission. A baseline calculation for CO_2 emission is shown in Section 5.A.2.

Assuming an emission rate of CO_2 at 64.2 t/TJ (terajoule) for natural gas (used as a primary fuel), and from the thermal and electricity consumption rates of Table 5.16, the corresponding CO_2 emission of the desalination processes of MSFD, MED, and RO plants can be computed. Detailed calculations of CO_2 emission are outlined in Section 5.A.2. The baseline emissions for thermaoelectrical energy consumption for the conventional desalination methods are compared to the emission obtained by the AD plant, as shown in Table 5.16, where the last (rightmost) column indicates the CO_2 savings.

The thermoeconomic viability of the adsorption–desalination (AD) process has been presented using a lifecycle approach. The specific-energy cost of AD plant is found to be 1.38 kWh/m^3, which is twice that of the thermodynamic limit: the lowest energy usage ever reported for a desalination method. At ≥1000 m^3/day, the total lifecycle cost asymptotic to a unit production cost of \$0.457/m^3 as compared

Table 5.16 CO_2 Emissions of Conventional and AD Desalination Methods for a Water Production Capacity of 1000 m^3/day

Desalination Method	$TB_{th,y}$ (tCO_2/year) (Thermal)	$EB_{elec,y}$ (tCO_2/year) (Electricity)	BE_y (tCO_2/year) (Combined)	Savings if AD Cycle Were Used for Water Production ER_y(tCO_2/year)
MSFD	1637	875	2512	2512–207 = 2305
MED	1383	640	2023	2023–207 = 1816
RO	0	1380	1380	1380–207 = 1173
AD	0	207	207	(Reference datum)

[a] In this table, $TB_{th,y}$ is the annual CO_2 emission from the burning of natural gas, $EB_{elec,y}$ is the emission from the generation of electricity consumed, BE_y is the baseline annual CO_2 emission, and ER_y is the annual CO_2 emission reduction by the AD plant for the same amount of desalting.

to \$0.944/m^3 of an equivalent RO plant. With low specific-energy consumption and simplicity of operation, the AD cycle is obviously a superior method for desalination when a low-temperature waste heat source is available. In addition, it could produce both the high-grade water (TDS $<$10 ppm) and cooling (as discussed previously) with almost no chemical usage for the pre/post-treatment of seawater. Owing to waste heat recovery, the AD cycle emits significantly lesser CO_2. When compared to a RO plant of the same water production capacity, a savings of at least 1172 t of CO_2 per year or 3.2 kg CO_2/m^3 of water can be realized with the AD cycle. Hence, the AD cycle is an excellent and practical solution for quenching the problem of global thirst by desalination.

5.10 OVERALL CONCLUSIONS

Theory and experimentation on the low-temperature heat-source-driven (waste heat or solar) adsorption–desalination (AD) cycles have been extensively explored in this chapter. The thermodynamic framework for silica gel–water systems as functions of pore widths and volumes has been developed for desalination. The mathematical modeling of the adsorption desalination plant has been described, and a prototype AD plant has been designed and constructed. Performance tests were conducted extensively, such as different hot water inlet temperatures, different cycle times, and different evaporation temperatures of the seawater. The results also showed that two-bed mode AD process was capable of producing a specific daily water production (SDWP) rate of 7.8 m^3 of potable water per metric ton of silica gel per day, while a SDWP of 10 was achieved by a four-bed mode AD plant. The lumped-parameter model was adapted since the current objective was to evaluate the performance of the AD plant. The optimal cycle times for a specific hot-water inlet temperature ranging from 60°C to 85°C for both two-bed and four-bed operation modes were experimentally evaluated. The experimental results showed that the operation cycle time for optimal performance of the AD plant varies with the regeneration temperature of the plant. A longer cycle time is required for lower heat source temperatures in order to completely regenerate, subsequently giving the higher water production rate. On extensive experimentation on the AD plant, the recovery ratio of the AD plant was found to be 70% without compromising the performance of the plant. This significantly high recovery ratio is attributed to (1) the low evaporation temperature of the seawater and (2) the fact that the evaporation is governed by the adsoprtion phenomenon of silica gel.

Economic analysis of the AD process was also briefly discussed in this chapter. The analysis was explored for different sizes of AD plant, and comparisons between the AD and reverse osmosis plants were also presented. The results showed the unit potable-water production cost of the AD plant is lower than that of an RO plant for larger-scale operation with implementation of the concrete bed design. Thus, AD plants not only are cost-effective but also produce water of high quality, and almost no chemical is used at the pre- and post-treatment stages of seawater or brackish-water. The AD processes emit lesser CO_2 by comparison to a RO plant of the same production capacity:- a reduction of $\geq$1172 t CO_2/year, or a savings of 3 kg of CO_2 per m^3 of water. Hence, the AD cycle is a promising and practical solution for quenching the global thirst by desalination as well as an excellent method of reducing global warming.

APPENDIX

5.A.1 Minimum-Energy Requirement for Desalting by Gibbs Free-Energy Approach

A typical desalting process is illustrated in Figure 5.A.1, where the feed seawater has concentration X_s and the rejected brine has concentration X_b, which is greater than X_s with the production of salt-free potable water.

A simple expression for calculating the free energy of separation per unit amount of potable water can be written assuming that the solution is dilute, the recovery approaches zero, and the product water is salt-free:

$$(dG)_{P_0,T_0} = \sum_i \mu_i \, dM_i = (\mu_w^0 dM_d)_{\text{distillate stream}} + \left(\mu_w dM_b + \sum_s \mu_s \, dM_s\right)_{\text{brine stream}} \tag{5.A.1}$$

Here, G is Gibbs free energy, μ_i is the chemical potential of species i, and M_i is the mass of species i; μ_w^0 represents pure water, M_d is mass of the distillate, and μ_w is the water in the brine, whereas μ_s refers to the salt species. It is assumed that the product is salt-free and thus, from mass balance $dM_d = -dM_b$ and $dM_s = 0$, Equation (5.A.1) becomes

$$(dG)_{P_0,T_0} = (\mu_w^0 - \mu_w) \, dM_d \tag{5.A.2}$$

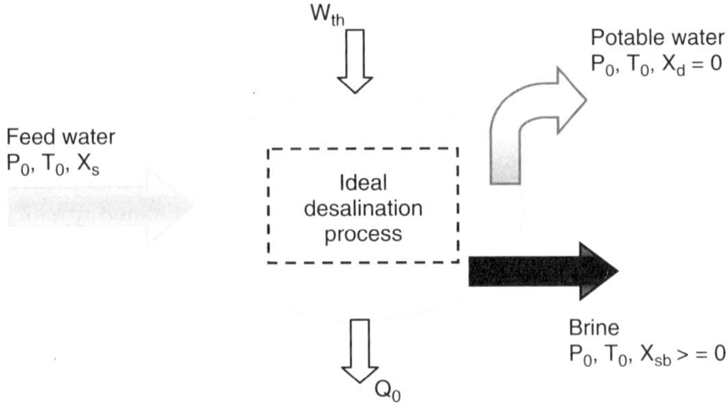

Figure 5.A.1 A typical desalination process.

For dilute solutions, $(\mu_w^0 - \mu_w) = -RT_0 \ln(X_w)$, and for large amounts of X_w, $-\ln(X_w) = X_s$, where X_s is the mole fraction of the total salts in the feed and X_w is the mole fraction of water. Thus, the theoretical minimum work per unit product for a small recovery approaching zero can be expressed as

$$W_{\text{th0}} = \left(\frac{\partial G}{\partial M_w}\right)_{P_0,T_0} = R\, T_0\, X_s \tag{5.A.3}$$

For the ionic composition, the factor φ is introduced into Equation (5.A.3), and then the theoretical minimum work for desalination is given by

$$W_{\text{th0}} = \left(\frac{\partial G}{\partial M_w}\right)_{P_0,T_0} = \varphi\, R\, T_0\, X_s \tag{5.A.4}$$

Using Equation (5.A.4), the theoretical minimum work for the separation for desalination process for different initial salt concentrations at constant solution temperature of 27°C is given in Figure 5.A.2. The theoretical minimum free energy for the desalination process varies linearly with the initial salt concentration, while the minimum free energy at salinity 35.4 g/kg of water is ~0.86 kWh/m^3.

For the finite recovery amount of potable water from the feed seawater, Equation (5.A.4) integrates the feed concentration X_s with the brine concentration X_b as:

$$W_{\text{thR}} = \frac{1}{M_d}\int_{X_s}^{X_b} R\, T_0\, X\, dM_d \tag{5.A.5}$$

For any X and M between X_s, M_f and X_b, M_b, the differential increase in salt content of M for pure water is $M\,X = (M - dM_w)(X + dX)$ and, therefore, $M\,dX =$

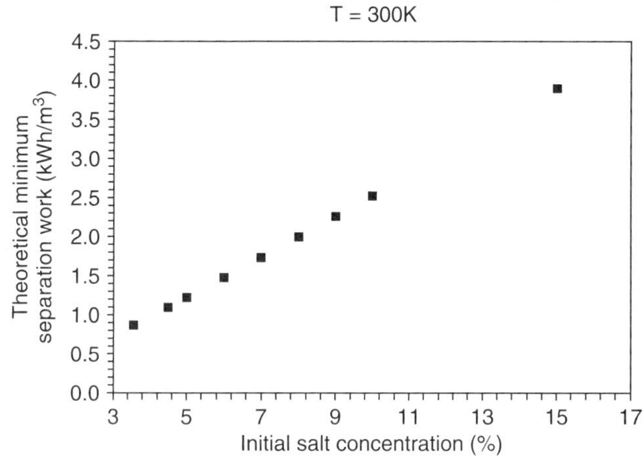

Figure 5.A.2 Theoretical minimum separation work for desalination with initial salt concentration.

$(X + dX)dM_w$. However, $dX\ dM_w$ is negligible, and thus $M\ dX = X\ dM_w$. Furthermore, $M\ X = M_f\ X_s$, and Equation (5.A.5) becomes

$$W_{thR} = \frac{M_f}{M_d} R\,T_0\,X_s \ln \frac{X_b}{X_s} \tag{5.A.6}$$

However, $\frac{M_d}{M_f}$ is the recovery ratio, using salt and mass balance. Then $M_f/M_d = [X_b/(X_b - X_s)]$, and finally the theoretical free energy for finite recovery becomes

$$W_{thR} = R\,T_0 \frac{X_s X_b}{X_b - X_s} \ln \frac{X_b}{X_s} \tag{5.A.7}$$

Using Equation (5.A.7), the free energy for different recovery ratios of the desalination process is presented in Figure 5.A.3 which shows the rates for recovery of potable water from seawater with different initial salt concentrations such as 4.5%, 3.5%, and 0.5% at constant temperature (27°C). It can be noted that for brackish-water desalination, the theoretical free energy required is <0.4 kWh/m^3.

5.A.2 Baseline Calculation for CO_2 Emission

The baseline CO_2 emissions in a desalination process can be estimated as the sum of CO_2 emissions from utilization of both thermal energy and electrical energy. For thermally activated systems, CO_2 emission emanates from the energy consumed in evaporating the seawater as well as the electricity consumption for moving the coolant or heat sources. On the other hand, the membrane desalination processes would theoretically consume electricity for pushing the saline solution and permeate. The following equations provide a method of calculation for the baseline emission for a desalination process: $BE_y(tCO_2/\text{year}) = TB_{th,y} + EB_{elec,y}$, where

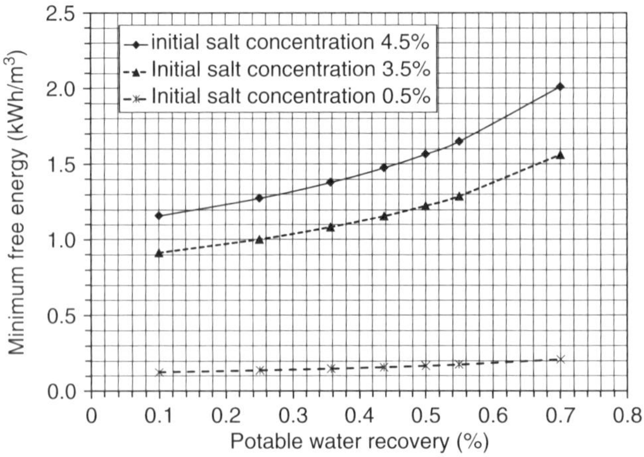

Figure 5.A.3 Theoretical free energy of desalting for potable-water recoveries of solutions with different initial salt concentrations.

BE_y is the baseline annual CO_2 emission, $EB_{elec,y}$ is the emission from generation of electricity consumed, and $TB_{th,y}$ is the annual CO_2 emission from burning of natural gas for desalting. Then $TB_{th,y}$ can be calculated using the equation $TB_{th,y} = Q_{elec,y}(\text{MJ/year}) \times EF_{gas}(\text{tCO}_2/\text{MJ})$, where $Q_{elec,y}$(MJ/year) is the annual thermal energy required for the desalination of a typical amount of potable water and $EF_{gas}(\text{tCO}_2/\text{MJ})$ is the emission factor for the burning of natural gas. The value of EF_{gas} is taken as 6.42×10^{-5} tCO_2/MJ.

The emission from the electricity consumed in the desalination process $EB_{elec,y}$ is calculated as $EB_{elec,y} = EG_y(\text{MWh/year}) \times CEF_{elec}(\text{tCO}_2/\text{MWh})$, where EG_y is the amount of electricity generated by the power plant and CEF_{elec} is the CO_2 emission factor for the generation of the electricity, and its value is taken as 0.4612.

The CO_2 emission reduction by AD process is given by $ER_y(\text{tCO}_2/\text{year}) = BE_y - ADE_y$, where ER_y and ADE_y are the annual CO_2 emission reduction and the annual CO_2 emission by the AD plant, respectively. The following expression is used to estimate the CO_2 emission by the AD plant: $ADE_y(\text{tCO}_2/\text{year}) = EB_{elec,y}$ The units here are similar, at metric ton of CO_2 per year.

NOMENCLATURE

Symbols and Abbreviations

A	area, m^2
BPE	boiling-point elevation, K (kelvins)
c	concentration, kg/kg
c^0	limiting uptake, kg/kg
c^*	equilibrium uptake, kg/kg

COP	coefficient of performance
c_p	specific-heat capacity, J kg^{-1} K^{-1}
D_{so}	a kinetic constant for silica gel–water system, m^2 s^{-1}
E_a	activation energy of surface diffusion, kJ/mol
h_{fg}	latent heat, kJ/kg
k	thermal conductivity, W m^{-1} K^{-1}
K_0	preexponential coefficient, k/Pa
L	length, m
m	mass, kg
$\dot{m}$	mass flow rate, kg/s
n	number of beds
N^*	number of adsorbent surface atoms in Henry region
N^0	number of adsorbent surface atoms in condensing phase
N_s	number of adsorbent surface atoms
OPR	overall performance ratio
P	pressure, Pa
P_0	reference pressure, Pa
PR	performance ratio,
q	fraction of refrigerant adsorbed by adsorbent, kg/kg
Q	total heat or energy, W or J
Q_{st}	isosteric heat of adsorption, kJ/kg
Q_{st}^*	isosteric heat of adsorption at zero surface coverage, kJ/mol
Q^0	least adsorption energy, kJ/mol
R	universal gas constant, J kg^{-1} K^{-1}
R_p	average radius of silica gel, m
SDWP	specific daily water production, m^3/t·day (where t = metric ton)
T	temperature, K
t	time/(t/t_{hcycle}), s
T_0	reference temperature, K
U	overall heat transfer coefficient, W m^{-2} K^{-1}
u	internal energy per unit mass, J kg^{-1} K^{-1}
v	specific volume, m^3/kg
v_p	pore volume, cm^3/g
$\gamma, \delta, theta$	flags governing mode of operation
λ	thermal conductivity, W m^{-1} K^{-1}
ρ	effective density, kg/m^3
τ	number of cycles per day

Subscripts

a	adsorbate
ads	adsorbent; adsorption
ch	chilled water

cond	condenser
cw	cooling-water
d	distillate
evap	evaporator
g	gaseous phase
hw	hot water
HX	heat exchanger
in	inlet
Mads	master adsorption
Mdes	master desorption
out	outlet
reg	regeneration
s	salt, adsorbent, or saturation
Sads	slave adsorber
Sdes	slave desorber
sg	silica gel
w	pure water

REFERENCES

1. WHO, *Guidelines for Drinking-Water Quality*, Vol. 1, *Recommendation*, World Health Organization, 2008, p. 218.
2. Shiklomanov IA, An introduction to global fresh water issues, in Gleick PH, ed., *Water in Crisis: A Guide to the World's Fresh Water Resources*, Oxford Univ. Press, New York, 1993. pp. 4– 5.
3. National Intelligence Council, Water scarcity map adapted from the original map and research published by the International Water Management Institute (IWMI) (Annual Report 2007–2008, p. 11), in *Global Trends 2025: A Transformed World*, Government Printing Office, Washington, DC, Nov. 2008, p. 55.
4. See http://atlas.nrcan.gc.ca/site/english/maps/freshwater/1 7.
5. Anonymous, A thirsty world, *UNESCO Courier* p. 21 (Sept. 2001).
6. Wagner L, *Water Desalination: Tap in to Liquid Gold*, Research Report, 2007.
7. US Dept. of Energy, *Desalination and Water Purification Technology Roadmap Report*, 2003.
8. Darwish MA, Al-Najem NM, Energy consumption by multi-stage flash and reverse osmosis desalters, *Appl. Thermal Eng.*, **20**:399 (2000).
9. Fiorinia P, Sciubbaa E, Sommarivab C, A new formulation for the non-equilibrium allowance in MSF processes, *Desalination* **136**:177–188 (2001).
10. Aly NH, El-Fiqi AK, Thermal performance of seawater desalination systems, *Desalination*, **158**:127–142 (2003).
11. A1-Shammiri M, Safar M, Multi-effect distillation plants: State of the art, *Desalination*, **126**:45–59 (1999).
12. Rahman H, Hawlader MNA, Malek A, An experiment with a single-effect submerged vertical tube evaporator in multi-effect desalination, *Desalination* **156**:91–100 (2003).

13. El-Dessouky HT, Ettouney HM, Multiple-effect evaporation desalination systems: Thermal analysis, *Desalination* **125**:259–276 (1999).
14. El-Dessouky HT, Ettouneya HM, Mandani F, Performance of parallel feed multiple effect evaporation system for seawater desalination, *Appl. Thermal Eng.*, **20**:1679–1706 (2000).
15. El-Nashar AM, Qamhiyeh AM, Simulation of the steady-state operation of a multi-effect stack seawater distillation plant, *Desalination*, **101**:231–243 (1995).
16. Van der Bruggen B, Vandecasteele C, Distillation vs. membrane filtration: Overview of process evolutions in seawater desalination, *Desalination* **143**:207–218 (2002).
17. Kershman SA, Rheinlander J, Gabler H, Seawater reverse osmosis powered by reneawable energy sources—hybrid wind/photovoltaic power/grid power supply for small-scale desalination in Libya, *Desalination*, **153**:17–23 (2002).
18. Ahmad GE, Schmid J, Feasibility study of brackish-water desalination in Egyptian deserts and rural regions using PV systems, *Energy Convers. Manage.* **43**:2641–2649 (2002).
19. Tzen E, Theofilloyianakos D, Sigalas M, Karamanis K, Design and development of a hybrid autonomous system for seawater desalination *Desalination* **166**:267–274 (2004).
20. Ng KC, Wang XL, Gao LZ, Chakraborty A, Saha BB, Koyama S, *Apparatus and Method for Desalination*, SG Patent Application 200503029-1 (2005) and WO Patent 121414A1 (2006).
21. Dabrowski A, Adsorption—from theory to practice, *Adv. Colloid Interface Sci.* **93**:135–224 (2001).
22. Chakraborty A, *Thermoelectric Cooling Devices: Thermodynamic Modelling and Their Applications in Adsorption Cooling Cycle*, PhD thesis, National Univ. Singapore, Nov. 2005.
23. IUPAC, Reporting physisorption data for gas/solid systems, *Pure Appl. Chem.* **57**(4): 603–619 (1985).
24. Sukuki M, *Adsorption Engineering*, Kodansha Ltd., Tokyo and Elsevier Science Publishers B. V., Amsterdam, 1990.
25. Ruthven DM, *Principles of Adsorption and Adsorption Processes*, Wiley, New York, 1984.
26. Do DD, *Adsorptrion Analysis: Equilibrium and Kinetics*, Series on Chemical Engineering, Vol. 2, Imperial College Press, 1998.
27. Authors' unpublished data on silica gel characteristics, 2011.
28. Cortés FB, Chejne F, Carrasco-Marín F, Moreno-Castilla C, Pérez-Cadenas AF, Water adsorption on zeolite 13X: Comparison of the two methods based on mass spectrometry and thermogravimetry, *Adsorption* **16**:141–146 (2010).
29. Simonot-Grange MH, Elm'Chaouri A, Weber G, Dufresne P, Raatz F, Joly J-F, Characterization of the dealumination effect into H faujasites by adsorption: Part 1. The water molecule as a structural aluminum ion selective probe, *Zeolites* **12**:155–159 (1992).
30. Aristov YI, Tokarev MM, Freni A, Glaznev IS, Restuccia G, Kinetics of water adsorption on silica Fuji Davison RD, *Micropor. Mesopor. Mater.* **96**:65–71 (2006).
31. Gordeeva LG, Tokarev NLM, Parmon VN, Aristov YmL, Selective water sorbents for multiple application. 6. Freshwater production from the atmosphere, *React. Kinet. Catal. Lett.* **65**(1):153–159 (1998).

32. Restuccia G, Cacciola G, Performances of adsorption systems for ambient heating and air conditioning, *Int. J. Refrig*. **22**:18–26 (1999).
33. Gavrilov KB, Okunev AG, Aristov YuI, Monolayer physical adsorption in narrow pores. Apparent surface dimension, *React. Kinet. Catal. Lett*. **58**(1):39–48 (1996).
34. Alefeld G, Bauer HC, Maier-Laxhuber P, Rothmeyer M, A zeolite heat pump, heat transformer and heat accumulator, *Proc. Int. Conf. Energy Storage*, Brighton, UK, 1981.
35. Henninger SK, Habib HA, Janiak C, MOFs as adsorbents for low temperature heating and cooling applications, *J. Am. Chem. Soc*. **131**:2776–2777 (2009).
36. Wang X, Ng KC, Experimental investigation of an adsorption desalination plant using low-temperature waste heat, *Appl. Thermal Eng*. **25**(17–18):2780–2789 (2005).
37. Zejli D, Benchrifa R, Bennouna A, Bouhelal OK, A solar adsorption desalination device: First simulation results, *Desalination* **168**(1–3):127–135 (2004).
38. Wang X, Chakarborty A, Ng KC, Saha BB, How heat and mass recovery strategies impact the performance of adsorption desalination plant: Theory and experiments, *Heat Transfer Eng*. **28**(2):147–153 (2007).
39. Ng KC, Saha BB, Chakraborty A, Koyama S, Adsorption desalination quenches global thirst, *Heat Transfer Eng*. **29**(10):845–848 (2008).
40. Thu K, Ng KC, Saha BB, Chakraborty A, Koyama S, Operational strategy of adsorption desalination systems, *Int. J. Heat Mass Transfer* **52**(7–8):1811–1816 (2009).
41. Thu K, Saha BB, Chakraborty A, Chun WG, Ng KC, Study on an advanced adsorption desalination cycle with evaporator-condenser heat recovery circuit, *Int. J. Heat Mass Transfer* **54**(1–3):43–51 (2011).
42. Chakraborty A, Saha BB, Ng KC, Koyama S, Srinivasan K, Theoretical insight of physical adsorption for a single-component adsorbent + adsorbate system: I. Thermodynamic property surfaces, *Langmuir* **25**:2204–2211 (2009).
43. Chakraborty A, Saha BB, Ng KC, Koyama S, Srinivasan K, Theoretical insight of physical adsorption for a single component adsorbent + adsorbate system: II. The Henry region, *Langmuir* **25**(13):7359–7367 (2009).
44. Chakraborty A, Saha BB, Koyama S, Ng KC, Srinivasan K, Adsorption thermodynamics of silica gel-water systems, *J. Chem. Eng. Data* **54**(2):448–452 (2009).
45. Mcenaney B, Estimation of the dimensions of micropores in activated carbons using the Dubinin-Raduskevich equation, Carbon, **25**:69–75 (1987).
46. Chakraborty A, Leong KC, Thu K, Saha BB, and Ng KC, Theoretical insight of adsorption cooling, *Applied Physics Letters* Vol. 98, No. 22. art no. 221910 (2011).
47. Steele WA, The interaction of gases with solid surface, *The International Encyclopedia of Physical Chemistry and Chemical Physics*, Vol. 3, Pergamon Press, Oxford, 1974.
48. Schindler JB, Levan MD, The theoretical maximum isosteric heat of adsorption in the Henry's law region for slit-shaped carbon nanopores, *Carbon* **46**:644–648 (2008).
49. Ravikovitch PI, Vishnyakov A, and Neimark AV, Density functional theories and molecular simulations of adsorption and phase transitions in nanopores, *Physical review E*, **64**: art no. 011602, (2001).
50. Thu K, *Adsorption Desalination: Theory and Experiments*, PhD thesis, National Univ. Singapore, 2010.
51. Chakraborty A, Saha BB, Koyama S, Ng KC, Specific heat capacity of a single component adsorbent-adsorbate system, *Applied Physics Letters* **90**(17): art. no. 171902, (2007).

52. Tien C, *Adsorption Calculation and Modeling*, Butterworh-Heinemann Series in Chemical Engineering, Boston, 1994.

53. Saha BB, Kashiwagi T, Experimental investigation of an advanced adsorptiosn refrigeration cycle, *ASHRAE* (Am. Soc. Heating, Refrigerating, and Air-Conditioning) *Trans*. **103**(2):50–58 (1997).

54. Boelman EC, Saha, BB, Kashiwagi T. Experimental investigation of a silica gel–water adsorption refrigeration cycle—the influence of operating conditions on cooling output and COP, *ASHRAE Trans*. **101**(2):358–366 (1995).

55. Chua HT, Ng KC, Chakraborty A, Oo NM, Othman MA, Adsorption characteristics of silica gel + water systems, *J. Chem. Eng. Data* **47**(5):1177–1181 (2002).

56. Aristov YI, Tokarev MM, Cacciola G, Restuccia G, Selective water sorbents for multiple applications: 1. $CaCl_2$ confined in mesopores of the silica gel: Sorption properties, *React. Kinet. Catal. Lett*. **59**(2):325–334 (1996).

57. Tokarev MM, Okunev BN, Safonov MS, Kheifets LI, Aristov YI, Approximation equations for describing the sorption equilibrium between water vapor and a SWS-1L composite sorbent, *Russ. J. Phys. Chem*. **79**(9):1490–1493 (2005).

58. Saha BB, Chakraborty A, Koyama S, Aristov YI, A new generation cooling device employing $CaCl_2$-in-silica gel-water system, *Int. J. Heat Mass Transfer* **52**(1–2):516–524 (2011).

59. Chihara K, Suzuki M, Air drying by pressure swing adsorption, *J. Chem. Eng*. (Japan) **16**:293–298 (1983).

60. Ng KC, Thu K, Saha BB, Chakraborty A, Study on a waste heat-driven adsorption cooling cum desalination cycle, *International Journal of Refrigeration* Vol. 35, No. 3, pp. 685–693, (2011).

61. Spiegler KS, El-Sayed YM, *A Desalination Primer* Balaban Desalination Publications, Santa Maria Imbaro, Italy, 1994.

62. Ng KC, Thu K, Chakraborty A, Saha BB, Chun WG, Solar-assisted dual-effect adsorption cycle for the production of cooling effect and potable water, *Int. J. Low-Carbon Technol*. **4**(2):61–67 (2009).

63. Thu K, Chakraborty A, Saha BB, Chun WG, Ng KC, Life-cycle cost analysis of adsorption cycles for desalination, *Desal. Water Treat*. **20**(1–3):1–10 (2010).

64. Atikol U, Aybar HS, Estimation of water production cost in the feasibility analysis of RO systems, *Desalination* **184**(1–3):253–258 (2005).

65. Ng KC, Thu K, Yanagi H, Saha BB, Chakraborty A, Tawfiq Y, Al-Tawfiq YG, *Apparatus and Method for Improved Desalination*, PCT/SG2009/000223, 2008.

66. Saha BB, Ng KC, Chakraborty A, Thu K, *Desalination System and Method*, PCT/IB2010/001757, publication date Jan. 27, 2011.

67. Atikol U, Aybar HS, Estimation of water production cost in the feasibility analysis of RO systems, *Desalination* **184**:253–258 (2005).

68. Fiorenza G, Sharma VK, Braccio G. Techno-economic evaluation of a solar powered water desalination plan, *Energy Convers. Manage*. **44**:2217–2240 (2003).

69. Spiegler KS, El-Sayed YM, The energetics of desalination processes, *Desalination* **134**:109–128 (2001).

70. California Coastal Commission, Energy use, Chap. 1 in *Seawater Desalination in California*, (available at http://www.coastal.ca.gov/index.html; accessed 1/15/07).

CHAPTER 6

Advanced Instrumentation, Measurement, Control, and Automation (IMCA) in Multistage Flash (MSF) and Reverse-Osmosis (RO) Water Desalination

NOAM LIOR*, ALI EL-NASHAR, and CORRADO SOMMARIVA

*Corresponding author.

Advances in Water Desalination, First Edition. Edited by Noam Lior.

6.1 INTRODUCTION AND CHAPTER OBJECTIVES

A paramount goal in the proper design and operation of any desalination plant is the reduction of its investment and operating costs to minimize the cost of the freshwater produced. Important methods for cost reduction include the minimization, per unit desalted product water, of the costs of the (1) capital investment, (2) energy, (3) chemical additives consumed by the plant, (4) labor (amount and skill level), and (5) plant environmental impact. Another goal, related but also independent, is to maximize production, and to that end increase reliability and decrease downtime. One potential tool for these optimization tasks is the use of adequate and economical control strategies that can stabilize the operation of the plant at high-efficiency operating points while effectively handling the different constraints on the process variables. Better instrumentation, measurement, control, and automation (IMCA), indeed, are important for meeting these optimization objectives. This chapter focuses on the status and improvement needs for MSF and RO (the predominant seawater desalting processes in the world) plant automation and operation optimization to reduce desalted water costs. It includes (1) a state-of-the-art review as reference material, (2) a presentation of cost–benefit analysis, and (3) a description of the recommended automation systems for RO and MSF plants. Modeling needs and recommendations are also included. The chapter systematically presents topic-related principles, equations, and applications; describes the IMCA systems in several representative MSF and RO desalination plants; and presents solved examples to assist both practitioners and researchers. Comprehensive lists of literature on IMCA as related to MSF and RO desalination are given in the Further Reading section, following the cited references list, at the end of the chapter.

The chapter is intended to assist both practitioners and researchers. It is outlined as follows:

- A review of the literature and of operating plants, including solicited expert opinions from automation system manufacturers and other experts to determine how automation has been and can be applied in desalination plants to meet the project objectives. This review also establishes which parameters in the RO and MSF desalination processes need to be measured for process automation and the conditions under which they need to be measured.
- Identification of the current and potentially future process automation methods for RO and MSF using all the publicly available information.
- Description of simple process models for the automation/control system cost-benefit analysis.
- Identification and calculation of the impact of the degree of automation and control precision, including performance optimizing controls, on product-water cost (automation cost–benefit ratio) in desalination plants.
- Recommendations for the most suitable automation system for RO and MSF plants based on the cost–benefit analysis.

The readers should keep in mind that costs, technology, and system requirements and designs change with time, and use the information in this report accordingly. The descriptions in this chapter of the plant configurations, instrumentation, and controls were valid when the plants were constructed, but may have changed in the meantime, and some of the plants may even have been shut down.

6.2 A REVIEW OF THE CURRENT STATUS OF IMCA IN MSF AND RO WATER DESALINATION

6.2.1 General

To establish some starting point of this state-of-the-art and literature review, we shall regard it as the book and reviews published by the first author [1–6], which define much of the published material to the year 2000. A paper by the authors [7] briefly reviewed the state of the art at that time. In addition to publications cited as part of this paper, IMCA sources are listed in Further Reading Section 6.F.1

for MSF; in Section 6.F.2, for RO; in Section 6.F.3 for general desalination and related topics; and in Section 6.F.4 for general control theory.

A dynamic source of information is the *Encyclopedia of Desalination and Water Resources (DESWARE)*, published in 2001, which has a fairly large chapter on process instrumentation, control, and automation. It contains brief general reviews on control theory and components, but not specific to desalination, in the articles by Bunzemeier and Krause [8], Bunzemeier and Litz [9], and two articles by Muroni and Borsani [10,11].

A brief review of the theory of fault diagnosis, a systematic method for determining (after detection of a fault) unacceptable deviation of a measured or calculated parameter in chemical processes, is given by Tarifa and Scenna [12].

6.2.2 Multistage Flash Desalination IMCA Literature

DESWARE contains a section on plant availability analysis of MSF distillation [13], using fault-tree logic, and contains methodology and examples and evaluations of some MSF plants in the UAE based on a somewhat small and old set of data. It concludes that availability is reasonably high (~80%), with pumps typically having the highest failure rate, followed by valves. Hence, it recommends consideration of the once-through rather than the commonly used recirculation MSF plant design, the use of variable-speed pumps (without flow control valves), and hydrostatic protection from boiling of the brine in the brine heater (which causes scaling) instead of a Δp control valve. In summary, it states: "Considering future MSF plants, the central message of this work for improved design and operation is to simplify the design substantially. This shows a much higher potential than any sophisticated modeling control and simulation effort." This conclusion is identical to ours. We note, however, that current operational experience shows that both MSF and MED plants operating in the Gulf have an availability in excess of 99.5% and have operated continuously without interruption for more than 5 years between acid cleanings, and the specific suggestion to implement once-through plants may not offer the advantage perceived from much older experiences.

A paper on brine flow hydraulics in MSF plants, [14] is useful for modeling improvement, and another expounds the need for multivariable control in MSF desalination plants [15].

Tarifa and Scenna [12] briefly review the theory of fault diagnosis using artificial intelligence, as applied to MSF plants. They followed this work up to apply the fault diagnostic system to an MSF plant, and met with some, albeit insufficient, success. In another paper Tarifa and Nicolas [16] used neural networks for developing the model, and stated that the training method they developed makes the system simpler yet more accurate than previous ones. More work remains to link this model with a qualitative one to generate appropriate explanations for each diagnostic.

Another article in DESWARE is on control of cogeneration power plants for MSF desalination [17]. Since combined cycles using gas turbines with bottoming steam turbines are being aggressively and successfully developed and used for raising the energy efficiency (now approaching 60%) and reducing emissions in power generation applications, this article discusses control schemes and strategies for some of these cycles in conjunction with MSF plants.

A brief review of hybrid modeling and control techniques with examples for the control of an MSF brine heater and of the TBT, as well as level control of the pretreatment stage in an RO plant, are contained in a publication by Gambier and Badreddin [18].

The application of neural network process simulation to a large-scale commercial MSF desalination plant (22,270 m^3/d, Az-Zour, Kuwait) and a RO desalination plant (56,800 m^3/d, Jeddah, Saudi Arabia) was compared with that obtained from process models developed by the combination of the ASPEN PLUS and SPEEDUP commercial systems. The approach to neural network modeling is described in some detail with recommendations for the best approach. Neural networks are stated to have been effective for performance modeling but seem to have produced larger errors than model-based process simulation.

Nonlinear model predictive control (NLMPC) was analyzed as a control strategy for MSF desalination plants [19], and was found to offer good control performance and substantial energy savings. At the same time, it was found in comparison with conventional proportional integral (PI) control that its tuning was difficult and computation long. A thorough search of desalination-related IMCA patents was not part of this study, although but a number of patents were issued on these topics. Krause [20] describes a new method for controlling MSF.

6.2.3 Reverse-Osmosis Desalination IMCA Literature

It was realized very early that careful plant supervision and maintenance, assisted in a major way by real-time computer-aided automatic controls, are of vital importance for the economical and reliable operation of RO desalination plants [21–23]. The early publications have described system design and methodology, but not field testing.

Conventional automatic control of RO uses the PI control strategy [proportional integral; the derivative mode is often disabled because of large signal–noise ratio (SNR) in the process]. In an attempt to examine the application of advanced control and optimization to RO processes [24,25], constrained model predictive control (CMPC) was applied to laboratory-scale systems. While only some of the relevant parameters were controlled, the application of CMPC was found to increase the plant output by up to 13.6% and slightly improve the product quality, as compared with the PI system. An important aspect of this improvement is that CMPC allows for some knowledge-based corrections. Further investigations of applying neural networks and fuzzy logic to desalination plant control were made but very superficially [26], and similarly to all the other proposals and studies for the application of advanced control systems to water desalination conducted in the past, none have included a cost–benefit analysis.

6.2.4 Major Actionable Conclusions from the Literature Review

One of the most critical issues for maintaining low product cost is the achievement of high plant availability, and to this end the following measures could be beneficial:

- Using fault-tree analysis in some form, identify the most vulnerable components, attempt to replace them by an improved design, and monitor them more carefully.
- Simplifying the design wherever possible.

Model-based predictive control seems to have the potential for improving control of both MSF and RO plants, but needs to be examined and tested in the field.

Artificial intelligence, neural networks, and fuzzy logic have been studied further for desalination plants, but haven't been proven (perhaps not yet) as viable approaches.

6.3 DESALINATION PROCESSES AUTOMATION AND OPERATION OPTIMIZATION

6.3.1 Introduction

Currently (as of early 2012), the majority of operating MSF and RO plants are controlled by simple proportional integral derivative (PID)-type controllers (a glossary of abbreviations and instrumentation, measurement, control, and automation terms is given in the Nomenclature section at the end of this chapter (before the References list). The use of simple PID controllers, which can maintain the stable operation of the plant, certainly overlooks the complex interactions that may exist within the plant. Both MSF and RO plants are nonlinear multivariable systems, and an in-depth investigation of the different interactions between the inputs and outputs of the plant is necessary to optimize the selection of the control structure.

The use of advanced model-based control schemes such as dynamic matrix control (DMC), and nonlinear model predictive control (NLMPC) can have the potential of providing better control of the plant, but their cost and added complexity must be justified by the cost–benefit ratio resulting from any improvements it may produce.

Both MSF and RO plants are particularly suited for constraint-accommodating control strategies for a number of reasons; for instance, limits on temperature in the brine heater (in MSF plants) and feedwater (in RO plants) must be observed, based on the scale control method used. These constraints can be easily incorporated within the framework of an advanced control strategy that can accommodate, or make adjustments for, such constraints.

While the most important goal in the past was to achieve the highest possible availability of plants, today, optimization of plant, performance becomes increasingly important. For that purpose, it is no longer sufficient to operate plants according to the operation manuals of the process manufacturer or to follow the experience of the operators. It now becomes necessary to use supplementary calculations to determine the optimal setpoints of the individual control loops based on a rigorous process model, which allows for improvement of efficiency under all modes of operation.

The rigorous plant model may be used in online mode for optimization of the plant under all modes of operation and changing boundary conditions (e.g., summer/winter mode, changing salinity or temperature of the seawater). As it is built using physical laws, the model may also be used offline to determine in advance the anticipated behavior of a plant by using the actually designed geometric and material data, including additional heuristic knowledge with which to determine, for instance, the brine level. This helps to examine the construction of the plant and to calculate and evaluate different alternatives. Dynamic simulations allow examination of the transient behavior of the plant and assurance of smooth and safe operation during load changes.

6.3.2 IMCA in the MSF Process

6.3.2.1 General Description A schematic diagram of an MSF process is shown in Figure 6.1. The system consists of flashing stages, a brine heater, vacuum system, control loops, and pumping units [27,28]. Each flashing stage includes condenser tubes, distillate trays, water boxes, demister, and venting connections. The flashing stages are divided into heat rejection stages and heat recovery stages. The number of stages in the heat rejection section is commonly set at 3, this is necessary to avoid a large increase in the specific heat transfer area [29]. However, the number of stages in the heat recovery section varies between 12 and 29. Common design in the Gulf countries adopts the 24-stage system with 21 stages in the heat recovery section.

The brine recycle stream enters the brine heater tubes, where a saturated heating steam is condensed on the outside surface of the tubes. The brine stream absorbs the latent heat of condensing steam, and its temperature increases to its maximum design value known as the *top brine temperature* (TBT). Such a value varies

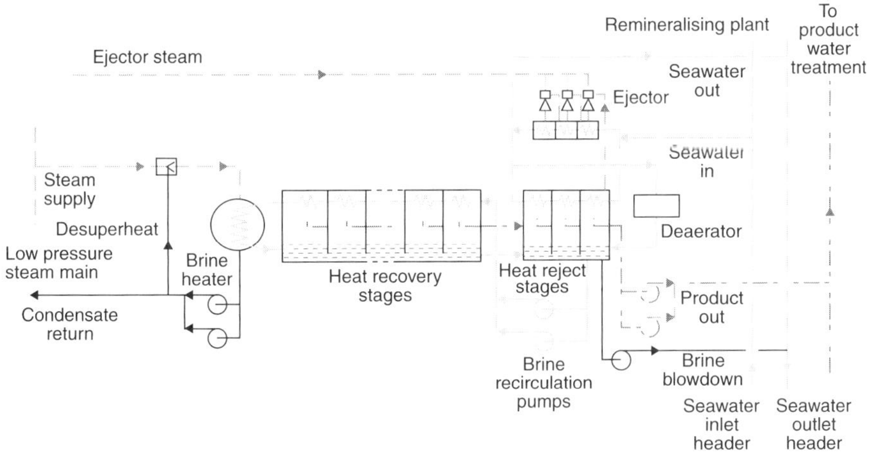

Figure 6.1 A typical MSF plant flow diagram.

between 90°C and 110°C, where the saturation temperature of the heating steam is approximately higher by 10°C. Low- or high-temperature operation is associated with the type of antiscalant material, where use of polyphosphate limits the TBT to 90°C, and use of a special polymer additive allows for a higher temperature of 110°C. The temperature difference between the TBT and brine recycle temperature is limited to 5°C, which is necessary to maintain a high-performance ratio for the system.

The hot brine enters the flashing stages in the heat recovery section and then enters the heat rejection section, where a small amount of freshwater vapor is formed by brine flashing in each stage. Simultaneously, the stage pressure decreases since it corresponds to the saturation temperature of the condensing vapor. Vapor formation results in an increase of the brine stream salinity, which is limited to a maximum value of ~70,000 ppm at the last stage. In each stage the flashed-off vapors flow through the demister pads, which are used to separate the entrained brine droplets. This is necessary to maintain low product conductivity and scale formation on the outside surface of the condenser tubes. The vapor condenses on the outside surface of the condenser tubes where the brine *recycle* flows inside the tube from the cold to the hot side of the plant.

The condensed freshwater vapor accumulates outside the condenser tubes, and accumulates across the stages, forming the distillate product stream, which is extracted from the last stage in the heat rejection section.

The intake seawater stream is introduced into the condenser tubes of the heat rejection section, where its temperature is increased to a higher level by absorption of the latent heat of the condensing water vapor. The warm discharge stream of intake seawater is divided into two parts:

1. Cooling seawater, which is rejected back into the sea
2. Feed seawater, which is chemically treated and then mixed in the brine pool of the last flashing stage in the heat rejection section

The brine recycle stream is extracted from the brine pool of the last stage in the heat rejection section and is introduced into the condenser tubes of the last stage in the heat recovery section. The remaining brine in the last stage of the heat rejection section is rejected into the sea.

Treatment of the intake seawater is limited to simple screening. However, treatment of the feed seawater stream is more extensive and includes deaeration and the addition of chemicals to control scaling, foaming, and corrosion.

A steam jet ejector is used for removal of the noncondensable gases that are released during the flashing process. Presence and accumulation of these gases results in corrosion of the condenser tubes and reduction of the overall heat transfer coefficient for the condenser tubes, which also reduces the process efficiency and system productivity.

6.3.2.2 Design Features of MSF Process The MSF process includes a number of control loops that operate to maintain high system performance and production rate. In addition, the system includes a number of design features, which cannot be defined as a control element; however, these units control the system characteristics:

1. *Demisters*. The function of the demisters is to separate entrained brine droplets within the flashed-off vapor. Proper design of these units also prevents reentrainment of the brine droplets within the demister mesh with the vapor flow. As a result, use of the demister results in low distillate product conductivity or salinity values with the design limits (<5 ppm).
2. *Interstage Weirs and Submerged Orifices*. These components control the brine flow rate per unit width of the flashing chamber and strongly affect the flashing rate of distillate product. Gates generate brine height between flashing stages to prevent vapor blowthrough and set the height of the brine pool. Gate height is adjusted to 0.2–0.5 m above the stage bottom; however, readjustment might be necessary to prevent blowthrough and to control the liquid level within the stages.
3. *Heat Transfer Area*. The heat transfer area of the condenser tubes is a major design feature that controls the temperature of the brine recycle entering the brine heater. This parameter significantly impacts the system performance ratio. If the heat transfer area is smaller than the thermal load of the condensing vapor, the stage pressure will increase as a result of accumulation of the noncondensed vapors. This pressure increase will reduce the amount of flashed-off vapor. Eventually, the system will reach a new steady state with lower flashing rates and a smaller flow rate of the distillate product. Also, the temperature of the brine recycle entering the brine heater will decrease, which will result in an increase of the amount of the required heating steam and subsequent reduction in the system performance ratio. Although the initial system design provides sufficient heat transfer within various stages, poor operation, and increase in fouling resistance, will reduce the effective heat transfer coefficient and create conditions where the thermal capacity of the condenser unit is lower than the thermal load of the condensing vapor. In addition, tube blockage may also result.

6.3.2.3 Overall MSF Plant Operation Philosophy and Matching with Power Plant In general, a quite consistent control and instrumentation philosophy is applied in all main MSF plants in the Middle East, with some small variation for different customers and manufacturers. For the majority of the MSF plants installed in the Gulf during normal operation, steam required for the brine heater is obtained from the low-pressure (LP) steam header. The LP steam header can, in turn, be associated to different thermal cycles. The possible combinations are summarized in Table 6.1.

In cases where the MSF plant is fed by steam from the steam turbine, there are always provisions for the desalination brine heaters to be automatically supplemented by steam from the HP steam headers via pressure-reducing desuperheating valves in case of a trip of the BPST or inadequate supply of LP steam. These

Table 6.1 Classification of Process Interfaces for Thermally Driven Desalination Plants

With associated power generator
Steam from heat recovery steam generation
Steam from backpressure steam turbine (BPST) combined cycle
Steam from condensing steam turbine
Without associated power generator
Steam from auxiliary boiler

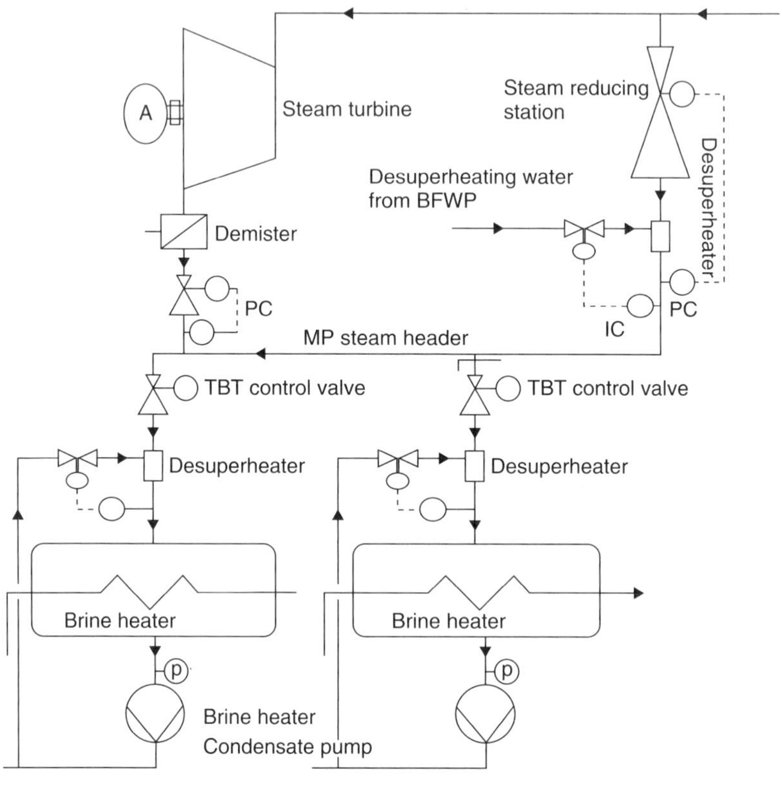

Figure 6.2 MSF–power yard process interface.

results in an arrangement as schematically indicated in Figure 6.2. Actual combined cycle operating requirement tend to achieve maximal power output from the BPST's by maintaining maximum steam flow through the BPST's while producing the required water demand.

6.3.2.3.1 Steam System: Top Brine Temperature Control Loop The steam system is related to the control of the top brine temperature of the distiller, which is one of the most critical process parameters to be controlled in an MSF plant.

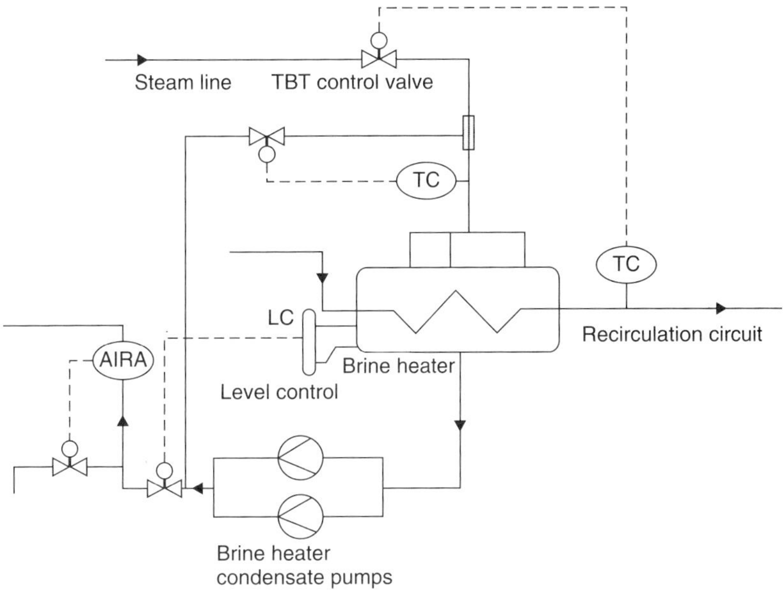

Figure 6.3 MSF TBT control system.

The top brine temperature (TBT) of the brine leaving the brine heater is controlled by the LP steam control valve (TBT control valve) in the steam pipe to the brine heater inlet. Therefore, the opening or closing of the steam control valve is governed by the TBT. MSF plants are normally equipped with a TBT controller with an operator input set point automatically regulating the quantity of steam feed to the brine heater. Figure 6.3 illustrates the principle of operation.

Fouling or scaling of the MSF plant brings about a decrease in the inlet temperature to the brine heater and a subsequent decrease in the TBT, which needs to be compensated for by a larger steam quantity to be provided to the brine heater. The measurement of the TBT is very delicate. In particular, this measurement is carried out on large-bore pipes (in some cases ND2000), where there are substantial differences between the brine bulk temperature and the outside temperature of the pipe. These differences may lead to substantial errors in evaluation of the overall MSF plant performance. In general, a minimum of three temperature measurements are used to measure the top brine temperature for large-bore pipes at various different points across the pipe section. This is carried out by installing thermowells of different lengths, with each thermowell measuring the brine temperature at different depths of the tube from the pipe wall and inside the fluid as shown in Figures 6.4 and 6.5.

6.3.2.3.2 Desuperheater Water Spray System: Steam–Brine Heater Temperature Control Loop The control loop is schematically represented in Figure 6.2. Its function is to control the temperature of the steam inlet to the brine heater to prevent the steam temperature from becoming too high

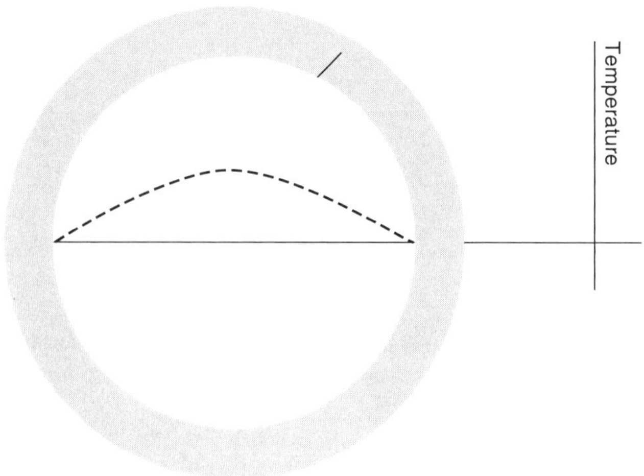

Figure 6.4 A typical temperature profile across a pipe section.

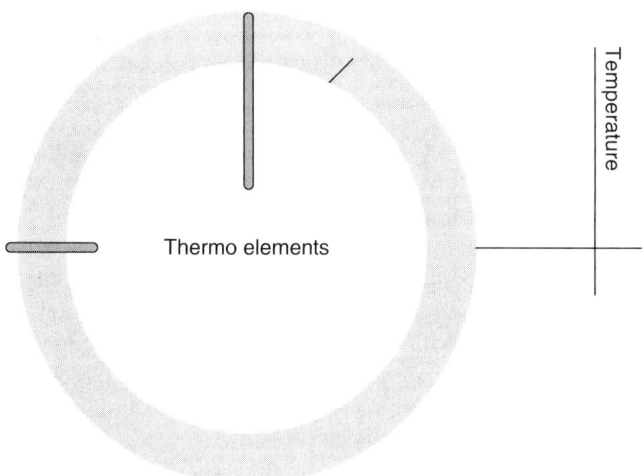

Figure 6.5 Typical temperature instrument location for large-bore pipe.

and causing localized tube scaling in the external side of the brine heater. As the steam temperature into the brine heater is close to saturation, a straight temperature control is not practical. The quantity of spray water is therefore controlled automatically on an open-loop ratio basis between the position of the pressure-reducing valve and the spray control valve. There is also an override control that increases the ratio between the valve positions if the steam temperature increases by more than 5°C above saturation.

The main issue related to this control loop is the reliability of the desuperheating system. Poor mixing and atomization between the overheated steam downstream of the control valve and the condensate used for temperature control often results

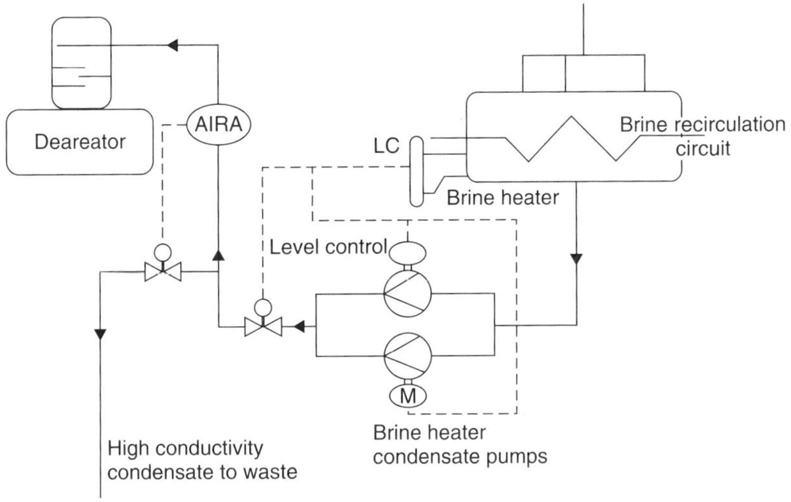

Figure 6.6 Brine heater level control in a condensate system.

in areas where steam is not properly de-superheated. Therefore, the condensate injected in the system tends to remain in the liquid phase and accumulate in the pipe, clogging the drainage system and increasing the possibility of erosion in the pipe.

6.3.2.3.3 Condensate System: Brine Heater Level Control The condensate level in the brine heater is controlled by regulation of the flow control valves at the outlet of the condensate extraction pumps. The condensate extraction pump is tripped on when a low condensate level is detected in the brine heater. In general, the current automation systems are designed in such a way that loss of both condensate pumps, or high–high[1] outlet brine temperature, for more than a few seconds, will result in tripping of the steam supply to the brine heater.

Normally, condensate is returned directly to the boiler system, however, its outflow conductivity should be continuously monitored, and if high conductivity is measured, the condensate flow should be automatically diverted to drain. A schematic diagram of the condensate system loop scheme is shown in Figure 6.6.

The brine level in the last stage is controlled by regulation of the brine blowdown flow. A low–low brine level in the last stage will trip both the brine blowdown and brine recirculation pumps (Fig. 6.6).

For the majority of plants in the Middle East, there is no standby facility for the brine blowdown pump. Therefore, in the event of failure of the single brine blowdown pump, operation of the distiller will continue by use of the emergency

[1]The terms *high–high* and *low–low* refer to two control thresholds for a measurand. For instance, for brine temperature there are two different thresholds in the control system: one for the alarm and the other one to activate a trip.

blowdown system. This system is designed to evacuate the blowdown through an emergency pipe downstream of the brine recirculation pumps. The emergency blowdown control valve is controlled manually from the central control room (CCR), and any necessary isolating valves to bring it into service are operable by remote control. The emergency blowdown system is not operated for more than a few hours because of the high differential pressure dissipated across the control valve, which induces both noise and cavitation. The purpose of the emergency blowdown circuit loop is to ensure sufficient time for a smooth shutdown of the distiller, or repair of the blowdown pump failure. In general, the emergency blowdown line is not intended for continuous operation.

6.3.2.3.4 Brine Recirculation System: Brine Recirculation and Evaporator Last Stage Level Control Figure 6.7 is a schematic diagram of the brine recirculation system. The brine recirculation flow rate is set by the DCS system according to the operating top brine temperature as indicated in Figures 6.7 and 6.8. A second control loop operating in parallel to the brine recirculation flow control loop is provided to protect the brine heater tubes and piping against the potential adverse effect of boiling brine that could be caused by a drop in brine pressure before entering the first stage. A differential pressure control valve actuated by the logic control is provided for that purpose and ensures that brine recirculation does not flash before entering the first-stage evaporator, which could result in erosion in the pipeline upstream the evaporator. A differential pressure controller is installed at the brine heater outlet to monitor the pressure difference between the first-stage evaporator and the upstream side of the valve. This controller is set to override the

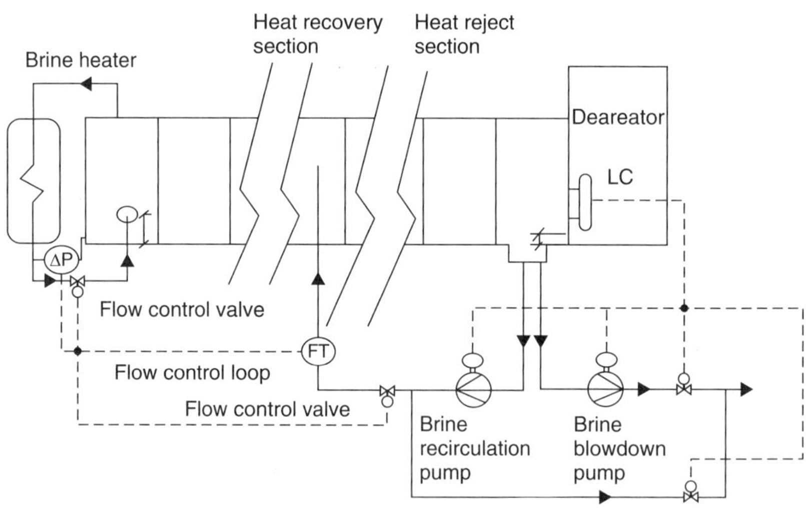

Figure 6.7 Brine recirculation system.

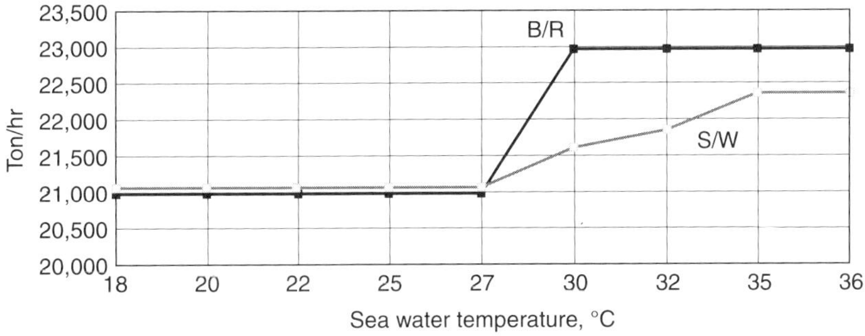

Figure 6.8 Correction curve for brine recirculation (B/R) flow rate and rate of seawater (S/W) flow to the heat reject section.

flow controller in the event that the pressure at the inlet fails within the required value of the outlet.

6.3.2.3.5 Seawater System: Makeup and Control of Seawater Flow Rate Through Heat Rejection Section The seawater flow rate through the heat rejection section is maintained at the desired value either by throttling of a control valve positioned in the drain circuit, or by speed variation. The setpoint of flow rate through the heat rejection section is given either by the central (or local) control room, or is automatically generated using algorithms that factor in the temperature of the seawater entering the heat rejection section and the required temperature at the outlet of the heat reject section. A thermodynamic model run by an online computer projects the flow rate. A typical correction curve for the brine recirculation and seawater flow rate through the heat rejection section is shown in Figure 6.8 for a 12-migd MSF plant.

The curve in Figure 6.8 is derived from the theoretical projection of the MSF thermodynamic simulator program. The thermodynamic simulation correlates the seawater temperature and flow rate to the heat reject section and brine recirculation flow rate to maintain the desired distillate production and performance ratio. These curves are used to generate, at different seawater temperatures, the setpoints for the rate of seawater flow through the heat reject section and brine recycle flow rate, which enables the plant to follow the required load variations.

As can be seen from Figure 6.8, the brine recirculation and seawater flow rates are maintained constant until the seawater temperature reaches 27°C. As the seawater temperature increases, it is also necessary to increase the seawater supply flow rate through the heat rejection section. This is done mainly to avoid excessive increase in the bottom brine temperature and therefore losses in production due to a decrease in the flashing range. Furthermore, the temperature difference between the seawater entering the evaporator and the outlet stream (discharged to the sea) is often a guaranteed parameter and must not exceed 10–12°C depending on the contract specification. However, despite the increase in the heat rejection flow rate, bottom brine temperature tends to increase. To compensate for the decrease in the flashing range, the brine recirculation flow rate must also increase as indicated in Figure 6.8.

To ensure that the optimum minimum flow rate passes through the seawater supply pump regardless of the load conditions of the evaporator, part of the seawater can be sent directly to the discharge culvert by means of a *minimum-flow line* and control loop. Centrifugal pumps are unstable if they operate below a certain flow rate, but during winter operation the seawater flow rate required is often below this threshold (called *minimum flow*). To ensure that the pumps operate without vibration, the flow rate through the seawater supply pump must be maintained above the threshold of minimum flow. Therefore, the difference between this threshold and the requirement of the distiller is recirculated back to the sea. Various interlock systems can be designed for the pump trip in case of a low level in the intake bay suction chamber.

Makeup flow is generally controlled in direct ratio to product flow with allowance for adjusting the ratio if required. The operator may manually adjust the ratio in the operation–monitoring station using the setpoint adjuster. Any seawater that exceeds the makeup flow requirements is discharged into the seawater discharge channel, carrying the excess heat with it. After cooling of the heat rejection section stages of the evaporator, a part of the warmed seawater flows to the deaerator as a makeup water supply for the brine recirculation system.

6.3.2.3.6 Distillate Product Water System: Distillate Level Last Stage The level of product water in the last stage is controlled by regulation of outlet product water flow. The product water pump is tripped in the event of low-level detection of product water in the last stage. In the event of high conductivity being detected in the product water, the product is automatically diverted to waste. Return to normal production in general requires manual initiation.

6.3.2.3.7 Cooling-Water Recirculation System: Seawater Temperature to Heat Rejection Section Multistage flash evaporators are designed for summer temperature but take advantage of the lower winter seawater temperature to increase the evaporator overall flash range and output. High-vacuum conditions in the MSF distiller caused by very low bottom brine temperature (BBT) in winter season may create excessive carryover in the demisters and subsequent high salinity of the product water. In some of the main MSF desalination plants a cooling-water recycle circuit is provided to maintain the last stage temperature at a level that prevents brine carryover or demister fouling at the lower winter seawater temperatures.

6.3.2.3.8 Antiscale–Antifoam and Sodium Sulfate Dosage The dosing rate of the chemicals is automatically controlled by means of control valves or displacement volume in a metering pump. The control system receives signals from

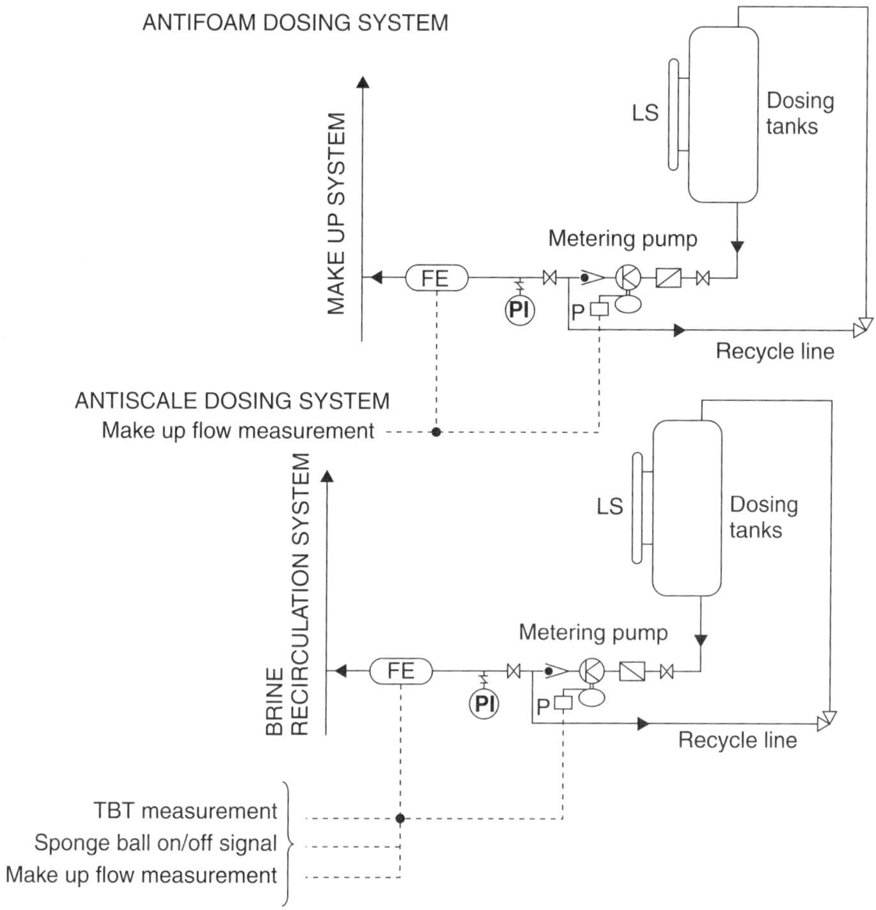

Figure 6.9 Typical antiscale–antifoam dosing system.

the instrumentation that measures the flow of the chemical additive and of the makeup water. Figure 6.9 is a flow diagram of a typical antiscale–antifoam dosing system. Antiscale is dosed via metering pumps in correlation with the makeup flow rate. However, the antiscale dosage controller takes into account different dosing curves according to the operating top brine temperature and status of the sponge ball cleaning system.

Typical dosing rates correlated with top brine temperature are illustrated in Figure 6.10 [30] Antifoam dosage, on the contrary, is simply controlled by the measurement of the make-up flow rate. With reference to Figure 6.10, antiscale in MSF plants is dosed in the brine recirculation line and not in the makeup line, to avoid chemicals losses in the deaerator and in the blowdown flow, while antifoam is dosed in the makeup line to prevent foaming in the deaerator and the last stage evaporator.

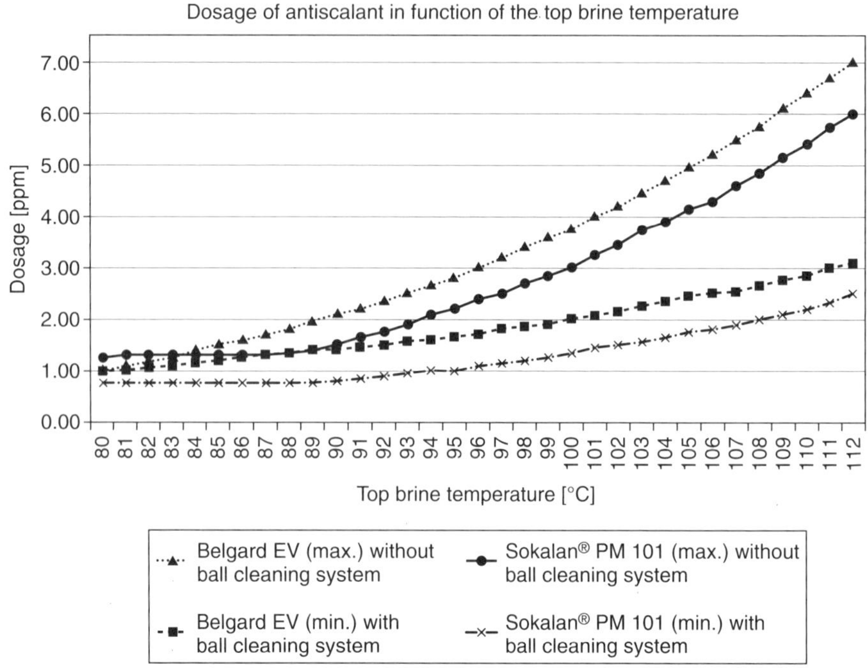

Figure 6.10 Typical antiscale dosage as a function of TBT.

6.3.2.4 Main Control Loops of MSF Process The main control loops of the MSF process are shown in Table 6.2 and Figure 6.11. As shown, all of the control loops involve a single-input single-output (SISO) control loop. In Figure 6.11, all control loops are referred to by numbers.

The Functions of these control loops are listed as follows:

Loop 1. Temperature of the cooling seawater entering the last (final) flash stage (controlled variable) by adjusting the circulation flow rate of the cooling seawater (manipulated variable). This loop operates only during the winter season when the intake seawater temperature drops to values close to 15°C. As a result, the control valve and pumping system of the control loop operate to adjust the temperature of the intake seawater. Setpoints for the intake seawater temperature vary between 25°C for winter operation and 32°C for summer conditions. This loop is not found in the once-through MSF layout [29].

Loop 2. TBT (controlled variable) by adjusting the heating steam flow rate (manipulated variable). An increase or decrease of the heating steam flow rate is necessary to control the TBT. This may be necessary to factor in an increase in the brine circulation rate, fouling and scaling effects, or decrease in

Table 6.2 List of Control Loops in a Typical MSF Plant

Loop Number	Controlled Variable	Manipulated Variable	Controller Type
1	SW inlet temperature	SW circulation flow	PID
2	Top brine temperature	Steam flow rate	PID
3	Pressure of heating steam	Throttling valve	PI
4	Heating steam temperature	Flow rate of condensate spray	PID
5	Condensate level in brine heater	Discharge valve adjustment	PI
6	Makeup flow rate	Feed valve	PI
7	Cooling SW reject temperature	Discharge valve	PID
8	Distillate level in final stage	Discharge valve of distillate	PI
9	Brine level in final flash stage	Brine blowdown flow rate	PI
10	Brine level in final flash stage	Brine recycle flow rate	PI
11	Dosing flow rate of chemical additives	Discharge valve	PI
12	Pressure of ejector motive steam	Throttling valve	PI
13	Temperature of ejector motive steam	Flow rate of condensate spray	PID

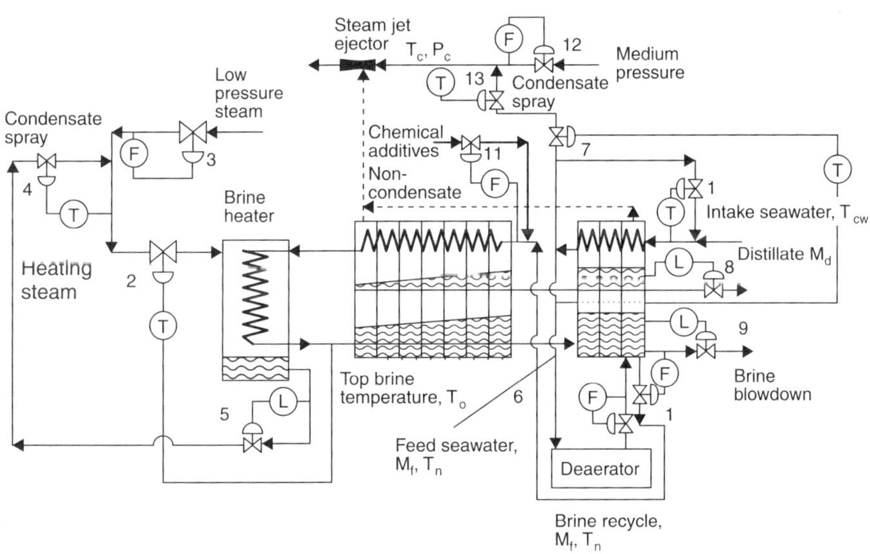

Figure 6.11 Flowsheet for Al Taweelah B plant layout showing main control loops (1996 testing protocol).

the temperature of the brine circulation. The setpoint for the TBT depends on the type of antiscalant used in the plant where 90°C is suitable for a polyphosphate type and the temperature range 100–110°C is used for polymer-type additives.

Loop 3. Pressure of the heating steam (controlled variable) and adjustment of the throttling valve (manipulated variable). Commonly, the low pressure steam has a pressure of 4–7 bar, and it is necessary to reduce the pressure to a value of ~2 bar, where it becomes superheated with a temperature close to 170°C.

Loop 4. Temperature of the heating steam (controlled variable) by adjusting the flow rate of the condensate spray (manipulated variable). This loop changes the steam quality from superheated to saturated, where its temperature drops from 170°C to 100–110°C. The steam temperature is controlled to be higher than the TBT by 5–10°C. This is necessary to prevent the formation of hotspots and scale formation.

Loop 5. Condensate level in the brine heater (controlled variable) and opening of the distillate discharge valve (manipulated variable). It is necessary to maintain a sufficient static head above the condensate pump to prevent condensate flashing within the discharge line or the pumping unit, which would result in violent vibrations, loosening of various connections, and severe erosion.

Loop 6. The flow rate of the makeup seawater entering the last stage (controlled variable) by adjusting the valve opening on the feedline (manipulated variable). The set-point for this control is adjusted according to the flow rate ratio of distillate and brine blowdown. This control loop maintains a constant *conversion ratio* (defined as the ratio between distillate and make-up flow rates), which is necessary to keep the salinity of the brine blowdown from the last stage at a design constant value of ~70,000 ppm. In the Gulf, the intake seawater salinity varies between 42,000–48,000 ppm and accordingly, the conversion ratio varies between 0.4 and 0.314. This sets the controller ratio (ratio of distillate and blowdown) at 0.46–0.67, measured as kg distillate per 1 kg brine blowdown.

Loop 7. Temperature of the reject flow rate of the cooling seawater (controlled variable) by adjusting the opening of the discharge valve (manipulated variable). The set-point of the controller is made in reference to the temperature of the last flash stage. For a constant plant capacity, an increase in the final-stage temperature results in a reduction in the flashing range and a consequent decrease in performance ratio. The final-stage temperature is normally set at ~40°C for summer operation and 32°C for winter operation.

Loop 8. Distillate level in the last stage (controlled variable) by adjusting the distillate product flow rate (manipulated variable). This control loop has a function similar to that of the level controller of the heating steam condensate where a decrease of the distillate product static head would cause operational problems in the associated pumping unit.

Loop 9. Brine level in the last flash stage (controlled variable) by adjusting the brine blowdown flow rate (manipulated variable). This is one of the most important control loops in the MSF plant since the level in the last stage affects the levels in the previous stages. Simultaneously, the flashing efficiency is affected in various stages as well as the amount of distillate product. The increase of brine level reduces the flashing rates as well as the amount of distillate product; consequently, the amount of heat released on condensation of the distillate vapor becomes lower. Eventually, the temperature of the brine recycle stream entering the brine heater decreases, which results in an increase of the amount of heating steam and reduction in the system performance ratio. However, reduction of the brine level may result in blowthrough of the flashed-off vapors across the stages, thus resulting in a reduction of the effective number of stages and a decrease in plant performance ratio.

Loop 10. Brine level in the last flash stage (controlled variable) by adjusting the brine recycle flow rate (manipulated variable). Function and effects on system performance for thus control loop are similar to those of loop 9.

Loop 11. Dosing flow rate of the chemical additives (controlled variable) by adjusting the opening of the discharge valve of the chemical additive (manipulated variable). Variations in the brine circulation flow rate necessitate adjustment of the dosing rate of the chemical additives. Such variations are caused by adjustments in the flow rate of the brine circulation stream, which might be necessary to adjust the brine level in the last flashing stage.

Loop 12. Pressure of the ejector motive steam (controlled variable) and the opening of the throttling valve (manipulated variable). Commonly, the medium-pressure steam has a pressure of 16 bar and it is necessary to reduce its value to 7 bar.

Loop 13. Temperature of the ejector motive steam (controlled variable) by adjusting the flow rate of the condensate spray (manipulated variable). This loop reduces the ejector motive steam temperature from $\sim$201°C to 165°C by spraying with steam condensate.

6.3.2.5 Identification of Major Process Parameters in MSF

6.3.2.5.1 Introduction In this section we will attempt to identify the major process parameters, which need to be measured for process automation of the MSF process. Table 6.3 represents the state of the art in a typical modern installation, typically an MSF crossflow plant with flow recirculation, where instrumentation has been optimized to take into account more recent process automation techniques. Table 6.3 also provides information on instrumentation in auxiliary systems (remineralization system and seawater intake system).

6.3.2.5.2 Typical Instrument List Table 6.3 is a typical instrument list for an MSF plant and auxiliaries. In an industrial plant, each instrument is identified in the field and in the control room by a label code. The label generally identifies the distiller unit, the type of instrument, and the serial number. A list of abbreviations used in Table 6.3 immediately follows the table.

Figure 6.12 is a typical recirculation MSF plant flowsheet showing instrumentation and controls.

6.3.2.5.3 Differences in Instrumentation Due to Plant Size and Manufacturer Design Multistage Flash crossflow with brine recirculation is the technology that has been applied in desalination plants since the 1980s. The level and extent of instrumentation are almost the same for each MSF plant of this type, regardless of unit size. There are few differences in the control philosophy and extent of instrumentation among various manufacturers. The number of instruments, however, depends on various factors, such as the number of stages and the unit size. An example of how unit size may influence instrumentation is the impact of an error in the measurement of the TBT. Suppose that the design top brine temperature for the plant is 110°C and the plant brine bottom temperature is 40.6°C at a given seawater temperature, then the actual distillate production can be evaluated by the equation

$$\text{Nominal net distillate production rate} = D_{\text{A}} \frac{(110 - 40.6)}{(\text{T}_{01\text{a}} - \text{T}_{02\text{a}})} \tag{6.1}$$

where $\text{T}_{01\text{a}}$ is the brine recirculation temperature measured in the main brine recirculation pipe interconnecting the brine heater with the first-stage evaporator, $\text{T}_{02\text{a}}$ is the brine bottom temperature usually measured in the evaporator, last stage and D_{A} represents the distillate flow rate actually measured during plant operation by the flowmeter installed. Using the formula above, the 0.5°C error in TBT ($\text{T}_{01\text{a}}$ in this example) produces a water production error of 0.7%. For a 100-migd MSF plant installation that translates to an error of 132 m^3/h, at the selling rate of US\$80/m^3, this brings about an uncertainty in water payment of ~US$1 M (million) per year.

A MSF plant of 12 migd capacity has a brine recirculation pipe of ND1800 (nominal diameter of 1800 mm) and the brine temperature across this large pipe is not uniform, tending to ward the distribution shown in Figure 6.4. This requires measurement of the TBT in such large pipes by several temperature detectors located at various depths in the pipe, as shown in Figure 6.5, and properly averaged at the DCS level. This sophistication is not necessary for smaller scale desalination plants.

Table 6.3 Standard Instrumentation Installed in MSF Systems[a]

	Table of Instrumentation																			
	Subsection : Desalination Unit																			
Tag No.	Description and Service		Central Control Room									Local/Local Panel or Board								
	E—Electrical Signal F—Flow L—Level P—Pressure T—Temperature Q—Quality/Analysis S—Speed/Frequency Y—Vibration/Shaft Displacement	Services	Indication	Integration	Recording	Closed-loop Control	Open-Loop Control	Alarm	Trip	Fault Printer	Redundant Transmitter	Indication	Integration	Recording	Closed-Loop Control	Open-Loop Control	Alarm	Trip	Test	
FIRCA	Heat Reject inlet SW	F	X		X	X		X		X	X									
FIRA	Condensers Cooling SW	F	X		X			X	X	X										
FIRCA	SW Makeup	F	X		X	X		X		X	X									E
FIRCA	Brine recycle	F	X		X	X		X	X	X	X									E
FIRA	Brine Blowdown	F	X		X			X		X										D
FIRQCA	Distillate	F	X	X	X	X		X		X										E
FIRQA	MPS to Ejectors	F	X	X	X			X		X										D
FIRQA	Condensate to power plant	F	X	X	X			X		X										E
FIR	Desuperheater inject water	F	X		X															E
FFC	Antifoam	F				X														R
FFC	Antiscale	F				X														E
FIA	Condensate to condensate cooler	F						X				X								R

(continued)

Table 6.3 (*Continued*)

FI	MPS to desuperheater	F											X								D
LIRA	Evap: stage brine	L	X		X		X		X												E
LICA	Evap: stage brine	L	X				X	X	X		X										E
LA	Evap: stage brine	L					X	X													E
LIRCA	Evap: stage distillate	L	X		X	X		X	X	X		X									E
LA	EVAP: stage distillate	L						X	X												E
LICA	Brineheater condensate	L	X			X		X		X		X									E
LA	Brine heater condensate	L						X	X												E
LA	Brine heater condensate	L						X	X												D
LI-	Brine in evap: all stages	L												X							D
LI-	Distillate in evap: all stages	L												X							E
LI	Distillate box	L												X							E
LI	Antifoam tank	L												X							E
LI	Antiscale tank	L												X							E
PIR	LP steam	P	X		X																D
PIRCA	Brine heater	P	X		X	X		X	X	X		X									D
PIA	Brine recycle	P	X					X		X											D
PI	Brine heater out	P																			D
PDICA	Brine heater out PCV	PD	X			X		X	X	X		X									E
PI	Brine blowdown	P	X																		D
PI	Evap: Stage 1 vapor	P	X																		D
PIR	Evap: stage vapor	P	X	X																	D
PI	BRP in	P												X							D
PI	BRP out	P												X							D

Table 6.3 (*Continued*)

PI	Brine recycle	P												X								D
PI	BDP in	P												X								E
PI	BDP out	P												X								E
PI	Heat rejection inlet SW	P												X								E
PI	Brine heater inlet	P												X								D
PI	Brine heater outlet	P												X								E
PI	Evap: stage	P												X								E
PI	Distillate pump inlet	P												X								D
PI	Distillate pump outlet	P												X								E
PI	MP steam	P												X								E
PI	Precondenser	P												X								D
PI	Intercondenser	P												X								D
PI	Condensate pump inlet	P												X								E
PI	Condensate pump outlet	P												X								E
PI	Antifoam dosing pump out	P												X								D
PI	Antiscale dosing pump out	P												X								E
PI	Antifoam	P												X								D
PI	Antiscale	P												X								E
PI	Ball circulation pump out	P												X								D
QIRA	Brine recycle conductivity	Q	X		X			X		X												E
QIRCA	Distillate conductivity	Q	X		X		X	X		X												E
QIRA	Deaerated SW	Q	X		X			X														E
QIRCA	Brine heater condensate conductivity	Q	X		X		X	X	X													E
TIR Thermowell	Heat reject inlet SW	T	X		X																X	FI

(*continued*)

Table 6.3 (*Continued*)

TIR Thermowell	Heat reject outlet SW	T	X		X																X	'FI
TIR Thermowell	Brine recycle	T	X		X																X	FI
TIR Thermowell	Brine Heater inlet	T	X		X																X	'FI
TIRCA	Brine heater outlet	T	X		X	X		X	X	X		X		X							X	D
TIR Thermowell	Brine in evap: 1st stage	T	X		X																X	D
TI	Brine in evap: stages	T	X																		X	D
TIR Thermowell	Brine in evap: stages	T	X		X																X	D
TIR	Distillate	T	X		X																	E
TIR	LP steam	T	X		X																X	E
TIR Thermowell	Brine heater inlet LPS	T	X			X		X		X		X									X	D
TIR Thermowell	Brine heater	T	X		X																X	D
TI	Brine heater condensate	T	X		X																X	E
TI	Condensate cooler out	T	X																			R
TI	Intercondenser out cooling SW	T	X					X														D
TI	Intercondenser out cooling SW	T	X					X														D
TI	After condenser out cooling SW	T	X					X														D
TI	Heat reject inlet SW	T												X								E
TI	Heater reject outlet SW	T												X								E
TI	Brine recycle	T												X								E
TI	Brine heater inlet	T												X								E
TI	Brine heater outlet	T												X								E
TI	Brine heater condensate	T												X								E
TI- Thermowell	Precondenser out cooling SW	T												X							X	D

Table 6.3 (*Continued*)

TI	Intercondenser out cooling SW	T												X								D
TI	Intercondenser out cooling SW	T												X								D
TI	After condenser cut cooling SW	T												X								D
TI	MPS from Power plant	T												X								D
TI- Thermowell	Distillate in evap: all stages	T																			x	D
	MV motor pump																					D
	Brine recycle pump																					E
	On		X				X							X				X				E
	Off		X				X							X				X				E
	Trip							x		x									X	X		E
	Current	E	X																			E
	Motor																					E
	Winding (3 points)	T	X					X	X	X												E
	Bearing	T	X					X	X	X												E
	Pump																					E
	Bearing	T						X	X	X												E
	LV motor pump																					E
	Distillate pump																					E
	Blowdown pump																					E
	Condensate pump																					E
	On		X				X							X				X				E
	Off		X				X							X				X				E
	Trip							X		X									X	X		E
	Current	E	X																			E

(*continued*)

Table 6.3 (*Continued*)

	LV motor pump																					E
	Antifoam dosing pump																					E
	Antiscale dosing pump out																					E
	On		X											X				X				E
	Off		X											X				X				E
	Trip																		X	X		E
	Ball cleaning system																					E
	System fault (common)							X		X												E
	All pumps													X				X	X	X		E
	All motor operated valves													X				X	X	X		D
	Ball collector													X				X	X	X		D
	Ball strainer													X				X	X	X		E
PDIA	No. 1 travelling band screen	PD						X						X					X			E
PDIA	No. 2 travelling band screen	PD						X						X					X			D
PDIA	No. 3 travelling band screen	PD						X						X					X			D
PA	Spray water to no. 1 TBS	P						X											X			D
PA	Spray water to no. 2 TBS	P						X											X			D
PA	Spray water to no. 3 TBS	P						X											X			D
PI	No. 1 spray water pump out	P												X								D
PI	No. 2 spray water pump out	P												X								D
PI	Sodium phosphate pump out	P												X								D
PI	Hypochlorite	P												X								D

Table 6.3 (*Continued*)

PI	Lime	P												X								D
PI	CO_2	P												X								D
PI	Sodium phosphate	P												X					X			D
FIA	Distillate water	F												X								D
FIC	Hypochlorite	F												X			X					D
FIC	Lime	F												X			X					D
FIC	CO_2	F												X			X					D
FIC	Sodium phosphate	F												X			X					E
QICA	Product water (hardness)	Q												X			X		X			E
QICA	Product water (conductivity)	Q												X			X		X			E
QICA	Product water (pH)	Q												X			X		X			E
	MV motor pump																					
	Mail cooling-water pump																					
	On		X				X							X				X				E
	Off		X				X							X				X				E
	Trip							X		X									X	X		E
	Current	E	X																			E
	Motor																					E
	Winding (3 points)	T	X					X	X	X												E
	Bearing	T	X					X	X	X												E
	Pump																					E

(*continued*)

Table 6.3 (*Continued*)

	Bearing	T						X	X	X												E
	LV motor pump																					E
	Auxiliary cooling-water Pump																					E
	On		X				X							X				X				E
	Off		X				X							X				X				E
	Trip							X		X									X	X		E
	Current	E	X																			E
	MV Breaker																					E
	Breaker for 11-kV incomer																					E
	BREAKER FOR 11/0.42-kV trans feeder																					E
	On		X				X							X				X				E
	Off		X				X							X				X				E
	Trip							X		X								X	X			E

[a]The last (rightmost) column in this table identifies the measurand/instrument categories as follows: E—essential; D—desirable; R—redundant; I—improvement needed.

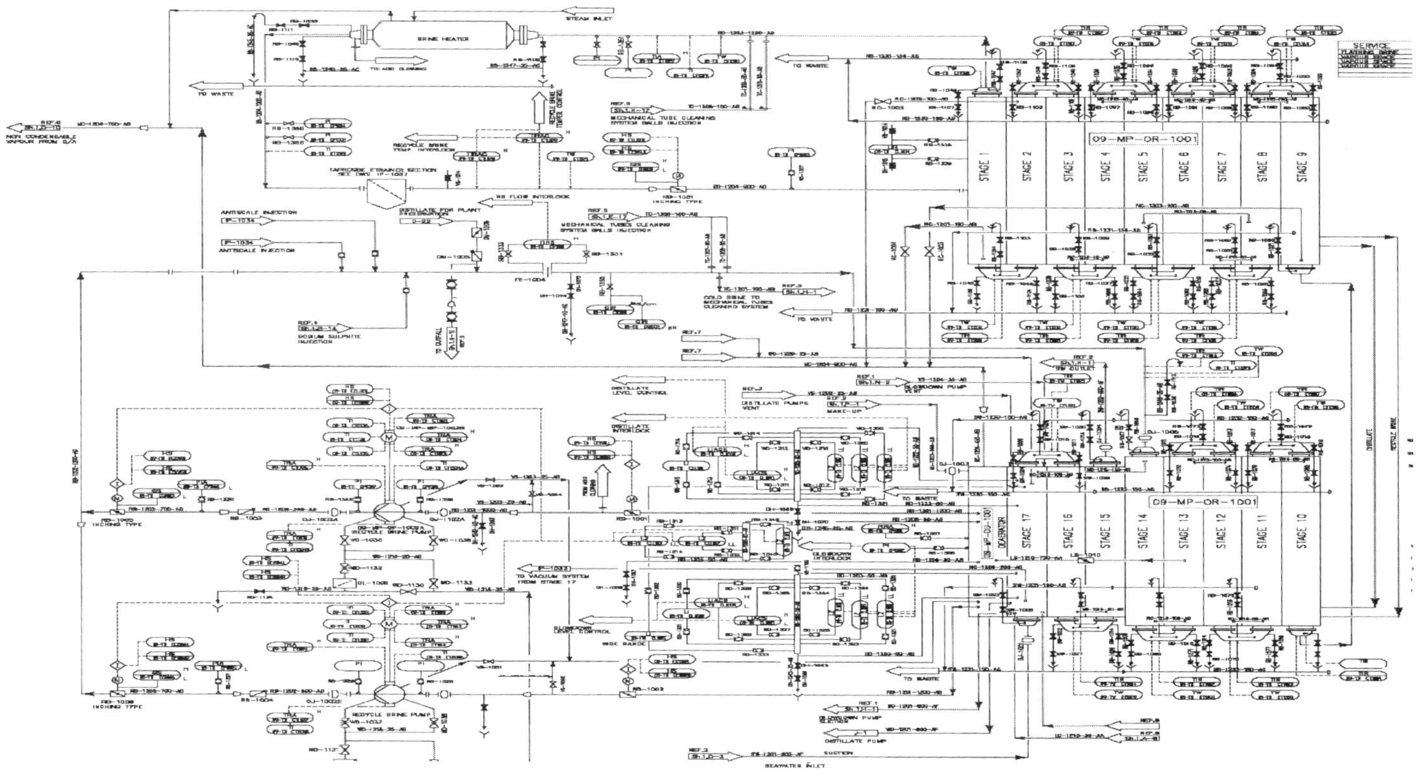

Figure 6.12 Proportional integral diagram of a typical recirculation MSF plant.

LIST OF ABBREIVATION IN TABLE 6.3

Flow Rate

FI	Flow indicator
FIC	Flow indicator controller
FIRCA	Flow indicator recorder controller alarm
FIRA	Flow indicator recorder alarm
FIRQCA	Flow indicator recorder totaliser controller alarm
FIRQA	Flow indicator recorder totaliser alarm
FIR	Flow indicator recorder
FIA	Flow indicator alarm

Level

LI	Level indicator
LIRA	Level indicator recorder alarm
LICA	Level indicator controller alarm
LA	Level alarm
LIRCA	Level indicator recorder controller alarm

Pressure

PI	Pressure indicator
PA	Pressure alarm
PIR	Pressure indicator recorder
PIRCA	Pressure indicator recorder controller alarm
PIA	Pressure indicator alarm

Differential Pressure

PDICA	Differential pressure indicator controller alarm
PDIA	Differential pressure indicator alarm

Conductivity or Oxygen Dissolved

QIRA	Quality (refers to analysis of conductivity or dissolved oxygen) indicator recorder alarm
QICA	Quality (refers to analysis of conductivity or dissolved oxygen) indicator controller alarm

Temperature

TIR	Temperature indicator recorded
TIRCA	Temperature indicator recorder controller alarm
TI	Temperature indicator

6.3.2.6 Control of Some Existing MSF Plants

6.3.2.6.1 The Umm Al Nar East (Ext.) MSF Plant [31] This section describes the computerized process control system of UANE (Ext.) MSF units 4, 5, and 6 at the Umm Al Nar power and desalination complex in Abu Dhabi, UAE. These units are an extension to the existing UANE units 1, 2, and 3, which have been in operation for a long period. The specifications for each UANE (Ext.) unit are given in Table 6.4.

At this plant, 2 × 160 t/h exhaust heat recovery boilers and 2 × 160 th auxiliary boilers supply the required steam to the units. The exhaust heat recovery boilers are installed on the exhaust ducts of two gas turbines. All the process parameters are measured and controlled by electronic analog instruments and the operating conditions of the distillers are automatically maintained at the specified values by regulating the flow rates, levels, pressures, temperatures, and chemical concentration of the process fluids. More than 20 main control loops allow the stable running of each desalination unit independently from the fluctuation of external parameters, such as

- Seawater temperature
- Seawater conductivity
- Main steam pressure
- Main steam temperature
- Distillate production

Table 6.4 Specifications of UANE (Ext.) Units

Technical Features		Evaporator Data	
Process	MSF	Recovery stages	15
Geometry	Crossflow	Rejection stages	3
Unit production	1140/1370 m^3/h	Tube bundle length	15.9 m
Total production	3420/4110 m^3/h	Tube material	CuNi 90/10 CuNi 70/30
Distillate quality	50 μS/cm	Tubesheet material	CuNi
Performance ratio	7 kg/kg	Shell material	Clad steel CuNi 90/10
Top brine temperature	90–106°C	—	—
Chemical treatment	High-temperature antiscalant	—	—

Source: Rebagliati et al. [31].

The process control system (PCS) provides supervision and control of the three MSF desalination units: recovery and auxiliary boilers and gas turbines. It consists of two VAX 11/750 computers operating in backup. The architecture of the PCS is shown in Figure 6.13.

The PCS is capable of performing the following functions:

- Data acquisition and validation
- Data transmission to main plant supervisor
- Supervisory control
- Event recording
- Data and event logging
- Failure analysis
- Maintenance logs
- Graphic displays of plant systems

The supervisory control function allows the processing of plant data either online or offline. The online supervisory program performs a running plant parameter estimation of the current values of different plant parameters (e.g., fouling factors, nonequilibrium allowance, ΔT across demister) using the current validated data. The mathematical simulation program is then activated, and the setpoints of all controlled variable are calculated for the optimal plant running condition.

The PCS allows several operational tasks to be carried out:

- Automatic distillate production change
- Automatic constant distillate production
- Plant trip analysis

The load variation is carried out automatically by changing the setpoints of the top brine temperature and brine circulation flow (recovery section), seawater flow and seawater inlet temperature (rejection section), antiscalant dosing rate, and brine heater inlet steam temperature. The first two variables are responsible for production, and performance ratio, seawater temperature, and flow are controlled to maintain correctly operated heat rejection sections, while the last two variables are adjusted to prevent tube fouling. When the PCS receives a request for load change, it checks for compatibility with the following constraints:

- Compatibility of type of antiscalant in use with the maximum TBT requested by the load variation
- Available steam flow rate compatible with the maximum steam flow rate requested by the new load
- Assurance that the constraints of the brine level in the stages are within allowable limits.

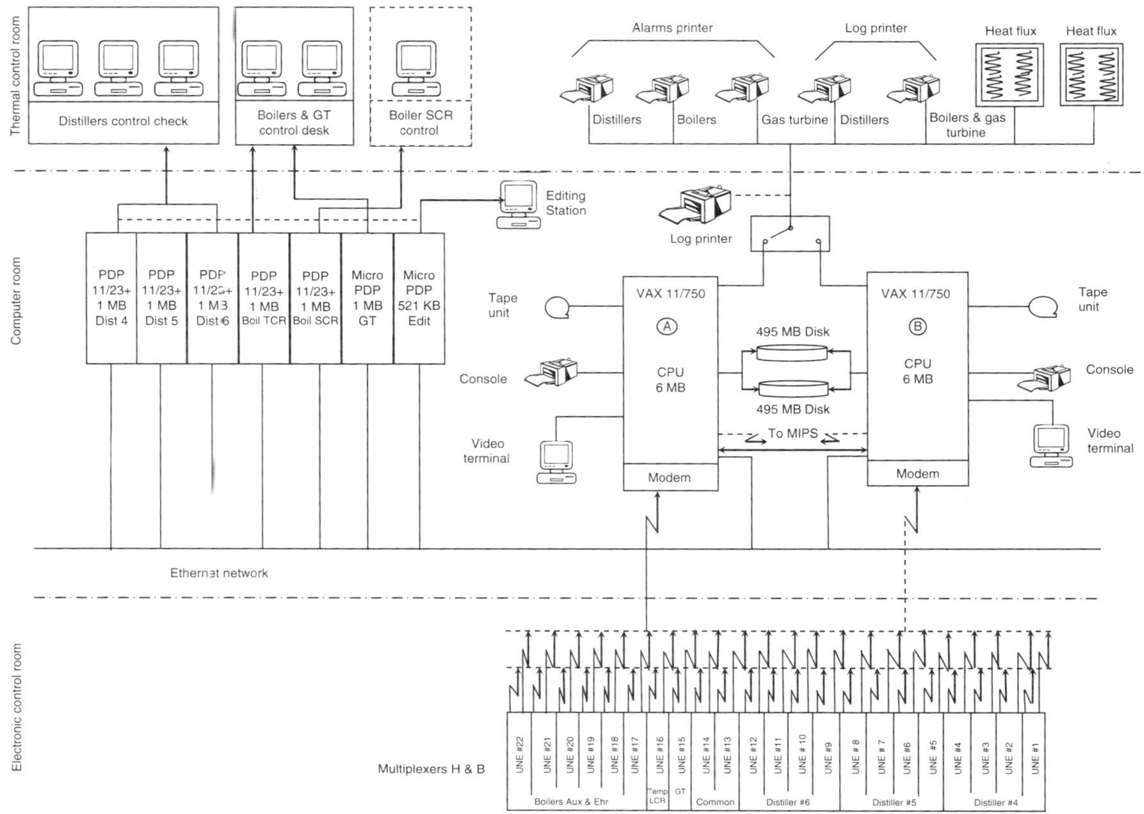

Figure 6.13 Architecture of process control system of UANE (Ext.) plant [31].

During operation under automatic distillate production, different external factors can interfere with maintaining a steady production. Typically, distillate production changes as a result of seawater temperature variation or excessive increase in fouling factor of the tube bundles. Such variations can be difficult to control manually, whereas in automatic operation, these variables are kept under constant control by the PCS, which takes immediate action to maintain the plant under optimum condition. The PCS can also carry out plant trip analysis so that in the case of a general plant trip, the cause can be identified easily. This procedure enables the operators to see all the recordings for last 10 min of the values for the main variables connected to the possible trip causes. Some of the possible trip causes are

- Very low flow of seawater
- Very low flow of recirculating brine
- Very high TBT
- Very low antiscalant flow

6.3.2.6.2 The Al-Taweelah B MSF Plant [32,33] The seawater desalination plant consists of six MSF evaporator units of cross-tube design with single-deck configuration. The nominal production rate of each unit is 45,500 t/day at 100°C top brine temperature and 32°C seawater temperature. The performance ratio at nominal production is 8.0. A maximum production of 57,610 t/day is reached at 112°C TBT and 25°C seawater temperature. In reality, the operational experience on the plant has shown that the design performances have been substantially exceeded because of excessively conservative fouling allowances specified at the design stage. The network structure for the control system of Al Taweelah B is shown in Figure 6.14. Two of the existing six desalination units have been combined for service by one process computer. In addition to the configuration stations necessary for configuration and modification work, two supervisor stations have been provided in the central control room (CCR). All process computers can exchange data among themselves. This ability also permits installation of further computing capacity, if required. It is possible via the coupling to the wide-area network (WAN) to have the plant condition displayed in a load dispatch centre (LDC) and to also have the production setpoints centrally specified.

The basic module for all activities in the plant management level is the plant data management (see Fig. 6.15). It handles the communication with the process control system and makes the acquired data available for further processing.

The core of plant data management is a staggered database structure. While an image of the current measured data is kept in the real-time database, these data can remain accessible in the *midrange database* for a selectable period before being transferred to the *archive database*. The data are compressed and preprocessed via various filter functions before being transferred to the database. This permits, in a very convenient manner, e.g., the user to form average values for the calculation of performance indicators or to generate minimum and maximum values for statistical evaluations. Important data information on the mode of operation of the plant,

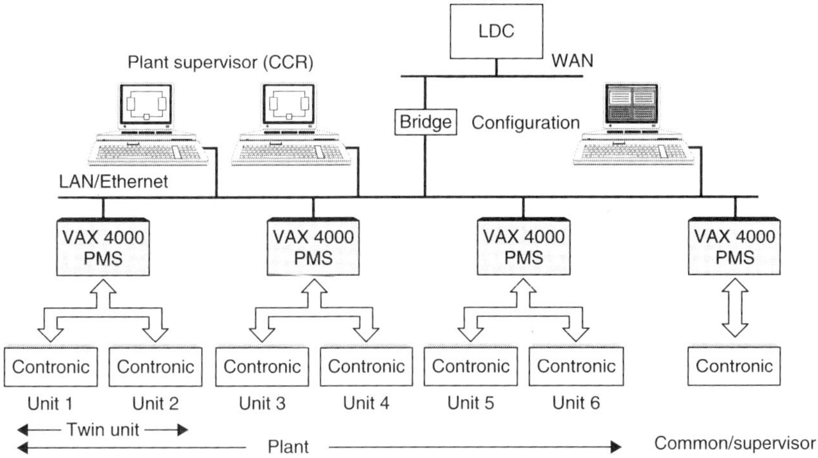

Figure 6.14 Control architecture of the Al Taweelah B plant (six MSF units) [32].

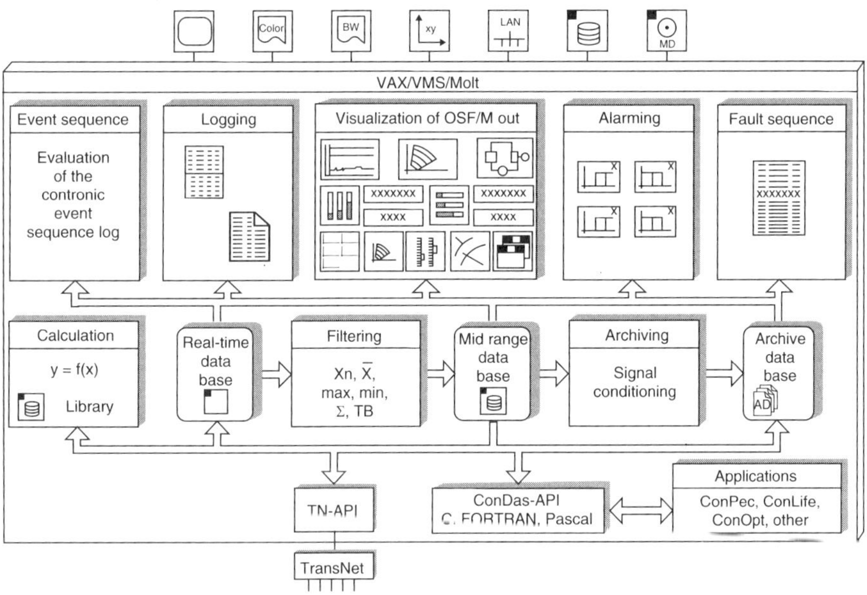

Figure 6.15 Plant data management for Al Taweelah B plant [32].

which may be required for the calculation of the service life of specially stressed components, can finally be retained in the archive database for practically any length of time. An archive automatically takes care of managing the data inventory.

In addition to other functions such as alarm generation, creation of logs, or the visualization of computed results, the application program interface (API) and the networking possibility of the various nodes must be particularly emphasized. The

API permits the coupling of any application to the database. It is possible thereby to include additional solution algorithms into the plant management level.

A *performance calculations package* aimed at improving the operation and availability of the plant is included. It consists in continuously monitoring the plant condition in order to detect malfunctions in time and to initiate counter-measures. The performance calculations package ConPec integrated into Contronic M is a modular, expandable program for the online determination of important operating parameters in power and desalination plants. On the basis of the process data transferred from the installed process control system, efficiency, setpoint deviations, plant condition, and other parameters helpful in assessment of the current operation are calculated and displayed to the supervisor.

Both the model formation and the configuration of ConPec take place modularly (see Fig. 6.15). Each component is described completely in terms of its thermodynamic behavior and is linked with its adjacent components. This procedure enables complex circuits to be described as well without great effort. New modules can be integrated easily according to the modular principle.

The optimization system performs the calculation of the main plant control setpoints in order to minimize a specific cost function calculated considering the operating costs of the operating unit. The values calculated allow the plant to keep the actual product on target and satisfy a defined set of operating constraints. The setpoints used in the optimization procedure are

- Top brine temperature
- Brine recirculation flow rate
- Seawater to reject temperature (winter operation)
- Seawater to reject flow rate
- Antiscalant dose level
- Makeup flow rate

The objective function is a cost relation representing the sum of the following parameters:

- LP steam to the brine heater
- Antiscalant chemical consumption
- Power consumption of the main pumps

A mathematical model is used to simulate the desalination plant behavior for different values of the setpoints and to calculate the corresponding values of the cost function. Since the operating conditions are affected by the fouling degree of the evaporator, the estimation of the fouling factor is necessary at predetermined time intervals in order to adapt the mathematical model to the plant behavior. Figure 6.16 is a block diagram showing the interconnection between system functions.

Measured values are brought from the plant and stored in the database at a predetermined frequency. The *data treatment module* reads the values from the

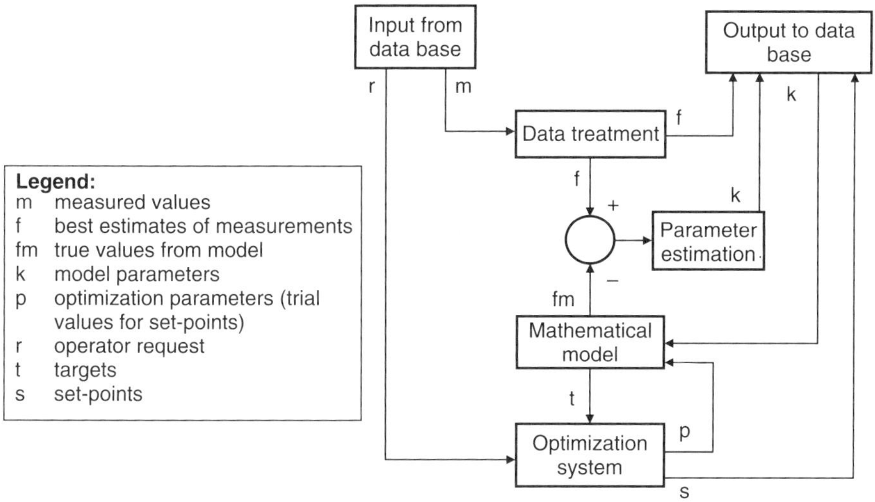

Figure 6.16 Block diagram showing the interconnection between system functions [32].

database and carries out validation checks in order to feed the optimization system with reliable values. These values are stored in the database for archiving and visualization.

Some of the reliable values are used to check the mathematical model in order to account for variations in the external parameters (fouling factors). The *parameters estimation module* evaluates the model parameters that allow the mathematical model to fit the real plant behavior. The updated model parameters are stored for archiving and visualization.

The mathematical model is used to simulate the plant operating conditions taking place for a predetermined set of input values. It is a steady-state model. The optimization module calculates the best set of control parameters that allow the plant to maintain the required distillate production. It uses the mathematical model to simulate various conditions in the operating range starting from the actual operating point. The calculated setpoints are stored in the database for downloading. The optimization module also operates in steady-state mode.

6.3.3 IMCA in the RO Process

6.3.3.1 General Description of the RO Process A schematic flowsheet of the membrane section of a typical seawater RO plant is shown in Figure 6.17a. Raw seawater is pumped from seawater intake to the pretreatment unit. At the intake, a disinfection of seawater can be conducted using chlorination in an intermittent mode, if necessary. Prior to media filters, seawater pH is adjusted to improve coagulation. As a coagulant, ferric salts are frequently used. In some seawater systems flocculation follows coagulation step. During flocculation, this lasts ~10 min, and size of flock particles increases, improving efficiency of removal of colloidal particles in the subsequent filtration step.

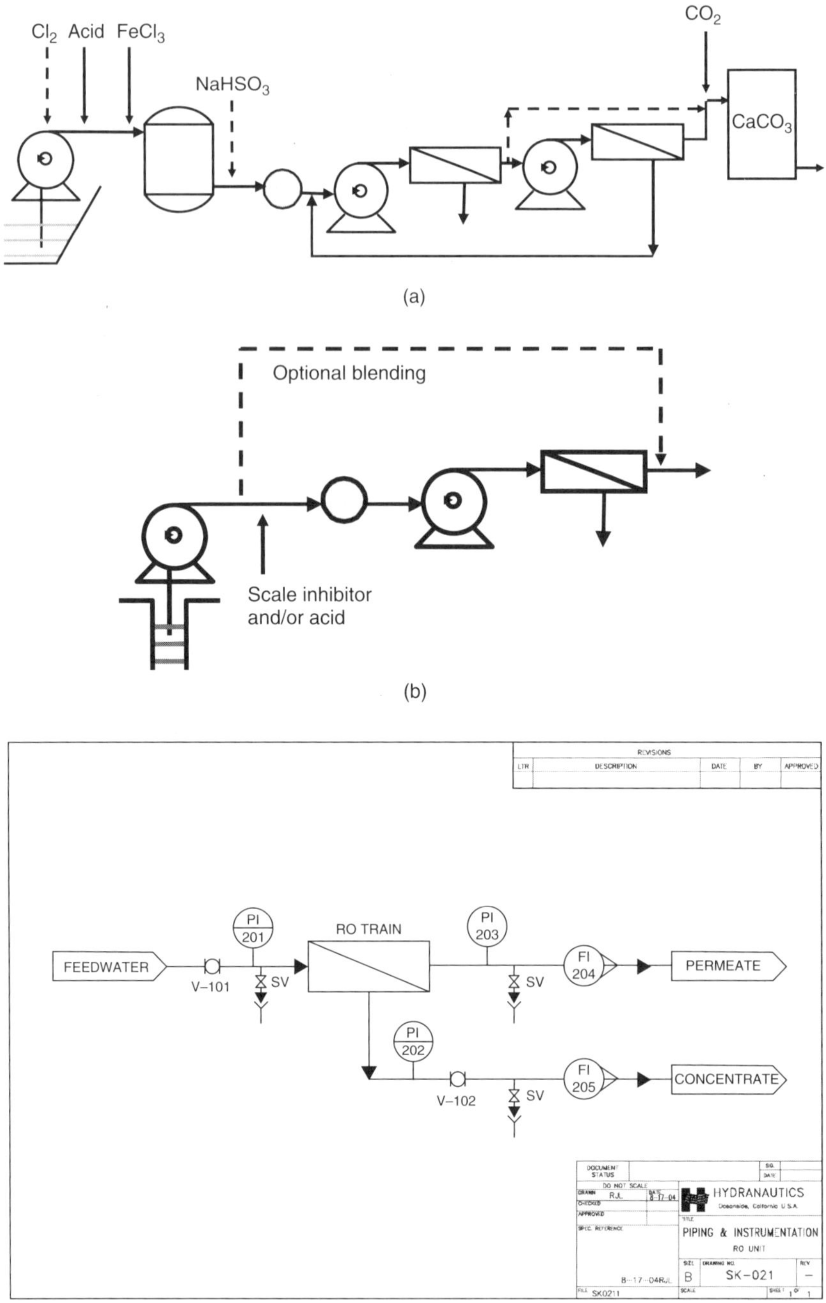

Figure 6.17 (a) Seawater RO operating on surface water with pressure media filters; (b) brackish-water RO unit operating from a wellwater source; (c) simplified flowsheet of an RO unit (Hydranautics).

After flocculation, seawater is pumped through gravity or pressure sand filters. Backwashing of the sand filters is carried out when they are loaded beyond the specified value indicated by differential pressure across the sand filter or according to volume of water filtrated. Specified doses of commercial hydrochloric or sulfuric acid is then added for pH correction and alkaline scale control. In some seawater systems, scale inhibitor is used as well. Filtrated seawater is then pumped through micron cartridge filters (5–10 μm, nominal pore size) before it is fed to the high-pressure pump of the RO system. The effluent from the pretreatment unit is sampled to check for SDI and pH values to safeguard the RO membrane systems.

The RO system shown in Figure 6.17a is configured as a two-pass unit. In the first pass, treated seawater is pumped at ~5–8 MPa into the spiral-wound or fine-hollow fibre RO modules arranged in parallel. Then 40–50% by volume of the feedwater is collected as permeate (product water) in an intermediate tank. All or part of the permeate from the first pass is processed by the second-pass RO unit. The second-pass RO operates at ~85–90% recovery rate. In the case of partial two-pass processing, first-pass permeate and second-pass permeate are blended together as a product water of required salinity, usually below 400 ppm TDS.

The reject from the first-pass membrane unit is sent to the energy recovery device and finally drained to the sea. With the present seawater membrane technology, two-pass processing is required only in case of very high feedwater salinity or high feedwater temperature (conditions at some Middle East locations). In most cases, a partial two-pass processing is usually sufficient. However, if the project specifications call for low boron concentration in the permeate (<0.5 ppm), a complete two-pass system, with interstage pH adjustment, is required.

The final product is sent to a posttreatment section to stabilize the water condition in respect to corrosion potential. The posttreatment process consists of adjustment of pH, increase of hardness, and sometimes increases in alkalinity as well. This could be achieved by adding lime slurry or passing product water through a limestone filter. Sometimes addition of CO_2 is required to dissolve sufficient quantity of solid $CaCO_3$ into HCO_3 ion. The posttreatment step results in overall increase of permeate salinity by ~30–60 ppm, therefore, RO systems have to be designed to account for this increase, and produce designed salinity of final product at the whole range of operating conditions. In case of a two-pass system, the reject from the second pass is mixed with the feed of the first stage to increase overall system recovery rate. The configuration of RO seawater systems is relatively uniform from site to site.

Figure 6.17b shows a configuration of brackish-water RO plant treating well water. In RO systems treating well water, the pretreatment is very limited, as well water usually contains very low concentrations of suspended particles. In most cases, the pretreatment consists of addition of scale inhibitor and cartridge filtration. The membrane unit is configured as a two- or three-stage system. Concentrate from each stage is sent as feedwater to the next stage, and permeates from all stages are collected together. A brackish system operates at feed pressure 1.5–2 MPa at recovery rate 65–90%. In brackish systems, power recovery equipment is not used frequently because of the relative low volume and pressure of the concentrate stream. Permeate from the brackish RO unit can be stabilized with respect to corrosion potential by blending with other water sources with sufficient hardness and alkalinity concentration, and/or by removal of CO_2 using a degasifier and pH adjustment with caustic.

Typical RO seawater commercial plant flow and instrumentation diagrams are shown in Figures 6.17c and 6.18. Seawater delivery for large system is almost exclusively based on seawater intake. However, there is growing tendency to locate seawater systems at the sites of power plants and utilize, the discharge stream from the steam condenser as feed. In currently built large RO systems, the pretreatment unit is still based on a conventional process of settling, flocculation, and media filtration. It is expected that in the future pretreatment in seawater plants will be based on submerged membrane technology. The membrane train consists of a single-stage array of pressure vessels housing seven or eight elements per vessel. The train capacity in large plants is in the range of 2–4 mgd. If required permeate salinity, or specific ion concentration, is lower than can be produced in a single pass, the first pass permeate is processed with the second-pass unit.

Depending on the condition, the second pass could be designed to process either full or partial first-pass flow. Usually, the second-pass concentrate is returned to the suction of the first-pass feed pump. The pumping system consists of high-pressure centrifugal pump and power recovery device. The power recovery device could be of Pelton wheel type, centrifugal reverse-running pump, or newer positive-displacement device (pressure exchanger). Seawater RO permeate is usually treated to reduce its corrosion tendency. This is accomplished by increasing the level of hardness and alkalinity. The control system is based on field sensors, transmitters, data storage, and process controlling PLC.

The RO system may consist of the following major units:

- Feedwater supply system
  - Seawater intake unit
  - Conveying and storage unit
- Pretreatment
  - Conventional technology (or)
  - Membrane technology
- RO membrane unit
  - Single-pass (or)
  - Two-pass (or)
  - Partial two-pass (or)
  - Two-pass, split partial (or)
- Permeate storage and treatment
- Membrane cleaning unit
- High-pressure pumping and power recovery unit
  - Centrifugal power recovery device (or)
  - Positive-displacement power recovery device (pressure exchanger)
- Control system

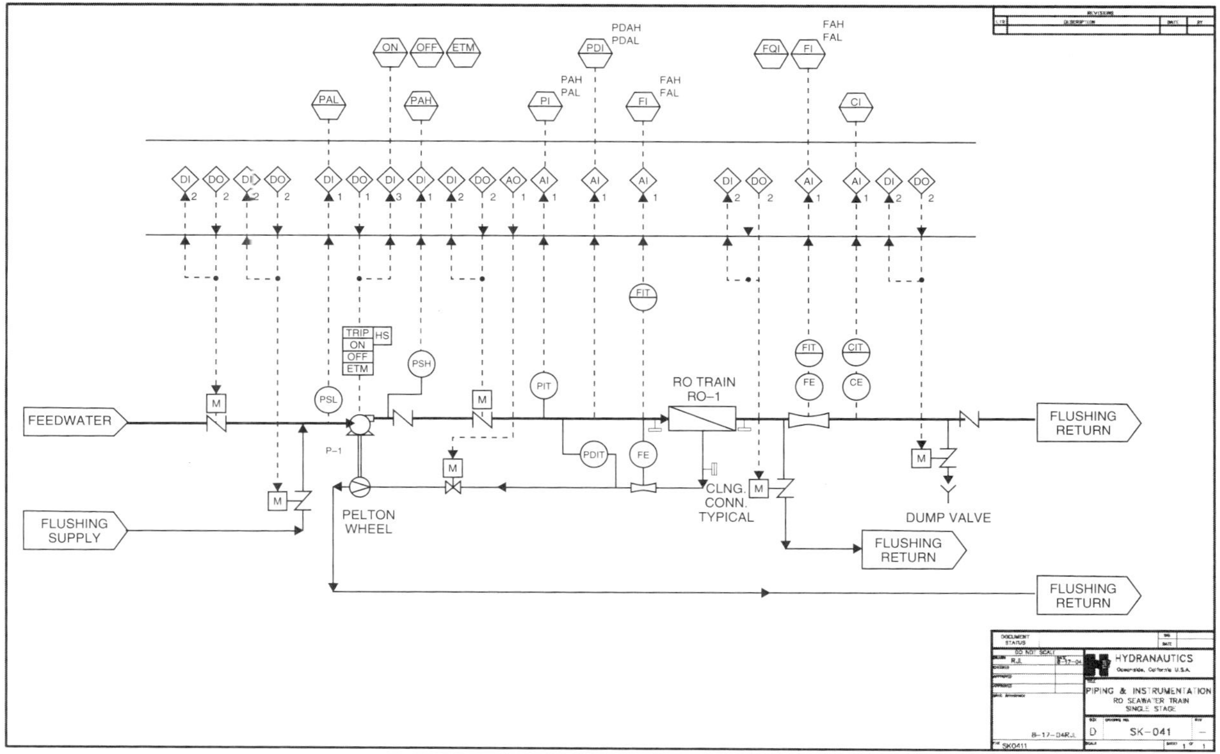

Figure 6.18 Flowsheet of a single-stage RO seawater desalination train (Hydranautics).

6.3.3.2 Identification of Major Process Parameters in a RO System

The process parameters are monitored to control performance of the plant to maintain operating parameters within the design range and to protect the integrity of the equipment. The following process parameters are monitored in RO plants:

- Raw-water conductivity
- Raw-water temperature
- Raw-water flow
- Raw-water pump suction and discharged pressure
- Raw-water turbidity
- Dosing rates of pretreatment chemicals
- Raw-water free (combined) chlorine
- Media filters head loss
- Filter effluent turbidity
- Filter effluent particle count
- Filter effluent SDI (MFI)
- Cartridge filter pressure drop
- High-pressure pump suction and discharged pressure
- Feedwater pressure
- Feedwater pH
- Feedwater free (combined) chlorine
- RO permeate flow
- RO permeate pressure
- RO permeate conductivity
- RO permeate temperature
- RO permeate pH
- RO concentrate flow
- RO concentrate pressure
- Dosing rate of posttreatment chemicals
- Product-water turbidity
- Product-water free (combined) chlorine
- RO permeate storage tank level

The monitoring activity conducted to protect plant equipment includes monitoring operating parameters of major equipment. These include setting alarms and shutoff switches to indicate off-limit conditions of the following parameters:

Levels in water storage tanks
Levels in chemical storage tanks
Flow of treatment chemicals
Water temperature

Water pH
Water turbidity
Free (combined) chlorine concentration
Pressure drop in cartridge filters
Pump suction pressure
Pump discharged pressure
Feed pressure
Permeate pH
Permeate conductivity
Permeate temperature
Permeate pressure
Concentrate flow
Concentrate pressure
Pressure drop in RO system
Temperature of electric motors

6.3.3.3 RO Automation System

6.3.3.3.1 Introduction Control of operation of the RO system can be accomplished with a very small amount of instrumentation. Figure 6.17c depicts a simplified process–instrumentation diagram showing only pressure sensors on the feed, concentrate, and permeate lines and flow sensors on permeate and concentrate streams. Such instrumentation would be sufficient to determine the capacity of the unit and to calculate permeate recovery ratio. In an extremely simplistic approach, the system could be operated and controlled even with instrumentation that is more limited. The only instrument required online would be a pressure gauge on the high-pressure pump discharge. Readings of discharge pressure and pressure–versus flow diagram of the pump would be sufficient to determine feed flow to the RO block. Then the permeate recovery rate (R) can be determine by measuring salinity (conductivities) of the feed, permeate and the concentrate streams, according to the following equation

$$R = \frac{C_{\mathrm{f}} - C_{\mathrm{p}}}{C_{\mathrm{c}} - C_{\mathrm{p}}} \tag{6.2}$$

where C_{f} = feed salinity
C_{P} = permeate salinity
C_{c} = concentrate salinity

Equation 6.2 is derived based on conservation of flow and concentration in the RO unit. Information of feed flow and recovery ratio enables calculation of permeate flow. The above approach is sometimes practiced in RO systems at the early stages of operation if the instrumentation–control system is fully functional. This method is also used to reconfirm readings of the flowmeters in the system.

In commercial units, the monitoring and control system is based on a large number of local gauges, sensors, and transmitters as shown in Figure 6.18.

6.3.3.3.2 Current Status of RO Process Control Process control and monitoring in commercial RO desalination systems utilizes an almost exclusively computer-based supervisory control and data acquisition (SCADA) system. The SCADA configuration includes central computer providing videodisplay, data storage, and reports is connected through a data highway with a distributed network of process monitoring and controlling microprocessors. The local microprocessors evaluate process parameters of designated system unit and control its operation within determined limits. Process control is achieved by evaluating the output of sensors installed in the plant and controlling operation of pumps and valves.

The current control system provides the following functions:

1. Protects the system from operating at conditions that may result in equipment damage. For example, equipment operation is started in a predetermined sequence. Pumps are protected from operation if the suction pressure is inadequate. At extreme pH or temperature feedwater is diverted to drain.
2. Maintains equipment operation within the design process limits. For example, operation is controlled to maintain design limits of feed temperature, pressure, flow and other parameters.
3. Maintains production of the design quantity and quality of product water. For example, feed pressure is adjusted to produce design output capacity. Permeate is diverted to drain if design quality is not met. Dosing pumps are controlled to maintain designed pH and hardness of the product water.
4. Stores operating data, generates reports, visual and hardcopy, from unit operation. For example, historical results and performance trends reports are generated. Membrane performance results are normalized. Operating cost data are calculated.

At present the application of process automation does not provide direct optimization of system performance to achieve the optimum product quality or minimum water cost. The only direct cost reduction measure applied currently is sequencing of operation desalting units according to predetermined cost parameters, for example, system capacity utilization according to variable power cost or priority of unit operations according to prior determined operating cost of individual units (unit required highest operating pressure will be activated last). The abovementioned cost reduction measures are applied on the basis of the offline operator's evaluation of the prevailing economic conditions.

6.3.3.4 The Major RO Process Parameters and Their Control According to the points mentioned above, the performance of RO membranes is affected by the following parameters:

- System recovery rate
- Design permeate flux

- Feed salinity
- Feed temperature
- Membrane surface condition

Effects of the individual process parameters on permeate salinity (salt passage) and feed pressure (water permeability) are listed in Table 6.5. However, in practice, the possibility of changing these parameters to optimize performance of an existing system is constrained by the system configuration and design.

6.3.3.4.1 Recovery Rate RO systems are designed to operate in the narrow range of two basic operating parameters: recovery rate and permeate flux.

In the majority of commercial systems, the pumping and pretreatment equipment available determines the maximum rate of feed flow. Therefore, for constant feed flow, reduction of recovery rate will result in lower permeate flow. Increase in recovery rate is possible as far as the pretreatment system is concerned. At conditions of constant feed water temperature, recovery increase requires higher feed pressure to maintain the same permeate output. Therefore, the motor of the high-pressure pump or booster pump has to be equipped with a variable-frequency drive (VFD) to enable increase of feed pressure. In the case of increase in feedwater temperature below 25°C, it would be possible to increase recovery rate in parallel to increase of membrane permeability. Depending on feedwater salinity, at temperatures above 25°C, usually the effects of the osmotic pressure increase are similar to or higher than those due to increase in water permeability with temperature.

In an existing system, the recovery rate may be adjusted by

1. Changing throttling position of the valve on the concentrate line
2. Changing the rpm of the pump's electric motor using VFD and, in parallel, adjusting throttling position of the concentrate valve
3. Allowing changes in feedwater temperature

Table 6.5 Effect of Process Parameters on Permeate Salinity and Feed Pressure

Parameter	Parameter Value Change	Permeate Salinity Value Change	Feed Pressure Value Change
Feed salinity	Increase	Increase	Increase
	Decrease	Decrease	Decrease
Recovery rate	Increase	Increase	Increase
	Decrease	Decrease	Decrease
Feedwater temperature	Increase	Increase	Decrease
	Decrease	Decrease	Increase
Permeate flux rate	Increase	Decrease	Increase
	Decrease	Increase	Decrease
Membrane age	Increase	Increase	Increase

6.3.3.4.2 Permeate Flux The permeate flux rate is determined by the net driving pressure, which is the function of the difference between applied feed pressure and osmotic pressure. An increase of flux rate would require operation at higher recovery rate, which would require an additional increase of feed pressure. Another option that would result in higher flux rate is reduction of membrane area in operation. Changing membrane area in operation is not a very practical approach in commercial systems, except for systems with extreme fluctuation of feed salinity.

In an existing system the permeate flux rate can be changed by

1. Adjusting feed pressure by changing rpm of pump's electric motor using VFD and by adjusting, in parallel, the throttling position of the concentrate valve
2. Allowing changes of feed water temperature

6.3.3.4.3 Feedwater Temperature Fluctuation of feedwater temperature is usually a result of the natural weather pattern. If the RO system is located adjacent to the sea, seawater at elevated temperature is available that can be used to increase the feedwater temperature. Fluctuation of feedwater temperature can be utilized to some extent to reduce power requirements by adjusting feed pressure and recovery rate. For example, increased feedwater temperature could allow decrease in the feed pressure or increase in recovery rate while maintaining the feed pressure constant—this is possible only if the permeate salinity at elevated temperature is still within design limits.

Fluctuations of feedwater temperature are natural phenomena. On sites where hot feedwater is available, some control of feedwater temperature may be managed by blending water from hot and cold sources.

6.3.3.4.4 Feedwater Salinity At a majority of seawater sites, feed salinity fluctuates within a narrow range not exceeding 5–10%. RO systems are designed to provide rated output at maximum feed salinity. During periods of lower salinity, the system can operate at higher recovery rate or lower feed pressure. At some locations, very wide salinity fluctuations are experienced, mainly because of rainwater influx. These conditions are addressed during the design stage to provide sufficient flexibility of pumping–power recovery equipment. In extreme cases hybrid membrane systems are considered, on which some limited pilot work has been conducted.

Control of feed water salinity can be executed to some extent in brackish systems receiving feedwater from multiple wells. In seawater systems, fluctuations of feedwater salinity are mainly a weather related phenomenon.

6.3.3.4.5 Water and Salt Transport Changes Related to Membrane Age and Fouling Increased membrane age almost universally results in higher salt passage and lower water permeability (need for higher feed pressure). A parallel process of membrane fouling has a very similar effect on membrane performance. The effect of compaction is not reversible. The effect of fouling can be reversed to some extent by membrane cleaning. This provides another dimension to the optimization process. For example, ability of increasing recovery rate is restricted to a narrower range with older membranes than with newer ones, due to increased salt passage.

6.3.3.4.6 High-Pressure Pumping Equipment and Energy Recovery Units

Currently, large commercial plants utilize, almost universally, centrifugal pumps for pumping feedwater to the membrane elements. In seawater systems, the newer pumping configuration consists of pressure centers. In this configuration, a small number of high-capacity high-pressure pumps are connected to a common feed manifold, connected to number of RO trains. In the same manner, the concentrate manifolds from all trains are connected together to the energy recovery devices. Use of high-capacity pumps provides higher pump efficiency than would be available in the conventional configuration, *i.e.*, a smaller pump and energy recovery device dedicated to a single train. Because of the very high efficiency of large-capacity pumps and new energy recovery devices (pressure exchangers), there is a very small energy penalty when some RO trains are offline due to maintenance work and when the RO system operates at lower recovery rate [34].

In brackish-water RO units, the energy recovery devices, which are gaining popularity, are turbochargers. A *turbocharger* consists of two pump impellers, directly connected on a common axis. The two impellers have blades configured in opposite directions. Therefore, the impeller that receives the concentrate stream acts as a driver for the second impeller that provides boost to the feedwater. Turbochargers are usually used to increase interstage pressure in two-stage brackish units. Pressure boost provided by turbocharger is proportional to the flow and pressure of the concentrate stream. Turbochargers are very compact and relatively inexpensive devices. Operation of turbocharger does not require any specific control equipment or operators intervention to adjust its operation, after the initial setting. Usually the turbocharger device is equipped with valves that enable redirect concentrate flow in order to bypass turbocharger partially or completely. This arrangement enables regulation of pressure boost provided by turbocharger, if required.

6.3.3.5 Comparison of Specific Requirements of Control Systems in Brackish and Seawater RO Plants Desalination processes in seawater and brackish-water (or nanofiltration) plants are very similar in most cases. The significant differences are related to feed salinity, operating pressure, recovery rate, and system configuration. Feed salinity and feed temperature in brackish and nanofiltration systems that mostly process wellwater are relatively constant. Stability of feed salinity and temperature results of operation at a narrow range of feed pressure. Feed pressure in nanofiltration systems is below 0.7 MPa and in brackish systems; in the range of 1.0–2.0 MPa. The tendency in nanofiltration and brackish systems is to maximize recovery rate, due to limited availability of feedwater and difficulty with disposal of the concentrate. Therefore, nanofiltration and brackish systems operate at conditions of oversaturation of the sparingly soluble salts in the concentrate stream. To maintain sparingly soluble salts in solution, acid and/or scale inhibitor is added to the feed stream. The control system plays an important role in assuring uninterrupted flow of acid and/or scale inhibitor. Acid addition is usually controlled through measurement of pH and flow rate of relevant streams (usually permeate and concentrate). Controlling addition of scale inhibitor is more difficult as direct measurements of concentration of scale inhibitor in feed or concentrate is not available. A range of indirect measurement approaches are used. One technique, which is close to direct measurement, is use of scale inhibitor that contains fluorescent marker additive and applying light sensor to measure marker concentration. The other method of control relays on measurement of quantity of scale inhibitor removed over time from the storage container. In case that scale inhibitor flow is interrupted, membrane system has to stop operation, otherwise irreversible scaling of membrane element could result.

Brackish and nanofitration systems have a multistage configuration. The number of desalination stages (concentrate staging) is usually two for brackish-water units and up to three for nanofiltration units. It is common in large nanofiltration plants to have a separate third-stage unit that processes combined concentrate from more than one first- or second-stage units. These third-stage units have a dedicated pump and require a separate control system. Potential for optimization of operation of nanofiltration or brackish systems is limited to adjusting permeate output according to fluctuation of potable water demand. Large systems are composed of a number of wells and RO units. The operation schedule will be based on the predetermined sequence of adjusting operation of water sources and RO units according to demand. Better quality water sources and lower energy consuming unit would be used as a base load and less efficient unit would be put online later to satisfy increasing demand. This operation schedule would be adjusted periodically according to the unit's performance evaluation, conducted offline.

Seawater systems operate at feed pressures of 5.5–7.5 MPa. Feed pressure in a seawater RO unit could vary over a wide range. With few exceptions, seawater is subject to moderate salinity fluctuations and relatively wide fluctuations of feedwater temperature. The normal fluctuation of seawater salinity is in the range of ~5%. This corresponds to a salinity change of ~2000 ppm TDS, which is equivalent to average osmotic pressure change of ~0.2 MPa. Feed pressure would have to change roughly by the same amount to compensate for a decrease of net driving pressure. In some locations, seawater salinity is affected by rivers discharge or low-salinity runoff during rainy seasons. Then the salinity fluctuation could be quite significant, sometimes in the range of 50%. Fluctuations of seawater temperatures are quite common (with some exceptions for locations close to the equator). Current trend of locating seawater RO plant at the site of electric power plants and utilizing condenser discharge as a RO feed, increases even more the potential range of feedwater temperatures. If one compares winter conditions of a power plant operating at low capacity and summer conditions of high-capacity power plant operation, the temperature difference could reach up to 20°C. This difference alone could change the specific water permeability of RO membranes by 40–60%. In addition, there is the effect of membrane compaction and fouling that could result in additional permeability changes of up to 20%. The other parameter that can affect system operation is salt passage. Salt passage is affected by temperature (see previous section for more detailed discussion) and condition of membrane surface. Higher salt passage may require increased operation of second-pass RO.

6.3.3.6 Control of Some Existing RO Plants

6.3.3.6.1 A Fully Automated 5000-gpd Seawater RO Test Unit in Kuwait [35]

Description of Plant Figure 6.19 shows a schematic diagram of the process flow and instrumentation of the fully automated RO plant. A beachwell seawater intake was utilized for seawater feed for the RO plant. The feedline from the beachwell was connected to a 110-mm inlet feedline from the RO unit with one secondary line for silt density index (SDI) measurements. The feed flow rate was set at 12 m^3/h at a pressure of 0.4 MPa. A 1000 liter intermediate tank was connected after the two cartridge filters. The tank was equipped with three level controls to ensure that the high pressure pump never ran dry and that constant water pressure was maintained on the feed line to the RO membrane. A 2000-L cleaning/flushing tank was used as a product tank and for the preparation of cleaning solutions, for the cleaning cycle, and for flushing of the RO membrane after stoppage and cleaning.

A 5000-gpd RO test unit, consisting of a high-pressure pump, a pulsation damper, and a high-pressure vessel, was used to complete the RO plant. The maximum capacity of the piston pump is 134 L/min running at 800 rpm at a pressure of 9.0 MPa. The diameter of the pump's pistons is 42 mm, and the pump has a maximum operating temperature of 60°C. The driving motor is a three-phase, star-delta, asynchronous type with a maximum power consumption of 23.8 kW. The motor is equipped with an overcurrent relay thermistor and fuse protection. The safety pulsation damper range is 4.0–11.0 MPa; the set point of the damper was set at 6.5 MPa to ensure protection of the membrane. An intermediate tank and a product tank were connected to the feed- and product lines, respectively.

Description of Control System A Siemens SU-115U PLC with relays, contactors, and supply voltage was installed on a frame located in the upper-floor control room (see Fig. 6.20). A cable tray was installed to carry all cables and wires to the R&D room where the unit was located. A three-phase 415-V AC supply voltage and a 24-V DC control voltage were connected from the low-tension cubicles to the PLC frame bus. The membrane used was the Fluid Systems spiral-wound, model 2822SS. The critical parameters that were identified for control purposes as follows:

Feed Flow Rate. The feed flow rate was set at 13 m^3/h through the intermediate tank. This flow *rate* ensured that the feed line and up to the intermediate tank was full at all times and that there was enough flow to the side stream feeding the automatic SDI unit. This was ensured by the proper setting of the minimum flow limit of the flow sensor controller.

Feed Temperature. One advantage of the beachwell intake is the steadiness of the feedwater temperature variations throughout the year (25–30°C). The operating temperature range was set to be within the allowable temperature

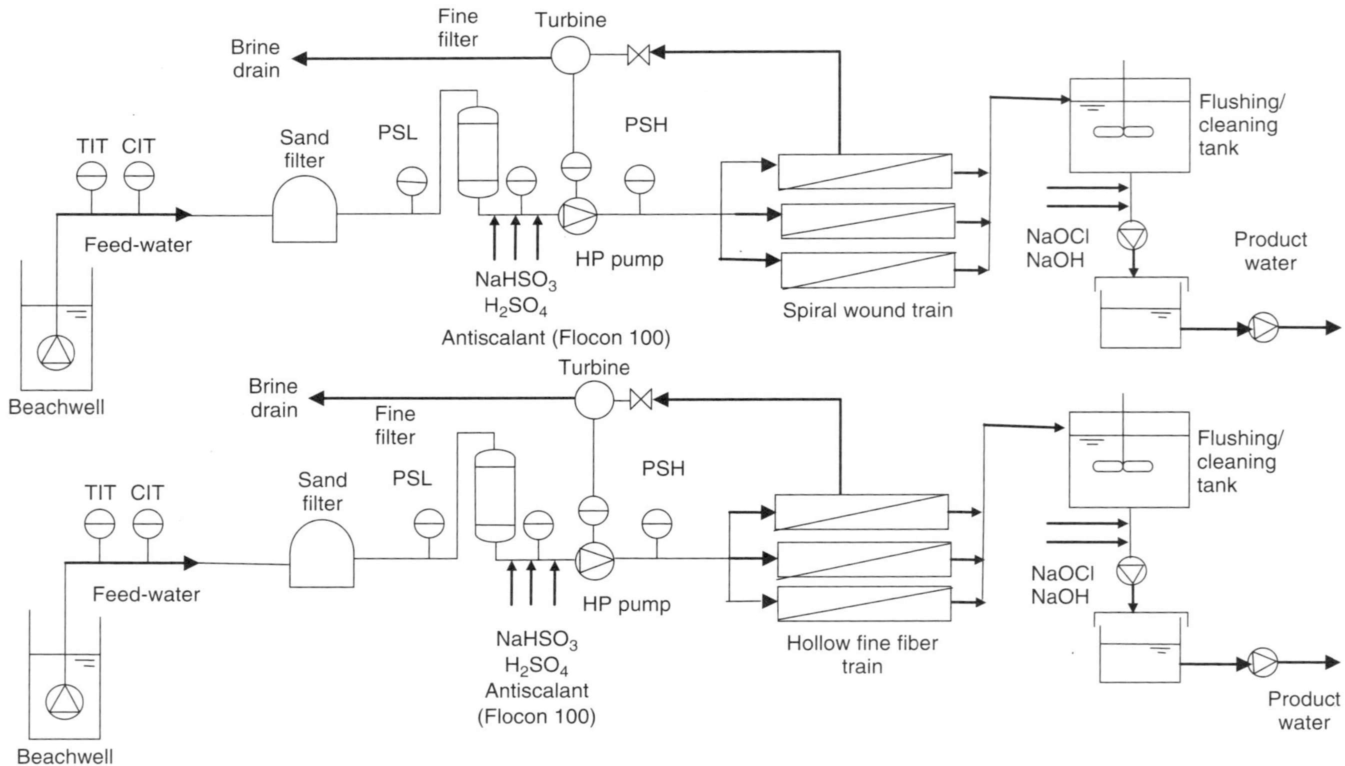

Figure 6.19 The fully automated seawater RO plant in Kuwait [35].

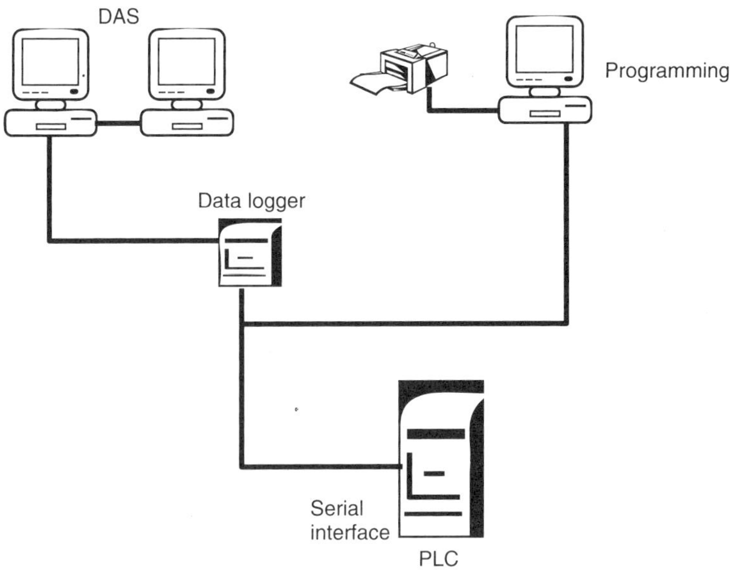

Figure 6.20 Control system structure for the Kuwait seawater RO test unit [35].

range of the membrane module. The lower and higher setpoints were set at 10°C and 40°C, respectively.

Feed Pressure. The high feed pressure is one of the most critical parameters for the RO membrane since the membrane element is located directly after the high-pressure pump. If the pressure exceeds the maximum allowable operating pressure of 12.0 MPa, the membrane can be severely damaged. Two devices, a pulsation damper and a pressure gauge, were installed to ensure that this would not occur. In this case, according to the manufacturer's recommendations, the maximum operating pressure was set at 6.5 MPa.

Feed pH. On the basis of the manufacturer's recommendations, the acceptable pH range for feed-water through the membrane was 4–11; therefore, the setpoint of the pH of the feed was 7.1. The lower and higher alarm limits were set at pH 5 and pH 10, respectively. Two automatic acid dosing units were connected and metered to control the dosing level and stroke length. An online pH-measuring instrument was installed, and the signal output was connected to the programmable logic controller (PLC) analog input card. An output signal of 4–20 mA, corresponding to a 2–12 pH value, was used as input to the PLC, which translated the signal to a 32-bit floating-point number and paced a prominent solenoid-driven acid dosing pump.

Silt Density Index. A Mabat automatic SDI 2000 was used for this project. The instrument has a PLC for controlling SDI measurement. The time interval for SDI measurements was set at one hour. The line was continuously flushed between tests to avoid any stagnation in the line. The SDI test duration time was the standard 15 min; the flushing time before the test was set at 10 min.

The values of the automatic SDI unit were cross-checked randomly with manual SDI measurements on the same line. The automatic SDI unit showed consistency with the manual sampling.

Chlorine Concentration. A chlorine measuring instrument (0–2 ppm) was installed and connected to the analogue input of the PLC. The manufacturer recommends that no chlorine pass through the membrane; therefore, the set point for the high Cl_2 level was set at zero. An automatic dosing pump was connected for the dosing of a deoxidizing agent, *i.e*. sodium bisulphite.

Cartridge Filter Differential Pressure. Two alternate 5-μm cartridge filters units, each with a capacity of 15 m^3/h, were installed in the feedline in front of the intermediate tank. Before installation, a test was conducted using the same beachwell feedwater to determine the size and number of filter elements required. Tests showed that a considerable drop in the flow rate (<8 m^3/h) resulted when the differential pressure across the cartridge filters reached 0.1 MPa. Therefore, the limit for the high differential pressure of the filters was set at 0.09 MPa.

Membrane Differential Pressure. A high differential pressure across the membrane is caused by salt deposits or fouling of the membrane on the feed side. On the basis of the manufacturer's recommendations, the setpoint for the membrane differential pressure was set at 0.5 MPa, at which point the system diverted the flow and an alarm was activated. The frequency of filter changes was 3–4 days.

High Product Conductivity and/or Low Product Flow. When either of the parameters' setpoints was reached, flow diversion and alarm signals were activated. In addition, a signal for membrane cleaning was activated. Cleaning normally resulted in a reduction in product conductivity to 800 μs/cm^{-1}, an increase in product flow to 1.2 m^3/h, and a drop in feed pressure to 4.3 MPa.

Low Chemicals Level. When the levels of both acid tanks or the level in the sodium tribisulfate tank was low, flow was diverted and an alarm was activated. When the operator refilled the tanks, the system restarted the unit automatically.

6.3.3.6.2 Larnaca 54,000 m^3/day Seawater RO Plant [36,37]

Description of Plant This plant, which has a state-of-the-art controls and monitoring system and is designed to treat seawater of salinity in the range of 38,000–40,000 mg/L with a minimum of operator interface.

Pretreatment consists in addition of sulfuric acid for pH adjustment prior to coagulation–flocculation, followed by gravity media filters. Media filter effluent flows into clear well. From a clear well, seawater is pumped using booster pumps through cartridge filters. The booster pump motors are equipped with variable-frequency drive converters (VFCs). Use of VFCs enable adjustment of suction pressure to the high-pressure feed pump, which enables convenient adjustment of feed pressure to the RO membranes. Use of VFC on a smaller motor of the booster pump, rather than on the large motor of the high-pressure pump, reduces the cost of

VFC and results in a small reduction in the electric efficiency. The cartridge filters have 20 μm porosity nominal rating and are of vertical configuration. Scale inhibitor is added to seawater upstream of the cartridge to protect against potential scale formation in the RO concentrate. From the cartridge filters, seawater feed is distributed to six high-pressure pumps operating in parallel. The pumps are of horizontal configuration, multistage, split case. Each pump is driven by an electric motor and an energy recovery turbine connected on the same axis. The energy recovery turbines are of the Pelton wheel type. Each pumping station is dedicated to operation of a single RO train. Train design capacity is 9000 m^3/day each. RO trains operate at 50% recovery rate. The seawater trains are configured as single-stage, with 120 pressure vessels per train. Each pressure vessel contains eight spiral-wound seawater elements, manufactured by Hydranautics. Permeate quality specifications include limit on boron concentration to be below 1 ppm. For this reason, the RO plant is configured as a partial two-pass system. Depending on concentration of boron in the feedwater and feedwater temperature, up to 15% of the first pass permeate can be processed by the second-pass RO units. The second-pass units are equipped with brackish membranes, configured in a 8–4 array configuration, eight elements per vessel. In order to improve boron rejection of the brackish membranes, the pH of feedwater to the second pass is increased to about 10 by addition of sodium hydroxide. Permeate from the second-pass RO units and unprocessed permeate from the first-pass RO are blended together and sent to the posttreatment system. In the posttreatment system, hardness and pH of the combined permeate are increased by passing permeate through a limestone filter with the addition of NaOH, if necessary.

Description of Control System The control system, at Larnaca, shown in Figure 6.21a, is of distributed configuration, divided into eight functional modules controlled through a central programmable logic controller (PLC):

1. Raw-water supply system
2. Pretreatment system
3. Feedwater transfer pumps
4. First-pass high-pressure pumps and seawater RO trains
5. Second-pass high-pressure pumps and brackish RO trains
6. Posttreatment system
7. Product delivery system
8. Auxiliaries (cleaning system, chemicals delivery, etc.)

Such a distributed control system is usually more expensive in terms of hardware cost and maintenance than the central system, but this control system configuration provides better flexibility and higher process availability.

The unique conditions of the electric tariff schedule in Cyprus provided strong incentives to reduce power consumption during the peak demand hours. To benefit from lower electric rates, the Larnaca plant has to reduce power consumption for 3 h during the peak electricity demand period. To accommodate this requirement, three

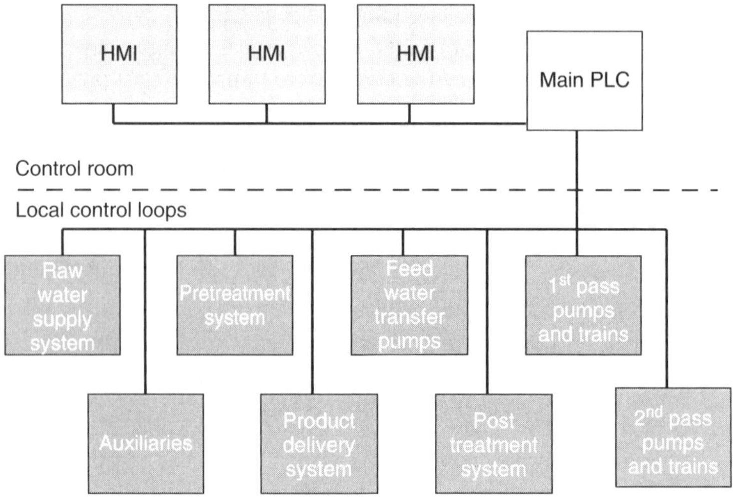

Figure 6.21 Configuration of the control system at RO seawater plant at Larnaca, Cyprus. (HMI: Human-Machine Interface System).

seawater trains (out of six installed) are taken offline for 3 h during the peak electricity demand period. Following the partial shutdown and corresponding capacity reduction, the plant recovers automatically and returns to the design capacity after a 3-h period, bringing all six trains seawater RO trains back in to operation. This mode of operation has been maintained for a number of years without any adverse effect on membrane performance.

The process control is implemented through central PLC with operator interaction taking place in the control room at the display and control stations. All system process information is displayed on a host of selectable screens. The process variable control is accomplished using loop control modules in the PLC, which collect data from locally mounted transmitters.

Systems are continuously monitored, started, or stopped from the central control room master computer PLC. A telemetry system provides the control via the master computer for the operation and monitoring of the offsite seawater intake pumping station. The RO trains are automatically controlled after operator selection and are brought online in a slow start manner (slow ramp) until full production is achieved. Constant permeate flow is achieved by flow transmitter–PLC/VDF interaction. Normalization of operational data is conducted a number of times per hour and compiled on a daily basis. Real-time and historical trends on all system process parameters are remotely accessible through the internet server, and hardcopies may be printed immediately for records.

6.3.3.6.3 Fujairah Seawater RO Plant [38]

Description of Plant The major western coast desalination plant in the UAE, the Fujairah seawater reverse-osmosis desalination plant, has a capacity of 9000 m^3/day or 2.4 million gpd. Figure 6.22 gives the flowsheet of this plant. This plant can also be considered as the major reference for a large-scale beachwell application in the Gulf, operating continuously for more than 4 years, with the beachwells providing excellent results in terms of seawater turbidity and SDI. The plant was commissioned in February 1991. The raw water is drawn from 10 beachwells located parallel to the shoreline behind the beach crest. The wells have been drilled to a depth of 60 m each and tested for yield and drawdown to provide a nominal capacity of 128 m^3/h each.

The main parameters of the raw water from the well field are as follows:

- Temperature 19°C average with a seasonal variation of ±1°C
- SDI 0.3–1.0
- TDS ~39,300–40,500 ppm
- pH 7.25

A major advantage of beachwells is the very stable temperature throughout the year as well as the low SDI factor. The TDS in this case is higher than the usual seawater TDS probably because of a higher salinity in the ground. However, it is certain that almost no potable water is abstracted from the landside aquifers; this confirms that the location of the well field is correct.

Chemical Conditioning Upstream of the sand filters, chlorine and H_2SO_4 are injected. Prechlorination and pH adjustment with sulfuric acid is carried out automatically. The chemistry of the wellwater and its rather low contamination does not require much chlorination; however, due to the presence of Fe^{2+}, chlorine is used as an oxidizing agent in order to convert Fe^{2+} into Fe^{3+}, and therefore 2 ppm are dosed.

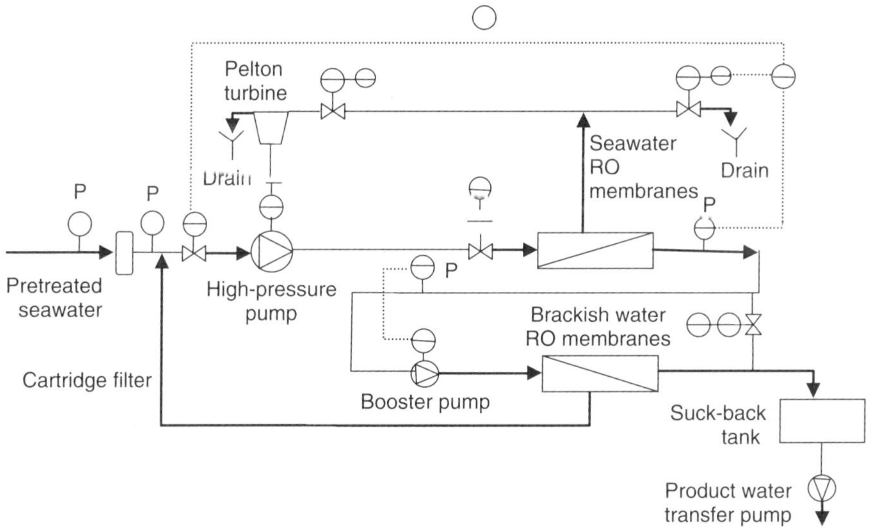

Figure 6.22 Flow diagram of the Fujairah seawater RO plant [38].

Acid is dosed at 15–20 ppm to maintain a feed pH in the range 6.9–7.0. Upstream of the cartridge filters, sodium bisulfite $NaHS0_3$ (7–8 ppm) is injected to eliminate the residual chlorine (~1.2 ppm) and to protect the membranes. Downstream of the cartridge filters, the redox (reduction–oxidation) potential is measured in the RO feedwater. If insufficient $NaHS0_3$ is present in the RO feedwater, the RO units are shut down automatically and the water is dumped back in to the sea via the dump valve.

The dump valve is also used for startup of the plant whereby the RO feedwater will be dumped automatically regardless of the water quality for 15 min before any RO stream can be started in order to ensure that the pretreatment and chemical dosing has been strictly established.

The Sand Filter Station This consists of nine pressure sand filters of 3.5 m diameter each, internally hard-rubber-lined with an incorporated nozzle floor. The sand filters are backwashed automatically in sequence with air scouring and water followed by an infiltration period. The filter medium composed of sand of different grain sizes (coarse and fine sand) as well as anthracite to enhance the filtration effect, eliminating suspended iron. The water out of the sand filtration station is led to an intermediate filtered-water tank providing storage capacity for the backwash water for the sand filters and surge capacity for the filtered water transfer pumps, which transfer the water through the cartridge, filters to the RO trains.

Desalination Membranes Three module racks are used, each RO stream equipped with 48 pressure vessels. Each pressure vessel can hold six spiral wound membranes. The maximum number of membranes that can be installed per train is 288.

The permeate is transferred to the surge suckback tank acting as an intermediate tank for the product-water transfer pumps and further to a 5000-m^3 product storage tank outside the main building. A postchlorination and a pH adjustment with sodium hydroxide is carried out automatically in front of the product-water tank.

At every shutdown of each RO train, the flushing pumps fed from the surge suckback tank start automatically replacing the brine solution within the system by low-salinity water, thus avoiding scaling and corrosion by long-term exposure to high-salinity brine.

Description of Control System Programmable logic controllers control the parameters and devices that are automated. They allow the control of all processes within the plant, as well as providing data to external systems via analog signals or serial strings. The controls of the individual plant sections are individually carried out by sub-PLC stations, which are all linked together by a digital bus system. The overall control is made by the master station, which communicates with the VDU system and the protocol system. A human–machine interface (HMI) system is used to indicate and record data gathered by the PLC.

Two levels of control are carried out by the PLCs:

Level 1 —involves sequencing operations such as manipulating valves or starting/stopping pumps, instrument verification, data acquisition, online maintenance, and fail-safe shutdown procedures

Level 2 —control parameters such as flow, pressure, temperature, and pH

The PLC can be connected via a digital communication link to an HMI computer. This computer may be programmed with graphical–equipment/process screens to enable the operator to view the system from a primary location and adjust the setpoint values of the control valves.

The process parameters and status indications are displayed on a backup process and power mimic. The automatic operation of the plant is accomplished using a process visualization system (VDU) with videoscreen, terminal, and mouse. A separate printer prints all incoming faults and system alarms for the record.

All process data are also stored in the system for 36 h in the form of graphs. The plant monitoring system (of the redox type) has a duplicate system for backup with a self-check function to identify sensor faults and any deviation from calibration.

The control system also provides alarm conditions for

- Pressure at high-pressure pump inlet
- Pressure at high-pressure pump discharge
- Pressure at membrane inlet
- Product (permeate) water pressure
- Reject output pressure

Pressure switches and transmitters monitor these pressures. To ensure long-term safe plant operation, pressure switches shut down the high-pressure pumps during low inlet pressure and high discharge pressure conditions. A relief valve in the product water line protects membrane elements from damage due to any backpressure downstream. Feedwater parameters (e.g., high pH, free chlorine, high temperature) that could adversely affect membrane longevity are indicated by an alarm condition. Automatic ON/OFF valves are provided to divert the feed or the permeate stream to drain when stream quality is not within the safe range. For example, a much higher than design product-water conductivity is an indication of membrane scaling, a breakdown in membrane integrity, or membrane damage. Hence, a "low membrane rejection" alarm is provided.

6.3.3.6.4 Sabha Brackish-Water RO Plant [21]

Description of Plant This plant was located 3 km north of the town of Eilat, Israel and desalts brackish-water having a salinity of 4600–6000 ppm TDS and supplies potable water to the town of Eilat. Plant capacity started with 700 m^3/day in 1978 using a single RO unit to 13,000 m^3/day in 1984 using four RO units. The plant is fed from an array of 11 wells, which differed from each other in flow rate and salinity. As shown in Figure 6.23, the plant consists of four RO units, each unit consisting of two stages of RO vessels connected such that the brine discharge from the first stage is used as the feed to the second stage (brine staging). The

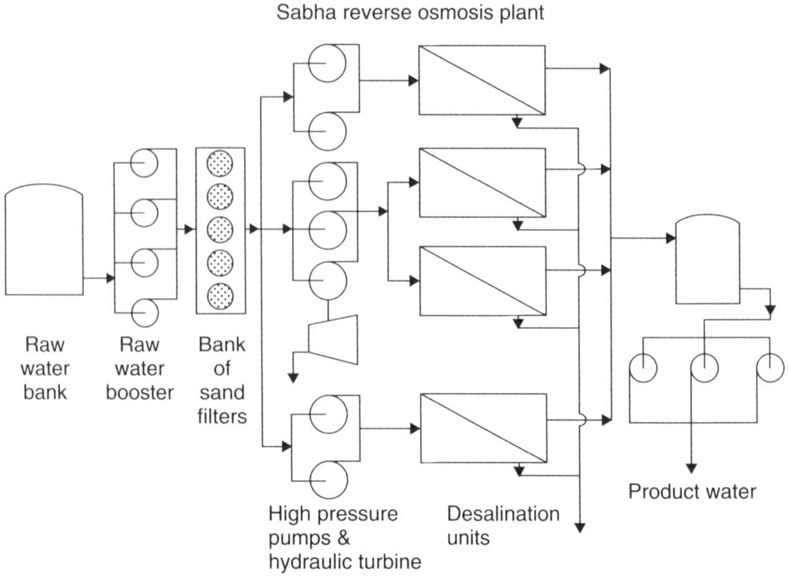

Figure 6.23 Flowsheet of Sabha, Israel brackish-water RO plant [21].

plant incorporates an energy recovery hydraulic turbine that is supplied by brine collected from all four units.

Seven high-pressure pumps connected in parallel were used to provide the feedwater at the required pressure to the first-stage pressure vessels. Pretreatment consisted of sand filtration and chemical dosing (not shown) of hydrochloric acid and scale inhibitor. A flow diagram of one of the units (unit 12) as produced by the data acquisition system is shown in Figure 6.24 for April 1984. The permeate recovery ratio of this unit was estimated at 60.3%, and the specific energy consumption, at 1.59 kWh/m^3.

Description of Control System This control system is based on a combination of programmable logic controller (PLC) and computer technology (Fig. 6.25). Two unit PLCs, one main PLC, and one communication PLC, one industrial computer coefficient of performance (COP), and one personal computer (PC) are used. A PLC is a special-purpose computer that reads input signals, runs control logic, and then writes output signals. The function of a PLC is to read discrete and analog inputs, run some control (conversion) logic, and then write discrete and analog outputs.

The PLCs are ISSC (Industrial Solid State Control) machines from the IPC620 family. The unit PLCs 1 and 2 receive discrete input signals such as start/stop commands, pressure switches, and overloads. The main PLC receives discrete output from the two unit PLCs as well as direct analog inputs from the sensors.

The data are processed by the main PLC and the COP computer, which provide online information to the monitor (screen) and printer, and to a communication

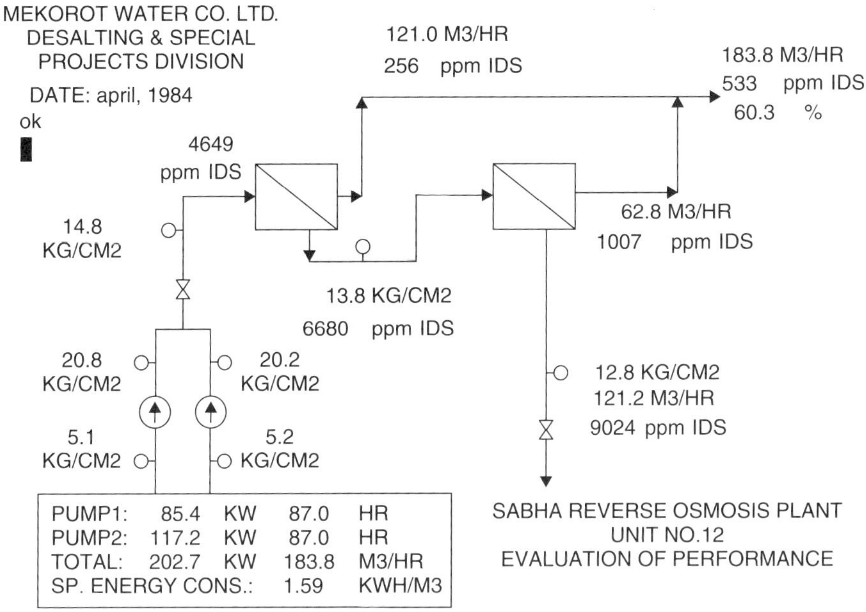

Figure 6.24 Flow diagram of unit 12 at the Sabha brackish-water RO plant [21].

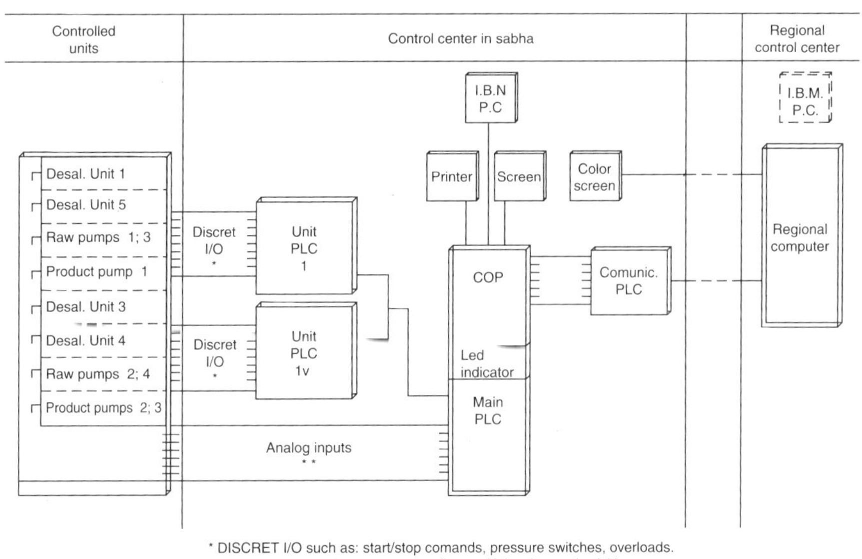

Figure 6.25 Control and data acquisition system of the Sabha brackish-water RO plant.

PLC that can provide information to a regional computer. The system is provided with software, which was developed for the following purposes:

- Online control of the RO plant
- Online processing of measured operational data for generating displays and performance reports
- Management of the historical database stored in memory
- Offline processing of the historical data for optimization of plant performance

6.3.4 Conclusions and Recommendations on Current and Potentially Future Desalination Process Automation Methods

This section reviews the MSF and RO processes and the conventional control methods used in the majority of existing plants and introduces a number of advanced control techniques that have been proven successful in other process industries and may prove economically feasible in desalination plants.

The following conclusions may be drawn from this section:

- Good process instrumentation and measurements are a vital prerequisite for control and automation.
- The majority of desalination plants currently in operation make use of single-loop (SISO) PI or PID controllers.
- New methods of advanced and intelligent control techniques have been proposed by a number of authors with promising technical performance results.
- The literature is almost devoid of investigations related to the economic feasibility of the use of advanced controllers in place of conventional ones in desalination plants.
- The use of computer-based data reconciliation is becoming standard in large plants.
- Model-based control and automation with optimization was introduced successfully in some large plants, and this trend is expected to continue and is likely to be applied to smaller plants.
- The use of techniques such as neural networks, fuzzy logic, and artificial intelligence in desalination plant control have been studied by some researchers without any clear beneficial results, and there are no plans to apply them in plants.
- There are several advanced control techniques currently available in the marketplace, and the most appropriate technology for a particular desalination process has to be investigated both technically and economically in pilot-scale tests
- To emphasize the obvious, any improvements in IMCA must reduce product-water cost, which includes not only the reduction of energy and capital costs,

but also the increase in plant availability. The latter requires, among other things, the use of simple and robust IMCA systems.

6.4 CURRENT AND POTENTIALLY FUTURE PROCESS AUTOMATION METHODS FOR RO AND MSF DESALINATION PROCESSES

6.4.1 Current Status and Projected Development of MSF Process Automation

6.4.1.1 Some Promising Future Control and Automation Options

6.4.1.1.1 Level of Automation The level of automation in the last generation of MSF plants is very high to an extent that it raises increasing concerns related to the "deskilling" of MSF plant operators. To explain, in the past operators had a minimal display panel containing the temperature and pressure gauges, and had to operate the control valves, manually, thus having to be completely aware of the operations taking place and requiring full knowledge of the machinery and associated performance actions/reactions. "Deskilling" is a phenomenon that takes place where there is a high level of automation; this brings about a disincentive to understand the plant and learn on the job.

There are, nevertheless, areas where improvements and modifications of the automation system, with particular regard to the process parameters to be controlled, could bring about savings in plant operation costs.

In standard MSF plants installed since 1990 in the Gulf, the regulatory control is fully automatic to maintain the plant at the setpoint without any further operator intervention for 70–100% load (distillate production capacity; normally 70% load is the first continuous guaranteed point occurring at minimum TBT; load is then increased to 100% by increasing TBT). Load variation within 70–100% is carried out using algorithms and correction curves as indicated schematically in Figure 6.26.

Curves similar to the one indicated in Figure 6.26 are derived from the theoretical projection of the MSF thermodynamic simulation program. The thermodynamic simulation correlates the required distillate output (indicated as *operation load*) to the seawater flow through the heat reject section, brine recirculation, makeup flow rate, and top brine temperature, needed to achieve the desired distillate production and performance ratio. These curves are used to generate the setpoints for the above mentioned process parameter, which allow the plant to follow the required load variations.

In standard MSF automation systems, if two brine recirculation pumps are operated in parallel, then a trip or failure of one pump results in an automatic adjustment of the brine top temperature setpoint. Otherwise, such brine recirculation flow rate halving makes it impossible to maintain the maximum top brine temperature while preventing instabilities due to blowthrough in the brine orifices. To overcome this problem, the pump trip generates a number of different setpoints for brine recirculation flow rate and top brine temperature to maintain the desalination plant operation at reduced load. The setpoints of the brine recirculation flow and

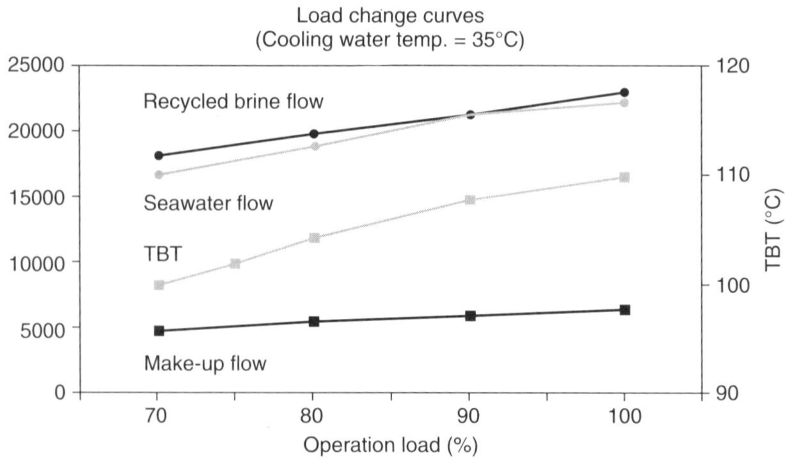

Figure 6.26 Typical load variation curve of an MSF plant.

the brine top temperature are usually automatically linked within the DCS such that both operate in a coordinated way to achieve the correct process plant response to an operator, or dispatch a request for a specific product-water production quantity.

Desalination plant protection is fully automatic, and it is not possible to disable the protection system from the central control room (CCR).

6.4.1.2 Future Automation Improvements

6.4.1.2.1 Full Automation of Startup– Shutdown and Flushing Sequence

It is envisaged that the implementation of fully automatic startup–shutdown sequence will be a feature of future plants.

However, not all control is automatic yet; there are still situations that should be avoided or correct or incorrect steps or procedures to be performed during plant operation that are not controlled by the DCS but by the operator in the present state of the art. In particular, shutdown is a very delicate procedure since the distiller is in a hot condition and aeration is likely to occur. In this situation, corrosion and scaling due to a wrong shutdown maneuver can develop very rapidly. There have been reports of problems arising from untimely plant shutdowns in the Middle East, stating that an unduly rapid shutdown sequence with insufficient brine recirculation flow rate in the heat recovery section had caused the formation of hard scales responsible for cutting and loss of the cleaning balls.

If the correct shutdown procedure is neglected, the consequences in the long term may be corrosion and maintenance problems. Before carrying out the plant shutdown, flushing the unit with product water is necessary to prevent formation of stagnant pockets of salty water, which can cause pitting and crevice corrosion.

Maintenance activities during major or minor overhaul or routine maintenance require that the unit be shut down and fully drained before releasing it for maintenance. During that time, the loss of water production and power generation is high.

In addition, units are not fully dry or protected from corrosion during the flushing time. The following steps are recommended for future systems:

- Modify existing manual procedures so as to reduce shutdown, flushing, and dryout time
- Investigate possibilities to implement an automatic procedure to reduce the down time
- Protect the unit from corrosion during flushing time

It is difficult to provide an economic evaluation of the financial advantages arising from the implementation of these procedures; we can, however, in general, highlight the fact that in IWPP (International Water and Power Project) loss in water production involves a loss in water revenues. For example, reduction of the downtime by 3 h per year in a 100 MIGD plant at the current cost of water (US$ 0.8/m^3) would allow a saving of US$ 45,000 per year in terms of water revenues. This implies that even this small reduction in downtime would pay back for the DCS modifications in just a few years.

6.4.1.2.2 Modification in Top Brine Temperature Control Scheme The arrangement shown in Figure 6.27 is proposed as a TBT control pattern alternative to the classical system which has been described in Figure 6.26. While allowing full temperature control of the desalination plant, this classical system has the disadvantage of incurring several heat and pressure losses due to control valve throttling.

The arrangement as indicated in Figure 6.27 shows direct control of top brine temperature both on HP (MP)-LP steam reducing station and/or steam turbine extraction, thus avoiding further throttling of steam in the TBT control valve as done in the currently used scheme (Fig. 6.26). The modified scheme has been implemented so far in only a few installations with steam reducing stations.

The main disadvantage of this control scheme is the uncertainty of energy input when steam that is extracted from the steam turbine is in wet condition. In particular, it is impossible to ascertain the degree of wetness of the steam and therefore precisely measure the enthalpy of the steam entering the desalination plant in this configuration.

The requirement of a total energy control at the brine heater inlet becomes redundant once the MSF performance ratio has been performance-tested. In general, the design of the steam generation plant allows sufficient flexibility to operate the plant with superheated steam during the performance test of the plant, for acceptance purposes. The actual value of the performance ratio during operation can be considered relatively unimportant.

It is actually more important to control the fouling trend and the variation over the time of the energy input, rather than ensure a precise absolute value. In this respect, implementation of control software matching the expansion curve of the steam turbine with the steam energy input to the brine heater can be considered. The

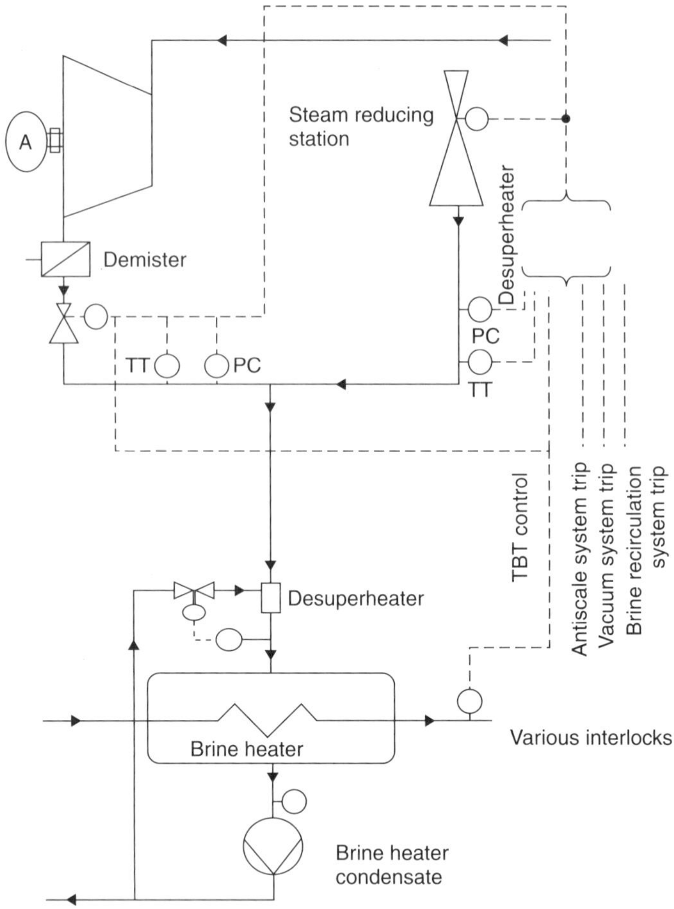

Figure 6.27 Alternative TBT control system (PC: pressure controller; TT: temperature transmitter).

economic advantages of the implementation of this control concept in retrofitting existing MSF–steam turbine plants are described in Section 6.4.1.2.3.

The proposed scheme is better because it allows

1. Reduction in plant CAPEX (capital expenditure) due to the deletion of one steam reducing station and related instrumentation, and deletion of one desuperheating system at steam reducing station.
2. More efficient steam desuperheating due to the higher pressure difference ΔP across the desuperheating system in the low-pressure steam manifold. This implies a more reliable measurement of the steam temperature to the brine heater.
3. Substantial simplification of the control system configuration and elimination of two control systems operating in series.

4. Reduction of pressure losses in the steam line to brine heater. This feature allows in lower steam extraction pressure from the steam turbine with a higher power output rendered at turbine shaft.

6.4.1.2.3 Automation in Control of Fouling Aspects

Introduction to Automation of Fouling Control. Experimental results on existing MSF desalination plants have demonstrated that the actual antiscale dosing rate can be up to 10% lower than the dosing rate recommended by the manufacturer for the same operating conditions [39]. These results can yield a substantial saving in plant O&M cost if fouling and scaling in the MSF desalination plant are kept under control.

Automation can provide a useful means of keeping scaling under control with optimized chemical dosing. The parameter to be controlled in this respect is the *fouling rate*.

Technical Background on Fouling in Existing MSF plants. In the operating conditions found in MSF plants today, the fouling rate increases quite rapidly during the first 2 months of operation (from startup).

Curve E in Figure 6.28 shows a typical trend of performance ratio decline in MSF plants due to increasing fouling of the evaporator. As can also be seen, the present operational procedures using a sponge ball cleaning system in continuous operation and antiscale dosing after 2 months of operation maintains a stable performance ratio, and no or little performance decline is experienced after the initial stabilization period. After this period, the fouling factor tends to become stable and no substantial variation in MSF plant performance ratio is observed.

As indicated in Figure 6.29, the fouling rate becomes constant after 6000–7000 h of operation. In general, there are safety margins in the assumptions of the fouling factor, whose effect is described in Figure 6.29. Feedback data from operating plants show that the actual fouling factors develop from a value of ~0.06 $m^2\ C^{-1}\ kW^{-1}$ to reach an asymptotic value of ~0.09 $m^2\ C^{-1}\ kW^{-1}$, while the fouling factors assumed in the design are more conservative, ranging from 0.106 to 0.152 $m^2\ C^{-1}\ kW^{-1}$. The situation is summarized in Figure 6.29, where the upper curve represents the EPC contractor contractual obligation, while the lower curve represents the real plant expectation during test results. The difference between the curves reflects the fact that the interval between two successive acid cleanings is generally greater than the stated value (usually 17,000 h).

In existing plants, this margin may be converted into operational cost savings by decreasing the antiscale dosage to an optimum level, and by controlling the fouling rate of the evaporator by fine-tuning the sponge ball cleaning system variables as indicated below.

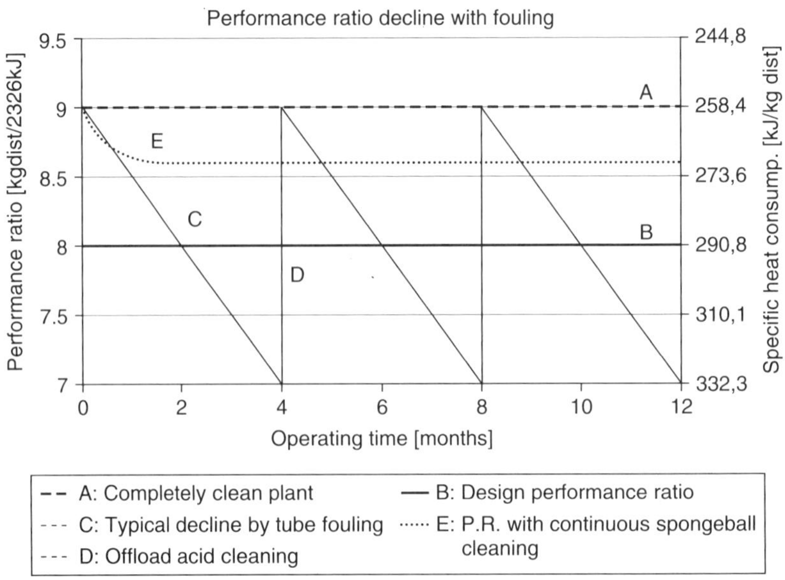

Figure 6.28 Typical MSF performance ratio behavior with time.

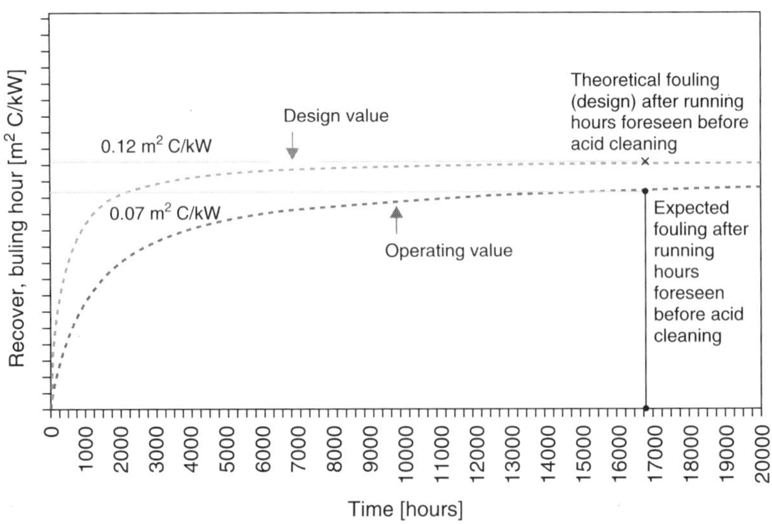

Figure 6.29 Projected fouling factor against time.

Implementation of Fouling Detection Automation System The fouling rate can be expressed as

$$\gamma = \frac{df}{dt} = \frac{d}{dt}\left(\frac{1}{h_\mathrm{f}} - \frac{1}{h_\mathrm{c}}\right) \tag{6.3}$$

where h_f is the heat transfer coefficient of a fouled heat exchanger, h_c is the heat transfer coefficient of a clean (unfouled) heat exchanger, and f is the fouling factor, generally defined as the ratio between the thickness of the scale (x_{sc}) and the thermal conductivity of the scale (k_{sc}) as indicated in the following formula:

$$f \equiv \frac{x_{sc}}{k_{sc}} \tag{6.4}$$

The scale growth asymptotic model was first studied by Kern and Seaton [40], who used the formula

$$\rho_{sc}\frac{dx_{sc}}{dt} = \dot{m}_{sc} - \frac{\tau}{B}\rho_{sc}x_{sc} \tag{6.5}$$

where ρ_{sc} is the scale density, t is time, $\dot{m}_{sc}$ is the gross scale deposition flux, τ is the rate of shear of the scale due to the flow across it, and B is a system-based coefficient that represents the strength (or adherence) of the scale.

In Equation (6.5), it is assumed that $\frac{dx_{sc}}{dt}$, the scale growth velocity, is a balance between gross scale deposition flux $\dot{m}_{sc}$ diminished by the rate of shear of the scale due to the flow across it.

The asymptotic value of the scale layer thickness is defined by

$$x_{sc,\infty} = \frac{\dot{m}_{sc}t_c}{\rho_{sc}} \tag{6.6}$$

where t_c is a time constant defined as

$$t_c \equiv \frac{\tau}{B} \tag{6.7}$$

The solution of these equations gives the time-dependent actual thickness value as follows:

$$x_{sc} = x_{sc,\infty}(1 - e^{-(t/t_c)}) \tag{6.8}$$

Consequently, the fouling factor evolution can be described as follows:

$$f = f_{\infty}(1 - e^{-(t/t_c)}) \tag{6.9}$$

It has been shown in operating results that a good sponge ball cleaning system can maintain a layer of scale within a thickness of 20–25 μm. By deriving the fouling factor from the thickness of the scale, it therefore can be concluded that the fouling factor during operating conditions should in general never exceed 0.08 $m^2\ C^{-1}\ W^{-1}$. This represents a substantial margin compared to the current design practice value, which ranges between 0.15 and 0.12 $m^2\ C^{-1}\ W^{-1}$.

Unfortunately, however, neither the fouling factor nor the heat transfer coefficient are directly measurable variables in MSF plants; hence, in order to estimate

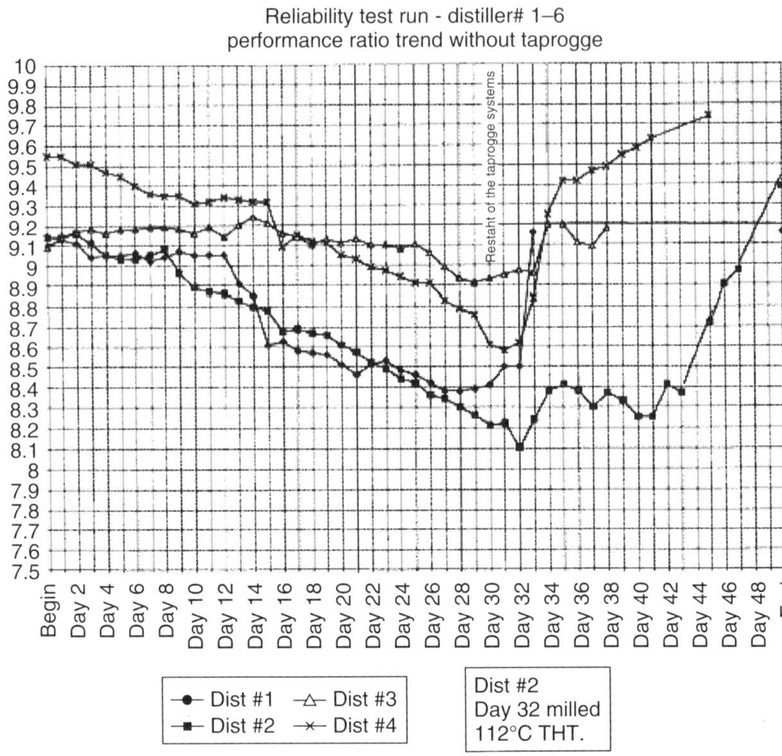

Figure 6.30 Plant performance ratio decline without sponge ball cleaning system in operation, and its restitution after operating the system (from the Al Taweelah B plant, 1996 testing protocol).

tube scaling in a running plant, it is necessary to infer it in some other way from field data.

Typical values for such parameters are presented in Table 6.6.

Figure 6.30 illustrates schematically how the sponge ball cleaning mode of operation can be related to fouling rate control.

The existing site instruments allow calculation of the plant performance ratio. From the curves indicated in Figure 6.30, the actual fouling factor can be calculated from Equation (6.3) as

$$f = \frac{1}{h_\mathrm{f}} - \frac{1}{h_\mathrm{c}} \tag{6.10}$$

Table 6.6 Sponge Ball Cleaning Anticipated Performance Parameters

Frequency of Operations of Ball Cleaning Equipment	50 Cycles/24 hours
Cycle time	30 Minutes

where the clean heat transfer coefficient h_c, is known from the design data and the actual fouled heat transfer coefficient h_f, can be measured from the plant instrumentation.

Antiscale dosage is set to the minimum level, and deviations in fouling factor from the observed operating conditions and progressive increase of the fouling rate due to the development of scales can be controlled by increasing the operation of the sponge ball cleaning system.

Conclusions Regarding Automation of Fouling Control The best operational pattern for fouling and scaling control in large MSF plants is the use of high-temperature improved antiscalants together with the use of a ball cleaning system at a relatively high frequency.

The frequency and duration can be fine-tuned and related to the control and monitoring of the fouling rate in order to minimize the chemical consumption. This method of operation definitely prevents tubesheet and tube clogging, making it possible to reduce additive dosage and maintain optimal heat consumption. Moreover, by allowing an almost complete elimination of manual and/or offline cleaning, fouling control leads to very high plant availability.

Figure 6.31 is a flow diagram of a proposed scheme for operational fouling rate control.

6.4.2 Status and Projected Development of RO Process Automation

The lack of utilization of automation as an online optimization tool is related to the intrinsic complexity of the membrane desalination processes and absence of mathematical models that could be applied to simulate system operation. These conditions are undergoing improvement, owing to better understanding of unit process operations conducted in RO systems, introduction of new measuring equipment

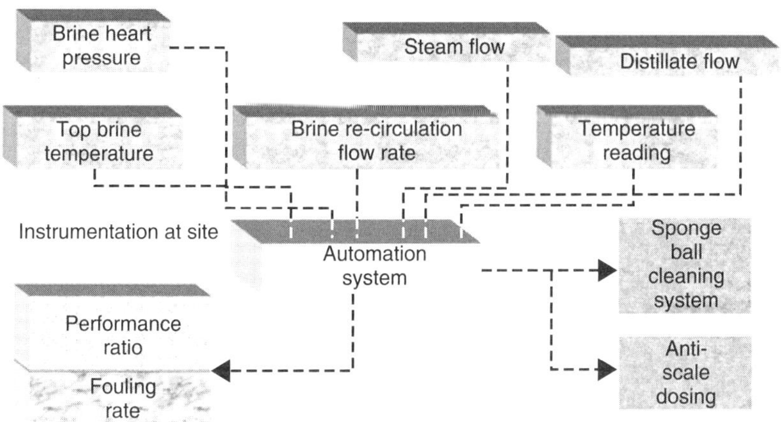

Figure 6.31 Operational patterns proposed for fouling rate control.

especially suitable for water treatment applications, and also the availability of control equipment that has "learn by example" capability. In parallel, the salt rejection of new RO membranes have been improved and therefore can operate within a wider range of operating parameters (recovery rate, flux rate, feed temperature) and still produce permeate water within potable quality limits.

It is expected that in the future, automation will be applied to achieve the following objectives in the RO desalination systems:

1. Improve performance of major process steps
2. Reduce operating cost through optimization of process parameters

The process step that has potential for significant improvement by applying 'smart' automation is the pretreatment of the feedwater. In both conventional and membrane-type pretreatment systems, advanced process control may result in better quality of effluent (RO feed) and lower operating cost. The objective of the RO pretreatment system is to reduce the concentration of colloidal and microbiological concentration in the feedwater. The current approach is to control dosing of treatment chemicals based on flow and according to dosing rates that were determined in the past from jar tests and/or operation of pilot unit. The only chemical whose dosing rate is controlled in a closed-loop mode is acid addition, paced according to the downstream pH. Availability of new measuring equipment (automatic SDI-MFI monitors, zeta potential, streaming current monitors) enables control of dosing rate of all chemicals according to resulting quality of the effluent. Another area in the pretreatment system that could benefit from more extensive control is backwash of media filters or UF/MF units. Optimization of backwash duration and frequency could result in more consistent effluent quality, higher online factor, and higher rate of water utilization.

Another segment of an RO system that could benefit from automation is the RO unit itself. The current approach is to operate RO units at constant permeate output (constant flux rate) and predetermined recovery rate. The above mentioned operating parameters are kept constant at the designed values, with little consideration to changing feedwater quality (salinity, temperature, SDI) or fluctuation of demand. The operators' concern with adjusting the recovery rate is that higher recovery rates may result in membrane scaling and increased permeate salinity. Reduction of recovery rate will result in lower operating pressure but will require greater flow of the feedwater. In both cases (higher and lower recovery), operation of the raw-water supply unit and the pretreatment system will have to be adjusted accordingly. There are also concerns with adjusting permeate flux rate that higher flux rates may increase membrane fouling rate and lower flux rates will result in lower output capacity and higher permeate salinity. However, lower flux rates result in lower feed pressure and lower power consumption. Therefore, the benefit of adjusting the recovery and flux rate is the potential reduction of the operating cost. The operating cost can be reduced through optimization of power consumption, usage of chemicals, frequency of membrane backwash, and cleaning.

Such optimization is possible with the existing sensors and control hardware. What is required is a suitable RO process control algorithm, development of which requires advances in understanding of the integrated RO desalination process. These processes include raw-water supply, pretreatment, RO assembly, posttreatment, and concentrate disposal. The cost optimization algorithm has to provide analytical solution for minimum water cost, considering all the above mentioned processes running concurrently and mutually dependent. The conditions of minimum cost should be achieved within constrains of maintaining safe and suitable operating conditions for the equipment and for maintaining capacity and quality of the product water. Optimization can be attempted for the complete process; however, considering the current level of technology, the most likely target for optimization will be one of product water cost components, which has a significant contribution to the water cost. A typical water cost distribution in RO seawater system is shown in Figure 6.32. It is evident that minimizing power consumption even by a small fraction can result in significant reduction of product-water cost.

As shown in Figure 6.33, the high-pressure pump unit utilizes the major fraction of the RO process power demand.

6.4.3 Prospects for Advanced Control Methods in MSF and RO Desalination Plants

6.4.3.1 Introduction In this section, the current state and prospects of advanced control methods in MSF and RO desalination plants are discussed based on the state-of-the-art. Conclusions and recommendations about future progress are given in Section 6.4.4.

Al-Gobaisi et al. [42] observed that the current practice of MSF and other desalination plant control, process design, and optimization conspicuously lags behind

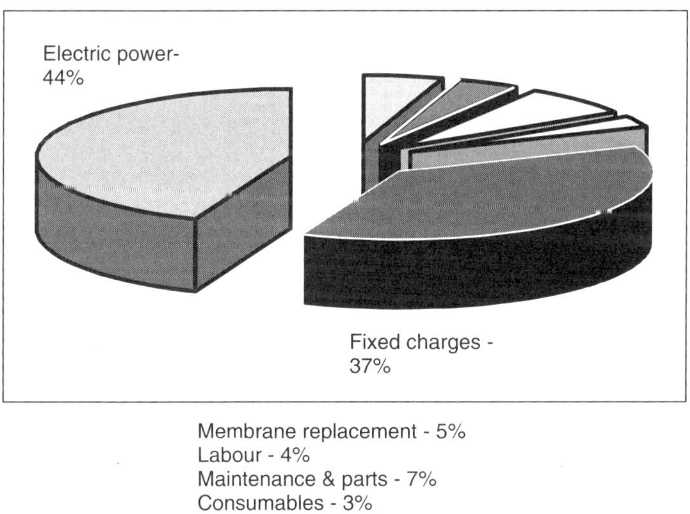

Figure 6.32 Water cost distribution in RO desalinaton.

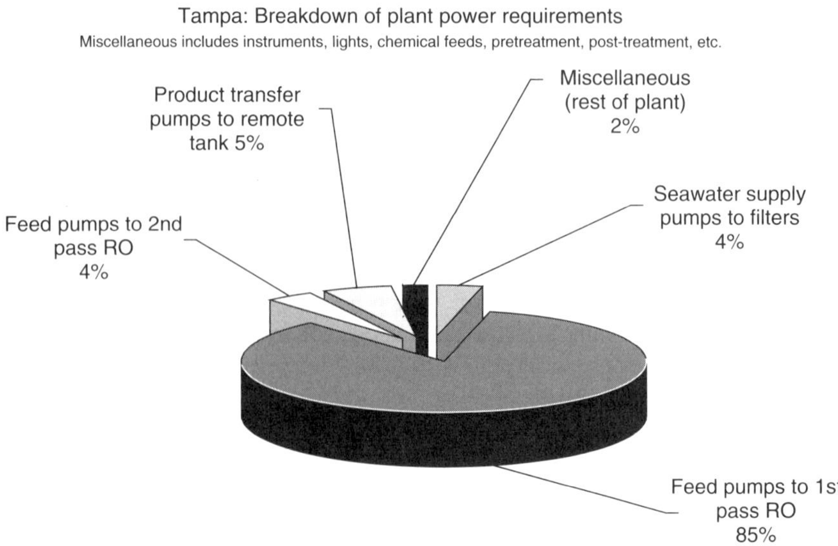

Figure 6.33 Breakdown of power requirement in the Tampa, Florida seawater RO desalination plant.

contemporary developments and advances in these fields. Conventional process control techniques do not ensure the safest and the most profitable operation when compared to advanced process control, which applies he knowledge of process economics and provides insight into the process to maintain essential variables at optimum setpoints.

We attempted to promote the use of advanced control techniques by providing answers to the following questions:

- What are advanced control strategies?
- Where and when can they be applicable?
- What can be expected of them?

It can be expected that advanced control cannot be economically and operationally feasible in all desalination plants. Small desalination plants may not be able to justify the expenses (both the capital and operating costs of the software and hardware required) of sophisticated control systems. Medium and large desalination plants and particularly rather new ones (with many years of operating life still remaining) are expected to benefit from the use of advanced control techniques, due to measurable savings in the operating expenses that can be achieved by running the plant at the optimum operating condition. The unavailability of skilled computer programming technicians at a particular site may be another impediment to the use of advanced control technology.

The main goal in using advanced control in desalination plants is to produce a product (potable water) that has a minimum cost all the time and during all phases of operation of the plant (full-load or part-load operation or during transient operation). This has to be achieved taking into consideration the normal performance deterioration of the different plants components such as fouling.

6.4.3.2 Prospects for Dynamic Matrix Control Method Computer simulations were performed by Robertson et al. [42] using the dynamic matrix control (DMC) algorithm for the control of an RO desalination plant, and compared the performance of the DMC controller with that of a conventional PI controller. It was found that with the selection of proper tuning constants, the DMC controller allows for substantial improvement over the PI control, based on the integral square error (ISE) performance criteria. For the same magnitude setpoint change in product flow rate, the ISE results for the flow rate were comparable between the two controllers; however, the ISE results for conductivity were substantially better using DMC control (for PI controller, ISE $= 1688.92$ and for DMC controller, ISE ≤ 1.0).

6.4.3.3 Prospects for Nonlinear Model Predictive Control (NMPC) Method Aly et al. [19] investigated the use of the nonlinear predictive control (NMPC) method as an advanced control strategy for the control of multistage flash desalination plants and compared the performance of the NMPC controller with a PI controller. For this purpose, a comprehensive nonlinear dynamic model was developed for the MSF plant and was validated with actual plant data (Khobar II, Saudi Arabia, with a capacity of 1184 t/h). A control structure for the plant was then developed from which two control objectives were selected. One control objective is to maximize the distillate product, and the other is to maximize the performance ratio while maintaining the manipulated variables in both cases under tight constraints.

The controlled variables are the top brine temperature (TBT) and the distillate production (M_d) and the manipulated variables are the recirculation flow rate (M_r), and the steam flow rate to the brine heater (M_s). The NLMPC algorithm was tested for a setpoint change of 4°C in the TBT and a step change of 1.8 t/min in M_d. For the first control objective (maximization of distillate flow), the controller succeeded in increasing M_d by 10% by increasing M_s by 12.5% and increasing M_r by 8.2%. For the second control objective (to maximize performance ratio), the controller was able to increase the performance ratio from its initial value by 2.8% (from 3.21 to 3.30) by reducing M_r by 8.5% (from 217.4 to 198.8 t/min) and reducing M_s by 2.8%, which translates to 35,250 tons of steam saved per year [19].

The NMPC model was tested for the rejection of plant internal disturbances by inserting changes in the heat transfer coefficient and the boiling point rise. These changes represent a realistic case in plant operation. The aim was to see whether the controller utilizing a simplified model can still maintain the performance ratio of the plant in light of these internal disturbances. Despite the large propagated disturbances introduced (20% decrease in the heat transfer coefficient and 20% increase in the boiling-point rise), the controller was still able to keep the system at its steady state despite slower transient responses.

6.4.3.4 Prospects for Constrained Model Predictive Control (CMPC) Method Deshpande et al. [43] have discussed prospects for advanced controls in desalination plants. Currently, most operating MSF plants are running under the command of single-loop PID-type controllers. Typically, nine control loops are employed in an MSF plant. The output of primary interest, namely distillate flow, is not explicitly controlled and operators must rely on design graphs showing how top brine temperature and recycle flow are related to distillate flow to achieve the desired distillate production. They suggested that CMPC methods appear to be ideally suited for MSF plants to

1. Control distillate flow explicitly
2. Hold brine levels within user-specified upper and lower limits to avoid blow-through and flooding
3. Maximize distillate production or minimize steam usage as desired without operator intervention

The CMPC strategy can be constructed to be one level higher than the existing controls, downloading the setpoints into the PID controllers. The advantage of this approach is that in the initial period when the familiarity with CMPC is not as high as it should be, the operator would simply be able to turn CMPC off, reverting to the familiar PID scheme in the event that it becomes necessary.

Maniar and Deshpande [44] conducted an in-depth study of an MSF desalination plant control system based a constrained model predictive control (CMPC) technique. The dynamic mathematical model developed by Husain et al. [45] was used in their study. A locally developed CMPC was designed to achieve a variety of operational objectives, such as maximizing distillate production or performance ratio or minimizing energy consumption. CMPC provides setpoints to the existing PID controllers: thus, integrity within the existing instrumentation is maintained. The performance of CMPC was tested utilizing the SPEEDUP dynamic simulation software, and the results were reported as excellent. For example, it was claimed that for an illustrative plant (no plant capacity was mentioned!), steam savings of $1.6 million per year is possible. The potential impact of CMPC for MSF plants was considered rather large considering that there are hundreds of such plants throughout the world that are currently on PID-type control.

Assef et al. [46] carried out an experimental investigation of constrained model predictive control (CMPC) for a RO desalination unit. For comparison purposes, results with traditional PID-type control have also been obtained. The experimental unit consists of a series of four cellulose acetate membranes. A PC is used as the data acquisition and control computer. It is interfaced to the experimental unit via analog-to-digital and digital-to-analog converter boards. The models required for CMPC- and PID-type controls are obtained by step testing. The RO system has four outputs (permeate flow rate; permeate conductivity, indicative of the salt content in the product; transmembrane pressure; and, inlet PH.) and two inputs (flow rate of reject water and inlet acid flow rate).

The production objectives are to produce the specified flow rate of permeate, having the desired salt content, subject to the constraints that the inlet pH and the trans-membrane pressure are within specified bounds. It is shown that CMPC can achieve these goals. It is also demonstrated that CMPC can maximize the throughput subject to the constraints on the other three outputs. A comparison of the results with CMPC and PI control reveals the excellent capability of CMPC for RO desalination plant operations. The features that CMPC offers are unique and could lead to significantly improved plant operations, extend membrane life, and result in less downtime. Small plants can be controlled with PCs, while CMPC implementation in large RO units would benefit from distributed control systems. Engineers with training in advanced control and specially trained operators would also be required for successful installations.

To demonstrate the throughput maximization capabilities of CMPC, the authors allowed the costs on the manipulated variables to be appropriately changed. The CMPC took the plant to a new production level where a higher quantity of permeate (3.5%) is produced while obeying the constraints on the other three outputs.

6.4.3.5 Prospects for Use of Neural Networks in MSF Plants It is well known that in the blowthrough and flooding regimes, the operational behavior of MSF plants can become highly nonlinear. A neural network may be used to identify the nonlinear relationships among the inputs and outputs over the entire operating region of the plant. The neural network model can then be used in conjunction with a nonlinear optimization algorithm to implement nonlinear model predictive control. This strategy (N-NMPC) could yield additional benefits over and beyond that possible with CMPC. A simulation study conducted confirms the feasibility, but to fully evaluate the benefits from N-NMPC, it is necessary to test an experimental plant over the entire operating range, including the non-linear regions, and carry out rigorous simulations. Full details on the application of neural networks to MSF plants may be found in Ramasamy et al. [47]. The main conclusion of this study was that neural networks could serve as a good model for the nonlinear behavior of the MSF process. However, no economic results were included.

6.4.3.6 Intelligent Control Approaches *Intelligent control* is concerned with algorithms, which are developed by emulating certain characteristics of intelligent biological systems [48]. This emerging area of control is fueled by advancements in computing technology, artificial intelligence tools, and the industry's need for simple and efficient controllers [49,50]. Expert controllers, based on expert systems development and validation tools, are being used to automate the actions of a human operator who controls the system. Fuzzy logic–based controllers operate, in a similar manner, to automate the perceptual, cognitive, and action-taking characteristics of humans who perform control tasks [51]. Artificial neural networks, emulating biological neural networks, are used to control systems from learning the way humans perform control actions. Genetic algorithms, an optimization technique based on the survival-of-the-fittest principle, are used to evolve controllers and select the best controller, which fits the systems [47].

These modern control tools are currently available either as a software package or embedded in computer chips. Using these technologies, control engineering can provide alternative strategies for controlling dynamical systems with more functionality and simpler implementation methods compared to similar model-based advanced control strategies. In fact, one of the more interesting trends is the integration of the functions of intelligent systems, such as those mentioned above, with conventional control systems to form *autonomous control systems*, which can perform complex control tasks independently with a high degree of success. The development of such complex controllers fits within the conventional system methodology for the construction of control systems where mathematical modeling along with heuristics is used. Some intelligent control tools, such as fuzzy logic, rely mainly on heuristics, but others, like artificial neural networks, utilize mathematical models, while others (neurofuzzy techniques and adaptive fuzzy control), use a combination of mathematical models and heuristics.

Kurdali et al. [52] introduced prospects of possible benefits of utilizing artificial intelligence techniques, in particular expert systems in MSF desalination plant problems. The proposed applications include online and offline computer-aided control system design, supervisory adaptive control, and alarm monitoring. The expert supervisor has two roles: a passive one in monitoring the system in the desalination plants control center that may include the following operations: electric utility failure, failure of seawater supply, failure of low pressure steam supply, failure of medium pressure steam supply, abnormal valve operating conditions, abnormal pump operating conditions, water contamination, and equipment failure. Fumagalli et al. [32] proposed the use of expert systems as a qualitative tool to aid the operator in managing the transient operation of desalination plants. The aim of the expert system is the supervision of the operation and the online control of the plant. The main functions of the utilized expert system are plant operation management, transients management, troubleshooting, and diagnostics. Abdulbary et al. [53] introduced the idea of using neural networks in identifying large MSF plants by manipulating collected input–output data. A plant consisting of 13 recovery plus three reject stages was modeled using a basic 20-neuron, single hidden layer backpropagation network. Preliminary results showed satisfactory agreement between the collected and estimated values.

A more complete work on MSF identification using neural networks was conducted by Selvaraj et al. [54]. Neural networks with a single hidden layer were used for both multiple-input/single-output and multiple-input/multiple-output-type identification. An error backpropagation learning rule with a momentum algorithm is used to adjust the neural network weights. The results obtained from comparing the predicted and actual outputs are almost identical. The process identification concerned the nonlinear relationship between three plant outputs (top brine temperature, distillate flow form the last stage, and the first-stage level), and two plant inputs (steam and recycle flow rates). In total, input–output data of 1500 points were used for network training, cross-validation, and testing. Zilouchian and Jafar [26] conducted a case study on the design and implementation of an intelligent system for direct seawater RO system located near the Atlantic Ocean at Boca Raton, Florida. The authors reported that the operation of the prototype plant demonstrated the effective and optimum performance of the design for two types of membrane module: spiral-wound (SW) and hollow fine fiber (HFF), under forced diverse operating conditions. The system has achieved a constant recovery of 30% and salt passage of 0.026%. The implementation of the proposed intelligent control methodology has achieved a 5% increase in system availability and a reduction in workforce requirements as well as a reduction in overall chemical consumption. The authors claim that the cost of producing water can be decreased using the proposed fully automatic control strategy but did not actually present typical savings.

Fuzzy logic technology was claimed to offer a revolutionary method by which to automate and increase the longevity of seawater desalination plants [55]. Methods of scale control, pump control, compressor control, valve control, liquid level control, and optimum management of production rate, all lend themselves to existing fuzzy logic control techniques. Development tools, software, and hardware resources are readily available, are more robust, and easier to use. In the future, fuzzy logic will allow systems to make automated and optimal decisions for the control of seawater desalination systems.

6.4.3.7 Prospects for Online Optimization Online optimization is the technology of maintaining a plant at its optimum operating condition at all times. The more competitive a specific market is, the more sophisticated the optimization and control strategies must be. To summarize the status of some industrial applications of on-line optimization, Lauks et al. [56] reviewed the literature from 1983 to 1991 and cited nine applications for five ethylene plants, a refinery, a gas plant, a crude unit, and a power station. Profits increased from 3% to 20%, and intangible improvements from a better understanding of the plant behavior were significant. Similar results were reported by Pierucci et al. [57] for an olefin plant, by Krist et al. [58] for a Dow Chemical benzene separations process, by Van Wijk and Pope [59] for a catalytic cracking complex at Shell's Stanlow refinery in the UK, by Bailey et al. [60] for a hydrocracker fractionation plant, by Lowery et al. [61] for GE Plastics's two biphenol plants, by Gott et al. [62] for Conoco's Billings refinery FCC units, and by Mullick [63] for a refinery crude unit. We will now focus on the developments in using advanced control methods in MSF and RO plants.

Alhumaizi [64] developed a dynamic model for the simulation and control of multi-stage flash desalination plants. The model was developed and validated with real plant data from an operating MSF plant in Jubail, Saudi Arabia. A multivariable control analysis was carried out to determine the optimal control structure. Both a conventional PI controller and an advanced nonlinear model predictive control (NLMPC) strategy were implemented for the control of the plant. Numerical simulations showed the superiority of the advanced control strategy as it allows the plant to operate at operating points of higher performance ratio, which has the potential of substantial energy savings for the plant. A robust control design was also carried out for the plant. Simulations show that the advanced control strategy allows good control performances of the plant in face of unavoidable model uncertainties.

Keyes et al. [65] described a multilevel control and optimization technique for improving product quality, productivity, and operation cost of MSF desalination plants. The proposed multilevel control strategy consists of three levels: (1) dedicated sequential and regulatory controls, (2) supervisory and optimizing controls, and (3) management information control. The major emphasis is on the awareness of system integration techniques related to energy management opportunities associated with desalination plants. With increased fuel cost, it is essential to ensure optimization of the integrated system, consisting of the electric energy system source and the desalination plant, rather than separate optimization for each system independently.

Optimum power plant operation provides minimized cost of steam for desalination thus providing maximum performance ratio of the desalination plant and minimum desalted-water cost. With the condensing turbine flexibility and freshwater storage possibility, an integrated plant optimization over a period provides an overall minimum energy cost, besides satisfying both freshwater and electrical energy demands. The integrated dual-purpose plant optimization is performed as follows. Electrical, steam, and freshwater production rate versus time are determined such that the energy cost is minimized and the demand for electricity and fresh water is satisfied. In the regulatory and sequential level, control loops, which directly affect plant performance, are regulated. The loops considered are top brine heater temperature, seawater feed flow rate, brine recirculating flow rate, and concentration ratio.

At the supervisory–optimizing control level, the setpoints of dedicated control loops are adjusted on the basis of a specified optimization performance index. These control loops include the previously mentioned desalination plant loops, besides the following control loops of the combined cycle power plant: gas turbine firing rate controls, steam bypass control, desuperheater pressure control, and steam turbine hotwell level control. An optimum load allocation based on minimization of the unit cost of distilled water is utilized. The implementation of the proposed method is achieved by utilizing distributed microprocessor-based control architecture.

Krause and Hassan [66] investigated the setpoint optimization of MSF plants in order to achieve the most economic operating conditions, by using a rigorous model that takes the changing plant behavior (e.g., fouling and changing environmental conditions) into consideration. As the model is based on physical effects, it allows the determination of the plant behavior from the physical and geometrical data of the plant, thus permitting it to modify and optimize the plant layout during the planning phase. Consultants and manufacturers will use the optimization strategy based on rigorous models online either during operation or offline during the planning phase. During operation, the geometric data of the plant remain constant, but the environmental conditions will change. These, together with the restrictions that are imposed on the process (maximum temperatures, chemical dosing, restricted steam production, etc.), are also taken into account by the model, and the optimal setpoints are calculated accordingly. In order to follow up the real plant behavior, it is necessary to determine the changes imposed by, for example, fouling. This is achieved by calculating the changed heat transfer coefficients. Another module accommodates of the measurement errors and tries to minimize their effects.

It was shown how the above-described tasks are implemented in the distributed control systems manufactured by Hartmann & Braun and which tools are available for consultants and manufacturers to support their evaluation work during the planning phase.

6.4.4 Conclusions and Recommendations on Potential of Future IMCA Systems

Conclusions and guidelines are as follows:

1. The level of automation in the last generation of MSF plants is very high to the extent that it brings up increasing concerns related to the deskilling of MSF plant operators.
2. Full automation of the startup–shutdown sequence is desirable but has so far been implemented in only one plant.
3. Modification in the control schemes that affect reductions in exergy (or just pressure) losses in the plant should to be explored, and an example of such a proposed scheme to control the top brine temperature in MSF plants was given.
4. Automation to control fouling is an important proposed approach.
5. The process step that has potential for significant improvement by applying smart automation is the pretreatment of the feed water, especially in RO.
6. In RO, implementation of automation of the recovery and flux rate combination, instead of the conventional method of maintaining constant flux, is recommended for potential reduction of the operating cost.
7. From the various advanced control methods considered, the use of constrained model predictive controllers (CMPCs) appears to be promising for implementation.
8. Automation as online optimization is highly desirable, but adequate mathematical dynamical models of the process are as yet unavailable to implement it.

6.5 CALCULATION OF IMPACT OF AUTOMATION ON PRODUCT WATER COST

6.5.1 Introduction

Automation is intended to simplify plant operation and optimize its performance, while maintaining safety. Performance optimization includes the minimization of operating expenses (energy, chemicals, labor, etc.) and maximization of availability. The general issue of instrumentation, measurements, control, and automation (IMCA) has three major impacts on produced water cost:

1. The sophistication level of automation that should be implemented to increase control accuracy and optimize plant performance
2. Some areas in which the IMCA philosophy should be changed
3. Process aspects that have not been automated yet but probably should be

It is obvious that automation is, from a fundamental viewpoint, not a vital component of a desalination plant since the plant can produce water even with purely manual control, so consequently any level of automation, or generally IMCA, must thus be economically justified. That justification must also include the consideration of IMCA system reliability, which would manifest itself in any changes in plant availability that may take place when the intended IMCA system would be implemented.

As opposed to the enormous amount of published information on control systems and their advancement, those on cost–benefit analysis of their implementation are very few in general [67,68], and, except for a paper by the authors [69], nearly nonexistent for desalination plants in particular. Some introductory comments on the subject including citations to available papers are therefore presented below.

A systematic approach to the problem was proposed by Martin et al. [70], presenting seven industrial sample cases. The first step in such a cost–benefit analysis is to identify the process variables, which affect the process parameters that carry the highest potential economic benefit [71], and that can, in fact, be changed in the desired direction by the application of improved controls. For example, in distillation desalination processes, raising the top brine temperature will reduce the specific-energy consumption, but should not be done because of scaling that takes place at the higher temperatures. Nevertheless, the constraints that might limit benefits must be carefully challenged before their acceptance. The base-case system and operation, existing before the IMCA system improvements are made, must also be clearly defined, based on existing data. It is important at this point to note that the conventional control scheme was to just aim at keeping a variable close to a target value, yet a more rewarding scheme is to keep the variables within certain limits that accomplish a desired final objective, such as, maximizing throughput, quality, or profit, under varying operating and economic circumstances.

The next step is to establish the relationship between the variation of this parameter and the resulting economic benefit. The latter is the product of this variation and its economic sensitivity factor. This is then repeated for each control function, taking care not to ignore possible parameter interactions.

The significant parameters and their relation to the process performance can be determined by proper mathematical modeling of the process, or when the modeling is insufficiently accurate, or absent altogether, by examining the actual operating data over a long enough period. Because of the inherent fluctuations of the data, statistical methods should be used to extract the information that is significant [70,72]. For some simple steady-state systems, Contreras-Dordelly and Martin [73] have developed a method for selecting the control system yielding the highest steady-state operation profit from among candidate multiloop designs, taking into account disturbances and model mismatch.

Historical operating data of a plant are akin to a gold mine, which can be mined for information useful for improving plant performance and safety. If examined and correlated carefully to show trends of the significant performance parameters, such as production rate, quality of product, energy and chemical consumption, and fouling, as a function of the process variables, they could teach us about the influences of process variables that perhaps could not have been predicted by intuition or even modeling. Analyzing the data can lead to conclusions about the need for additional or different instrumentation, for different control methods, or changes in the process. In some examples from various process plants, it was shown [74] that such analysis has had an excellent cost–benefit ratio, where in one example it resulted in savings of $35,000 per month at the one-time expense of only $21,000 for the addition of certain instruments and control procedures. Since larger plants use computers anyway, all the relevant data can be stored for statistical analysis and correlation. Such programs are relatively easy to implement and run automatically at desired periods.

Theoretical predictions of utility of IMCA schemes must eventually be put to the test in an operating plant. The test should assess the system functionality, as well as its economic promise. The latter is especially difficult in view of fluctuations and changes in operation inherent with any real plant and control scheme, which may be larger than the examined sought outcome, but good statistical analysis is able to extract the desired information, as shown in the literature [67,70,75–77], presenting an approach that attempts to permit the calculation of the important issue of simultaneous impact of process dynamics, model uncertainty, and controller complexity on the process economics. A benchmarking questionnaire, including a simple statistical tool for evaluation of benefits, sent to plant owners/operators to determine whether the plant has good potential for benefits from improved process control ("Should I invest in improved process control"?) was developed and tested by Lant and Steffens [78].

The above-described analysis of cost–benefit ratios of IMCA systems consider only the direct economic gains resulting from operating closer to the optimal point that gives the desired product yield and quality. It is important to note, however, that operating with tighter tolerances near the optimal point, for example, by reducing fluctuations near the setpoint and/or moving the setpoint closer to the specifications, while often considered unnecessary trouble, may have benefits that even exceed these direct ones. These benefits, often called 'intangibles' [76], include customer perceptions of plant owner reliability and speed of response, and owner sales costs, advertising effectiveness, discounts, refunds, legal penalties, insurance ratings, and fraud and neglect, all of which are very complex to quantify ahead of time, yet in fact become very tangible and may often exceed the direct costs of say operating at a lower energy efficiency or product yield. The concept of robust design addresses this issue by reducing variation beyond the conventional tangibly set limits, and is usually known as the *Taguchi method* [79]. Application of this method to the design of desalination plants would impose tighter tolerance limits on fluctuations in the important parameters, requiring usually improved instrumentation and controls, yet the benefits will also be higher than those computed conventionally.

Among several ways to analyze economic viability of investments in IMCA systems is breakeven analysis. *Breakeven analysis* allows comparison of the additional costs related to the introduction of more advances in automation, with the annual savings in O&M. It identifies clearly the commercial advantages, from an investor's point of view, that result from an additional investment in automation. In particular, automation is aiming at optimizing the O&M costs by both a reduction of the personnel and a better tuning of the operating parameters, that allows, in turn, savings in chemicals, optimization of the performance ratio, and so on.

Breakeven analysis is an application of marginal costing techniques normally used in standard accounting practice. In the decision-making process, if additional IMCA investment is required, it is necessary to ascertain the profit made as a consequence of a given investment, as well as the level of distillate production at which there is no profit or loss. This level of production is called the *breakeven point*. This can be represented by the following equation:

$$\text{Breakeven point} = \frac{\text{additional investment costs}}{\text{savings achieved per unit of production}} \tag{6.11}$$

when it is equal to 1. Figure 6.34 shows schematically the approach to be used for an IMCA breakeven analysis. In this example it is assumed that the investment capital cost of the additional (or more sophisticated) IMCA system is 5000 price units, and that 0.5 price unit per unit of production is the saving achieved by this investment. The graph shows that the breakeven point is achieved when 10,000 units are produced, and the investment produces profit when the production is above 10,000 units. The breakeven analysis concept can be applied to both the introduction of more advances in desalination plants of new design and construction, and to retrofits of an existing plant. More information on costing and economic analysis methods for desalination plants can be found in the book by Sommariva [80], and a proposal for the more appropriate net present-value-analysis, which takes into account the time value of money, was performed by Bawden and MacLeod [72].

It probably is understood that any improvements must consider the human interface with the IMCA system, to support the operators so that they can be successful in charge of the system [81]. Failure to consider these aspects is bound to render any system improvement futile.

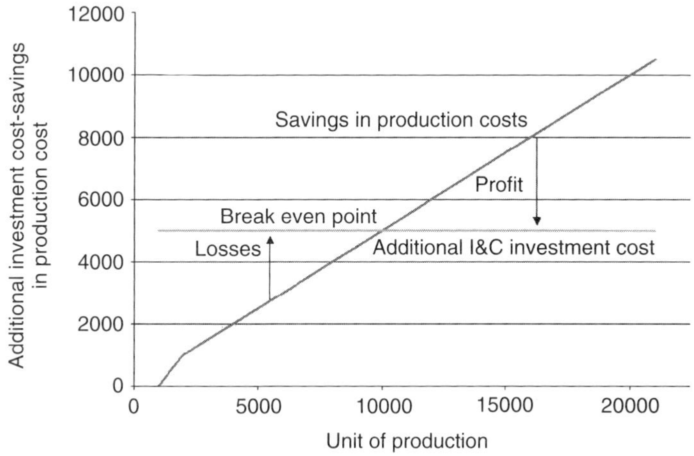

Figure 6.34 Breakeven analysis example for I&C investment cost–benefit analysis, as a function of system capacity.

Specifically now for desalination applications, multistage flash (MSF) and RO desalination plants are the most widely used seawater desalting systems in the world. An important goal in the design and operation of any desalination plant should be the reduction of its operating costs in order to minimize the cost of the freshwater produced. One potential area for cost reduction is minimization of the energy and chemicals consumed by the plant. A MSF plant consumes substantial amounts of thermal energy in the form of low-pressure (LP) steam provided by boilers or turbines, as well as considerable amounts of electrical energy as pumping power. RO plants use large amounts of pumping electrical energy. Both processes consume considerable amounts of chemicals for feedwater pretreatment and product (distillate) posttreatment. One potential tool for the optimization task is the use of adequate control strategies that can stabilize the operation of the plant at high-efficiency operating points while handling efficiently the different constraints on the process variables.

As stated earlier, currently, most MSF and RO plants are controlled by simple proportional integral derivative (PID) controllers. The selection of the control structure is governed by engineering and an intuitive approach. The PID controllers are tuned adequately to keep the process-controlled variables within tolerable limits. These controllers do not aim to achieve optimum plant operation despite the fact that they have being able to stabilize the plants adequately. The use of simple PID controllers, which can maintain the stable operation of the plant, certainly overlooks the complex interactions that may exist within the plant [82]. MSF and RO plants are nonlinear multivariable systems, and an in-depth investigation of the different interactions between the inputs and outputs of the plant is necessary to optimize the selection of the control structure. The use, on the other hand, of advanced model-based control schemes such as model predictive controllers (MPCs) can have the potential of providing better control of the plant as well as cost savings.

An advanced process control project in the desalination industry (MSF and RO processes) typically includes the installation of microprocessor-based distributed control system instrumentation, a supervisory computer, and implementation of sophisticated advanced control software for process control and optimization. In an oil refinery or petrochemical plant, such a project would cost somewhere in the range \$1,000,000–\$5,000,000 in 1991 dollars [70]. An economic justification is therefore essential for the approval of such a project, which is the IMCA cost–benefit ratio introduced at the beginning of this section.

The current study includes the first and predominant step in an economic feasibility of using advanced control to retrofit existing conventional PID controllers. The aim of this study is to evaluate the potential savings in the major operating components, namely, electrical energy, heating steam and chemical additives. Obviously, the operating cost involves other components such as administration, spare parts, operating and maintenance personnel, and insurance. These components are not included in the present study and should be included in a larger investigation.

All of the economics-related results shown in this chapter are based on costs of materials, energy, equipment, labor, etc., that existed at the time the studies were conducted, and some well before the date of publication of this chapter. Since these costs vary with time, sometimes significantly, readers of this chapter who wish to conduct economic analyses would naturally have to update the costs. The emphasis in our analysis is on methodology, not on actual value magnitudes, which are time- and case-dependent.

6.5.2 Rigorous and Detailed Estimation of Economic Benefits from Advanced Process Control

6.5.2.1 General Approach *Advanced process control* is a service available to plant operators, helping them improve the performance of the process. The primary objective of an advanced control system is to maximize profits by operating the process at its optimum point while satisfying criteria such as safety, environmental regulations, product specifications, and operational constraints. Advanced process control, therefore, aims to deliver higher plant performance than conventional control. Most desalination processes must follow a given target during operation and they have to follow this target as closely as possible despite the continued presence of disturbances. These targets are seen as the optimal values of the controlled variables. Every deviation from the optimal values on either side results in some kind of economic loss or cost penalty. The variance of the variable in question has to be reduced. Performance is related to the magnitude of the variations of a certain variable or property around the target. The more we can reduce the variance, the smaller is the cost penalty. It is important to note that performance optimization also includes the management of transient conditions and load changes as well as startup sequences to maintain and increase plant life time and reduce corrosion.

Estimating the quantitative economic benefit from advanced process control begins by determining the base operation. This operation is characterized by process data that are collected during a period of typical closed-loop operation with the existing (conventional) control system. From these data, the mean value and variance for the key economic controlled variables are determined. The use of advanced process control is expected to reduce the variation in the controlled variables. Because of this reduction in the controlled variables, the mean operating value can be moved closer to a product specification or operating constraint without increasing the frequency of violating a constraint [83]. This process is referred to as the *improved control operation*, as shown in Figure 6.35.

The economic benefit from improved control operation stems from two sources:

1. Operation with variance reduction
2. Operation at optimum value of controlled variable

6.5.2.2 Savings Due to Reduction in Variance of Controlled Variables

The improved control mean operation is determined by shifting the base operating mean value toward the optimum value without increasing the constraint violations. This is achieved due to the reduction in the controlled variable variance.

The change in the mean operation is calculated by specifying a fractional violation of the limit and determining the mean value that results in this violation. The mean value and standard deviation of a controlled variable can be expressed as

$$\bar{x} = \frac{1}{n}\sum x \tag{6.12}$$

$$s = \left[\frac{1}{n(n-1)}\left(n\sum x^2 - \left(\sum x\right)^2\right)\right]^{1/2} \tag{6.13}$$

The basis for estimating benefits described in Martin et al. [70] is to calculate by how much the mean value of the process variable can be moved, by installing an

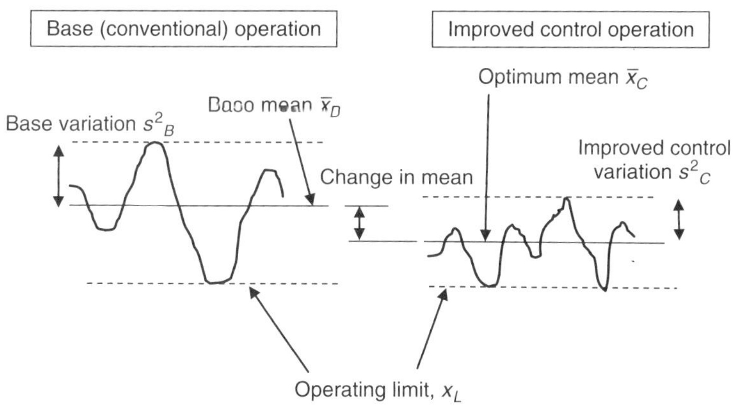

Figure 6.35 Variation of a controlled variable in base operation and improved control operation [83].

advanced control device, with that movement defined as

$$\Delta\bar{x} = \bar{x}_c - \bar{x}_B \tag{6.14}$$

The method of estimating $\Delta\bar{x}$ depends on the assumption that a normal distribution of the data occurs. Based on normal distribution, the probability $p(m)$ that $m\%$ of the data fall beyond the limit x_L is given by

$$p(m) = \frac{x_L - \bar{x}}{s} \tag{6.15}$$

The values of $p(m)$ for different values of m can be found in statistical tables as follows:

$m\%$	$p(m)$
25.0	0.675
15.86	1.000
5.00	1.650
2.27	2.000
0.14	3.000

The advanced control data are assumed to violate the limit by the same percentage as for the case of base (conventional) controllers. This means that both base and advanced controller data have the same value of m. In this case, the change in the mean process variable can be calculated as:

$$\Delta\bar{x} = p(m)(s_B - s_C) \tag{6.16}$$

A typical assumption is $s_C = s_B/2$ [70], and in this case, $\Delta\bar{x} = p(m)/2$. For $m = 5\%$, $p(m) = 1.650$ and $\Delta\bar{x} = 0.825 s_B$. The unit operating cost of water can be expressed as a function of $\bar{x}$ as $\Phi = f(\bar{x})$, where f is a general cost function depending on how the process economics varies with the operating point or values of the controlled variables. The cost saving or cost penalty resulting from this move in the average process variable $\bar{x}$ can be estimated from

$$\Delta\Phi = \frac{\partial f}{\partial x}\Delta\bar{x} \tag{6.17}$$

When the objective function (operating cost in \$/m^3 of product water) is dependent on a number of independent variables, $x_1, x_2, \ldots\ldots x_n$, the effect of changes in the mean values of these variables on the saving in the unit cost of water can be calculated from

$$\Delta\Phi = \sum_{i=1}^{n} \frac{\partial f}{\partial \bar{x}_i}\Delta\bar{x}_i \tag{6.18}$$

The partial differential parameters in this equation are calculated from the cost function

$$\Phi = f(\overline{x}_1, \overline{x}_2, \ldots, \overline{x}_n) \tag{6.19}$$

which is obtained from the process model for the MSF and RO processes.

6.5.2.3 *Savings Due to Running at Optimum Condition* The optimal operating condition is defined as the one that results in a minimal operating cost of water. We express the optimal cost as Φ^* and the operating parameters that correspond to this cost are $\overline{x}_1^*, \overline{x}_2^*, \ldots, \overline{x}_n^*$;

$$\Phi^* = f(\overline{x}_1^*, \overline{x}_2^*, \ldots, \overline{x}_n^*) \tag{6.20}$$

If the actual operating conditions using conventional PID controllers are $\overline{x}_1, \overline{x}_2, \ldots, \overline{x}_n$, then the savings in operating costs that can be achieved due to the use of advanced process control can be estimated as:

$$\Delta\Phi = \Phi - \Phi^* = \frac{\partial f}{\partial \overline{x}_1}(\overline{x}_1^* - \overline{x}_1) + \frac{\partial f}{\partial \overline{x}_2}(\overline{x}_2^* - \overline{x}_2) + \cdots + \frac{\partial f}{\partial \overline{x}_n}(\overline{x}_n^* - \overline{x}_n) \tag{6.21}$$

6.5.3 Process Control Structure

6.5.3.1 *Architecture of Automatic Setpoint Control System* Figure 6.36 illustrates how an automatic set-point adjustment system can be applied to a desalination plant for process operation optimization. The online computer hardware that is used for process optimization is often connected to the actual control system. The process measurements are supplied to the computer and form the inputs to its optimization program, which carries out a search to find the optimum setpoint values, which maximize or minimize a defined objective function subject to a system of constraints that must be maintained. The output of the computer will be the optimal values of the set points that are directly implemented by the control loops. An interface terminal allows the plant operator or supervisor to communicate with the computer and change the value of some of the input parameters or change the objective function.

In Figure 6.36 it has been assumed that the process control computer is a single hardware unit, that is, one computer performing all the tasks required for optimal set-point control. The recent rapid development of solid-state and digital technology has led to a very different approach to the hardware configuration of computer control system. Figure 6.37 illustrates a typical arrangement that might now be appropriate for the control of modern desalination plants.

The tasks of measurements, DDC (direct digital controls), operator communications and sequence control, etc. are distributed among a number of separate processing units, each of which will incorporate a microcomputer of one form or another. The microcomputers are linked together *via* a common serial communications highway and are configured in a hierarchical command structure [32].

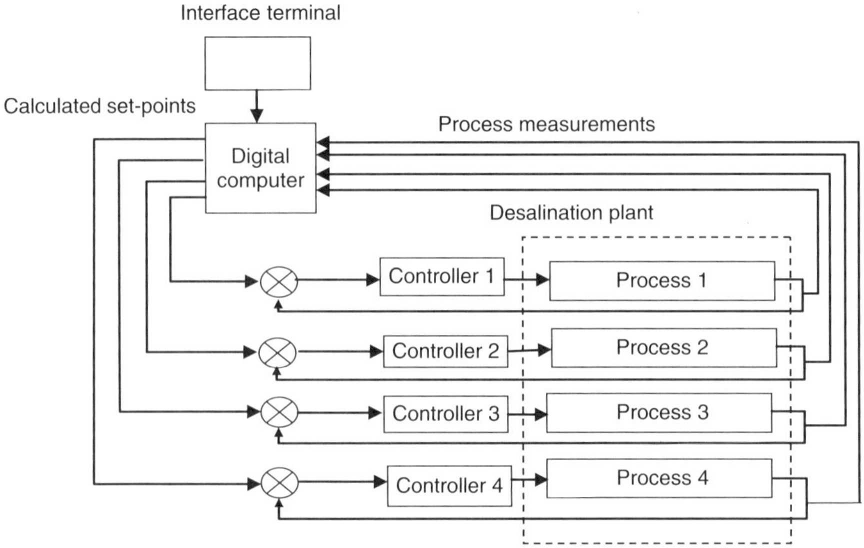

Figure 6.36 An optimal setpoint control system for a desalination plant.

6.5.3.2 Main Components of Setpoint Optimization System

Figure 6.38 represents the main components of a setpoint optimization system for a desalination plant. The desalination process exchanges material flows (seawater inflow, brine blowdown, distillate (product water), chemicals, etc.), energy flows and information with the environment. An advanced control system such as the setpoint optimization system controls information flows to ensure optimal functioning of the process with respect to a fixed objective. The process supplies measurements via sensors (output variables) and receives orders generated by the controllers (input variables).

The heart of the system is a mathematical model that allows satisfactory representation of the desalination process at least in the steady-state condition. The equations developed in this model use the fundamental laws of thermodynamics such as the conservation of mass and energy and the phase equilibrium equations. The model usually takes the form of a system of nonlinear algebraic equations

$$\Psi(\boldsymbol{s}\ \boldsymbol{p},\ \boldsymbol{u}) = 0 \qquad (6.22)$$

where Ψ is a function, $\boldsymbol{s}$ is the vector of state variables defining the process (flow rates, compositions, temperatures, pressures, concentrations, etc.), $\boldsymbol{p}$ is the vector of the model's parameters (heat transfer coefficients, fouling factors, etc.), and $\boldsymbol{u}$ is the vector of control variables. On the basis of this model, the main functions that such an advanced control system must fulfill are

- Reconciliation of the measurements
- Identification of the model's parameters
- Analysis of sensitivity and optimization

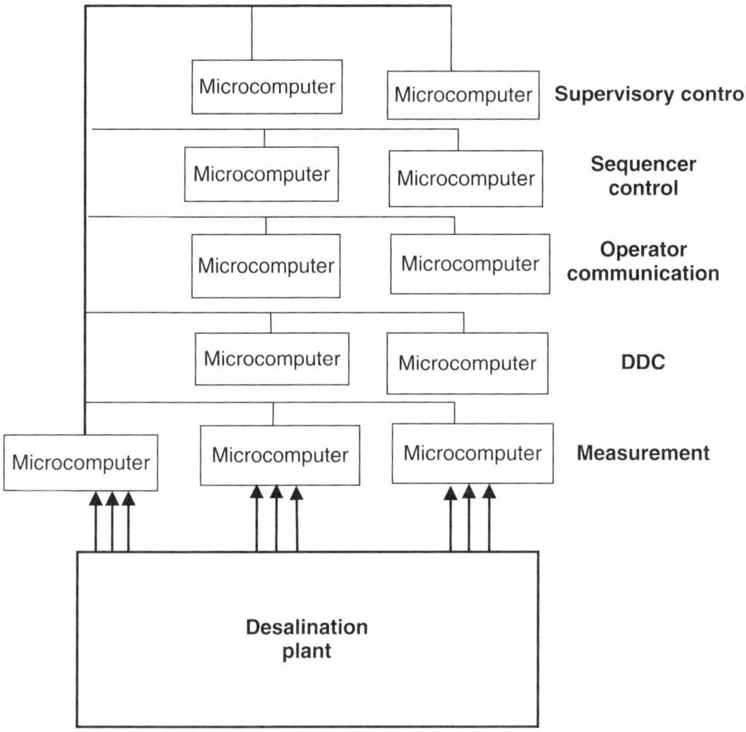

Figure 6.37 Possible arrangement of a modern control system for a desalination plant.

6.5.3.3 Reconciliation of Measurements The measurements carried out on the process are usually tainted with errors and therefore incoherent. The validation (or reconciliation) of the measurements by exploiting the redundancy created by the relationships, which exist between the measurements (material and heat balances, phase equilibrium relationships, etc.) allow the following objectives to be met:

- Obtain a coherent set of data
- Calculate the unmeasured state variables of the process
- Carry out a diagnosis on the plant to reveal faulty sensors

6.5.3.4 Identification of Model Parameters This step aims at evaluating the values of the model's parameters from measured data, thus allowing minimization of the deviations between the values calculated by the model s and the values obtained from measurements s_m. It should be realized that some of the model's parameters such as the fouling factors in condenser tubes evolve slowly over the course of time, and it is therefore important to identify such change in order to adjust the model to reflect the evolution of the process.

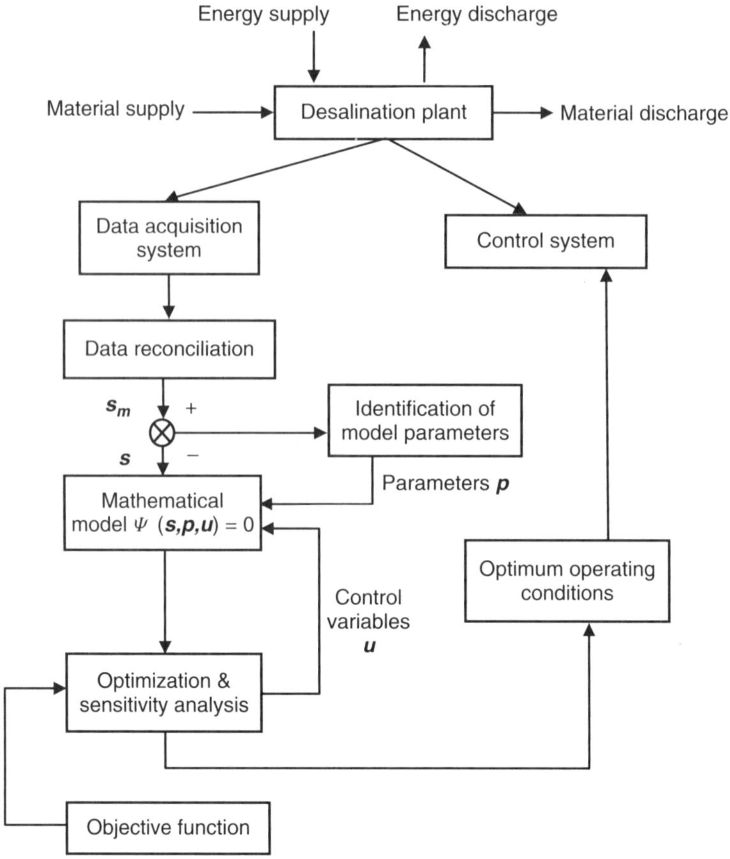

Figure 6.38 Main components of a setpoint optimization system.

6.5.3.5 Optimization and Sensitivity Analysis Having defined a criterion (objective function), it is possible to optimize this criterion, that is, to adjust the operating variables of the plant in order to determine the values of the controller's setpoints that ensure optimal process functioning at anytime. This is a conventional nonlinear optimization problem with equality and inequality constraints. These constraints, in fact, represent the mathematical model of the process and define the field of feasibility of some variables.

6.5.3.6 Mathematical Simulation Model *Mathematical modeling* refers to formulating a set of equations that describes the different processes taking place in the MSF plant. In the simulation phase, the formulated model is solved using a suitable solution procedure, as well as entering the values of input process parameters. Obviously, this has to be done using a computer, leading to a function termed *computer-aided simulation*. The simulation model used in setpoint optimization is a steady-state model. This model is nonlinear in nature mainly because of the dependence of the physical properties of the streams (brine, distillate, vapor, etc.) on temperature, pressure, and salinity. In addition, relations for the heat transfer coefficients also contribute to the nonlinearity of the model. Because of to the complexity of the model, the simulation procedure can be based on a simple model, a rigorous model, or a combination of both.

6.5.4 MSF Cost-Benefit Ratio of New IMCA Systems

6.5.4.1 General Comments The cost fraction of the IMCA component relative to the desalination plant turnkey cost is typically very small, and thus any investment in it is likely to have small impacts, probably <1%, on the overall water tariff. Looking at a typical MSF plant cost breakdown shown in Figure 6.39, for example, an increase of say 10% in the advanced IMCA costs would increase the turnkey plant costs by only ~0.7%, which would have created a relatively negligible increase of the water tariff. The resulting reductions in O&M costs could be much more significant as detailed further below. From Figure 6.39, the investment costs for the automation and control system for an MSF plant are quite low compared with the overall MSF plant capital costs.

6.5.4.2 Modeling and Simulating the Performance of the MSF Process
6.5.4.2.1 Process Description The flow diagram of a typical MSF process is shown in Figure 6.40. Modeling of the process in steady-state operation has been carried out by many researchers [33,84–90], and there are many proprietary commercial programs available. It is noteworthy that there hardly exists any comparison between them, and many show no validation. Many publications also include models without referring to prior work, leaving it unclear whether they advance the state of the art.

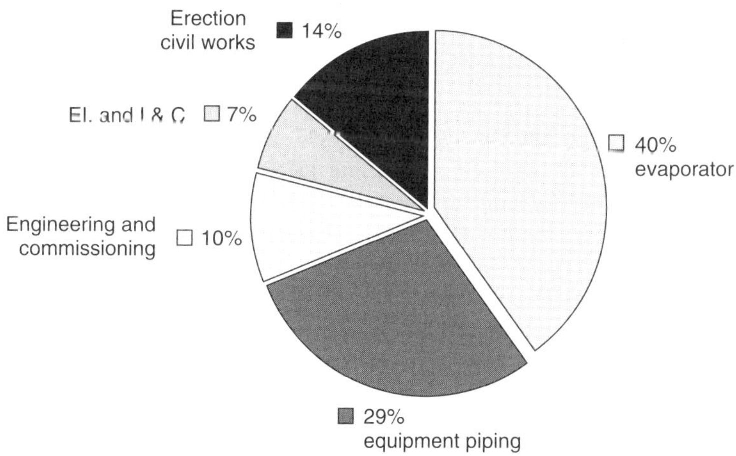

Figure 6.39 Example of an MSF desalination plant capital cost breakdown.

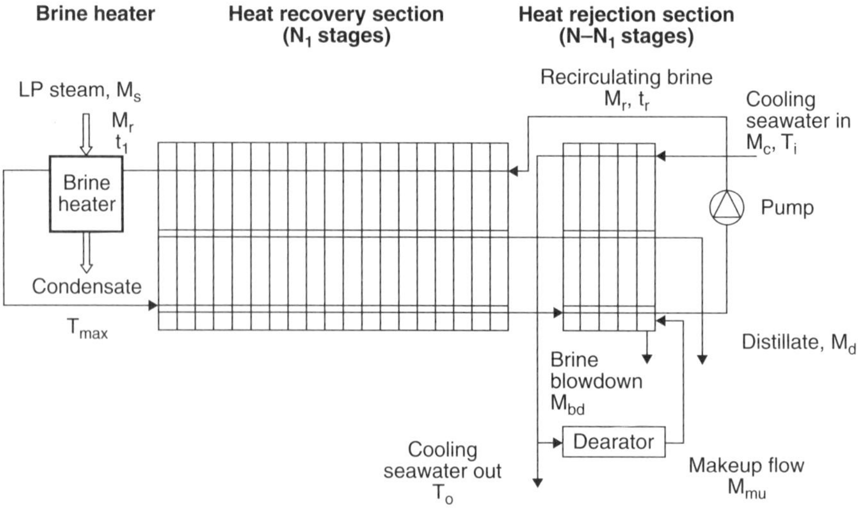

Figure 6.40 Simplified flow diagram of an MSF process.

Such plant models relate system design variables, such as performance ratio, top brine temperature (T_{max}), recirculation flow rate, and a number of stages with the design and operating parameters such as temperature profile, flow profile, heat transfer area, and heat transfer coefficients. Each stage is divided into four spaces: flash chamber, vapor space, tube bundle, and product tray. Applying the mass–energy equation for each of these spaces as well as the heat transfer equation for the tube bundle, we obtain a system of equations for each stage and the brine heater. The model is usually developed under the following assumptions:

- The distillate from each stage is salt-free.
- There are no heat losses.
- There is no entrainment of salt mist by the flashed vapor.

6.5.4.2.2 Mathematical Simulation Model *Mathematical modeling* refers to formulating a set of equations that describe the different processes taking place in the MSF plant. In the simulation phase, a computer using a suitable solution procedure, as well as entering the values of input process parameters, solves the formulated model.

The simulation model used here in setpoint optimization is steady-state, and is non-linear mainly because of the dependence of the physical properties of the streams (brine, distillate, vapor, etc.) on temperature, pressure, and salinity, and because of the relations for the heat transfer coefficients. Owing to the complexity of the model, the simulation procedure is sometimes based on a simple model, adequate for approximate results, and on a rigorous model or a combination of both when better accuracy is sought.

The simple model shown and used here is based on El-Nashar [33], and has the following simplifying assumptions:

1. The brine properties such as the specific heat and latent heat are constant over the temperature and concentration ranges of the heat recovery and heat rejection sections, and their average values can be used for each section.
2. Temperature drop per stage is identical for each stage of the recovery section and rejection section.
3. The overall heat transfer coefficients for the different stages of the heat recovery section are assumed constant, as they are for the heat rejection section.
4. Temperature loss per stage, which consists of the boiling-point elevation, nonequilibration loss, and demister loss, is constant in each stage for the whole plant.
5. Reflashing and recondensation of the distillate in each stage, as implemented in most plants for higher heat recovery, are ignored for model simplicity, making these model results more conservative by predicting performance ratios that are lower by at most 8% than those where the distillate heat balance is included.

Using mass and heat balances for the brine heater, heat recovery section, and heat rejection section, the following equations for an MSF plant having N stages can be written (symbols are defined in the Nomenclature section at end of the chapter) as follows:

Heat balance equations:

Brine heater:
$$Q = M_r C_{p1}(T_{max} - t_1) \tag{6.23}$$

Recovery section:
$$M_r C_{p1}(t_1 - t_r) = M_{d1} L_1 \tag{6.24}$$

Rejection section:
$$\begin{aligned} M_c C_{p2}(T_o - T_i) &= (M_d - M_{d1}) L_2 \\ (M_r - M_{d1}) C_{p2} T_{N1} &= M_c C_{p2}(T_o - T_i) \\ &\quad + M_d C_{pd}(T_N - \varepsilon) - M_{d1} C_{pd}(T_{N1} - \varepsilon) \end{aligned} \tag{6.25}$$

Total plant:
$$\begin{aligned} Q &= (M_c - M_{mu}) C_{p2} T_o + M_d C_{p1} \\ &\quad (T_N - \varepsilon) + M_{bd} C_{p2} T_N - M_c C_{p2} T_i \end{aligned} \tag{6.26}$$

Brine mixing in stage N:
$$\begin{aligned} &M_{mu} C_{p2} T_o + (M_r - M_{mu}) C_{p2}(T_N - \varepsilon) \\ &= M_r C_{p2} t_r \end{aligned} \tag{6.27}$$

Heat transfer equations:

Recovery section:
$$t_1 - t_r = N_1(T_{N1} - t_r - \varepsilon)\left\{1 - \exp\left[\frac{-(U_1A_1)}{M_rC_{p1}}\right]\right\} \tag{6.28}$$

Rejection section:
$$T_o - T_i = (N - N_1)(T_N - T_i - \varepsilon)\left\{1 - \exp\left[\frac{-(U_2A_2)}{(M_cC_{p1})}\right]\right\} \tag{6.29}$$

Material balance equations:

$$M_{bd} = \frac{M_d}{r - 1} \tag{6.30}$$

$$M_{mu} = \frac{r}{r - 1}M_d \tag{6.31}$$

Quantities with subscript 1 refer to the average value for the heat recovery section; subscript 2 refers to the heat rejection section. Appropriate average property values for the specific heat and latent heat and for the overall heat transfer coefficients and stage temperature loss are to be used in the above set of equations. Fifteen operating variables appear in Equations (6.23)–(6.31), namely, $Q, t_o, t_r, T_i, T_o, T_{max}, T_{N1}, T_N, M_c, M_r, M_{mu}, M_{bd}, M_d, M_{d1}$, and r. Five control (decision) variables must be selected in order to effect a solution of the problem: T_{max}, T_i, M_c, M_r, and r. The model was run using the design specification data of the UANE (Ext.) plant, and the results are shown. This program was developed on a Mathematica platform and was validated against a number of MSF plants at the Abu Dhabi Water and Electricity Authority. We note that the results are shown primarily to demonstrate modeling and trends, and while the trends shown are correct, the absolute values depend on the plant, the specific operating conditions, and on the model assumptions and rigor.

The effect of the top brine temperature, T_{max} and brine recirculating flow rate, M_r, on the distillate production and LP steam consumption of this evaporator is shown in Figures 6.41 and 6.42 for a constant inlet cooling-water temperature to the heat rejection section, $T_i = 30°C$, and assuming that $M_c = M_r$. As can be seen, increasing T_{max} and M_r causes an increase in distillate production but the steam consumption also increases correspondingly. These figures demonstrate that these two control variables, T_{max} and M_r, have a significant influence on the performance of the evaporator.

This simple model can thus be used to predict the performance of the evaporator for given values of the control parameters. As a result of the calculations, the following data will become available: distillate production, performance ratio, LP steam flow rate, temperature profile inside the evaporator, brine blowdown flow

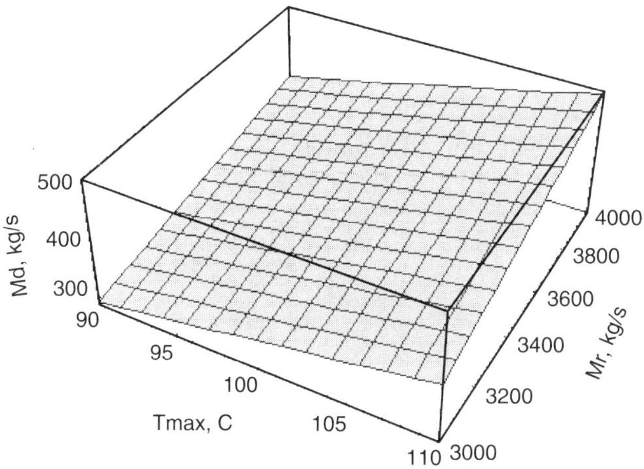

Figure 6.41 Distillate production (M_d) versus T_{max} and M_r for a constant inlet cooling-water temperature to the heat rejection section ($T_i = 30°C$, $M_c = 3500$ kg/s).

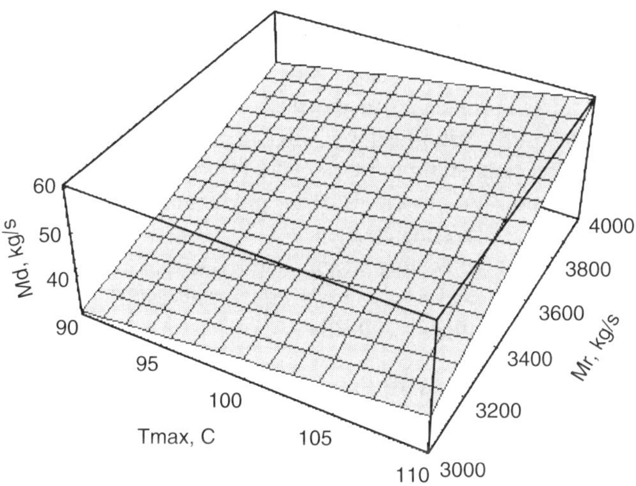

Figure 6.42 Heating steam consumption M_s versus T_{max} and M_r for a constant inlet cooling-water temperature to the heat rejection section ($T_i = 30°C$, $M_c = 3500$ kg/s).

rate, seawater recycle flow rate, and blowdown salt concentration. This model may be used to provide initial values for the iteration parameters to be used to start the rigorous model. The major performance index used in MSF plants is the performance ratio (PR), defined as

$$\text{PR} \equiv \frac{M_d \; \Delta H_d}{M_s \; \Delta H_s} \tag{6.32}$$

where ΔH_d is the enthalpy increase of the seawater when converted to vapor (the latent heat of evaporation) and ΔH_s is the enthalpy drop of the heating steam in the brine heater, from its entrance to the condensate that it forms. Often $\Delta H_d \approx \Delta H_s$, as assumed in this simplified analysis, and then a simplified PR [sometimes denoted as GOR (gain output ratio)] is defined simply as the ratio of distillate flow rate M_d, to the steam flow rate M_s, to the brine heater, as follows:

$$\mathrm{PR} = \frac{M_d}{M_s} \tag{6.33}$$

Maintaining T_{max}, M_c, and M_r constant and increasing the steam temperature T_s causes the overall heat transfer coefficient of all stages to increase, thus causing the distillate production M_d to increase, and a slight increase in performance ratio PR, as demonstrated in Figure 6.43. In this figure, M_d and PR are expressed relative to the design values M_{d0} and PR_0.

When the seawater temperature T_i at the inlet to the rejection section changes (as may occur at different geographic locations and seasons) while the other variables remain constant, this can have a great effect on the distillate production but very little effect on the performance ratio as shown in Figure 6.44. The reason is that increasing the seawater inlet temperature while maintaining T_{max}, M_c, and M_r at constant values causes an increase in the brine temperature of the last stage, thus reducing the flash range of the plant. This reduces the distillate production rate because it is in direct proportion to the flash range. Because of the increase in T_i, the heat added in the brine heater is also slightly reduced owing to the small increase in the brine temperature entering the brine heater. Therefore, increasing T_i results in a reduction in both the distillate production and the heat added in the brine heater. Since the performance ratio PR is defined as the ratio between the distillate production and the heating steam supplied to the brine heater, and because the reduction in both M_d and M_s is nearly equal, the result is that the increase in T_i causes has very little influence on PR.

6.5.4.2.3 A Case Study (UANE Ext. MSF Plant [33] To demonstrate how the mathematical model may be used to optimize the operation of an MSF unit, select

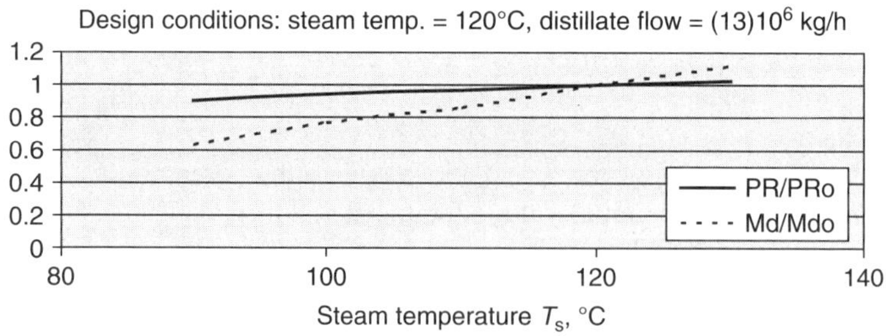

Figure 6.43 Effect of steam temperature T_s on distillate production and performance ratio for constant: $T_{max} = 105°C$, $M_c = 3500$ kg/s, $M_r = 3500$ kg/s.

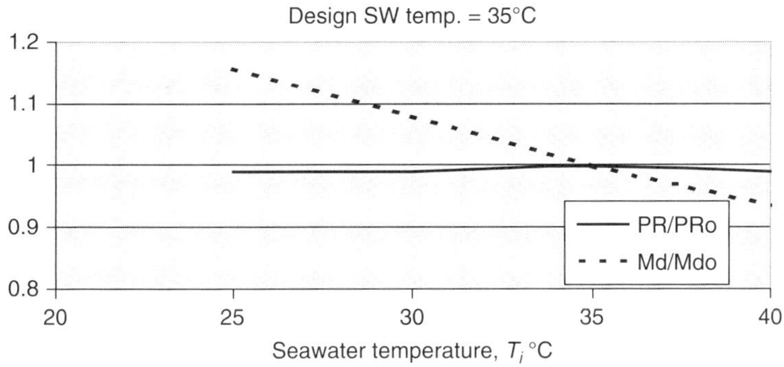

Figure 6.44 Effect of seawater inlet temperature T_i on the performance ratio, PR/PR_0, and distillate production, $M_d = 380$ kg/s, $M_c = 3500$ kg/s, $T_{max} = 105°C$.

one of the three identical MSF units at the UANE (Ext.) desalination plant. The design specifications of these units are given below:

Design rated capacity, t/d	32,700
Performance ratio	7.6
Design top brine temperature, °C	109.6
Number of stages	18
Heat rejection stages	3
Recovery stage condenser surface, m^2	4537
Rejection stage condenser surface, m^2	4304
OHTC (overall heat transfer coefficient) of recovery section (fouled condition), kW m^{-2} C^{-1}	2.85
OHTC of rejection section (fouled condition), kW m^{-2} C^{-1}	2.45

The simple model was used to predict the performance ratio PR. The distillate flow rate M_d and steam flow rate M_s for given values of the input (control) parameters: were also estimated the top brine temperature T_{max}, the recirculating brine flow rate M_r, the cooling seawater flow rate M_c, the inlet seawater temperature to heat rejection section T_i, and the concentration ratio r, defined as the ratio of the blowdown to seawater concentration. A computer program based on the simple model was developed, assuming that the plant was in a fouled condition. The program was run at different values of the input parameters that covered the permissible operating range of the plant, and the program output was used to establish correlations between the output and input parameters. The correlations are based on least-squares fit of the values of the output parameters and are given below:

$$PR = 15.483 T_{max}^{0.5005} M_r^{-0.2780} M_c^{-0.1102} T_i^{0.1105} r^{-0.2297} \tag{6.34}$$

$$M_d = 3.1(10^{-4}) T_{max}^{1.5840} M_r^{0.9270} M_c^{0.0617} T_i^{0.3864} r^{0.0525} \tag{6.35}$$

$$M_s = 7.463(10^{-6}) T_{max}^{1.1293} M_r^{1.2387} M_c^{0.1992} T_i^{0.4130} r^{0.2623} \tag{6.36}$$

where T_{max} and T_i are in °C and M_r and M_c are in kg/s. PR and r are dimensionless. The operating range of each variable is as follows:

$$90 \leq T_{max}\ (^\circ C) \leq 110$$
$$3000 \leq M_r\ (kg/s) \leq 4000$$
$$3000 \leq M_c\ (kg/s) \leq 4000$$
$$15 \leq T_i\ (^\circ C) \leq 35$$
$$1.2 \leq r \leq 1.4$$

It should be noted that these correlations are applicable only during winter months when the seawater recirculation pump is in operation. This is obviously so since T_i is one of the control parameters used as input in this model. In summer operation, the seawater recirculation pump is normally shut down, and the value of T_i, will be identical to the ambient seawater temperature; thus this parameter will not be incorporated into the model, and there will be only four control parameters instead of five.

The economic parameters used in this case study are

Cost of LP steam c_s = \$3.5/t (\$0.0035/kg)
Cost of electricity c_{el} = \$0.06/kWh
Cost of antiscalant c_{as} = \$1.0/kg

The dosing rate in ppm d_{as} for the antiscalant is assumed to be dependent only on the top brine temperature according to a second-degree polynomial given by [32]

$$d_{as} = 32.25 - 0.808T_{max} + 0.00523T_{max}^2 \qquad (6.37)$$

The pumping power requirements for the brine recirculating pump and the seawater pump are assumed to constitute the bulk of the pumping power requirement. The power requirement for these two pumps depends on the flow rates M_r and M_c, according to the operating curves for each pump. In order to simplify the analysis, an average specific pumping power in kWh/m^3 was estimated for each pump [33] using these curves, and the following values were obtained:

Brine recirculation pump 0.286 kWh/m^3
Cooling seawater pump 0.057 kWh/m^3

The objective function represents the operating cost (\$/kg distillate) and is given as follows:

$$\Phi = \frac{M_s c_s}{M_d} + \frac{M_r(0.286)c_{el}}{M_d} + \frac{M_c(0.057)c_{el}}{M_d} + \frac{r}{r-1} d_{as} \cdot 10^{-6} c_{as} \qquad (6.38)$$

The aim is to minimize this function subject to the plant operating constraints as shown above, and given again below:

$$90 \leq T_{max}\ (^\circ C) \leq 109.6$$

$$3000 \leq M_r\ (kg/s) \leq 4000$$

$$3000 \leq M_c\ (kg/s) \leq 4000$$

$$15 \leq T_i\ (^\circ C) \leq 35$$

$$1.2 \leq r \leq 1.4$$

This is a typical constrained nonlinear optimization problem that can be solved by a number of alternative methods (e.g.), steepest descent [33], to obtain the minimum operating cost Φ^* (\$/m^3 distillate) and the associated values of the control parameters, $x_1^*, x_2^*, x_3^*, x_4^*, x_5^*$ corresponding to each distillate production level. The results are shown in Table 6.7, giving the optimum values of the control parameters for different levels of distillate production. The optimum top brine temperature for a given load can be seen to increase as the load increases. The cooling seawater inlet temperature to the rejection section exhibits the opposite trend; that is, the inlet temperature drops as the load increases. In general, one can observe that the optimum values of the control parameters experience wide variation for different levels of distillate production.

The optimum performance ratio for different distillate production levels is shown in Figure 6.45, and the corresponding minimum operating costs are shown in Figure 6.46. Operating the plant at parameters different from the optimum will obviously result in a cost penalty. It would be convenient to define a *cost penalty* as the percentage increase in the specific production cost of water due to the shift in the operating point from the optimum, thus to define the percent cost penalty δ (%) as

$$\delta = \frac{\Phi - \Phi^*}{\Phi^*} \times 100 \qquad (6.39)$$

Table 6.7 Values of Control Parameters Corresponding to Minimum Distillate Production Cost

Distillate M_d kg/s	Top Brine Temperature T_{max}, °C	Recirculating Brine M_r, kg/s	Cooling Seawater M_c, kg.s^{-1}	Inlet SW Temperature to Reject Section T_i, °C	Concentration Ratio r
250	76.0	3510	3581	33.6	1.2
300	83.0	3662	3624	33.4	1.21
350	105.3	3253	3949	34.9	1.35
400	105.6	3000	3000	27.5	1.36
450	109.3	3259	3026	25.9	1.39
500	109.6	3574	3087	25.1	1.39

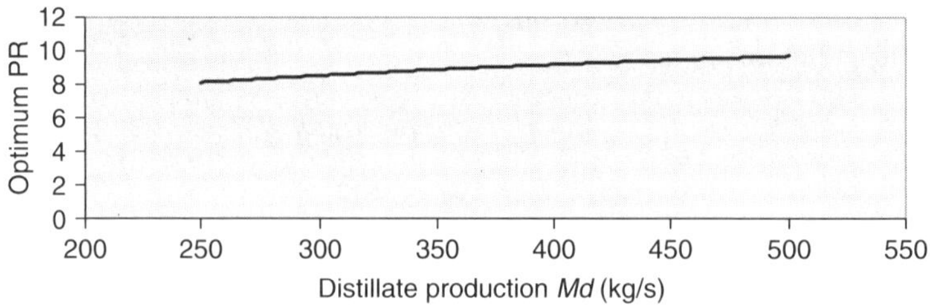

Figure 6.45 Optimal performance ratio PR corresponding to different levels of distillate production M_d.

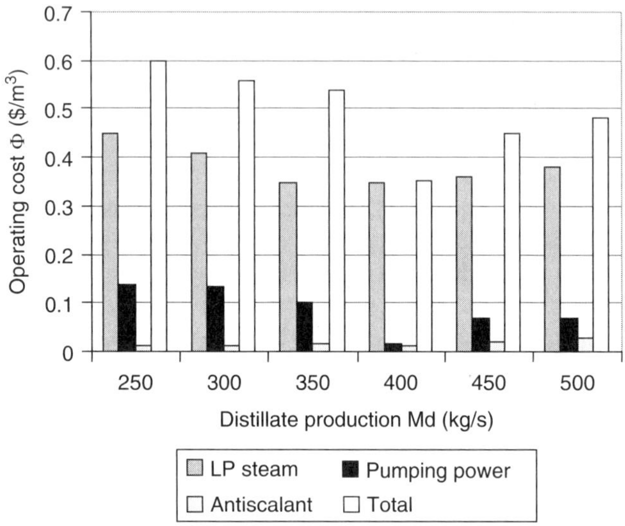

Figure 6.46 Minimum operating costs at different production levels.

where Φ^* is the specific production cost of water at the optimum operating condition and Φ is the cost at an operating condition deviated from the optimum.

The cost penalty for not operating the plant at the optimum T_{max}, for example, was estimated for a production level of 500 kg/s, and the results are shown in Figure 6.47. This represents the percentage increase in operating cost ($/m^3 distillate) incurred because of operation at a top brine temperature different from the optimum value. At this production level, the minimum operating cost is $0.48/t with the optimum $T_{max} = 109.6°C$. Operating the plant at 100°C instead of 109.6°C is shown to result in an increase of 6.9% in operating cost, which becomes $0.51/t (an increase of $0.033/t), thus incurring an additional operating expense of $0.51 M per year.

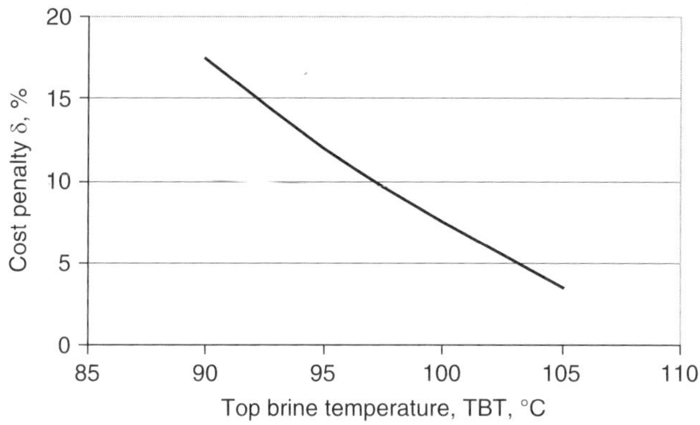

Figure 6.47 Cost penalty δ for deviation from optimum TBT ($M_d = 500$ kg/s, $\delta = 0$ at $T_{max} = 109.6^\circ$ C, Table 6.7).

6.5.4.3 Optimization of Operation of the MSF Process

6.5.4.3.1 The Optimization Method This follows the methodology presented in general in Section 6.5.2. The product-water cost function to be minimized has the following form

$$\Phi(\mathrm{vi}_1 \cdots \mathrm{vi}_n, \mathrm{vd}_1 \cdots \mathrm{vd}_m) = \sum c_i x_i \qquad (6.40)$$

where Φ = cost function
$\mathrm{vi}_1 \cdots \mathrm{vi}_n$ = independent variables
$\mathrm{vd}_1 \cdots \mathrm{vi}_m$ = dependent variables–
$c_1 \cdots c_k$ = unit costs
$x_1 \cdots x_k$ = specific consumption (energy and chemicals) terms

The dependent variables are functions of the independent variables vi_i, and the relations are described by the mathematical model equations:

$$\mathrm{vd}_j = f_j(\mathrm{vi}_1 \cdots vi_n) \qquad (j = 1 \cdots m) \qquad (6.41)$$

These are the system constraints. Other constraints can be added to them for specific applications.

The minimization of Φ using the simplex method requires that the expressions of the cost function and the constraints be linear in the variables. In order to approach the problem with this method, it is necessary to linearize the equations around the operating point. Starting from an initial point, we can write

$$\Phi(vi_1 \cdots \mathrm{vi}_n, vd_1 \cdots \mathrm{vd}_m) = \Phi(\mathrm{vi}_1^0 \cdots \mathrm{vi}_n^0, vd_1^0 \cdots vd_m^0) + \sum_{i=1}^{n} \frac{\partial \Phi}{\partial vi_i} d(\mathrm{vi}_i) + \sum_{j=1}^{m} \frac{\partial \Phi}{\partial vd_j} d(\mathrm{vd}_j) \qquad (6.42)$$

For our case, the cost function has the form

$$\Phi = c_s \cdot 10^{-6} \frac{M_r C_{pr}(T_{max} - t_1)}{M_d} + c_{as} d_{as} \cdot 10^{-6} \frac{M_{mu}}{M_d} + c_{el} \frac{\sum P_i}{M_d} \tag{6.43}$$

where T_{max} = top brine temperature
M_r = brine recirculation flow rate
T_i = seawater to reject section inlet temperature
M_c = cooling seawater flow rate
M_{mu} = makeup flow rate

are the independent variable, while

M_d = distillate product flow rate
M_s = steam to brine heater flow rate
M_{swr} = seawater recycle flow rate
M_{bd} = brine blowdown flow rate
t_l = brine temperature at brine heater inlet

are the dependent values. The unit costs are as follows:

c_s = unit cost of steam ($/GJ of heat)
c_{as} = unit cost of antiscalant ($/kg of antiscalant)
c_{el} = unit cost of electricity ($/kWh)

The specific consumption of steam, antiscalant, and electricity are calculated from:

Steam heat consumption/kg distillate = $M_r C_{pr}(T_{max} - t_1)/M_d$
Antiscalant consumption/kg distillate = $d_{as} \cdot 10^{-6} \cdot M_{mu}/M_d$
Pump power consumption/kg distillate = $\Sigma P_i/M_d$

The pump power consumption for a particular pump P_i is related to the corresponding fluid flow rates Q_i according to the following relation:

$$P_i(Q_i) = \rho g(H_f + aQ_i^2)\frac{Q_i}{\eta p} \tag{6.44}$$

where H_f is the contribution to the required head by the pressure and geodetical head, and aQ_i^2 is the friction head loss due to the irreversibility induced by the flow circuit.

The mathematical model describes all these relations in the following way:

$$M_d = f_d(T_s, M_r, T_i, M_c, M_{mu}) \tag{6.45}$$

$$M_c = f_c(T_s, M_r, T_i, M_c, M_{mu}) \tag{6.46}$$

$$M_{swr} = f_{swr}(T_s, M_r, T_i, M_c, M_{mu}) \tag{6.47}$$

$$M_{bd} = f_{bd}(T_s, M_r, T_i, M_c, M_{mu}) \tag{6.48}$$

$$t_1 = f_T(T_s, M_r, T_i, M_c, M_{mu}) \tag{6.49}$$

where the functions f_j represent the dependent variables resulting from the application of the model with the set of independent variable $(T_s, M_r, T_i, M_c, M_{mu})$. The minimization of the cost function is subject to a series of operating constraints, as follows:

Operating limits of control parameter:

$$T_s^{min} \le T_s \le T_s^{max} \tag{6.50}$$

$$M_r^{min} \le M_r \le M_r^{max} \tag{6.51}$$

$$T_i^{min} \le T_i \le T_i^{max} \tag{6.52}$$

$$M_c^{min} \le M_c \le M_c^{max} \tag{6.53}$$

$$M_{mu}^{min} \le M_{mu} \le M_{mu}^{max} \tag{6.54}$$

Congruence relations, such as balance of flow rates at reject section inlet and outlet

$$M_{sw} - M_{mu} - M_{swr} \ge 0 \tag{6.55}$$

Limit for brine recycle concentration

$$r = \frac{M_{mu}}{M_r} \cdot \frac{M_r - M_d}{M_{mu} - M_d} \le r_{max} \tag{6.56}$$

Product distillate target

$$M_d = M_d^o \tag{6.57}$$

Operating limits of steam supply to brine heater

$$M_s \le M_s^{max} \tag{6.58}$$

Operating limit of flashing brine temperature in last stage:

$$T_N = f_{T_N}(T_s, M_r, T_i, M_c, r) \le T_N^{max} \tag{6.59}$$

The starting point of the calculation is the actual operating point. At each step of the calculation, a linearization of the cost function and of the constraints is realized around the working point in order to allow the application of the simplex method. The linearization is performed through the calculation of the derivatives of the cost function versus both the independent and dependent variables, and of the derivatives of the dependent variables versus the independent ones. The derivatives that are not explicit are numerically calculated with the use of the mathematical model.

6.5.4.3.2 Optimization Results It was shown that the use of advanced controllers such as MPC can result in substantial reduction in the variance of controlled variables. Muske [83] showed measurements of operation temperature data with conventional controller and MPC (advanced controller), and his data are depicted in Figures 6.48 and 6.49, which also show the temperature constraint (limit). These figures show the fluctuations in a process fluid temperature (controlled variable) in a heat exchanger and demonstrate the sizable reduction in variance by using advanced controllers compared with a conventional PID controller. As was explained above, two types of savings can be achieved by using advanced control: savings due to reduction in variable variance $\Delta\Phi_1$ and savings due to operating the plant close to optimum conditions $\Delta\Phi_2$. The savings that can be achieved by the reduction of variable variance can be estimated from the following relation

$$\Delta\Phi_1 = \sum_{i=1}^{n} \frac{\partial f}{\partial \overline{x}_i} \Delta\overline{x}_i \tag{6.60}$$

where $\Delta\overline{x}_i$ is estimated from the relation $\Delta\overline{x}_i = P_i(m_i)(s_{\mathrm{B}i} - s_{\mathrm{C}i})$. The assumptions made in the following calculations are $m = 5\%$, $s_{\mathrm{C}i} = s_{\mathrm{B}i}/2$ and $s_{\mathrm{B}} = 0.02\,\overline{x}$ (i.e., the standard deviation in the base case is 2% of mean value).

The savings that can be achieved from operating close to the optimum condition can be estimated from

$$\Delta\Phi_2 = \Phi - \Phi^* = \sum_{i=1}^{n} \frac{\partial f}{\partial \overline{x}_i} (\overline{x}_i^* - \overline{x}_i) \tag{6.61}$$

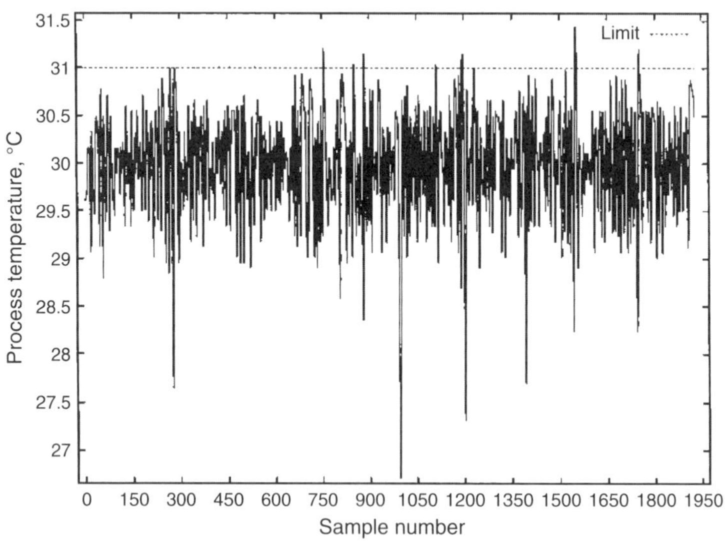

Figure 6.48 Temperature measurements with conventional PID controller [83].

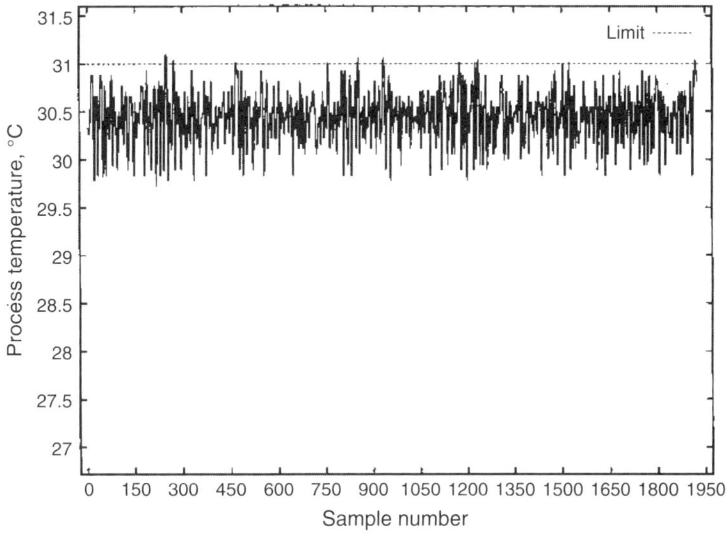

Figure 6.49 Temperature measurements with advanced MPC controller [83].

The extent of this saving depends on how close the current operating condition using conventional control devices is to the optimum condition; that is, it depends on the approach from the optimum operating condition.

The total savings that can be accrued from using advanced controllers in place of the conventional PID controllers in MSF plants were estimated by calculating both parts of the savings, namely, the savings due to reduction in the variance of the controlled variables, and the savings due to operating close to the optimum. $\Delta\Phi_1$ was calculated from Equation (6.60) with $\Delta\overline{x}$ obtained from Equation (6.14), and the terms $\partial f/\partial \overline{x}_i$ were calculated from the cost function of Equation (6.19). $\Delta\Phi_2$ was calculated from Equation (6.61).

Thus, the total savings can be written as follows:

$$\Delta\Phi = \Delta\Phi_1 + \Delta\Phi_2 \tag{6.62}$$

Let us define the potential percentage saving that can be achieved by replacing PID controllers with advanced controllers by

$$\text{Saving } \% = \frac{\Phi - \Phi^*}{\Phi^*} \times 100 \tag{6.63}$$

The saving that can be achieved is expected to depend on the approach of the value of the actual plant controlled variables from the optimum operating condition. Let us define this *percent approach* as:

$$\text{Approach } \% = \frac{x - x^*}{x^*} \times 100 \tag{6.64}$$

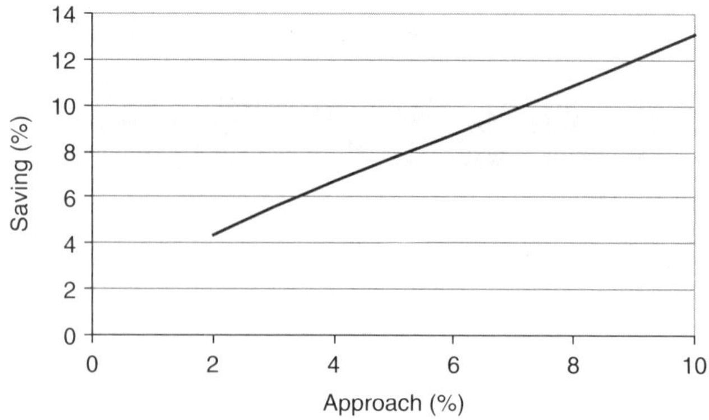

Figure 6.50 Savings in operating cost resulting from using advanced controllers in MSF plants based on the assumed economic factors.

where x is the value of a controlled variable and x^* is the corresponding optimum value.

On the basis of the specific economic cost factors presented above, the saving that can be achieved in an MSF plant that currently uses PID control is shown in Figure 6.50. This figure shows the dependence of this saving (as a percentage of the operating expenses at the optimum operating condition) on the approach of the current operating condition from the optimum. As can be seen, the savings range from 4.5% for a 2% approach to a saving of 13% for a 10% approach. In a typical MSF plant, a 5% approach appears to be normal. In such a case, the saving is ~\$0.04/m^3, which corresponds to ~\$66,000 per year for each 1 migd of installed capacity. This saving is achieved by maintaining the operating parameters of the plant close to its optimum values, which are estimated from the mathematical model as well as from the reduction in the variance of the controlled variables resulting from the use of advanced controllers. For a hypothetical MSF plant having specific operating cost and specific total water cost of \$0.75 and \$1.5 per cubic meter of water, respectively, this saving represents 5.3% in specific operating cost and 2.6% in specific total water cost. Actual savings for a particular plant depend on its capital and operating costs as well as its design and operating conditions.

6.5.4.4 An Additional Way to Improve: Change of IMCA Philosophy

So far we have discussed only the consequence of improving IMCA systems for an existing plant flowsheet, without any attempt to improve the plant configuration itself, and/or the automation control philosophy. In some cases such changes and/or improvements may actually bring about a reduction in the capital cost accompanied by reduction in the O&M costs at the same time. This is, for example, the case of the modification of the control system for the MSF brine top temperature, where a more sophisticated control algorithm can allow a substantial reduction in hardware.

In this example, we address modifications to the top brine temperature (TBT) control scheme in a 12-migd MSF desalination plant, which uses conventional steam turbine extraction as its heat input.

A retrofit of an existing plant is considered, where the extraction steam is obtained at 2.17 bar from a 120-MW steam turbine. The extraction pressure is controlled by a valve that ensures that a constant pressure of 1.8 bar is maintained at the MSF desalination plant battery limits[2] at various power plant running conditions. The steam turbine heat–mass balance is shown in Figure 6.51. This flowsheet shows a typical steam turbine configuration with extraction of high-pressure steam

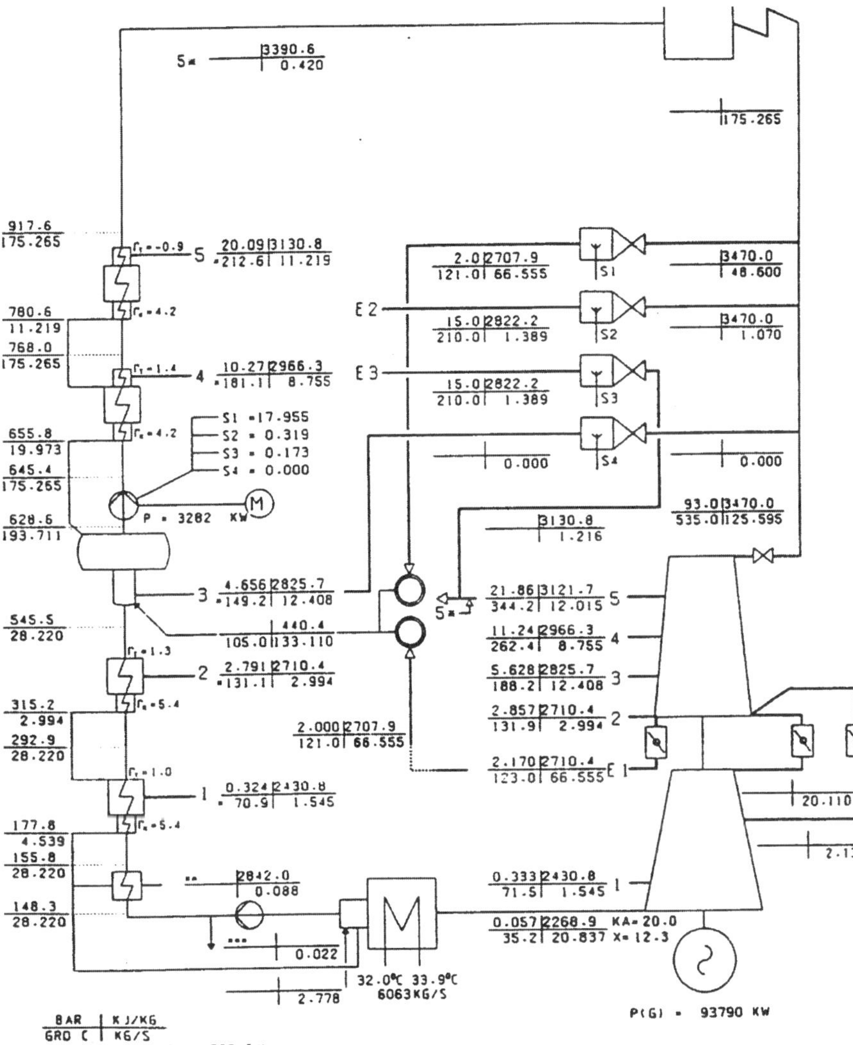

Figure 6.51 Typical steam turbine configuration with extraction to an MSF brine heater.

[2]*Battery limits* is a term used in the desalination industry to indicate the connection interface between two yards, in this case the power and the desalination.

and low-pressure steam to MSF (at 2.17. bar). The MSF desalination plant brine heater, indicated by a circle in the flowsheet, receives steam from the extraction point located in the LP part of the first turbine casing.

The pressure is further decreased at desalination battery limits by the TBT control valves, which throttles the steam flow to the brine heater to maintain a constant TBT. The current arrangement differs from the previous control scheme because it is designed to include a single control loop.

Accordingly, TBT is directly controlled at turbine extraction without the intermediate control loop, which is normally installed to control the pressure in the steam extraction manifold. This arrangement brings about additional and unnecessary pressure losses in the system, resulting in a loss of power that could have been produced by the turbine. This phenomenon is described in Figure 6.52, which shows the steam expansion curve in a Mollier diagram for a typical 120-MW steam turbine. The modification proposed to the TBT control scheme involves control of the TBT directly from the steam turbine extraction, without the further pressure reduction upstream of the brine heater.

The present arrangement allows substantial reduction in energy loss due to drops in valve pressure. Consequently, the power rendered at steam turbine shaft is up to ~1% higher.

The breakeven analysis for this kind of modification would involve:

1. Labor requirement for
   a. Piping modification
   b. Control system modification
   c. Wiring
   d. Insulation replacement
   e. Engineering
   f. Modification of setpoint
2. Fixed (capital) cost estimation for the equipment = US$1,200,000
3. Savings with modified automation/control scheme:
   a. Savings in spare parts due to deletion of one item of equipment neglected (conservative assumption)
   b. Energy savings due to deletion of one desuperheating station and booster pumps calculated at the current UAE energy price of 0.12 AED (UAE dihram) per kWh)
      (1) Energy savings = US$ 12/3.67/kWh saved = US$ 0.032/kWh
      (2) 50 kW saving achieved by implementation of system
   c. Additional power output of steam turbine shaft: 800 kW

The solution is shown graphically in Figure 6.53.

The total savings are just under $200,000 per year, and the analysis shows that the breakeven point occurs at just less than 6 years of operation. The proposed retrofit continues to produce these savings until the end of its functional life.

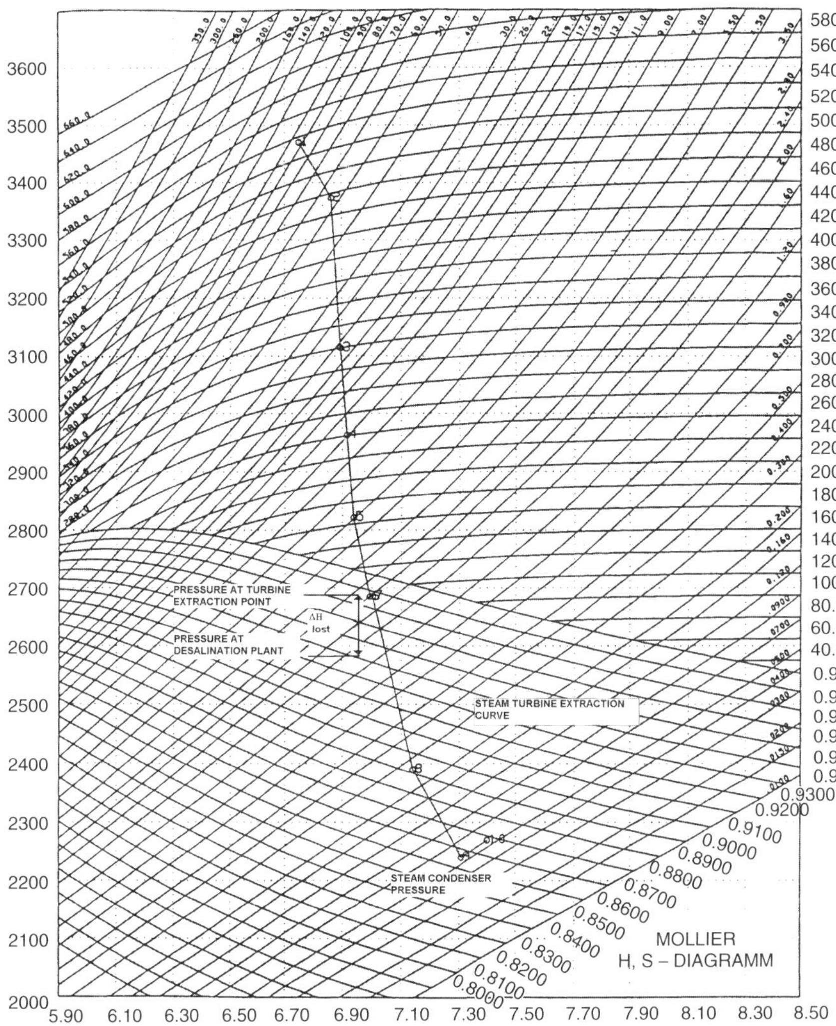

Figure 6.52 Typical steam expansion path in a turbine.

6.5.5 RO Cost–Benefit Ratio of New IMCA Systems

6.5.5.1 RO Process A commercial RO desalination plant consists of four major sections:

1. Seawater intake system
2. Pretreatment section
3. RO membrane elements
4. Posttreatment section

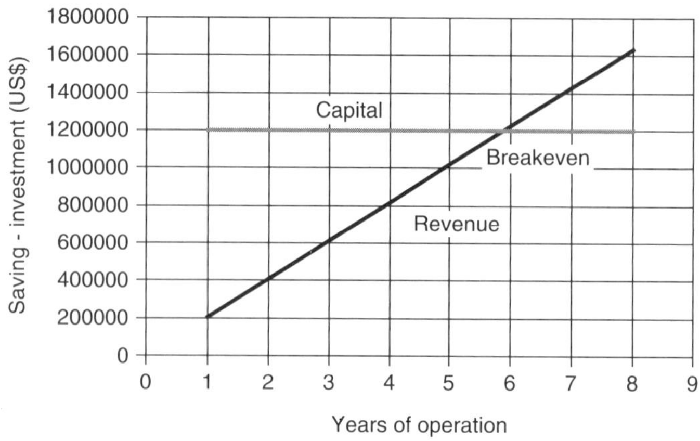

Figure 6.53 Breakeven analysis of the proposed MSF retrofit example.

These elements, as discussed below, constitute the bulk of the capital cost of the plant as well as the operating costs:

1. Seawater intake and pretreatment system
2. High-pressure pumps
3. RO membrane elements
4. Energy recovery turbines

A simplified flowchart of a single-stage seawater RO plant is shown in Figure 6.54. Due to the excellent solute separation obtainable with current membranes (average ~99%), it is theoretically not necessary to include a second stage for the product stream in order to obtain a product concentration below 500 ppm if the feed concentration is less than 50,000 ppm [91]. However, in practice, a small number of second-stage membrane elements may need to be installed in order to ensure uninterrupted supply of product water well within specification. This is particularly beneficial in the unlikely event of excessive membrane fouling or damage. In some situations, it is beneficial, instead, to feed the reject stream from the first stage to a second set of RO membrane elements to obtain an additional product.

6.5.5.2 ***Model of an RO Process*** The model used in this study is the same as that used by Taniguchi et. al. [92] for a spiral-type RO element, designated SU-820, manufactured by Toray Industries of Japan. The transport equations of the water flux J_V and salt flux J_S through RO membranes are given as follows, based on nonequilibrium thermodynamics

$$J_V = L_P[\Delta p - \sigma\{\pi(C_M) - \pi(C_P)\}] \tag{6.65}$$

$$J_S = P(C_M - C_P) + (1 - \sigma)\overline{C}_S J_V \tag{6.66}$$

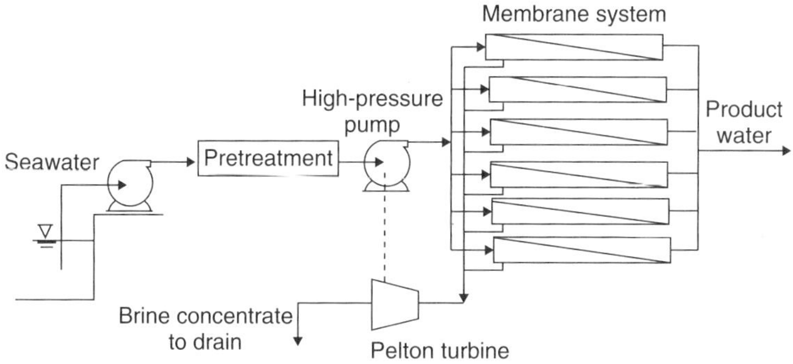

Figure 6.54 Simplified flow diagram of a single-stage seawater RO plant.

where L_P is a solution permeability, P a salt permeability, σ a reflection coefficient, π is the osmotic pressure, C_M is the salt concentration at the membrane surface, C_P is the permeate (product concentration), and $\overline{C}_S$ is an average concentration; L_P, P, and σ are membrane transport parameters that specify membrane properties.

For high-rejection membranes (rejections are >99%), σ is ~1 and Equation (6.66) can be simplified as follows:

$$J_S = P(C_M - C_P) \tag{6.67}$$

C_P is also given by

$$C_P = \frac{J_S}{J_V} \tag{6.68}$$

The balance of mass transfer across the membrane gives rise to the following equation

$$\frac{C_M - C_P}{C_B - C_P} = \exp\frac{J_V}{k}, \qquad k = \frac{D}{\delta} \tag{6.69}$$

where k is a mass transfer coefficient, C_B is the bulk concentration, D is the diffusivity, and δ is a boundary layer thickness. The correlation used for evaluating the mass transfer coefficient k is given by the following equation

$$\text{Sh} = a\text{Re}^{0.875}\text{Sc}^{0.25} \tag{6.70}$$

where $\text{Sh} = kd/D$ is the Sherwood number, $\text{Re} = \rho ud/\eta$ is the Reynolds number, and $\text{Sc} = \eta D$ is the Schmidt number. Here, d is a hydraulic diameter of the feedwater channel, defined as

$$d = \frac{4 \times \text{cross-sectional flow area}}{\text{wetted perimeter}} \tag{6.71}$$

and u is the flow velocity.

A membrane element is divided into N sections along the feed flow direction. Mass balance equations at each section ($\Delta h = h/N$) are given as

$$\Delta Q_{\mathrm{P}i} = JVi\,\Delta h \tag{6.72}$$

$$C_{\mathrm{P}i} = \frac{J_{\mathrm{S}i}}{J_{\mathrm{V}i}} = \frac{PC_{\mathrm{M}i}}{J_{\mathrm{V}i} + P} \tag{6.73}$$

$$Q_{\mathrm{Po}} = \sum_{i=1}^{N} \Delta Q_{\mathrm{P}i} = \sum_{i=1}^{N} J_{\mathrm{V}i}\,\Delta h W \tag{6.74}$$

$$C_{\mathrm{Po}} = \frac{\sum_{i=1}^{N} C_{\mathrm{P}i}\,\Delta Q_{\mathrm{P}i}}{Q_{\mathrm{Po}}} = \frac{\sum (PC_{\mathrm{M}i}/J_{\mathrm{V}i} + P)J_{\mathrm{V}i}\,\Delta h W}{Q_{\mathrm{Po}}} \tag{6.75}$$

where subscript i is the ith section, subscript "o" refers to the outlet of the membrane, W is the width of the membrane, and Q is the volume flow rate.

The properties of seawater are calculated from the correlations used by Taniguchi et al. [92]. For the osmotic pressure (Pa), density (kg/m^3), viscosity (Pa·s), and diffusivity (m^2/s), the correlations are

$$\pi(c, T) = (0.6955 + 0.0025T) \times 10^{8}\frac{c}{\rho} \tag{6.76}$$

where C is the salt concentration in kg/m^3 and T is the temperature in °C.

For the density of seawater:

$$\rho = 498.4m + \sqrt{248400m^2 + 752.4mC} \tag{6.77}$$

where

$$m = 1.0069 - 2.757 \times 10^{-4}T \tag{6.78}$$

For the viscosity of seawater:

$$\eta = 1.234 \times 10^{-6}\exp\left(0.00212C + \frac{1965}{273.15 + T}\right) \tag{6.79}$$

For the diffusivity:

$$D = 6.725 \times 10^{-6}\exp\left(0.1546 \times 10^{-3}C - \frac{2513}{273.15 + T}\right) \tag{6.80}$$

A computer program was developed to estimate the properties of the RO membrane elements at different operating conditions. The program was written using Mathematica software. Using a very similar analysis, Taniguchi et al. [92] compared their results with those of actual experimental results obtained from the Ehime test plant in Japan under different operating conditions, and the agreement between his simulation program and the test results was excellent.

With the membrane parameters L_P and P given and seawater conditions (seawater concentration C_f and seawater temperature T_f) known and the operating condition (feed pressure P_f) fixed, the program estimates the fluxes of salt and water for each section of membrane and evaluates the product concentration C_p and product recovery R_c.

A membrane made of crosslinked fully aromatic polyamide material identical to the material used by Toray Industries of Japan is assumed [92]. For this material, the membrane parameters at the standard operating conditions (25°C temperature and 6.5 MPa pressure) are

$$L_P = (2.52)10^{-12}\ \mathrm{m^3 m^{-2} (Pa \cdot s)^{-1}} \text{ and } P = (12.1)10^{-9}\ \mathrm{m/s}$$

The membrane element used is of the spiral-wound type having the dimensions: $W = 1$ m and $h = 6$ m, $d = 2$ mm. The inlet Reynolds number is Re $= 100$. Several program runs were made at different feed pressures, feed concentrations, and feed temperatures, and the results are shown as three-dimensional graphs in Figures 6.55 and 6.56. As demonstrated in these figures, the product concentration is quite sensitive to both temperature and concentration of the feedwater; increasing either of them results in a substantial increase in C_p. The feed pressure, however, has the opposite effect; increasing the feed pressure decreases the product concentration.

The effects of T_f, C_f, and P_f on recovery R [defined as the ratio between permeate (product) water flow rate and feedwater flow rate] of the RO membrane element is demonstrated in the three-dimensional graphs of Figures 6.57 and 6.58. It can be seen that the increase in feedwater temperature for a given value of feed concentration causes an increase in water recovery. This increase is more pronounced at the high feed concentration than at the low feed concentration. It can also be seen that the recovery levels are higher for lower feed concentration than for higher concentration. The effect of the feed pressure is such that increasing the pressure results in an increase in the recovery. Such an increase is more appreciable at lower feed concentrations than at higher ones.

6.5.5.3 Optimization of RO Process Operation In the same way as for the MSF process (Section 6.5.4), the cost function to be minimized has the following form

$$\Phi(\mathrm{vi}_1 \cdots \mathrm{vi}_n, \mathrm{vd}_1 \cdots \mathrm{vd}_m) = \sum_{i=1}^{k} c_i \cdot x_i \tag{6.81}$$

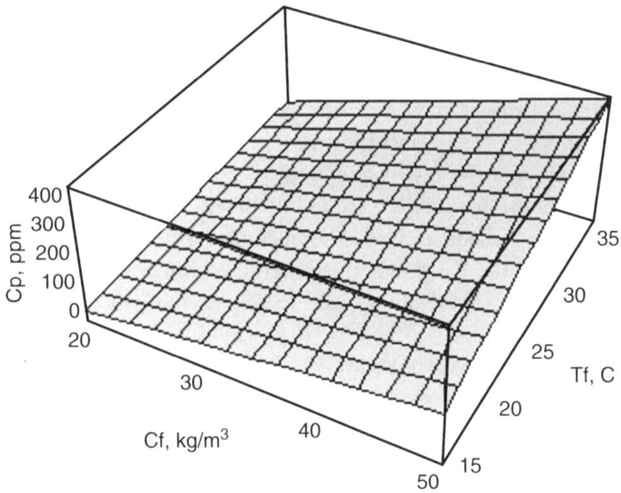

Figure 6.55 Effect of feed concentration C_f and feed temperature T_f on the product concentration $C_p(P_f = 7$ MPa).

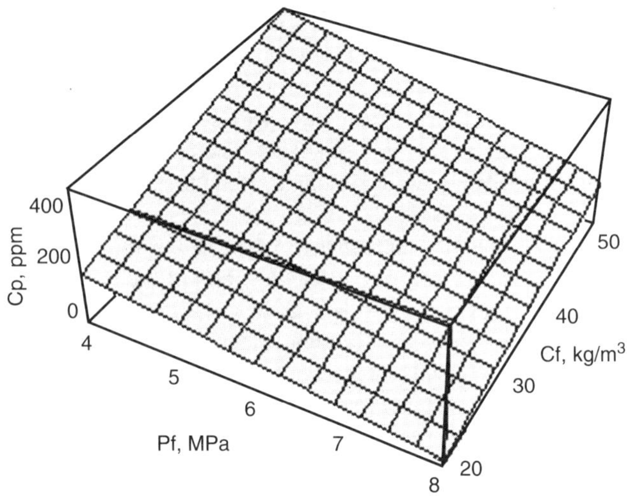

Figure 6.56 Effect of feed pressure P_f and feed concentration C_f on product concentration $C_p(T_f = 25°C)$.

where Φ = cost function
$\text{vi}_1 \cdots \text{vi}_n$ = independent variables
$\text{vd}_1 \cdots \text{vd}_m$ = dependent variables
$c_1 \cdots c_k$ = unit costs
$x_1 \cdots x_k$ = specific consumption terms

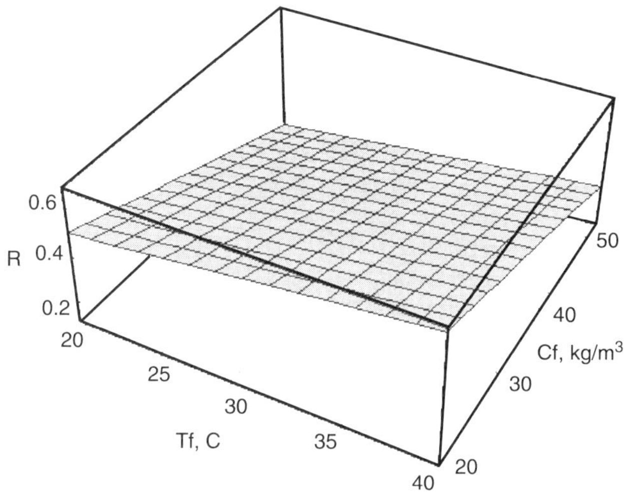

Figure 6.57 Effect of feed temperature T_f and feed concentration C_f on recovery R ($P_f = 7$ MPa).

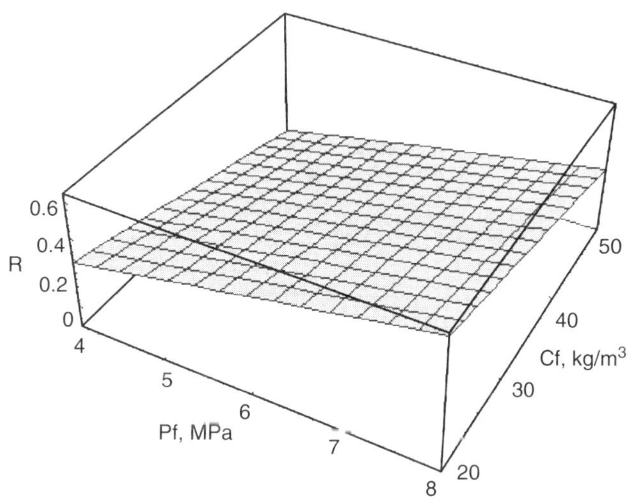

Figure 6.58 Effect of feed pressure P_f and feed concentration C_f on recovery R ($T_f = 25°C$).

The dependent variables vd are functions of independent variables *vi*, and the relations are described by the mathematical model equations:

$$\mathrm{vd}_i = f_j(\mathrm{vi}_1 \cdots \mathrm{vi}_n) \qquad (j = 1 \cdots m) \tag{6.82}$$

For our case, the cost function can be approximated as

$$\Phi = \sum (c_{\mathrm{ch}} d_{\mathrm{ch}} \cdot 10^{-6}) \frac{M_{\mathrm{f}}}{M_{\mathrm{d}}} + c_{\mathrm{el}} \frac{P_{\mathrm{HPP}} - P_{\mathrm{T}} + P_{\mathrm{PT}}}{M_{\mathrm{d}}} \tag{6.83}$$

where
c_{ch} = cost of chemicals used in pretreatment, \$/kg
d_{ch} = dose of chemicals, ppm
M_{f} = feedwater flow rate, kg/s
M_{d} = product-water flow rate, kg/s
P_{HPP} = hp pump power requirement, kW
P_{T} = power generated by recovery turbine (Pelton turbine), kW
P_{PT} = pumping power requirement of seawater intake and pretreatment section, kW

The pumping power requirement (in kW) is evaluated from

$$P_i(Q_i) = \rho g (H_{\mathrm{f}} + a Q_i^{\,2}) \frac{Q_i}{\eta_{\mathrm{p}}} \tag{6.84}$$

where η_{p} is the pump efficiency. An equation developed for pump efficiency based on the *GPSA Engineering Data Book* [93] is

$$\eta_{\mathrm{P}} = 80 - 0.2855\ h + 0.000378\ hQ - 0.000000238\ hQ^2 + 0.000539\ h^2 - 0.000000639\ (h^2)Q + 0.0000000004\ (h^2)(Q^2) \tag{6.85}$$

where h is developed head in feet of water and Q is the flow in gpm. Ranges of applicability are $h = 50$–300 ft (15.24–91.44 m) and $Q = 100$–1000 gpm, error documented at 3.5%.

The power generated by the recovery turbine (in kW) is calculated from

$$P_{\mathrm{T}} = (1.67)(\text{flow, m}^3/\text{min})(\text{pressure drop, bar})\eta_{\mathrm{t}} \tag{6.86}$$

where η_{t} is the recovery turbine efficiency. For a Pelton turbine, an efficiency of 87% was selected (Fig. 6.59). The constant 1.67 in this equation is a unit conversion factor that converts the unit of P_{T} from m^3 bar^{-1} min^{-1} to kW. The amount of recovery power generated will obviously depend on the recovery of the membrane system; the higher the recovery, the lower will be the amount of recovered power due to a lower flow rate through the recovery turbine.

The operating variables are assumed to be feedwater pressure P_{f}, feedwater temperature T_{f}, and feedwater concentration C_{f}. The feedwater flow rate was assumed constant. Using the least-squares technique on the results of the RO model described above, the product recovery and product concentration of the RO system may be expressed in terms of these three variables as follows

$$R = -6.609\ 10^{-2} - 1.239\ 10^{-2}\ C_{\mathrm{f}} + 9.209\ 10^{-2}\ P_{\mathrm{f}} + 8.38\ 10^{-3}\ T_{\mathrm{f}} \tag{6.87}$$

$$C_{\mathrm{p}} = 156.728 + 8.27541\ C_{\mathrm{f}} - 72.3046\ P_{\mathrm{f}} + 8.36\ T_{\mathrm{f}} \tag{6.88}$$

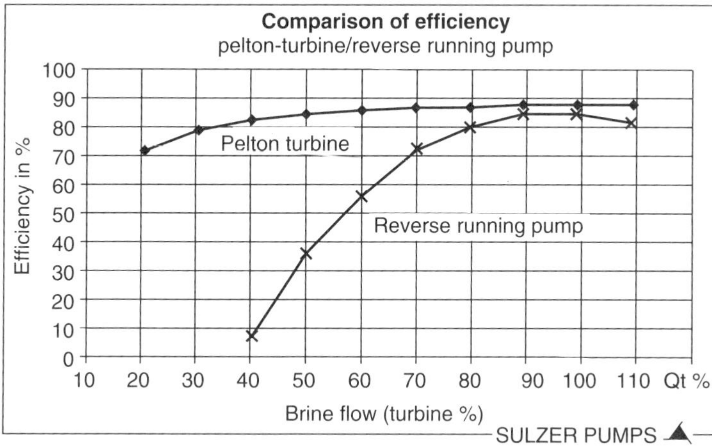

Figure 6.59 Efficiency of Sulzer–Pelton turbine and reverse-running pumps.

where P_f is the feedwater pressure (MPa), T_f is the feedwater temperature (°C), and C_f is the feedwater salt concentration (kg/m).

The constraints on these variables are as follows:

$$4 \le P_f \le 8 \quad \text{MPa}$$
$$15 \le T_f \le 40 \quad {}^{\circ}C$$
$$35 \le C_f \le 50 \quad \text{kg/m}^3 \tag{6.89}$$

The specific cost of electricity and chemicals per unit product water ($/m^{-3}) was expressed in terms of the three main operating variables mentioned above:

$$\Phi = 0.643392 + 3.96 \times 10^{-2}\ C_f - 0.206595\ P_f - 7.89\ 10^{-2}\ T_f$$
$$+6.826\ 10^{-4}\ C_f\ T_f + 1.41 \times 10^{-2}\ P_f\ T_f - 2.12\ 10^{-4}\ C_f\ P_f\ T_f \tag{6.90}$$

where Φ is the specific water cost in \$/m^3. Equations (6.88)–(6.91) are obtained by least-squares fit of the results obtained by the computer model of the RO system. Several runs of the model were made at different values of operating parameters that spanned the range given in Equation (6.90).

The effect of feedwater pressure and concentration on the specific water cost of an RO plant, calculated according to the model above, is shown in Figure 6.60 for an operating feedwater temperature of 25°C. It can be seen that at high feedwater concentration, increasing the feedwater pressure results in a decrease in the specific water cost. At low feed concentration, increasing the pressure causes a smaller cost decrease. Increasing feedwater concentration while maintaining a constant pressure results in an increase in cost, with the increase more pronounced at lower pressure than at higher pressure. This trend stems from the fact that for a particular

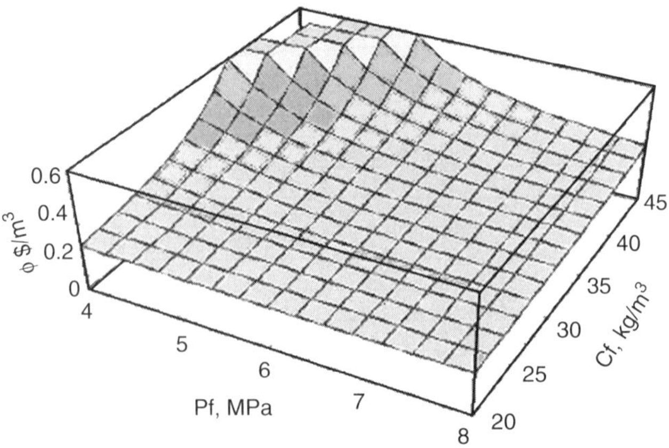

Figure 6.60 Specific water cost Φ versus feedwater pressure, P_f, and feedwater concentration C_f ($T_f = 25°C$).

feedwater concentration and membrane area installed in the system, increasing the feedwater pressure results in a higher product water recovery. The resulting effect on product-water cost will depend on the balance between two opposite cost factors. One is reduction of feed flow and therefore smaller processing capacity of pretreatment system and pumps, and the second is an increase of required feed pressure due to higher average feed salinity and therefore higher osmotic pressure. The net water cost result is a function of system operating parameters and local economic conditions. The system operating parameters are influenced by type of membrane elements used and its water permeability and salt rejection. The economic conditions include power cost, cost of treatment chemicals, membrane replacement cost, and cost of any equipment that its size depends on volume of water processed. Comparing calculated results with performance of commercial membrane elements, it seems that the model above predicts membrane performance with sufficient accuracy. The ability to project economics of RO systems is still approximate, and additional adjustments of parameters used in calculations are required.

Figure 6.61 shows the influence of feedwater temperature and concentration on the specific water cost for a constant operating pressure of 6.0 MPa. It can be seen that increasing the feedwater temperature results in a drop in the specific water cost due to the increase in membrane flux and the associated increase in the plant water recovery.

Based on the specific economic cost factors presented above, the percentage saving that can be achieved by the SWRO process for the particular assumed unit costs for electricity and chemicals is shown in Figure 6.62. For a 5% approach, the potential saving in the operating cost of water is ~2%, which corresponds to ~$13,000 per year for each 1 migd of desalination capacity. In a seawater RO plant, the salinity and temperature of the seawater vary seasonally and the optimum feed pressure entering the RO modules therefore has to be adjusted to match this variation in order to maintain at optimum operating condition at all times. If the

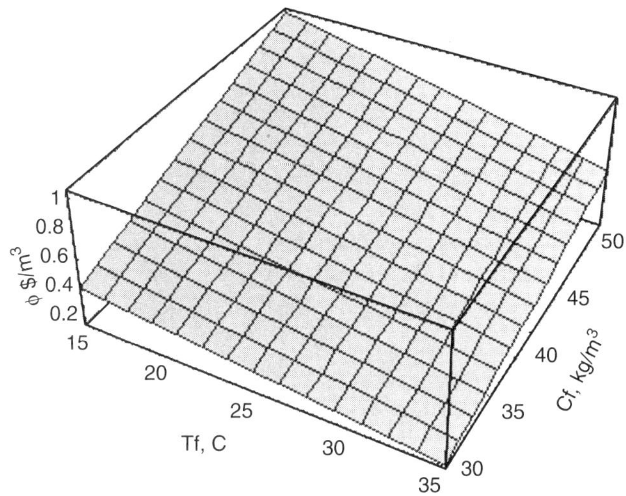

Figure 6.61 Specific water cost Φ versus feedwater temperature T_f and feedwater concentration C_f ($P_f = 6$ MPa).

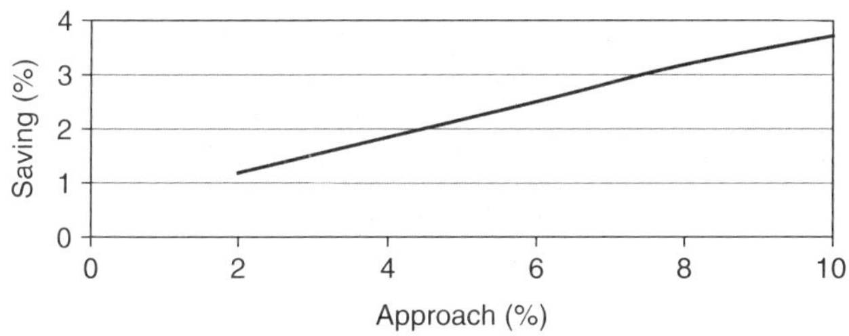

Figure 6.62 Savings in operating cost resulting from using advanced controllers in RO plants based on the assumed economic factors ($T_f = 25°C, C_f = 40$ kg/m^3).

actual feed pressure is maintained constant throughout the year, the approach from the optimum will vary according to the season of the year. The use of advanced controllers would enable the plant to run at optimum condition throughout the year, thus allowing a reduction in operating expenses.

It should be borne in mind that the cost saving potential depicted in Figure 6.62 is based on the assumed values of the unit costs of heating steam, electricity, and antiscalant chemical and does not include any potential saving in operating cost due to any saving in labor requirement.

6.5.5.4 Promising RO Automation Methods and Their Potential Cost–Benefit Ratios In conventional RO plant operation, the control system tends to maintain the flows and thus the recovery constant at the design value. Any change in the membrane flux, such as by temperature or fouling, is compensated by adjusting the feed pressure. However, the maximum specified feed pressure must not be exceeded, nor should too much fouling be tolerated. If the feedwater analysis changes such that the scaling potential increases, the system recovery has to be decreased. When using advanced control, all operating parameters such as pressure, flows, recovery, and pH are allowed to vary as long as they do not exceed the limits specified by the module manufacturer. Therefore, the operating point is allowed to move within a small range of operating parameters to maintain optimum operation.

Automation can contribute to reduction of water cost in RO desalination systems through optimization of the operation process. The areas of operation where advanced automation can be especially effective are described below.

6.5.5.4.1 Power Consumption Optimization of power consumption can be achieved by adjusting the recovery rate and flux rate according to fluctuation of feedwater temperature and salinity, product-water demands, variable power rate, and variability of power consumption among parallel RO units. In RO plants where blending of different quality-products is practiced, adjusting operating conditions to produce highest salinity, within required limits, usually results in lower power consumption. An example of optimization of power consumption of seawater RO process is shown in Figure 6.63.

The results demonstrate power consumption in RO system processing to be 35,000 ppm TDS seawater with a permeate recovery range of 35–55%. The power consumption values shown in the figure include feedwater supply, pretreatment, a high-pressure pumping system, permeate pumping, and auxiliary equipment. The calculations were made assuming, a high-efficiency (94%) pressure-exchange-type energy recovery device. The relations shown in Figure 6.64 will differ for different system configurations and site conditions. For example, long feedwater supply lines and pretreatment pressure losses will favor higher recovery rates. High feed salinity will tend to shift the optimum to lower recovery. The optimization will be more complicated in case of a two-pass system, and then optimization of both passes in respect of operating parameters must be considered. It is evident that optimization of operating parameters, using a smart control system, to consistently

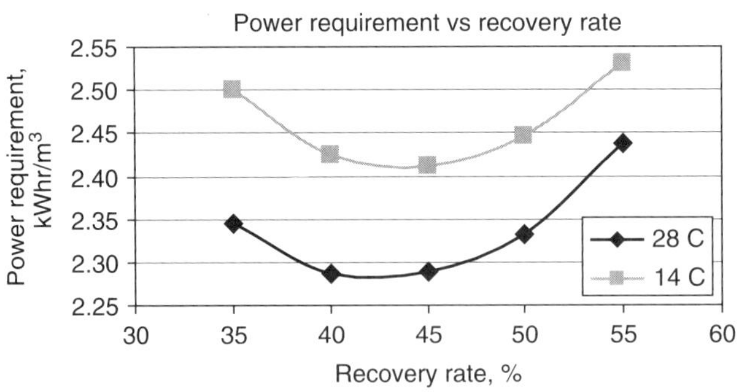

Figure 6.63 RO power requirement as a function of recovery ratio for two feedwater temperatures.

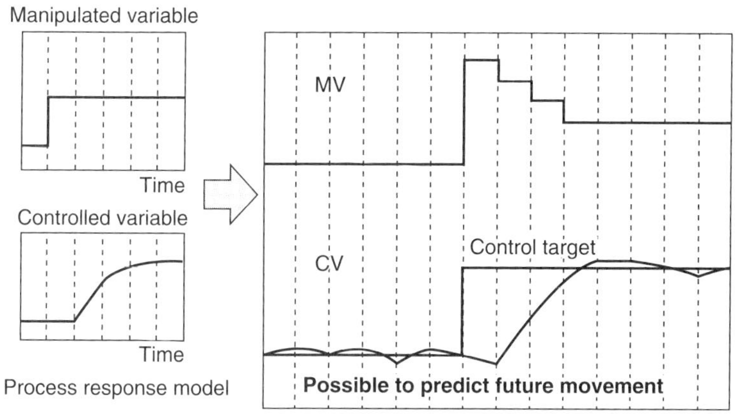

Figure 6.64 Action of an MPC on a controlled variable using a model.

operate at lowest power consumption is more important in large-capacity desalination plants. Expected power savings in RO system utilizing high-efficiency pumps and power recovery systems could be in the range of 0.05–0.20 kWh/m^3. For a system capacity of 200,000 m^3/day, reduction of power consumption by 0.1 kWh/m^3 could result in a \$400,000 per year saving of the operating expenses. The estimated cost of automation including instruments, PLC, control valves, programming, and installation is ~\$2,500,000. The additional cost for developing and implementation of a smart control program should not exceed \$150,000–\$200,000. Therefore, the return on investment would be very short. As it is expected that the optimization of operating parameters will be within a narrow range, it should not adversely affect membrane longevity or other operating expenses (frequency of cleaning or other maintenance operations).

6.5.5.4.2 Chemicals Consumption Optimization of chemicals consumption can be achieved by adjusting dosing rate according to conditions of the raw water and operating parameters of the RO unit. For example, the dosing rate of flocculant and polymer can be reduced during periods when concentration of suspended solids in the raw water is lower. The dosing rate of flocculant pumps can be paced according to raw-water parameters that affect the filtration/clarification process. The relevant parameters that are conveniently measurable online are raw-water turbidity, TOC, pH, and temperature. Correlation between those parameters, flocculant dosing rate, and quality of pretreatment effluent can be developed in the framework of a pilot study or during initial operating period of the RO plant. In addition, dosing rate of acid or scale inhibitor can be adjusted according to recovery rate and permeate flux rate.

6.5.5.4.3 Membrane replacement Frequency (or rate) of membrane replacement can be reduced by analyzing performance trends of membrane elements in operation and initiating preventive maintenance flushing or cleaning procedures.

Indirectly, automation can reduce the membrane replacement rate by helping to maintain more consistent feedwater quality and therefore ensure consistent membrane performance.

In this area of application, the computerized automatic control system is used to store operating data, calculate performance trends, and calculate projected performance at various operating conditions and ranges of performance of membrane elements. The current approach to correction of system performance through membrane replacement includes replacement of old elements with a new one in a whole train or stage. The replaced elements are being individually tested. Poorly performed elements are being discarded; the better ones are stored for future replacement. Utilization of a computerized system to calculate system performance, based on performance of elements in the system and utilization of the old elements in the inventory, can result in reduced use of new elements for replacement while maintaining required plant performance.

6.5.5.4.4 Labor and Maintenance Labor requirements can be reduced by automation. Currently, even some large RO systems, are remotely controlled. However, extent of labor reduction due to automation strongly depends on local regulations. At some locations, local regulations require the presence of operators even in fully automated plants.

The water cost in a typical large (~200,000 m^3/day) seawater RO system is ~\$0.70/m^3. The contributions to the water cost are as follows:

Product-Water Cost Component	Cost, \$/m^3
Capital cost, including land fee (25 years at 6.0% interest)	0.320
Electric power (\$0.060 $kWhr^1$)	0.210
RO membrane replacement (5 years membrane life)	0.035
Membrane replacement (7 years membrane life)	0.030
Chemicals	0.025
Maintenance and spare parts	0.036
Labor	0.030
Total cost	0.686

The cost contribution parameters that potentially could be optimized by more advanced automations are electric power, RO membrane replacement, chemicals, and possibly maintenance (frequency of membrane cleaning). The total contribution of these components in the example above is \$0.336. Assuming that it would be possible to reduce contribution of these components by 10%, this would result in a 200,000-m^3/day capacity system and a water cost saving of ~\$61 M over a period of 25 years of plant operation. By comparison, the total budgetary cost of such a project is in the range of \$200M. Comparing this to potential reduction of operating cost due to better automation provides justification for a more sophisticated automation equipment to be implemented.

6.5.6 Conclusions

The following major conclusions of this section are:

- A basic premise of this study is that any level of automation or IMCA in general must be economically justified, and consideration of IMCA system reliability must also be included.
- As opposed to the enormous amount of published information on control systems and their advancement, those on cost–benefit analysis of their implementation are very few in general, and nearly nonexistent for desalination plants in particular.
- A critical review of the status of cost–benefit analysis of IMCA systems with emphasis on their use in MSF and RO desalination, was performed and presented.
- An explicit methodology for such analysis was developed and presented.
- To implement this methodology, simplified models for predicting the performance of MSF and RO plants were developed and successfully implemented.
- Specific examples of IMCA system cost–benefit ratios for large RO and MSF plants were calculated.
- Suggestions were made for some promising ways where automation could be implemented in MSF and RO plant design and operation to reduce product water cost.
- Proper saving and analysis of historical plant data can lead to product cost reduction in ways not always possible to discover by conventional modeling.
- The use of advanced controllers in MSF and RO plants may result in economic saving in the operating expenses [steam consumption (MSF), electrical energy consumption (MSF and RO), and chemical consumption (MSF and RO)].
- In the examples analyzed, based on unit cost assumptions, the following values were estimated: steam cost = \$3.5/t, electricity cost = \$0.06 kWh, average cost of chemicals \$1.0/kg^{-1}, direct savings of \$66,000 per year (MSF) and \$13,000 per year (RO). These values can be achieved for each migd of plant installed capacity if the advanced IMCA system results in a deviation from optimum that is at most 5%. The real savings depend on the actual operation strategy of the existing plant as well as the extent of seasonal variation in seawater temperature and salinity in plant site.
- The economic feasibility of replacing conventional PID controllers with advanced controllers depends on the results of a cost–benefit analysis to be carried out on the particular plant under consideration. This analysis should take into consideration both the operating condition of the plant as well as the purchase and installation costs of the controller hardware and software.
- The direct cost–benefit ratio for large plants (say, 200,000 m^3/day) is estimated to be ~30 for MSF and ~23 for RO processes, indicating that advanced IMCA systems have a very desirable financial outcome. That ratio becomes even higher when the indirect benefits are added.

- It is well recognized in process IMCA system design and implementation that there are many intangible benefits to IMCA system improvement that may actually exceed those simply calculated from the resulting higher product output and quality; more attention to such robust design is thus recommended.

6.6 IMCA SYSTEMS RECOMMENDED FOR RO AND MSF DESALINATION PLANTS

6.6.1 Introduction

Based on the critical review, analysis of the information and the cost–benefit analysis described in Section 6.5, we make in the following several general recommendations for IMCA for RO and MSF plants. We note, however, that concrete recommendations for specific plants can be made only after the plants have been analyzed individually with respect ot the prevailing technology, prices, and plant use at that time. The report provides adequate guidance for performing such analyses.

Because of concern for producing water at the lowest possible cost, it is becoming increasingly important to operate the desalination plants as close as possible to the optimum at all times.

In many application areas of automation, such as power stations, chemical, steel, and pulp-and-paper industries, modern control schemes were implemented together with new microprocessor-based I&C equipment, this development is in a very early stage for desalination plants. Modern control modules, however, will have a measurable effect on the economy of desalination plants.

As stated, most desalination plants currently in operation use simple conventional PID controllers. The use of simple PID controllers, which can maintain the stable operation of the plant, certainly overlooks the complex interactions that may exist within the plant. MSF and SWRO plants are nonlinear multivariable systems, and an in-depth investigation of the different interactions between inputs and outputs of the plant is necessary to optimize the selection of the control structure.

In this section, we identify the desirability or prospects of advanced control systems for desalination plants of different capacities by investigating their economic feasibility through a cost–benefit analysis of replacing existing PID controllers with advanced ones to existing desalination plant under operation.

6.6.2 Model Predictive Advanced Controller (MPC) a Recommended Advanced Controller

6.6.2.1 Rationale for Selecting MPC Controllers Because of the interactive nature of MSF and SWRO process variables where a change in one affects more than one related variable, it would be ideal to have a controller that is able to combine the operation of a number of single-loop controllers. At the same time, this controller should be able to choose, intelligently, a comfortable selective group of those variables whose manipulation will drive the object variable(s) to its (their) optimum targets.

Interaction among process variables is a very common situation encountered in many process plants. Often the selection of variables to be driven to their limits and extent of their manipulation is left to the subjective judgment consequent of experience of the operator-in-charge. The selection is essentially a tradeoff between variables to be driven to their limits. This is largely due to the complexity of interactions among variables. This judgment, while not wrong from a process or operational viewpoint, may not be in line with the company's objectives or market demands. Every operator recognizes the interaction among process variables. However, it is the impracticality of negotiating these variables to maintain an optimum condition at all times that forces operators to maintain the variables at a "comfortable" location, away from their constraints. A direct result of this is that the operation is never at its optimum point. This is where multivariable controllers becomes useful.

6.6.2.2 Brief Introduction to MPC To drive a process to its operating target, we have to know how the process or plant will respond to each step change. A behavioral pattern of the process is therefore necessary to predict "process status" for each change. This pattern is represented by a set of mathematical equations and is known as the *process model*. Using this model, the controller makes a decision in real time, the extent of move required for a manipulated variable that will move the process as close as possible to a reference trajectory. This is done at each control step and for the entire horizon. It is this predictive nature of the controller that enables it to handle constraints, and feedforward and allows us to incorporate abilities to handle noise *etc*.

Model predictive control algorithms are algorithms that compute a sequence of manipulated variable adjustments to optimize future behavior of a plant. They use a model to evaluate how control strategies will affect the future behavior. They can deal with explicit constraints. After finding a good strategy, MPC pursues a strategy for one control step and then reevaluates its strategy based on the plant's response.

Simultaneously, MPC copes with amplitude constraints on inputs, outputs, and states. It is in effect a controller having a process response model inside, which enables it to predict the optimum manipulating output to elicit the desired process response. With this information, the controller drives a manipulated variable into steps that will ultimately drive a controlled variable to its target. All the while, it checks for constraints, both on manipulated and controlled variables, and any others that the operator may choose to define and specify. Figure 6.64 shows how a model predictive controller is able to drive a controlled variable to its target by using a process model. A model predictive controller is also able to handle processes with long dead time, or inverse or integral response. A feedback mechanism enables the controller to constantly update its prediction model (Figure 6.65). It is typically suited to processes where large interactions among controlled and manipulated variables are witnessed. Figure 6.66 shows how an MPC moves a manipulated variable

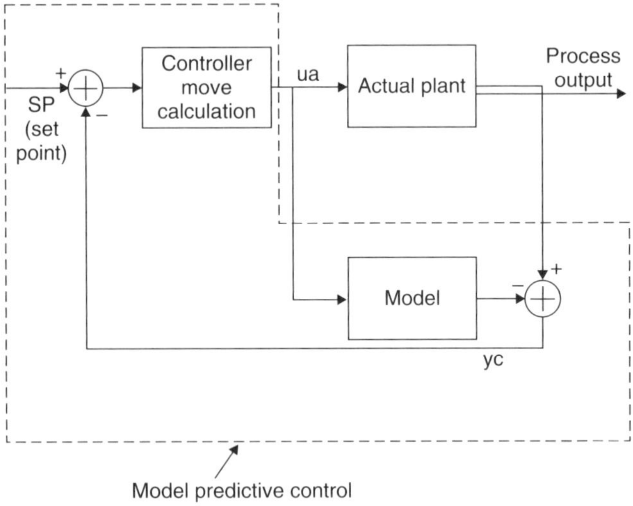

Figure 6.65 Block diagram of a model predictive controller.

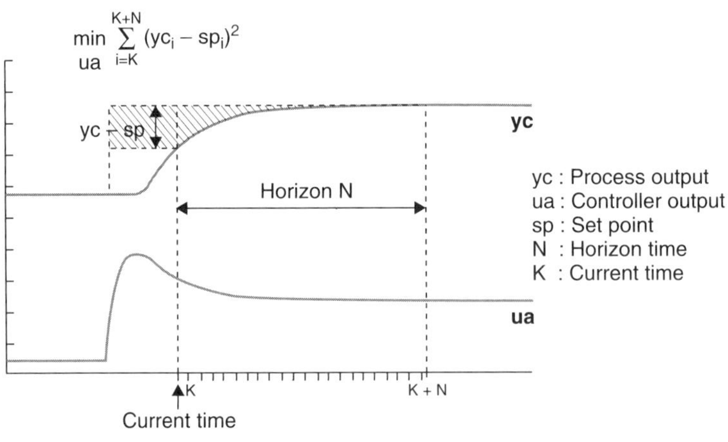

Figure 6.66 Aim of a model predictive controller.

so that the controlled variable moves along a reference trajectory to minimize the error between the target and actual value, at each execution step.

The principal aims of an MPC are to

- Drive variables in a process to their optimum targets, keeping in mind the interaction among variables
- Effectively deal with constraints
- Respond quickly to changes in optimum operating conditions
- Achieve economic objectives

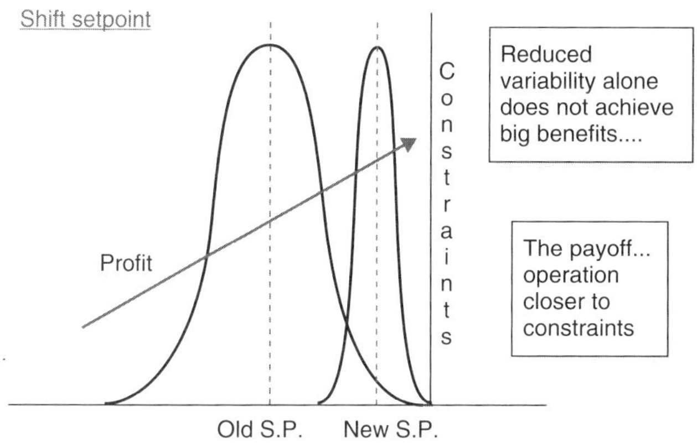

Figure 6.67 Reduced variability allows operation closer to constraints.

What are the advantages of using an MPC besides simply driving multiple variables to a set target? What is the monetary attraction to implementing MPC?

The real benefit from MPC is obtained not only by reducing variations but also by operating closer to constraints, as illustrated in Figure 6.67.

The purpose of MPC and an optimization feasibility study is to determine the envelope of plant operability and control strategies that will drive the plant to run at the optimum, maximum yield, or conversion and/or maximum throughput, while simultaneously honoring process, equipment, and operating constraints.

In a crude oil unit, for example, high feed rates are commonly encountered. An incremental change in the yield of profitable streams often generates very large annual benefits. However, the benefits achieved on a particular unit depend on that unit's operating objectives and the plant environment.

The major benefits will come from some or all of the following general areas:

- Increased throughput, to be achieved by better control of process parameters
- Enhancing the unit performance by
  - *Reduced energy costs*—lower fuel, steam, and electricity usage and improved unit efficiencies.
  - *Easier and smoother operation*—less operator intervention. This will reduce their stress while more rapidly returning the process to normal operations and provide a faster and seamless transition between different feed rates.
  - *Increased operating safety*—this is in direct proportion to uptime of APC. Higher uptimes signify less operator intervention in the process. This reduces chances of error, making the process inherently safer.

6.6.2.3 APC Selection Considerations To obtain the full benefits from APC, one should pay special attention to aspects of an APC package that provide for high uptime and easy maintainability. One aspect of APC that is repeatedly ignored is the requirement to maintain it [94]. It should be borne in mind that advanced controllers, once they are designed and implemented, also need to be maintained, just like any other piece of equipment such as a pump or compressor [94]. How easy a controller is to handle, maintain, or alter is determined by important factors such as

- Robustness
- How easily a controller adapts itself to changing process conditions, which is directly influences extent of intervention required
- How easily it can be maintained, or the model updated, or returned
- How easily it can be remodelled, if necessary, for gross excursions in process conditions from what it is designed for
- How one validates a model developed
- The peripherals that are required
- How easily it can be integrated with existing equipment

6.6.2.4 System Configuration of a Typical MPC The principal components of an overall integrated system (Fig. 6.68) will be

- One PC running on MS Windows for model building
- One PC running on MS Windows that will act as a control station

The difference in system configuration will be due to the operating system of the existing DCS, the communication network that is used by DCS, and the corresponding communication protocol that needs to be established between DCS and the control station (see Fig. 6.68) [94].

To solve this problem, the control package usually provides an interface that is compliant with an OPC (OLE for process control systems) standard interface. OPC is a server running on Windows, which can be connected to a variety of process control systems and provides an OPC client with process data via an OPC interface. The OPC client can acquire and define process data to and from process control and the control station. While the control station is connected to an Ethernet bus, OPC is connected to both Ethernet via Ethernet card and to the control bus.

6.6.2.5 Advanced Identification and Analysis (AIDA) The first and foremost task in designing a controller is to develop a model. An *AIDA Model* is the behavioral representation of a process. It represents process behavior to specific process inputs. AIDA establishes, mathematically, the relationship between input (manipulated) and output (controlled) variables. This is crucial in defining the scope of the control algorithm. It should represent adequately all aspects of system behavior. It should also be robust with sufficient flexibility to face uncertainties in the process.

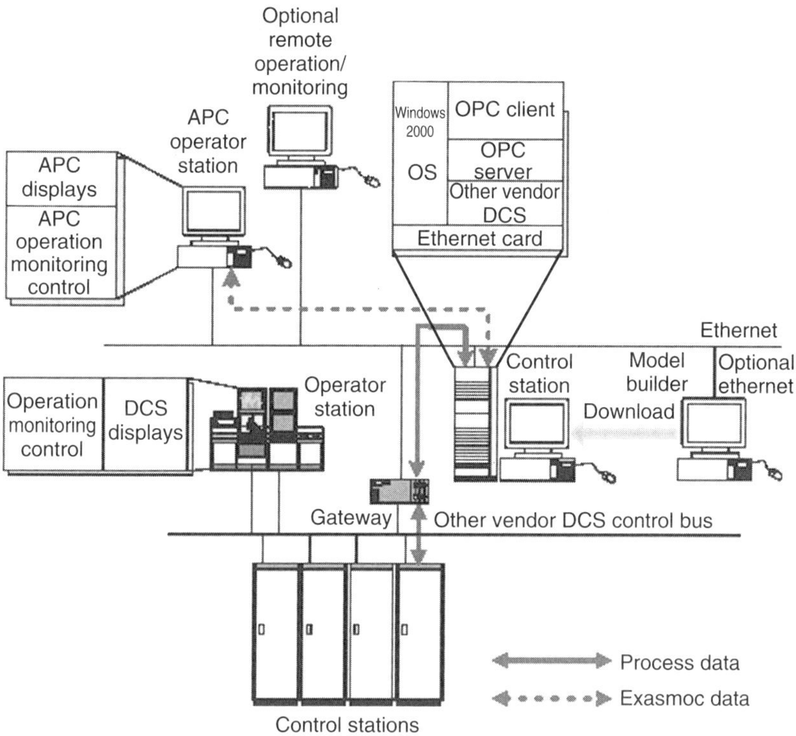

Figure 6.68 Typical system configuration with other vendor DCS [94].

AIDA is a software package used to estimate a linear dynamic model for a process unit. Deriving a dynamic model for a desalination plant on the basis of plant data is called *identification*. The process data are collected at a given sampling period, typically one minute. Collected data include the setpoint of the plant operating "handles" (the process inputs, or manipulated variables). These include the plant operating variables that are affected by changes in the process inputs and for which there are regulation or optimization objectives in the future advanced control scheme (the process outputs, or dependent variables), and a number of associated variables that provide information on the current plant conditions and helps obtain a more accurate model (the measured disturbance variables).

AIDA generates a linear, dynamic, parametric model that could be first-order, second-order, or ramp; that is, plant data are used to establish the relationship between various interacting variables. It fits a discrete-time transfer function model and then converts it into a continuous (Laplace) transfer function. It then uses white noise to reflect unmeasured disturbances.

A plant model is developed by observing a plant's response and behavior to a step or impulse change made to the input. The response and mathematical equation

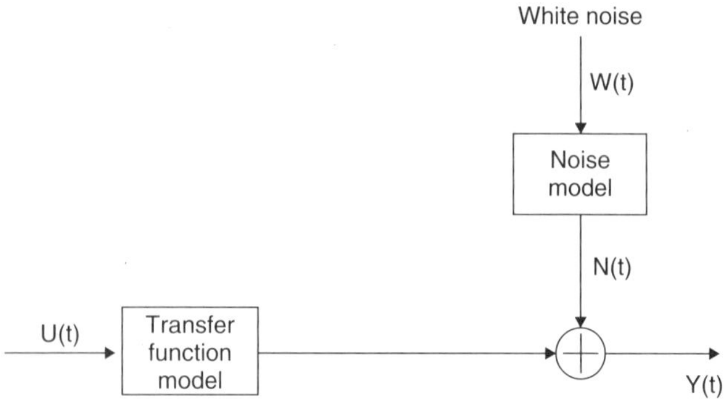

Figure 6.69 Block diagram of AIDA model generation [94].

representing this response becomes the process model. Data collected from the plant to indicate response are never free from noise. Therefore, a mathematical equation striving to represent this response inadvertently incorporates noise in its formula as if it were a response. The source of this noise could be any process noise or measurement noise (Fig. 6.69).

In addition, in a multivariable interacting environment, it is practically impossible to model the effect of all influencing input components. Therefore, we need a model that is able to adequately represent process response while simultaneously rejecting uncertain components from data collected.

6.6.2.6 MPC Controller The MPC controller requires a PC, which is an offline design–simulation package. It is a computer-based package designed for synthesis and simulation of linear, multivariable, and model predictive control. It allows the process control engineer to design and test the MPC controller prior to implementation on an on-line system. The other computer is the online package.

The different steps of MPC controller development are

- Configure process models
- Configure controller objectives
- Set tuning factors
- Generate.def (DCS tag definition file) and. con file (controller definition file)
- Test controller by simulating
- If performance is adequate, download to the online computer.
- In the online system, it is possible for the user to configure the controller by setting up the process model. As opposed to the usual "blackbox representation of a process for other MPC's, this system visualizes the process as a gray box". This concept allows the user to develop the process block by block. The user has the option of allowing the system to select the model from AIDA or develop it themselves.
- It is possible to define, in addition to CVs, MVs, and DVs, another variable called the *intermediate variable*. This variable is used when a measure of

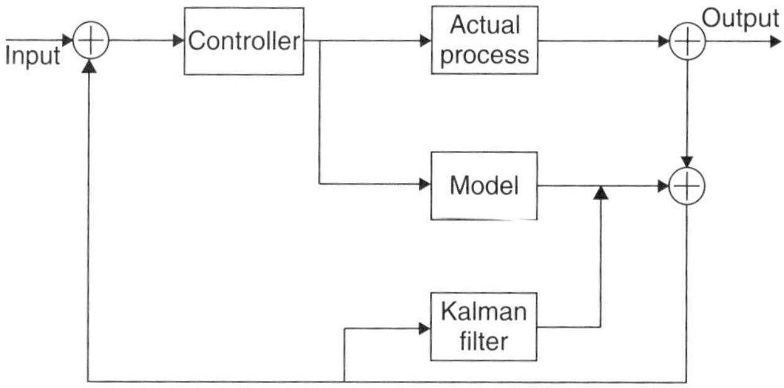

Figure 6.70 Block diagram of Kalman filter in control scheme.

that variable is indicative of a change of controlled variable in advance (e.g., column top temperature against top quality)

- Controller objectives can be easily set as *maximize*, or *minimize*. The optimizer within the online computer will automatically drive the variables toward this target without violating any constraints [94].
- Update the model online using a Kalman filter. This updates online the error model, called an *observer model*, providing for more robust control Figure 6.70 is a block diagram of a Kalman filter incorporated in a control scheme.

6.6.2.7 Cost Estimation for Advanced Controller This section has been prepared to serve as a general guide for estimating the total cost of the selected MPC control system, realizing, of course, that actual prices change with time. The section begins with a discussion of the cost of hardware components, followed by guidelines for estimating the costs of software and other nonhardware expenses [95].

Equipment making up the control system is separated by function and is installed in two different work areas of a processing installation switch as a desalination plant. Equipment that the operator uses to monitor process conditions and to manipulate the setpoints of the process operation is located in a central control room. From this location, the operator can view information transmitted from the processing area and displayed on a cathode ray tube (CRT), and change control conditions using a keyboard. The controlling portions of the system, which are distributed at various locations throughout the process area, perform two functions at each location:

1. Measurement of analogue variable
2. Discrete inputs and generation of output signals to actuators that can change process conditions

Input and output signals can be both analog and discrete. By means of electrical transmission, information is communicated between the central location and the remotely located controller locations. The communication path is either a cable from each remote location to the central station, or a single-cable data highway interfacing all the remote stations. The parts of the system are always sold as a package. Suppliers do not sell only the remote portions or only the centrally located portion. Since the parts function together as a system, they must be completely integrated and tested as a system. Because the components of the system communicate over a shared data highway, no change is required to the wiring when process modifications result in revisions to the control system of the plant (Fig. 6.68).

6.6.2.8 Distributed Control System (DCS) Configuration As shown in Figure 6.71, the operators' console in the control room [high-level operator interface (HLOI)] can be connected through a shared communications facility (data highway) to several distributed system components. These components can be located either in rooms adjacent to the control room or out in the field. Such distributed local control units (LCUs) can also be provided with some limited amount of display capability (low-level operator's interface, LLOI) [94].

A specific DCS for a particular plant is configured from standard building blocks that are marketed by most DCS suppliers. Figure 6.71 illustrates the categories of components that are available for configuring various DCS systems. These components include the CRTs and the associated console in the central control room, the interface packages serving the interconnections with other digital systems such

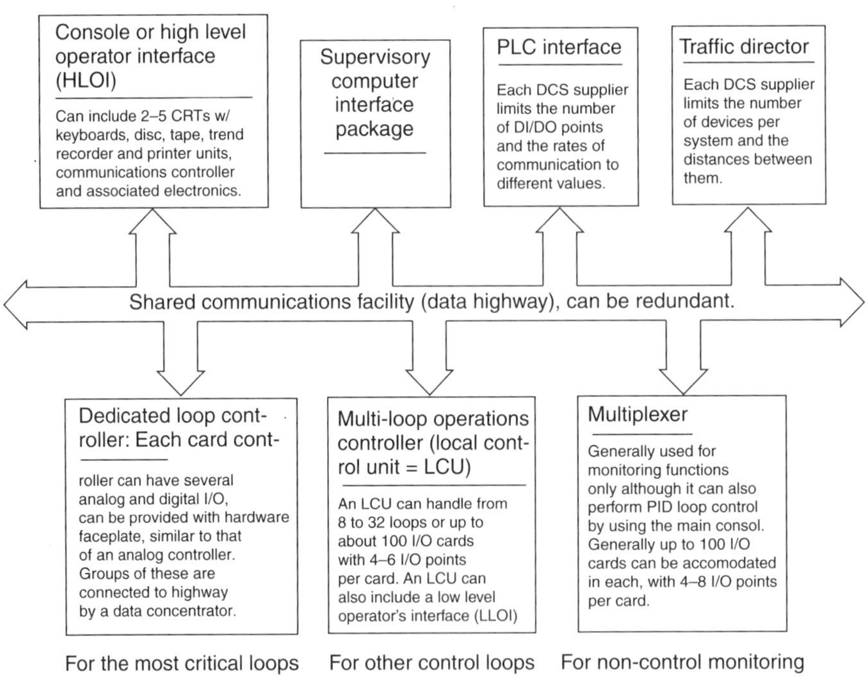

Figure 6.71 The main components of a distributed control SWRO system (DCS) and the functions and limitations of these building blocks.

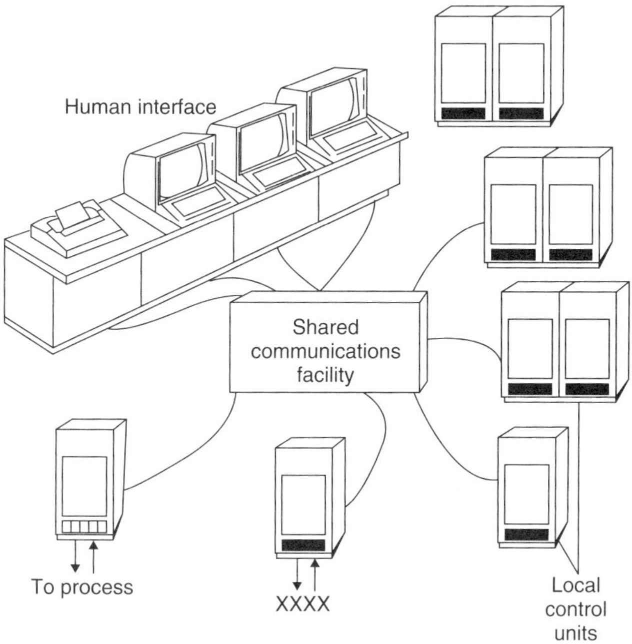

Figure 6.72 In a typical DCS, the panel board and consoles are eliminated, and communications occur over a shared data highway.

as PLCs and supervisory computers, and a data highway that connects the various parts of the system and is operated by the traffic director.

The process I/O signals are connected to logic control units (LCUs), which can be of various levels of cost and reliability. The I/O for the most critical control loops, the one that should continue functioning even if the central processor or the data highway fails, should be connected to dedicated card controllers (Figure 6.72).

6.6.2.9 Hardware Components The hardware components of a DCS can be grouped into

- Items related to the operator's console (CRTs), archiving disk/tape units, printers, engineering station
- Items related to the data highway (communication traffic director (T), interface for PLC, MIS computer
- Field-mounted items (I/O files, dedicated loop controllers, multiloop unit controllers (UC), multiplexers/data acquisition units (ACQ)
- Field instrumentation, if required

6.6.2.10 Software Components Nonhardware expenses may include

- Software development costs for configuration, special displays, etc.

- System integration, factory acceptance testing (FAT), field installation and checkout, site acceptance testing (SAT), documentation, and support devices, including project management, training, and maintenance

6.6.2.11 DCS Configuration Definition The first step in the preparation of a DCS estimate is to define and size the main components of the system. To serve that goal, a block diagram representation of the DCS is prepared (see Fig. 6.73). A single operator console with three CRTs, two printers, and one disk/tape unit for archiving is used. Three multiloop control units (CUs), three data acquisition multiplexers (ACQ), one management information system (IMS) computer, and one interface for communication (COMM) are used. A data highway is used to interconnect the CUs and the ACQs. Three process areas are depicted:

- The desalination process
- The seawater intake and pretreatment
- The posttreatment

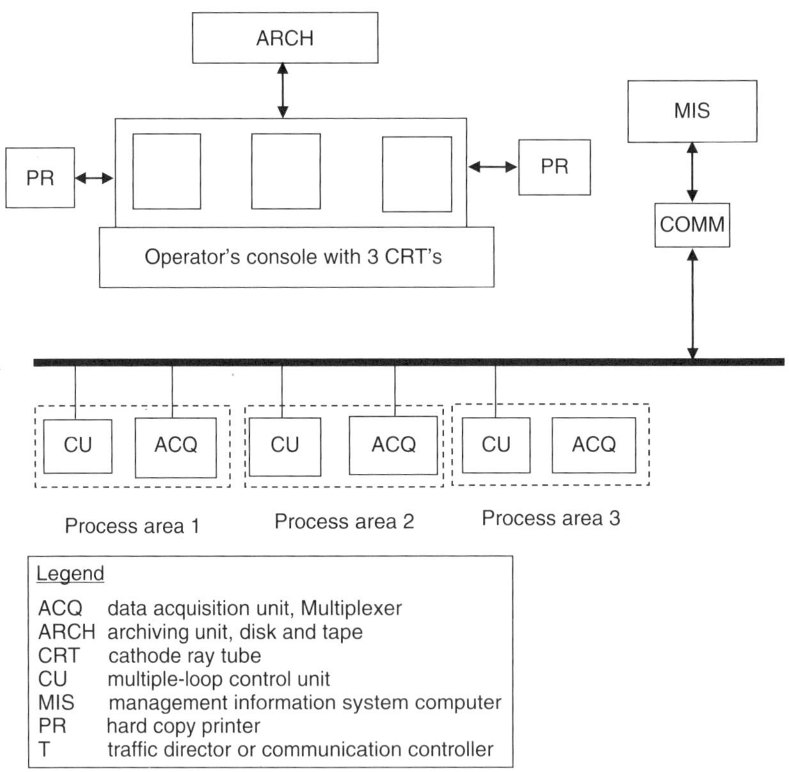

Figure 6.73 Block diagram representation of the DCS for MSF and SWRO plants.

Table 6.8 I/O Requirements for MSF and Seawater SWRO Plants

	MSF Plant					SWRO Plant				
A/D, I/O Type	PSWR Process Area 1	PSWR Process Area 2	PSWR Process Area 3	Total	Cost, $	PSWR Process Area 1	PSWR Process Area 2	PSWR Process Area 3	Total	Cost, $
Analog input	32	32	32	96	14,000	32	32	32	96	14,000
Analog output	32	32	32	96	23,000	32	32	32	96	23,000
Digital input	80	80	80	240	21,000	30	30	30	90	8,000
Digital output	80	80	80	240	21,000	30	30	30	90	8,000
Total	—	—	—	—	97,000	—	—	—	—	53,000

Source: Hertanu and Lipták [95].

Its own control and a data acquisition unit (CU, ACQ) serve each processing area. The I/O requirements for each processing area are assumed to be as listed in Table 6.8.

6.6.2.12 Cost Estimate The next step in the preparation of the cost estimate is to obtain price lists from a number of DCS suppliers. An example of one price list is given by Hertanu and Lipták [95] (see Table 6.9).

From this price list, we made estimates from Tables 6.10 and 6.11, for the MSF and SWRO plants of large size.

6.6.2.12.1 Hardware Three basic groups of hardware and related items are included in the price list:

Operator's Console and Related Items. The following items are listed:

Operator's console with three real-time clocks (RTCs) and keyboards: $70,000

Archiving unit with 40 GB hard disk and tape drive: $30,000 each

High-speed hardcopy printer: $3000 each

An engineering station with Intel Pentium 4 processor, 40 GB hard disk, CD-SWROM drive, colour monitor, keyboard, laser printer, provided with all software needed for offline configuration and programming: $30,000.

Communications-Related Items. The next step in the preparation of the estimate is to determine the type of communication network required. The traffic director (T in Fig. 6.71) and the interface devices (COMM in Fig. 6.71) appropriate for the number and type of devices supported by the data highway have to be specified. For our case, the following estimates can be obtained from the price list in Tables 6.10 and 6.11:

Trafic communications director (T): $8000

Interface unit (COMM): $9000

Table 6.9 Sample Price List for DCS Components

Description	Price Range, $
1.0 *Hardware*	
1.1 ContSWROl unit, multiloop (CU)	25,000–72,000
1.2 Data acquisition unit, multiplexer (ACQ)	14,000–18,000
1.3 Operator's console	55,000–115,000
1.4 Archiving unit, disk, and tape (ARCH)	30,000–40,000
1.5 Hardcopy devices, printers (P)	1,000–3,000
1.6 Interface unit for MIS (COMM)	8,000–1,000
Communication unit, traffic director (T)	8,000–11,000
1.8 I/O file for 6 I/O cards	2,500
1.9 Analog input card (16 points)	1,400
1.10 Analog output card (8 points)	1,400
1.11 Digital I/O card (16 points)	700
1.12 Engineering station	30,000
2.0 *Software*	
2.1 Configuration—MSF (database, displays, alarms, control loops)	165,000
2.2 Configuration—SWRO (database, displays, alarms, control loops)	165,000
2.3 Software development (MSF—MPC)	255,000
2.4 Software development (SWRO- MPC)	200,000
3.0 *Integration factory acceptance test*	
3.1 Space and utilities	800–1200/day
3.2 Technical support	120–160 h^{-1}
4.0 *Site acceptance test* (travel and living expenses; not included)	120–160 h^{-1}
5.0 *Documentation* (full set of printed manuals or CD-SWROM)	1500–2500
6.0 *Support Services*	
6.1 Project management	150 h
6.2 Training (can be computer-based)	
6.2.1 At manufacturer facility (priced per student per day)	400–800
6.2.2 Instructor at site	1000–1200/day^{-1}
6.3 Maintenance (varies with DCS size and complexity)	5000–7500/month
6.4 Software upgrading or troubleshooting	—[a]
6.5 Hardware warranties	—[a]
6.6 System installation	—[a]
6.7 System startup and loop checkout	—[a]

[a] Varies from project to project, but can amount to 20–50% of total cost.
Source: Hertanu and Lipták [95].

Table 6.10 Cost Estimate for Model Predictive Controller (MPC) for Large MSF Unit

Description	Material			Labor			Total, $
	Quantity	Unit Price, $	Total, $	Hours	$/h	Total, $	
1.0 Hardware							
1.1 Contol unit	3	50,000	150,000				150,000
1.2 Data acquisition unit	3	15,000	45,000				45,000
1.3 Operator's console	1	70,000	70,000				70,000
1.4 Archiving unit	1	30,000	30,000				30,000
1.5 Hardcopy drive	2	3,000	6,000				6,000
1.6 Interface unit	1	9,000	9,000				9,000
1.7 Communications	1	8,000	8,000				8,000
1.8 I/Os			97,000				97,000
1.10 Engineering station	1	30,000	30,000				30,000
Total hardware cost							*445,000*
2.0 Software							
2.1 Configuration			15,000	1,500	100	150,000	165,000
2.2 Software development			5,000	2,500	100	250,000	255,000
Total software cost							*420,000*
3.0 Other expenses							
3.1 Integration technical support				80	150	12,000	12,000
3.2 Factory acceptance test				120	150	18,000	18,000
4.0 Field site testing							
4.1 Field installation & checkout				80	120	9,600	9,600
4.2 Site acceptance test				80	150	12,000	12,000
5.0 Documentation							6,000
6.0 Support services							
6.1 Project management				200	150	30,000	30,000
6.2 Training (10 people)							10,000
6.3 Maintenance (12 months)							60,000
Total MSF project cost							*1,022,600*

Field Mounted Items. The control units (CUs) and the multiplexers (ACQ) are mounted in the field. The CUs are assumed to be fully packaged and wired. The ACQs were specified to be provided with integral power supplies and integral I/O in the same cabinet. The unit costs of the components are as follows:

Control unit (CU): $50,000

I/O requirement of CU: $133,000

Data acquisition/multiplexer: $15,000

6.6.2.12.2 Software Cost Estimate Estimating the software development time required for the project is probably the most difficult task. The person-hour

Table 6.11 Cost Estimate for Model Predictive Controller (MPC) for Large SWRO Unit

	Material			Labor			
Description	Quantity	Unit Price, $	Total, $	Hours	$/h	Total, $	Total, $
1.0 Hardware							
1.1 Control unit	3	50,000	150,000				150,000
1.2 Data acquisition unit	3	15,000	45,000				45,000
1.3 Operator's console	1	70,000	70,000				70,000
1.4 Archiving unit	1	30,000	30,000				30,000
1.5 Hardcopy drive	2	3,000	6,000				6,000
1.6 Interface unit	1	9,000	9,000				9,000
1.7 Communications	1	8,000	8,000				8,000
1.8 I/Os			53,000				53,000
1.10 Engineering station	1	30,000	30,000				30,000
Total hardware cost							*401,000*
2.0 Software							
2.1 Configuration			15,000	1,500	100	150,000	165,000
2.2 Software development			5,000	1,500	100	150,000	200,000
Total software cost							*365,000*
3.0 Other expenses							
3.1 Integration technical support				80	150	12,000	12,000
3.2 Factory acceptance test				120	150	18,000	18,000
4.0 Field site testing							
4.1 Field installation and checkout				80	120	9,600	9,600
4.2 Site acceptance test				80	150	12,000	12,000
5.0 Documentation							6,000
6.0 Support services							
6.1 Project management				200	150	30,000	30,000
6.2 Training (10 people)							10,000
6.3 Maintenance (12 months)							60,000
Total SWRO project cost							*923,600*

requirements of software preparation are a function not only of the number of I/O and the complexity of operations but also of the experience of the programming personnel. The more experience the engineering firm has with similar past projects, the more accurately can the required software effort be projected. It has to be broken down into specific programming task units related to the preparation of regulatory, interlock; alarm, etc. software packages, and each unit must be multiplied by the associated person-hour requirement.

In the case of large MSF and SWRO plants, the total software preparation cost was estimated at $420,000 and $365,000, respectively.

6.6.2.12.3 Other Expenses After the hardware components of the DCS have been procured, the system must be integrated and made operational before it can be used. The location, space, utility, and labor requirements of the integration effort vary with project size and complexity. The cost of integration is usually estimated based on a specific integration plan, listing all the tasks that are to be performed during the integration process. For our case, it was estimated that this task would cost $12,000 for the MSF and SWRO cases.

The factory acceptance test (FAT) seldom represents an additional cost for the project, as the same engineering firm, which prepared the software, performs it and the person-hours are charged against the software preparation budget. The FAT tests should be started by testing the integrity of the hardware components and their connection paths. This is followed by loading both the standard and application software packages and verifying the functional operation using the application programs and their database. Field checkout and installation costs for this project were estimated based on providing supervision only at $9600 and consisted of supervising the receiving, unpacking, installing, wiring, and powerup of the DCS components. The material costs for cables, wires, and mounting hardware were included in the DCS hardware estimate.

The *site acceptance test* (SAT) follows installation and precedes the actual plant startup. During SAT, all required DCS functions are verified. The cost of this task was estimated in Tables 6.10 and 6.11 as $12,000.

The estimate also includes $6000 for documentation. The project documentation should include manufacturer's cataloges and manuals for the relevant hardware, software, installation, site preparation, and maintenance guidance. It also includes all engineering drawings, validation documents, and test plans, procedures, and results that were specifically prepared for the project.

The total cost estimate in Tables 6.10 and 6.11 also includes $31,200 for project management, $6000 for training operators, and $60,000 for the first year of maintenance.

The additional cost of water due to the estimated capital and operating expenses of a newly installed DCS system on an existing desalination plant depends on the remaining lifetime of the plant, the capacity of the plant, and the technology used (whether MSF or SWRO). Using the data in the tables above, the additional cost of water can be estimated as shown in the following section.

6.6.2.13 Added Cost of Water Due to DCS Installation Installing an advanced controller is associated with capital and operating expenses as shown in Table 6.12. This results in an added cost to the original water cost. The capacities of the small, medium, and large MSF and SWRO plants are defined in Table 6.13.

The amount of added cost depends on a number of factors:

- The remaining lifetime of the desalination plant
- The actual cost of the DCS system equipment and installation

Table 6.12 Estimated Capital and Operating Costs of DCS Systems for Different Sizes of MSF and SWSWRO Plants

	Capital Cost, \$		Operating Cost, \$/year	
Plant Size	MSF	SWSWRO	MSF	SWSWRO
Large	962,600	862,600	60,000	52,000
Medium	721,000	647,000	43,000	39,000
Small	480,000	430,000	29,000	26,000

Table 6.13 Capacities of Small, Medium, and Large MSF and SWRO Plants

Plant Size	MSF, (m^3/day^{-1})	Seawater SWRO, m^3/day
Small	400	400
Medium	4,000	4,000
Large	40,000	10,000

- The cost of maintenance, upgrading, troubleshooting, etc. of the hardware and software

The added cost $\Delta\Phi$ can be estimated from the following relation:

$$\Delta\Phi = \frac{i(1+i)^n}{(1+i)^n - 1}\frac{\text{CAP}}{M_d(0.9)(365)} + \frac{\text{OPER}}{M_d(0.9)(365)} \tag{6.91}$$

where CAP = capital cost of advanced control system, \$
OPER = operating cost of advanced control system, \$ per year
i = interest rate
n = number of years remaining in lifetime of desalination plant, years
M_d = nominal capacity of desalination plant, m^3/day
(0.9) = plant availability
(365) = number of days per year

The added cost of water $\Delta\Phi$ is calculated for MSF and SWRO plants with large, medium, and small capacities using Equation (6.91). The results of the calculations are plotted in Figures 6.74–6.76 as the added water cost versus the number of years remaining in the lifetime of the plant. For large and medium MSF plants, Figure 6.74 shows that the added cost of water decreases with the number of remaining years and that the cost increase is substantially lower for plants of large capacity compared with plants of medium capacity. The reason is that the cost of DCS equipment does not increase proportionally with plant capacity but increases

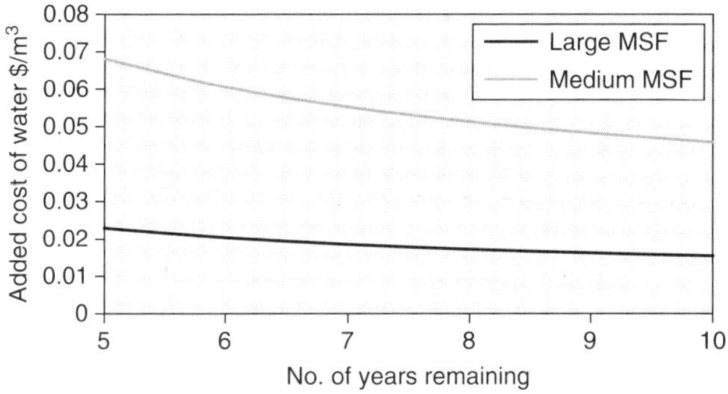

Figure 6.74 Added cost of water for large and medium MSF plants.

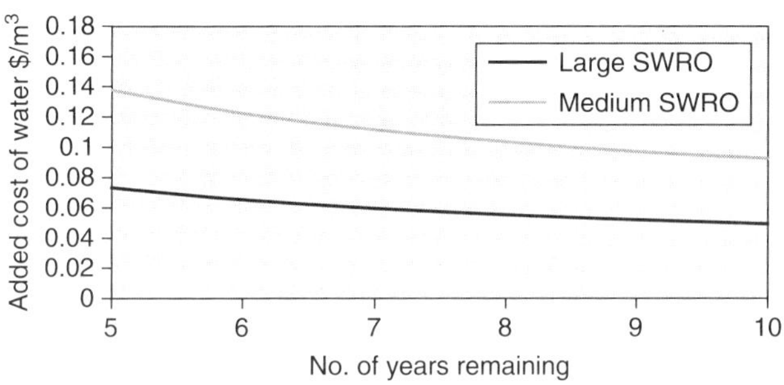

Figure 6.75 Added cost of water for large and medium SWRO plants.

rather slowly. For large MSF plants, the added cost varies between \$0.018 and \$0.022/m^3, and for medium plants it varies within \$0.05–\$0.07/m^3.

For large and medium SWRO plants (Table 6.13), the trend is similar to that of the MSF plants except that the added cost for SWRO plants is larger than for MSF plants. For large SWRO plants, the added cost ranges between 0.05 and 0.08 \$/m^{-3} and for medium plants, between 0.1 and 0.15 \$/m^3.

The added cost for small MSF and SWRO plants is shown in Figure 6.76, where it can be seen that the added cost for both types of plant can be substantial, ranging from \$0.8 to \$1.15/m^3 for small MSF plants and from \$0.7 to \$1.2/m^3 for small SWRO plants.

6.6.2.14 Benefit Due to DCS Installation As explained in detail earlier, the economic benefit resulting from operation with improved control stems from two sources:

1. Operation with variance reduction
2. Operation at optimum values of controlled variables

The economic benefits from each of these sources are estimated below.

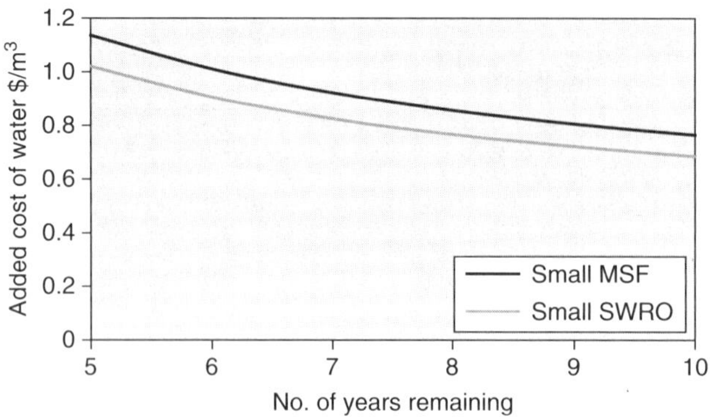

Figure 6.76 Added cost of water for small MSF and SWRO plants.

The total savings that can be accrued from using advanced controllers in place of the conventional PID controllers in MSF plants were estimated by calculating both parts of the savings, namely, the savings due to the reduction in the variance of controlled variables and the savings due to operating close to the optimum. Thus, the total savings can be expressed as

$$\Delta\Phi = \Delta\Phi_1 + \Delta\Phi_2 \tag{6.92}$$

Let us define the potential percentage saving that can be achieved by replacing PID controllers with advanced controllers by

$$\text{Saving }\% = \frac{\Phi - \Phi^*}{\Phi^*} \times 100 \tag{6.93}$$

The saving that can be achieved is expected to depend on the approach of the value of the actual plant controlled variables from the optimum operating condition. Let us define this *percent approach* as

$$\text{Approach }\% = \frac{x - x^*}{x^*} \times 100 \tag{6.94}$$

where x is the value of a controlled variable and x^* is the corresponding optimum value.

The potential saving that can be achieved in a typical MSF plant that currently uses PID control systems is shown in Figure 6.77 and for RO plants, in Figure 6.78.

6.6.2.15 Results and Discussion The index for measuring the economic feasibility of DCS in MSF and SWRO plants is the *net added cost of water* (NACW), which is defined as the difference between the *added cost of water* (ACW) and the *saving in water cost* (SWC). ACW accounts for the capital

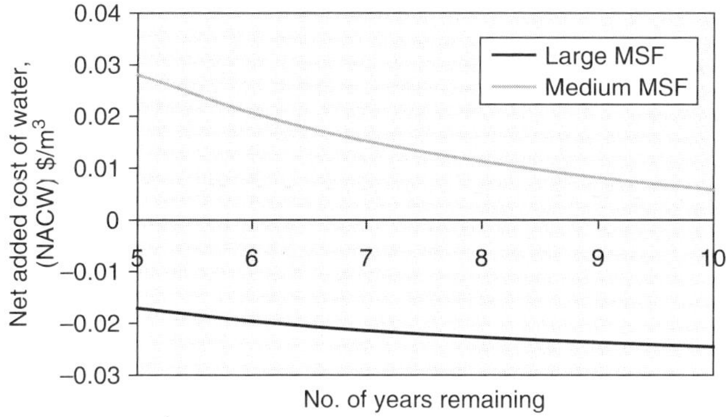

Figure 6.77 Net added cost of water for large and medium MSF plants.

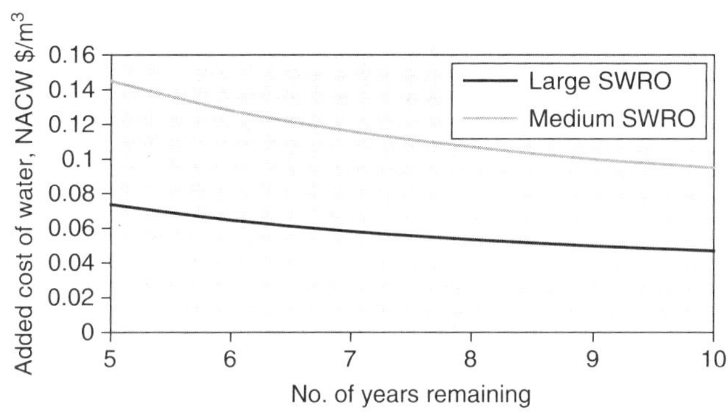

Figure 6.78 Net added cost of water for large and medium SWRO plants.

and operating costs of purchase, installation, operation, and maintenance of the advanced control system, while SWC is the savings in operating cost of water (basically cost of energy and chemicals) that can be achieved by operating the plant near optimum condition. A positive NACW indicates a non-feasible situation, while a negative NACW shows a feasible one:

$$\text{NACW} = \text{ACW} - \text{SWC} \tag{6.95}$$

Figure 6.77 depicts the variation in NACW with the number of remaining years for large and medium MSF plants. Large MSF plants exhibit a negative NACW, indicating that the installation of DCS in these plants can result in a net reduction in the cost of water; medium MSF plants, on the other hand, exhibit a positive NACW, thus displaying a "no feasibility" situation. For large MSF plants, the benefit that can be accrued from using DCS ranges from \$0.017 to \$0.024/m^3, depending on the remaining life time of the plant.

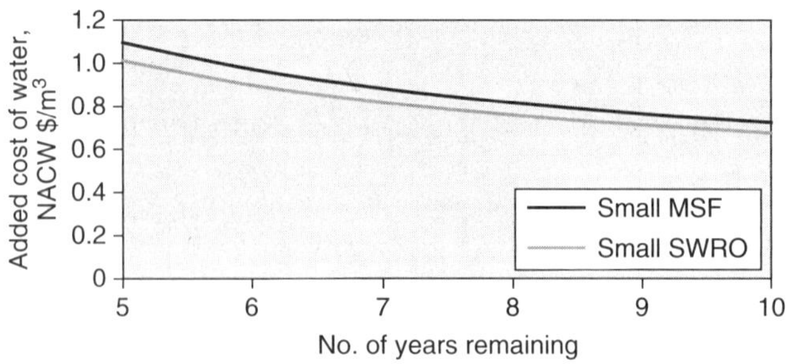

Figure 6.79 Net added cost of water for small MSF and SWRO plants.

Figure 6.78 shows the NACW for large and medium-size SWRO plants and indicates a no-feasibility situation for both plant capacity ranges. This means that from an economic perspective, and according to the assumptions made, it is not feasible to install DCS on this type of plant. Figure 6.79 shows the NACW for small MSF and SWRO plants and indicates a no-feasibility situation for both types of plants.

6.6.2.16 Conclusions and Recommendations The following conclusions can be drawn from this study:

- Installing DCS on large MSF plants can result in a reduction in the resulting cost of water and thus represents an economically feasible situation.
- On the basis of the economic assumptions made, SWRO plants of all capacities are not expected to produce water at a lower cost if equipped with DCS.
- The economic feasibility of DCS improves with increase in the remaining years of plant life time.

These conclusions depend on the economic assumptions made and on the current market prices for software and hardware as well as on the current operating conditions of the existing plant. In making these estimates for the potential saving due to the advanced controller, it was assumed that the operating condition of the plant deviates from the absolute optimum condition by 5%, which seems to be quite reasonable for many plants. Plants that operate at >5% away from target optimum are expected to have a potential for larger saving in water cost when equipped with advanced controllers. However, plants that operates closer than 5% from target optimum would achieve a smaller saving in water cost.

These conclusions cannot be considered as absolute and can vary with market conditions and price levels that are continuously changing with technology development in the field of control engineering.

6.6.3 Recommendations for the Most Suitable Automation Systems: MSFD Plants

The recommendations have been based on the cost–benefit analysis described earlier.

The actual level of automation introduced in modern MSF plants is the result of several years of optimization and improvement.

Table 6.14 summarizes the relative importance/and desirability of the automation control loops actually implemented in MSF plants. It shows assessments of the existing main control loops implemented in thermal desalination systems and their suitability.

As indicated in Table 6.14, the present level of automation is covers the main process control loops; however, as the design trend tends to move toward higher top brine temperature, there is a need to increase the MSF response to scale formation.

Plans to introduce an automation system in the typical nonoperating activities such as distillate flushing and routing maintenance are not envisaged at the present time; however, with the introduction of private projects, there is increased research to find a more rapid load change and startup–shutdown procedure to maximize revenues from water production. Both these targets can be achieved with increased automation and modeling under unsteady-state conditions.

It can be seen that most of the essential control loops are already implemented and are an integral part of the MSF control schemes in industrial plants. However, some of these control schemes appear to be redundant and ineffective. Some improvements can be carried out by introducing new control schemes as indicated in Section 6.3.2.3.

6.6.4 RO: Suitable and Recommended IMCA System

6.6.4.1 Control System Architecture While the analysis in this section is specific to RO plants, the general recommendations are also largely applicable to MSF plants. The recommended control system architecture should be based on the programmable logic controller (PLC).

The PLC should be a modular system that can be configured for future expansion, enable addition of control loops, and/or increase the number of sensor inputs and control outputs.

It should support multiple communications protocols, including RS-232, data highway, universal I/O links, analog and digital inputs, and outputs. The high-speed communication protocol should be of a type widely adopted by major equipment manufacturers.

The PLC should be suitable for centralized or distributed processing (depending on end-user/design engineer preference) and high-speed communications, and have Ethernet capability.

Table 6.14 MSF Plant Main Control Loops Assessment Table

Item	Description	Judgment				
		Actual Status	Essential	Desirable	Redundant	Improvement
1	Top brine temperature (TBT)	Available	√			
2	Steam to brine heater temperature	Available		√		Revision as indicated in Section 6.3.2
3	Brine heat condensate level	Available	√			
4	Brine recirculation flow rate	Available	√			
5	Final stage level	Available	√			
6	Makeup flow rate	Available	√			
7	Seawater to reject (or drain)	Available	√			
8	Distillate level final stage	Available		√		
9	Seawater temperature water to heat reject section	Almost always available			√	
10	Antiscalant flow	Available	√			
11	Antifoam flow	Available			√	
12	Sodium sulfite dosing	Available		√		
13	Fouling rate control	Unavailable				√

The main PLC should be configured for operation without interference from harsh environmental or industrial conditions such as electric noise, fluctuation of ambient temperature, and vibrations. At present, noise-free communication is achieved mainly by connecting different system segments through fiberoptic lines.

Sufficient redundancy should be provided to ensure continuous operation while maintenance is conducted on part of the control system (independent operation of control modules).

Operation of the control system should be supported by an uninterrupted power supply (UPS). The UPS should provide sufficient power capacity to maintain the control system operational during the time lapse between the time of failure of main power supply and the time that emergency power generator is online and operational. The control system should be divided into a functional sections (control loops) according to logic of plant operation and functions performed by individual sections of the plant.

In the following discussion, it is assumed that some of the control loops are controlled by a separate PLC, which communicates with the main PLC through a dedicated Ethernet network (distributed control configuration). However, it is possible that separate software modules that reside on the main PLC (centralized control configuration) could manage all individual control loops. The selection of hardware or software specific for distributed control loops will be dictated by consideration of cost, logistics, and reliability. If the main PLC is capable of handling all inputs and outputs and all system components are located on a single site, then most likely a software solution of managing and control of connected subsystems will be preferred.

The local control loops will include sensors, activators, control logic software, and setpoints and will provide output to the local or central PLC.

6.6.4.2 Control Room The control room will include a number of workstations, connected through a dedicated Ethernet network, that provide a display of operating parameters and enable process control. The workstation should be interchangeable and configured to provide five levels of accessibility according to authorization level of operator:

Level 1—anyone; view any display screen except for setpoint values

Level 2—operator level; view any display screen except for setpoint values; ability to stop any equipment, print reports, and enter data

Level 3—plant engineer level; access to all operator functions and ability to modify setpoints and all process parameters

Level 4—plant supervisor level; access to all functions of plant engineer and ability to add/delete users, change access level, and change password

Level 5—PLC program developer; unrestricted access and modification authority

In addition, the control room should include a server with an industrial database for plant data management. The database will provide information for evaluation of system performance and conducting calculations for performance projections and optimization. The server should have the ability to record information on external media (CD or other mass data storage). Printers are also located in the control room.

6.6.4.3 *Intake Control Loop* The equipment for the intake control loop includes

- PLC with required power supply and adapters
- Enclosures for instrumentation and sensor input
- Communication modem for connection to RO control loop

6.6.4.4 *Pretreatment Control Loop* Equipment for the pretreatment control loop includes

- PLC with required power supply and adapters
- Enclosures for instrumentation and sensor input
- Communication modem for connection to RO control loop

6.6.4.5 *Main RO Control Loop* Equipment for the pretreatment control loop includes

- PLC with required power supply and adapters
- Enclosures for instrumentation and sensor input
- Communication modem for connection to intake and pretreatment control loops

6.6.4.6 *RO Train Control Loops* Equipment for the RO train control loops includes

- Enclosures for sensors and instrumentations input and local control
- Remote nodes with required power supplies and adapters
- Analog input–output modules

6.6.4.7 *Motor Control Center*

- Enclosures for sensors and instrumentation input and local control
- Remote nodes with required power supplies and adapters
- Analog input and output modules

6.6.4.8 *Electrical Circuits and VFD*

- Enclosure for sensors and instrumentations input and local control
- Remote nodes with required power supplies and adapters
- Analog input and output modules

6.6.4.9 Permeate Posttreatment and Storage

- Enclosures for sensors and instrumentation input and local control
- Remote nodes with required power supplies and adapters
- Analog input–output modules

6.6.4.10 Wastewater Neutralization and Discharge

- Enclosure for sensors and instrumentations input and local control
- Remote node with required power supplier and adapters
- Analog input–output modules

6.6.4.11 Operator–Computer Interface (HMI), Alarms They will be managed through integrated software that provides capability of management of plant operation, display process parameters, process control, data logging, and data management. The integrated software will manage plant operation by maintaining operating parameters within limits included in the setup parameter database. According to plant conditions or process parameters, alarms could be initiated. Alarms are initiated according to the following conditions:

1. Failure of equipment to operate (motors, valves, etc.)
2. Operating parameters outside limits (temperature, pH, pressure, etc.)

Software will initiate equipment shutdown if operating conditions are in a range that could result in equipment damage (high free chlorine, low suction pressure, etc.).

Both alarms and shutdown procedure will start after some delay, length of which will be defined during the detail design process.

Some of the alarms will be designed to clear when conditions change; others may require acknowledgment of operator or plant supervisor.

6.7 MODELING AND ANALYSIS RECOMMENDATIONS FOR MSF AND RO PLANTS

6.7.1 Modeling Requirements

It is obvious that at least for the larger plants, future automation systems will be based on line optimization where the measuring instruments, the control system, and the computer model of the process will interact in real time for the optimum performance of the plant. This requires the establishment of models that describe the process properly under both steady-state and transient conditions. Transient "dynamic" modeling of both RO and MSF plants is still in its infancy (representative state-of-the art references are shown in the listed in the Further Reading section below). This is especially true for MSF plants, where the flashing brine flow is still not adequately modeled. In our earlier studies that we cite in this chapter, we specifically identify the modeling needs for the implementation of improved automation and control.

Many companies have recognized that computer modeling is a strategic technology. Process modeling has become the enabling technology to drive a revolution in the chemical-based industry. The desalination industry has been lagging behind other chemical industries and should catch up, especially in view of the rapidly increasing demand for affordable clean freshwater.

Multistage flash (MSF) and RO desalination plants currently provide the most popular processes producing large quantities of potable water. While in the past the most important goal was to achieve the highest possible availability of the plants, nowadays the best performance of the plants becomes increasingly important. For that purpose, it is no longer sufficient to operate plants according to the operating manuals of the process manufacturer or by following the experience of the operators. It now becomes necessary to use intelligent equipment associated with supplementary computation to determine the optimal setpoints of the individual control loops on the basis of a rigorous process model, which allows an improvement of the efficiency under all modes of operation in an unprecedented way.

6.7.2 Steady-State or Dynamic Modeling

6.7.2.1 Steady-State Modeling Steady-state analysis is used to define the operating regions overall which maximization of the performance of MSF and RO plants can occur and also those regions where undesirable operating conditions may occur. This is achieved by studying the effect of each operation parameter on the overall performance of the plant. Further, the combined effect of two or more operation parameters on the process behavior at steady state can be examined. Steady-state analysis is helpful in identifying the good operating regions of maximum production or highest performance parameters and in isolating unfavorable operating conditions, such as, the brine levels that result in blowthrough or carryover in MSF. However, favorable operating conditions can be determined for each operation parameter separately. It is of interest to locate the optimum value for all operation parameters that maximize a certain performance index. In addition, it is important to consider the physical bounds imposed on the plant parameters and to satisfy certain constraints such as those induced by requiring safe brine level conditions. In this case, the optimum operating condition is cast as a multivariable constrained optimization problem and solved by available numerical techniques [14,85,96–98].

Steady-state models for MSF and RO processes can be developed with different grades of accuracy depending on the task of the model. It is a nonlinear model with several tens or even hundreds of equations. For online setpoint optimization, normally a less rigorous model is sufficient. Simplified models can be achieved by assuming constant physical properties, heat transfer coefficients and temperature drop in all stages, usage of less rigorous equations, etc. Supervisory means have to be provided to determine steady state and valid plant conditions for the measured values used.

Two groups of methods have been established for the solution of steady-state MSF and RO models: simultaneous (equation-based) methods and stage-by-stage (sequential) methods.

Equation-based methods use special programming languages for coding process models. These are high-level languages that specialize in the task of specifying equation-based models. The languages are declarative rather than procedural; the purpose is to specify the functional form of the model (e.g., a system of equations) and is not a series of statements to be executed sequentially. This high-level representation can be automatically translated into FORTRAN subroutines to interface with numerical solvers. Although highly flexible, equation-based methods can be used only by experts because they are not user-friendly and require expert knowledge. Examples of commercial equation-based software are ACSL, CSMP, DYNAMO, and SIMUSOLV.

The *stage-by-stage method* is sequential in nature. Each unit operation model (blackbox model) is coded with its own numerical integration routine and can be solved for the time variation of its outputs, given the time variation of its inputs. Therefore, given the input trajectories, each unit model can be integrated independently over a given time horizon. A coordinator algorithm has to be developed to coordinate the solution of individual models, particularly if there is feedback of material or information in the flowsheet. Examples of commercial stage-by-stage software are SPEEDUP, DIVA, DYNSIM, and ASCEND.

The stage-by-stage approach has been widely used because of the following advantages:

- The equation system of each unit is calculated separately. Therefore, the need to solve an equation system with a huge, unwieldy number of equations is avoided.
- It is move user-friendly and most users have some experience with it. It solves the problems the way the engineer would solve the problem by manual calculation.

However, the stage-by-stage method can introduce convergence and stability problems, particularly in flowsheets with feedback loops.

In the *equation-based simultaneous* methods, a set of nonlinear equations describing the MSF and RO processes are solved simultaneously. These equations are interdependent but sparse in nature. The latter fact reduces the calculation efforts considerably and permits the use of special methods. A wide variety of methods for simultaneous solution are presented in the literature. The most important of them for MSF plants are the global Newton–Raphson method and the linearization method used by Helal et al. [85]. The Newton–Raphson method converges very rapidly when using start values close to the solution, otherwise it does not converge properly, or stability problems are encountered. The Helal method [85] is based on decomposition of the equations after linearization. The enthalpy balance equations and the heat transfer relations are linearized using data from previous iterations as well as from the present calculations. The equation system is decomposed into subsets by appropriate choice of the iteration variables; in this way, the equations are grouped by type rather than by stage. The enthalpy balance equations are formulated into a tridiagonal matrix form, which is solved by the Thomas algorithm. The subsets are solved iteratively in sequence. The importance of simultaneous methods has increased more recently with the advances in calculation speed of computers and the development of sophisticated resolution methods and have been used in the optimization package presented in this chapter. The model was extensively validated by comparison of the calculated results with measured data.

Since the 1980s or so, several applications of setpoint optimization have been developed for the process industry, such as for economic consumption of energy in chemical and petrochemical plants [99], for an olefin plant with linkage of an economic model with specific parameters of the plant [100], or for fluidized catalytic crackers [101]. A further publication on setpoint optimization for the process industry is by Perregard et al. [102]. For the setpoint optimization, an objective function must be set up and maximized or minimized depending on the nature of the objective function. At the same time, the constraints imposed to ensure a safe and stable operation of the plant must be satisfied [87,103,104]. In cogeneration plants, for example, the availability of steam may be restricted because of a predetermined electrical load that must be produced and that has a higher priority than water generation. The actual use of setpoint optimization in MSF and RO plants will depend to a large extent on its economic feasibility, which is affected by the local economic environment of the plant as well as its design and performance parameters.

6.7.2.2 Dynamic Modeling The rigorous dynamic plant model may be used in online mode for optimization of the plant under all modes of operation and changing boundary conditions (e.g., summer/winter mode, changing salinity or temperature of the seawater). As it is constructed using physical laws, it may also be used offline to determine in advance the anticipated behavior of a plant by using the actually designed geometric and material data, including additional heuristic knowledge (e.g., determining the brine level). This helps to examine the construction of the plant and to calculate and evaluate different alternatives. Dynamic simulations enable the examination of the transient behavior of the plant so as to ensure smooth and safe operation during load changes. During online operation of the optimization package, it is very important that the rigorous model used always track the behavior of the real plant. Different measures, such as measured data validation and determination of the effects of fouling and scaling are to be applied to fulfill this task [64,105].

During a transient condition, the MSF control system normally receives the input from the load dispatch center to increase or decrease the production capacity from a given setpoint to another value as indicated in Figure 6.80.

Generally, the system has a limited time to respond to the load variation, and with the increase in unit size of the MSF evaporator, the reaction time becomes slower as the system thermal inertia increases.

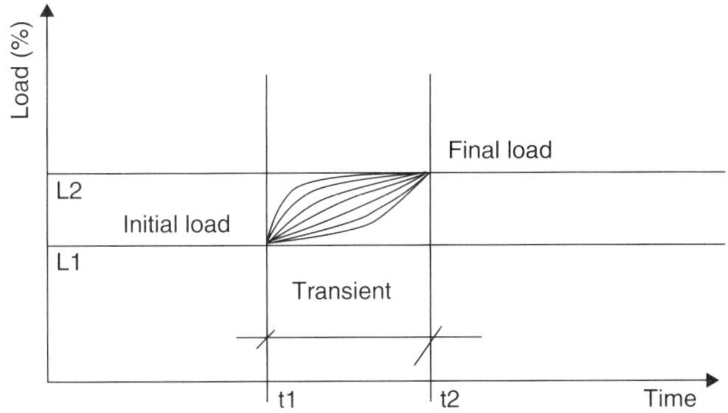

Figure 6.80 Dynamic responses of control systems to a load change.

In modern international water projects (IWPs) and international water and power projects (IWPPs) projects it is essential that the plant remain available to the plant load required by the load dispatch center, as substantial deductions in the final settlement are foreseen in case of unavailability.

The deduction for unavailability of the desalination facility is calculated by the following formulas:

$$\mathrm{WCDO}_m = \sum_{i=1}^{N_m} \mathrm{WCDO}_i \tag{6.96}$$

$$\mathrm{WCDO}_i = \max(\mathrm{OD}_i - \mathrm{SO}_i, 0) \times (1 + \mathrm{WF}_i) \times \mathrm{WCCR}_m$$

where $WCDO_m$ = deduction for unavailability of plant for billing period
WCDO_i = deduction for unavailability of plant for hour i in US\$
OD_i = output delivered by plant for hour i in, m^3/h
SO_i = scheduled output for hour i, m^3/h
N_m = number of hours in billing period
WCCR_m = water capacity related charge rate for billing period m

On the basis of this formula, the dynamic model and control may be used to reduce this availability deduction. Generally, the MSF system reacts to the load dispatch center's request by a variation in the following process parameters:

1. Steam flow rate
2. Make up flow rate
3. Brine recirculation flow rate

Variation in these values normally affects the system top brine temperature, which tends to increase or decrease according to both brine recirculation and steam flow rate. The distillate flow rate during transient conditions can be expressed as

$$\dot{m}_{\mathrm{d}}(t) = \varphi(t) \tag{6.97}$$

$$\dot{m}_{\mathrm{d}}(t) = \psi[\dot{m}_{\mathrm{br}}(t), \dot{m}_{\mathrm{st}}(t), \dot{m}_{\mathrm{mu}}(t)] \tag{6.98}$$

where t = time
$\dot{m}_{\mathrm{d}}(t)$ = distillate flow rate
$\dot{m}_{\mathrm{br}}(t)$ = brine recirculation flow rate
$\dot{m}_{\mathrm{st}}(t)$ = makeup flow rate
$\dot{m}_{\mathrm{mu}}(t)$ = steam flow rate

The basic relationship governing the heat transfer and distillate production in the desalination brine heater section is the definition of performance ratio as indicated in equation (100) below:

$$\eta(t) = \frac{\dot{m}_{\mathrm{d}}(t)}{\dot{m}_{\mathrm{st}}(t)} \frac{\Delta H_{\mathrm{ref}}}{\Delta H_{\mathrm{bh}}} \tag{6.99}$$

where $\eta(t)$ = performance ratio
ΔH_{ref} = reference enthalpy (normally 2,326 kJ/kg)
ΔH_{bh} = difference between the steam and condensate enthalpies in the brine heater

These sets of relationships may be simplified through the introduction of the distillate production recirculation ratio parameter λ, defined as

$$\lambda \equiv \frac{\dot{m}_{\mathrm{d}}}{\dot{m}_{\mathrm{br}}} \tag{6.100}$$

where λ can be expressed using Silver's simplified equation as

$$\lambda = \frac{1}{\left(1 - e^{-[(C_{\mathrm{v}}\Delta T)/\Delta H_{\mathrm{bh}}]}\right)^{ns}} \tag{6.101}$$

where ΔT = flashing range of the evaporator (difference between highest and lowest brine temperatures)
ns = number of MSF stages
C_{v} = brine specific heat

Ideally, the transient conditions should be managed in such a way that the plant performance ratio remains constant (at, say, value K) until the new steady state

$$\eta(t) = K \tag{6.102}$$

is obtained. Thus

$$\frac{d[\eta(t)]}{dt} = \frac{d}{dt}\left[\frac{\dot{m}_{\mathrm{d}}(t)}{\dot{m}_{\mathrm{st}}(t)}\frac{\Delta H_{\mathrm{ref}}}{\Delta H_{\mathrm{bh}}}\right] = 0 \tag{6.103}$$

Improved thermodynamic performance of the plant can also be sought by minimizing entropy production. Entropy changes in the heat transfer process in the brine heater can be expressed as

$$\Delta S_{\mathrm{s}} = \Delta S_{\mathrm{bh}} + \frac{\Delta \mathrm{H}_{\mathrm{bh}}}{\mathrm{T}} \tag{6.104}$$

where ΔH_{bh} and ΔS_{bh} represent the enthalpy drop and entropy rise of the steam in the brine heater, and ΔS_{s} is the total entropy production. By combining Equations (6.10) and (6.106), we obtained the following equation:

$$\Delta S_{\mathrm{s}} = \Delta S_{\mathrm{bh}} + \frac{\dot{m}_{\mathrm{d}}}{\dot{m}_{\mathrm{s}}}\frac{\Delta H_{\mathrm{ref}}}{\eta T} \tag{6.105}$$

To achieve an efficient process, the overall entropy production ΔS_{s} must be minimized, and therefore the operating conditions that satisfy Equation (6.106) should be selected for that purpose:

$$\frac{d\,\Delta S_{\mathrm{s}}}{dt} = 0 \tag{6.106}$$

The power cycle simulation model for the MSF desalination plant can be used to find the optimal settings for the abovementioned controllable parameters inside each MSF. The simulation and monitoring of the MSF desalination allows the operator to find deviations from the designed operation and to define optimized setpoints for operation of the MSF desalination. It will be possible to analyze

- The behavior of the process
- The production costs
  - Steam costs
  - Electrical costs

Further optimization can be related to optimization of the antiscaling costs. In fact, the antiscale dosing rate depends on the top brine temperature and the feedwater flow according to the MSF evaporator antiscale dosing schedule. Its influence on the transient costs is calculated on the basis of this function with a given price per quantity. The following inputs are used for the model:

- Top brine temperature
- Pressure of LP steam to brine heater
- Temperature of LP steam to brine heater

- Flow of LP steam to brine heater
- Brine recirculation flow rate
- Makeup flow rate
- Distillate flow rate setpoint at time 1
- Distillate flow rate setpoint at time 2

6.7.3 Mathematical Modeling: Description and needs in Desalination IMCA

6.7.3.1 The Modeling in General To analyze the behavior of desalination processes and to answer some of the questions raised regarding its control, we need a mathematical representation of the physical phenomena taking place in it. Such a mathematical representation constitutes the model of the system, while the activities leading to the construction of the "model" will be referred to "as modeling."

Modeling the MSF and RO desalination processes is a very synthetic activity, requiring the use of all the basic principles of thermal/fluid engineering science, such as thermodynamics, heat transfer, fluid mechanics, and transport phenomena. For the design of controllers for MSF and RO processes, modeling is a very critical step. It should be approached with care and thoughtfulness. The goal is to develop a control system that will guarantee that the operational objectives of the process are satisfied in the presence of ever-changing disturbances. Then, why do we need to develop an accurate mathematical description (model) for the MSF and RO processes that we want to control? Often, the physical equipment we want to control has not been constructed. Consequently, we cannot experiment to determine how the process reacts to various inputs and therefore cannot design the appropriate control system. Nevertheless, even if the process equipment is available for experimentation, the procedure is usually very costly. Therefore, we need a simple description of how the process reacts to various inputs, and this is what the mathematical models can provide to the control engineer.

In general, processes can be represented by dynamic as well as steady-state models. The former are based on differential equations and/or state machines, and they are required for solving problems in the transient phase, process interactions, troubleshooting, etc. However, they are necessary to implement advanced control, supervision, fault detection, and recovery strategies and dependable systems. A steady-state model consists of algebraic equations, and it is a useful tool at the design stage, helping in the appropriate dimensioning of the plant, optimization aspects, and calculation of the set points. They may also be used to determine fixed parameters and to validate dynamical models. However, they are not useful for control design and analysis purposes. The modeling process is usually integrated with other types of engineering analysis, such as steady-state and dynamic process simulation and online optimization. Dynamic modeling is normally needed for control system design, while steady-state models are sufficient for optimization. The focus on dynamic modeling is on determining first-principle-based mathematical models with emphasis on lumped-parameter (one-dimensional) systems with the aim of predicting the evolution of the system behavior with respect to time. In terms of control, for a mathematical model to be useful it has to represent the dynamic behavior of the system inside its operation range. This means that the dominant time constants and the gains of both plant and model must match with adequate precision.

It should be noted that mathematical modeling of a process is only an abstraction of the true system behavior. The model should mimic the system's behavior with sufficient accuracy and should be verified against system behavior whenever possible. There are many models for a single system, each representing a different level of abstraction, and the key task is to select the appropriate level of abstraction.

6.7.3.2 State Variables and State Equations To characterize a desalination process (or generally any system) and its behavior, we need

- A set of fundamental dependent quantities whose values will describe the natural state of the system
- A set of equations including the variables presented above that will describe how the natural state of the system changes with time

For desalination plants, the processes of interest have four fundamental quantities: mass, species, energy, and momentum. Quite often, the fundamental dependent variables cannot be measured directly and conveniently. In such cases, we select other variables that can be measured conveniently, and when grouped appropriately, they determine the value of the fundamental variables. Thus, mass, energy, and momentum can be characterized by variables such as density, concentration, temperature, pressure, and flow rate. These characterizing variables are called *state variables* and their values, the *state* of a processing system.

The equations that relate the *state variables* (dependent variables) to the various independent variables are derived from application of the conservation principles on the fundamental quantities and are called *state equations*.

6.7.3.3 Conservation Equations The principle of conservation of a quantity S states that

$$\frac{\text{Accumulation of } s \text{ within system}}{\text{Time period}} = \frac{\text{flow of } S \text{ in system}}{\text{time period}} - \frac{\text{flow of } S \text{ out of system}}{\text{time period}} + \frac{\text{amount of } S \text{ generated within system}}{\text{time period}} - \frac{\text{amount of } S \text{ consumed within system}}{\text{time period}} \quad (6.107)$$

The quantity S can be any of the following fundamental quantities:

- Total mass
- Mass of individual species

- Total energy
- Momentum

We now review the specific forms most often used for the balance equations with reference to the general system shown in Figures 6.1 and 6.11; therefore

Total mass balance:

$$\frac{d(\rho V)}{dt} = \sum_{i=1}^{N} \rho_i m_i - \sum_{j=1}^{M} \rho_j m_j \tag{6.108}$$

Mass balance on component A:

$$\frac{d(n_A)}{dt} = \frac{d(C_A V)}{dt} = \sum_{i=1}^{N} C_{Ai} m_i - \sum_{j=1}^{M} C_{Aj} m_j \tag{6.109}$$

Total energy balance:

$$\frac{dE}{dt} = \frac{d(U + K + P)}{dt} = \sum_{i=1}^{N} \rho_i m_i h_i - \sum_{j=1}^{M} \rho_j m_j h_j \pm Q \pm W_{\mathrm{s}} \tag{6.110}$$

Momentum balance:

$$\sum \underline{F} = \frac{d(m\underline{v})}{dt} = \frac{d}{dt} \int_{cv} \rho \underline{v}\, dV + \int_{cs} \rho \underline{v}\underline{v}\, d\underline{A} \tag{6.111}$$

For a general stage j of an MSF plant, for example, expressed as a lumped system (no spatial dependence), the following ordinary differential equations (ODEs) can be derived from the conservation of brine, salt, and energy:

$$\begin{aligned} \frac{dM_{\mathrm{b}_j}}{dt} &= m_{\mathrm{b}_{j-1}} - m_{\mathrm{b}_j} - m_{\mathrm{v}_j} \\ \frac{dM_{\mathrm{s}_j}}{dt} &= m_{\mathrm{b}_{j-1}} C_{j-1} - m_{\mathrm{b}_j} C_j \\ M_{\mathrm{b}_j} C_{\mathrm{p}_j} \frac{dT}{dt} &= m_{\mathrm{b}_{j-1}} (T_{j-1} - T_j) - m_{\mathrm{v}_j} L_j \end{aligned} \tag{6.112}$$

where $M_{\mathrm{b}j}$ and $M_{\mathrm{s}j}$ are brine and salt holdup, respectively, in stage j, m_{b} is the brine mass flow rate, C is the salt mass fraction, m_{v} is the mass flow rate of vapor, T is the temperature, and L is the latent heat of vaporization. These are three ordinary differential equations (ODEs).

6.7.3.4 Additional (Algebraic) Equations In addition to the conservation equations, we need other relationships to express thermodynamic equilibria, such as transport rates for heat, mass, momentum, property relationships, equation of state, and nonequilibrium allowance. Examples of these additional equations used in modeling MSF and RO desalination plants are given below.

6.7.3.4.1 Non-equilibrium Loss in MSF When flashing brine enters a flash chamber, it flashes down (i.e., loses temperature) to a temperature that is slightly higher than the vapor temperature above the flashing brine. This temperature difference consists of the boiling-point elevation (BPE) and the non-equilibrium loss (ΔT_{NE}). ΔT_{NE} is dependent on a number of stages, dimensional and flow parameters such as the stage length l, stage vapor temperature T_{v}, stage flashdown, ΔT_{fd}, brine mass flow rate per unit chamber width ω, brine depth BD, and boiling-point elevation. Several experimental correlations have been developed, one of which is shown below [106]:

$$T_{\mathrm{b}} - T_{\mathrm{v}} = \mathrm{BPE} + \Delta T_{\mathrm{NE}} = BPE + \alpha\left(\frac{\Phi}{\alpha}\right)^{0.3281l} \tag{6.113}$$

Here, α and Φ are given by

$$\Phi = 0.9784^{T\mathrm{v}}15.7378^{BD}1.3777^{\omega\times10^{-6}} \tag{6.114}$$

$$\alpha = 0.5\Delta T_{\mathrm{fd}} + \Phi \tag{6.115}$$

where T_{b} is the brine temperature, °C; T_{v} is the vapor temperature, °C; ΔT_{fd} is the stage flashdown, °C; l is chamber length, ω is the chamber load, kg/(h·m)$^{-1}$, and BD is the brine depth, m.

6.7.3.4.2 Heat Transfer in Heat Exchangers (condensers) of MSF Plant The equation for heat transfer in tube bundles of MSF plant is calculated using the following equation:

$$Q = UA\,\Delta T_{\mathrm{lm}} \tag{6.116}$$

where U is overall heat transfer coefficient, A is the area of heat transfer, and ΔT_{lm} is the logarithmic mean temperature difference. The overall heat transfer coefficient U for use in Equation (6.116), can be calculated from individual resistances using the equation

$$\frac{1}{U} = \frac{1}{h_{\mathrm{o}}} + \frac{d_{\mathrm{o}}}{h_{\mathrm{i}}d_{\mathrm{i}}} + \frac{td_{\mathrm{o}}}{k_{\mathrm{w}}d_{\mathrm{lm}}} + f \tag{6.117}$$

where h = individual heat transfer coefficient, h_{o} for outside and h_{i} for inside
d = tube diameter, d_{o} for outside, d_{i} for inside
k_{w} = thermal conductivity of tube material

f = fouling factor to allow for scale and dirt both inside and outside
t = tube thickness

Numerous correlations are available in the literature to estimate the heat transfer coefficients h_o and h_i for a condenser horizontal tube bundle where steam condensation occurs outside the tubes and cold brine flows inside. Many of them are applicable only to specific geometries and conditions, and also depend on the presence of noncondensable gases, and on other effects. Typical correlations used in the design of existing plants [107] are:

$$h_o = (6.189 \times 10^{-3}) F_n \left(\frac{k_f^3 \rho_f g L^2}{n \mu_f d_o \Delta T_f} \right)^{0.25} \tag{6.118}$$

$$h_i = E_i (0.01222) \left(\frac{k_b}{d_i} \right) \left(\frac{d_i G}{\mu_b} \right)^{0.8} \left(\frac{C_p \mu_b}{k_b} \right)^{0.333} \left(\frac{\mu_b}{\mu_w} \right)^{0.14} \tag{6.119}$$

where C_p = specific heat of brine at bulk brine temperature, kcal $kg^{-1}C^{-1}$
E_i = tube inner surface enhancement factor (this = 1 for smooth surface)
F_n = empirical correction factor
k_f = thermal conductivity of condensate film, kcal $hr^{-1}m^{-1}C^{-1}$
ρ_f = density of condensate film, kg/m^3
g = acceleration of gravity, m/hr^2
G = mass velocity of brine, kg $h^{-1}m^{-2}$
h_i = internal convective heat transfer coefficient; here in kcal $h^{-1}m^{-2}C^{-1}$
h_o = outer convective heat transfer coefficient; here in kcal $h^{-1}m^{-2}C^{-1}$
L = latent heat of vaporization, kcal/kg
m = mass flow rate
n = number of tubes in a vertical row
μ_f = viscosity of condensate film, $(N \cdot h)/m^2$
μ_b = viscosity of bulk fluid, $(N \cdot h)/m^2$
μ_w = viscosity of fluid at wall temperature, $(N \cdot h)/m^2$
ΔT_f = temperature drop across condensate film, °C
d_i = tube inner diameter, m
d_o = tube outer diameter, m

The wall temperature for determining μ_w can be found from

$$T_w = T_b + \frac{U(LMTD)}{h_i} \tag{6.120}$$

where U is the overall heat transfer coefficient (kcal h^{-1} m^{-2}·C^{-1}) and LMTD is the logarithmic mean temperature difference (°C), which is defined as

$$\text{LMTD} = \frac{\Delta T_{\text{large}} - \Delta T_{\text{small}}}{\ln \dfrac{\Delta T_{\text{large}}}{\Delta T_{\text{small}}}} \tag{6.121}$$

where ΔT_{large} and ΔT_{small} are the greater and smaller temperature differencees between the condensing vapor outside the tube bundle and the brine inside, respectively.

The value of F_n can be calculated from the formula:

$$F_n = 1.238 + (0.035n) - (0.00157n^2) \tag{6.122}$$

where $n = 10$ for tube bundles having more than 10 tubes in a vertical row. n can be estimated from the following formula:

$$n = \left(\frac{\text{NT}}{\text{NVR}}\right)^{0.5} + 0.9 \tag{6.123}$$

where NT is the total number of tubes in a bundle and NVR is the number of vertical rows in a bundle.

The fouling factor f has a value close to zero when the plant is new with the tube surfaces absolutely clean and increases as the plant continues operation and scale and dirt accumulate on the tube surfaces. As f increases, plant production and performance ratio decrease and acid cleaning becomes necessary when they drop to a certain level (to be decided by plant operators).

The presence of noncondensable gases with the condensing vapor on the outside of the tube bundle surface can result in a gas blanket on the surface that results in values of h_0 smaller than that determined from Equation (6.123). Noncondensable gases inhibit mass transfer by imposing a diffusional resistance. The efficiency of the venting system is therefore important for ensuring the removal of these gases from the tube bundle.

6.7.4 Modeling Requirements in RO Desalination

6.7.4.1 Modeling of RO Process A conventional model, relatively simplified but widely used in practical modeling of RO systems [107], can be used to indicate where improvements are warranted for better prediction. With some approximation it is assumed in these models that performance of RO membranes follows the simple relations of a net driving pressure model for water transport and diffusion, concentration sought, and salt transport.

6.7.4.1.1 Osmotic Process *Osmosis* is a natural process involving fluid flow across a semipermeable membrane barrier. It is selective in the sense that the solvent passes through the membrane at a higher rate than the passage of dissolved solids. The difference of passage rate results in solvent–solids separation. The direction of solvent flow is determined by its chemical potential, which is a function of pressure, temperature, and concentration of dissolved solids. Pure water in contact with both sides of an ideal semipermeable membrane at equal pressure and temperature has no net flow across the membrane because the chemical potential is equal on both sides. If a soluble salt is added to water on one side of the membrane, the chemical potential of this salt solution is reduced. Osmotic flow from the pure-water side across the membrane to the salt solution side will occur until the equilibrium of chemical potential is restored. Equilibrium occurs when the hydrostatic pressure differential resulting from the volume changes on both sides is equal to the osmotic pressure. This is a solution property independent of the membrane. Application of an external pressure to the salt solution side, which is equal to the osmotic pressure, will also cause equilibrium. Additional pressure will raise the chemical potential of the water in the salt solution and cause a solvent flow to the pure-water side, because it now has a lower chemical potential. This phenomenon is called *reverse osmosis*. The osmotic pressure P_{osm} of a solution can be approximated indirectly by measuring the concentration of dissolved salts in solution

$$P_{\text{osm}} = RT \sum_{i} m_i \tag{6.124}$$

where P_{osm} = osmotic pressure (bar), R is the universal gas constant (0.082 1 atm mol^{-1} K^{-1}), T is the temperature (K), and Σm_i is the sum of molar concentration of all constituents in a solution. An approximation for P_{osm} can be made by assuming that 1000 ppm concentration of total dissolved solids (TDS) equals ~0.77 bar (11 psi) of osmotic pressure. For example, in a RO unit, operating at 75% recovery rate, feed salinity is 3000 ppm TDS and concentrate salinity is ~11,500 ppm TDS. Accordingly, the osmotic pressure of the feed is 2.3 bar (33 psi) and of the concentrate, it is 8.7 bar (126 psi). Equation (6.124) only holds for dilute salt solutions and temperatures close to 25°C. Under significantly different conditions a more rigorous calculation, which takes into consideration ion activities rather than concentrations, has to be applied.

6.7.4.1.2 Permeate Recovery Rate (Conversion) Permeate recovery is one of more important parameters in the design and operation of RO systems. Recovery or conversion rate of feedwater to product (permeate) is defined (in %) by

$$R \equiv \frac{m_{\text{p}}}{m_{\text{f}}} 100 \tag{6.125}$$

$$R \equiv \frac{m_{\text{p}}}{m_{\text{p}} + m_{\text{c}}} 100 \tag{6.126}$$

where R is recovery ratio (%), m_p is the product-water mass flow rate, m_f is the feedwater mass flow rate, and m_c is the concentrate mass flow rate. The recovery rate affects salt passage and product flow. As the recovery rate increases, the salt concentration on the feed–brine side of the membrane increases, which causes an increase in salt flow rate across the membrane. In addition, a higher salt concentration in the feed–brine solution increases the osmotic pressure, reducing the net driving pressure available and consequently reducing the product-water flow rate. Using the definition of R as a fraction from Equation (6.127) and species mass balances; we obtain

$$R \equiv \frac{m_p}{m_p + m_c}$$

$$m_f c_f = m_p c_p + m_c c_c$$

$$m_f c_f = m_p c_p + (m_f - m_p) c_c$$

$$m_f (c_f - c_c) = m_p (c_p - c_c) \tag{6.127}$$

We then obtain a relation between the recovery ratio and the concentration of various streams

$$R \equiv \frac{m_p}{m_f} = \frac{c_c - c_f}{c_c - c_p} \tag{6.128}$$

where c_f is the feed concentration, c_p is the permeate concentration, and c_c is the concentrate concentration.

Equation (6.128) can be applied to determine recovery rate from concentration values of ions in the feed, permeate, and concentrate stream. Usually, these calculations are based on concentrations of chloride or calcium ions, which can be determined easily and with a high degree of accuracy.

6.7.4.1.3 Average Feed Salinity *Average feed salinity* (AFS) is a representative value of feed concentration used for calculation of performance of a membrane element or RO system. AFS accounts for the phenomenon of salinity increase in the RO system from the salinity of feedwater at the entrance to the RO device to the final salinity of concentrate leaving the system. The AFS is calculated as an arithmetic [Eq. (6.129)] or logarithmic mean [Eq. (6.130)]. AFS can be expressed as a function of recovery rate, assuming at first approximation that ions are totally rejected by RO membranes:

$$\text{AFS} = 0.5 c_f \left(1 + \frac{1}{1 - R}\right) \tag{6.129}$$

$$\text{AFS} = c_f \ln \left[\frac{1/(1 - R)}{R}\right] \tag{6.130}$$

The arithmetic mean is usually applied for calculations in cases of low recovery (single-element calculations). In cases of high recovery rates, the logarithmic mean is used for performance calculations.

6.7.4.1.4 Net Driving Pressure The *net driving pressure* (NDP) is the driving force of the water transport through the semipermeable membrane. The value of NDP decreases along the RO unit. Therefore, for the purpose of membrane performance calculations, it is defined as an average NDP. NDP is the fraction of the applied pressure in excess of average osmotic pressure of the feed and any pressure losses in the system according to the following equation

$$\mathrm{NDP} = p_{\mathrm{f}} - p_{\mathrm{os}} - p_{\mathrm{p}} - 0.5p_{\mathrm{d}} + p_{\mathrm{osp}} \tag{6.131}$$

where p_{f} = feed pressure
p_{os} = average feed osmotic pressure
p_{p} = permeate pressure
p_{d} = pressure drop across RO elements
p_{osp} = osmotic pressure of permeate

In regular RO applications the osmotic pressure of permeate is negligible; however, in nanofiltration (NF) applications, where salt rejection is relatively low, permeate salinity is significant compared to the feed concentration. Therefore, osmotic pressure of permeate has to be considered in calculation of NDP in NF systems.

6.7.4.1.5 Water Transport The rate of water passage through a semipermeable membrane is defined by

$$m_{\mathrm{w}} = K_{\mathrm{w}}\frac{S}{d}(\Delta p - \Delta p_{\mathrm{osm}}) \tag{6.132}$$

where m_{w} is the mass water flow rate through the membrane, Δp is the hydraulic pressure differential across the membrane, Δp_{osm} is the osmotic pressure differential across the membrane, K_{w} is the membrane permeability coefficient for water, S is the membrane area, and d is the membrane thickness.

This equation can be simplified to

$$m_{\mathrm{w}} = AS(\mathrm{NDP}) \tag{6.133}$$

where A is the water transport coefficient, which represents a unique constant for each membrane material type, and NDP is the net driving pressure or net driving force for the mass transfer of water across the membrane.

6.7.4.1.6 Salt Transport The rate of salt flow through the membrane is defined by

$$m_{\mathrm{s}} = K_{\mathrm{s}}\frac{S}{d}\Delta c \tag{6.134}$$

where m_{s} is the mass flow rate of salt through the membrane, K_{s} is the membrane permeability coefficient for salt, Δc is the salt concentration differential across the membrane, S is the membrane area, and d is the membrane thickness.

This equation can be simplified to

$$m_s = BS\ \Delta c \tag{6.135}$$

where B is the salt transport coefficient and represents a unique constant for each membrane type; and Δc is the concentration difference, which is the driving force for the transfer of dissolved ions through the membrane

Equations (6.133) and (6.135) show that for a given membrane

- The rate of water flow through a membrane is proportional to the net driving pressure differential (NDP) across the membrane
- The rate of salt flow is proportional to the concentration differential across the membrane and is independent of applied pressure

The salinity of the permeate c_p depends on the relative rates of water and salt transport through the reverse-osmosis membrane:

$$c_p = \frac{m_s}{m_w} \tag{6.136}$$

The fact that water and salt have different mass transfer rates through a given membrane creates the phenomena of water–salt separation and salt rejection. No membrane is ideal in the sense that it absolutely rejects salts; rather, the different transport rates create an apparent rejection. Equations (6.133)–(6.136) explain important design considerations in RO systems. For example, an increase in operating pressure will increase water flow without significantly affecting salt flow, thus resulting in lower permeate salinity. However, higher recovery rates will increase the concentration difference (gradient) and result in higher permeate salinity.

6.7.4.1.7 Salt Passage and Salt Rejection *Salt passage* is defined as the ratio of concentration of salt on the permeate side of the membrane relative to the average feed concentration. Mathematically, it is expressed as

$$\mathrm{SP} = \frac{c_p}{c_{fm}} 100 \tag{6.137}$$

where SP is the salt passage (%), c_p is the salt concentration in the permeate, and c_{fm} is the mean salt concentration in the feed stream.

Applying the fundamental equations of water flow and salt flow illustrates some of the basic principles of RO membranes. For example, salt passage is an inverse function of pressure; that is, salt passage increases as applied pressure decreases. This is because reduced pressure decreases permeate flow rate, and hence, dilution of salt (the salt flows at a constant rate through the membrane as its flow rate is independent of pressure).

Salt rejection is the opposite of salt passage, and is defined by

$$\mathrm{SR} = 100 - \mathrm{SP} \tag{6.138}$$

where SR is the salt rejection (%) and SP is the salt passage as defined in Equation (6.137). Salt rejection is an important product quality parameter of RO membranes, determining suitability of a given membrane for various applications. The equations given above for water and salt transport imply constant values of a transport rates. However, they are strongly affected by temperature, increasing at similar rates with temperature rise.

6.7.4.1.8 Temperature Effect on Transport Rate Feedwater temperature affects the rate of diffusive flow through the membrane. For RO calculations, the following equation is used to calculate the temperature correction factor (TCF), applied for calculation of water permeability

$$TCF = e^{-C[(1/T)-(1/298)]} \qquad (6.139)$$

where T is temperature, K; and C is a constant, characteristic of membrane barrier material; for instance, C values of 2500–3000 are being used for polyamide membranes.

For RO applications, the reference temperature is 25°C, for which TCF = 1.0. The water and salt transport increases ~3%/°C. Figure 6.81 shows values of TCF in the temperature range of 5–50°C. Relative values of water viscosity are included for comparison. There is a striking similarity between both curves, suggesting that changes in water permeability with temperature are result of viscosity changes. The results in Figure 6.81 suggest that, as a result of increased permeability with temperature increase, the operating feed pressure should be lower. This is the situation in the case of low-salinity feed processing (brackish applications). This is also the case for RO seawater applications in the low range of feedwater temperatures. However, at feedwater temperatures of >30°C the subsequent decrease of required feed pressure is insignificant. The effect of increased permeability is reduced by increased osmotic pressure of the seawater feed. In addition, increase of salt passage and potential need for partial second-pass processing may actually result in higher overall power consumption at the high end of feed temperature (Fig. 6.82).

6.7.4.1.9 Average Permeate Flux *Average permeate flux* (APF), is defined as the permeate flow divided by the total membrane area in the RO unit [typical units: L m^{-2} h^{-1} or gfd (gal ft^{-2}/day^{-1})]

$$\mathrm{APF} \equiv \frac{Q_\mathrm{p}}{\mathrm{EN(MA)}} \qquad (6.140)$$

where Q_p = permeate flow rate
EN = number of elements in system
MA = membrane area per element

Conversely, the design APF is used to determine the required number of membrane elements in RO systems, for a required permeate capacity.

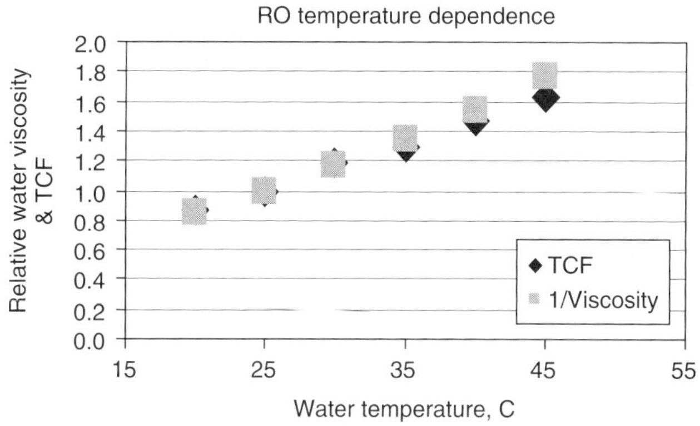

Figure 6.81 Water permeability and viscosity dependence on temperature.

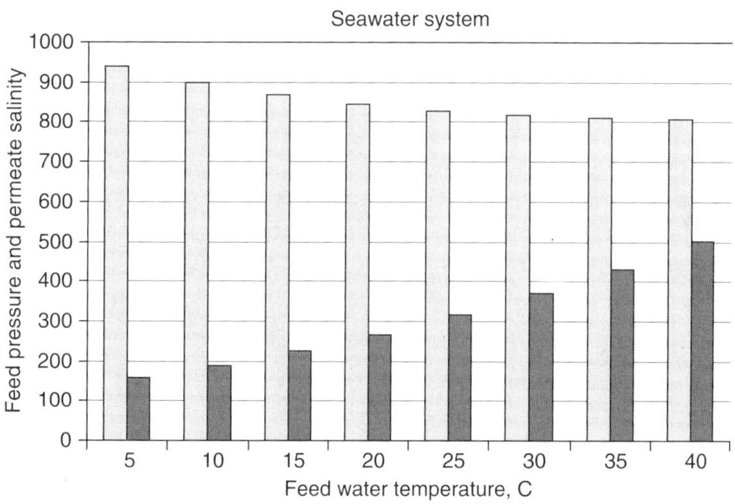

Figure 6.82 Feed pressure (light columns, psi) and permeate salinity (dark columns, ppm) changes with temperature in RO seawater systems.

6.7.4.1.10 Specific Permeability of a Membrane *Specific permeability*, or *specific flux* (SF), characterizes the membrane material in terms of water flux rate driven by the gradient of applied net driving pressure:

$$SF = \frac{APF}{NDP} \tag{6.141}$$

Specific permeability depends on the resistance of a membrane to water flow. This resistance is a composite of flow resistance of the membrane barrier, support layers, and any foulant layer on the membrane surface. It is usually calculated for the feedwater temperature of 25°C. Specific flux unit are L m^{-2} h^{-1} bar^{-1} (or gal ft^{-2} day^{-1} psi^{-1}).

6.7.4.1.11 Concentration Polarization As water flows through the membrane and salts are rejected by the membrane, a boundary layer is formed near the membrane surface. In this layer the salt concentration exceeds the salt concentration in the bulk solution. This increase of salt concentration at the membrane surface is called *concentration polarization*. As shown on Figure 6.83, during the RO process there is a convective flow of water and ions toward the membrane surface. Ions rejected by the membrane diffuse back to the bulk, due to the concentration gradient. The observed effect of concentration polarization is reduction of actual product water flow rate and salt rejection versus theoretical estimates.

Concentration polarization has the following effects on the RO process:

1. A higher osmotic pressure at the membrane surface than in the bulk feed solution Δp_{osm} and reduced net driving pressure differential across the membrane $(\Delta p - \Delta p_{\mathrm{osm}})$
2. Reduced water flow across membrane (Q_{w})
3. Increased salt flow across membrane (Q_{s})
4. Increased probability of exceeding solubility of sparingly soluble salts at the membrane surface, and distinct possibility of precipitation causing membrane scaling

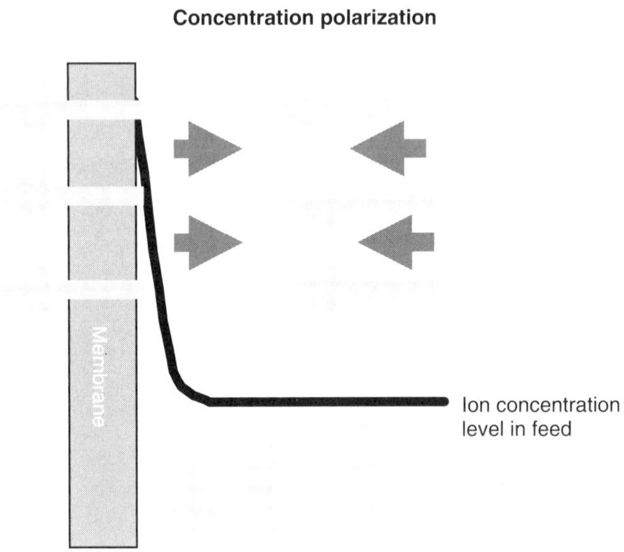

Figure 6.83 Concentration polarization (salt is shown in dark arrows; water, in light arrows).

The *concentration polarization factor* (CPF) is defined as a ratio of salt concentration at the membrane surface (c_s) to bulk concentration (c_b):

$$\mathrm{CPF} = \frac{c_s}{c_b} \tag{6.142}$$

An increase in permeate flux will increase the delivery rate of ions to the membrane surface and increase c_s. An increase of feed flow, parallel to the membrane surface, increases turbulence and reduces the thickness of the high-concentration layer near the membrane surface. Therefore, the CPF will increase with increase in permeate flow (Q_p), due to increase of permeate flux, and decrease with an increase of the average feed flow ($Q_{f\ avg}$),

$$\mathrm{CPF} = K_p e^{Q_p/Q_{f\ avg}} \tag{6.143}$$

where K_p is a constant depending on membrane element geometry, namely, configuration and dimensions of feed channels and feed spacer.

Using the arithmetic average of feed and concentrate flow as an average feed flow, we can express CPF as a function of the permeate recovery rate of a membrane element (R_i):

$$\mathrm{CPF} = K_p e^{2R_i/(2-R_i)} \tag{6.144}$$

The value of the concentration polarization factor of 1.2, which is the recommended limit by some of the membrane manufacturers, corresponds to 18% permeate recovery for a 1 m (~40″) long membrane element as shown in Figure 6.84. The value of CPF is applied in calculations of RO element performance to express excess concentration adjacent to the membrane surface.

6.7.4.1.12 Membrane Age The performance of membrane elements changes with operating time. Usually permeability declines and salt passage increases. The membrane barrier, made of aromatic polyamide, is very robust. However, formation of a fouling layer on the membrane surface, abrasive effect of particles in the

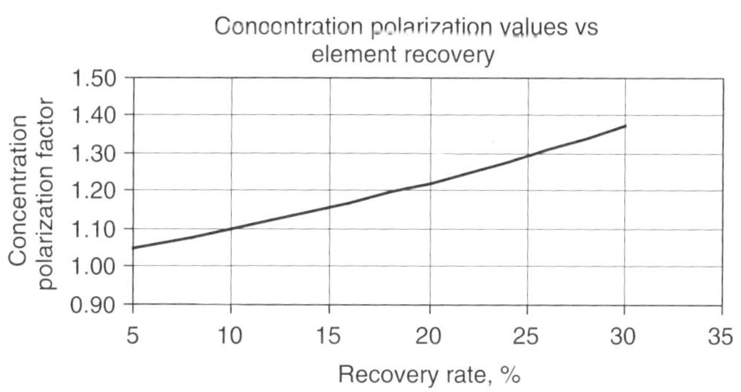

Figure 6.84 Effect of recovery rate on concentration polarization.

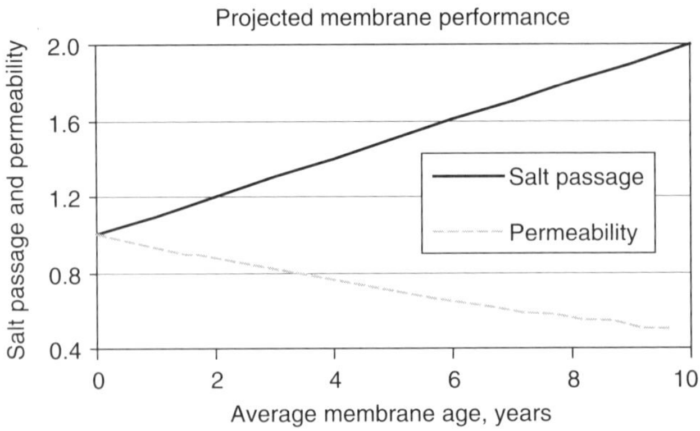

Figure 6.85 Projected change of membrane performance with operating time.

feedwater, and exposure to extreme pH cleaning chemicals will eventually change the properties of the membrane surface and result in performance deterioration. These expected changes in performance are accounted for in calculation of projected performance of an RO system by assuming annual increase of salt passage and decrease of permeability. The approach varies among different membrane manufacturers, but it is generally assumed that salt passage will increase by up to 10% per year and permeability will irreversibly decline by ~7% per year. In Figure 6.85, it is assumed that the salt passage increase is linear, and the permeability decline is a compound function of operating time.

$$\mathrm{SP}_y = \mathrm{SP}_0(1 + \mathrm{SPI})Y \tag{6.145}$$

$$\mathrm{Perm}_y = \mathrm{Perm}_0(1 - \mathrm{Decl})^Y \tag{6.146}$$

where SP_y is the salt passage at the year y, SP_0 is the initial salt passage, and SPI is the *salt passage increase factor*, expressed as a decimal fraction. For the permeability "Perm" the same designation applies, where "Decl" is the permeability annual declination rate.

Accordingly, for year 3 of operation

$$\mathrm{SP}_3 = \mathrm{SP}_0(1 + 0.1)3 = 1.3\ \mathrm{SP}_0$$

$$\mathrm{Perm}_3 = \mathrm{Perm}_0(1 - 0.07)^3 = 0.80\ \mathrm{Perm}_0 \tag{6.147}$$

6.7.4.2 Power Requirement Modeling of RO Process

Energy consumption in an RO unit is the sum of energies used in all segments of the RO process: energy used for water supply (E_{ws}), energy used in pretreatment unit (E_{pt}), net energy spent by the high-pressure feed pumping unit (E_{pu}), energy for permeate treatment and pumping (E_{pe}), and energy used by auxiliary equipment (E_{ax}):

$$E = E_{ws} + E_{pt} + E_{pu} + E_{pe} + E_{ax} \tag{6.148}$$

In a RO plant, except for a small quantity of energy that is used for lighting and to operate the control system, the rest is used to power the water pumping equipment. Energy of pumping (E_p) is a direct function of flow rate (Q_f), head differential (Δp_f) developed by the pump, and specific gravity (ρ) of the fluid being pumped. Energy usage for the pumping is inversely affected by hydraulic efficiency of the pump (η_p) and transformation efficiency of the electric motor (η_m):

$$E_p = k \frac{Q_f \Delta p_f \rho}{\eta_p \eta_m} \qquad (6.149)$$

where k is a unit conversion factor.

The energy recovered (E_r) in the energy recovery device is a function of concentrate flow rate (Q_c), head differential (Δp_c) between inlet and outlet of the energy recovery device and its efficiency (η_t), and of the specific gravity of the concentrate stream (ρ). It is expressed as

$$E_r = k \rho Q_c \Delta p_c \eta_t \qquad (6.150)$$

Accordingly, the pumping energy is

$$E_{pu} = E_p - E_r \qquad (6.151)$$

and Equation (6.148) can be modified as follows:

$$E = E_{ws} + E_{pt} + E_{pe} + E_{ax} + E_p - E_r \qquad (6.152)$$

6.7.4.3 Practical Approach to RO System Performance Optimization

It is intuitively obvious what the ideal result of process optimization would be: a set of operating parameters that would result in operating conditions of lowest power consumption and/or lowest consumption of pretreatment chemicals. This is under conditions of producing permeate salinity within the design limits. Due to a large number of major process parameters that are interrelated, it is not very likely that a simple analytical model can be developed that could be universally applied to commercial membrane systems. Conditions of the RO system, including intrinsic properties of RO membranes, are not stationary. They change with time, fluctuating within the range of parameters discussed above. Still there are calculations algorithms, based on the NDP model and equations listed above, that enable calculations of RO membrane performance on the basis of element performance at reference conditions. The calculations are conducted using state-of-the-art computers in an iterative mode, providing results that are close to actual field performance.

The calculated performances are close to results seen in field operation for conditions of new membranes. There is no analytical model that could be used to predict, with sufficient accuracy, performance of membrane systems as a function of time and possible membrane performance deterioration. The assumptions made during plant design stage are based on a collective industrial experience with given membrane type and water source. As expected in such situations, a good design is considered one with sufficient safety margins. Such design ensures that the plant will provide the required performance, but it is not a design that results in optimum efficiency of operation. Therefore, a simplified practical approach would be to adopt the algorithms of commercial computer programs for process optimization. In a first step, performance of the RO system will be used to calculate current membrane performance to establish a reference point for calculations. This reference point will provide an average performance of elements in the RO system that is adjusted for fouling and compaction. Such performance of an *average element* will be used to project system performance in the range of operating parameters that can be controlled. The result will be a matrix of discrete values of feed pressures and permeate salinity. Then, on this matrix, limits of designed permeate salinity and capacity of pumping equipment (water flow and head) will be imposed. For values that are allowable in terms of these limits, the operating cost will be calculated on the basis of the operating parameters. The net results will be a set of operating parameters that would result in a minimum operating cost. With the state-of -the-art computers, the calculation process will be very fast and operation of the plant could be adjusted instantaneously in response to change of process parameters. Such a control system requires the continuous intervention of the system operator, who would adjust parameters of membrane elements, stored in the database, according to conditions of the membrane elements. These can be derived from plant performance records and calculated using normalization programs. All the components of such a system, including algorithms of calculations programs, are currently available. What is required is a demonstration in a small- to medium-scale RO system to prove functionality and benefits of such approach.

A more correct way of controlling performance of RO systems would be through the utilization of dynamic models. These could be developed based on artificial neural networks and fuzzy logic control. This technology, used to some extent in the electronics industry, is being proposed as a control approach in the desalination industry.

6.7.5 Mathematical Formulation of the Dynamic Modeling Problem

Because of the importance of applying dynamic modeling to desalination processes, we offer a brief introduction to the approach here.

6.7.5.1 *Model Equations* The dynamic model equations for the MSF and RO processes normally give rise to a set of equations that consist of a group of first-order differential equations with time as an independent parameter, and another group of algebraic equations. This set of equations, called *differential algebraic equations* (DAE), has the form

$$\begin{aligned} \underline{f}(\dot{\underline{x}}, \underline{x}, \underline{y}, \underline{u}, t) &= \underline{0} \\ \underline{g}(\underline{x}, \underline{y}, \underline{u}, t) &= \underline{0} \\ \underline{u} &= \underline{u}(t) \end{aligned} \tag{6.153}$$

The group of equations $\underline{f}$ contains the differential equations, and the group $\underline{g}$ contains the algebraic equations. In these equations $\underline{x}$ is the vector of state variables, $\underline{y}$ is the vector of output variables, $\underline{u}$ is the vector of input (manipulating) variables, and t is the time. Since $\underline{u}$ is a known function of time, it can be substituted in the equations $\underline{f}$, $\underline{g}$, and the model equations can be simplified to the form

$$\underline{f}(\dot{\underline{x}}, \underline{x}, \underline{y}, t) = \underline{0}$$

$$\underline{g}(\underline{x}, \underline{y}, t) = \underline{0} \tag{6.154}$$

Therefore, if we have n differential variables $\underline{x}$, m algebraic variables $\underline{y}$, and l input (given) variables $\underline{u}$, we have a total of $(n + m)$ unknown variables.

The following form of the differential group of equations is more appropriate for MSF and RO desalination processes since all differential equations can be cast in this form:

$$\dot{\underline{x}} = \underline{f}(\underline{x}, \underline{y}, \underline{t}) \tag{6.155}$$

To reduce this form to the fully determined ordinary first-order differential equation (ODE) system $\dot{\underline{x}} = \underline{f}(\underline{x}, t)$, it is necessary to cast the group of algebraic equations in the following form:

$$\begin{aligned} y_1 &= g_1(\underline{x}) \\ y_2 &= g_2(\underline{x}, y_1) \\ &\cdots \\ &\cdots \\ y_m &= g_m(\underline{x}, y_1, y_2, \ldots, y_{m-1}) \end{aligned} \tag{6.156}$$

Each algebraic variable $y_1, y_2, \ldots$ can be calculated explicitly given $\underline{x}$ and any previously calculated y_i; that is, the algebraic variables can be eliminated by direct substitution of the algebraic equations into the ordinary differential equations. Note that most of the models do not conform directly to this formulation without some form of manipulation (in general, differentiation [108]. Substituting the y vector into the vector function f, we obtain the fully determined form:

$$\dot{\underline{x}} = \underline{f}(\underline{x}, t) \tag{6.157}$$

This is the most natural form in which to derive the component models. Direct integration of this set of ODE's ensures that the algebraic equations are satisfied in each time step, hence the algebraic variables will remain consistent with the differential variables, [108].

A very simple example of the above-described procedure follows. Assume that the system of equations is composed of ODEs and the algebraic equation shown below:

$$\dot{x}_1 = 2x_1 + x_2 + x_3 \tag{6.158}$$

$$\dot{x}_2 = x_1 - 2x_2 - 3x_3 \tag{6.159}$$

$$0 = x_1 + x_2 - x_3 \tag{6.160}$$

For consistent initialization at $t = 0$; we specify $x_1(0)$, $x_2(0)$:

$$x_3(0) = x_1(0) + x_2(0) \tag{6.161}$$

$$\dot{x}_1(0) = 2x_1(0) + x_2(0) + x_3(0) \tag{6.162}$$

$$\dot{x}_2(0) = x_1(0) - 2x_2(0) - 3x_3(0) \tag{6.163}$$

Differentiating the algebraic equation [Eq. (6.160)], we obtain

$$0 = \dot{x}_1 + \dot{x}_2 - \dot{x}_3 \tag{6.164}$$

and now Equations (6.158), (6.159), and (6.164) form a system of ODEs in x_1, x_2, x_3 that can be easily solved using Runge–Kutta methods.

6.7.5.2 Solution for the Dynamic Simulation Model Solving this initial value problem over the interval $[t_0, t_\mathrm{f}]$ requires solution of two subproblems:

1. Defining and solving for the initial state of the system at t_0
2. Numerical integration of the equations: $\underline{\dot{x}} = \underline{f}(\underline{x}, t)$ from t_0 to t_f

To define the initial state of a fully determined system of DAEs, we need to find a set of consistent initial values for the variables $\underline{x}(t_0), \underline{\dot{x}}(t_0)$, and $\underline{y}(t)$ at $t = t_0$. Clearly, a necessary condition for a set of consistent initial values is that the model equations be satisfied at $t = t_0$:

$$\underline{\dot{x}}(t_0) = \underline{f}[\underline{x}(t_0), t_0] \tag{6.165}$$

Degree-of-freedom analysis gives n equations and n unknowns, so the initial state of the system is fully defined. The initial state of the system is usually chosen as the steady state for which all the initial differential variables vanish: $\underline{\dot{x}}(t_0) = \underline{0}$.

6.7.5.3 Solution algorithm The fourth-order Runge–Kutta algorithm was found to be the most appropriate method for solving the system of first-order differential equations [109]. The method has the advantage of one-step integration across the ith time interval for each equation. For a system of n equations, this one-step integration is used to estimate the values of the state variables $\underline{x}$ at the end of interval i if the values at the beginning of the interval are known:

$$x_{j,i+1} = x_{ji} + h\varphi_j = y_{ji} + \frac{h(k_{j1} + 2k_{j2} + 2k_{j3} + k_{j4})}{6} \tag{6.166a}$$

$$k_{j1} = f_j(t_i, x_{1i}, x_{2i}, \ldots, x_{ni}) \tag{6.166b}$$

$$x^*_{ji} = x_{ji} + \frac{1}{2}hk_{j1} \tag{6.166c}$$

$$k_{j2} = f_j\left(t_i + \frac{1}{2}h, x_{1i}{}^*, x_{2i}{}^*, \ldots, x_{ni}{}^*\right) \tag{6.166d}$$

$$\overline{x}_{ji} = x_{ji} + \frac{1}{2}hk_{j2} \tag{6.166e}$$

$$k_{j3} = f_j\left(t_i + \frac{1}{2}h, \overline{x}_{1i}, \overline{x}_{2i}, \ldots, \overline{x}_{ni}\right) \tag{6.166f}$$

$$\overline{x}^*_{ji} = x_{ji} + hk_{j3} \tag{6.166g}$$

$$k_{j4} = f_j(t_i + h, \overline{x}_{1i}{}^*, \overline{x}_{2i}{}^*, \ldots, \overline{x}_{ni}{}^*). \tag{6.166h}$$

The function φ_j is termed the *increment* function and is simply a chosen approximation to $f_j(t, x_1, x_2, \ldots x_n)$ on the interval $t_i \leq t \leq t_{i+1}$; h is the timestep size $(t_{i+1} - t_i), j$ is the equation index number, and i is the index number for the state variables, $x_1, x_2, \ldots, x_n$.

These relationships are applied in parallel at each point in the algorithm for all n equations, that is, for $j = 1, 2, \ldots, n$. φ_j is a weighted average of four derivative evaluations k_1, k_2, k_3, k_4 as shown in Equations (6.166a)–(6.166h). The solution algorithm proceeds as follows:

1. Calculate the values of $f_j, j = 1, 2, \ldots, n$, using the current t_i and $x_{1i}, x_{2i}, \ldots, x_{ni}$ values. These are equivalent to the values $k_1, j = 1, 2, \ldots, n$ of Equation (6.166b).
2. Compute the values x^*_{ji} from Equation (6.166c).
3. Increment t to the value required in Equation (6.166d) namely $(t + h/2)$.
4. Calculate k_2 from Equation (6.166d) using the new value of t and the previously calculated values of x^*_{ji}.
5. Calculate $\overline{x}_1, x_2, \ldots x_n$ for $j = 1, 2, \ldots, n$ from Equation (6.166e), given the values of k_2.
6. Calculate the values of k_3 from the known values of $\overline{x}_1, \overline{x}_2, \ldots \overline{x}_n$ for $j = 1, 2, \ldots, n$ using Equation (6.166f).
7. Calculate $\overline{x}_1{}^*, \overline{x}_2{}^*, \ldots, \overline{x}_n{}^*$ for $j = 1, 2, \ldots, n$ from Equation (6.166g) using the previously calculated values of $\overline{x}_1, \overline{x}_2, \ldots, \overline{x}_n$ for $j = 1, 2, \ldots, n$.
8. Calculate k_4 from Equation (6.166h) using the values of $\overline{x}_1{}^*, \overline{x}_2{}^*, \ldots, \overline{x}_n{}^*$.
9. Calculate the values of x at the end of interval i (x_{i+1}) from Equation (6.166a) using the previously calculated values of k_1, k_2, k_3, k_4.
10. Repeat the calculations for interval $i+1$ and continue until the last interval

Shivayyanamath and Tewari [110] adopted the following procedure to solve the dynamic simulation problem of MSF plant:

1. Assign the initial values for all the differential variables.
2. Choose proper timestep size.
3. Calculate the brine heater outlet temperature by solving the energy and heat transfer equations for the brine heater.
4. Calculate the flashing brine temperature using the energy equation applied to the flashing brine.
5. Calculate the stagewise recirculating brine temperatures and tube wall temperatures using the energy equation applied to the recirculating brine and tube material of the condenser tube bundles in each stage.
6. Calculate the flashing brine flow rate in each stage using a mass balance on the flashing brine in each stage.
7. Calculate the salinity of flashing brine leaving each stage using using a salt mass balance in each stage.

The predicted values are used to estimate the variables for the next period using the Runge–Kutta method. Calculations are continued until the steady state is reached or until the end of the specified period.

6.7.6 Summary of Modeling Needs for the Future

It appears that the current modeling capabilities and experience are in general sufficient for integration with advanced control systems for MSF and RO plants, and the main improvement recommended at this time is to implement them more widely and economically, especially for larger plants, where appropriate. At the same time, two major targets of improvement should be sought for even more effective implementation and use of IMCA systems that would reduce water costs:

1. Better physical/chemical modeling of the basic fluid dynamics and transport processes in MSF and RO plants that would reduce the predictive error under the varying conditions and compositions in the plant, especially under transient conditions.
2. More rapid computational algorithms, enabling models to produce results closer to the real process time.

6.8 CONCLUSIONS AND RECOMMENDATIONS

- Good process instrumentation and measurements are a vital prerequisite for control and automation.
- The majority of desalination plants currently in operation make use of single-loop (SISO) PI or PID controllers.
- New methods of advanced and intelligent control techniques have been proposed by a number of authors with promising technical performance results:

  - Model-based predictive control seems to have the potential for improving control of both MSF and RO plants, but needs to be examined and tested in the field.
  - Artificial intelligence, neural networks, and fuzzy logic have been studied further for desalination plants, but have not been proven as viable approaches.

- One of the most critical issues for maintaining low product cost is the achievement of high plant availability, and to this end:
  - Using fault-tree analysis in some form (after component reliability statistics such as failure and repair rates were collected), identify the most vulnerable components, attempt to replace them by an improved design, and monitor them more carefully.
  - Simplify the design wherever possible.

- A basic premise of this study is that any level of automation or IMCA in general must be economically justified, to also include consideration of IMCA system reliability.
- As opposed to the enormous amount of published information on control systems and their advancement, those on cost–benefit analysis of their implementation are very few in general, and nearly nonexistent for desalination plants in particular.
- The literature is almost devoid of investigations related to the economic feasibility of the use of advanced controllers in lieu of conventional ones in desalination plants.
- The use of computer-based data reconciliation is becoming standard in large plants.
- Model-based control and automation with optimization was introduced successfully in some large plants, and this trend is expected to continue and is likely to be applied to smaller plants as well.
- The use of techniques such as neural networks, fuzzy logic, and artificial intelligence in desalination plant control have been studied by some researchers, without any clear beneficial results, and there are no plans known to us to apply them in plants.
- There are several advanced control techniques currently available in the marketplace, and the most appropriate technology for a particular desalination process has to be investigated both technically and economically in pilot-scale tests.
- To emphasize the obvious, any improvements in IMCA must reduce product-water cost, which includes not only a reduction in energy and capital costs but also an increase of plant availability. The latter requires, among other things, the use of simple and robust IMCA systems.
- The level of automation in the previous generation of MSF plants is very high, to the extent that it brings up increasing concerns related to the deskilling of

MSF plant operators. Training simulators could prove very useful in improving the skill level of plant operators.

- Full automation of the startup–shutdown sequence is desirable but has thus far been implemented in only one plant.
- Modification of the control schemes that affect reductions in exergy (or simply pressure) losses in the plant need to be explored, and an example of such a proposed scheme to control the top brine temperature in MSF plants was presented.
- Automation to control fouling is an important proposed approach.
- The process step that has potential for significant improvement by applying "smart" automation is the pretreatment of the feedwater, especially in RO.
- In RO, automation of the recovery–flux rate combination, instead of the conventional method of maintaining constant flux, is proposed for potential reduction of operating cost.
- From the various advanced control methods considered, the use of constrained model predictive controllers (CMPCs) appears to be promising for implementation.
- Automation as online optimization is highly desirable, but adequate mathematical dynamical models of the process are not yet available to implement it.
- A critical review of the status of cost–benefit analysis of IMCA systems, with emphasis on their use in MSF and RO desalination, was performed and presented.
- An explicit methodology for such analysis was developed and presented.
- To implement this methodology, simplified models for predicting the performance of MSF and RO plants were developed and successfully implemented.
- Specific examples of IMCA system cost–benefit ratios for large RO and MSF plants were calculated.
- Some promising ways in which automation could be implemented to improve MSF and RO plant design and operation to reduce product-water cost were suggested.
- Proper saving and analysis of historical plant data can lead to product cost reduction in ways not always obviour with conventional modeling.
- The use of advanced controllers in MSF and RO plants can result in economic saving in the operating expenses [steam consumption (MSF), electrical energy consumption (MSF and RO), and chemical consumption (MSF and RO)].
- In the examples analyzed, on the basis of unit cost assumptions (steam cost = \$3.5/t, electricity cost = \$0.06/kWh, average cost of chemicals \$1.0./kg), it was concluded that *direct* savings of \$66,000 per year (MSF) and \$13,000 per year (RO) can be achieved for each migd of plant installed capacity if the advanced IMCA system ensures a departure from optimum that is at most

5%. The real savings depend on the actual operation strategy of the existing plant as well as the extent of seasonal variation in seawater temperature and salinity in the plant site.

- The economic feasibility of replacing conventional PID controllers with advanced ones depends on the results of a cost–benefit analysis to be carried out on the particular plant under consideration. This analysis should factor in the operating condition of the plant as well as the purchase and installation costs of the controller's hardware and software.
- The direct benefit–cost ratio for large plants (say, 200,000 m^3/day) is estimated to be ~30 for MSF and 23 for RO processes, indicating that advanced IMCA systems have a very desirable financial outcome. That ratio becomes even higher when the indirect benefits are added.
- It is well recognized in process IMCA system design and implementation that there are many intangible benefits to IMCA system improvement that may actually exceed those simply calculated from the resulting higher product output and quality; more attention to such "robust design" is thus recommended
- Design life of the automation system is much shorter than that of the desalination plant, and better ways to match the automation system with the overall plant lifetime should be found.
- Automatic control procedures and systems should be developed for optimizing the operation of hybrid desalination plants.
- It appears that the current modeling capabilities and experience are qualitatively sufficient for integration with advanced control systems for MSF and RO plants, and the immediate improvement recommended at this time is to implement them more widely and economically, especially for larger plants, where appropriate.
- At the same time two major targets of improvement should be sought for even more effective implementation and use of IMCA systems that would reduce water costs:
  - Better physical/chemical modeling of the basic fluid dynamics and transport processes in MSF and RO plants that would reduce the predictive error under the varying conditions and compositions in the plant, especially under transient conditions
  - Faster computational algorithms that enable the models to produce results closer to the real process time

ACKNOWLEDGMENT

The authors are grateful to the Middle East Desalination Research Center for partial support of this study. Dr. Mark Wilf has made major contributions to this work, writing most of Sections 6.3.3, 6.4.2, 6.5.5.4 and 6.7.4 and participating in conclusions and recommendations of most sections and in Section 6.8.

NOMENCLATURE

Symbols

a	constant
A	heat transfer area, m^2
$\underline{A}$	area vector
A_m	total membrane area in operation, m^2
APF	average permeate flux, m^3 (permeate, m^{-2}; membrane, s^{-1})
B	adhesion parameter, $(N{\cdot}s)\ m^2$
BPE	boiling-point elevation, K
C	salt concentration, kg/m^3
c	unit cost, $\$/m^3$ water
c_p	specific heat, $kJ/(kg{\cdot}^\circ C)$
CAP	capital cost of advanced control system, \$
C_B	bulk concentration, kg/m^3
C_c	concentrate concentration
C_f	feed concentration
C_M	concentration at membrane surface, kg/m^3
C_P	concentration of product water (permeate), kg/m^3
$\mathbf{d}$	disturbance vector
D	diffusivity, m^2/s
d	hydraulic diameter, m
$D(s)$	disturbance in Laplace domain
d_{as}	dose rate of antiscalant, ppm
$e(t)$	error in time domain
E	error function; energy
$E(s)$	error in Laplace domain
f	fouling factor, $m^2/W{\cdot}^\circ C$
F	pump head, ft or m
$\underline{F}$	force vector
g	gravitational acceleration, m/s^2
$\boldsymbol{G}$	matrix of transfer functions
G	pump flow, gpm
$G_c(s)$	transfer function of controller in Laplace domain
$\boldsymbol{G}_d$	matrix of disturbances
$G_p(s)$	transfer function of process in Laplace domain
h	length of membrane, m; heat transfer coefficient, $W\ m^{-2}\ {}^\circ C^{-1}$; specific enthalpy
H_f	friction head, m
i	interest rate; index of time period
I	input; input node index
ID	inner tube diameter, m
j	output node index; index of equation number; interest rate

J	flux
J_S	salt flux, kg/(m^2· s)
J_V	water flux, m^3/(m^2· s)
K	current time, s; kinetic energy
k	mass transfer coefficient
k_{sc}	thermal conductivity of scale
k_1, k_2, k_3, k_4	Runge–Kutta parameters
L	latent heat of evaporation, kJ/kg; MSF stage length, m
L_P	solution permeability, m^3/(m^2·Pa·s)
m	percent of data falling beyond limit; or m = gross scale deposition flux, kg/(s·m^2); mass flow rate, kg/s
$m(t)$	manipulated variable in time domain
$\dot{m}$	mass flow rate, kg/s
$\dot{m}_{sc}$	gross scale deposition flux, kg s^{-1} m^{-2}
M	mass, kg
$M(s)$	manipulated variable in Laplace domain
M_{bd}	blowdown flow rate, kg/s
M_c	seawater flow rate to rejection section, kg/s
M_d	flow rate of distillate, kg/s; nominal capacity of desalination plant, m^3/day
M_{dl}	distillate flow rate from recovery section, kg/s
M_{mu}	makeup flow rate, kg/s
M_r	brine recirculating flow rate (MSF process) or recirculating flow rate, kg/s
M_s	flow rate of steam (MSF process); mass flow rate of LP steam to brine heater, kg/s
n	number of nodes; number of years remaining in plant lifetime; number of moles; number of tubes in a vertical row; number of equations
N	horizon time; number of stages in MSF plant; also number of sections in feed direction along RO membrane
N_l	number of stages in heat recovery section in MSF plant
N_o	number of outputs of the network
N_P	number of input/output pairs of data
O	output
OPER	operating cost of advanced process controller, \$/year
$\mathbf{p}$	vector of parameters
$p(m)$	probability that $m\%$ of a controlled variable value falls beyond limit x_L or probability that $m\%$ of data fall outside limit; pump flow rate, gpm or m^3/s
P	pumping power, kW; membrane permeability, m/s; pressure, MPa; potential energy, J
Q	heat transfer rate, kJ/s
Q_P	total permeate flow rate, m^3/s

r	concentration ratio
$r(t)$	setpoint in time domain
R	RO plant recovery ratio; membrane rejection
$R(s)$	setpoint in Laplace domain
Re	Reynolds number
s	Laplace transform variable; standard deviation; standard deviation of a controlled variable
$\boldsymbol{s}$	vector of state variable
Sc	Schmidt number, dimensionless
Sh	Sherwood number, dimensionless
sp	setpoint
t	time, s
T	temperature,°C
t_1	brine inlet temperature to brine heater,°C
T_i	seawater inlet temperature to heat rejection section,°C
t_i	temperature of recirculating brine or cooling seawater,°C
T_{max}	top brine temperature,°C
T_N	flashing brine temperature in the last stage,°C
T_{N1}	flashing brine temperature out of heat recovery section,°C
T_o	outlet temperature from heat rejection section,°C
t_r	temperature of recirculating brine entering last recovery stage,°C
$\mathbf{u}$	vector of manipulated variables
$\boldsymbol{u}$	vector of independent parameters
u	velocity, m/s
U	overall heat transfer coefficient, kW/(m^2·°C); internal energy, J
$U(t)$	input signal to controller
ua	controller output
vc	calculated variable
vd	dependent variable
vi	independent variable
V	volume, m^3
w	weighting factor
W	width of membrane, m; work, J
$W(t)$	white noise
x	controlled variable, also, specific consumption terms
$\mathbf{x}$	vector of state variables
$\overline{x}$	mean value of a controlled variable
$\overline{x}_B$	mean value of a controlled variable with conventional controller
$\overline{x}_C$	mean value of a controlled variable with advanced controller
x_L	limit value of a controlled variable
$\boldsymbol{x}_{sc}$	scale layer thickness, m
$x_{sc,\infty}$	asymptotic scale layer thickness, m
$\boldsymbol{y}$	vector of measured data
y_c	process output

$Y(s)$	output (controlled variable) in Laplace domain
$y(t)$	output (controlled variable) in time domain
β	bias neuron weight
γ	fouling rate, $m^2\ W^{-1}\ ^\circ C^{-1}\ s^{-1}$
δ	boundary-layer thickness, m
Δc	total saving in water cost achievable by using advanced controllers, $\$/m^3$
Δh	length of a membrane section, m
ΔP	pressure difference, bar
ΔQ_{Pi}	permeate volume flow at ith section, m^3/s
$\Delta \overline{x}$	amount that mean value is shifted on installation of advanced controllers; change in mean value of process variable
$\Delta \Phi$	added cost of water, $\$/m^3$
$\Delta \Phi_1$	saving due to reduction in variance of controlled variable
$\Delta \Phi_2$	saving due to operation close to optimum value of controlled variable
ϵ	stage temperature loss, (BPE+nonequilibrium loss + demister loss), $^\circ$C
ζ	damping ratio
η	efficiency; also viscosity, kg/(m·s)
κ_{sc}	scale thermal conductivity, $kW\ m^{-2}\ ^\circ C^{-1}$
μ	dynamic viscosity
π	osmotic pressure, Pa
ρ	density, kg/m^3
σ	membrane reflection coefficient
τ	time constant, s
τ_f	flowing shear stress, N/m^2
Φ	cost of water, $\$/m^3$
Φ^*	cost of water at optimum operating point, $\$/m^3$
ω_n	natural angular frequency, s^{-1} (reciprocal seconds)

Subscripts

1	recovery section
2	rejection section
as	antiscalant
b	brine
bd	blowdown
B	base (conventional)
c	clean condition
C	advanced (improved) controller
ch	chemicals
d	distillate (product)

el	electricity
f	feed; or refers to fouling; film
fo	fouled condition
HPP	high-pressure pump
i	stage number or index
j	index
L	limit
m	membrane
mu	makeup
o	cooling-water out
p	product
p(m)	probability that $m\%$ of data fall outside limit
P	pump
PT	pretreatment
s	steam
sc	scale
swr	seawater recycle
T	turbine
v	vapor
w	wall
∞	infinity

Superscripts

–	average
*	optimum condition

Abbreviations

ACW	added cost of water, $/m^3
ANN	artificial neural network
APC	advanced process control(ler)
API	application program interface
BPST	backpressure steam turbine
CAPEX	capital expense (cost)
CCR	central control room
COMM	communication
CRT	cathode ray tube
CU	control unit
CV	controlled variable
DCS	distributed control system
DMC	dynamic matrix control

DV	disturbance variable
FAT	factory acceptance testing
HFF	hollow fine fiber
HLOI	high-level operating interface
HP	high pressure
IMCA	instrumentation, measurement, control and automation
ISE	integral square error
LAN	local-area network
LCU	local control unit
LDC	load dispatch center
LLOI	low-level operating interface
LP	low pressure
MIMO	multi-input multi-output
MP	medium pressure
MPC	model predictive control; model predictive controller
MSF	multistage flash
MSFD	multistage flash desalination
MV	manipulated variable
NACW	net added cost of water, $/m^3
NB	negative big
Neg	negative
NLP	nonlinear programming
NM	negative medium
NMPC	nonlinear model predictive control
N-NMPC	neural nonlinear model predictive control
O&M	operation and maintenance
OLE	object linking and embedding
OPEX	operating expense
PI	proportional integral
PID	proportional integral derivative
PLC	programmable logic(al) controller
PR	performance ratio
RO	reverse osmosis
SAT	site acceptance testing
SCADA	system control and data acquisition
SDI	silt density index
SISO	single-input single-output
SW	spiral wound
SWRO	seawater reverse osmosis
TBT	top brine temperature in MSF process
TDS	total dissolved solids
WAN	wide-area network

REFERENCES

1. Lior N, ed., *Measurements and Control in Water Desalination*, Elsevier Science, Amsterdam, 1986.
2. Lior N, Measurements and instrumentation in water desalination, keynote paper, *Proc. IDA World Congress Desalination and Water Sciences*, Abu Dhabi, 1995, Vol. 1, pp. 229–253.
3. Lior N, The status of measurements, instrumentation & control in water desalination: Part 1, *Int. Desal. Water Reuse Q*. **8**(2):46–51.
4. Lior N, The status of measurements, instrumentation & control in water desalination: Part 2, *Int. Desal. Water Reuse Q*. **8**(3):24–32 (1998).
5. Lior N, The status of measurements, instrumentation & control in water desalination: Part 3, *Int. Desal. Water Reuse Q*. **9**(1):19–26 (1999).
6. Lior N, The status of measurements, instrumentation & control in water desalination: Part 4, *Int. Desal. Water Reuse Q*. **9**(4):31–35 (2000).
7. Lior N, El-Nashar A, Sommariva C, Wilf M, An update on the state of instrumentation, measurements, control and automation in water desalination, *Proc. IDA World Congress Desalination and Water Reuse*, Singapore, Sept. 2005, International Desalination Association, Topsfield, MA, 2005.
8. Bunzemeier A, Krause H, Introduction to process control, in Darwish MK, Al-Gobaisi DMK, eds., *Encyclopedia of Desalination and Water Resources* (DESWARE), *Encyclopedia of Life Support Systems* (EOLSS) Publishers, 2001 (www.desware.net).
9. Bunzemeier A, Litz L, Process control systems, in Darwish MK. Al-Gobaisi DMK, eds., *Encyclopedia of Desalination and Water Resources (DESWARE)*, Eolss Publishers, 2001 (www.desware.net).
10. Muroni O, Borsani R, Control valve positioners, in Darwish MK, Al-Gobaisi DMK, eds., *Encyclopedia of Desalination and Water Resources (DESWARE)*, EOLSS Publishers (www.desware.net), 2001.
11. Muroni O, Borsani R, Control valve actuators, in Darwish MK, Al-Gobaisi DMK, eds., *Encyclopedia of Desalination and Water Resources (DESWARE)*, EOLSS Publishers (www.desware.net), 2001.
12. Tarifa EE, Scenna NJ, Fault diagnosis in chemical processes, its relation to thermal desalination systems, in Darwish MK, Al-Gobaisi DMK, eds., *Encyclopedia of Desalination and Water Resources (DESWARE)*, EOLSS Publishers (www.desware.net), 2001.
13. Rautenbach R, Schafer S, Availability analysis of MSF distillers using fault tree logic, in Darwish MK, Al-Gobaisi DMK, eds., *Encyclopedia of Desalination and Water Resources (DESWARE)*, EOLSS Publishers, (www.desware.net), 2001.
14. Flehmig F, Watzdoff RV, Marquardt W, Reddy KV, Dynamic MSF plant modeling and simulation: Brine flow hydraulics, in Darwish MK, Al-Gobaisi DMK, eds., *Encyclopedia of Desalination and Water Resources* (DESWARE), EOLSS Publishers, Oxford, 2000.
15. Blum J, Marquardt W, The need for multivariable control in MSF desalination plants, in Darwish MK, Al-Gobaisi DMK, eds., *Encyclopedia of Desalination and Water Resources (DESWARE)*, EOLSS publishers, 2000.

16. Tarifa EE, Nicolas J, Scenna fault diagnosis for a MSF using a SDG and fuzzy logic, *Desalination* **152**:207–214 (2002).

17. Dastych J, Pickhardt R, Unbehauen, H, Control schemes of cogenerating power plants for desalination, in Darwish MK, Al-Gobaisi DMK, eds., *Encyclopedia of Desalination and Water Resources (DESWARE)*, EOLSS Publishers, 2001 (www.desware.net).

18. Gambier A, Badreddin E, Application of hybrid modeling and control techniques to desalination plants, *Desalination* **152**:175–184 (2002).

19. Aly E, Alhumaizi K, Ajbar A, Model reduction and robust control of multi-stage flash (MSF) desalination plants, *Desalination* **121**:65–85 (1999).

20. Krause H, Hartmann & Braun, GmbH, (*Methods for Controlling a Nonlinear Technical Process*), German Patent DE4434294C2,29.4, 1999.

21. Glueckstern P, Wilf M, Etgar J, Ricklis J, Use of microprocessor for control, data acquisition and on line performance evaluation in a reverse osmosis plant, *Proc. 2nd Congress Desalination and Water Reuse*, Bermuda, 1985.

22. Fredkin E, New concepts in energy savings and process control for desalination plants using reverse osmosis, *Desalination* **65**:321–330 (1987).

23. Fredkin E, Banks R, Computerized instrumentation and control for reverse osmosis systems, *Desalination* **75**:141–148 (1989).

24. Assef JZ, Watters JC, Deshpande PB, Imad M, Alatiqi IM, Advanced control of a reverse osmosis desalination unit, *J. Process Control* **7**(4):283–289 (1997).

25. Burden AC, Deshpande PB, Watters JC, Advanced process control of a B-9 Permasep permeator desalination pilot plant, *Desalination* **133**:271–283 (2001).

26. Zilouchian A, Jafar M, Automation and process control of reverse osmosis plants using soft computing methodologies, *Desalination* **135**:51–59 (2001).

27. El-Saie M, Hafez M, Selecting and tuning the control loops of MSF desalination for robustness, *Desalination* **97**:529–540 (1994).

28. Ayoub A, Abushama F, El-Hares H, Control loops in the MSF desalination plant of west Tripoli, *Proc. 1st Al-Fateh/IFAC Workshop, Automatic Control in Desalintion and the Oil Industry*, Tripoli, SPLAJ, May 12–16, 1980.

29. Alatiqi I., Ettouny H., El-Dessouky H. Process control in water desalination industry: An overview, *Desalination* **126**:15–32 (1999).

30. Hamed OA, Al Washmi H, Mustafa GM, Bamardouf K, Al-Shamri FA, Al-Bunain RE, Al-Rowely E, *Antiscalant Testing at Low Dose Rate in Al Jubail Phase II MSF Desalination Plant*, IDA Report BAH03-036, 2002.

31. Rebagliati S, Ghiazza E, Abueida KS, One year operational experience on the process control system at UANE MSF desalination plant, *Proc. 4th World Congress Desalination and Water Reuse*, Kuwait, 1989, Vol. III, pp. 149–169.

32. Fumagalli B, Loddo R, Odicino G, *Taweelah Desalination Plant Extension Stage B—Plant Process Optimization System*, Report WED/PS/AD-165/91, 1991.

33. El-Nashar A, Optimization of operating parameters of MSF plants through automatic setpoint control, *Desalination* **116**:89–107 (1998).

34. Wilf M, Bartels C, Optimization of seawater RO system design, *Desalination* **173**:1–12 (2005).

35. Jafar M, Abdel-Jawad M, Design & evaluation of a fully automated reverse osmosis plant, Proc. *IDA World Congress Desalination and Water Reuse*, Madrid, 1997.

36. Faigon M, Process control of Larnaca 51,000 m^3/d seawater RO plant, *Desalination* **138**(1–3):297 (Sept. 2001).
37. Liberman B, Larnaca desalination plant, *Desalination* **138**:293–295 (2001).
38. Schierach M, Seawater RO plant in Fujairah-UAE, Proc. *IDA World Congress Desalination and Water Reuse*, Abu Dhabi, 1995.
39. Hamed OA, Al-Washmi H, Power/water cogeneration cycle analysis Part 1: Design data theoretical analysis, *Proc. IDA World Congress Desalination and Water Reuse*, Manama, Bahrain, Oct. 26–31, 2002.
40. Kern DQ, Seaton RE, Surface fouling—how to calculate limits, *Chem. Eng. Prog.* **55**(6):71–73 (1959).
41. Al-Gobaisi DMK, Hassan A, Rao GP, Sattar A, Woldai A, Borsani R, Towards improved automation for desalination processes, Part I: Advanced control, *Desalination* **97**:469–506 (1994).
42. Robertson MW, Watters JC, Deshpande PB, Assef JZ, Alatiqi IM, Model based control for reverse osmosis desalination processes, *Desalination* **104**:59–68 (1996).
43. Deshpande PB, Prospects for advanced controls in desalination plants, *Proc. IDA World Congress, Desalination and Water Reuse*, Abu Dhabi, 1995, Vol. VII.
44. Maniar VM, Deshpande PB, Advanced control for multi-stage flash (MSF). Desalination plant optimization, *J. Process Control* **6**(1):49–66 (1996).
45. Husain A, Woldai A, Al-Radif A, Kesou A, Borsani R, Sultan H, Deshpande PB, Modeling and simulation of a multistage flash (MSF) desalination plant, *Desalination* **97**:555–586 (1994).
46. Assef JZ, Watters JC, Deshpande P, Alatiqi I, Advanced control of a reverse osmosis desalination unit, *Proc. IDA World Congress Desalination and Water Reuse*, Vol. VII, Abu Dhabi, 1995.
47. Ramasamy S, Deshpande PB, Tambe SS, Kulkami BD, Identification and advanced controls of MSF desalination plants with neural networks, paper presented at *IDA World Congress Desalination and Water Reuse*, Abu Dhabi, Nov. 18–24, 1995.
48. Armstrong P et. al., Controls, in Kreider JF, ed., *Handbook of Heating, Ventilation, and Air Conditioning*, Boca Raton, FL,CRC Press, 2001.
49. Curtiss PS, Brandemuehl MJ, Kreider JF, Energy management in central HVAC plants using neural networks, *ASHRAE Trans*. Paper 3784, **100**:(1):476–493(1994).
50. Curtiss PS, Kreider JF, Brandemuehl MJ, Adaptive control of HVAC processes using predictive neural networks, *ASHRAE Trans*. Paper 3671, **99**(1):496–504(1993).
51. Narendra KS, Parthasarathy K, Identification and control of dynamical systems using neural networks, *IEEE Trans. Neural Networks* **1**:4–27(1990).
52. Kurdali A, Woldai A, A1-Gobaisi DMK, Knowledge based systems for desalination processes, *Desalination* **92**:295–307 (1993).
53. Abdulbary AF, Lai LL, Al-Gobaisi DMK, Husain A, Experience of using the neural network approach for identification of MSF desalination plants, *Desalination* **92**:323–331 (1993).
54. Selvaraj R, Deshpande PB, Tambe SS, Neural networks for the identification of MSF desalination plants, *Desalination* **101**:185–193 (1995).
55. Campbell RL, The future of fuzzy logic and desalination: Making clear operator decisions in the control of seawater desalination systems, *Proc. IDA World Congress Desalination and Water Reuse*, Abu Dhabi, 1995, Vol. II, pp. 407–416.

56. Lauks UE, Vanbinder RJ, Valkenburg PJ, van Leeuwen C, On-line optimization of an ethylene plant, *Proc. European Symp. Computer Aided Process Engineering* (ESCAPE-l); *Supplement to Comput. Chem. Eng*. **16**(Suppl.) S213–S220 (1992).

57. Pierucci SM, Faravelli T, Brandani P, A project for on-line reconciliation and optimization of an olefin plant, *Comput. Chem. Eng*. **18**(Suppl.): S241–S246 (1994).

58. Krist JHA, Lapare MR, Wassink SG, Neyts R, Koolen JLA, Generic system for on-line optimization and the implementation in a benzene plant, *Comput. Thermal Eng*. **18**(Suppl. P): S517–S524 (1994).

59. Van Wijk RA, Pope MR, Advanced process control and line optimization in shell refineries, *Proc. European Symp. Computer Aided Process Engineering*, ESCAPE-l; supplement to *Comput. Chem. Eng*. **16** (Suppl.) S69–S580 (1992).

60. Bailey JK, Hrymak AN, Treiber SS, Hawkins RB, Non-linear optimimtion of hydrocracker fraction plant, *Comput. Chem. Eng*. **17**(2):123–128 (1993).

61. Lowery PR, McConville B, Yocum FH, Hendon SR, Closed-loop-real-time optimization of two biphenol-A plants, *Proc. AIChE Spring National Meeting*, Paper 39g, Houston, TX, 1993.

62. Gott J, Roubidoux C, Heersink R, On-line optimization for smart FCC controls, *Proc. Natl. Petroleum Refiners Association (NPRA) Computer Confe.*, Paper CC-91-130, Houston, TX, 1991.

63. Mullick S, Rigorous on-line model (ROM) for crude unit planning, scheduling, engineering and optimization, *Proc. AIChE Natl. Meeting*, Houston, TX, 1993.

64. Alhumaizi K, Modeling, simulation and control of multistage flash desalination plants, *Proc. IDA World Congress Desalination and Water Reuse*, Madrid, 1997.

65. Keyes M, Schieb T, Kaya A, Multi-level control and optimization methods for desalination plants, *Proc. IFAC Workshop on Automatic Control in Petroleum, Petrochemical, and Desalination Industries*, Kuwait, 1986.

66. Krause H, Hassan A, Optimization of MSF plants as integral part of a distributed control system and as a tool for consultants, *Proc. IDA World Conf. Desalination and Water Reuse*, Abu Dhabi, Madrid, 1997.

67. Craig IK, Henning RGD, Evaluation of advanced industrial control projects: A framework for determining economic benefits, *Control Eng. Pract*. **8**:769–780 (2000).

68. Oosthuizen DJ, Craig IK, Pistorius PC, Economic evaluation and design of an electric arc furnace controller based on economic objectives, *Control Eng. Pract*. **12**:253–265 (2004).

69. El-Nashar A, Sommariva C, Lior N, *Proc. cost benefit ratio of advanced control and automation in water desalination*, Singapore, Sept. 2005; International Desalination Assoc. Topsfield, MA, 2005.

70. Martin GD, Turpin LE, Cline RP, Estimating control function benefits, *Hydrocarb. Process* **70**:68–73 (1991).

71. Tolfo F, A methodology to access the economic returns of advanced control projects, *Proc. Americam Control Conference*, paper FP1-3:30, pp. 1141–1146, (1983).

72. Bawden NM, MacLeod IM, Benefits of advanced process control, *Elektron* **12**:13–17 (1995).

73. Contreras-Dordelly JL, Marlin TE, Control design for increased profit, *Comput. Chem. Eng*. **24**:267–272 (2000).

74. Stanton BD, Using historical data to justify controls, *Hydrocarb. Process* **69**:57–60 (1990).
75. Bozenhardt H, Dybeck M, Estimating savings from upgrading process control, *Chem. Eng*. **93**(3):99–102 (1986).
76. Latour PR, Quantify quality control's intangible benefits, *Hydrocarb. Process* **71**:61–68 (1992).
77. Lear JB, Barton GW, Perkins JD, Interaction between process design and process control: The impact of disturbances and uncertainty on estimates of achievable economic performance, *J. Process Control* **5**(1):49–62 (1995).
78. Lant P, Steffens M, Benchmarking for process control: Should I invest in improved process control? *Water Sci. Technol*. **37**(12):49–54 (1998).
79. The American Supplier Institute (trademark owner of Taguchi Methods) (http://www.amsup.com/).
80. Sommariva C, *Desalination Management and Economics*, Surrey, UK, 2004.
81. Martin T, (18–23 July). Considering social effects in control system design: A summary, *Proc. 12th World Congress International Federation of Automatic Control*; Sydney, Australia, July 18–23, 1993, Vol. 7, pp. 325–330.
82. Ali E, Understanding the operation of industrial MSF plants, Part I: Stability and steady state analysis, *Desalination* **143**:53–72 (2002).
83. Muske K, Estimating the economic benefits from improved control, *Industr. Eng. Chem. Res*. **42**:4535–4544 (2003).
84. Aly N, El-Fiqi A, Thermal performance of seawater desalination plants, *Desalination* **158**:127–142 (2003).
85. Helal AM, Medani MS, Soliman MA, Flower JR, A tridiagonal matrix model for multistage flash desalination plants, *Comput. Chem. Eng*. **10**(4):327–342 (1986).
86. Mazzotti M, Rosso M, Beltramini A, Morbidelli M, Dynamic modeling of multistage flash desalination plants, *Desalination* **127**:207–218 (2000).
87. Husain A, Hassan A, Al-Gobaisi DMK, Al-Radif A, Woldai A, Sommariva C, Modeling, simulation, optimization and control of multistage flashing (MSF) desalination plants, Part I: Modeling and simulation, *Desalination* **92**:21–41 (1993).
88. Rosso M, Beltramini A, Mazzotti M, Morbidelli M, Modeling multistage flash desalination plants, *Desalination* **108**:365–374 (1996).
89. Bourouis M, Pibouleau L, Floquet P, Domenech S, Al-Gobaisi DMK, Simulation and data validation in multistage flash desalination plants, *Desalination* **115**:1–14 (1998).
90. El-Dessouky HT, Ettouney HM, *Fundamentals of Salt Water Desalination*, Elsevier, Amsterdam, 2002.
91. Malek A, Hawlader MNA, Ho JC, Design and economics of RO seawater desalination, *Desalination* **105**:245–261 (1996).
92. Taniguchi M, Kurihara M, Kimura S, Behavior of a reverse osmosis plant adopting a brine conversion two-stage process and its computer simulation, *J. Membrane Sci*. **183**:249–257 (2001).
93. Gas Processors Assoc., *GPSA data book*, 12th ed., Tulsa, OK (USA), 2004.
94. Yokogawa Electric Corp., *Advanced Process Control Solutions—Exasmoc*, Report TI36J06D10-01E, Oct. 2001.
95. Hertanu H, Lipták B, DCS-cost estimating, in Lipták BG, ed., *Instrument Engineers' Handbook*, 3rd ed., Vol. 2, *Process Control*, Chilton Book Co., Radnor, PA, 1995.

96. Glueck AR, Bradshaw RW, A mathematical model for a multistage flash distillation plant, *Proc. 3rd Int. Symp. Fresh Water from the Sea*, 1970, Vol. 1, pp. 95–108.

97. Delene JG, Ball SJ, *A digital computer code for simulating large MSF Evaporator Desalting Plant Dynamics*, Oak Ridge National Laboratory Report ORNL-TM-2933, 1971.

98. Ettouney HM, El-Dessouky H, A simulator for thermal desalination processes, *Desalination* **125**:277–291 (2001).

99. Poje JB, Smart AM, On-line energy optimization in a chemical complex, *Chem. Eng. Progr*. 39–41 (May 1988).

100. Prichard PC, *Computer control of an olefin plant using an on-line optimisor*, Inst. Chem. Eng. Symp. Series, 1988, Vol. 100, pp. 277–283.

101. Moro LFL, Odloak D, Constrained multivariable control of fluid catalytic cracking converters, *J. Process Control* **5**(1):29–39 (1995).

102. Perregard J, Genovese F, Gani R, Efficient integration of simulation and optimization for study of complex chemical processes, *Proc. Computer-Oriented Process Engineering* (COPE-91), Barcelona, Oct. 14–16, 1991, pp. 139–144.

103. Rimawi MA, Ettouney HM, Aly GS, Transient model of multistage flash desalination, *Desalination* **74**:327–338 (1989).

104. Reddy KV, Husain A, Woldai A, Dynamic modeling of the MSF desalination process, *Proc. IDA and WRPC World Congress on Desalination and Water Treatment*, Al-Gobaisi DMK, ed., Abu Dhabi, 1995, pp. 227–242.

105. Thomas PJ, Bhattacharyya S, Patra A, Rao GP, Steady state and dynamic simulation of multistage ash desalination plants: A case study, *Comput. Chem. Eng*. **22**:1515–1529 (1998).

106. Lior N, Formulas for calculating the approach to equilibrium in open channel flash evaporators for saline water, *Desalination* **60**:223 (1986).

107. Morin OJ, Hempstead TD, for DSS Engineers, Inc., *27,486 m^3/d MSF Desalination Plant—Design Verification*, Vol. 1, report submitted to Government of Abu Dhabi Water & Electricity Dept., 1984.

108. Barton Paul I, *Dynamic Simulation*, Sec. 109. II, Special Summer Program, MIT, Cambridge, MA, Aug. 1–10, 1994.

109. Carnahan B, Luther HA, Wilkes JO, *Applied Numerical Methods*, Wiley, New York, 1969.

110. Shivayyanamath S, Tewari PK, Simulation of startup characteristics of multi-stage desalination plants, *Desalination* **155**:277–286 (2003).

FURTHER READING

6.F.1 Literature on MFSD-Related IMCA

Aly E, Ajbar A, Alhumaizi K, Robust control of industrial multi-stage desalination plants, *Desalination* **114**:289–302 (1997).

Wangnick K, *IDA Worldwide Desalting Plants Inventory*, Wangnick Consulting for International Desalination Assoc., 2004.

Kirstaedter K, Measurement, modeling, identification and optimized controller-parameter settings of MSF-desalination units, *Proc. IDA World Congress Desalination and Water Reuse*, Abu Dhabi, 1995.

Akbarzadeh-T, M-R, Kumbla KK, Jamshidi M, Al-Gobaisi DMK, Evolutionary PID fuzzy control of MSF desalination processes, *Proc. IDA World Congress Desalination and Water Reuse*, Madrid, Oct. 6-9, 1997, Vol. III, pp. 35–45.

Al-Aamry N, Saudi Arabia shutdown and flushing procedure modification for MSF units, *Proc. IDA World Congress Desalination and Water Reuse*, Bahrain 2002, pp. 78–79.

Alatiqi I, Ettouney H, El-Dessouki H, Al-Hajri K, Measurement of dynamic behavior of a multistage flash water desalination system, *Desalination* **160**:233–251 (2004).

Al-Gobaisi DMK, Barakzai AS, EI-Nashar AM, An overview of modem control strategies for optimizing thermal desalination plants, *Desalination* **84**:3–43 (1991).

Alhumaizi K, Modeling, simulation and control of multistage flash desalination plants, *Proc. IDA World Congress Desalination and Water Reuse*, Madrid, 1997, Vol. III, pp. 111–129.

Ali E, Alhumaizi K, Ajbar A, Model reduction and robust control of multi-stage flash (MSF) desalination plants, *Desalination* **121**:65–85. (1999).

Aly E, Ajbar A, Alhumaizi K, Robust control of industrial multi-stage desalination plants, *Desalination* **114**:289–302 (1997).

Ayoub A, Abushama F, El-Hares H, Control loops in the MSF desalination plant of West Tripoli, Proc. *1st Al-Fateh/IFAC Workshop, Automatic Control in Desalintion and the Oil Industry*, Tripoli, SPLAJ, May 12–16, 1980.

Barba D, Luizzo G, Tagliferri G, Mathematical model for multiflash desalting plant control, *Proc. 4th Int. Symp. Fresh Water from the Sea*, 1973, pp. 153–168.

Borsani R, Barone A, Four years operation of the largest single train MSF desalination plant, *Proc. IDA World Congress Desalination and Water Reuse*, San Diego, CA, 1999.

Borsani R, Claps B, *First Year Operation of the Largest MSF Units at Jebel Ali K2 Site* IDA BAH03-026, 2002.

Bourouis M, Pibouleau L, Floquet P, Domenech S, Al-Gobaisi DMK, Data reconciliation and gross error detection in multistage flash desalination plants, *Proc. IDA World Congress Desalination and Water Reuse*, Madrid, Oct. 6–9, 1997, Vol. III, pp. 167–180.

Bunzemeier A. A new control concept for an improved level control in MSF desalination plants, *Desalination* **114**:13–22, (1997).

Burns and Roe, Inc., *Universal Design—Report and User's Manual on Design of 2.5 Million Gallon per Day Universal Desalting Plant*, Vols. I–V (especially Vol. V), US Dept. Interior, OSW Contract 14-01-0001-955, Washington, DC, 1969.

Chidambaram M, Al-Gobaisi DMK, Controller tuning for a MSF desalination plant, *Proc. IDA World Congress Desalination and Water Reuse*, Abu Dhabi, Nov. 18–24, 1995, Vol. III, pp. 263–278.

Dastych J, Pickhardt PC, Unbehauen H, Control schemes of cogenerating power plants for desalination, in Darwish MK, Al-Gobaisi DMK, eds., *Encyclopedia of Desalination and Water Resources* (DESWARE), EOLSS Publishers (178H178H, www.desware.net), 2001.

Drake FA, Measurements and control in flash evaporation plants, in Lior N, ed., *Measurements and Control in Water Desalination*, Elsevier, Amsterdam, 1986, Sec. 21, pp. 241–262.

El-Hares H, Control loops in MSF desalination plants, *Proc. 7th Symp. on Fresh Water from the Sea*, 1980, Vol. 2, pp. 417–432.

El-Nashar AM, Optimization of operating parameters of MSF plants through automatic setpoint control, *Desalination* **116**:89–107 (1998).

El-Saie Ali AH, Ramadan AH, Gomaa SS. Multi stage flash desalination process control system, *Proc. 1st Al-Fateh/IFAC Workshop, Automatic Control in Desalination and the Oil Industry*, Tripoli, SPLAJ, May 12–16, 1980.

Fertig M, *Modeling and Control of a Brine Heater on the Basis of Hybrid Systems*, diploma thesis, Automation Laboratory, Univ. Mannheim (Germany), 2001.

Flehmig F, Watzdoff RV, Marquardt W, Reddy KV, Dynamic MSF plant modeling and simulation: Brine flow hydraulics, in Al-Gobaisi, DMK, ed., *Encyclopedia of Desalination and Water Resources* (DESWARE), ELOSS publishers, 2000.

Fortuna L, Graziani S, Manganaro G, Muscato G, The CNN's as innovative computing paradigm for modeling, *Proc. of IDA World Congress Desalination and Water Sciences*, Abu Dhabi, Nov. 18–24, 1995, Vol. IV, pp. 399–409.

Fumagalli B, Loddo R, Odicino G, *Taweelah Desalination Plant Extension-Plant Process Optimization System*, MEI/TIM/AUT/ BF, WED Report WED/PS/AD-16/91, 1991.

Gambier A, Fabricius SMO, Badreddin E, Simulation using block-oriented and object-oriented software for desalination plants, *Proc. Asian Control Conf*., Singapore, 2002, pp. 1991–1996; *Proc. 13th IASTED International Conf., Modeling Identification and Control*, Anaheim, CA, 1994, pp. 256–259 (paper presented at both confs.)

Gambier A, Fabricius SMO, Badreddin E, Simulation using block-oriented and object-oriented software for desalination plants. *Proc. Asian Control Conf*., Singapore, 2002, pp. 1991–1996.

Gambier A, Fertig M, Badreddin E, Hybrid modeling for supervisory control purposes for the brine heater of a multi-stage flash desalination plant. *Proc. Am. Control Conf*., Anchorage, 2002, pp. 5060–5065.

Ghiazza E, Fumagalli B, Odicino G, Drops: The first *d*ata *r*econciliation and *o*ptimization *s*ystem for MSF distillers has been installed and Al-Taweelah B, First feedback data and results after some months of operation on the world's largest desalination plant, *Proc. IDA World Congress Desalination and Water Reuse*, Madrid, Oct. 6–9, 1997, Vol. III 79–90.

Hamed, Osman A, Al Washmi H, Mustafa GM, Bamardouf K, Al-Shamri FA, Al-Bunain E, Rashid E, Al-Rowely E, *Antiscalant Testing at Low Dose Rate in Al Jubail Phase II MSF Desalination Plant*, IDA BAH03—036, 2002.

Hornburg CD, Watson BM, Operational optimization of MSF systems, *Desalination* **92**:333–351 (1993).

Husain A, Woldai A, Al-Radif A, Kesou A, Borsani R, Sultan H, Deshpandey PB,. Modeling and simulation of a multistage flash (MSF) desalination plant, *Desalination* **97**:555–586 (1994).

Ismail A, Oct. 6–9, (1997). Fuzzy model reference learning control of multi-stage flash desalination plants, *Proc. IDA World Congress Desalination and Water Reuse*, Madrid, Oct. 6–9, 1997, Vol. III, pp. 3–18.

Ismail A, Control of multi stage flash (MSF) desalination plants; a survey, *Proc. IDA Congress Desalination and Water Reuse*, Madrid, 1997, Vol. II, pp. 1–38.

Ismail A. Control of multi-stage flash desalination plants, a survey, *Desalination* **116**:145–156 (1998).

Jernqvist A, Jernqvist M, Aly G, Simulation of thermal desalination processes, *Desalination* **134**:187–193, (2001).

Kirstaedter K, Measurement modeling, identification and optimised controller-parameter settings of MSF-desalination units, *Proc. IDA World Congress Desalination and Water Reuse*, Abu Dhabi, Nov. 18–24, 1995, Vol. VII, pp. 52–81.

Krause H, Hartmann & Braun GmbH, (Methods for controlling a nonlinear technical process), German Patent DE4434294C2,29.4, 1999.

Krause H, Hassan A, Training simulators for MSF plants, *Proc. IDA World Congress Desalination and Water Reuse*, Abu Dhabi, Nov. 18–24, 1995, Vol. IV, pp. 185–201.

Kurdali A, Woldai A, A1- Gobaisi DMK, Knowledge based systems for desalination processes, *Desalination* **92**:295–307 (1993).

Maniar VM, Deshpande PB, Advanced control for multi-stage flash (MSF) desalination plant optimization, *J. Proc. Control* **6**(1):49–66 (1996).

Mazzotti M, Rosso M, Beltramini A, Morbidelli M, Dynamic modeling of multistage flash desalination plants, *Desalination* **127**:207–218 (2000).

Omar AM, Simulation of MSF desalination plants, *Desalination* **45**:65–76 (1983).

Parenti SB, Masarani A, Industrial application of real time neural networks in multistage desalination plant, *Proc. IDA World Congress, Desalination and Water Reuse*, Abu Dhabi, 1995, Vol. II, pp. 457–467.

Ramaswamy S, Deshpande PB, Tambe SS, Kulkami BD,. Identification and advanced controls of MSF desalination plants with neural networks, *Proc. IDA World Congress Desalination and Water Reuse*, Abu Dhabi, Nov. 18–24, 1995.

Rautenbach R, Schafer S, Availability analysis of MSF distillers using fault tree logic, in Darwish MK, Al-Gobaisi DMK, eds., *Encyclopedia of Desalination and Water Resources* (DESWARE), EOLSS Publishers (179H179H www.desware.net), 2001.

Reddy BC, Chidambaram M, Al-Gobaisi DMK, Design of centralized controllers for a MSF plant, *Desalination* **113**:27–38, (1997).

Reddy KV, Husain A, Woldai A, Al-Gobaisi DMK, Dynamic modeling of the MSF desalination process, *Proc. IDA World Congress Desalination and Water Reuse*, Abu Dhabi, Nov. 18–24, 1995, Vol. IV, pp. 227–242.

Rosso M, Beltramini A, Mazzotti M, Morbidelli M, Modeling multistage flash desalination plants, *Desalination* **108**:365–374, (1996).

Rost M,. Design and Simulation of Adaptive Multivariable Control for the model of MSF Plant, graduation thesis (in German), Univ. Dresden, Germany, 1995.

Shivayyanamath S, Tewari PK Simulation of startup characteristics of multi-stage desalination plants, *Desalination* **155**:277–286 (2003).

Sommariva C, A new milestone: The 72 MIGD multi stage flash distillation plant at Al Taweelah, Abu Dhabi U.A.E, *Proc. Int. Desalination and Water Re-use Conf*., quarterly May/June 1996, pp. 30–36.

Tarifa EE, Scenna NJ, A dynamic simulator for MSF plants, *Desalination*, **138**:349–364 (2001).

Tarifa EE, Nicolas J, Scenna fault diagnosis for a MSF using a SDG and fuzzy logic, *Desalination* **152**:207–214 (2002).

Tarifa EE, Demetrio, H, Samuel F, Sergio L, Martinez AF, Nunez NJ, Scenna fault diagnosis for a MSF using neural networks, *Desalination* **152**:215–222 (2002).

Uche J, Serra L, Valero A, Correas L, On-line thermoeconomic diagnosis of a dual purpose power and desalination plant, *Proc. IDA World Congress* on *Desalination and Water Reuse*, March 8–13, 2002, pp. 78–79.

Woldai A, Al-Gobaisi DMK, Dunn RW, Kurdali A, Rao GP, An adaptive scheme with an optimally tuned PID controller for a large MSF desalination plant, *Control. Eng. Pract.* **4**(5):721–734 (1996).

Woldai A, Al-Gobaisi DMK, Dunn RW, Kurdali A, Rao GP, Simulation aided design and development of an adaptive scheme with optimally tuned PID controller for a large multistage flash seawater desalination plant. *Proc. 5th IFAC Symp. Adaptive Systems in Control and Signal Processing*, Parts I–III, Budapest, 1995, pp. 95–112.

Zhang H, Wang S, Studies on the dynamic behavior of MSF systems, *Proc. IDA and WRPC World Congress Desalination and Water Reuse*, Madrid, 1997, Vol. 3, pp. 103–109.

6.F.2 Literature on RO-Related IMCA

Assef JZ, Watters JC, Deshpande PB, Imad M, Alatiqi IM Advanced control of a reverse osmosis desalination unit, *J. Process Control* **7**(4):283–289 (1997).

Bartman AR, Zhu A, Christofides PD, Cohen Y, Minimizing energy consumption in reverse osmosis membrane desalination using optimization-based control, *Proc. Am. Control Conf.*, Baltimore, June 30–July 2, 2010, paper ThB14.6 2010, pp. 3629–3635.

Bou-Hamad S, Abdel-Jawad M, Al-Tabtabaei M, Al-Shammari S, Comparative performance analysis of two seawater reverse osmosis plants: Twin hollow fine fibre and spiral wound membranes, *Desalination* **120**:95–106 (1998).

Burden AC, Deshpande PB, Watters JC, Advanced process control of a B-9 Permasep permeator desalination pilot plant, *Desalination* **133**:271–283 (2001).

Caldwell Membrane pre-treatment and operation monitoring and control, *Proc. Ultrapure Water Expo*, Orlando, FL, April 2002.

Fredkin E, New concepts in energy savings and process control for desalination plants using reverse osmosis, *Desalination* **65**:321–330 (1987).

Fredkin E, Banks R, Computerized instrumentation and control for reverse osmosis systems, *Desalination* **75**:141–148 (1989).

Glueckstern P, Wilf M, Etgar J, Ricklis J, Use of microprocessor for control, data acquisition and on line performance evaluation in a reverse osmosis plant, *Proc. 2nd Congress Desalination and Water Reuse*, Bermuda, 1985.

Goswami G, Srivastava VK, Hanra MS, Misra BM, Sadhukhan HK, Design of a 1800 m^3/day two stage RO plant for drinking water supply in water scarce area, *Proc. IDA World Congress, Desalination and Water Reuse*, Abu Dhabi, 1995.

Jafar MM, Abdel-Jawad M, Design, implementation, and evaluation of a fully automated reverse osmosis plant, *Proc. IDA World Congress Desalination and Water Reuse*, Madrid, Oct. 6–9, 1997, Vol. III, pp. 189–217.

Kadaj R, McMillan D, Losch J, The 4 MGD RO plant in Marco Island, Florida, *Desalination* **102**:25–26 (1995).

Malik A, Hawlader MNA, Ho JC, Design and economics of RO seawater desalination, *Desalination* **105**:245–261 (1996).

Matsuda M, Kamizawa C, Precise measurement of membrane constants of cellulose acetate membranes by direct osmosis tests, *Desalination* **49**:367–378 (1984).

Mindler AB, Epstein AC, Measurements and control in reverse osmosis desalination, in Lior N, ed., *Measurements and Control in Water Desalination*, Elsevier, Amsterdam, 1986, Sec. 2.6, pp. 343–379.

Palacin LG, Tadeo F, Elfil H, de Prada C, Salazar J, New dynamic library of reverse osmosis plants with fault simulation, *Desal. Water Treat*. **25**:127–132 (2011).

Poullikkas A, Optimization algorithm for reverse osmosis desalination economics, *Desalination* **133**:75–81 (2001).

Robertson MW, Watters JB, Deshpande PB, Assef JZ, Alatiqi IM, Model base control for reverse osmosis desalination processes, *Desalination* **104**:59–68 (1996).

Schierach M, Seawater RO plant in Fujairah-UAE, *Proc. IDA World Congress Desalination and Water Reuse*, Abu Dhabi, 1995.

Senthilmurugan S, Ahluwalia A, Gupta SK, Modelling of a spiral-wound module and estimation of model parameters using numerical techniques, *Desalination* **173**:269–286 (2005).

Singh R, Tembrock J, Effectively control reverse osmosis systems, *Chem. Eng. Prog*. **95**: 9 (Sept. 1999).

Starov VM, Smart J, Lloyd DR Performance optimization of hollow fibre reverse osmosis membranes, Part I. Development of theory,. *Membrane Sci*. **103**:257–270 (1995).

Taniguchi M, Kurihara M, Kimura S, Behaviour of a reverse osmosis plant adopting a brine conversion two-stage process and its computer simulation, *J. Membrane Sci*. **183**:249–257 (2011).

Tanny S, Ricklis WM, An automated modified fouling index device for monitoring and controlling pretreatment process, *Proc. Symp. Progress of Reverse Osmosis Desalination*, Ashkelon, Israel, 1983.

Zilouchian A, Mutaz J, Automation and process control of reverse osmosis plants using soft computing methodologies, *Desalination* **135**:51–59 (2001).

6.F.3 Literature on IMCA References Related to Desalination in General (Mostly Excluding MSF and RO)

Abdulbary AF, Lai LL, Reddy KV, Al-Gobaisi DMK, Artificial neural networks as efficient tools of simulation, *Proc. IDA World Congress Desalination and Water Reuse*, Abu Dhabi, Nov. 18–24, 1995, Vol, IV, pp. 361–374..

Abdul-Majid S, Applications of neutron back-diffusion technique for wall thickness and scale measurements at desalination or chemical plants, *Desalination* **94**:101–107 (1993).

Albuquerque JS, Biegler LT, Data reconciliation and gross-error detection for dynamic systems, *AIChE J*. **42**:2841–2856 (1996).

Ackers P, White WR, Perkins JA, Harrison AJM, *Weirs and Flumes for Flow Measurement*, Wiley, New York, 1978.

AIChE, *Testing Procedure for Evaporators*, American Institute of Chemical Engineers, New York, 1961.

Al-Shayji K, Liu YA, Neural networks for predictive modeling and optimization of large-scale commercial water desalination plants, *Proc. IDA World Congress Desalination and Water Reuse*, Madrid, Oct. 6–9, 1997 Vol. III, pp. 35–45.

Al-Sum EA, Satter A, Aziz MA, Automation of water treatment plants and its application in the power and desalination plants, *Proc. Arabian Gulf Regulation Water Desalination Symp*. (DESAL'92), Al Ain, UAE, 1992.

Altmann T, DPMS—desalination plant management system, *Proc. IDA World Congress Desalination and Water Resuse*, Abu Dhabi, 1995, Vol. V, pp. 43–50.

Aly NH, Marwan MA, Dynamic response of multi-effect evaporators, *Desalination* **114**:189–196 (1997).

APHA, Standard Methods for the Examination of Water and Wastewater, *Proc. 15th Am. Public Health Assoc. Conf.*, Washington, DC, 1981.

ASME, *Plant Performance*, PT 46, American Society of Mechanical Engineers, NY, 1996.

ASME, *Fluid Meters—Their Theory and Application*, American Society of Mechanical Engineers, New York, 1971.

Atherton Derek P, Plant models for controller design, *Proc. IDA Congress Desalination and Water Reuse*, Abu Dhabi, 1995, Vol. I, 175–194 pp. 175–194.

Bailey JK, Hrymak AN, Treiber SS, Hawkins RB, Non-linear optimization of hydrocracker fraction plant, *Comput. Chem. Eng*. **17**(2):123–128 (1993).

Bunzemeier A, A new approach to the modulating control of processes without self-regulation—the SPI(D) controller, *Proc. IDA Congress Desalination and Water Reuse*, Madrid, 1997, Vol. III, pp. 61–77.

Cadet C, Beteau JF, Carlos Hernandes S, Multicriteria control strategy for cost/quality compromiso in wastewater treatment plants, *Control Eng. Pract*. **12**:335–347 (2004).

Campbell RL, The future of fuzzy logic and desalination: Making clear operator decisions in the control of seawater desalination systems, *Proc. IDA World Congress Desalination and Water Reuse*, Abu Dhabi, Nov. 18–24, 1995, Vol. II, pp. 407–415.

Craig IK, Henning RGD, Evaluation of advanced industrial control projects: A framework for determining economic benefits, *Control Eng. Pract*. **8**:769–780 (2000).

Deshpande PB, Prospects for advanced controls in desalination plants, *Proc. IDA World Congress Desalination and Water Reuse*, Abu Dhabi, Nov. 18–24, 1995, Vol. VII, pp. 189–203.

Effelsberg H, Hirschfeld D, Sampling and analysis in seawater desalination in Heitmann, H-G, ed., *Saline Water Processing*, VCH, Weinheim, Germany, 1990, pp. 17–24.

Forbes JF, Marlin TE, Design cost: A systematic approach to technology selection for model-based real-time optimization systems, Energy 717–734 (1996).

Gambier A, Wolf, M, Miksch T, Wellenreuther A, Badreddin E, Optimal systems engineering and control co-design for water and energy production: A European project, Desali. Water Treat. 10:192–199 (2009).

Hahn WJ, Measurements and control in freeze-desalination plants, in Lior N, ed, *Measurements and Control in Water Desalination*, Elsevier, Amsterdam 1986, Sec. 2.5, pp. 321–341.

Halversen IJ, Skogestad S, Optimal operation of Petlyuk distillation: Steady state behavior, *J. Process Control* **9**:407–424 (1999).

Hewitt GF, Lovegrove PC, *Experimental Methods for Two-Phase Flow Studies*, Report EPRI NP-118, Electric Power Research Inst. Palo Alto, CA. 1975.

Hovd M, Michaelsen R, Montin T, Model predictive control of a crude oil distillation column, *Comput. Chem. Eng*. **21** (Suppl.): S893–S897 (1997).

Howe ED, Measurements and control in solar distillation plants, in Lior N, ed., *Measurements and control in Water Desalination*, Elsevier, Amsterdam, 1986, Sec. 2.4 pp. 307–320.

Jamieson DT, Experimental methods for the determination of the properties of seawater, in Lior N, ed., *Measurements and Control in Water Desalination*, Elsevier, Amsterdam, 1986, Sec. 1.6, pp. 219–240.

Jones DA, *Principles and Prevention of Corrosion*, 2nd ed., Prentice-Hall, Englewood Cliffs, NJ, 1996.

Lauks UE, Vanbinder RJ, Valkenburg PJ, van Leeuwen C, Online Optimization of an ethylene plant, *Proc. European Symp. Computer Aided Process Engineering, ESCAPE-l; Supplement to Comput. Chem. Eng*. **16** (Supp.): S213–S220 (1992).

Lausterer GK, Aiwanger G, Improving the economics of desalination plants by advanced control methods, *Proc. IDA World Congress*, Abu Dhabi, Vol. **III**:279–296.

Lear JB, Barton GW, Perkins JD, Interaction between process design and process control: The impact of disturbances and uncertainty on estimates of achievable economic performance, *J. Process Control* **5**(1):49–62 (1995).

Leitz FB, Measurements and control in electrodialysis, in Lior N, ed., *Measurements and Control in Water Desalination*, Elsevier, Amsterdam, 1986, Sec. 2.7, pp. 383–401.

Lior N, Instrumentation principles for performance measurement of solar heating systems, *Energy* **4**:561–573 (1979).

Lior N, *Measurements and Control in Water Desalination*, Elsevier, Amsterdam, 1986 (this book is out of print at the publishers, but a few remaining copies are available for purchase from the author).

Lior N, Measurements and instrumentation in water desalination, keynote paper, *Proc. IDA World Congress Desalination and Water Reuse*, Abu Dhabi, 1995, Vol. I, pp. 229–253.

Lior N, Measurement of small temperature and pressure differences, in Lior N, ed., *Measurements and Control in Water Desalination*, Elsevier, Amsterdam, 1986, Sec. 1.2, pp. 19–60.

Lior N, The status of measurements, instrumentation & control in water desalination: Part 1, *Int. Desal. Water Reuse Q*. **8**(2):46–51 (1998).

Lior N, The status of measurements, instrumentation & control in water desalination: Part 2, *Int. Desal. Water Reuse Q*. **8**(3):24–32 (1998).

Lior N, The status of measurements, instrumentation & control in water desalination: Part 3, *Int. Desal. Water Reuse Q*. **9**(1):19–26 (1999).

Lior N, The status of measurements, instrumentation & control in water desalination: Part 4, *Int. Desal. Water Reuse Q*. **9**(4):31–35 (2000).

Lowery PR, McConville B, Yocum FH, Hendon SR, Closed-loop-real-time optimization of two biphenol-A plants, *Proc. AIChE Spring 1993 Natl. Meeting*, paper 39g, 1993, Houston, TX.

Lu J, Challenging control problems and emerging technologies in enterprise optimization, *Control Eng. Pract*. **11**:847–858 (2003).

MacDonald RJ, Data reconciliation and parameter estimation in plant performance analysis, AIChE J. **34**:1–8 (1988).

Martin GD, Turpin LE, Cline RP, Estimating control function benefits, *Hydrocar. Process*.:–(June 1991).

Marquardt K, Continuous measuring of colloidal substances in the treatment of fresh water and wastewater, in Heitmann HG, ed., Saline Water Processing, VCH, Weinheim, Germany, 1990, pp. 25–36.

Marquardt K, Greiner G, Colloidal index. Determining the colloidal index (fouling index; silt density index) of various types of water, in Heitmann HG, ed., Saline Water Processing, VCH, Weinheim, Germany, 1990, pp. 37–43.

McGarvey FX, Fisher SA. Measurements and control in ion exchange installations, in Lior N, ed. *Measurements and Control in Water Desalination*, Elsevier, Amsterdam, 1986, Sec. 2.8, pp. 404–424.

McIlhenny WF, Measurements and control of feed water and product water composition, in *Measurements and Control in Water Desalination*, Elsevier, Amsterdam, 1986, Sec. 2.10, pp. 445–460.

Moore FE, Today's status of pH measurement and control in desalination plants, *Desalination* **39**:23–31 (1981).

Moro LFL, Odloak D, Constrained multivariable control of fluid catalytic cracking converters, *J. Process Control* **5**(1):29–39 (1995).

Mullick S, Rigorous on-line model (ROM) for crude unit planning, scheduling, engineering and optimization, *Proc. AIChE Natl. Meeting*, Houston, TX, 1993.

Muske K, Estimating the economic benefits from improved control, *Industr. Eng. Chem. Res*. **42**:4535–4544. (2003).

Nath R, Alzein Z, On-line dynamic optimization of olefins plants, *Comput. Chem. Eng*. **24**:533–538 (2000).

Oak Ridge National Laboratory (ORNL), *Instrumentation Guide for Seawater Distillation Plants*, Report OWRT/S-76/28, NTIS Publication PB-252–693, Office of Water Research and Technology, US Dept. Interior, Washington, DC, 1972.

Oosthuizen DJ, Craig IK, Pistorius PC, Economic evaluation and design of an electric arc furnace controller based on economic objectives, Control Eng. Pract. **12**:253–265, (2004).

Pierucci SM, Faravelli T, Brandani P, A project for on-line reconciliation and optimization of an olefin plant, *Comput. Chem. Eng*. **18**(Suppl.): S241–S246 (1994).

Pusch W, Measurement techniques of transport through membranes, in Lior N, ed., *Measurements and Control in Water Desalination*, Elsevier, Amsterdam, 1986, Sec. 1.4, pp. 105–198.

Quinn MF, Lo J-M, Electromagnetic flow meter errors caused by incorrect installation procedures, *Proc. IDA World Congress Desalination and Water Reuse*, Madrid, Oct. 6–9, 1997, Vol. III, pp. 241–248.

Raymond C, Titli A, Boverie S, About the fuzzy control of complex systems, *Proc. IDA World Congress Desalination and Water Reuse*, Abu Dhabi, 1995, Vol. II, pp. 429–438.

Reddy GP, Husain A, Al-Gobaisi DMK, Modeling and optimization of multi-effect horizontal tube falling film evaporators, *Proc. IDA World Congress Desalination and Water Reuse*, Madrid, Oct. 6–9, 1997, Vol. III, pp. 131–149.

Riley JP, Skirrow G, *Chemical Oceanography*, 2nd ed., Academic Press, New York, 1975.

Schreiber CF, Measurements and control related to corrosion and scale in water desalination installations, in Lior N, ed.,. *Measurements and Control in Water Desalination*, Elsevier, Amsterdam, 1986, See. 2.9, pp. 425–444.

Shinskey FG, *pH and pIon Control in Process and Waste Streams*, Wiley, New York, 1973.

Simon GP, Calmon C, Experimental methods for the determination of non-transport properties of membranes, in Lior N, ed., *Measurements and Control in Water desalination*, Elsevier, Amsterdam, 1986, Sec. 13, pp. 61–103.

Standiford FS, Control in multiple effect evaporator plants, in Lior N, ed., *Measurements and Control in Water Desalination*, 1986, Elsevier, Amsterdam, 2.2, pp. 263–292.

Standiford FS, Control in vapor compression evaporators, in Lior N, ed., (1986). *Measurements and Control in Water Desalination*, Elsevier, Amsterdam, 1986, Sec. 2.3, pp. 293–305.

Unbehauen H, Supervisory control systems and problems for imbedding advanced control strategies, *Proc. IDA World Congress, Desalination and Water Reuse*, Abu Dhabi, 1995, Vol. I, pp. 105–116.

Vanderbroucke EJ, SCADA System design: Meeting the owner requirements, *Proc. Computer Conf. Productivity through Innovation*, Am. Water Works Assoc., 1966.

van der Leeden F, Troise FL, Todd DK, *The Water Encyclopedia*, 2nd ed., Lewis Publishers, Chelsea, MI, 1990.

VDI, Bestimmung der mittleren temperatur in strömenden flüssigkeiten, VDI/VDE 2640 Blatt 4, in *VDI/VDE-Handbuch Mess- und Regelungstechnik*, VDI-Verlag GmbH, Dusseldorf, Oct. 1983.

Wangnick K, How incorrectly determined physical and constructional properties in the seawater and brine regimes influence the design and size of an MSF desalination plant—stimulus for further thought, *Proc. IDA World Congress Desalination and Water Reuse*, Abu Dhabi, Nov. 18–24, 1995 Vol. II, pp. 201–218.

WHO, *International Standards for Drinking Water*, World Health Organization, Geneva, 1971.

Wilson JR, (1960). *Demineralization by Electrodialysis*, Butterworths, London, 1960.

Zahedi G, Saba S, al-Otaibi M, Mohd-Yusof K, Troubleshooting of crude oil desalination plant using fuzzy expert system, *Desalination* **266**:162–170. (2011).

Zhang Z, Pike RW, Hertwig TA, An approach to on-line optimization of chemical plants, *Comput. Chem. Eng*. **19**, (Suppl.): S305–S310 (1995).

6.F.4 Literature on Some General Controls

Allgöwer F, Badgwell TA, Qin JS, Rawlings JB, Wright SJ, Nonlinear predictive control and moving horizon estimation—an introductory overview, in Frank PM, ed., *Advances in Control, Highlights of ECC'99*, Springer, 1999, pp. 391–449.

Amin M, Womack BF, Masada GY, (1985). Application of the Multi-Input Multi-Output One-Step-Ahead Adaptive Controller to a Power Plant Boiler. Paper presented at Proceedings of the 1985 American Control Conference Boston Marriott Hotel, Copley Place, Boston, MA., June 19–21, 1985, (pp. 620–625).

Armstrong P et. al. Controls, *Handbook of Heating, Ventilation, and Air* in Kreider JF, ed., Conditioning CRC Press, Boca Raton, FL, 2001.

Åstrom KJ, Hägglund T, *Automatic Tuning of PID Controllers*, ISA, Research Triangle Park, NC (USA), 1998.

Barney GC, Florez J, *Proc. IFAC Symp. Identification and System Parameter Estimation*, York, UK, 1985.

Beale R, Jackson T, *Neural Computing: An Introduction*, Adam Hilger, Bristol, UK, 1990.

Bhat NV, McAvoy TJ, Use of neural nets for dynamic modeling and control of chemical process systems, Comput. Chem. Eng. **14**:573–582, (1990).

Box GEP, Jenkins GM, *Time Series Analysis*, revised ed. Holden Day, Oakland, CA, 1976.

Chen H., Allgöwer F, Nonlinear Model Predictive Control Schemes with guaranteed stability, in Berber R, Kravaris C, eds., *Nonlinear Model Based Process Control*, Kluwer Academic, Dondrecht, 1998, pp. 465–494.

Curtiss PS, Kreider JF, Brandemuehl MJ, Adaptive control of HVAC processes using predictive neural networks, *ASHRAE Trans*. **99**(Pt. 1): –(1993).

Curtiss PS, Brandemuehl MJ, Kreider JF, Energy management in central HVAC plants using neural networks, *ASHRAE Trans*. **100**, (Pt. 1): –(1993).

Goldberg, DE, *Genetic Algorithms in Search, Optimization, and Machine Learning*, Addison-Wesley, 1989.

Graettinger TJ, Bhat NV, Heckendorn K, Buck JS, *Model predictive Control Using Neural networks, Proc. AIChE Spring Natl. Meeting*, Paper 100b, 1994.

Gupta M, Sinha N, eds., *Intelligent Control, Theory and Practice*, IEEE Press, Piscataway, NJ, 1995.

Harold D, Process controller tuning guidelines, *Control Eng*. v. 46, no. 8 pg 22 (Aug. 1999).

Harris CJ, Moore CG, Brown M, *Intelligent Control—Aspects of Fuzzy Logic and Neural Nets*, World Scientific, Singapore, 1992.

Hegglid GJ, *J. Model. Ident. Control* **4**:43–61(1983).

Huang YL, Lou HH, Gong JP, Edgar TF, Fuzzy model predictive control, *IEEE Trans. Fuzzy Syst*. **8**:665–677, (2000).

Hunt KJ, Sbarbaro D, Zbikowski R, Gawthrop, PJ, Neural networks for control systems—a survey, *Automatica* **28**:1083–1112 (1992).

Jantzen J, *Design of Fuzzy Controllers*, Technical report 98-E 864 (design), Technic,al Univ. Denmark, Dept. Automation, Lyngby, Denmark, 1998.

Kreider JF, Curtiss PS, Sheridan TB, Huang SH, Nelson RM, Mechanical systems control, in Kreith ed., *The CRC Handbook of Mechanical Engineering*, CRC Press, Boca Raton, FL, 1997, pp. 6-1–6-57.

Luyben WL, *Process Modeling, Simulation and Control for Chemical Engineers*, McGraw-Hill, New York, 1990.

Mah RSH, *Chemical Process Structures and Information Flows*, Butterworths, Stoneham, MA, 1990.

Ochiai, Calculating process control parameters from steady state operating data, *Instrument Society of America Trans*. **36**(4):313–320 (1998).

Passino K., Yurkovich S, *Fuzzy Control: Theory and Applications*, Addison-Wesley, 1997.

Pollard JF, Broussard MR, Garrison DB, San KY, Process identification using neural networks, Comput. Chem. Eng. **16**:253–270 (1992).

Psichogios DC, Ungar LH, Direct and indirect model based control using artificial neural networks, Industr. Eng. Chem. Res. **30**:2564–2573 (1991).

Riggs JB, *Chemical Process Control*, Ferret Publishing, Lubbock, TX, (1999).

Smith CA, Corripio AB, *Principles and Practice of Automatic Process Control*, Wiley, New York, 1985.

Tanttu JT, Lieslehto J, A comparative study of some multivariable PI controller tuning methods, in Devanathan R, ed, *Intelligent Tuning and Adaptive Control*, Pergamon Press, New York, 1991.

Wang L-X, *Adaptive Fuzzy Systems and Control: Design and Stability Analysis*, Prentice-Hall, Englewood Cliffs, NJ, 1994.

Zanin AC, Tvrzska de Gouvea M, Odloak D, Integrating real-time optimization into the model predictive controller of the FCC system, Control Eng. Pract. **10**:819–831 (2002).

INDEX

Advances in Water Desalination, First Edition. Edited by Noam Lior.